S0-BAC-947

Terrigenous Clastic Depositional Systems

W.E. Galloway
David K. Hobday

Terrigenous Clastic Depositional Systems

Applications to Petroleum, Coal, and Uranium Exploration

With 237 Figures

Springer-Verlag
New York Berlin Heidelberg Tokyo

W.E. GALLOWAY Bureau of Economic Geology, The University of Texas at Austin, Austin, Texas 78712 U.S.A.
D.K. HOBDAY Bridge Oil Limited, Level 33 CBA Centre, Sydney, N.S.W. 2000 Australia

Library of Congress Cataloging in Publication Data
Galloway, W. E.
Terrigenous clastic depositional systems.
Bibliography: p.
1. Sedimentation and deposition. I. Hobday, David K.
II. Title.
QE571.G27 1983 553.2 83-668

Typeset by Ampersand Inc., Rutland, Vermont.
Printed and bound by Edwards Brothers, Ann Arbor Michigan.
Printed in the United States of America.

9 8 7 6 5 4 3 2

ISBN 0-387-90827-7 Springer-Verlag New York Berlin Heidelberg Tokyo
ISBN 3-540-90827-7 Springer-Verlag Berlin Heidelberg New York Tokyo

We dedicate this book to our teachers and friends,
Frank Brown
John Ferm
Bill Fisher

Preface

The reserves, or extractable fraction, of the fuel–mineral endowment are sufficient to supply the bulk of the world's energy requirements for the immediately forseeable future—well into the next century according to even the most pessimistic predictions. But increasingly sophisticated exploration concepts and technology must be employed to maintain and, if possible, add to the reserve base. Most of the world's fuel-mineral resources are in sedimentary rocks. Any procedure or concept that helps describe, understand, and predict the external geometry and internal attributes of major sedimentary units can therefore contribute to discovery and recovery of coal, uranium, and petroleum.

While conceding the desirability of renewable and nonpolluting energy supply from gravitational, wind, or solar sources, the widespread deployment of these systems lies far in the future—thus the continued commercial emphasis on conventional nonrenewable fuel mineral resources, even though their relative significance will fluctuate with time. For example, a decade ago the prognostications for uranium were uniformly optimistic. But in the early 1980s the uranium picture is quite sombre, although unlikely to remain permanently depressed. Whether uranium soars to the heights of early expectations remains to be seen. Problems of waste disposal and public acceptance persist. Fusion reactors may ultimately eliminate the need for uranium in power generation, but for the next few decades there will be continued demand for uranium to fuel existing power plants and those that come on stream.

This book is, to some extent, a hybrid. It is directed toward the practicing exploration and development geologist who is, of necessity, something of a generalist. However, the stress on process and principle may also make this a suitable text for courses in resource geology.

Our grouping of coal, uranium, and petroleum may appear to be incongruous and artificial. However, our basic premise is that there are common genetic attributes shared by all three, and that the sedimentological principles governing their distribution are fundamentally similar. We have both had geologic careers divided among all three of the fuel minerals. Factors that we have found to be important include depositional processes and environments and their resultant genetic facies, interrelationships of genetic facies within depositional systems, early postdepositional modifications by circulating ground water, and, finally, the changes that take place at depth as sedimentary basins evolve in response to tectonic and regional hydrologic controls.

In many instances the paleoenvironmental factor is preeminent in controlling the distribution of fuel minerals. The origins of peat and both syngenetic and placer uranium are directly related to depositional environment. Peat is subsequently modified to coal by burial and heating during the normal sequence of basin evolution.

However, many attempts to relate fuel-mineral deposits to genetic facies associations alone have met with mixed success. Sedimentary facies with apparently all of the necessary attributes for hosting fuel minerals commonly prove to be singularly barren, whereas some rich deposits in ostensibly unfavorable host facies defy conventional explanation. These exceptions indicate the need to consider additional factors, some of which may not be reflected in static facies elements. For example, it was recognized 30 years ago that the role of postdepositional ground-water flow is crucial in sandstone-type uranium mineralization. Hydrologic setting is important in peat genesis, and critical to its preservation as coal; it may even have influenced the distribution of placer uranium in early Precambrian Witwatersrand-type algal mats. Thus, differences in ground-water circulation arising from topographic, structural, or climatic controls explain differences in uranium mineralization in sandstones of similar origin. They may also explain mutually exclusive distributions of coal and epigenetic uranium in identical, coeval facies in different parts of a sedimentary basin. For these reasons, we summarize principles of ground-water flow in large sedimentary basins and explore implications for fuel-mineral genesis.

Numerous excellent textbooks and other compilations are devoted to sedimentary facies, environments, and processes, reflecting the burgeoning interest and involvement of geologists in these fields. There has been a corresponding recent proliferation of literature on fuel minerals from the standpoint of their geographic distribution, regional geologic setting, host rock associations, and economic and engineering aspects of their exploitation. This book attempts to bridge the gap between process-related studies of sedimentary rocks and the more traditional economic geology of commercial deposits of coal, uranium, and petroleum. Due attention is paid to subsurface techniques which, integrated with outcrop data, enable the most realistic reconstructions of genetic stratigraphy, and offer the greatest application in exploration. After reviewing depositional systems and their component genetic facies with emphasis on field and subsurface recognition, we examine ground-water flow systems—how they evolve in relation to changing structural configuration, consolidation, climatic regime, and topography in the recharge area. This sets the stage for an account of the associated fuel minerals in terms of their paleoenvironmental setting, emplacement, and subsequent transformations.

Our views are necessarily prejudiced by our own experience, but we attempt to do justice to the prevailing state of the art in basin analysis. Prodigious volumes have been published on the relationship of petroleum to clastic depositional systems, so only an overview is possible here. However, we document important studies in mature hydrocarbon provinces that provide excellent models for exploration in less-explored basins. In contrast, with a few conspicuous exceptions, coal and sedimentary uranium have only recently attracted the same level of detailed attention from sedimentologists. This stems in part from the early dominance of petroleum as a fuel, the temporary eclipse of coal, and the relatively recent emergence of uranium; and probably also from an overemphasis on descriptive stratigraphy, particularly in coal basins. The burgeoning studies of sedimentary uranium have presently reached a plateau, which permits a fairly comprehensive synthesis. Although general environments of coal formation have been known since the last century, it was only with detailed studies of modern fluvial and deltaic environments, starting with the Mississippi, that predictive models were developed. These coal models are currently undergoing considerable refinement. Those that we describe have all shown economic application in exploration and mine development.

The importance of sedimentary facies in affecting the quality and extraction of fuel minerals is also being more widely appreciated. For example, the roof and floor properties of coal mines are largely determined by subfacies characteristics. Knowledge of the depositional framework and associated fluid flow and engineering properties has long been important in hydrocarbon production. Progressively more sophisticated geological input is used in genetic-predictive modeling, and this trend is likely to increase as reserves become depleted.

Compilation of a book which focuses on the geology and mineral deposits of many parts of the world brings one face to face with the problem of units of measurement. There is no ready solution to the complexity of English and metric units applied in different countries, or in the same country at different times, or for different commodities. We have attempted to cite measurements in their original units and to provide equivalencies in parentheses. Where original figures are rounded off, conversions are similarly rounded. In reality, the resource geologist must remain, for some time to come, conversant in both English and metric.

Acknowledgments

A book of this scope inherently transcends the personal experience of two authors. We have each been blessed with stimulating environments and co-workers who have contributed not only to the evolution to our ideas, but more directly, to the completion of this manuscript.

Friends and associates who suggested ideas, provided figures and data, and most importantly, served as sounding boards and reviewers for particular topics, include Sam Adams, John Barwis, Knut Bjørlykke, Pat Eriksson, Frank Ethridge, John Hunt, Brian Jones, Rufus LeBlanc, Evan Leitch, Robert Morton, Marjorie Muir, Jonathon Price, Ray Rahmani, Stan Schumm, Greg Skilbeck, Michelle Smyth, Ron Steel, Tony Tankard, Bruce Thom, Jŏzsef Tóth, Bill van Rensburg, Vic Von Brunn, Roger Walker, and John Ward. To each of these we extend our gratitude.

Additional associates who contributed to the preparation of the manuscript include Sheila Binns, Dawn Garbler, Marjorie Starr, Avril Troth, and Doris Tyler (typing); Micheline Davis, Len Hay, and John Roberts (drafting); and Lucille Harrell, Jim Morgan, and Dan Scranton.The index was compiled by Amanda Masterson.

The Department of Geology and Geophysics, University of Oklahoma, provided the much needed time and research support for one of us (WEG) during a tenure as Visiting Professor. Additional time and support was made available by our employers, the Bureau of Economic Geology, University of Texas at Austin, and the Department of Geology and Geophysics, University of Sydney. To each of these organizations we express our thanks.

Finally, we wish to acknowledge the many hours of time and effort spent by Martin Jackson in reviewing the entire manuscript. His thoughtful suggestions and comments have undoubtedly rounded off many of the rough edges endemic to joint authorship and aided clarity of the text.

To our wives, who seem always to welcome us back from our philanderings with mistress geology, we can only repeat "thanks again."

Contents

Chapter 6 Clastic Shore-Zone Systems 115

Chapter 7 Terrigenous Shelf Systems 143

Chapter 8 Terrigenous Slope and Basin Systems 167

Chapter 9 Lacustrine Systems 185

Chapter 10 Eolian Systems 203

Chapter 11 Depositional Systems and Basin Hydrology 223

Chapter 1

The Fuel-Mineral Resource Base

Introduction

Nonrenewable energy resources available in very large quantities are limited to heavy hydrogen and dry geothermal energy. Large-scale renewable resources are solar and atmospheric electricity (Moody, 1978). These four may be regarded as the ultimate energy sources, but they are unlikely to contribute significantly to the total energy budget for at least the next few decades. Synthetic fuels are already in production, but widespread conversion is being held back by economic and environmental considerations. Wind, water, and biomass conversion will play an enlarged, but still relatively minor, role. This leaves coal, uranium, and petroleum to provide the bulk of the immediately foreseeable energy requirements.

The use of coal goes back thousands of years to its combustion in Bronze Age funeral pyres. It was used by the ancient Greeks and Romans, and subsequently by the American Indians, Chinese, and European nations, where it gradually supplanted animals, wind, water, and wood as the main energy source, fueling the Industrial Revolution. Although superseded this century by petroleum, coal has made a strong comeback since the early 1970s and is on the ascent. Petroleum, too, has a long history of human utilization, being employed in warfare and embalming prior to 500 BC, and, subsequently, in medicines and street lamps. A petroleum field was established as early as 211 BC in Szechwan, China (Halbouty, 1980), but only with the 1859 Titusville discovery was the modern petroleum industry presaged. The degree to which Western nations became reliant on petroleum in motive and stationary power sources was made apparent by the 1973 OPEC oil embargo. Uranium, in contrast, has been used for less than 50 years, with consumption accelerated by World War II weapons research and, from 1968 to 1973, by fuel-mineral demand.

Exploitation of these three fuel minerals is subject to obstacles of different kinds (Clarke, 1978). In the case of petroleum, the bottleneck is in conceptual and technological considerations at the exploration stage, and in efficient extraction from the reservoir; in the case of uranium it is environmental apprehension at the energy-conversion stage; and in coal it is at the recovery stage.

Coal reserves are generally regarded as sufficient for at least the next 300 years at present rates of consumption, but current use is likely to increase severalfold. There is remarkably little consensus regarding the volume of coal resources

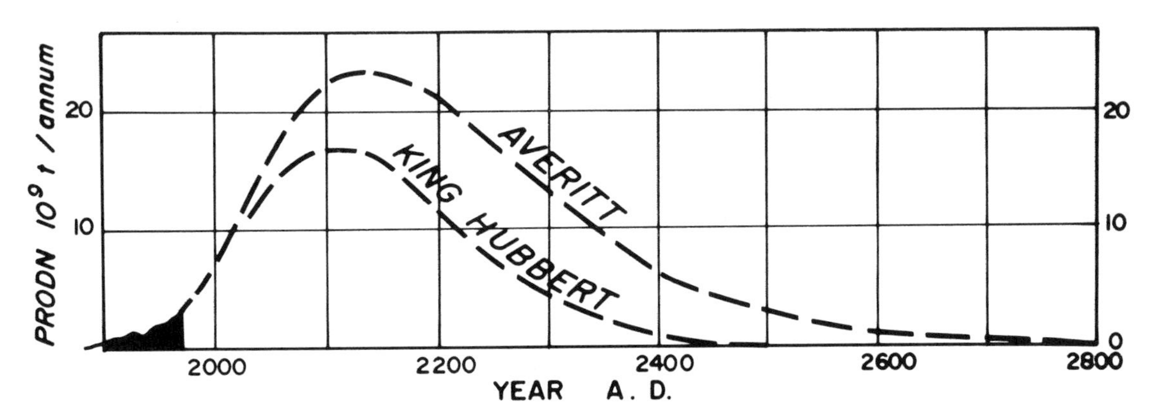

Figure 1-1. Projections of global coal production based on estimates by Averitt (1969) and King Hubbert (1969). (After Fettweis, 1979.)

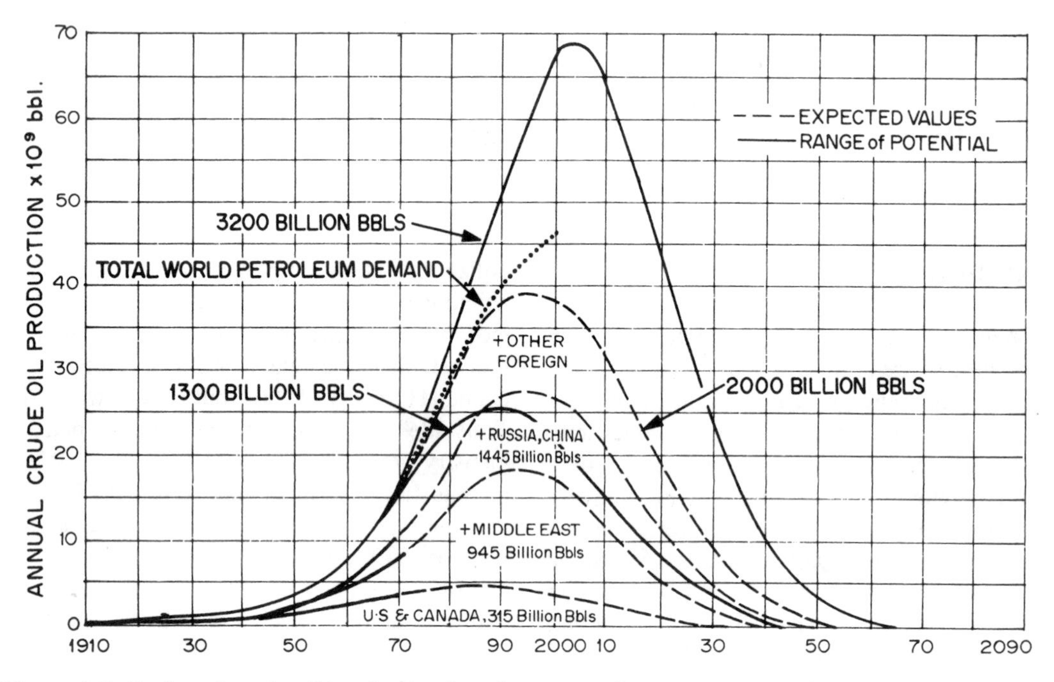

Figure 1-2. Projected crude oil production based on an estimated reserve of 2000 billion barrels. (After Moody, 1978.)

or reserves, with different countries frequently employing different criteria in their estimates. Based on King Hubbert's (1969) curve (Fig. 1-1), coal should represent a major energy source for the next 500 years. Averitt's (1969) figures are even more optimistic. Coal use is expected to peak between the years 2100 and 2200 at between six and nine times current annual production, declining to an insignificant level around the year 2800. The 1974 World Energy Conference estimated the world's coal resources at 11×10^{12}t. However, only 0.6×10^{12}t were known in detail and regarded as extractable under then-existing economic and technological constraints, thus constituting reserves (Fettweis, 1979, p. 19). The 1977 estimate was 12.9 $\times 10^{12}$t, of which 6 percent was regarded as reserves. Clarke's (1978) prediction of an increase in the world's coal reserves to $5–6 \times 10^{12}$t was based on expectations of increased demand, coupled with increased prices as other bulk energy supplies dwindle, leading to increased recovery.

Crude oil reserves from conventional sources are estimated at around 2000×10^9 bbl (Fig. 1-2). Of this, some 375×10^9 bbl have already been extracted, leaving about 725×10^9 bbl of proved reserves and some 900×10^9 bbl total undiscovered recoverable reserves, or potential reserves. World production is expected to peak during the next decade, but some additional 24×10^9 bbl need to be discovered each year in order to replace depleted reserves (Grivetti, 1981).

Natural gas resource estimates are in the range of $5–12 \times 10^{12}$ ft^3 ($0.14–0.37 \times 10^{12}$ m^3), of which a large proportion is in Middle Eastern and Eastern Bloc countries. Even in the United States, however, very large volumes of gas remain, and exploration for gas has accelerated since 1978. The price factor is critical in hydrocarbon reserve determinations. For example, price increases over the past few years instantly increased the reserves in some mature fields. Over a slightly longer term, price incentives add to reserves by increasing exploration activity.

Uranium reserve figures fluctuate even more markedly in response to price fluctuations. The size of uranium resources is limited only by cost factors because the element is widely distributed at low concentrations. Even so, and in spite of predictions for reduced nuclear generating capacity, temporary shortfalls in uranium supply are likely to result from the existing decline in exploration.

Table 1-1. Rounded Estimates of the Present World Distribution of Mineral Energy Resources[a]

	Oil and NGL	Gas	Tar	Oil Shale	Coal	Uranium	Total
Russia, China, etc.	1	—	1	—	26	NA	29
United States	—	—	—	2	18	15	35
Canada	—	—	1	—	—	9	10
Middle East	1	—	NA	—	—	—	1
Other	1	—	—	1	4	18	24
World	4	1	2	3	49	41	100

[a] After Moody (1978).

World energy resource distributions, as summarized in Table 1-1, show some surprises. The oil fields of the Middle East have about one percent of the world energy resources, with the total for oil amounting to a mere 4 percent. Gas, tar, and oil shale similarly represent only a few percent. The lion's share is coal and uranium, which together constitute 90 percent. Total energy resources of the United States amount to over one-third of the world figure.

Chapter 2

Approaches to Genetic Stratigraphic Analysis

The real world is immensely complex [and] continuous. Isolated structures are therefore subjective and artificial portions of reality, and the biggest initial problem is the identification and separation of meaningful sections of the real world. On the one hand, every section or structure must be sufficiently complex . . . so that its study will yield significant and useful results; on the other, every section must be simple enough for comprehension and investigation.

Chorley and Kennedy (1971, p. 1)

Introduction

One of the most difficult tasks in the application of genetic facies interpretation in resourc exploration, appraisal, and development is th delineation of depositional units of sufficient extent and appropriate scale for analysis. The depositional basin defines the boundaries and general conditions of the accumulation of a sediment pile. Depositional systems, as described in subsequent chapters, provide "meaningful sections" of the basin fill. Their recognition and delineation establish a framework for facies differentiation and mapping, using appropriate process–response models. It is commonly at the facies level that source units, fluid-migration pathways (the basin plumbing), potential hosts or reservoirs, and trapping configurations are sought and dissected.

In most sedimentary basins, exploration and development of energy minerals relies increasingly on generation and analysis of subsurface data. Detailed description of sedimentologic attributes requiring outcrop exposure as a basis for genetic stratigraphic interpretation becomes, at best, of limited use. Similarly, whole diamond core is a rare luxury that is typically available only in areally and stratigraphically restricted portions of the basin fill. However, the concept of a depositional system implies that component facies are spatially related, three-dimensional units, which may be readily described by commonly available types of subsurface data, augmented where possible with descriptions of core or outcrop sections. This approach to facies analysis relies heavily on reconstruction of basin morphology and bedding architecture, determination of gross lithology, quantitative delineation of the geometry of framework sandstones, and recognition of vertical and lateral successions and common facies associations. The following sections discuss approaches to three-dimensional facies analysis, with emphasis on subsurface data, and review basic sedimentologic concepts that are fundamental to development of flexible process–response facies models.

Depositional Architecture

Bedding geometry and spatial relationships within and among lithologic units is a fundamental property of genetic stratigraphic sequences constituting a basin fill. Delineation of "bedding style" or "depositional architecture" on both regional and local scales provides much information on depositional processes and probable depositional systems or environments.

Sedimentation within a basin of any size can occur at the margin or bottom. *Aggradation* is the process of vertical filling of the basin. Infilling from the margin is either by *progradation*, if sediment is washed into the basin, or by *lateral accretion*, if sediment moving within the basin preferentially accumulates against the margin (Fig. 2-1). Each of these three mechanisms produces a characteristic bedding architecture, and is typified by a general textural profile (Fig. 2-1). Aggradational bedding produces no inherent systematic textural trends; rather, each bed may display varying texture and composition. Progradation and lateral accretion both produce depositional units having a sigmoidal cross section. They are readily differentiated, however, by contrasting textural sequences: progradational sequences coarsen upward, whereas lateral accretion produces an upward-fining sequence. In

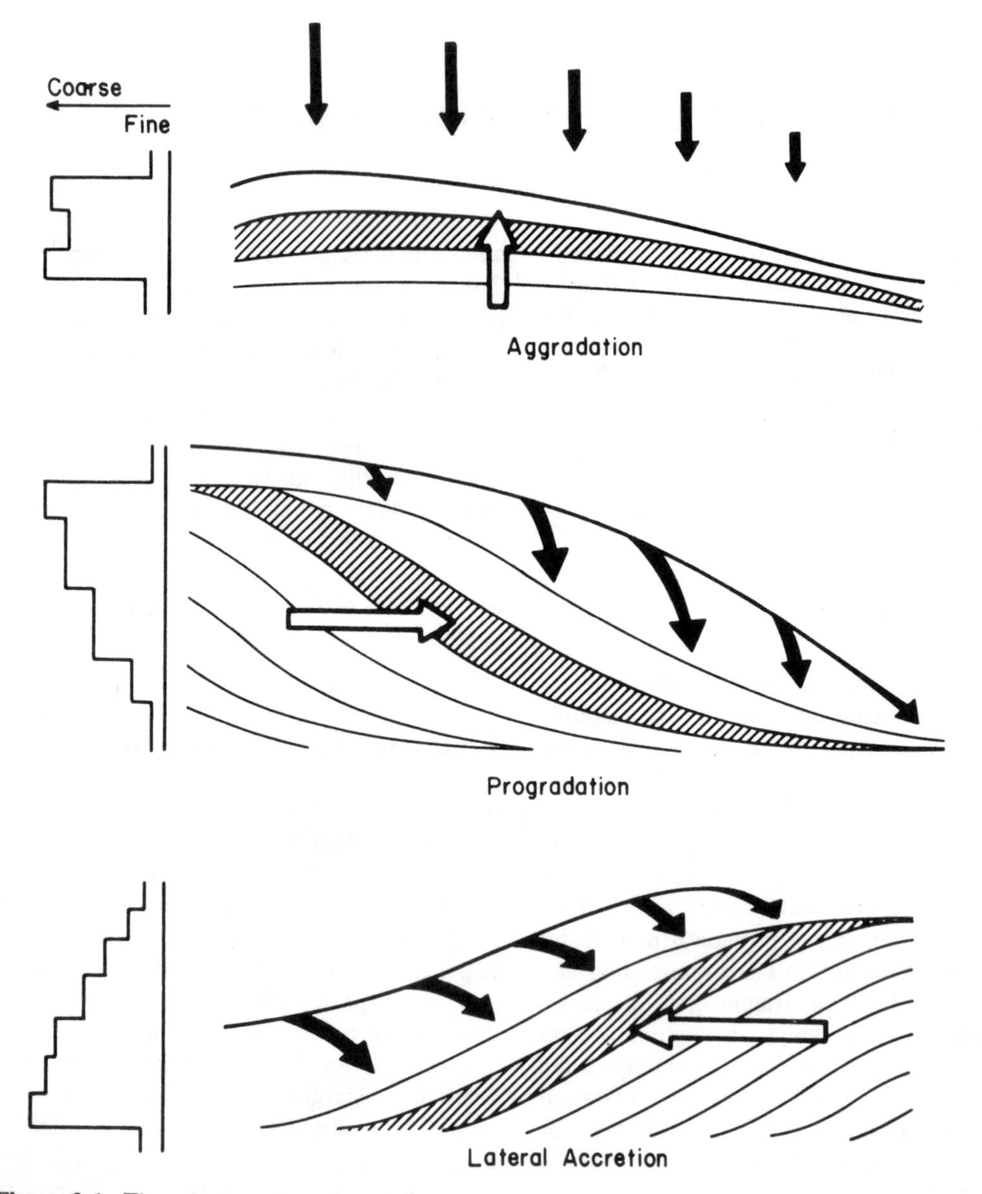

Figure 2-1. Three basic styles of basin-filling and their resultant bedding geometries and vertical textural sequences. (Modified from Galloway *et al.*, 1979.)

both, the sequence is reproduced laterally within a single genetic increment, and vertically as successive increments are stacked one on the other.

Accretionary, aggradational, and progradational depositional settings may exist side by side in the same depositional system. For example, an abandoned fluvial channel may fill by progradation along one side of sediment washed in during floods from an adjacent, active channel in which point bars are growing by lateral accretion. At the same time, overbank sediment deposited by flood waters causes the floodplain to aggrade, or build vertically. Whole depositional systems may be dominantly progradational (a delta system) or aggradational (an alluvial fan system); lateral accretion is more typical of local environments within larger systems.

In addition to bedding styles, most clastic depositional systems are characterized by specific geometries and processes of sediment dispersal. Almost all bed-load transport processes leave a depositional record of the path of the sediment dispersal system. The principal exceptions to this rule are systems in which down-slope gravitational remobilization of coarse sediment produces significant zones of bed-load sediment bypass. Definition of the geometry of the bed-load (sand) framework, or depositional skeleton of the larger genetic units is basic to unraveling sediment dispersal pathways, and, in turn, provides much useful information about depositional processes and possible environments.

A primary distinction may be made between dip-fed and strike-fed sediment dispersal systems. A dip-fed system, such as a fluvial system, primarily transports sediment down-slope toward the depositional basin. In contrast, a strike-fed system, such as a barrier bar system, moves bed-load sediment parallel to the basin margin. Many depositional systems contain both dip- and strike-fed elements. Relative volume, vertical and areal distribution, and cross-sectional geometry of the dip-fed and strike-fed elements are some of the most powerful guides for genetic stratigraphic interpretation. Further, these parameters are readily determined from subsurface data.

Depositional Episodes

Frazier (1974) developed a conceptual model based on extensive three-dimensional stratigraphic studies of Quaternary depositional systems of the Gulf Coast Basin which integrated the principal components of the basin fill. This model, with due recognition of geometric variability introduced in basins of differing tectonic setting and bathymetric configuration, provides a basis for recognition of genetic stratigraphic units within large marine or lacustrine basin fills.

Several sedimentologic principles form the basis for the model (Frazier, 1974). (1) Terrigenous clastic sediments are allochthonous, and must therefore be transported to the basin margin, primarily by rivers. (2) Basins are filled by clastic sediment through a repetitive alternation of depositional and nondepositional intervals. At any one time, active deposition is concentrated in specific areas of the basin, although infinitesimal amounts of sediment accumulate elsewhere. Consequently, essential nondepositional interludes or hiatuses separate depositional units. (3) The time interval represented by a hiatus and its resultant stratigraphic surface varies from place to place; however, at least one time line extends throughout the entire extent of the surface. Thus, the hiatal surface upon which a progradational series of beds is deposited represents a progressively longer time interval in the basinward direction. Conversely, the subaerial surface that forms over the progradational interval represents increasing amounts of time in the landward direction. (4) A simple depositional event, which is a localized pulse of deposition separated by hiatal intervals from underlying and overlying strata, consists of three phases. Deposits of the *progradational phase* progressively fill the basin, forming a wedge of sediment that commonly thickens basinward. Contemporaneous *aggradational* phase deposits cap the progradational platform, commonly thickening in the landward direction. Termination or decrease in sediment output and ongoing basin subsidence results in deposition of a veneer of reworked, *transgressive phase* sediments across the basinward portion of the depositional unit. Thus each depositional event produces a facies sequence recording initial progradation, penecontemporaneous aggradation, and terminal transgression.

As shown in the schematic time–distance diagram in Figure 2-2, multiple depositional events combine to produce a major physical, genetic stratigraphic unit called a *depositional episode*. The depositional episode is a complex of facies sequences derived from common sources along the basin margin, and deposited in a period of relative base-level or tectonic stability (Frazier, 1974). Each depositional episode is bounded basinward by major transgressive events and hiatal intervals (and their resultant surfaces) that have regional or world-wide significance (Fig. 2-2). Further, the depositional episode contains an extensive subaerial hiatal surface, which increases in temporal significance landward. Bounding transgressions may be a product of tectonic or isotatic subsidence or of eustatic changes in base level. Boundaries between depositional episodes are ill-defined landward of the shoreline of maximum transgression. As pointed out by Frazier, conventional stratigraphic units and depositional episodes may coincide. However, the transgressive facies, which are genetic components of the stratigraphic sequence produced by a depositional episode, are commonly given individual formational status or are com-

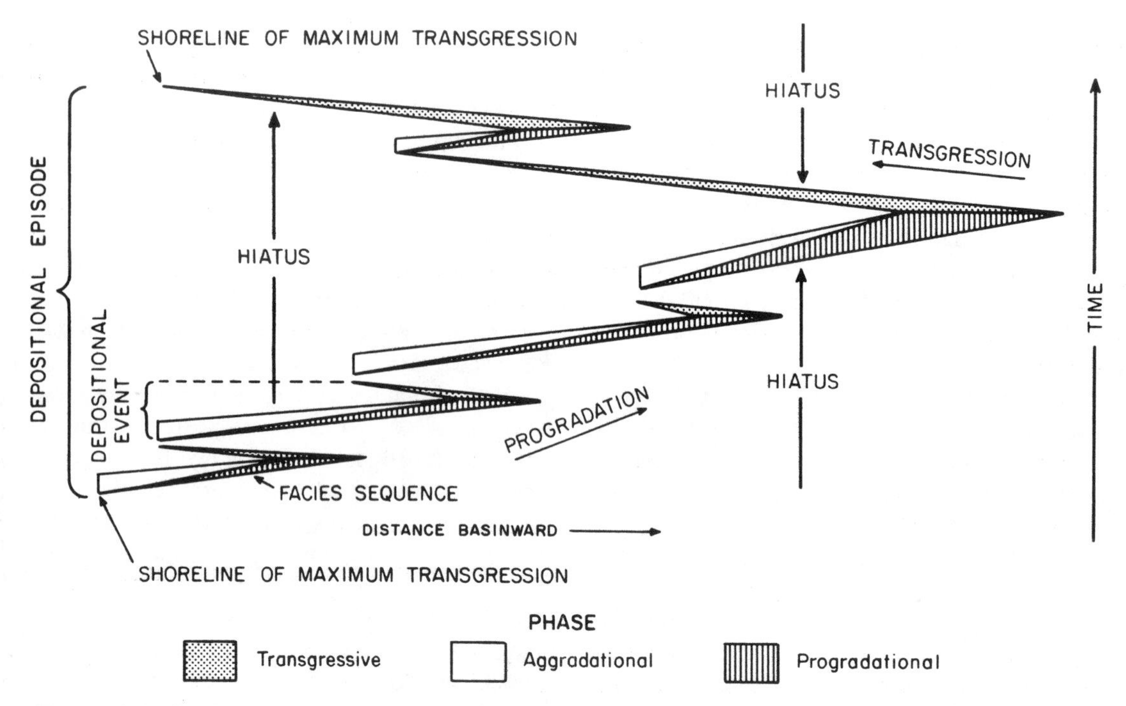

Figure 2-2. Schematic time–distance diagram illustrating the temporal and spatial relationships of a depositional episode and the phases of its component depositional events. Only patterned areas represent intervals of active terrigenous deposition in any one portion of the basin. The depositional episode is bounded by external hiatal surfaces and encompasses a single major internal hiatal surface. (Modified from Frazier, 1974.)

bined with the strata of the overlying depositional episode.

Recognition of depositional episodes and their component depositional events has two applications. First, these constitute both regional and local genetic units that must be recognized and correlated if quantitative facies mapping is to produce meaningful patterns. Secondly, the model relates the various bedding styles to preferred positions within the basin fill.

Both transgressive facies and hiatal surfaces provide physical stratigraphic correlation markers that can be used to define the boundaries of genetic units. Unlike hiatal surfaces, however, transgressive units do not necessarily incorporate a time line (Fig. 2-2). External hiatal surfaces, which represent long intervals of essentially no deposition of terrigenous clastic sediment, may be indicated by a variety of thin, laterally continuous beds or horizons, including: (1) marl and limestone beds, (2) richly glauconitic or phosphatic sand and mud, (3) fossiliferous or burrow-churned pelagic mud, and (4) veneers of marine-reworked, relict sand, silt, or mud. Presence of carbonate and other chemical constituents in these veneers reflects the long contact times between surficial sediments and the overlying water column. Chemical or biogenic materials may be indicated on subsurface well logs as thin zones of relatively dense, low-porosity, high-resistivity, and/or radioactive material. In contrast, nearly pure pelagic mud units are indicated by zones of minimum resistivity on electric logs. Thus, the most useful genetic boundary horizons are commonly distinguished by their log signature, as well as by their lateral continuity.

Alternatively, the hiatal surface may develop as a surface of marine erosion (Dietz, 1963). Resulting dissection, canyon cutting, and planation and concomitant aggradation of the basin floor are readily apparent in reflection seismic data (Brown and Fisher, 1980).

Internal hiatal horizons may be erosional or pedogenic surfaces, or may be recorded by deposition of peat. Though less useful for correlation or isolating genetic stratigraphic units, the chemical and physical attributes of such horizons may also produce distinctive log responses. However, lateral continuity is typically less well displayed than in their basinal counterparts, which form under more uniform subaqueous conditions.

In summary, interpretation of depositional architecture of the basin fill—determination of the major depositional episodes and the nature and extent of their bounding horizons, their contained bedding styles, and the overall geometry of framework sand facies—can be an adequate basis for early recognition of principal depositional systems and associated facies.

Quantitative Facies Mapping

Quantitative geologic mapping is standard procedure in energy resource geology. In addition to basic structure contour and interval isopach maps, isolith maps, such as a net sandstone or coal isopach, and proportion maps, including sandstone percentage and sand/shale ratio maps, define the areal extent and expected thickness of reservoirs or other economically important lithologies.

In terrigenous clastic depositional systems, a combination of genetic interval isopach map, net sand isopach map, and, if the interval thickness changes markedly, a sand percentage or ratio map is particularly useful for genetic stratigraphic interpretation. Such a map suite outlines principal depocenters for both total sediment and for the bed-load fraction, and displays the distribution, trends, and areal patterns of the framework sand facies. Further, the distribution of both framework and nonframework facies can be related to basement and intra-formational structure, and to basin morphology.

The detailed geometry of a specific sand body is typically obscured because the facies sequence of even a single depositional event consists of several partially superimposed sand units deposited in different environments. However, depositional grain of the framework sand facies dominates contour patterns. Even in thick sequences that incorporate many tens or hundreds of individual depositional cycles, vertical persistence of depositional environments characteristic of rapidly subsiding basins results in stacking of similar facies and preservation of framework trends. Although such stacking inherently decreases map resolution and causes loss of details of framework geometry, major attributes, such as positions of depocenters, relative abundance of dip- or strike-fed sand bodies, areal and stratigraphic distributions of framework sands within the genetic package, and impacts of contemporaneous structures on sand distribution, remain visible.

Utility of maps is further increased if basic sedimentologic concepts and genetic models are incorporated in contouring style. For example, with the exception of gravitational remobilization in subaqueous slope settings, total bed-load sediment bypass is rare in terrigenous clastic systems. Consequently, elements of each portion of the sediment dispersal system are likely to be preserved as a series of interconnected sand bodies whose trends reflect directions of sediment transport. Thick, isolated pods or lobes of sand are rare.

Contouring more appropriately emphasizes continuity rather than isolation of sand deposits. Exceptions have strong implications for interpretation of depositional processes and depositional systems. Similarly, contouring should attempt to recognize and emphasize emerging patterns such as discrete belts, radiating distributary aprons, or subparallel pods. Systematic areal changes in pattern, dimensions, or trend of contours likely reflect significant facies changes.

In addition to the basic interval, isolith, and sand percentage maps, several derivative facies maps may also be useful in facies delineation and interpretation (Forgotson, 1960). Maps showing the thickness of the thickest sand body or the number of discrete sand bodies within the interval provide rapidly derived overviews of facies trends, as well as adding information about facies distribution. Three-component maps outline proportional content of multiple lithologies such as sand, mudstone, and limestone. Vertical position of sediments within a genetic sequence can be quantified and displayed using a center-of-gravity mapping technique (Forgotson, 1960). If data are computerized, a mathematically derived gradient or trend surface may be generated and used to remove regional gradients, so that details of framework geometry can be better resolved from the background trends (Wermund and Jenkins, 1970).

Wire-Line Logs

Various wire-line logs are the most common type of geologic data available for subsurface geologic analysis. Together with drill cuttings, such logs provide a basic suite of information about the lithology, petrophysical properties, and pore-fluid content of the strata penetrated.

No wire-line log determines lithology or grain size directly. Consequently, lithologic and textural interpretation are based on calibration of log response with core or other independent lithologic data, use of assumed correlations between lithology and the property actually measured, or comparison of several log types. Details of log interpretation lie beyond the objectives of this book. The mechanics of well logging, and the assumptions, techniques, and theories of log interpretation are discussed in numerous petroleum engineering texts and reference manuals. However, logs can be readily used for qualitative lithologic information, and provide a three-dimensional data base for facies recognition and mapping.

Log Types

Two types of logs are commonly utilized for lithofacies interpretation: the electric log and the natural gamma log. The electric log, which comes under many different names, including resistivity log, induction log, and laterlog, typically displays two basic traces, an S.P. (Spontaneous Potential) and a resistivity curve. The S.P. curve, which lies along the left side of the log, measures the relative electrical potential developed between the fluid within the bore hole and the formation, referenced to the fixed potential of an electrode at the surface. Indirectly, S.P. measures permeability, but the direction and magnitude of the electrical potential, and consequently of the deflection of the log trace, are also a function of the electrochemical contrast between bore-hole and formation fluids. In deep wells, where bore-hole fluids are typically less saline than formation waters, the S.P. curve deflects to the left from the base line (indicating negative current flux) within porous, permeable lithologies such as sand. S.P. response is less stable at shallower depths where freshwater aquifers are encountered.

The resistivity trace is a direct or calculated measurement of the resistivity of the rock matrix and its contained pore fluids. Several types of resistivity measurements are commonly recorded on the same log. Because resistivity of sediment or rock matrix is high compared to that of saline or even brackish water, measured resistivity is primarily a function of pore-fluid chemistry rather than of lithology. However, if porosity and permeability are low, as in a tightly cemented or highly compacted, texturally immature lithology, the resistivity curve may register the high matrix resistivity by a deflection to the right from the base line. The resistivity curve may thus be used to determine and measure thickness of sand bodies in fresh water zones or in facies sequences characterized by very low intergranular porosity.

The gamma log measures natural gamma radiation of the subsurface formations. Such radiation is primarily emitted by radiogenic potassium contained in clay minerals. In mixed siliciclastic sequences, the gamma curve can be readily used to distinguish between sand and shale. Further, the degree of the deflection is an index of "shaliness" of the interval. In addition, gamma logs may be particularly useful for recognition and correlation of highly organic marine shales. Such black shales commonly contain anomalous amounts of uranium, making them readily apparent on the highly sensitive oil well gamma log. Problems in use of the gamma log may occur if small amounts of other radioactive materials, such as uraniferous heavy minerals, phosphatic or glauconitic grains, or detrital mica, are present in the sands. Similarly, low gamma counts also characterize relatively pure carbonate units, which might be interbedded with sands and shales.

For quantitative lithologic techniques, such as summation of the net sand within a genetic unit, logs are internally calibrated by defining typical log responses of end-member lithologies within an interval known or assumed to contain thick, clean sand and mud units. Connecting deflections produced by the end-member lithologies define sand and shale baselines that bracket the log trace. Intermediate deflections indicate interbedded or texturally mixed lithologies. An operational definition of sand can be established by adopting a minimal proportion deflection (such as ½ or ⅔) from the shale base line as the cut off point. The addition of core and sample data may allow even more accurate definition of lithology using electric or gamma logs.

Both electrical and gamma logging tools have finite lower limits to their resolution of thin beds. Trace deflections are subdued when bed thickness is less than the minimum. Intervals of thinly interbedded sand and mud, for example, may produce a trace deflection midway between shale and sand base lines. In progradational sequences, in which upward-coarsening is the result of the increasing number and thickness of sand beds relative to interbedded mud, both S.P. and gamma log traces ideally display a progressive deflection from the shale base line to the sand base line, reflecting the increasing proportion of permeable sand and decreasing proportion of radioactive clay. Similarly, vertical change in the textural maturity of a sand may be readily reflected by both log types. However, textural changes exclusively within sand-sized sediment are least likely to be indicated by the logs, although in theory the S.P. log would respond to the changes in permeability that accompany changes in grain size.

A much less common wire-line log, the dipmeter, appears to have potential for subsurface facies interpretation. Though originally designed as a tool to measure structural dip, highly specialized processing of dipmeter readings may produce information on the direction and vertical variation of depositional dip (bedding architecture) within genetic sequences (Selley, 1978b). Application of dipmeter interpretation to depositional system analysis offers a promising area for further research and experimentation.

Interpretation of Log Pattern

Log patterns may be used at three levels of interpretation: (1) determination of vertical sequence and bedding architecture, (2) recognition and mapping of log facies, and (3) interpretation of depositional environment. One of the most obvious and earliest uses of wire-line logs, beyond the determination of basic lithology, is the interpretation of vertical sequence. The characteristic erratic, upward-coarsening, and upward-fining textural patterns of aggradational, progradational, and lateral accretion bedding geometries are readily recognized on electric and gamma logs. The extended electric log segment shown in Figure 2-3 illustrates a succession of progradational upward-coarsening patterns reflected on both S.P. and resistivity curves (interval a), or the S.P. curve only (interval c). Blocky deflections characterize a sand-rich aggradational interval beginning at "d" on the log. A mud-dominated aggradational sequence (e) punctuated by scattered sandstone units, many of which have poorly developed upward-fining tops (suggesting lateral accretion bedding within the sand units), constitutes the upper portion of the logged interval. In addition, an interval of blocky, aggradational sand units (b) is seen to cap the lowermost, thick progradational interval.

Besides recognition of simple vertical sequences, comparison of many logs through a genetic stratigraphic unit will likely lead to recognition of recurrent patterns or log motifs. Such motifs or log facies may be characterized by vertical sequence, scale of units, dominant lithologic composition, or comparative response of different curves on the same or different logs, and their areal, lateral, and vertical distributions may be examined for systematic patterns. For example, in Figure 2-3, the lower half of the stratigraphic succession is seen to consist dominantly of a series of thick, simple-to-compound progradational units. The upper half of the succession consists of a lower aggradational sandy facies, a middle thick aggradational or nondescript muddy unit, and an upper mixed aggradational sand and mud sequence. Overall, the sequence suggests a systematic change upward from progradational to aggradational depositional architecture, accompanied by improved porosity and permeability (decreasing resistivity and increasing S.P. deflection in sand units) and increasing lithologic homogeneity (indicated by the increasingly less serrate character of the curves).

Beyond simple interpretation of vertical sequence and bedding style and descriptive classification of log facies, log patterns can be correlated directly with specific depositional facies (Jageler and Matuszak, 1972). Such interpretation attempts to make use of the fact that logs convey much information about vertical textural changes, degree of bedding, thickness, and nature of sand body contacts with underlying strata. These attributes, in turn, reflect the processes active during sand body deposition. Representative

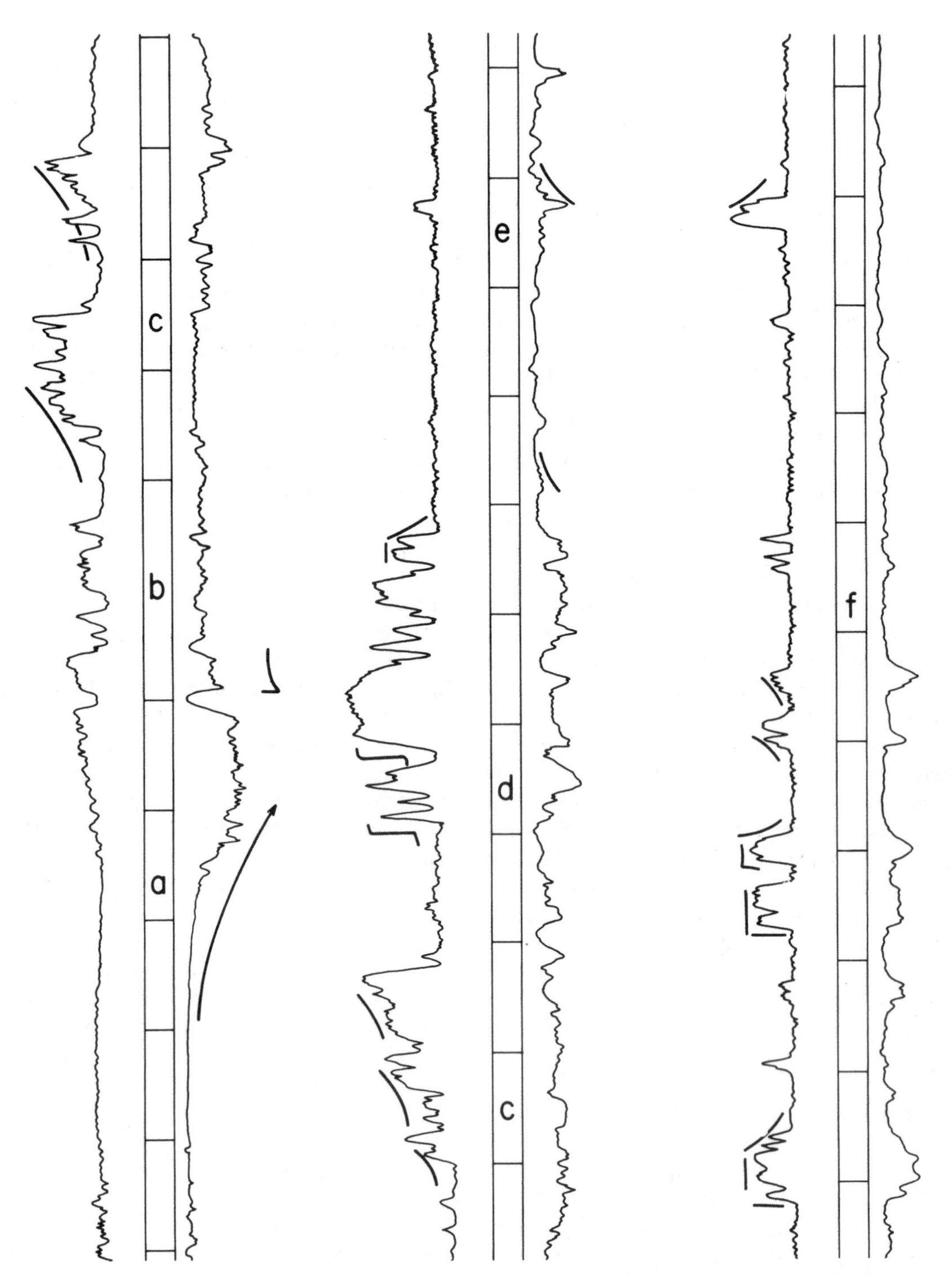

Figure 2-3. Extended electric log segment through the Frio Formation (Oligocene–Miocene, Texas Gulf Coast basin) containing a systematic vertical distribution of progradational, upward-coarsening units (intervals a and c) and aggradational units (intervals d and f). Individual sand bodies in the aggradational portion of the sequence may display upward-fining sequences (interval e).

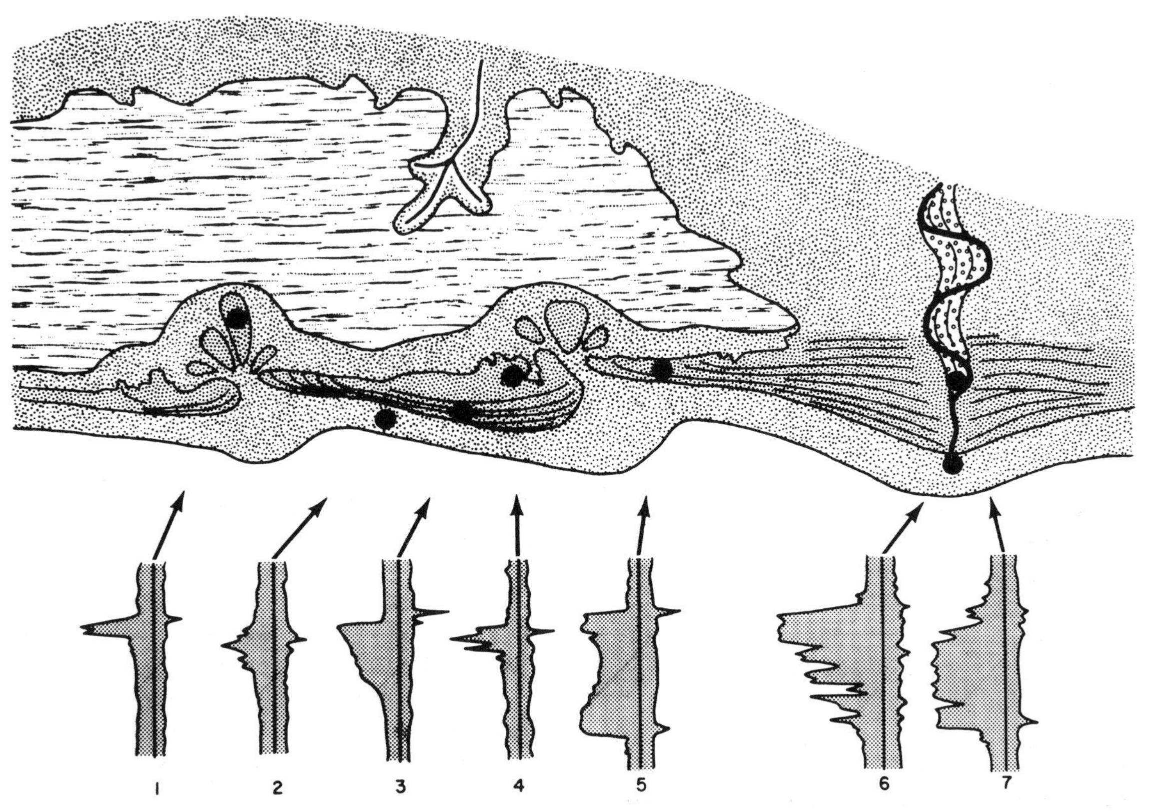

Figure 2-4. Idealized correspondence between log-pattern facies and specific depositional environments in a suite of coastal sand bodies. Thickness, nature of the vertical boundaries, and degree of internal bedding reflect processes active in each environment and are displayed by the log traces. Suggested environments include: (1) flood-tidal delta, (2) distal shoreface, (3) barrier core, (4) washover fan, (5) inlet fill, (6) channel mouth bar, and (7) meanderbelt.

examples of interpretation and delineation of depositional facies on the basis of log patterns are reviewed by Saitta and Visher (1968), Fisher *et al.* (1969), and Kruger (1968). A unique, enlightening suite of S.P. logs recorded in shallow drill holes in Holocene coastal plain environments is given by Bernard *et al.* (1970).

A log trace is, however, at best a generalized and incomplete description of facies characteristics. Genetic interpretation of log facies is most successful where applied within the context of a known depositional system, and improves as the details of log signature are calibrated with specific lithological attributes. Figure 2-4 illustrates an idealized example in which contemporaneous coastal environments, including fluvial, deltaic, barrier-bar, and tidal-inlet settings, are each characterized by a unique log facies. Differences include vertical sequence, thickness, and degree of bedding within the sand body. Vagaries of deposition and preservation preclude such idealized fidelity in real log shapes. However, examination of many logs, combined with emphasis on areal associations and recurrent patterns, could lead to considerable sophistication in facies interpretation within such a facies sequence—provided the depositional system, and consequently, the suite of likely sand bodies and their three-dimensional relations, are already known from regional study.

Subsequent chapters contain examples of vertical sequences and generalized S.P. or gamma log profiles characteristic of a variety of sand facies. In addition, many subsurface cross sections illustrate the use of wire-line log response in detailed as well as regional facies interpretation. Comparison of the example reveals both the potential of log facies interpretation as well as the limitations of the tool to uniquely identify depositional environments without supporting information on the regional depositional setting and calibration of log responses within individual depositional systems.

Seismic Stratigraphic Analysis

Reflection seismology provides a powerful tool for delineation of the depositional architecture of basin fills. Reflection patterns outline genetic sequences of conformable strata bounded by regional unconformities or marked changes in bedding style (Mitchum *et al.*, 1977; Vail *et al.*, 1977). Seismic facies analysis delineates internal features and bedding geometries within such major stratigraphic units (Sangree and Widmier, 1977; Brown and Fisher, 1980). Analysis of individual wave forms contained on record sections through manipulation of synthetic seismograms or seismic modeling may detail the geometry of individual lithologic units or define facies transitions (Meckel and Nath, 1977; Galloway *et al.*, 1977). In addition, seismically derived interval velocities of successive increments of the subsurface section can be used to determine gross lithology.

Acoustic waves reflect at interfaces between media with contrasting impedance, which is in turn a function of velocity and density. Both the rock matrix and pore fluid composition affect the impedance of earth materials. Consequently, lithologic boundaries are commonly reflective interfaces, and on a local scale, reflection patterns are indicative of facies variations. Regionally primary seismic reflections tend to follow time-stratigraphic horizons, such as regional bedding surfaces or disconformities, and are useful for defining bed architecture and boundaries of depositional episodes (Vail *et al.*, 1977).

Figure 2-5 shows a segment of a seismic field record section containing well-defined progradational, or offlapping reflectors bracketed by parallel, horizontal events. Above the field section is an interpretation of the equivalent stratigraphy, translated into acoustic stratigraphic units, and showing inferred lateral relationships of major rock units. The interpretative section was based on scattered well control in the general area of the section. Sonic logs provided the interval velocity data for major sandstone, shale, and limestone units that were used to convert the lithostratigraphic section into the acoustic stratigraphic model. The velocity model was then processed with a computer seismic simulation program, and the depth scale converted to a two-way sonic travel time scale. The computer plot displays the impulse response (in which the spikes show magnitude and polarity of the acoustic impedance contrast between adjacent beds) and a synthetic seismic record section that reproduces with reasonable fidelity the distribution, polarity, and trace spacing of each of the numbered seismic events. Comparison of the model section with the field record shows that the original inferred stratigraphic model is a reasonable interpretation of the subsurface bed configuration. In addition, the seismic model shows that some traces may indicate details of bed thickness by variations in wave-form amplitude. In this example, the combination of geologic information and seismic data allowed development and testing of a stratigraphic correlation between equivalent basin and shelf depositional sequences in a cratonic basin.

By providing information in the form of continuous cross sections or three-dimensional grids, reflection seismic records offer a powerful tool for recognition of genetic sequences, determination of their internal composition and bedding geometry, and delineation of the distribution of framework sand units. Continuous improvements in recoverable bandwidth, and thus resolution, combined with new processing techniques and display formats, presage an increasing role for seismic stratigraphic studies in energy resource geology.

Recognition of Depositional Systems—An Example

Integration of various subsurface methods in regional genetic stratigraphic analysis can be illustrated by a case history (Galloway *et al.*, 1982, 1983). The Frio Formation (Oligocene through Early Miocene) comprises a major wedge of sediment deposited as part of the Tertiary fill of the Gulf Coast Basin. The Frio depocenter is centered in the Rio Grande Trough, a broad, subtle structural downwarp that intersects the Gulf Coast Basin in South Texas.

Subsurface correlation of electrical logs and seismic lines along dip profiles shows that the Frio is a progradational or offlapping sequence of interbedded sandstone and mudstone. Regionally,

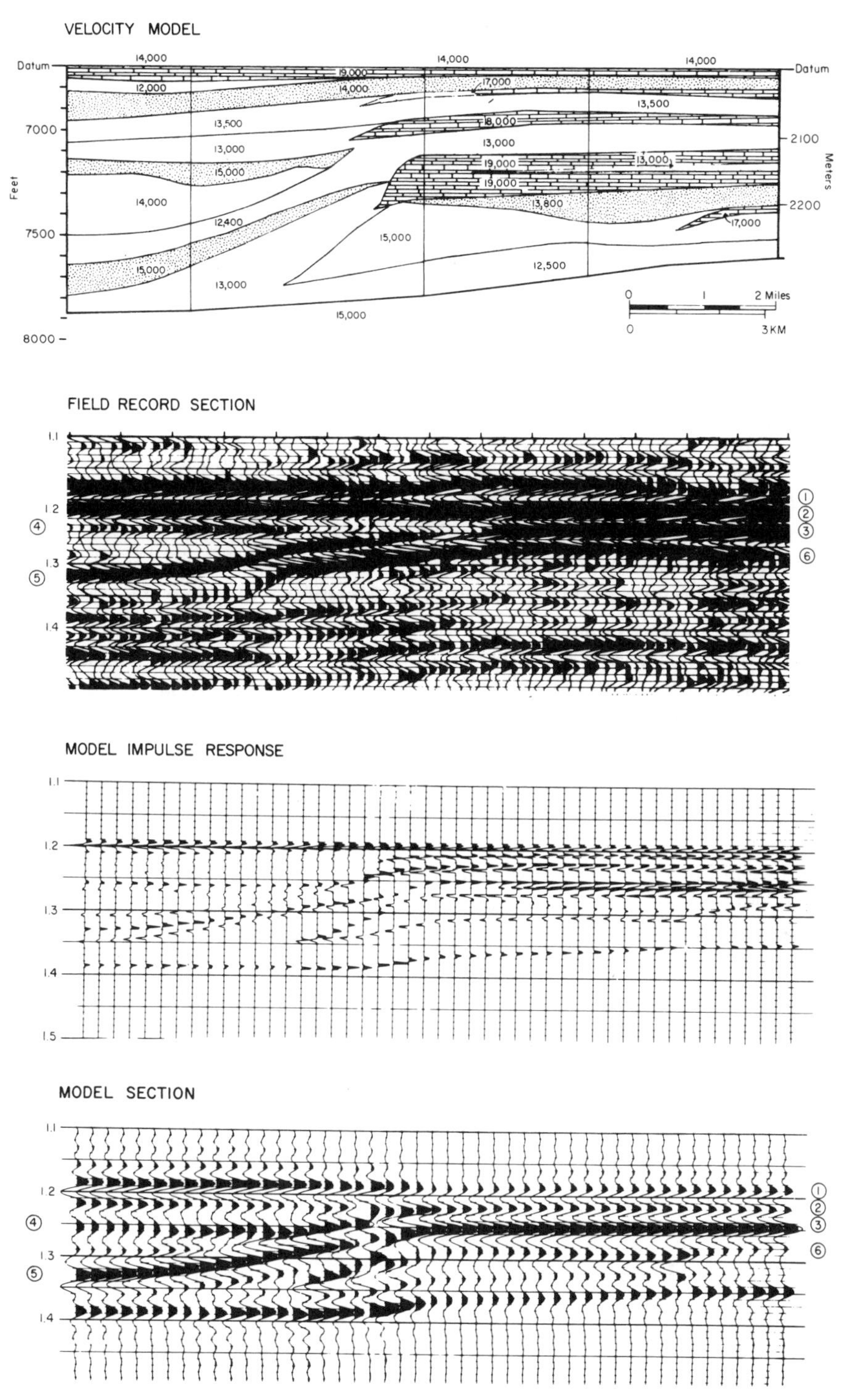

Figure 2-5. Interpreted acoustic stratigraphic, or velocity model of a shelf edge (Pennsylvanian of the Anadarko Basin, Oklahoma), an equivalent segment of a seismic field-record section, the impulse–response plot of the velocity model shown in the time domain, and a computer generated synthetic seismogram of the model. Comparison of the number, amplitude, spacing, and geometry of the model traces with the field record suggests the validity of the stratigraphic interpretation. (From Galloway *et al.*, 1977.)

the Frio is bounded by transgressive marine mudstones, and constitutes one of the major depositional episodes of the Gulf Coast Basin. The simple pattern of progradational basin-filling is interrupted by a succession of large, strike-parallel growth fault zones, which produce a series of depositional subbasins (Galloway *et al.*, 1983). Lithostratigraphic and paleontologic correlation within the Frio depositional episode provides a basis for subdivision of the total sequence, which may exceed 10,000 ft (3,000 m) in thickness, into operational map units.

Vertical persistence of facies over hundreds to thousands of feet, which is a common attribute of depositional sequences in rapidly subsiding basins, is apparent in wire-line logs penetrating the Frio section (see Fig. 2-3). This vertical persistence suggests that facies maps for the thick operational units, though incorporating genetic units of many successive depositional events, will reveal basic patterns and trends of framework sand distribution. In fact, the combination of net sandstone and percentage sandstone maps (Figs. 2-6 and 2-7) provides considerable information about depositional style. Total sandstone within the lower Frio operational unit (Fig. 2-6) shows an irregular, lobate sandstone depocenter centered along the gulfward margin of the Frio wedge (near the present shoreline). The depocenter is broken into a number of local sandstone thicks and intervening thins, reflecting the concentration of sand in fault-bounded subbasins. Updip, sandstone distribution is defined by a broad belt dominated by dip-oriented contours, which is centered around the axis of the Rio Grande Trough. To the northeast along depositional and structural strike, sandstone is concentrated within a relatively narrow strike-parallel belt that is bounded both basinward and landward by a mudstone-dominated Frio section. The strike alignment of the contours is offset across faults, indicating contemporaneous structural subsidence and deposition. Near the outcrop belt the Frio is thin, and regional subdivision is impractical. Here, a map displayed total sandstone contained within the entire Frio section. Interweaving, dip-oriented patterns dominate.

Electrical logs used in preparation of the correlation sections and quantitative facies maps exhibit recurrent curve patterns or motifs, which can be grouped either as progradational, aggradational, or mixed types (Fig. 2-8). Based on unit thickness, relative sand content, and suggested degree of bed uniformity, eleven log facies were identified. Using representative log segments for guidance, logs through each operational unit were categorized as to dominant log facies, and the information transferred to cross-section and maps (Fig. 2-9). Cross sections showed that an upward succession from progradational to aggradational sequences, similar to that shown in Figure 2-3, is typical. In map view, the Frio depocenter displays a systematic basinward change from aggradational to mixed, and finally to progradational log facies assemblages. To the northeast, sand-rich progradational patterns grade laterally into an area of sand-rich aggradational patterns. Thick, sand-poor progradational sequences front the strike-oriented sandstone belt; aggradational patterns lie updip.

Examined together, the three maps and supporting cross sections: (1) both distinguish and outline regional depositional elements of the Frio and suggest the probable trends of their framework sand facies, (2) document systematic vertical and lateral patterns in depositional style consistent with the ideal framework of a depositional episode, (3) demonstrate relationships between lithologic composition and contemporaneous structural features, and (4) suggest consistent lateral and vertical facies changes that reflect depositional processes.

These interpretations, combined with knowledge of common depositional styles and component systems of the Gulf Coast Basin, are the basis for the interpretation of a large Frio delta system and updip fluvial system centered along the Rio Grande Trough. Further, the interdeltaic strandline to the northeast is logically interpreted as a barrier bar system, bounded updip by lagoonal and coastal plain facies, and downdip by marine shelf and slope facies (Galloway *et al.*, 1983). Detailed local studies, using maps of genetic sand units as well as core examination, are consistent with, and add detail to, the regional interpretations.

Facies Models—Reference Points

Within the framework of a depositional system, more detailed facies interpretation commonly relies to a greater or lesser extent on the use of sedimentary models. Models may be largely

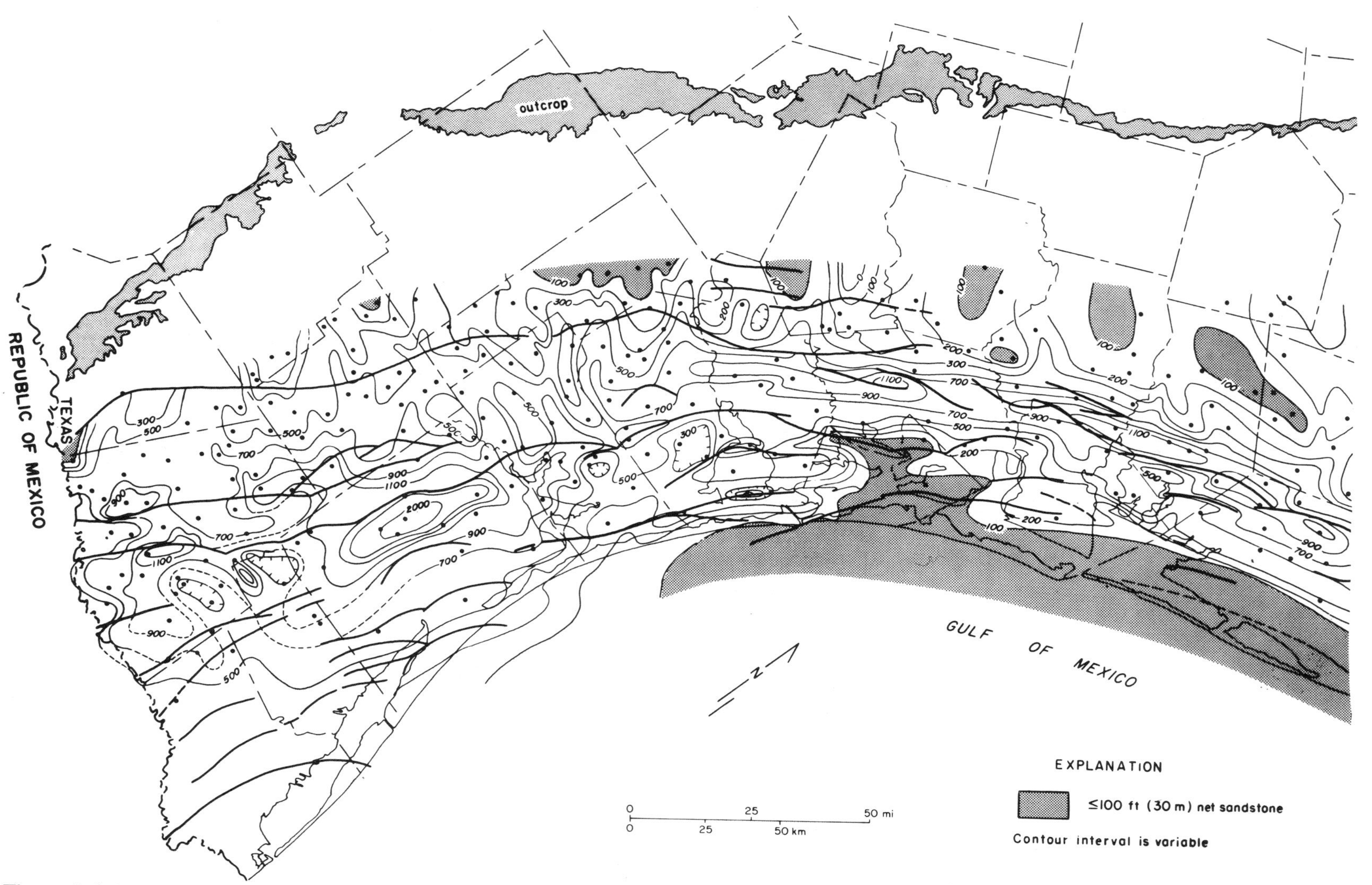

Figure 2-6. Net sandstone isopach map of the lower portion of the Frio Formation, the Rio Grande Trough area, northern Gulf Coast Basin. Solid heavy lines outline major growth faults. Density of control is indicated by well spots. Contour interval is in feet. (From Galloway *et al.*, 1982b.)

Figure 2-7. Sandstone percentage of map of the lower portion of the Frio Formation, Rio Grande Trough area. Three distinct sand-rich provinces are evidenced by sand content and contour trends. (From Galloway *et al.*, 1982b.)

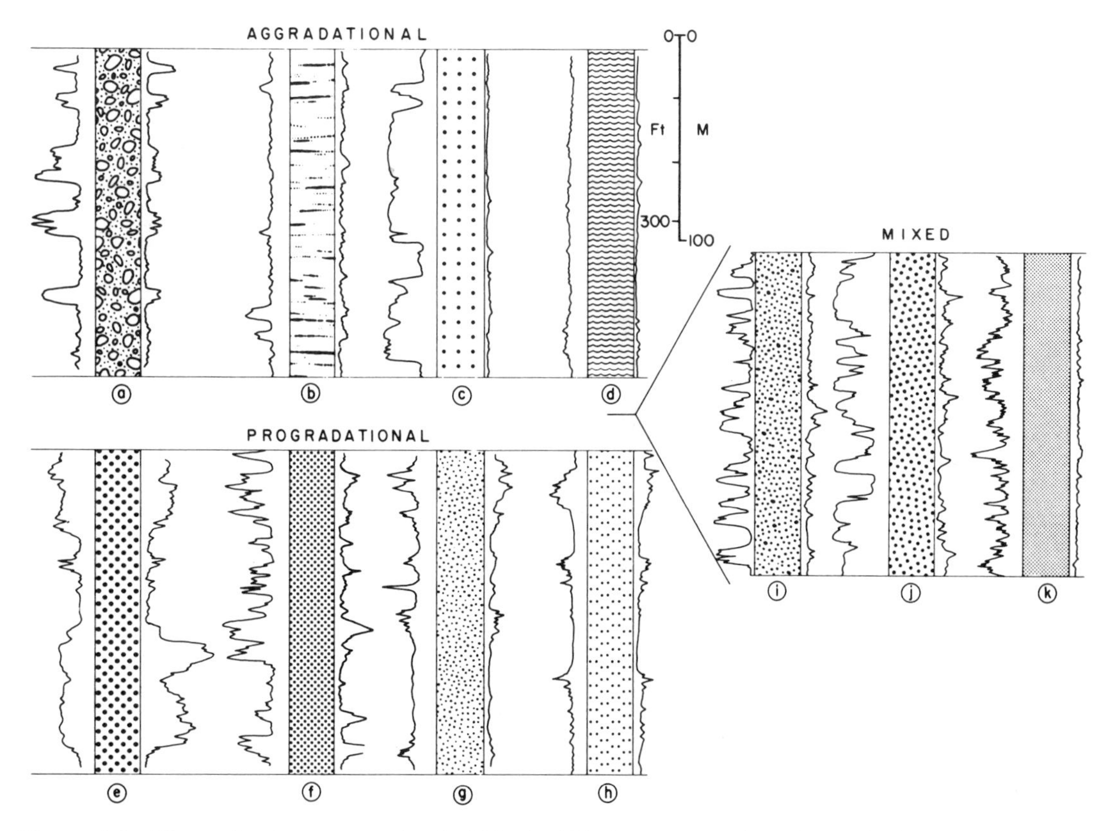

Figure 2-8. Electric log facies recognized within the Frio Formation of the Texas Gulf Coast Basin. (From Galloway *et al.*, 1982b.)

descriptive, attempting to generalize the physical attributes of the deposits of a particular depositional environment. Although appealing because of the relative simplicity of using idealized standards for comparison, physical models have real limitations in their potential for application to genetic stratigraphic problems. Few if any environments define a unique interaction between processes, changing base levels, and available sediment. Thus, great variation in the characteristics of the resultant facies is the rule rather than the exception. Likewise, systematic surveys of modern depositional basins show that many environmental settings produce facies potentially capable of serving as reservoirs, conduits, or hosts for mineral fuel resources. Thus, the fluvial channel model of the early 1960s (Visher, 1965) evolved into a braided channel, a distributary channel, and a point bar model. Currently, six or more models for braided streams alone have been suggested (Miall, 1978). The alternative approach, and the one emphasized in subsequent chapters, relies on development of process–response models. Fundamental to this approach is the qualitative understanding of processes responsible for the erosion, transportation, and deposition of sediment, and of the small- and large-scale features that record those processes. Process–response models result from the examination of the range of process inherent in each depositional environment and the range of products that may result. Environmental reconstruction is thus based on process interpretation in the context of the three-dimensional geometry and stratigraphic context of genetic facies.

The Record of Depositional Process

Process–response models of depositional environments and resultant deposits are founded on several key sedimentologic concepts. Although well known, they are worth restatement, both as

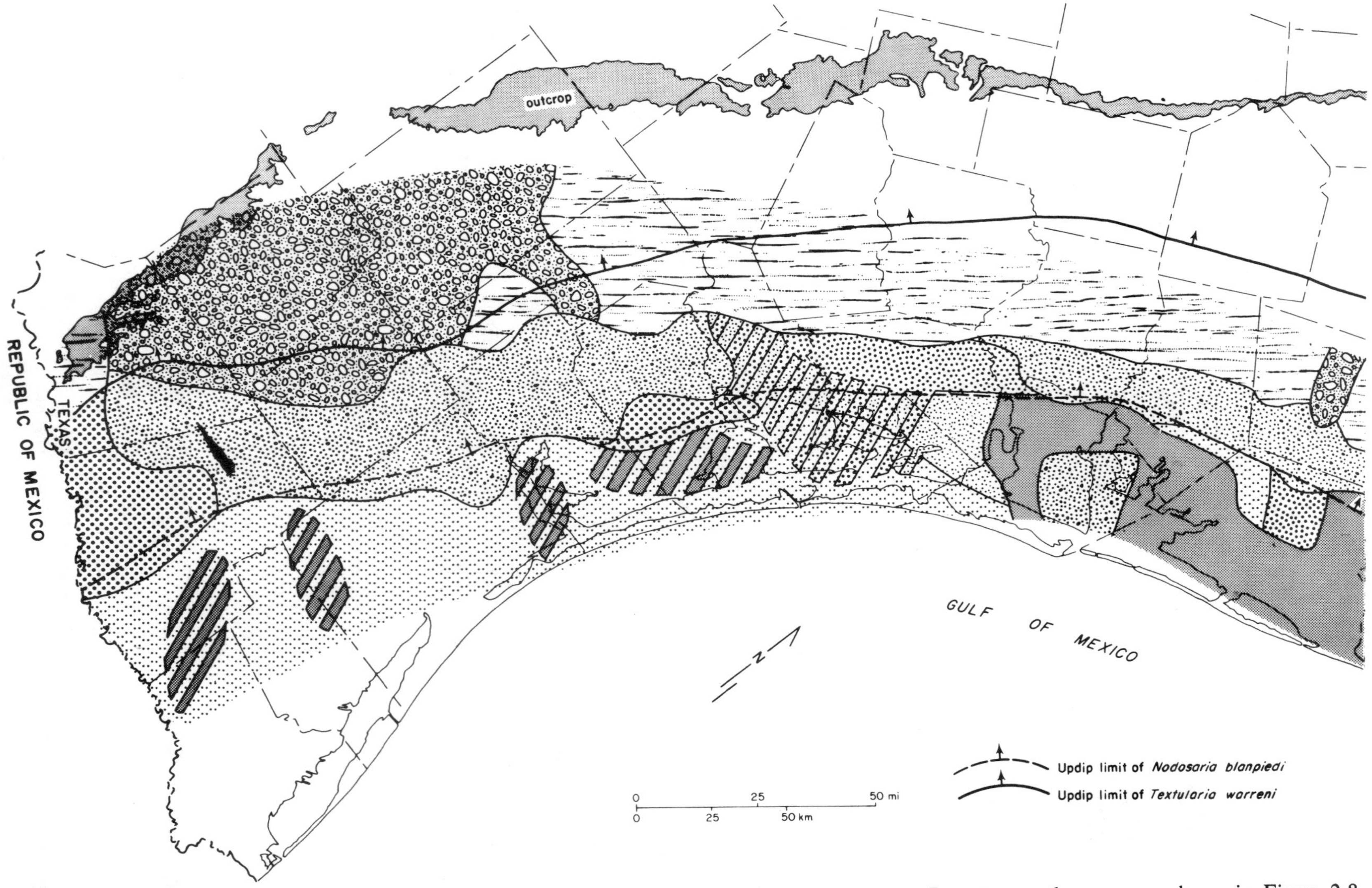

Figure 2-9. Electric log facies map of the lower portion of the Frio Formation, Rio Grande Trough area. Patterns are the same as shown in Figure 2-8. Aggradational log facies grade systematically basinward into progradational log facies. (Modified from Galloway *et al.*, 1982b.)

review and for clarification of the oftimes confusing uses of terminology.

Erosion, transport, sorting, and deposition of sediment are accomplished by several process agents, including:

Subaerial agents

(1) Gravitational potential energy
(2) Confined fluid flow
(3) Unconfined fluid flow (sheetflow; wind)

Reservoir agents

(1) Wave energy flux
(2) Tidal energy flux
(3) Permanent and barometric currents
(4) Gravitational potential energy

Differing depositional environments, or systems, are characterized by the interplay or dominance of these process agents, producing, in turn, specific erosional or depositional features.

In environments characterized by active fluid flow, significant sorting of an initially heterogeneous sediment mix occurs. Clay and sand are efficiently separated and deposited at different sites. Two sorting mechanisms operate in most depositional settings.

1. The ability of moving water to erode loose sediment is a direct function of current velocity, and is commonly greater than the velocity required to transport sediment of the same grain size. This velocity inequality was expressed conceptually by Hjulstrom in 1939. For fine sediments, such as clay and silt, the difference between the velocity necessary to erode a particle and that necessary to transport it once it is moving is more than an order of magnitude. Sand, on the other hand, responds readily to changing flow conditions and can be washed through the system in the course of several cycles of erosion, transport, and deposition.

2. The vertical concentration gradient of sediment within flowing water is dependent on grain size. Although any dividing line is arbitrary, sediment load is conventionally subdivided into bed load and suspended load. Bed-load sediment, which in normal-flow conditions includes sediment coarser than very fine sand, moves near the base of the channel. Suspended-load sediment (very fine sand, silt, and clay) is more uniformly distributed throughout the water column. The significance of this distinction is twofold. First, any process that either separates the moving water mass from the channel bottom, or bleeds off only a portion of the total water column, will separate bed load and suspended load. For example, if flow separation occurs at the channel base, sand-sized sediment is rapidly deposited but suspended load remains within the moving water column. Second, as flow velocity wanes, fine sediment, which is mixed throughout the water column and has very low settling velocities, may be transported a considerable distance before deposition. The bed-load sediment, which is moving near the bed and has a high settling velocity, will be quickly deposited before most of the suspended load settles out. Thus settling lag, combined with waning flow, can efficiently separate suspended-load and bed-load sediment.

Primary and biogenic sedimentary structures are inherent attributes of most clastic sedimentary sequences. Although their application in construction of process models is limited to situations where rock material is available, they provide a wealth of information on depositional processes, rates, and transport directions. Thorough reviews of structures and their interpretation are contained in Harms and Fahnestock (1965), Harms *et al.* (1975), and Potter and Pettijohn (1977).

Moving fluids interact with loose bed-load sediment, producing a predictable hierarchy of bedforms. These bedforms provide a record of sediment transport under varying conditions of depth, velocity, and turbulence, which is preserved in the form of familiar sedimentary structures.

The qualitative dynamics of fluid flow is expressed by the concept of flow regime. Upper flow regime occurs when water velocity is relatively high or the flow depth is relatively shallow. Fluid and sediment particle motion, as well as bedforms, evolve with increasing flow regime. In a shallow, medium-grained sand-bed channel the following sequence of bedforms appears as flow velocity increases:

Ripples
Sand waves (dunes with superimposed ripples)
Dunes
Washed-out dunes
Plane bed
Standing waves and antidunes

The sequence varies only if coarser or finer grain sizes dominate. Lower flow regime bedforms in silt consist only of ripples and plane bed. In contrast, ripples are absent from the lower flow regime sequence that evolves on a coarse sand bed. Migration of bed forms constitutes a mechanism of transport for bedload sediment, and generates internal sedimentary structures which, in turn, provide a partial record of the flow conditions at the time of deposition. Principal types of sedimentary structures and their parent bedforms include:

1. Ripple lamination (small-scale trough cross stratification) produced by the migrating ripples. Ripple structure is highly variable, depending on relative rates of deposition and bedform migration (climbing ripples), grain size, and variability of flow (oscillation ripples).

2. Tabular cross stratification, characterized by uniform bed sets and high dip angles, produced by migrating sand waves.

3. Large-scale trough cross stratification, showing spoon-shaped bed sets and high- to low-angle tangential foresets, formed by migrating dunes.

4. Horizontal stratification and parting lineation reflecting development of a plane bed at the transition from lower to upper flow regime.

Cross stratification produced by migration of antidunes during upper-flow regime is rarely preserved because receding flow reworks the earlier-formed structures. Larger-scale cross stratification, including steep avalanche foresets produced when migrating dunes spill over the top of an advancing bar, and accretionary cross stratification produced by down-current bar migration are indicative of development of larger, more permanent depositional features.

The proportion of the complete bedform preserved as a sedimentary structure provides an index of the rate of sedimentation relative to rate of bedform migration. For example, preservation of the truncated toes of cross-bed sets indicates dominance of sediment transport through the system. In contrast, more complete preservation of bedforms indicates rapid sedimentation rates. Ripple lamination, which reflects relatively tranquil, lower flow regime conditions, commonly displays ripple drift or climbing-ripple cross lamination. Increasing angle of climb, and concomitant increasing preservation of the ripple form is a distinctive product of waning velocity of a sediment-laden current in a variety of environments.

Once formed, primary structures are subject to partial or complete destruction by burrowing organisms or root penetration. Lateral or vertical variation in density of biogenic structures within a genetic sequence may provide information on comparative rates of sediment accumulation and intensity of reworking by physical processes.

Few, if any, primary structures are unique to a specific depositional environment. Rather, the structural suite is a response to the spatial and temporal evolution of depositional processes. For example, a common process sequence in many environments consists of: (1) a rapid rise in fluid flow velocity and associated turbulence, followed by (2) a period of transitional or upper regime flow, then by (3) an extended period of waning flow and ultimate stagnation, and culminating in (4) an extended interval of depositional inactivity. The depositional record of such an event, which might be a flood or storm, would include a basal scour surface reflecting increasing velocity and turbulence, and consequent erosion overlain by an interval of planar stratification, capped in turn by an interval of ripple cross stratification grading up into climbing-ripple stratification and covered by a mud drape. Subsequent reworking of the depositional unit might include sorting and reworking of the upper surface by oscillatory currents, producing wave ripples, or reworking by burrowing organisms or root penetration.

Base-Level Change and Vertical Sequence

Most process-response models imply ranges for both vertical and lateral facies dimensions. For example, concentration of sand along a marine shoreline would likely be limited to the interval between the high-tide beach and active wave base. This interval varies from a few feet in low-energy coasts to typical values of a few tens of feet on moderate energy coasts, to as much as 100 ft (30 m) on very high energy coasts. Simple progradation of the shoreline into a basin containing several hundred feet of water would produce sand bodies of comparable thickness. However, as discussed in detail by Matthews (1974), thickness of depositional sequences, as well as vertical superposition of successive facies, is affected by dynamic changes in base level inherent to all depositional basins of even moderate size.

External factors that determine the evolution of base level are: (1) isostatic subsidence or uplift, (2) tectonic deformation, and (3) eustatic sea level (or local lake level) changes. Isostatic subsidence is induced whenever an increment of sediment is added on a subaqueous or subaerial depositional surface. Documented subsidence rates exceed 10^1 m/1000 years. Tectonic deformation induced by crustal shortening, extension, loading, or cooling have demonstrated rates ranging from 10^{-1} to 10^1 m/1000 years (Matthews, 1974). Eustatic changes in sea level during the Pleistocene occurred at rates of 10^1 m/1000 years. Such rates approximate or exceed all but the most rapid rates of sediment accumulation within depositional systems. For example, the Holocene Mississippi delta system has deposited sediment at rates of 10^1–10^2 m/1000 years.

Thus, isostatic and tectonic subsidence as well as eustatic base level rise are capable of measurably expanding the thickness of a single depositional sequence produced in most environments. Isostatic subsidence is, of course, self limiting, and rates are proportional to sediment input. Tectonic or eustatic rises in base level may exceed rates of sediment accumulation and lead to shoreline retrogradation and inverted facies successions.

Integrated Genetic Stratigraphy

The tremendous variability of basin settings and stratigraphic sequences and of type and completeness of data bases precludes the use of a single approach to facies analysis. However, our thesis is that a hierarchical approach that first defines regional attributes of genetic sequences before attempting site-specific interpretation affords the greatest potential for successful interpretation of depositional sequences and application of that interpretation to resource discovery and development.

Thus, systematic stratigraphic analysis should describe and interpret several aspects of basin geology, including:

1. Framework of the depositional basin, including its size, shape, probable range of water depths, and major sources of sediment.
2. Principal genetic stratigraphic units or depositional episodes, and analysis of their depositional architecture. Both regional well log or seismic sections are particularly useful.
3. The three-dimensional geometry of framework facies outlining major depositional elements or systems and their component facies. Isolith and sandstone percent and various quantitative facies maps, cross sections, and fence diagrams are all basic tools at this level of analysis. Regional facies distribution may be determined by seismic facies mapping and wave-form modeling.
4. Recurrent vertical sequences, both within the larger genetic units as well as within specific facies. Wire-line logs are an especially effective tool for revealing vertical changes in lithology or bed thickness.
5. Lateral and vertical facies associations and cross-sectional geometry. Representative closely spaced logs and measured sections can be selected for detailed correlation and mapping.
6. Internal bedding and sedimentary structures, texture, and bedding features of main facies. This final step requires outcrop or core material. In subsurface settings the limited amounts of core typically available may be used to supplement sample logs in calibrating wire-line log response so that detailed facies interpretations can be extrapolated using log data.

The more completely each level of description and interpretation can be concluded, the more likely a unique, predictive interpretation can be made at the next-lower level.

Chapter 3

Alluvial-Fan Systems

Introduction

Alluvial fans are conical, lobate, or arcuate accumulations of predominantly coarse-grained clastics extending from a mountain front or escarpment across an adjacent lowland. They represent the coarsest, most poorly sorted, proximal unit in the range of subaerial depositional systems, and commonly merge downdip into finer grained, lower gradient fluvial systems. Some fans, however, terminate directly in lakes or ocean basins as fan deltas, which generally show some degree of distal modification by currents or waves (Fig. 3-1).

Fans form in response to an abrupt reduction in gradient and stream competence, commonly accompanied by gravitational instability. Consequently, fans are most numerous in tectonically active areas, such as rift- and strike-slip-related pull-apart basins, block-faulted terrane, foreland basins, and along the margins of foredeeps. Many such basins undergo prodigious aggradation (Steel *et al.*, 1977b). Fan growth may follow climatic changes, for example, the rapid liberation of moraine by ice wasting. These meltwater fans are most active seasonally, as are those affected by tropical storms. In some cases, sedimentation is controlled by the frequency of major volcanic eruptions, with the ash being redistributed during the next rainy season (Kuenzi *et al.*, 1979). Other causes of fan development may involve destruction of vegetation on steep slopes and lowering of base level due to eustatic sea-level changes (McGowen, 1970) or falling water levels in closed lacustrine basins (McGowen *et al.*, 1979).

Although commonly associated with arid or semiarid regions, fans are not restricted to particular climatic zones. Climate does, however, account for the basic distinction between the steep, relatively small fans of arid regions and the low-gradient, commonly larger fans of humid regions. Other factors influencing fan size are drainage basin area, discharge, relief, and source-rock lithology. Contrasting depositional styles in climatically similar regions may bring about significant differences in alluvial-fan size and geometry (Gloppen and Steel, 1981). Fans formed from erosion of shale and mudstone terrane tend to be significantly larger than those derived from a sandstone source (Bull, 1962). Fan size is modified by loss of sediment from the fan terminus. This is particularly true of fan deltas subject to high-energy, shore-zone sediment dispersion, but some small interior fans are truncated by axial trunk streams flowing at right angles to the fan axis. Fan area is significantly reduced by subsidence and encroachment of shallow lakes and marshes (Schramm, 1981).

Adjacent or opposed fans in narrow basins may be separated by playas and dunefields. Lateral coalescence of fans along a mountain front produces bajadas, common features of semiarid landscapes; these may grade into pediments of denudational origin.

Much of the world's gold production is from placers in ancient fan systems of the Witwatersrand Supergroup, South Africa (Pretorius, 1976; Tankard *et al.*, 1982). The Witwatersrand and related basins also contain large quantities of placer uranium, which was more stable in the Early Proterozoic atmosphere. The geometry, high ground-water transmissivity, and geochemical and permeability gradients of wet alluvial-fan systems make them one of the most favorable hosts for epigenetic uranium deposits (Galloway *et al.*, 1979). The marine-reworked fringes of fan deltas are excellent oil and gas reservoirs, with reversal of depositional dip by basin-margin downfaulting or folding providing stratigraphic pinchout and structural traps. Although many alluvial-fan deposits are oxidized (Bull, 1972), vast coal resources are associated with Carboniferous and younger systems.

Processes Acting on Alluvial Fans

Given the prerequisite active tectonic setting, two sedimentary processes dominate in response to the steep gradients and abundant coarse clastic

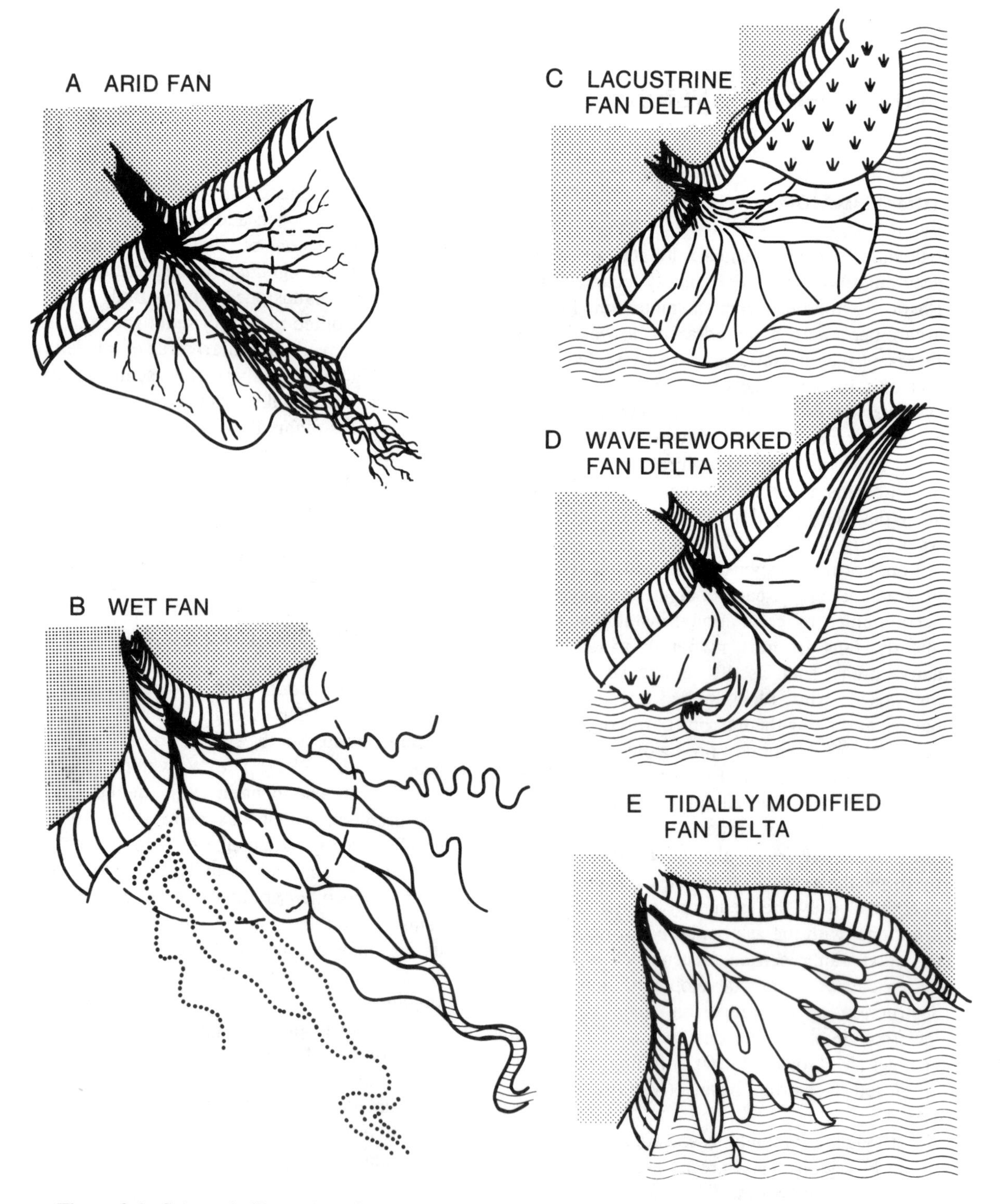

Figure 3-1. Schematic illustration of various fan and fan-delta systems reflecting differences in climate and degree of distal modification.

sediments. These are stream flow and debris flow, which vary in proportion according to factors such as sediment texture and climate. There is nothing unique about the processes acting on fans; it is the fan geometry that is distinctive (Miall, 1977). Some wet alluvial fans, ranging from the giant Kosi Fan (Fig. 3-2) of India and Nepal which spans 150 km (100 mi) from apex to toe, to the small tropical Honduras fans (Schramm, 1981), are a product of stream processes alone. As such, they are synonymous with Rust's (1978) category of braided alluvial plains. At the other end of the spectrum are fans produced almost entirely by debris flow. These fans are formed preferentially in semiarid areas, or in less-arid regions with abundant clays, for example, from glacial or volcaniclastic sources. Most fans, however, reflect a combination of debris-flow and stream processes. Subordinate processes, such as illuviation, infiltration, and sieve deposition, operate on gravel fans.

Stream Flow

Stream flow dominates fans with an effective vegetation cover and year-round rainfall which gives rise to perennial streams. In areas of exceptionally coarse sediment, stream transport may be confined to major flood events associated with tropical storms occurring at intervals as long as 20–25 years (Schramm, 1981, p. 135). Elsewhere, for example, on the Kosi Fan, stream reworking is continuous as one or two perennial channels shift sideways across the fan. This results from aggradation of the upstream channel floors causing diversion of flow in a consistent lateral direction (Miall, 1977). The main Kosi

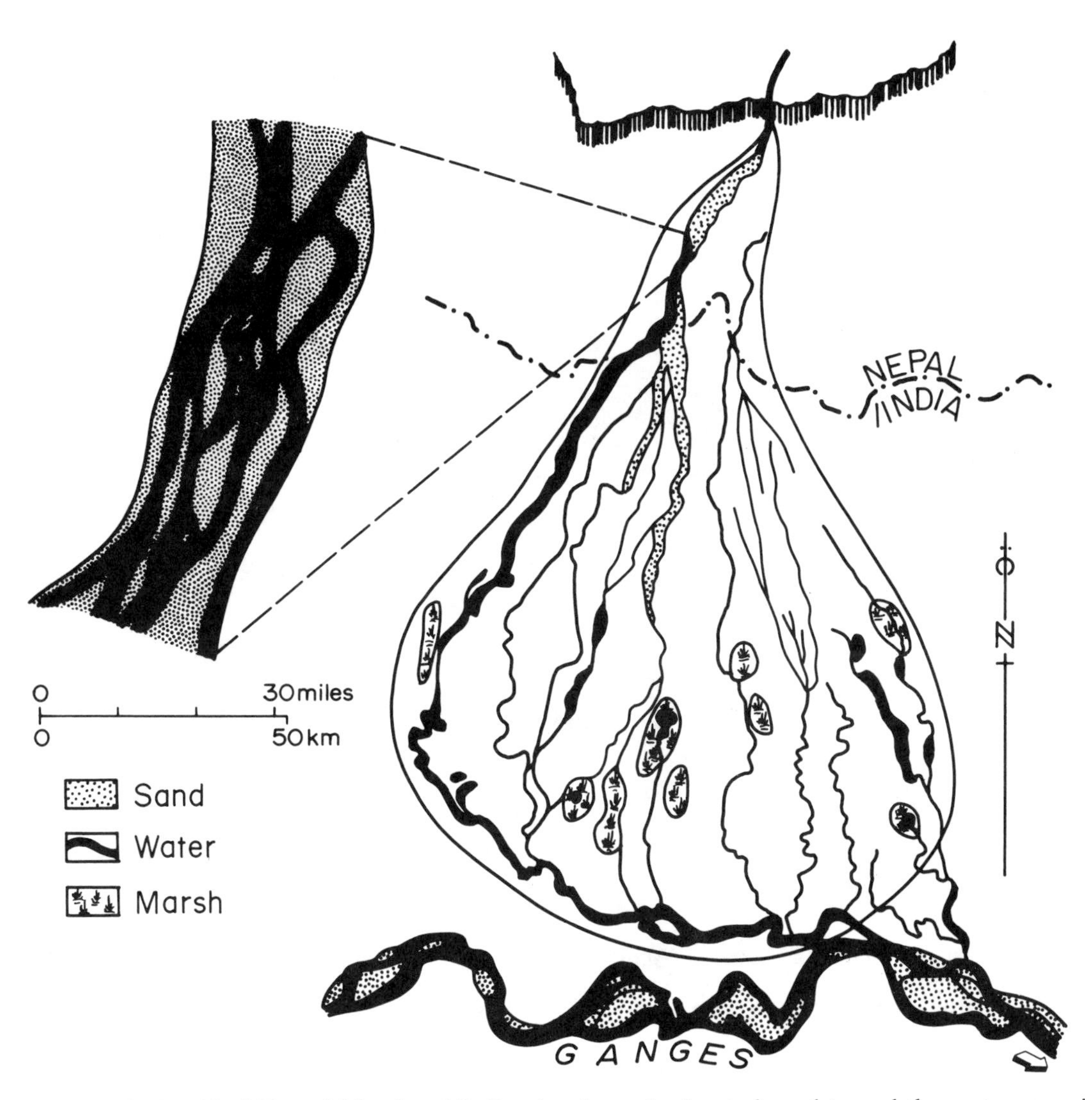

Figure 3-2. The Kosi Fan of Nepal and India, showing a dominant channel toward the western margin, enlarged. (Modified after Seni, 1980.)

channel has moved 112 km (70 mi) westward in 228 years.

Flow, therefore, may be confined to channels, as on the Kosi Fan, or may spread across the entire surface as sheet-flow, as during the Honduras floods. Most fan surfaces are only partly inundated even during flood, but over a geological timespan all parts of these fan surfaces are subject to stream flow (Rust, 1978). Some of the main channels on a small fan delta studied by McGowen (1970) are erosional features formed only during waning flood stages.

Streams emerging onto a fan from a mountain canyon are incised, but bifurcate and assume radiating braided patterns, with a few major channels dominating over smaller ephemeral channels. Stream competency is rapidly reduced by percolation into the gravelly deposits of older fans, and by a flattening gradient. Upper-fan channels may be a feature of low-discharge stage only, flowing between subdued, diamond-shaped sheet bars (Fig. 3-3). Bed relief increases downstream as the bars become higher. When flood depths are substantial, for example, as a result of glacier-burst flooding, flow separation on the gravel bars produces well-defined gravel foresets (Boothroyd and Ashley, 1975; Boothroyd and Nummedal, 1978). When flow is more evenly distributed, there is less tendency for foresets to form.

Along with a predictable pattern of downstream grain-size reduction, there is a change from longitudinal to transverse or linguoid bars on the sandy distal fan (Smith, 1974; Boothroyd and

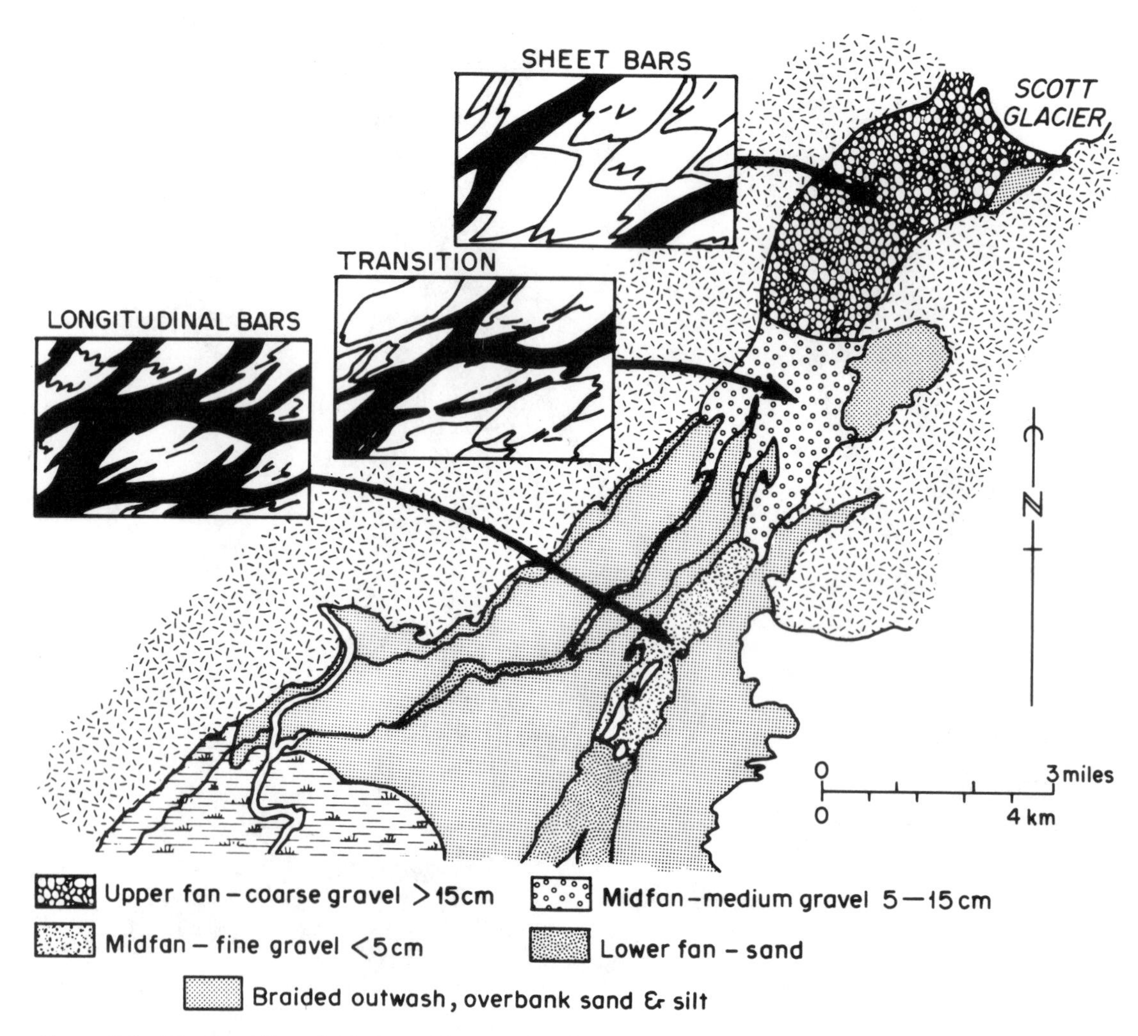

Figure 3-3. The Scott Fan of Alaska, with insets showing downfan change from sheet bars to longitudinal bars. (Modified after Boothroyd, 1972.)

Nummedal, 1978). Some bars form during high-river stage and are distinct from the more complex dissected bars that form at low-river stage.

On many small fans the distal sandy facies is absent (Boothroyd and Nummedal, 1978), and vegetation plays an important role in developing extensive meandering alluvial floodplains, or swamps and marshes with anastomosing channels. The number of channels is limited, and fan drainage may converge into a single meandering channel or narrow braided tract (Fig. 3-2). Where anastomosing patterns prevail, channel-floor aggradation accompanies peat growth, with the channels maintaining a constant position (Smith and Smith, 1980).

On steep proximal fan surfaces, particularly in arid climates, stream flow may occur only after the bulk of the previously accumulated clayey material has been removed by debris flows. Where discharge from a catchment basin is deficient in sediment as a result of prior flushing, erosion in the upper fan produces a fanhead trench (Beaty, 1970). The trench generally heads in a mountain stream and cuts through debris flow and sieve deposits to a depth of as much as 12 m (40 ft). Channel depth decreases downstream and the trench gradually loses its identity, merging with the fan surface (Hooke, 1967). Beyond this intersection point the fan surface is predominantly sandy and is molded by intermittent fluvial processes and wind. The smaller ephemeral braided streams show a uniform radial pattern, which is a response to frequent switching of active channels across the distal fan. Channels are blocked by their own sediments or by eolian sands, and are diverted around sedimentary obstructions during the next flood, a process which leads to branching (Glennie, 1970). Deflation of the finer fraction of abandoned channel sediments leaves a residue of coarse sand and gravel. The channels become poorly defined distally where they merge into a flat sandy sheet. During maximum discharge, sediment-laden sheet floods spread across the entire surface. Sheet floods also arise from direct precipitation on a fan surface (Wasson, 1977). Even on sandy fans with well-defined channels and bars, sheet-flood processes may dominate the sedimentary record.

Water depths are very shallow, commonly only a few centimeters, so upper flow regime conditions prevail, with plane-bed and antidune bedforms. Individual flood events are marked by rapid introduction of the coarsest debris, followed by progressively finer sediments as the flow wanes (Hardie *et al.*, 1978). Impounded waters flood the distal sandflat. Waves produced by wind stress on the shallow water result in a complex array of ripples reflecting variable wind directions and water depths (Hardie *et al.*, 1978). Clay drapes are desiccated as the lakes retreat, and mudclasts are largely redistributed by wind.

Secondary fans grow around the periphery of some major alluvial fans. They may result from headward retreat of small gullies which deposit small lobes at their downstream ends, or from continuation of a fanhead trench to the foot of the primary fan (Denny, 1965).

Stream-flow processes are strongly influenced by fluctuating local base levels created by damming through landslides or by interfering fan lobes, and by changes in lake or sea level (McGowen *et al.*, 1979). Impounding of water promotes aggradation, whereas a lowering of base level causes stream incision. Continued fan progradation reduces fan gradient, which promotes flow diversion by stream piracy of fan headwaters taking advantage of steeper gradients. Temporarily more vigorous flow is directed toward interlobe depressions.

Debris Flows

Debris flows result where clay and water provide a low-viscosity medium of high yield strength capable of transporting larger particles under gravity. The competence of debris flows is largely determined by clay matrix content, clay composition, and duration of flow (Hampton, 1975). Flow is laminar rather than turbulent. According to Hampton, debris flows commonly operate in conjunction with other mechanisms such as grain flow. The content of silt and clay in debris-flow deposits characteristically ranges from 10 percent to over 30 percent, but may be less than 1 percent in some instances (Miall, 1970). Debris flows of low clay content, and consequently low matrix strength, are referred to by Miall as debris floods.

Debris flows are most common in semiarid regions or where abundant glaciogenic or volcaniclastic material is available. Debris flows on low-gradient, wet alluvial fans have low preservation potential. The flows are often triggered by thunderstorms and are of brief duration. The frequency of storms, together with the relief of the drainage basin, source-rock lithology, and rate of weathering, determine the volume and caliber of sediment; this in turn governs the dominant processes, particularly in the most proximal parts of arid fans. A source terrane of rapidly weathering phyllite or shale yields voluminous fine sediments conducive to debris flows.

Many debris flows commence by landslides and slope wash. Prior accumulation of as much as 6 m (20 ft) of unconsolidated material in catchment depressions may be a prerequisite for debris flows (Beaty, 1970). Loose material accumulates as a result of mass movement and colluviation; flow is triggered by the addition of water (Wasson, 1977). The flow observed by Sharp and Nobles (1953) in the arid southwestern United States commenced on a steep slope of 32 degrees, which flattened abruptly downstream. Nevertheless, boulders 18 in (45 cm) in diameter were carried distances of almost 12 mi (20 km). In volcanic terrane of central Africa, a tropical thunderstorm initiated a debris flow in which great blocks of rock were transported like corks (Lester King, personal communication, 1964). In a desert environment, debris-stream wadis may be activated by flash floods as a slurry of sand, mud, and boulders (Glennie, 1970, p. 29). The debris flows tend to terminate as elongate lobes (Fig. 3-4A).

Viscosity of debris flows varies even within a particular flow. Examples studied by Bull (1972) comprised between 40 and 90 percent mud. Some portions of the flow may be penetratively sheared while others behave as semirigid bodies. Rigid zones along the margins may form levees (Fig. 3-4A). Differences in grading and degree of clast alignment result from this segregation (Bull, 1972; Harms *et al.*, 1975). In the most viscous flows, the larger clasts are distributed throughout the deposit, but with lower viscosities the deposits may show inverse, normal, and inverse-to-normal grading with stronger clast alignment. Inverse grading results from processes such as dispersive pressuring, kinetic sieving, and basal assimilation,

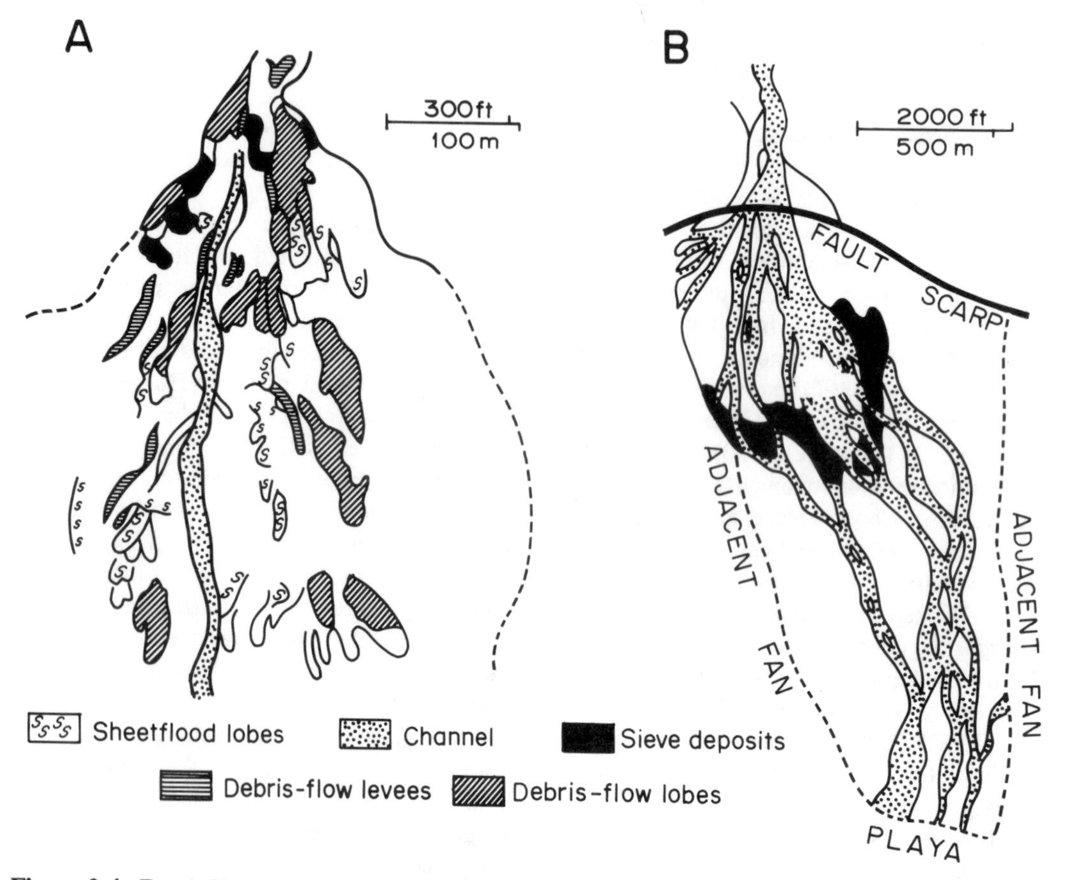

Figure 3-4. Death Valley fans showing varying proportions of debris-flow, sieve, sheetflood, and channel deposits typical of arid regions. (After Hooke, 1967.)

but loss of clay strength in the strongly sheared lower layers of décollement may be the most important mechanism in subaerial debris flows of soil, talus, and colluvium (Naylor, 1980).

Progressive downslope loss of moisture to the substrate leads to an increase in matrix strength (Rust, 1978). According to this mechanism, updip debris-flow deposits should show better developed textural arrangements and bedding characteristics than their downdip counterparts. However, alluvial-fan debris flows commonly extend from the subaerial fan into lakes, where the fluidity is increased. Here the debris flows are succeeded by more dilute density currents (Nemec *et al.*, 1980).

Textural inversion causes a decrease in sorting toward the foot of some fans. This anomalous pattern results from mixing of fan sediments with fluvial systems occupying the basin axes. Loading, slumping, and sliding of fan gravel into finer floodbasin sediments produces a pebbly mudstone (Larsen and Steel, 1978). Further disruption results from injection, liquifaction, and dewatering.

Eolian Processes

Redistribution of sediment by wind is important in poorly vegetated fan environments. Small dunes form on the distal sandflat, and may coalesce and migrate upslope to overlie even the coarsest deposit of the upper fan. This situation is particularly common in interior sand seas, where protruding highlands receive a higher rainfall and are flanked by a belt of "wadi" fan sediments (Glennie, 1970) which merges with eolian dunes. Pebbly deflation lags may show evidence of wind etching, and develop a coating of desert varnish produced by in-place weathering. In southern Iceland extensive dune fields are developed on the distal fan (Boothroyd and Nummedal, 1978). This area receives 1000 mm (40 in) precipitation a year but is subject to strong winds which exceed 10 m/s (68 mph) for over 30 percent of the time (Nummedal *et al.*, 1974). Dunes include longitudinal and smaller transverse forms which generally lack slip faces as a result of variable wind direction. Some dunes are stabilized by grass.

Sieve Deposition

Sieve deposition involves the stranding of lobes of coarse sediment on the surface of a fan as a result of water loss to the permeable gravelly substrate. Sieve deposits are concentrated around midfan (Fig. 3-4B) and are related to distinct channel systems, so they are commonly reworked during subsequent floods (Hooke, 1967). Stranding of an initial barrier of debris is followed by backfilling. Therefore, lenticular sieve deposits are commonly banked against larger boulders which formed the initial obstruction (Gloppen and Steel, 1981). After the channel has regraded its course, another debris barrier is formed upfan (Hooke, 1967). Sorting is initially good but illuviation, or invasion of openwork gravel by subsequent flow, may provide smaller particle sizes, significantly decreasing the sorting (Wasson, 1977). A granitoid source weathers to coarse gravel and sand, which are most conducive to infiltration.

Ground-Water Processes

Ground-water flow through wet alluvial-fan systems is extremely important because of their high transmissivities (permeability multiplied by thickness). Recharge on the upper fan occurs by direct precipitation or infiltration of stream discharge. Ground-water flows downslope below a water table that mimics the depositional surface (Galloway *et al.*, 1979, p. 121) and discharges in the lower fan. As the deposits are buried by younger fans, flow evolves from unconfined to semiconfined. The chemistry of the ground water changes as it migrates and encounters soluble salts, and discharging ground water tends to be slightly reducing. In arid climates, evaporite minerals such as gypsum and calcite are precipitated. Travertine and tufa may encrust springs emerging from the toe of the fan. Evaporative pumping of ground water results in caliche crusts, nodules, and calcite pore cements, particularly in the braided fan channel deposits. Plants on the distal fan have large root systems which disrupt bedding, tap ground water by evapotranspiration, and provide conduits for ground-water movement (Hardie *et al.*, 1978). In humid climates, peatswamps commonly form where ground water

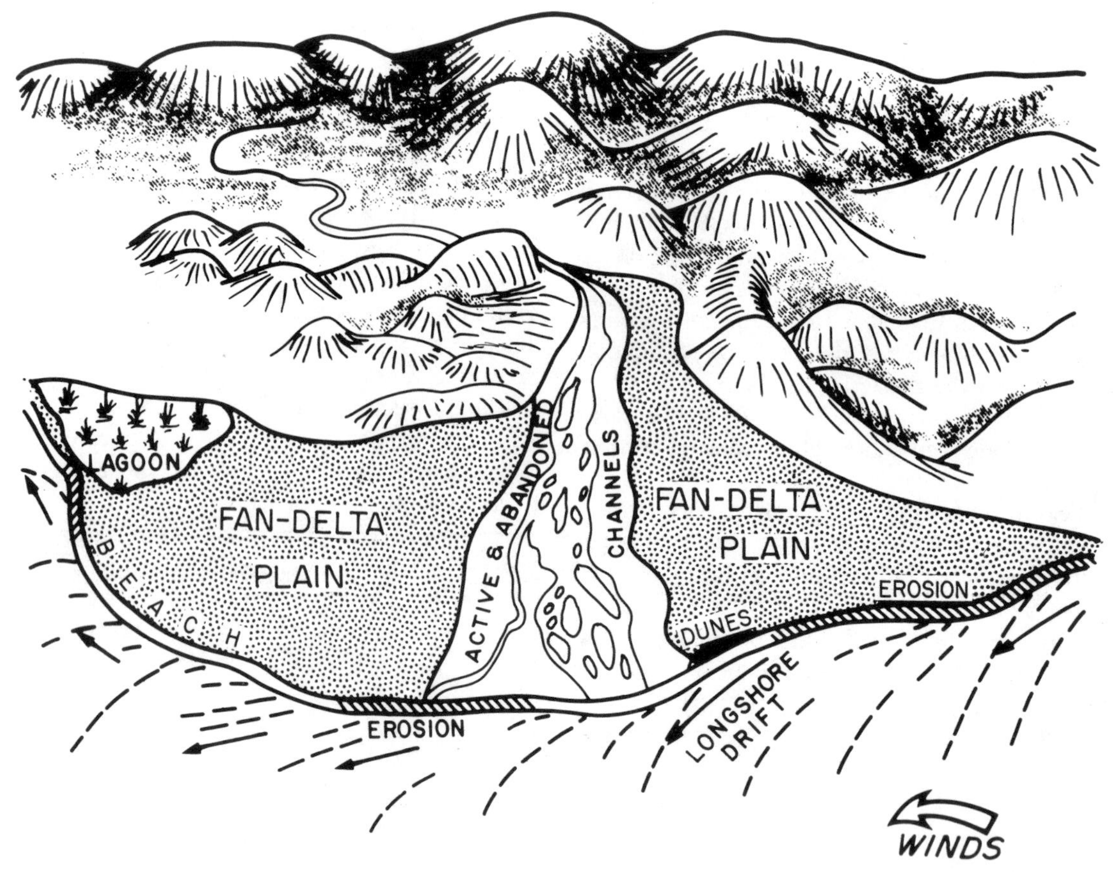

Figure 3-5. The Yallahs Fan of Jamaica with a marine-modified fan-delta margin. (After Wescott and Ethridge, 1980.)

discharges, and extend to the shores of lakes, or persist across the alluvial floodplain.

Basinal Processes

Waves and tides modify the distal terminations of alluvial fans (fan deltas) building into the open ocean, and even lacustrine fan deltas may show significant wave effects. Wave reworking effectively improves sorting and leads to better rounding of clasts with a unique closely packed, interlocking fabric (Gloppen and Steel, 1981). Some reworked conglomerates show low-angle deltaic foresets. Flooding by wind tides may affect large areas of the fan-delta plain (Boothroyd and Nummedal, 1978). Tides are normally ineffective in redistributing coarse sediment, but waves and associated longshore currents are important agencies. The Yallahs Fan (Fig. 3-5) of southeastern Jamaica (Wescott and Ethridge, 1980) and the small fan deltas of coastal Honduras (Schramm, 1981) are wave dominated, with spits, barriers, and beaches. For much of the year no drainage from the Yallahs River reaches the sea because the distributaries are sealed by wave-built berms. The Honduras fans prograde dramatically during extreme floods, and are subject to wave attack, coastal retreat, and downdrift spit growth during the intervening periods.

Wave and tidal processes may operate in conjunction, as on the Copper River fan delta (Fig. 3-6). A tidal range of several meters generates currents of up to 200 cm/s (6.6 fps) in inlets (Reinson, 1979). Ebb currents reinforced by river discharge are most effective in funnelling sand across the lagoon and out of tidal inlets between stunted barriers. Strong oceanic swell associated with winter storms results in powerful westward longshore drift which produces an asymmetrical sediment distribution (Galloway, 1976).

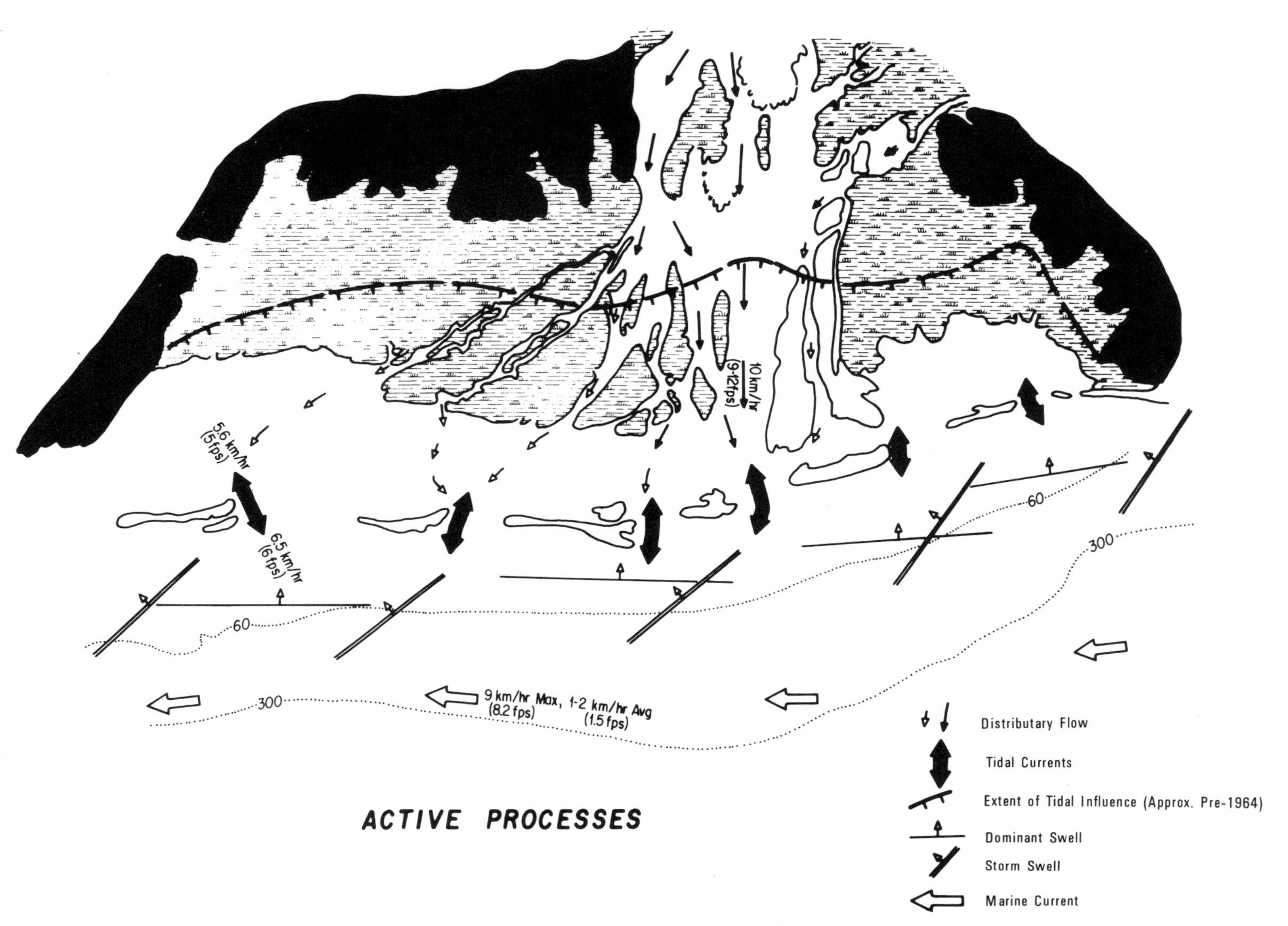

Figure 3-6. The Copper River Fan Delta of Alaska showing modification by tidal currents and waves. (After Galloway, 1976.)

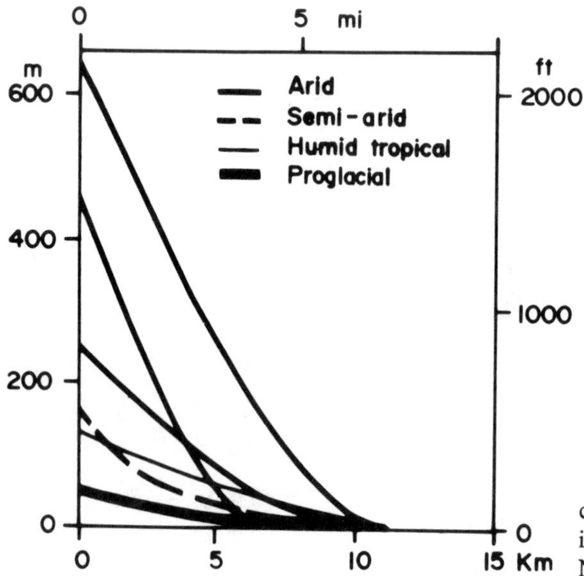

Figure 3-7. Typical fan gradients in different climatic settings. [Modified after Schramm (1981), including data from Hooke (1967) and Boothroyd and Nummedal (1978).]

Wet (Stream-Dominated) Alluvial Fans

The best-documented wet alluvial fans are in Nepal and India (Gole and Chitale, 1966), southeast Alaska (Boothroyd, 1972), tropical Honduras (Schramm, 1981), and Iceland (Boothroyd and Nummedal, 1978). Wet fans are active seasonally in response to monsoonal floods or ice melting in the catchment area; between 60 in (1500 mm) and 100 in (2500 mm) of annual precipitation is typical. The fans are normally characterized by perennial discharge (Schumm, 1977), but in some cases these perennial streams have little influence on fan sedimentation, which is dominated by rare exceptional floods (Schramm, 1981). Surface areas attain many hundreds of times the size of arid fans, with a maximum of around 62 mi^2 (16,000 km^2). The gradient of the fan surface is typically low (Fig. 3-7), sloping between 5 and 21 ft/mi (1 and 4 m/km) in the upper regions and flattening downfan to 1.6–7 ft/mi (McGowen, 1979). Sedimentation rates may be extremely high, with as much as 50 ft (15 m) of local aggradation in 2 to 3 years documented by Kuenzi *et al.* (1979). Fluvial processes tend to dominate almost the entire surface of humid fans (Fig. 3-2 and 3-3), but debris flows may be important in the upper fan. Debris-flow deposits of humid fans resemble those of arid fans, except that they may contain a mixture of rounded and angular clasts in muddy matrix. In some examples the matrix is sandy (Miall, 1978).

Vegetation is generally effective in stabilizing the fan surface and maintaining steep slopes in the catchment area, but the oversteepened deposits may collapse and slide during prolonged or violent floods (Schramm, 1981).

Wet fans generally show regular downstream reduction in maximum clast size (Fig. 3-8). Changes in sphericity are irregular, but an increase in rod-shaped or discoidal particles is brought about by selective traction and by selective suspension processes, respectively (Bluck, 1964). Gravels, commonly of boulder size, dominate the proximal fan. Clasts are imbricated, transversely aligned, and well rounded, the largest boulders being abraded in place. Most boulders and pebbles are in contact with one another, and markedly finer interstitial sediment is introduced subsequently.

The proximal fan is dominated by storm discharge. Crude horizontal bedding is developed in the low gravel bars, whereas trough cross bedding reflecting dune migration characterizes the in-channel gravels and sands. Two bar types are distinguished in the midfan area. The upper midfan or transition (Fig. 3-3) is composed largely of coarse gravel, with rhomboid bars

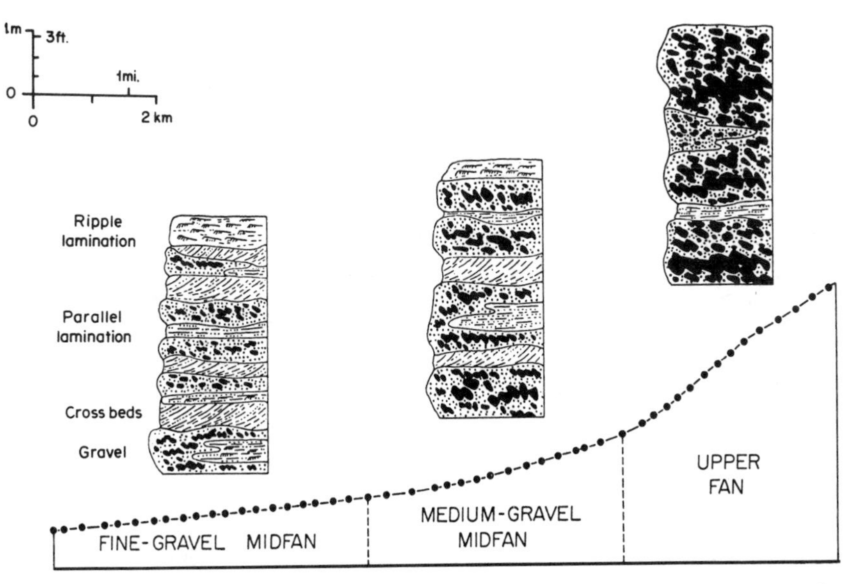

Figure 3-8. Downfan reduction of gradient accompanying reduction in particle size of a wet fan. (After Boothroyd, 1972.)

intermediate in form between the sheet bars upstream and longitudinal bars of the lower midfan (Boothroyd and Ashley, 1975; Boothroyd and Nummedal, 1978). These rhomboid bars and adjacent channels show greater relief than the sheet bars, and distinct slip faces are present on some, becoming better developed downstream. The largest clasts are concentrated at the upstream end of the bars, with planar cross-bedded sand wedges (Rust, 1972a) along lateral and downstream margins. Bar-top aggradation and slip-face accretion result in a predominance of horizontal bedding and planar gravelly foresets. The proportion of foresets increases downstream, and there is a progressive improvement in sorting. Trough cross beds form in the in-channel sandy gravels. Lower midfan longitudinal bars are composed of finer gravels and are separated by low-flow channels floored with dunes. Slip faces are well developed, and in some places accrete sandy bar-margin wedges. The bar gravels show horizontal bedding and planar foresets in approximately equal proportion.

The sandy distal fan is traversed by braided channels that become more numerous and shallower downstream, in some cases merging into a flat sandy surface. Bars include longitudinal, linguoid, and transverse forms. The most common sedimentary structures are planar cross beds, in places overlain by bar-top plane beds or ripples, and trough cross beds deposited by channel dunes. Elsewhere the entire distal sand deposit is plane bedded. Where the gradient flattens sufficiently and fine-grained sediments dominate, the sedimentary record will comprise upward-fining cycles of mixed-load channel type (Chapter 4). Some of these cycles terminate upward in a coal seam. In rare cases, wet alluvial fans may be composed entirely of unusually fine material such as organic-rich silt (Legget *et al.*, 1966).

Stream-Dominated Fan Systems in the Geologic Record

Despite the disproportionate emphasis of studies on modern fans in arid or semiarid regions, wet fans have been widely recognized in the geologic record. Individual fan systems tend to be large and show a range of fluvially produced textures and structures which generally permit ready identification of proximal, midfan, and distal fan components (Figs. 3-8 and 3-9). Humid-region fan deposits of pre-Carboniferous age are likely to differ from younger deposits when highland areas had been colonized by vegetation.

The areal extent of a fan system, its thickness, and preserved cross-sectional geometry, are largely dependent on the structural setting. The vast single fan system of the lower Paleozoic Table Mountain Group of southern Africa was

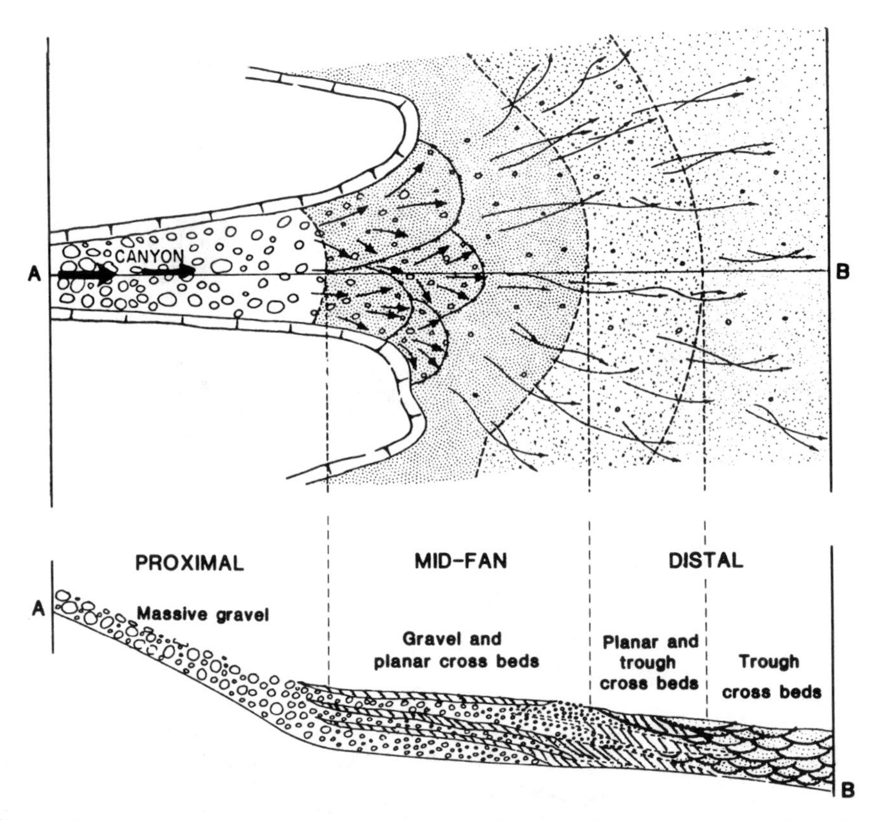

Figure 3-9. Idealized cross section and plan view through the wet alluvial-fan system of the Cambrian Van Horn Sandstone. (After McGowen and Groat, 1971.)

deposited in an early rift, the Natal Embayment, which was a precursor of Mesozoic Gondwana breakup (Hobday and von Brunn, 1979). Consequently only the western half of the basin remains in Africa. Fan-head conglomerates occupy paleovalleys up to 300 m (1000 ft) deep. Despite a heterogeneous Precambrian basement, only well-rounded, near-spherical quartzite boulders are present at the base of the fan deposits, attesting to their durability in a warm, humid climate that contributed to the disintegration of other lithologies. Progressive upward decrease in maximum clast size from 2 m (7 ft) to around 25 cm (10 in) is accompanied by horizontal bedding, shallow scour surfaces, and pods of conglomerate. This probably represents the downfan increase in bar and channel relief as observed in the Alaskan fans. Imbrication and transverse alignment of long axes are conspicuous throughout these deposits. In marked contrast are subordinate unsorted lenses interpreted as debris-flow deposits. In these, sandstone-supported pebbles and boulders show a weak tendency toward downstream alignment of long axes, and no bedding or imbrication are apparent. Most of these debris-flow deposits show evidence of partial reworking by stream flow.

Abrupt downslope interfingering of upper-fan conglomerates with arkosic sandstones coincides with reduction in basement relief. The same lithologic change took place vertically as aggradation proceeded. The arkoses show a suite of sedimentary structures, including planar foresets of variable direction on the margins of exhumed transverse bars, and scour and truncation surfaces reflecting low-water dissection and modification, identical to that of modern sandy braided streams.

Wet alluvial-fan systems of the Cambrian Van Horn Sandstone of Texas (McGowen and Groat, 1971) are very similar to those of the Table Mountain Group, but contain more clearly defined midfan facies (Fig. 3-9). These comprise conglomerate-floored channels containing trough and planar cross-bedded sandstone, and longitudinal bar conglomerates showing horizontal bedding and foresets ranging from low-angle to high-angle. Updip, massive boulder beds represent proximal surge deposits, and grade laterally and upward into alternating conglomerates and thin sandstones signifying growth of sheet bars. Distal-fan sandstones show a downfan decrease in scale of sedimentation units, an increasing proportion of trough to planar cross beds, and increasing mudstone content. These braided-stream deposits reflect a trunk system of large paleochannels and a network of smaller radiating channels. The most distal facies are not preserved.

Other wet-fan systems of Paleozoic age are broadly similar to the above, with differences attributed mainly to relief and provenance variation. A complex of eleven discrete fans was recognized as contributing to the deposition of poorly sorted Devonian conglomerates in Arctic Canada (Miall, 1970). The prevalence of debris-flow deposits, with subordinate fluvial sandstones, was attributed by Miall to the general absence of Early- to Mid-Devonian vegetation in mountainous terrain. Wet-fan systems of the Lower Proterozoic Witwatersrand Supergroup comprise several large individual lobes which emerged from canyons along the tectonically active margins of the basin (Pretorius, 1976; Tankard *et al.*, 1982). A further example is provided by basal Paleozoic fan systems of the southwestern Cape Province of South Africa. Overlying the irregular, highly faulted basement are poorly bedded, immature conglomerates which give way upward to well-bedded, texturally and lithologically more mature conglomerates corresponding with the blanketing of local basement highs (Vos and Tankard, 1981).

Carboniferous and younger wet-fan systems in northern Spain (Heward, 1978b), southeastern Australia (Davis, 1974), southern Africa (Hiller and Stavrakis, 1979), and western Canada (Long, 1981) show similar characteristics despite a range in tectonic setting from fault-bounded intermontane valleys to forelands and foredeep basins. Debris-flow deposits are volumetrically more significant than in older fan systems. The proportion of proximal and midfan conglomerates to distal-fan sandstones, siltstones, and mudstones varies in response to differences in climate, sediment supply, and degree of contemporaneous tectonism. In the Permian of the northeastern Sydney Basin, Australia, deglaciated highlands supplied sediment-charged floodwaters, which, in concert with a superabundance of contemporaneous pyroclastic sedimentation produced surges of ill-sorted debris in viscous suspension (Davis, 1974). These deposits were largely reworked by braided streams. Inactive fan surfaces were rapidly vegetated, with trees preserved in vertical position by ashfall debris. Finer-grained organic material is widely preserved in close association with the conglomerates as a consequence of the consistently high water table. The Triassic fans of the Karoo Basin, South Africa, occupy a similar foredeep-plain environment, and show a classical downstream fluvial progression from coarse, longitudinal bar to sandy transverse bar facies.

Mesozoic and Tertiary fans of the Canadian Cordillera were laid down during, or immediately following, a phase of major strike-slip faulting (Long, 1981). Landslides and scree were rapidly reworked into fan systems whose extent and internal character were governed to a large degree by the width and gradient of the basin. In the narrower basins, sedimentation patterns were strongly influenced by damming and local base-level changes caused by older fans building across the valleys. Carboniferous fans of northern Spain were similarly restricted to small, elongate, tectonically active basins (Heward, 1978a and b), and show comparable fluvially dominated conglomeratic successions. Organic material was abundantly preserved in both areas.

Wet alluvial fans of the Neogene Ogallala Formation of Texas (Fig. 3-10) were arranged in three major lobes along the eastern flank of the Rocky Mountains. Midfan deposits comprise broad sand and gravel sheets, which merge into digitate belts of the distal fan (Seni, 1980). The distribution of these relatively thin fan systems was strongly influenced by preexisting topography.

Arid-Region Fans

Fans in arid or semiarid regions tend to be small, cone-shaped piles with convex cross-fan profiles and concave radial profiles (Bull, 1972). Surface areas are generally between 2 and 100 km^2, with steep gradients (Fig. 3-7) of up to 100 m/km (McGowen, 1979). Typical examples are illustrated in Figure 3-4. Fans are also common in arid rift basins such as the Gulf of Aqaba

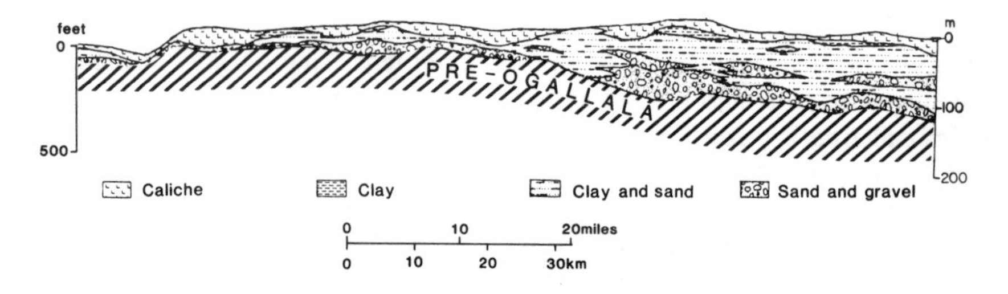

Figure 3-10. Dip section through the thin but widespread wet alluvial-fan system of the Neogene Ogallala Formation. (After Seni, 1980.)

(Friedman, 1968), the Dead Sea (Sneh, 1979), and Baja California (Thompson, 1968).

Arid-region fans are made up of the deposits of debris flows, sieve lobes, braided streams, and sheet floods, but their proportions vary from one fan to another. Of the Arizona fans studied by Blissenbach (1954), between 5 and 40 percent of the sediment was of debris-flow origin.

Some fans in arid regions show a dominance of stream-flow processes. For example the Kern River fan of the San Joaquin Valley of California (Davis *et al.*, 1964) is in an area that receives only 6 in (150 mm) of rain a year. But a higher rainfall, augmented by snow melt, in the nearby mountains provides a high level of stream activity. As a result the characteristics of the Kern River fan are intermediate between wet and dry fans (Seni, 1980).

Debris-flow deposits on arid fans may comprise steep-sided lobes several meters thick, or thin sheets inclined subparallel to the fan surface at angles of 1 to 8 degrees (Wasson, 1977). Successive flows occupy different areas of the fan surface so that the deposits overlap or coalesce. Some debris flows blanket topographic irregularities, so that they are irregularly based but flat-topped (Wasson, 1977); others occupy steep-sided channels, to the erosion of which they probably contributed. Sharp and Nobles (1953) noted a marked downslope reduction in thickness of some flows. The percentage of clay remains relatively constant downstream, and only the very largest clasts are deposited selectively (Hooke, 1967). But in contrast to wet fans, some arid fans show no clear relationship between slope and maximum clast size, with the downstream decrease in clast size tending to be irregular and unpredictable (Boothroyd and Nummedal, 1978). On the other hand, some arid fan deposits show clearly defined downfan reduction in bedding thickness and maximum clast size, although not as evenly as in stream-dominated fans (R.J. Steel, personal communication, 1982). Debris-flow deposits are characteristically unstratified and matrix-supported, with angular clasts.

Although matrix support is the most commonly used criterion in the recognition of ancient debris flows, care has to be exercised because matrix may be substantially reduced by compaction or flushing. Conversely, matrix content may be anomalously high in sediments subject to textural inversion or illuviation of fines.

Sieve deposits characteristically comprise clast-supported gravel showing a gradational contact with underlying units. They follow narrow, impersistent dip-oriented belts with a rounded cross section and steep downdip termination as high as 35 ft (10 m) (Hooke, 1967); but because of early modification these geomorphic characteristics are seldom preserved in the rock record. The proportion of sieve deposits is usually small, but where there is preponderance of gravel and very little silt and clay, as on Shadow Rock Fan (Fig. 3-4B), the fan may be composed largely of sieve deposits (Hooke, 1967).

The sandy distal deposits of arid-region fans show planar and trough cross beds, and ripple lamination. Upward-fining sequences result from rapid filling of flood-incised channels (Glennie, 1970). Sheet-flood deposits of the distal fan comprise parallel-laminated sands traversed by subdued channels (Bull, 1972). Pebble layers representing a wind deflation lag may include ventifacts and clasts with desert varnish.

These sandflat deposits show a distinctive association of upper-flow regime structures, mainly plane beds but including wavy antidune forms, together with wave ripples representing subsequent reworking. Each flood event deposits a sequence up to several tens of centimeters thick comprising pebbly sand overlain by parallel-laminated sand and finer ripple cross-laminated sand (Hardie *et al.,* 1978). Associated eolian sands show a typical suite of small dune structures. Structures in the distal fan are commonly disrupted by chemical precipitation and mineral growth, by plants with exceptionally large root systems, and burrowing.

Secondary fan lobes downslope of the primary fans produce thin upward-coarsening sequences. These sediments tend to be finer grained and better sorted than the primary fan deposits (Heward, 1978a).

Arid-Region Fan Deposits in the Geologic Record

Arid-region fan deposits can generally be distinguished in the stratigraphic record by their limited individual areal extent, poor sorting, textural and compositional immaturity, and paucity of current-produced structures in their proximal parts. The climatic distinction is more difficult to apply to the pre-Carboniferous record.

Direct evidence of an arid fan environment is provided by caliche-type paleosols, abundance of evaporites in the distal fan, absence of plant material, desiccation-cracked argillaceous interlobe facies, playa-lake deposits, abundant eolian sandstone intercalations, deflation residues including wind-faceted pebbles, and, rarely, a vertebrate fauna showing adaption to aridity. Arid alluvial-fan systems commonly comprise localized wedges of ill-sorted conglomerates and arkosic sandstones, or "granite wash" adjacent to steeply faulted basin margins (Fig. 3-13).

The New Red Sandstone fan deposits of Scotland include debris-flow, braided-stream, and streamflood facies, with convincing evidence of deposition in an arid climate (Steel, 1974). Sequences resulting from progradation of individual fans comprise debris-flow deposits overlain by streamflood and braided-stream facies. Debris-flow units are laterally extensive and consist of very poorly sorted conglomerates with generally nonerosive bases, attributed by Steel to highly viscous flow with little turbulence. Sandstones capping many of these debris-flow conglomerates are interpreted as deposits of winnowed sand carried during the high floodwater levels that characteristically follow debris flows on modern fans. Associated streamflood conglomerates, in contrast, are erosively based, relatively well sorted, and lenticular, with large-scale cross-beds. The braided-stream facies consist of finer grained, very well sorted conglomerates showing trough cross bedding and abrupt lateral variation in texture. These are interpreted as a reworked product of earlier debris-flow and streamflood deposits. Playa-lake siltstones and red mudstones with desiccation cracks are associated with the distal streamflood and debris-flow deposits, and caliche is common in adjacent floodplain facies.

Other cited examples of ancient arid-region fans are in the Permian Peranera Formation of Spain, which abuts the Hercynian core of the Pyrenees, and the widespread Lower Triassic Bunter Formation (Nagtegaal, 1973). Both formations are redbeds, with the Peranera showing evidence, in the form of abundant caliche and high labile content, of greater aridity. The Peranera redbeds are largely primary whereas the Bunter shows evidence of diagenetic oxidation. Multiple conglomeratic bar and channel-fill sequences of the Peranera Formation contain mudstone intercalations with desiccation cracks. The Permian Rotliegendes were deposited in a desert environment extending across northeastern Europe (Glennie, 1972); bajadas along the flank of the Variscan highlands merged northward into dunes, sabkhas, and a large lake. Porous wadi sediments were subject to rapid infiltration, with the sandy sieve deposits later redistributed by wind. Fan lobes in the mid-Proterozoic Waterberg Supergroup of South Africa are up to 1000 m (3280 ft) thick, and show evidence of distinct episodes of progradation and abandonment (Vos and Eriksson, 1977). Although cross bedding and pebble imbrication attest to the dominance of traction current deposition, there are numerous mudstone interbeds showing evidence of prolonged desiccation. In some areas, fan conglomerates are overlain by eolian sandstones which merge basinward into red lacustrine mudstone. The very large-scale foresets and other bedding characteristics indicate deposition by transverse and barchan-type dunes (Meinster and Tickell, 1975), representing reworked sand from the underlying braided-stream facies.

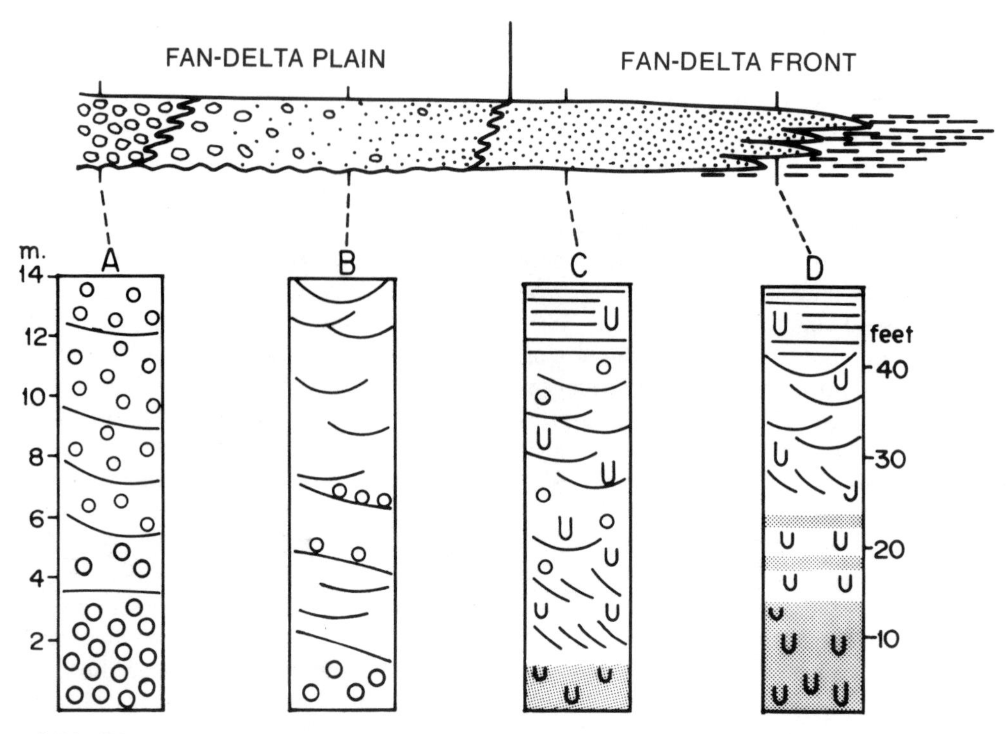

Figure 3-11. Proximal-to-distal change in vertical sequences from a channelized fan-delta plain to a fan-delta front. (After Ricci-Lucci *et al.*, 1981.)

Fan Deltas

Fan deltas form where braided rivers build into a standing body of water, and show a variable degree of peripheral modification. They are present along a variety of coasts with a range of climates and energy conditions, and are common along the shores of intermontane lakes in paraglacial areas; but large fan deltas are best developed along coasts of convergent plate margins in both continental and island-arc settings (Wescott and Ethridge, 1980). Some such as the Yallahs fan delta have a subaerial extent of only 10.5 km^2 (4.1 mi^2) but have a large submarine component. Others such as the Copper River fan delta (Fig. 3-6) are largely subaerial.

The distal facies of fan deltas form in shorezone, subaqueous marine, or lacustrine environments, with fluvial influence reflected only in the high rate of coarse sediment influx. The shorezone sector experiences a complex, fluctuating interplay between fluvial and wave or tidal processes. The marine influence may be reflected in the formation of beaches, spits, or tidal bars (Fig. 3-1), or it may be a more subtle biological manifestation in the form of marginal-marine trace fossils or benthic forams. Ricci-Lucci *et al.* (1981) distinguish between proximal, channellized fan delta (delta plain) with a dominance of upward-fining sequences, and distal, nonchannellized fan delta (delta front) with upward-coarsening or coarsening-then-fining textural patterns (Fig. 3-11).

The delta plain may build rapidly seaward during floods, subsequently retreating due to marine reworking (Schramm, 1981). During low discharge, marine conditions commonly extend up the distributary channels as a salt-water wedge.

Like alluvial fans, fan deltas show significant tectonic control over sedimentation patterns, and in addition are subjected to eustatic and lake-level fluctuations. These external controls have a dramatic influence on fan deltas, causing incision, progradation, or aggradation as the base level changes. Many of these base-level controls are accompanied by climatic changes that intensify the effects. Ricci-Lucci *et al.* (1981) are able to separate the effects of external controls, which generate a widespread sedimentary response, from the effects of inherent sedimentary controls which produce laterally interfingering and crosscutting relationships.

The Copper River fan delta receives a voluminous but intermittent supply of coarse sediment and is building onto a very high-energy, tectonically active shelf. Gravel is deposited in the lower reaches of the fluvial system and in the upper distributaries, while sand and mud are carried onto the delta plain and beyond. Today lagoon and shoreface facies are volumetrically most important, forming a wedge of well-sorted sand up to 115 ft (35 m) thick, 10 mi (16 km) wide, and 30 mi (50 km) along shore (Galloway, 1976).

This slowly prograding wedge is truncated by the advancing inlets and distributaries. The distributaries are of two types, braided and sinuous, which deposit tabular and lenticular sand bodies, respectively. The aggradational delta plain consists of marsh-covered silt and clay. Sporadic eolian dunes, supplied by violent winds from the interior, attain heights of 200 ft (65 m). Continued progradation would superimpose thick, gravelly alluvial-fan deposits on this coastal and offshore succession.

The distributaries of the Yallahs Fan (Fig. 3-5) are braided, with low longitudinal bars. Aggradation results in crudely bedded, imbricated gravels and sands, with mud accumulating in abandoned channels. Adjacent to the narrow active depositional tract is a broad, flat, seaward-dipping, fan-delta plain. The surface is largely vegetated and is underlain by 30 m (100 ft) of fluvial gravel and sand. Beaches along this microtidal coast are broad and sandy, but substantial tracts are erosional and gravelly. Shoreface and shallow offshore sediments are highly variable, and small patch reefs surrounded by skeletal sands parallel the shore. Submarine canyons, characterized by frequent slumping, carry nearshore sediments to the deep Yallahs Basin. Continued progradation of the Yallahs fan delta would produce a very thick succession dominated by resedimented sands and conglomerate overlain by progradational offshore, shoreface, and beach gravels, sands and silts, and aggradational fluvial gravels and sands.

Like the Yallahs, the late-Pleistocene to Holocene fan deltas along the Dead Sea Rift were constructed by flashy flood discharge and were subject to negligible reworking (Sneh, 1979). In contrast, however, the basinward components are lacustrine chemical precipitates and fine-grained clastics. The largest of these fan deltas has a surface area of 6 km^2 (2.3 mi^2), with a steep 2.5 degree slope. Typical arid fan-head gravels comprising angular sand- to boulder-sized clasts grade into upper-fan deposits of alternating cross-bedded gravels and coarser, crudely bedded gravels. These grade distally into complexly bedded sands of variable grain size, including poorly sorted, coarse-grained beds with mudballs, and finer sands with climbing ripple lamination. The outermost subaerial fan is composed of regularly interlayered sand and mud reflecting rapid, periodic changes in fluvial current regime. Prodelta lacustrine chalk shows varve-like interlamination of chemically precipitated aragonite needles, gypsum, fine detrital calcite, quartz, and clay.

One of the numerous intermontane fan deltas of western Canada is described by Long (1981). A small fan delta with a subaerial area of about 2 km^2 is supplied by a steep braided stream. The coarse gravel fan surface is largely inactive, and supports a mature stand of conifers. A narrow beach ridge along the seaward margin encloses small tracts of delta-plain marsh. The delta foresets are steep and have a sharp basal contact with finer grained lacustrine sediments.

Small fan deltas which are most active during floods were described by McGowen (1970) in a back-barrier setting in Texas, and by Schramm (1981) on the open Honduras coast. The Texas fan progrades by braided stream processes during episodes of moderate discharge, aggrades by sheet-flood deposition during hurricanes, and is partially reworked into destructional bars and eolian dunes during episodes of reduced flow. Vertical sequence resulting from prolonged progradation would coarsen upward from bioturbated bay facies into parallel-laminated, ripple cross-laminated, and burrowed prodelta silt and mud followed by sands of the distal fan and fan-delta plain. Fan-delta plain sands show small trough cosets in braided channel fill, and parallel lamination with subordinate planar foresets.

Honduras fan deltas (Fig. 3-12) vary according to differences in storm frequency, storm-related sea-level changes, wave direction and energy, and sediment volume and caliber (Schramm, 1981, p. 144). Maximum aggradation and progradation occur when storm surge raises sea level by several meters. Shoreline erosion occurs during fair weather, leaving a lag of large boulders. The overlying sediments of the shoreface and fan margins comprise sand with sporadic boulders

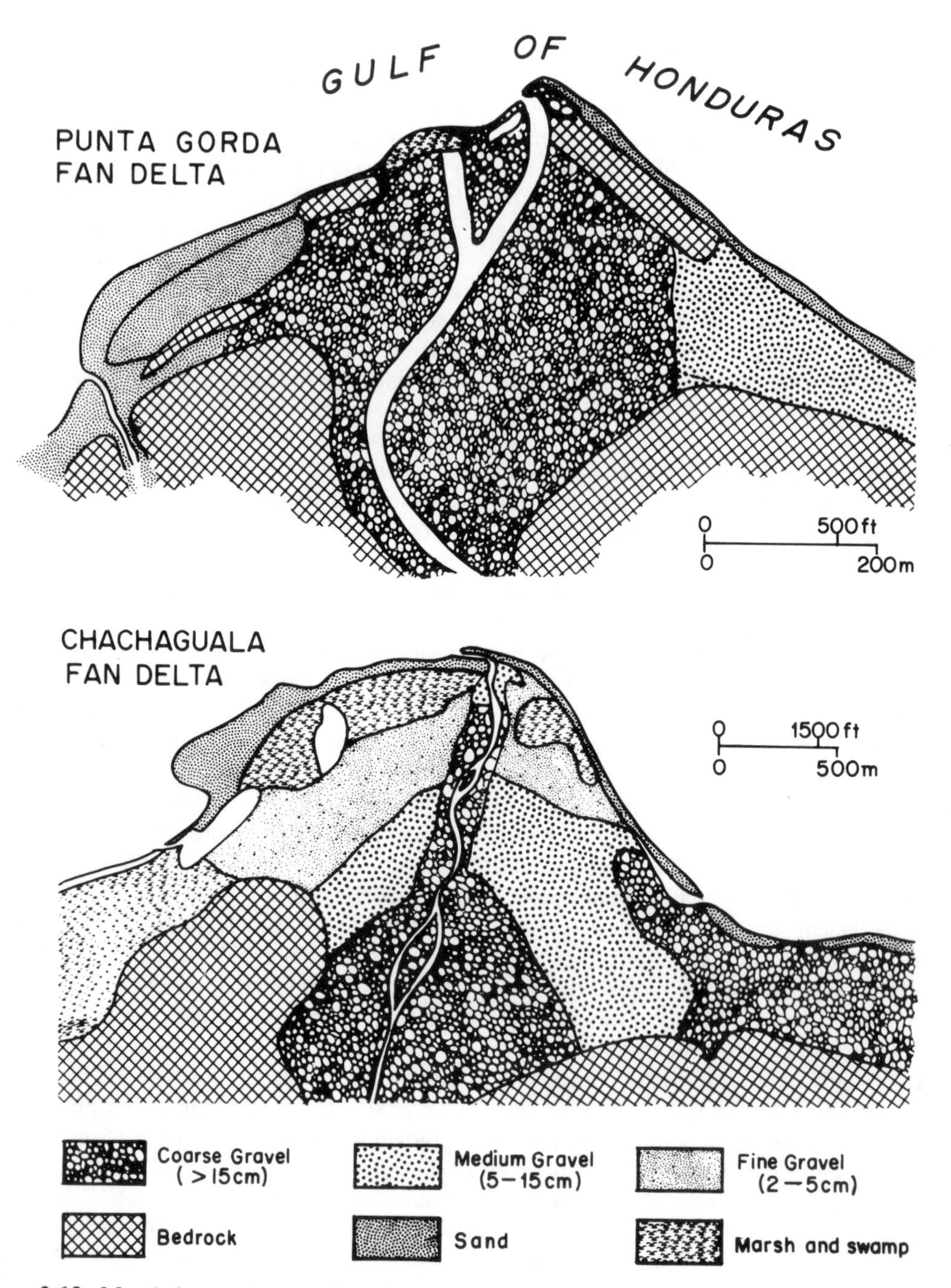

Figure 3-12. Morphology and textural variation in small fan deltas of coastal Honduras. (Modified after Schramm, 1981.)

and pebbles, and better sorted sandy spit deposits. These spits locally enclose small marshy ponds. Fan-delta plain deposits comprise stratified pebbly sand and sandy gravel on the lower fan, grading upfan into very poorly sorted, crudely bedded, coarser gravel. Progradation would generate an irregularly upward-coarsening sequence from fine-grained shelf deposits to coarse proximal delta-plain gravels, but some of the coarsest boulder conglomerates would be related to the wave-winnowed lag toward the base of the fan sequence.

Ancient Fan-Delta Systems

Patterns of peripheral reworking in ancient alluvial fans reflect a range from low-energy, lacustrine processes to powerful oceanic waves and tides. The fan-delta front is commonly represented by large conglomeratic foresets which preserve the successive profiles of progradation and provide an indication of the depth of water. Foresets are not developed, however, in very shallow basins.

Spectacular fan-delta foresets 30 m (100 ft) high in the northern Sydney Basin (Fig. 9-7) indicate continuous progradation along a broadly lobate front. Complex internal structures, including planar and trough cross beds, indicate bedform migration down the slope of the major foresets and obliquely across slope. Long (1981) recognized two types of fan-delta foresets in the Canadian Cordillera, both having abrupt but nonerosive contacts on finer grained lacustrine sediments. In one type, large planar conglomeratic foresets terminate in horizontally bedded sandstone bottomsets, reflecting simple avalanching down the delta front. The second type consists of graded conglomerates which extend beyond the fan-delta front, and interfinger with lacustrine facies. Long interprets these as originating from sediment gravity processes related to density underflows from fan-delta distributaries.

Prograding gravel beach facies preserved along the fringes of Pleistocene fan deltas in Italy comprise a tabular body 10 m (33 ft) thick containing offlapping gravel sheets inclined at angles of 10 degrees (Rainone *et al.*, 1981). Size-shape sorting characteristic of wave action shows concentrations of seaward-imbricated platy and bladed pebbles in the thin topset gravels and more spherical pebbles in the foresets. The foresets are gradational into bottomsets consisting of wavy and parallel-laminated sand with hummocky cross stratification reflecting storm-wave processes.

Fans that terminate in large, shallow lakes typically show a distal belt of shallow or intermittently emergent mudflat deposits. This peripheral zone generally contains evidence of sporadic introduction of coarser sheet-flow deposits from the landward sandflats. Plant rootlets are common, and some units are intensely bioturbated. Microstructures such as sandy streaks, lenses, and wave lamination indicate weak, wind-generated currents and waves. Abrupt changes in current direction, shallowing, and periodic emergence are indicated by complex ripples and desiccation features. These are reminiscent of tidal processes, but they are frequently present along the margins of intracratonic fresh-water basins (Van Dijk *et al.*, 1978). Muddy fan-delta deposits can contain abundant fungal spores (Teichmuller and Teichmuller, 1968a) and locally grade into coals.

Lacustrine fan-delta deposits are very responsive to fluctuations in water level. These events are preserved in the geologic record by abrupt changes from mixed-load fluvial deposition to channel entrenchment, and the development of coarse fan-delta foresets (McGowen *et al.*, 1979) stemming from a lowering of lake level. Evidence of major variations in lake level is contained in Pleistocene fan-delta deposits of the Dead Sea (Sneh, 1979). Continued rise in lake level during deposition produced marked stacking and transgression of facies. The resulting upward-fining deposits are quite different from the upward-coarsening progradational patterns normally associated with fan deltas. Prodelta chalks were deposited over the top of, and between, fan deltas rendered inactive during maximum transgression. Quite different facies relationships are present in Permian fan-delta systems of Texas (Fig. 3-13), where coarse arkosic sediments show abrupt lateral gradation into carbonate-bank and shelf-margin complexes.

Marine fan-delta deposits show a complete gradation from wave-modified to tide-modified forms, with intermediate types being widely recognized in the stratigraphic record. Ordovician fan-delta systems in Libya (Fig. 3-14) are represented by burrowed delta-front sandstones composed of distributary channel-fill and intervening shoal and swash-bar facies. The overlying fan-delta plain conglomerates are erosively based complexes of braided channel-fill origin (Vos, 1981). Examples in the Proterozoic Waterberg Supergroup of South Africa contain a combination of beaches and tidal flats (Vos and Eriksson, 1977). The beach facies consist of prominent quartzose sandstones dipping gently seaward in contrast to the cross-bedded immature alluvial facies. Wave reworking was concentrated on fan salients, whereas interlobe embayments over abandoned fans were dominated by tidal processes. A comparable association of fringing

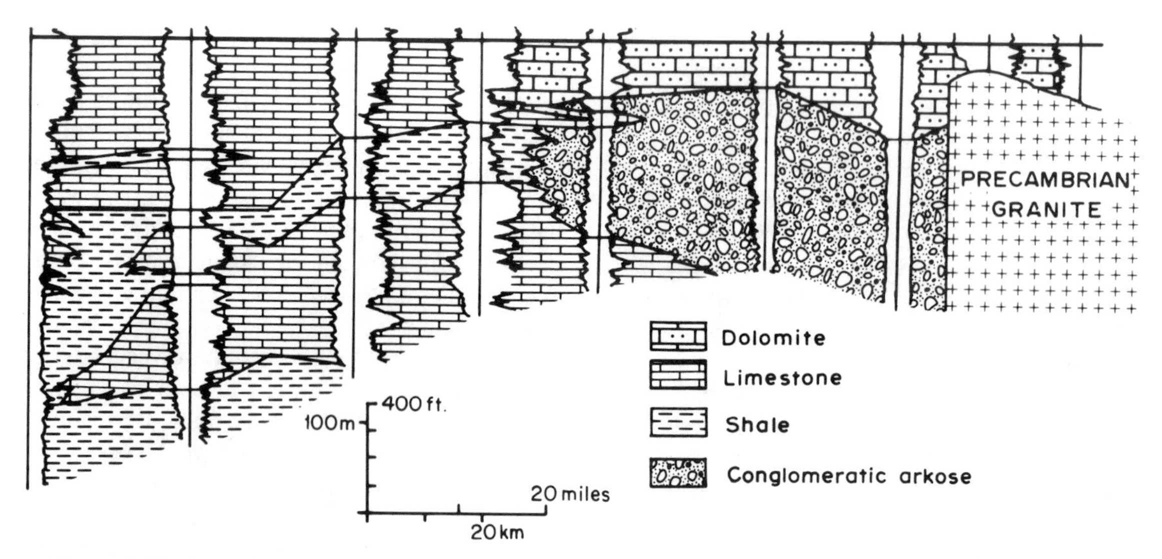

Figure 3-13. Lateral gradation of fault-bounded fan-delta wedges with shelf shales and carbonates, northern Palo Duro Basin, Texas. (After Handford, 1980.)

beaches and tidal-flat facies is present along fan-delta margins in the uppermost Precambrian of Arctic Norway (Hobday, 1974) and in Lower Paleozoic deposits at the southern Cape of Africa (Tankard and Hobday, 1977). Alluvial-fan systems in the Varangerfjord area of Norway display evidence of pronounced marine reworking during episodes of reduced fluvial energy that signalled the abandonment of a particular lobe and the initiation of a new fan of contrasting composition, texture, and direction of progradation. The southern African example is interpreted as a major, southward-building fan system grading into spectacular aggradational tidal-flat facies enclosed by stunted mesotidal barrier-island facies, a situation not unlike the Copper River Delta of Alaska (Fig.

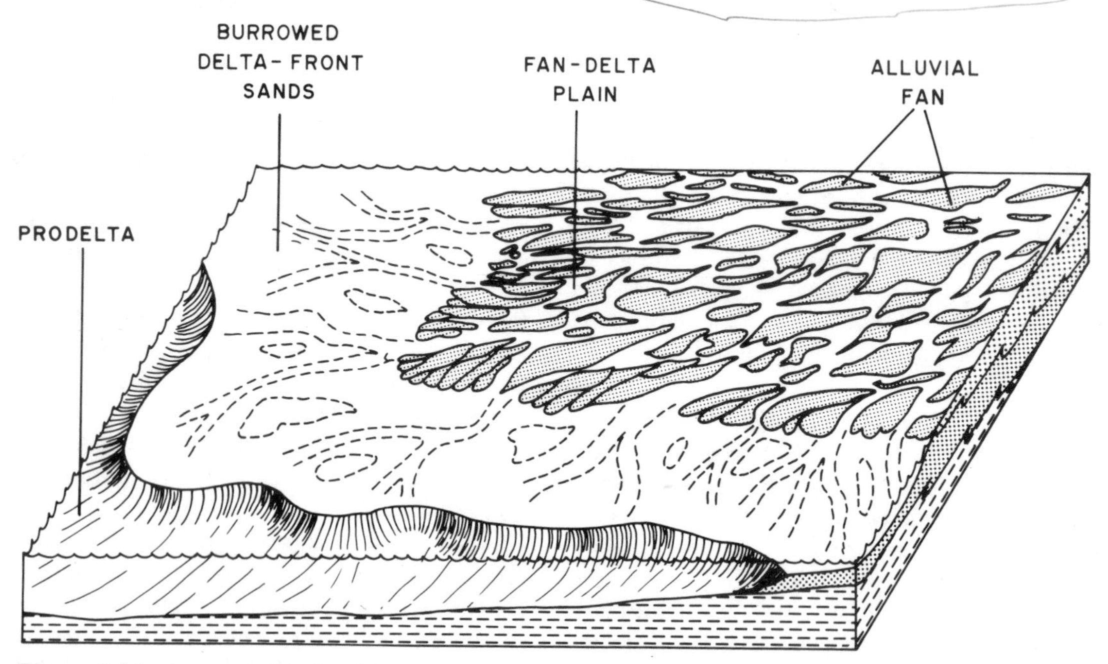

Figure 3-14. An extensive fan-delta plain gradational into shallow, delta-front sands with distributary channels, shoals, and swash bars, Ordovician of Libya. (Modified after Vos, 1981.)

3-6). Barrier sands were probably supplied by reworking of older fan deposits to the east.

More severe tidal modification of fan-delta deposits is illustrated by the Table Mountain Group along the east coast of southern Africa (Hobday and von Brunn, 1979). The great fan system that built down the rift axis terminated in a macrotidal embayment, where the coastal configuration served to augment the tidal range. The irregular fan margins were fashioned into a complex pattern of bars and channels, all aligned roughly perpendicular to the shoreline. Temporarily stable bars were oxidized in their uppermost few centimeters, where tracks and resting impressions of marine organisms are commonly preserved.

Basin-Fill Geometry

The geometry and facies interrelationships of alluvial-fan systems reflect a hierarchy of processes (Heward, 1978a) ranging from individual depositional events through short-term fan behavior to long-term fan behavior, in arrangements that depend on the tectonic evolution of the basin. Fans range from sheetlike to wedge-shaped.

Vertical Sequences Reflecting Short-Term Fan Behavior

Relatively thin sedimentary packages 200 ft (60 m) or less that show distinct vertical changes in grain size and bedding character, but which change laterally into facies showing different vertical patterns, may represent inherent sedimentary responses. Upward-fining conglomeratic debris-flow deposits thicker than 50 ft (15 m) may, therefore, be a response to gradual abandonment of proximal depositional lobes following episodes of fanhead entrenchment (Heward, 1978a). These sequences are laterally equivalent to distal fan mudstones, siltstones, sandstones, rootlet beds, and coals, which coarsen upward, fine upward, or coarsen then fine. Repetitive upward-coarsening sequences 30 to 80 ft (10 to 25 m) thick, which together make up thicker sequences, are interpreted by Steel *et al.* (1977b) as reflecting progradation in response to individual phases of vertical tectonism, and to lateral movement of the fan (Gloppen and Steel, 1981). Minor upward-coarsening sequences may also result from secondary fan growth. In certain areas the boundaries between sequences are accentuated by pedogenic horizons. Some basins show an almost random alternation of upward-coarsening and upward-fining sequences.

Additional complexity is introduced in narrow, fault-bounded basins through the blocking of axial valleys by fans prograding transversely (Smith and Smith, 1980). These base-level changes affect the deposits of adjacent fans (Long, 1981).

Thicker Tectonically Controlled Sequences

Upward-fining and upward-coarsening sequences which attain thicknesses of hundreds or even thousands of meters are widely recognized in alluvial-fan deposits. The gross upward-fining pattern that characterizes many fan systems represents initial vigorous influx of coarse gravel in response to tectonism, followed by gradually waning flow energy and finer sediment deposition as the source area is lowered or retreats. The coarsest fanhead and canyon-fill gravels commonly overlie a high-relief unconformity. Upward-fining alluvial-fan sequences of this type have been recognized by Steel and Wilson (1975), Turner (1978), Hobday and von Brunn (1979), and Ricci-Lucci *et al.* (1981). A Lower Proterozoic example 1100 m (3600 ft) thick consists of chaotic debris-flow conglomerates overlain by current-deposited conglomeratic sandstones, sandstones, and graded siltstones and shales with mudcracks (Tyler, 1979). The Pliocene succession represented in Figure 3-15 reflects an upward gradation from aggradational deposits of longitudinal or sheet bars at the base, to vertical and lateral-accretion deposits of braided and meandering bed-load channels, and lateral-accretion deposits of finer-grained channel-fill at the top.

Upward-coarsening patterns are equally characteristic, representing stacking of successively more proximal facies in response to major, persistent faulting. The most dramatic example (Fig. 3-16) is the prodigious 25,000 m (82,000 ft) thickness of regularly repetitive, basinwide, upward-coarsening cycles of alluvial-fan and floodplain origin in the Devonian Hornelen Basin of Norway (Steel *et al.*, 1977b). Each 100 m thick upward-coarsening cycle is a response to

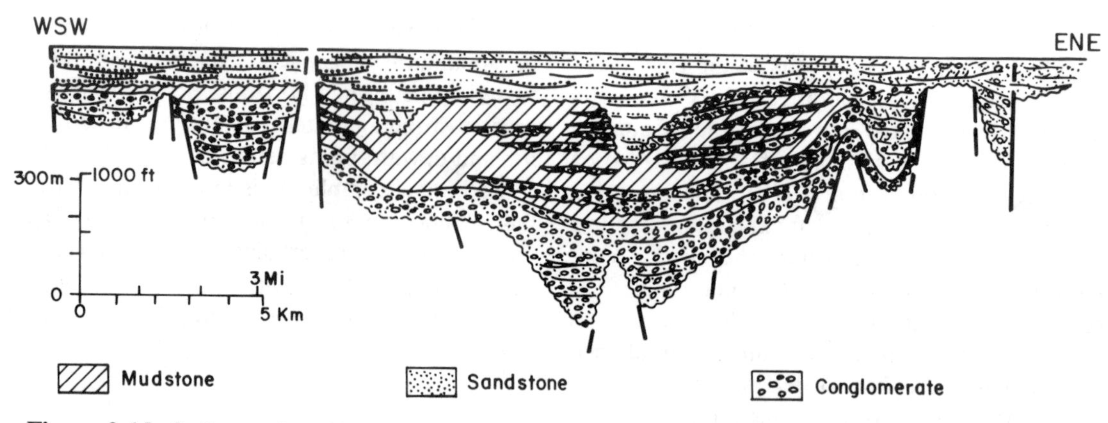

Figure 3-15. Strike section through a thick, Pliocene alluvial-fan and fan-delta succession of the Intra-Apenninic Basin, Bologna, Italy, showing marked vertical reduction in grain size. (After Ricci-Lucci *et al.*, 1981.)

basinal subsidence which accentuated relief in the bounding highlands and caused the influx of progressively coarser sediments as the more proximal facies migrated basinward. Each major progradational episode was followed by migration of the hingeline of subsidence away from the basin, resulting in overlap of the basement during deposition of the succeeding cycle. The upward-coarsening trends persist laterally through adjacent depositional systems, confirming a tectonic control.

Rust (1978) pointed out that the ideal sedimentary response to basin tectonism would be an upward-coarsening sequence representing fan progradation, followed by an upward-fining pattern representing gradual return to equilibrium. An ancient example (Fig. 3-17) is provided by Gloppen and Steel (1981). This coarsening-then-

Figure 3-16. Successive, laterally extensive, upward-coarsening alluvial-fan sequences in the Hornelen Basin, Norway. (Photo courtesy of Ron Steel, reproduced by permission of Fjellanger Widerøe).

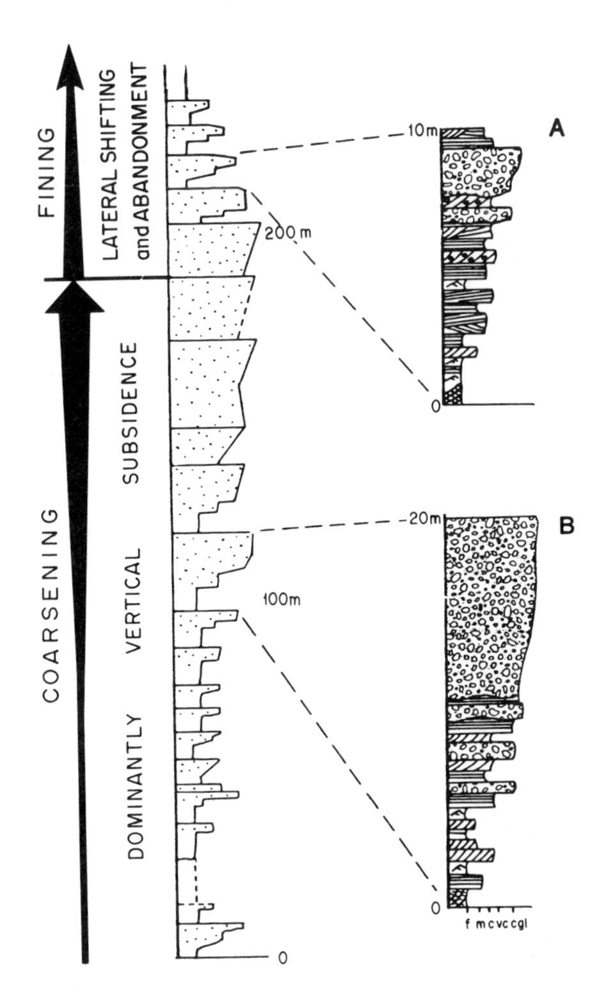

Figure 3-17. Vertical succession through a Devonian fan showing upward coarsening at two distinct scales. The thick upward-coarsening sequence is an overall response to tectonic subsidence; thinner upward-coarsening sequences were constructed by prograding fan lobes. Upward fining at the top reflects decay of the fan systems. (After Gloppen and Steel, 1981.)

fining pattern will only be attained where sufficient time elapses between tectonic pulses. Successive upward-coarsening cycles would therefore suggest relatively frequent faulting, possibly in small increments, whereas thick upward-fining successions suggest major tectonism followed by relative quiescence. Less active tectonism followed by stability results in progradation followed by dissection as the geomorphic threshold is exceeded. The recycled sediments build new lobes and bifurcating channel-fill deposits (Seni, 1981). These deposits are characteristically thin (Fig. 3-10).

Basin-Margin Fault Patterns and Their Effects

The cross-sectional geometry, too, of alluvial-fan systems is determined by contemporaneous tectonism, with the thickest, coarsest deposits commonly adjacent to the basin-margin fault system. It is not uncommon, however, for thickening to occur toward midfan, or even in the distal fan. In basin-and-range topography, the thickness of the fan deposits may vary as the prograding wedge traverses successive fault blocks and basins, which may be contemporaneously active.

Heward (1978a) recognized three contrasting basin-margin fault configurations which control the geometry and extent of basin-filling alluvial-fan systems. Persistent faulting in the same location is frequently associated with major strike-slip zones. This leads to pronounced aggradation along a relatively narrow tract (Fig. 3-18A). Limited backfaulting in a graben or half-graben situation produces moderate stacking of alluvial fan systems (Fig. 3-18B). Continued uplift and dissection may produce a pronounced

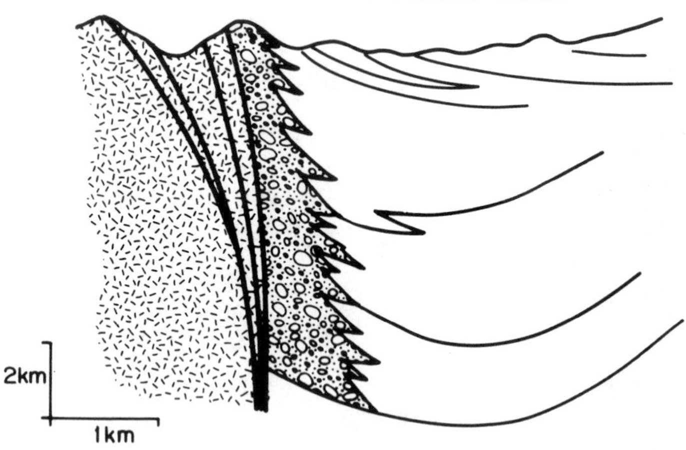

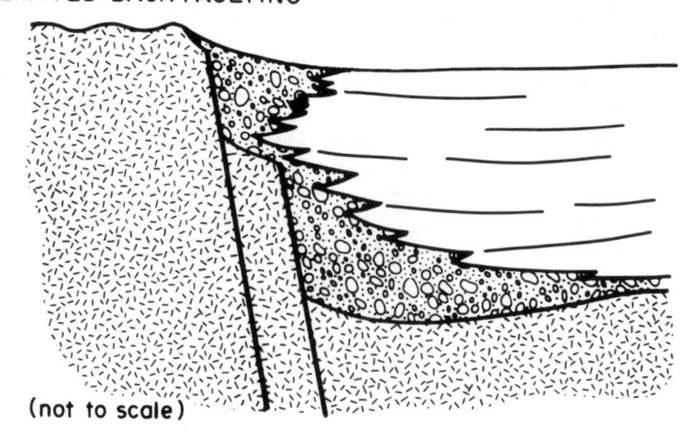

Figure 3-18. Contrasting alluvial-fan cross-sectional geometries arising from differences in basin-margin fault control. (After Heward 1978a.)

"inverted stratigraphy" of clast types. Repetitive backfaulting produces widespread alluvial-fan conglomerates of irregular thickness (Fig. 3-18C). The most proximal facies become progressively younger as backfaulting proceeds. Angular unconformities between deposits represent successive pulses of tectonic activity (Miall, 1970). Landward onlap of the 100–200 m (330–660 ft) thick, upward-coarsening increments in Hornelen Basin is attributed by Gloppen and

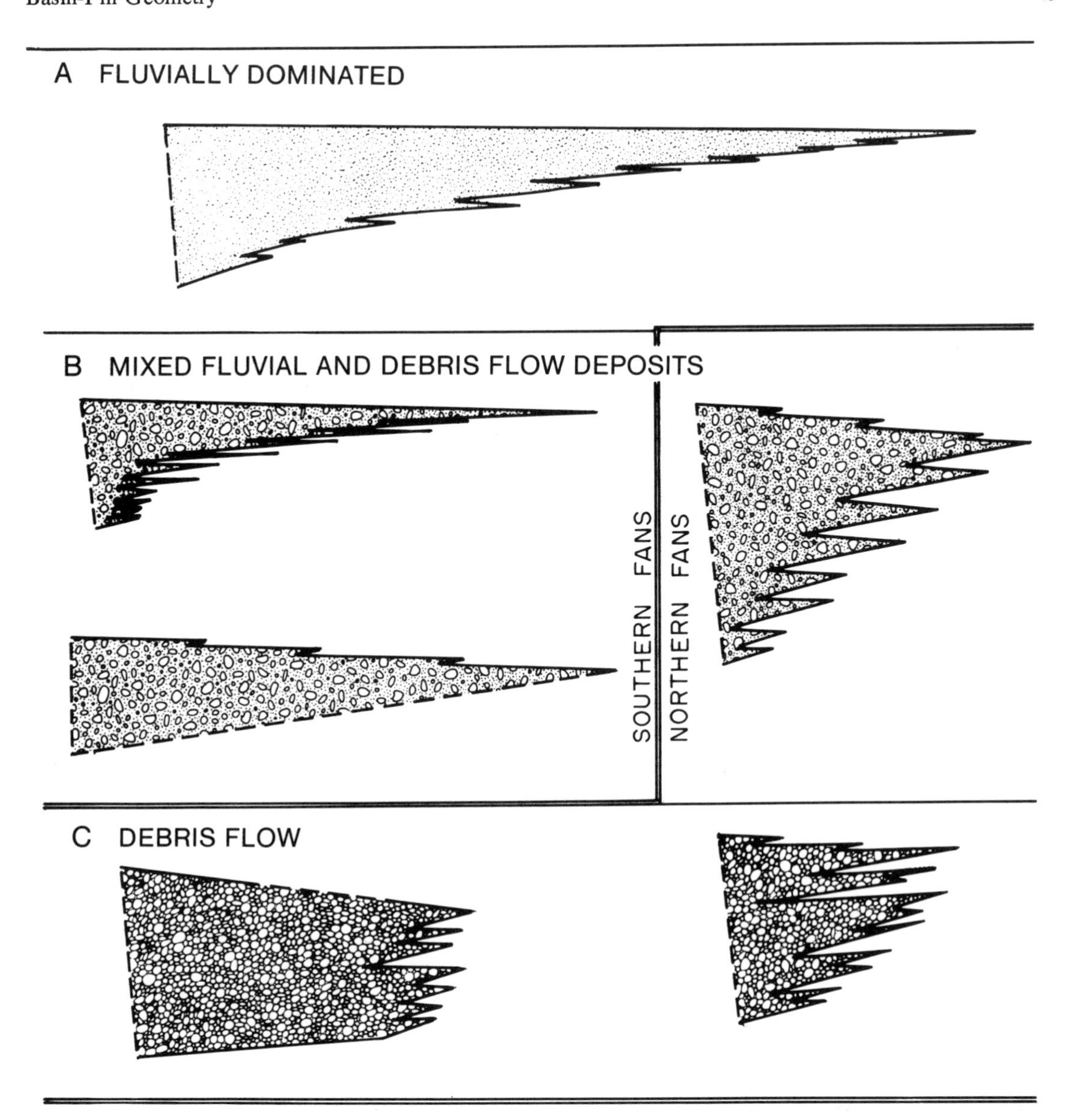

Figure 3-19. Variation in geometry of fluvially dominated, mixed, and debris-flow dominated fans in the Devonian of Norway. The former have a large radius of 4 km (2.5 mi) or more and a low-angle interface with more distal facies. Fans involving debris flow have a radius of 1-2 km and generally steeper interfaces with adjacent facies. The northern and southern margins of the same basin were dominated by debris flows and fluvial processes, respectively. (After Gloppen and Steel, 1981.)

Steel (1981) to intervals of significant strike-slip displacement superposed on steady subsidence.

Influence of the Catchment Basin

As pointed out by Steel (1976), the size and topography of a drainage basin strongly influences runoff distribution, so that fan systems supplied by large catchments may show characteristics of humid-region fans even in areas of relatively low rainfall. Similarly fans in areas of high rainfall, but with small catchments, may show features regarded as typical of arid-region fans. For example, tropical Honduran fans (Fig.

3-12) in a high rainfall area are comparable in size to Anstey's (1966) modal arid fan size of 1.5 to 8 km dimensions. Their slopes (Fig. 3-7) are intermediate between arid-region fans and large wet fans (Schramm, 1981). Schramm ascribes this steep profile to stabilizing vegetation in the source area and on the fan surface, and to the fixed apex. The small catchment basins are probably also an important factor.

In some basins, alluvial fans containing a preponderance of debris-flow or fluvial deposits, respectively, are present in different segments or on opposite sides (Steel, 1976; Gloppen and Steel, 1981). Where the basins, such as those of Devonian age in Norway, are small and relatively uniform in terms of source rock and climate, Steel attributes the contrasted fan types to differences in the fault-controlled paleotopography. An analogous modern situation is provided in Death Valley, California (Denny, 1965; Hooke, 1972), where fans associated with the dipslope of up-faulted blocks are substantially larger than those associated with fault scarps. In Norway (Fig. 3-19) small, steep fans consisting largely of debris-flow deposits accumulated along the northern, most actively subsiding faulted margin; thinner but more extensive fans consisting mainly of braided-stream deposits accumulated along the more subdued southern margin, characterized by larger drainage basins and more evenly distributed runoff. The distal interface between debris-flow-dominated fan deposits and adjacent basinal facies tends to be very steep in contrast to fans with a fluvial component (Gloppen and Steel, 1981).

Evolution of Alluvial-Fan Systems through Geologic Time

Prior to the advent of effective vegetation cover in mid-Paleozoic time, fans were probably the prevalent depositional system wherever sufficient relief existed. Alluvial-fan deposits are an important component of some of the oldest known sedimentary rocks, and were particularly abundant in the vast, unvegetated interior basins of Proterozoic and early Paleozoic age (Tankard *et al.,* 1982). Many such basins displayed a series of alluvial fans converging from the mountainous rim toward the lacustrine center. Others involved numerous transversely opposed fans between which a major channel system drained longitudinally.

The larger extent of ancient alluvial-fan systems in comparison with those of today is attributed by Schumm (1977) to the greater proportion of stream and sheet-flood deposits in Precambrian and lower Paleozoic fans. Prior to late Paleozoic development of terrestrial vegetation, greater sediment yield and deeper and more frequent floods deposited extensive tracts of braided alluvium (Rust, 1978). Debris flows are nevertheless a significant component of some alluvial fans deposited in prevegetation times. Rust pointed out that absence of vegetation could have enhanced debris flows, because the necessary clays were abundant in many areas. It is generally accepted, however, that the early record is biased toward stream and sheetflow deposits. Since vegetation became established, fans dominated by debris-flow processes have become more prevalent (Schumm, 1977).

Chapter 4

Fluvial Systems

Introduction

Fluvial systems serve primarily to collect and transport sediment into major lacustrine or marine basins. However, in certain settings favoring subaerial accumulation of sediment, such as subsiding coastal plains, intermontane basins, and tectonic forelands, fluvial depositional systems may become a major, or even dominant component of the basin's fill. Even where their volumes are minor, fluvial facies are often disproportionately important as hosts for mineral fuel deposits, particularly petroleum and uranium.

Fluvial depositional systems are primarily aggradational. Localized progradation and lateral accretion occur within specific environments. The locus of deposition and most important feature of the aggrading alluvial surface is the channel. In plan view, single channel segments form a continuum ranging from low sinuosity to high-sinuosity patterns (Fig. 4-1). Braided channel patterns are a well-known manifestation of sand-rich, low-sinuosity channels, in which flow during low-river stage weaves between multiple channel bars. The channel bars are submerged and become active, large-scale bed forms during high flow.

At the opposite end of the continuum, mud-rich, low-sinuosity channels are typified by submerged, migrating sand waves or alternating, side-attached bars (Fig. 4-1). True straight-channel segments of great length are rare. A broad, intermediate assemblage of channel segments that are moderately to highly sinuous are usually described as meandering channels.

Two multiple channel patterns, anastomosing and distributary, are common in aggrading fluvial systems. Anastomosing patterns occur where contemporaneous branches of a single river weave around permanent, commonly vegetated islands or disconnected segments of floodplain (Fig. 4-1). Anastomosing patterns are most common on extremely low-gradient alluvial plains (Schumm, 1968a). Individual channel segments within the anastomosing network may be either straight or highly sinuous. Distributary patterns are characteristic of certain types of rapidly aggrading alluvial surfaces, including deltaic plains and large alluvial fans. In both anastomosing and distributary channel networks, flow separates into two or more channels. Presence of contemporaneous multiple channels, if it can be documented in an ancient fluvial system, has considerable paleogeographic and sedimentologic significance.

Depositional Processes

Though commonly not considered as such, the fluvial system is an efficient sorter of the heterogeneous sediment load it receives. Several processes combine to segregate bed load and suspended load into specific depositional environments and to preferentially trap and retain the bed-load fraction.

Channel Flow

Flow within a channel, and its effect on the erosion, transport, and deposition of sediment, is determined by the distributions of velocity and turbulence. Areas of maximum velocity and turbulence are likely to be sites of erosion and bypass of sediment; conversely, areas of relatively low velocity and turbulence are likely to be areas of bed stability or deposition.

In a sinuous channel during low-flow conditions, the thread of maximum velocity hugs the outer or concave banks and cuts diagonally across the intervening straight reaches (Fig. 4-2). The swing of the thread of flow across the channel and its location near the water surface produce a second-order helical flow that moves across the channel base and up the sloping inner bank. Thus, sediment in transport tends to move across the channel, up the inner bank, and into a relatively low-velocity area, where it may be deposited. At

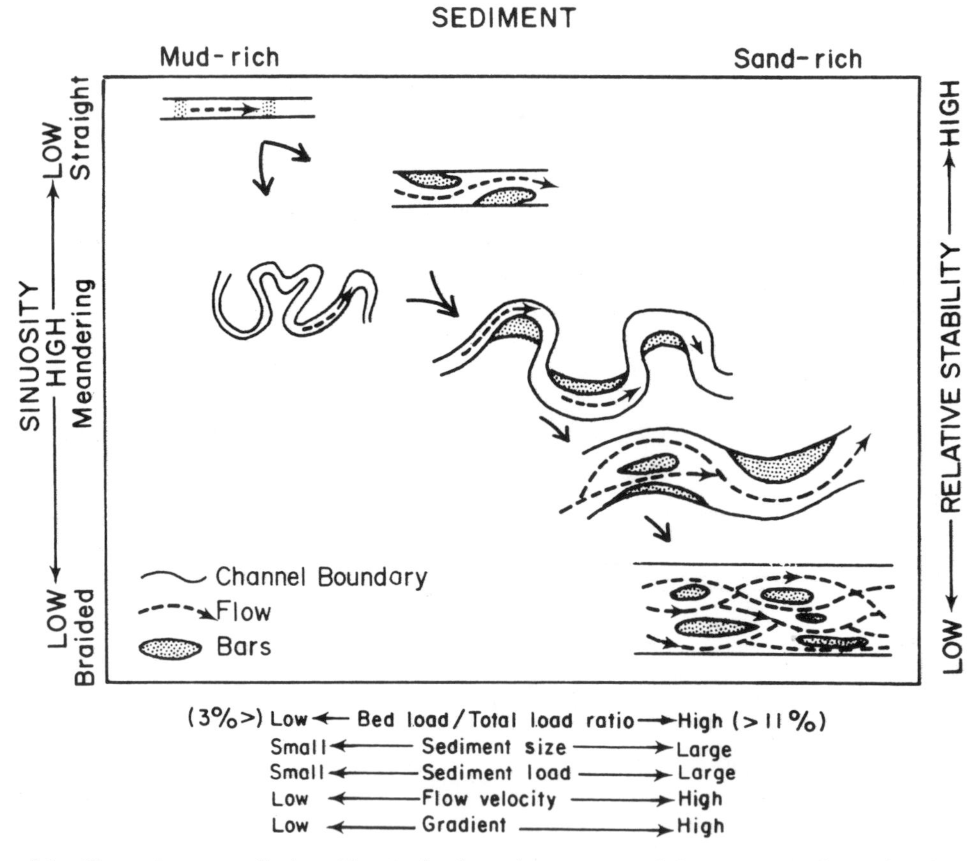

Figure 4-1. Channel patterns displayed by single-channel segments and the spectrum of associated variables. (Modified from Schumm, 1981.)

high-flow levels, the water takes a straighter path; the thread of maximum velocity moves toward the inner bank (Fig. 4-2) and may shoot across the bar that has been deposited there.

In the sinuous portions of the channel, maximum turbulence occurs near the base of the channel against the outer bank; in the straight reaches, it occurs along both channel margins (Fig. 4-2). Erosion is greatest in turbulent areas, and undercutting with concomitant bank slumping is a common mechanism of stream widening.

Stream meandering, which is an ongoing accentuation of channel sinuosity, is the natural consequence of the asymmetrical distribution of flow velocity and turbulence within a channel bend. The concentration of bed and bank erosion against the outer, concave bank and deposition of sediment on the low velocity, less turbulent, gently sloping inner bank cause a cumulative lateral migration of the channel bend. Sediment eroded from one cut bank moves downchannel to be redeposited on the laterally accreting convex bar of another meander loop.

In straight channel segments the thread of maximum velocity lies near the top center of the channel, and erosion likely occurs along both channel banks where turbulence is high.

The volume of water moving through a channel segment is a function of the cross-sectional area of the channel and the average flow velocity. Average velocity, in turn, is a function of hydraulic head, or the actual slope of the water surface (measured in change of elevation per unit distance). Thus, increase of flow during flooding has several predictable consequences. Straightening of the flow lines increases head by shortening the distance between successive points of lower elevation. Maximum efficiency is achieved by a straight channel, which has the same head gradient as the slope of the alluvial surface. Scouring

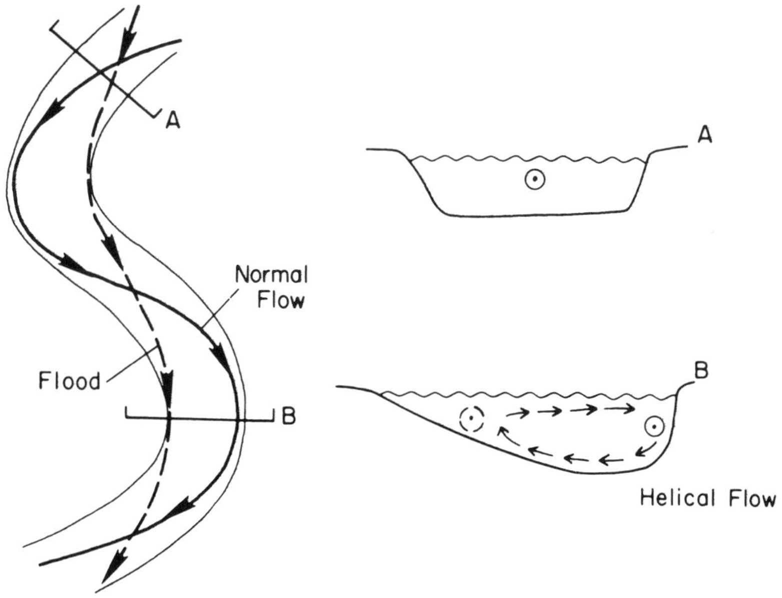

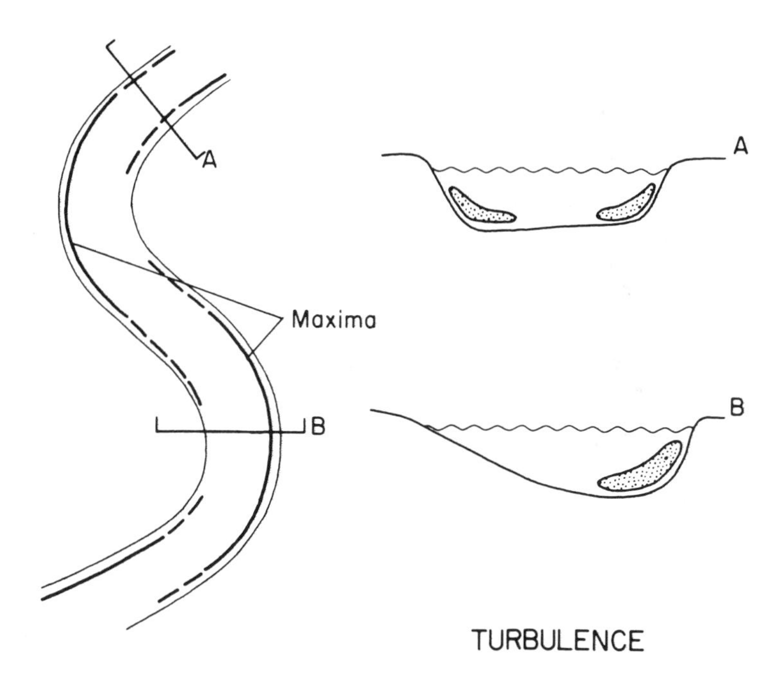

Figure 4-2. Distributions of velocity and turbulence within a sinuous channel segment.

of the bed and banks increases channel cross-sectional area. Bank scour and consequent widening of the channel is accentuated in low-sinuosity channel segments; conversely, channel deepening tends to characterize highly sinuous channels. If scouring and straightening are inadequate to handle river flow, water tops the channel banks and spills out into the surrounding valley floor or alluvial plain.

Overbank Flow

Floodbasin aggradation occurs when the sediment-laden flood waters overflow the river banks

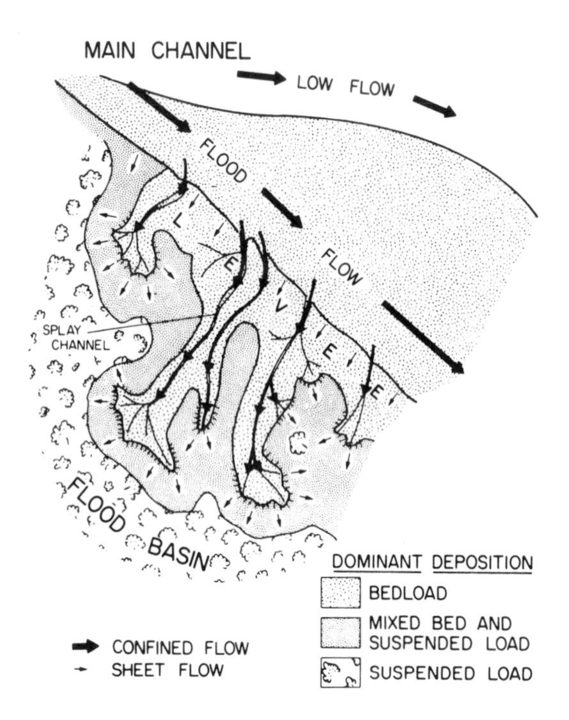

Figure 4-3. Processes and depositional framework of crevasse splays produced by flow through breaches in the natural levee of a main channel. Example based on splays of the Brahmaputra River described by Coleman (1969).

and spill across interchannel areas. The flood water, which is tapped from the uppermost portion of the river's water column, contains mostly suspended-load sediment. Upon topping the bank, flow is no longer confined, and, perhaps aided by the baffling effect of vegetation, flow velocity decreases abruptly. Sediment settles out rapidly—sand and silt near the channel margin, fine silt and clay farther away. The net result is the accumulation of sediment along the channel margin, forming a stable ridge, or natural levee, and a slow layer-by-layer aggradation of the interchannel floodbasin surface.

Crevassing occurs when flood waters pour through localized breaches or swales along the channel levees. Channeling of flow through the levee may cause scouring and deepening of the crevasse channel; consequently, crevasse channels may tap deep into the water column and funnel large volumes of water and sediment out of the main channel and onto the floodplain (Fig. 4-3). However, the sediment load of the crevasse is typically dominated by suspended-load and fine bed-load material in all but sandiest streams. Crevasse flow rapidly dissipates into distributaries or sheet wash across the splay surface, and sediment soon drops out (Fig. 4-3). Grain size decreases systematically from the crevasse axis, but locally complex internal bedding results from the multiple scour-and-fill episodes that occur within the crevasse splay channels.

In smaller streams that are characterized by short bursts of extremely high discharge, differentiation of channel and overbank flow, and consequent deposits, may be arbitrary. Examples of sand sheets and gravel sheets and bars deposited upon floodplain muds and soil zones of small ephemeral and flashy streams of subarid regions have been described by McKee *et al.* (1967) and Gustavson (1978).

Levee and splay deposits accumulate close to the margins of active channels. Floodplain deposition, in contrast, may spread up to several kilometers across the alluvial plain. However, rates of floodplain accretion are typically slow, commonly averaging a few centimeters per year in depositionally active systems. Thus, much of the alluvial surface is depositionally inactive at any point in its history.

Channel Abandonment

Aggrading alluvial plains are characterized by repeated shifts of active channels. On a local scale, meander loops may be cut off as flow diverts into an open chute channel or cuts a short channel segment across the neck of a meander loop. On a larger scale, entire channel segments can be gradually or abruptly abandoned during avulsion or channel diversion. With successive flood cycles the active channel builds up its levees and adjacent floodplain. Stable channels, in effect, become perched above the surrounding alluvial plain. Ultimately, levees are permanently breached and the channel establishes (or reoccupies) a more favorable course across a topographically lower portion of the floodbasin. Unlike progressive lateral migration associated with meandering, avulsion is a geologically abrupt process that repeatedly punctuates the history of an aggrading fluvial system. Leeder (1978) estimated that large-scale avulsion of many large rivers occurs on the order of 10^3 year intervals.

For various reasons, which are sometimes quite obscure, newly formed channels typically reoccupy previous channel axes. Controls favoring such reoccupation, and consequent channel stacking and amalgamation, include subtle tectonic influence on local subsidence rates, the comparative ease of erosion of sandy deposits of previous channel axes relative to floodplain muds and soils, and the presence of imperfectly developed levees at the nodes where earlier avulsions occurred. Vertical stacking of Holocene channel axes has been well described in studies by Bernard *et al.* (1970) and Winker (1979) of the Brazos alluvial plain, and it is apparent in many ancient fluvial sequences.

Upon avulsion, the abandoned channel segment becomes a topographic low within the relatively featuresless alluvial plain. It may then become the site of a local underfit stream, or it may form a series of isolated lakes. Except at the upstream end adjacent to the point of avulsion, most or all of the abandoned channel is removed from sources of terrigenous sediment other than floodbasin muds that may be washed in through minor tributaries or by major floods of the trunk stream. Consequently, the abandoned channel likely fills with suspended sediment, *in situ* organic debris, or small prograding lacustrine deltas, particularly if the newly occupied channel lies many miles distant.

Fluvial Facies

Fluvial depositional systems consist of a mosaic of genetic facies, including various combinations of channel fill, channel margin, and flood-basin deposits. Channel facies are, of course, the most diagnostic component, and their recognition is the key to interpretation and mapping of fluvial systems.

Channel-Fill Facies

Channel deposits contain most of the bed-load sediment retained within the fluvial system, and thus form the skeletal framework of the system. They include both aggradational and lateral accretion depositional units. Internal structure of the channel fill is determined primarily by the geometry of the channel. Bed accretion dominates within low-sinuosity sandy channels. Lateral accretion characterizes high-sinuosity channels.

Low-Sinuosity Channels. Low-sinuosity channels occur in both sand-rich and mud-rich fluvial systems, each characterized by a very different type of channel-fill facies. Bed-load or sand-rich, low-sinuosity channels (Fig. 4-4) include a variety of specific depositional features including lateral, transverse, and longitudinal bars.

Lateral bars or alternating sand bars form along the margins of low-sinuosity channel segments. Such bars are exposed during low flow and submerged during flood, when coarse material can be washed across the bar surface and deposited on the downstream margin. Primary structures include planar and low-angle accretionary foreset bedding (Fig. 4-4, sequence A). Transverse bars are downstream-migrating sand bars oriented transverse to flow, and are typical of sand-bed braided channels. During flood, sediment moves up the sloping, upstream bar flank and cascades down the lee side, producing avalanche or tabular cross bedding within the bar (Fig. 4-4, sequence B). The bar crest displays planar stratification, though subsequent erosion during bar migration may remove bar crest deposits. Transverse bars may be dissected during low-flow conditions. In contrast, longitudinal bars display long axes parallel to the flow, and are common in many braided streams (Fig. 4-4). During flood, shallow water flowing across the bar surface creates abundant horizontal stratification. Accretion along bar margins and in the downstream direction produces low- to moderate-angle cross stratification (Fig. 4-4, sequence A).

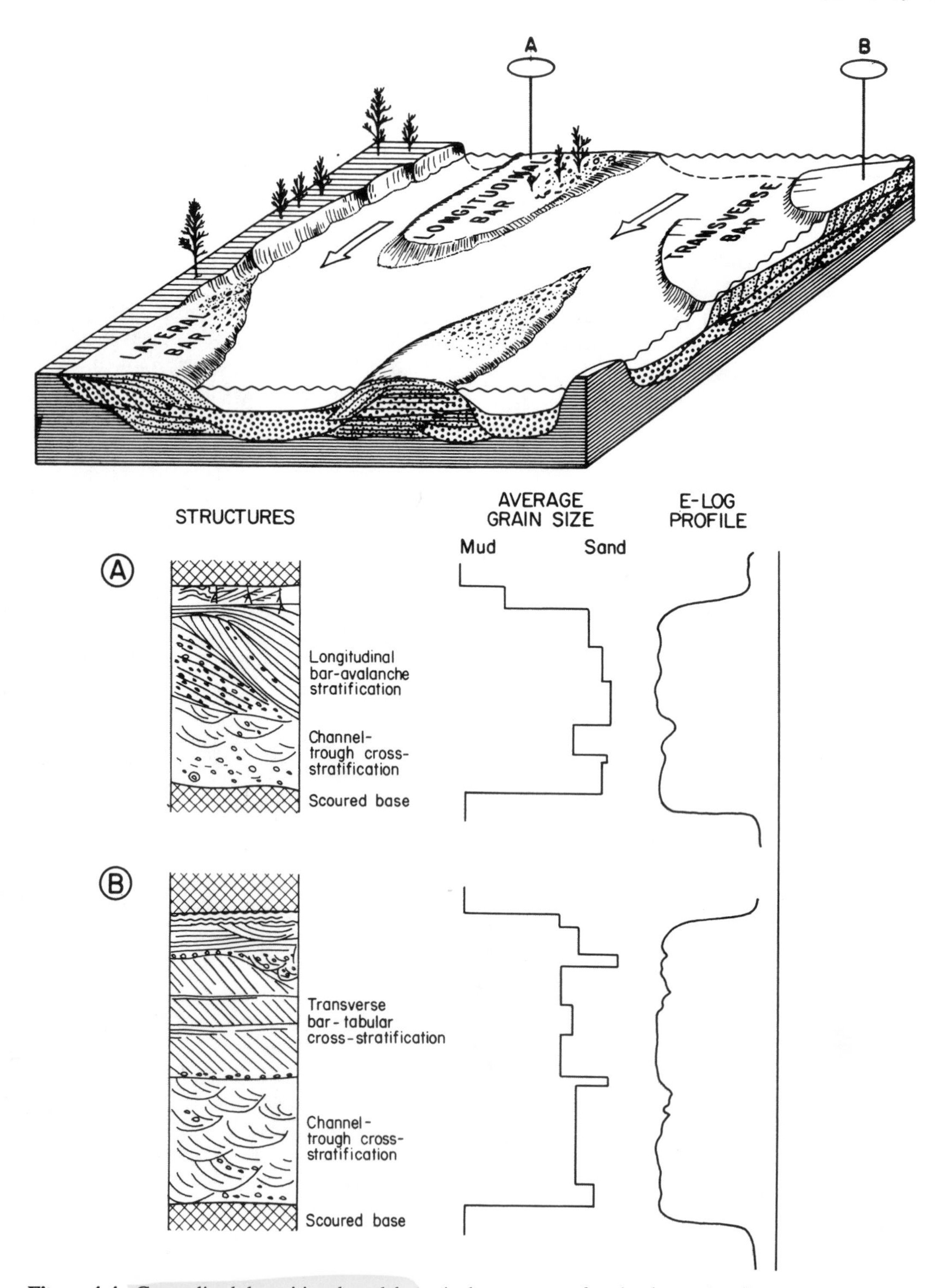

Figure 4-4. Generalized depositional model, vertical sequences of grain size and sedimentary structures, and electric (S.P.) log profiles produced by a low-sinuosity, braided channel. Sequence A is dominated by migration of a gravelly longitudinal bar. Sequence B records deposition of successive transverse bar cross-bed sets upon a braid channel fill.

During low flow, multiple braid channels interweave between the bars, locally eroding bar margins.

Channel-fill lenses occupy interbar areas, and accumulate as lenses of bed-load sediment that may be preserved in aggrading channels. Lenses may be structureless or display trough cross stratification, where water depths were adequate for formation of migrating subaqueous dunes (Fig. 4-4, sequences A and B). Channel plugs are volumetrically minor in most sand-rich low-sinuosity channel sequences, but do form local, thin, discontinuous lenses of sandy mud and clay that fill abandoned braid channels. Organic debris is a minor component of many such plugs.

Braided-stream models based on modern braided or low-sinuosity channels have been described by several authors including Coleman (1969), Cant and Walker (1978), Doeglas (1962), McKee *et al.* (1967), Rust (1972a), Smith (1970) and Williams and Rust (1969) and were reviewed by Miall (1977a, 1978). A generalized braided stream model, illustrated in Fig. 4-4 and summarized below, exhibits typical aspects of a low-sinuosity, sandy, channel-fill sequence.

Channel fills form broad, tabular, dip-oriented, multilateral sand belts with high width/thickness ratios. Basal scour surfaces are typically flat and display proportionally low erosional relief (Chitale, 1973). Channel-fill sequences consist dominantly of sand and are commonly conglomeratic. Mud and clay are volumetrically minor; sorting ranges from poor to good.

Depositional units include interbedded transverse or longitudinal bar lenses, or both, laced with abundant small- to medium-sized sand- to mud-filled scour channels. Internal structure is complex in detail, but thorough amalgamation of individual depositional units produces thick, widespread units of uniform composition.

Vertical sequences are commonly poorly developed within the bed-load channel facies. Minor textural fining-upward occurs within some scour channel fills and at the top of the composite sand body. Coarse sediment is typical throughout the sequence, which may consist of numerous incomplete cycles of channel and bar deposition. Multiple, thin fining-up and some coarsening-up packages occur within the sand body. A relatively simple suite of internal structures is dominated by horizontal stratification and planar, low-angle avalanche and tabular cross stratification. Trough cross stratification ranges from rare to common. Various types of ripple lamination may occur if sufficient fine sediment is deposited and preserved, but are generally a minor component of preserved sequences. A multiplicity of local scour surfaces, discontinuous lags, and diffuse pebble sheets occurs within the sand body.

Mud-Rich Low-Sinuosity Channels. Mud-rich low-sinuosity channel (Fig. 4-5) deposits differ greatly from their coarser counterparts. Channel cross sections are commonly highly convex and symmetrical. Lateral bars may form, but during waning flow or channel abandonment, simultaneous bed and bank accretion produces a symmetrically bedded channel-fill unit. The delta-plain distributary channel model (Brown *et al.*, 1973) provides one good example of a fine-grained low-sinuosity channel fill. Figure 4-5, section B illustrates a generalized depositional model of an anastomosing low-sinuosity channel described in Schumm (1968a), Fisher *et al.* (1969), and Smith and Smith (1980).

Channel-fill sequences form dip-elongate, narrow, lenticular units exhibiting a high-relief basal scour surface (Fig. 4-5). A low width/thickness ratio and vertical stacking of channel-fill units are characteristic. Average sand-body trend generally parallels the depositional slope, but distributing or anastomosing channel patterns are common, and may cause considerable directional dispersion. The channel fill consists dominantly of very fine sand with abundant silt and clay. Coarse material is typically sparse, but may include gravel, intraclasts, and plant debris, and typically accumulates near the base of the sequence as a lag.

Aggradational channel fill sediment consists of bed and suspended-load material deposited along the bed and banks. The pronounced bank accretion is indicated by well-developed assymetrical bedding produced in slightly sinuous segments (Fig. 4-5, sequence A), or symmetrical bedding produced in straight segments (Fig. 4-5, sequence B). Mud plugs are common and well developed. Prominant natural levees flank the channel fill. Fine-grained low-sinuosity channel fills commonly fine upward, but the vertical trend may be obscured, particularly in the subsurface, by the limited range of grain sizes available (Fig. 4-5, sequence B). Massive lower channel-fill sands

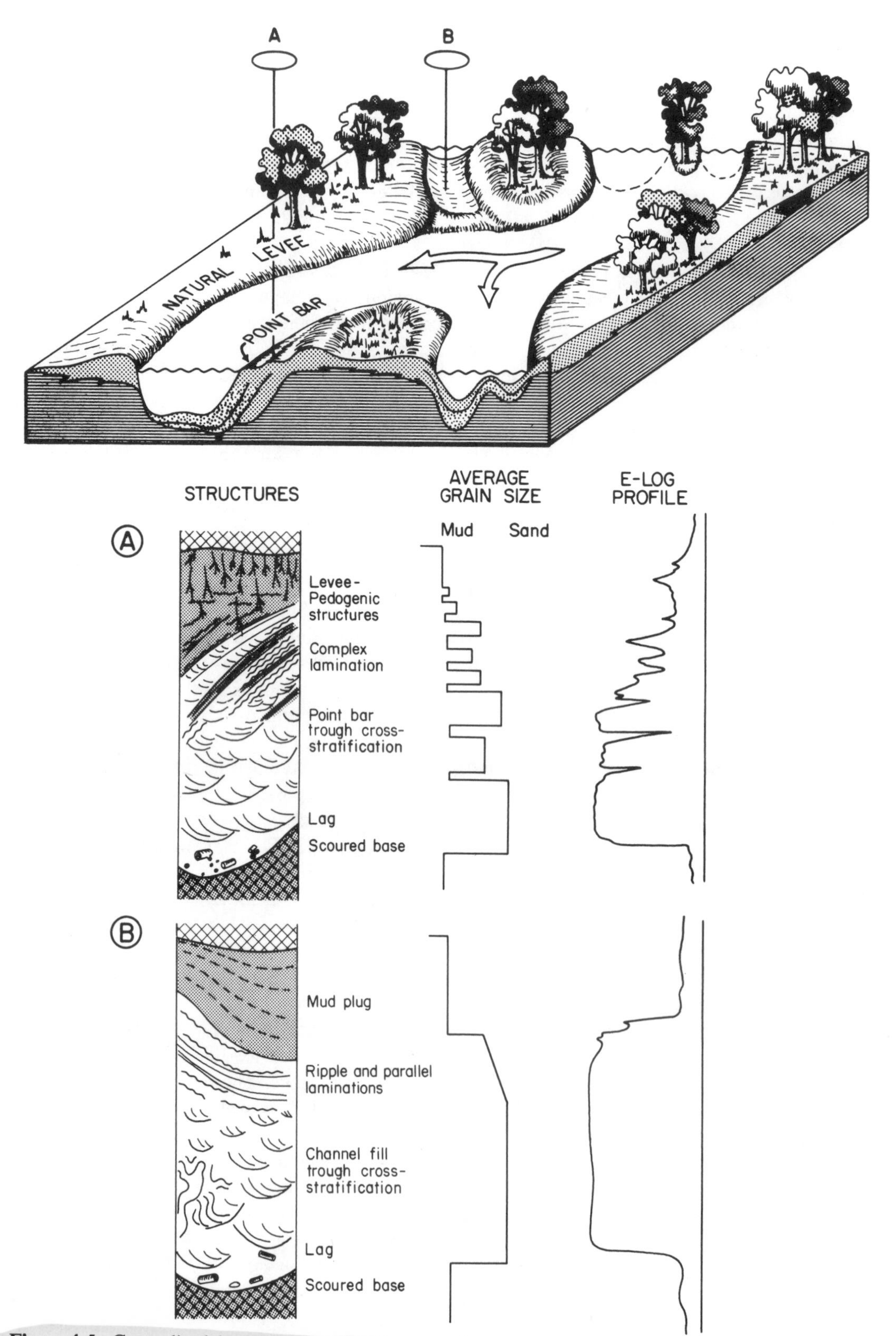

Figure 4-5. Generalized depositional model, representative vertical sequences, and idealized electric (S.P.) log profiles through laterally accreting (A) and symmetrically-filling channel segments (B) of an anastomosed channel system.

may grade up into or be abruptly overlain by a mud plug.

Large to small-scale trough cross stratification dominates internal structures; soft-sediment deformation is common to pervasive. Structures commonly show a poorly developed upward decrease in size, especially near the top of the channel-fill unit (Fig. 4-5, sequence B). Mud plugs contain wavy, ripple, and parallel lamination; local bioturbation may be evident. Root disturbance of primary structures is common and may be pervasive.

High-Sinuosity Channels. Meandering channels produce the features most popularly associated with fluvial sedimentation. Local environments, which produce diagnostic genetic subfacies, include the channel floor, point bars (Fig. 4-6), chutes and chute bars (Fig. 4-7), and the abandoned channel plug.

The channel floor, or thalweg, is the deepest part of the channel, and is the site of deposition of the coarsest material transported by the river. This channel lag, which lies on or just above the basal erosion surface, consists of locally derived material such as mudclasts and blocks eroded from the banks and bottom, water-logged plant debris, and coarse bedload gravel and sand (Figs. 4-6 and 7, all sequences). Coarsest lag characterizes scour pockets produced in areas of maximum velocity and turbulence. In addition, significant amounts of bed-load sediment accumulates and may be preserved in an aggrading channel. Migrating subaqueous dunes cover the active channel floor, thus large to medium-scale trough cross bedding is the predominant internal structure (Fig. 4-6, sequences A and B).

Point bars are produced as sediment eroded from upstream channel banks or washed into the channel is transported up gently sloping inner stream banks into areas of comparatively low velocity and turbulence (Fig. 4-2), where it is deposited on laterally accreting point bars (Fig. 4-6).

Cross-sectional area of the channel is maintained by concomitant erosion of the convex or cut bank. Thus, the curvature of the meander tends to become increasingly exaggerated. Because sediment is moving up and out of the channel onto the bar, a vertical decrease in grain size characterizes the point-bar sequence. The accretionary architecture of the point bar is commonly reflected by the development of ridge and swale topography (also known as meander scrolls) on the point bar surface (Fig. 4-6) and poor- to well-defined sigmoidal lateral accretion bedding (the epsilon cross bedding of Allen, 1970). Fine sediment washed across the bar surface during floods may pond in the swales, forming local muddy lenses and plugs. Older portions of the point bar are commonly vegetated and capped by fine-grained levee and floodplain sediment (Fig. 4-6), completing the upward-fining cycle that begins with the coarse channel lag. Sand transport across lower- and midbar surfaces is dominantly by dune migration, and medium- to large-scale trough cross stratification characterizes this part of the sand body (Fig. 4-6, sequence A). Size of bed-sets decreases upward. Ripple, climbing-ripple, and tabular and planar stratification occur in the finer-grained upper point bar sequence where water depths are shallow, or flow velocities are lower, or both. The bar surface may be modified by sheet-wash and gullying as well as by rooting and burrowing during periods of subaerial exposure. Upstream portions of the point bar may form an upstream-sloping ramp onto which high-velocity channel flow spreads during peak discharge. Thus, systematic grain size variation between channel floor and lower to mid-bar is modified, and unusually coarse-grained point bar deposits may be abruptly overlain by fine-grained levee deposits (Fig. 4-6, sequence B). The fining-upward textural trend is truncated, and planar or tabular stratification are likely structures in the upper part of the sand body.

Chutes and chute bars form during flood stage as a part of the river flow cuts directly across the surface of the point bar. When this happens, significant amounts of coarse bed-load sediment move up out of the main channel and are funneled through one or more chutes or channels which have been scoured into the upstream end of the point bar (Fig. 4-7). As flow spreads across the bar surface, bed-load material is deposited, forming a chute bar. The processes, physiography, and sediments of chute-modified coarse-grained point bars have been described by McGowen and Garner (1970) and Levey (1978). The chute contains coarse lag material typically found in the main channel. Chute bars consist of relatively coarse sediment deposited by flow separation at the lee side of the bar crest. Chute-bar units are thickest where the flow reenters the main channel and the bar progrades into the

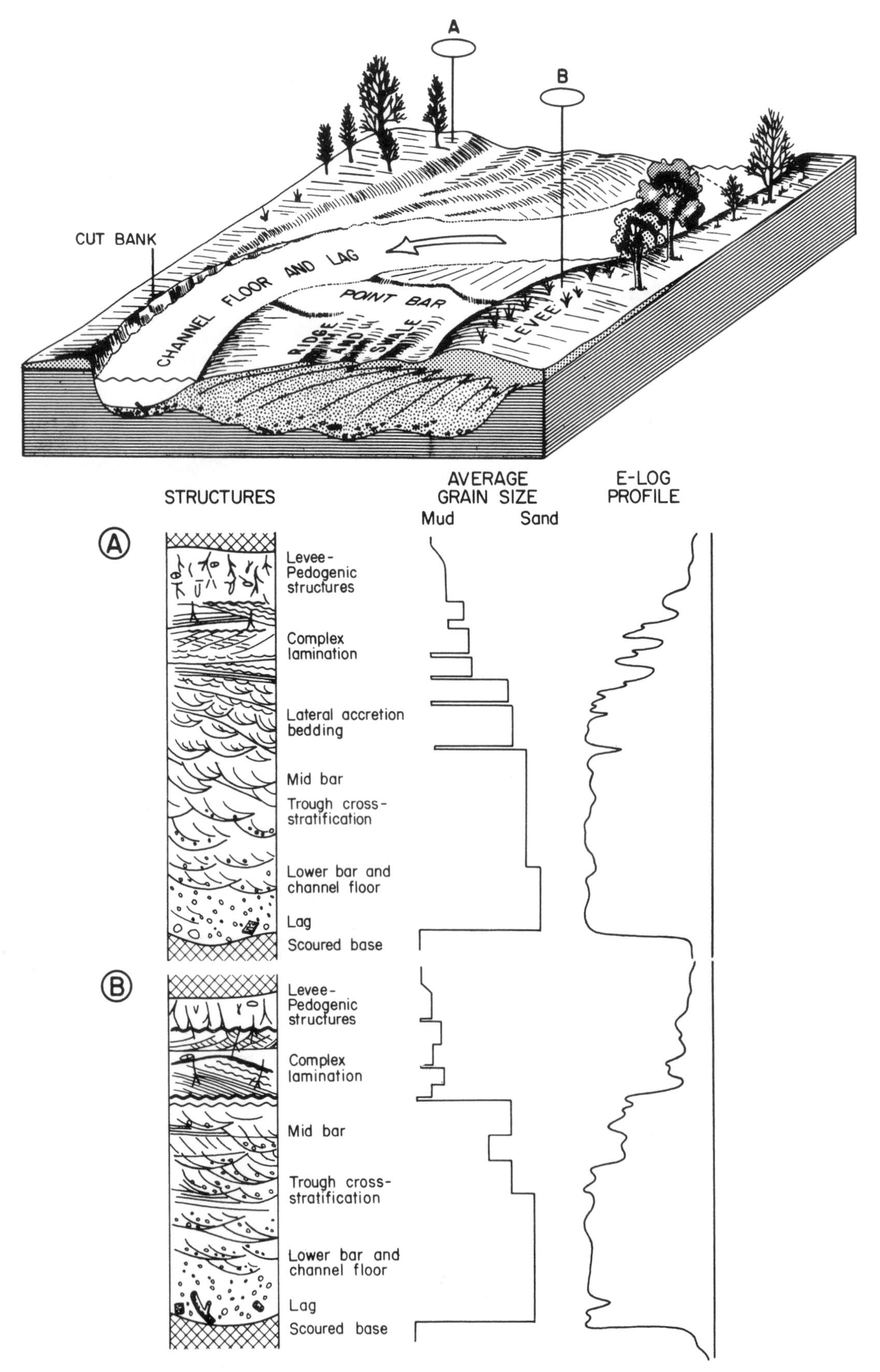

Figure 4-6. Generalized depositional model, vertical sequences, and electric (S.P.) log profiles of a meanderbelt sand body produced by a high-sinuosity channel. Sequence A illustrates a complete fining-upward sequence typical of the mid- or downstream point bar. Section B illustrates the truncated vertical sequence commonly found in the upstream end of the bar.

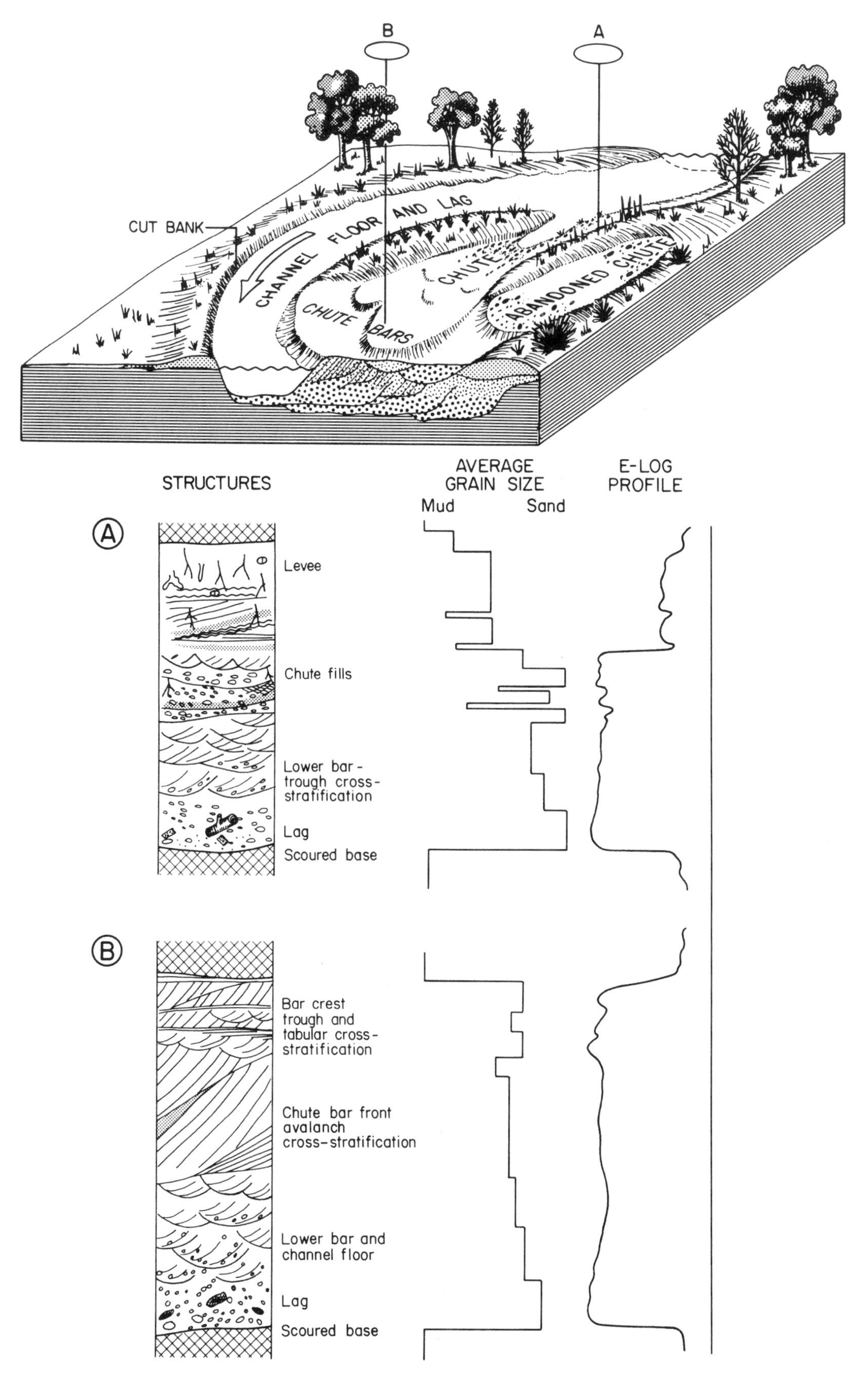

Figure 4-7. Generalized depositional model, vertical sequences, and electric log (S.P.) profiles of a chute-modified point bar. Upstream portions of the point bar are capped by chute-channel deposits (sequence A). Downstream, the channel and lower point bar deposits are capped by chute-bar sediments (sequence B).

relatively deep water of the channel. Dominant structures of the chute complex include imbricated pebble sheets, planar lamination, and mud lenses in upstream portions of the chute channel (Fig. 4-7, sequence A), trough cross stratification in distal and marginal portions of the chute, and planar and avalanche cross stratification in the chute bars (Fig. 4-7, sequence B). Cut-and-fill, associated with scour around vegetation or other obstructions to flow, is also a common feature of the chute-modified point bar. The principal result of chute development is the deposition of coarse bed-load sediment and large-scale sedimentary structures at the top of the point-bar sequence.

Two significantly different variations of the depositional model for sinuous channel-fill sequences can be summarized from the numerous studies of modern fluvial systems (Brown *et al.*, 1973; Collinson, 1978). Such models are distillations of common features found in many modern rivers; consequently, actual channel-fill sequences may be expected to vary from the idealized models presented here. As pointed out by Jackson (1975; 1978), considerable variation in composition, vertical sequence, and internal structure exists within a single point bar, and many exceptions to the generalized models exist. Nonetheless, the fine-grained meander point-bar model and the chute-modified point-bar model summarize salient attributes of many sinuous channel fills.

The *fine-grained point bar* sequence (Fig. 4-6) is the single best known fluvial model. It characterizes many highly sinuous channel systems and is readily recognized by its highly ordered internal structure. Well-described modern examples include the Brazos River (Bernard *et al.*, 1970), Mississippi River (Davies, 1966; Fisk, 1944; Frazier and Osanik, 1961), Red River (Harms *et al.*, 1963), and Wabash River (Jackson, 1975, 1976).

Belts or "beaded" chains of erosionally based meanderbelt sand lenses are dip oriented. The upper boundary is commonly, but not always, transitional into overlying sediments; multistory and vertically amalgamated sand bodies are common. Width/thickness ratios of meanderbelt sand bodies are moderate to high. Medium to very fine sand dominates the depositional unit, but a broad range of grain sizes, including gravel and mud, may occur within the complete sequence. Mud clasts and blocks, and woody plant debris are concentrated near the base.

The channel-fill deposit consists of channel floor and lag sediments, point-bar sand and silt (composed of a succession of lateral-accretion units), and topstratum materials (Fig. 4-6). Lateral accretion beds characterize these fluvial point bar sequences (Fig. 4-8). However, the best development and outcrop expression of this bedding style occurs in fine-grained, mud-rich channel fills (Jackson, 1978, 1981). Lower dip angles, complex point-bar morphology, and limited outcrop size make its recognition difficult in larger and coarser-grained meanderbelt deposits. Thin swale-fill mud lenses may locally cap the point-bar units. Fine-grained channel plugs record the final filling of the channel following abandonment. Point-bar sands are characterized by medium- to large-scale trough cross stratification with local zones of tabular, planar, and ripple stratification (Fig. 4-6, sequences A and B). Finer-grained sediments at the top of the sequence contain abundant small-scale structures, including ripple and climbing-ripple lamination, mud drapes, and root traces. The average current vector of internal directional features parallels the sand-body trend, but considerable variability exists locally.

Channel fill units of highly sinuous streams are characterized by a general upward decrease in both grain size and scale of sedimentary structures, but vertical trends may be gradational or quite erratic in detail, depending on flow variability of the river, range of sediment sizes available, position on the point bar, or other factors.

The *chute-modified point bar* model (Fig. 4-7) illustrates the impact of increasing discharge variability, and is commonly applicable to relatively coarse-grained fluvial systems (consequently, it has been described as the coarse-grained point bar model by most authors).

Sand belts are broad, laterally and vertically amalgamated, and generally dip oriented. The width/thickness ratio of the sand body is moderate to high, and development of coalesced multilateral channel fills is common. Channel fill is composed chiefly of fine to coarse sand; gravel zones are common, especially near the base and top of the sequence. Mud clasts and large, fragmental to finely macerated plant debris are minor constituents. Fine-grained sediments are limited to thin topstratum and channel-plug deposits.

Channel lag is overlain by channel floor and

Figure 4-8. Lateral-accretion bedding characteristic of point-bar deposition within a high-sinuosity channel segment.

lower point-bar sands and gravels (Fig. 4-7, sequences A and B). The upper bar sequence consists of interlensing, scour-based chute fill (Fig. 4-7, sequence A), and accretionary chute bar sands (Fig. 4-7, sequence B). Large- to medium-scale trough cross stratification with some tabular cross stratification dominates the lower portion of the channel fill. Large-scale avalanche and tabular cross stratification, as well as troughs, scour-and-fill, discontinuous gravel lags, and minor planar and ripple laminations characterize the upper part of the sequence. The average vector of paleocurrent indicators parallels the channel trend, but considerable local variability results from the complex depositional patterns.

In contrast to the trend characteristic of the fine-grained meanderbelt model, grain size exhibits little vertical fining. In fact, some of the coarsest sediment and largest structures may occur at the top of the sand body in chute and chute-bar units (Fig. 4-7).

Both the simple and chute-modified point bar sand bodies can occur along a single fluvial segment; however, discharge characteristics and nature of the sediment load usually favor one or the other depositional style. In addition, a plethora of intermediate and hybrid combinations exists. Together with the sand-rich and mud-rich straight channel examples, the models illustrate the potential complexity of channel-fill facies and, at the same time, provide idealized examples of four important members of the fluvial spectrum.

Channel Margin Facies

During flood stage, some bed-load and considerable suspended-load sediment is deposited along channel margins as waters overflow the confining banks or funnel through local breaches. Such extra-channel flow is mostly unconfined; consequently, velocity decreases rapidly away from the channel, and most of the entrained sediment is quickly deposited. Only the finest suspended sediment is transported into the interchannel floodbasins. Two distinct channel-margin environments occur: natural levees, which bound the channel, and crevasse splays, which extend into the floodbasin from breaches or low areas of the levees (Fig. 4-3).

Natural Levee. Fine sand, silt, and some clay is deposited along the margins of the channels as decelerating, suspended-load rich waters spill over the banks. Increments of sediment are built up with each successive flood, so that levees become major topographic features on the otherwise featureless alluvial plains of large rivers. The internal sedimentary features of levees reflect rapid deposition, multiple waning flow cycles, shallow flow depths, and intermittent periods of subaerial exposure. Sedimentary structures are dominated by fine ripple, climbing-ripple, wavy and planar lamination, abundant clay drapes, laminated mud layers, and root-disturbed zones (Fig. 4-6, upper portion of sequences A and B, for example). Local soft-sediment deformation and scour-and-fill structures also occur. Levees are subjected to repeated wetting and drying; thus, the sediments are compacted, oxidized, and highly leached. Pedogenic carbonate and iron oxide nodules and concretions are common.

Crevasse Splay. Local breaches in the levees funnel the flow from the channel, tapping lower portions of the water column and providing conduits for suspended- and bed-load sediment dispersal into near-channel portions of the floodbasin (Fig. 4-3). Small distributary or braided channels extend across the splay surface, and both channelized and unconfined flow occurs during flood events. Splays build partly by progradation into the floodbasin (which may be covered by standing water during floods when the crevasse splays are active) and by aggradation as bed load and suspended sediment is deposited as flow spreads across the splay surface (Fig. 4-9).

In flood-prone river systems, splays may become quite large, covering several square kilometers and coalescing into broad aprons flanking the main channel (Fig. 4-10). Sand isolith maps of such depositional systems generally depict both the channel-fill sands and the marginal, genetically associated splays.

The internal structure of crevasse splays is characterized by heterogeneity, and reflects their origin by multiple flood events, shallow flow conditions, and rapid sedimentation rates. Where the floodbasin consists of permanent interchannel lakes, progradation of the splay produces a vertical sequence analogous to that of a small lacustrine, or Gilbert delta (Fig. 4-9). Multiple, coalescing depositional units include small-to-

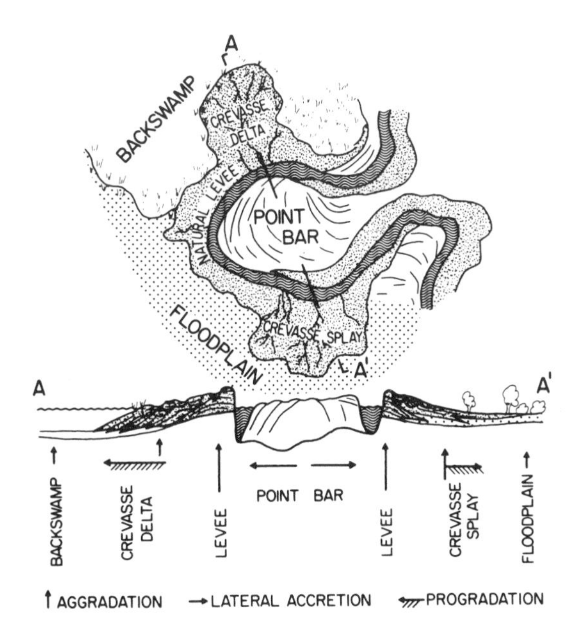

Figure 4-9. Channel margin and interchannel flood basin depositional environments include crevasse splays, which may prograde into flooded backswamps or interchannel lakes or aggrade on the subaerial floodplain, and the levees that flank fluvial channels. Geometry of internal depositional units reflects the aggradational or progradational history of the facies.

large splay channels and scours, interchannel sheet-flood remnants, mud drapes, and paleosols. Ripple, climbing-ripple, planar, wavy, and medium-scale trough cross-lamination, mud drapes, graded beds, and local scour-and-fill structures are common. In addition, splays are sedimentary "garbage piles", accumulating large amounts of plant debris and mud clasts. Grain sizes and unit thickness are generally less than that of associated channel sequences, but otherwise can be quite variable, depending on rate of floodplain aggradation, magnitude of flooding, effectiveness of the confining levees, and water depths in the floodbasin. Upper portions of splays may be highly compacted, oxidized, and leached where they stand above the local water table. Paleosols and root-disturbed zones are commonly interspersed through the splay sequence.

Interchannel Floodbasin Facies

Fine suspended-load sediment reaches the broad interchannel areas during floods. Here sedimentation rates are low, and reworking by burrowing, plant growth, and pedogenic processes typically destroys primary structures. In dry climates, where the water table is low, the floodbasin is a dry floodplain which may be vegetated by trees or grasses, or locally veneered by migrating eolian dune fields. Wet climates with their characteristically shallow water tables produce backswamp or interchannel lacustrine environments in the floodbasin. As will be discussed in Chapter 11, location within the ground-water basin may override climate in determining the nature of the floodbasin. High organic productivity, low rates of terrigenous sediment influx, and the shallow water table make certain backswamp environments ideal for deposition and preservation of plant debris. Major peat deposits may therefore accumulate in backswamp-lacustrine environments.

Abandoned channel segments lying within the floodbasin produce a volumetrically limited but important and distinctive facies typically referred to as the mud plug (Fig. 4-11). Channel plugs may in fact consist of muddy sand to nearly pure clay, but they are characterized by fine grain size relative to the channel-fill deposits in which they are embedded. Muddy channel plugs provide a true picture of actual channel dimensions, thus the plug is narrow (tens to a few hundred meters) and elongate. In complex meanderbelt sequences, channel plugs intricately compartmentalize upper portions of otherwise laterally extensive meanderbelt sand bodies (Fig. 4-11). Their geometry, position within or at the top of channel-fill sand units, and contrasting grain size provide diagnostic criteria that may be readily apparent in subsurface geophysical logs (Fig. 4-12). Dense

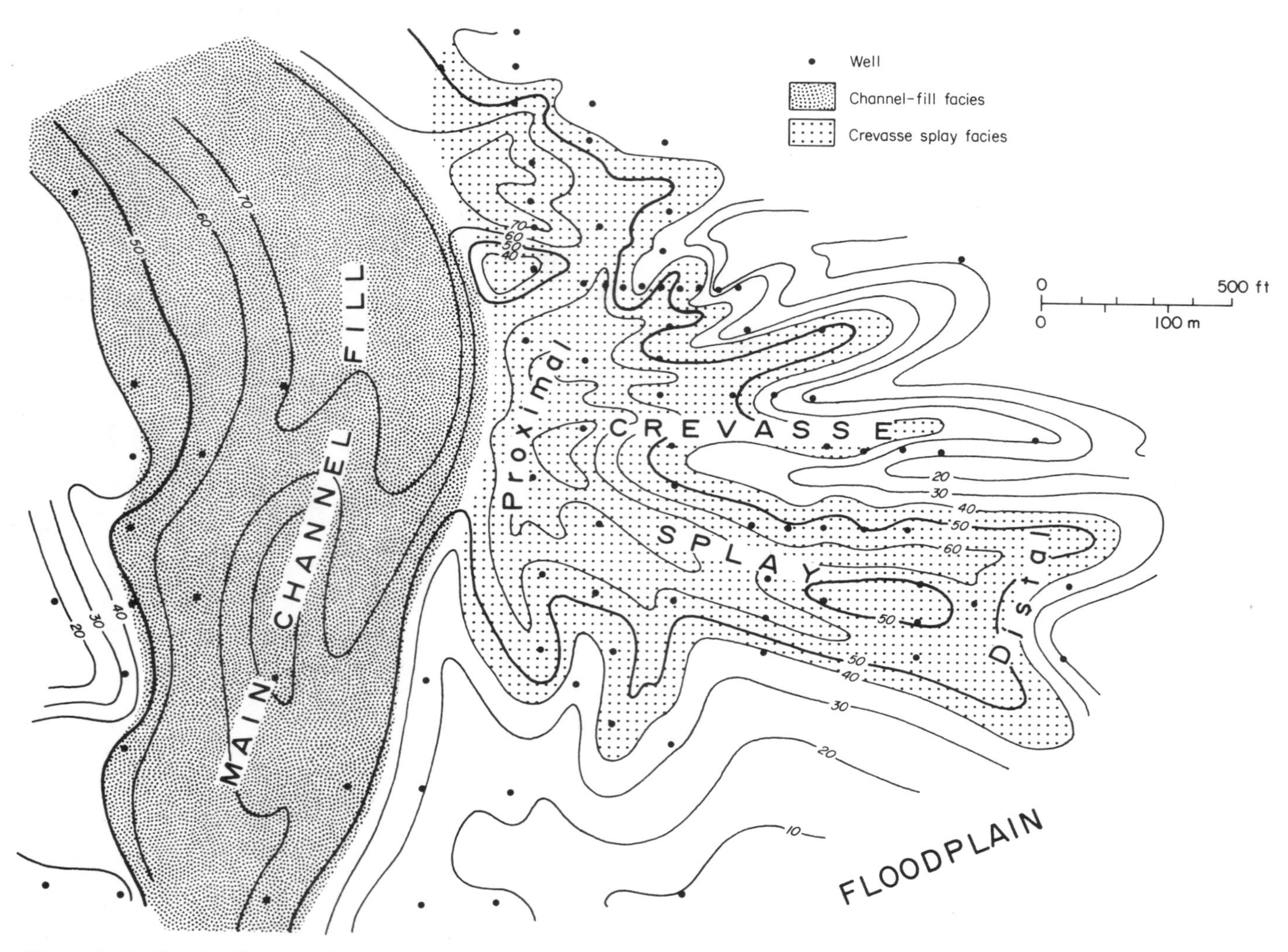

Figure 4-10. Sand isolith map of a large crevasse splay marginal to a sinuous channel fill. Splay thickness approaches that of the main channel. The splay deposits are a heterogeneous mix of fine to coarse sand and lenses of mud and gravel; Catahoula Formation, Oligo-Miocene, Texas Coastal Plain. (Modified from Galloway and Kaiser, 1980.)

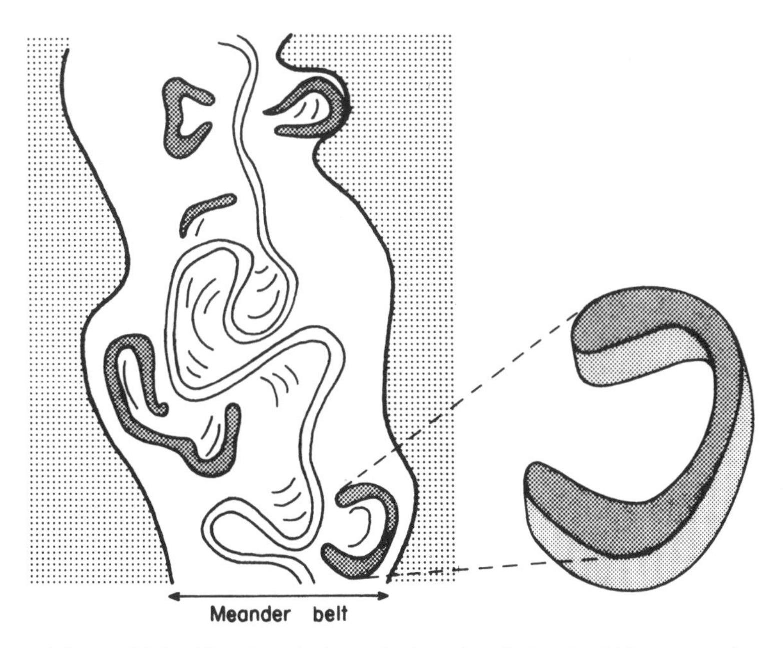

Figure 4-11. Abandoned channel plugs (mud plugs) within a meander-belt sand body.

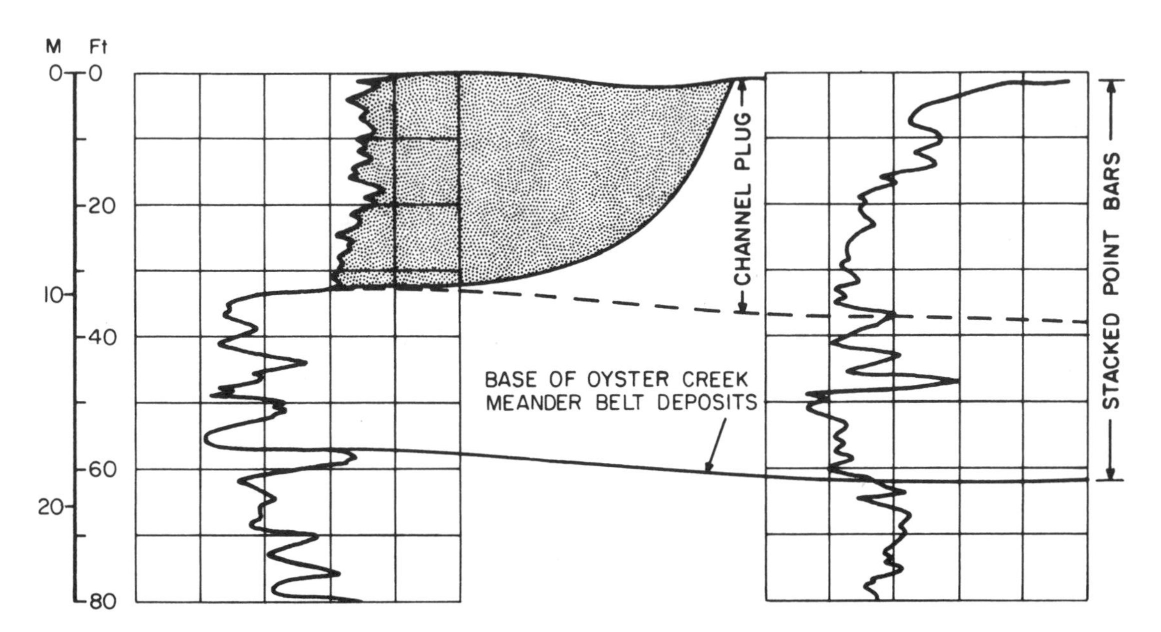

Figure 4-12. Electric logs of core holes drilled through a point-bar sand body and associated channel plug in the Brazos River alluvial plain (Texas Gulf Coast, U.S.A.). The exaggerated thickness of the total sand sequence is due to stacking of the youngest meanderbelt sequence (approximately 60 ft or 20 m thick) on the older point-bar sands of a rising sea-level-stage channel. (Modified from Bernard *et al.*, 1970.)

well data are required to map individual plugs accurately.

Depending upon climate and depth to the water table, abandoned channel segments may form permanent or seasonal interchannel lakes which fill with locally derived floodplain sediment, suspended river sediment washed in during floods, or unusual constituents such as air-fall volcanic ash or peat. Sedimentary structures include types common in fresh-water lakes, including steeply dipping delta foresets, load and fluid-escape structures, fine-scale lamination, and burrowing and root-mottling.

The Spectrum of Fluvial Depositional Systems

A fluvial system consists of a skeleton of fluvial channel-fill facies and closely associated but diverse splay and levee facies within a matrix of floodbasin muds and organics. Fluvial systems vary widely in such basic parameters as average proportion of sand to mud, and dimensions and geometry of sand bodies. Consequently, they also vary in their capacity to collect and transmit ground water or hydrocarbons. Further, within a fluvial system these same parameters vary systematically both parallel and transverse to the sediment dispersal axis.

Large fluvial complexes tend to produce integrated drainage networks containing one or more trunk streams of the same type (e.g., meandering, braided) for large parts of the network. Depositional characteristics of these trunk streams provide a logical basis for differentiating significant portions of a fluvial system. In a series of papers Schumm (1960, 1972, 1977) described relationships in modern streams among sediment load transported by the channel, channel geometry, and sediment type deposited by the channel. These relationships can be quantified for modern river segments, and indicate qualitative trends that can be applied to the interpretation and classification of ancient fluvial depositional systems.

The basis of Schumm's classification lies in the empirical observation of a fundamental correlation between the ratio of bed load to suspended load transported by a perennial stream and the cross-sectional and areal geometry of the channel (expressed as width-to-depth ratio and sinuosity). This relationship is independent of other variables such as slope, discharge, or, within broad limits, periodicity of flow. Thus, alluvial channels can be classed as bed-load, mixed-load, or suspended-load types, depending on the sediment load transported by the channel (Table 4-1). For each channel type, modern streams are characterized by consistent patterns of erosion and deposition, ranges of sinuosity, and proportion of suspended load deposited within the channel perimeter (Table 4-1). Significantly, Allen (1965a) graphically illustrated a similar relationship between channel sinuosity and proportion of suspended load. Although quantitative aspects of Schumm's classification have been questioned, particularly for very large rivers and for some tropical streams (Baker, 1978), the qualitative relationships seem to apply over a wide range of geologically significant channel sizes and climatic regimes and prove useful in the interpretation of ancient fluvial systems (Ethridge and Schumm, 1978).

Bed-load and suspended-load channels can be considered end members of a spectrum of possible fluvial channel types into which any channel segment can be fitted. At one end of the spectrum are suspended-load dominated channels, characterized by low width/depth ratios, high sinuosity, dominance of vertical erosion and deposition, and a relatively fine-grained channel fill containing abundant suspended-load sediment. At the other end are bed-load-dominated channels, characterized by very high width/depth ratios, straight to slightly sinuous channel patterns, lateral erosion, bed deposition, and coarse-grained channel fill containing little suspended load material. Mixed-load channels occupy the middle range of this spectrum. Further, bed-load channels tend to have steep gradients, and suspended-load channels have lower gradients. Thus, the commonly observed decrease in stream gradient along its length is paralleled by a systematic, downchannel change in channel geometry and types of fluvial deposits.

Absolute dimensions of the channel (and its resultant channel-fill deposits) are determined by stream discharge. Because the average width/depth ratio of bed-load channels has been found to exceed 40:1 (Schumm, 1977), two bed-load channels 2 ft and 20 ft deep will be at least 80 ft and 800 ft wide, respectively.

Additional factors influence deposition within

Table 4-1. Classification of Alluvial Channels[a]

Dominant mode of sediment transport and channel type	Channel-fill sediment (percent silt clay)	Bedload (percentage of total load)	Channel stability: Stable (graded stream)	Channel stability: Depositing (excess load)	Channel stability: Eroding (deficiency of load)
Suspended load	>20	<3	Stable suspended-load channel. Major deposition on banks; initial streambed deposition minor.	Depositing suspended-load channel. Major deposition on banks; initial streambed deposition minor.	Eroding suspended-load channel. Streambed erosion predominant; initial channel widening minor.
Mixed-load	5–20	3–11	Stable mixed-load channel. Width/depth ratio 10 to 40; sinuosity usually <2.0, >1.3; moderate gradient	Depositing mixed-load channel. Initial major deposition on banks followed by streambed deposition	Eroding mixed-load channel. Initial streambed erosion followed by channel widening
Bed load	<5	>11	Stable bed-load channel. Width/depth ratio >40; sinuosity usually <1.3; steep gradient	Depositing bed-load channel. Streambed deposition and island formation	Eroding bed-load channel. Little streambed erosion; channel widening predominant

[a]Modified from Schumm (1977).

each of the broad classes of channel type. Climate affects the flow periodicity and the presence (and type) of vegetation. Vegetation itself is a factor, as it stabilizes channel banks and reduces bank erosion, thus favoring bed erosion during rising water levels. Despite these and other variables, considerable data from modern channels support the basic interrelationships between sediment transport, sediment deposition, and channel geometry.

Interpretation of Paleochannels

Determination of cross-sectional dimensions or original bed-load to suspended-load ratio of a paleochannel is extremely difficult even under optimum conditions of exposure. However, channel-fill composition, geometry, and internal structure provide a basis for modifying Schumm's classification of modern channels for application to the stratigraphic record. The modified classification utilizes fundamental attributes of any ancient depositional system—composition and geometry of the framework elements—that can be defined from outcrop or subsurface data, or both. The well-known fluvial models can be placed within a spectrum of fluvial depositional systems analogous to that described by Schumm's classification. Importantly, fluvial sequences that do not readily conform to idealized models can also be interpreted and placed within a predictive framework. The composition and geometry of aggrading bed-load, mixed-load, and suspended-load paleochannel fills are discussed below and are summarized in Figure 4-13.

Bed-Load Channel. The channel-fill sequence is dominated by sand. Coarse sand and gravel are commonly present, but are not necessary components; the sequence could consist exclusively of fine to medium sand. A bed-load channel is characterized by a high width/depth ratio and a tendency to erode its banks; these factors produce a tabular or belted sand body. Development of multilateral channel fills may produce a fluvial sheet sand (Campbell, 1976). Straight channels are characterized by relatively uniform depths of scour along the base, and this may be reflected in the low relief on the base of the sand body and by preservation of laterally continuous sheet or tabular, remnant floodplain mud units. Amalgamated, multilateral sand bodies result from rapid lateral migration of bed-load channels (Fig. 4-14A). Internally, bed-load channel-fill sequences reflect the dominance of bed accretion units, such as longitudinal, transverse, lateral, and chute bars, which produce a channel fill consisting of multiple, interlensed depositional units (Figs. 4-14A and 4-4) of varying grain size, and displaying complex textural sequences. Braided models and many coarse-grained, chute-modified meanderbelt models are well-documented types of bed-load channel-fill sequences.

Mixed-Load Channel. Mixed-load paleochannel deposits consist of sand with subordinate silt and clay. That channels were sinuous, with variable depths of basal scour along the thalweg, is reflected in the development of irregular or beaded belts of sand (Fig. 4-13), and in the preservation of lenticular, discontinuous flood-basin remnants between channel-fill sequences. A record of mixed bank- and bed-accretion (point-bar and channel lag) deposits is preserved in the channel fill, and a repetitive fining-upward sequence typifies most vertical sections through the meanderbelt deposit (Fig. 4-6). Channel meandering and consequent point bar accretion, combined with the stacking and amalgamation of successive channel fills, commonly produces a composite sand body that is much larger than the original channel (Fig. 4-14B). The channel plug provides the best record of true channel dimensions.

As in modern mixed-load channels, the abundance of fine, suspended-load sediment (89–97 percent of total load) in ancient mixed-load fluvial systems favored deposition and preservation of extensive floodbasin muds (Fig. 4-13). Mixed-load channel fills are typified by the well-known fine-grained point-bar model and its many variations.

Suspended-Load Channel. Suspended-load paleochannels were narrow and confined; erosion occurred primarily at the base, and surrounding clay-rich sediments formed stable, steep channel banks. Deposition of channel fill was dominantly along the banks, and may have been one-sided in highly sinuous channel segments, or symmetrical in straight channel segments (Figs. 4-13, 4-5). The dominant channel-fill sediment ranges from very fine sand to silt and clay, but some very coarse sediment may be present. Sand body geometry is highly lenticular in cross section (Fig.

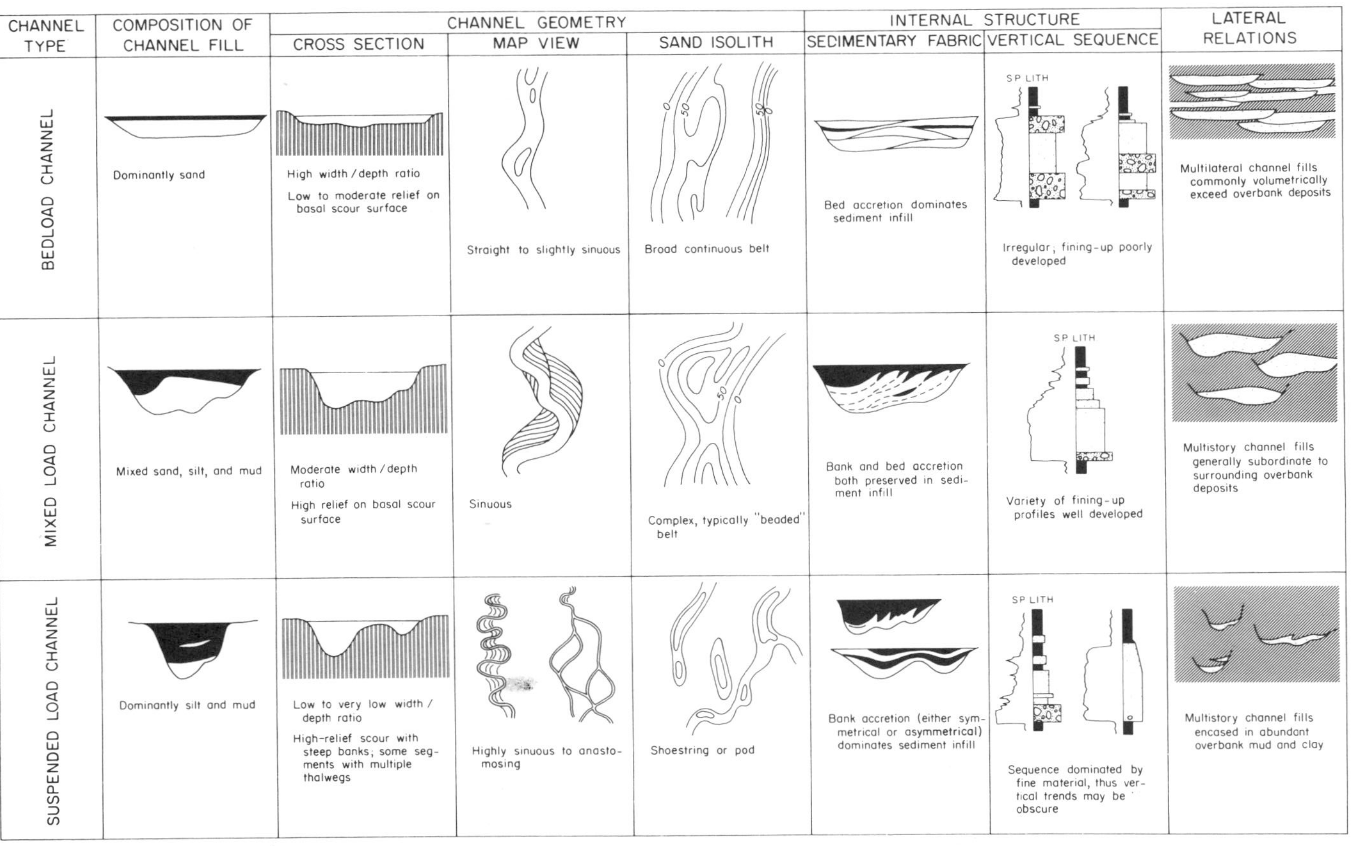

Figure 4-13. Geomorphic and sedimentary characteristics of bed-load, mixed-load, and suspended-load channel segments and their deposits. (From Galloway, 1977.)

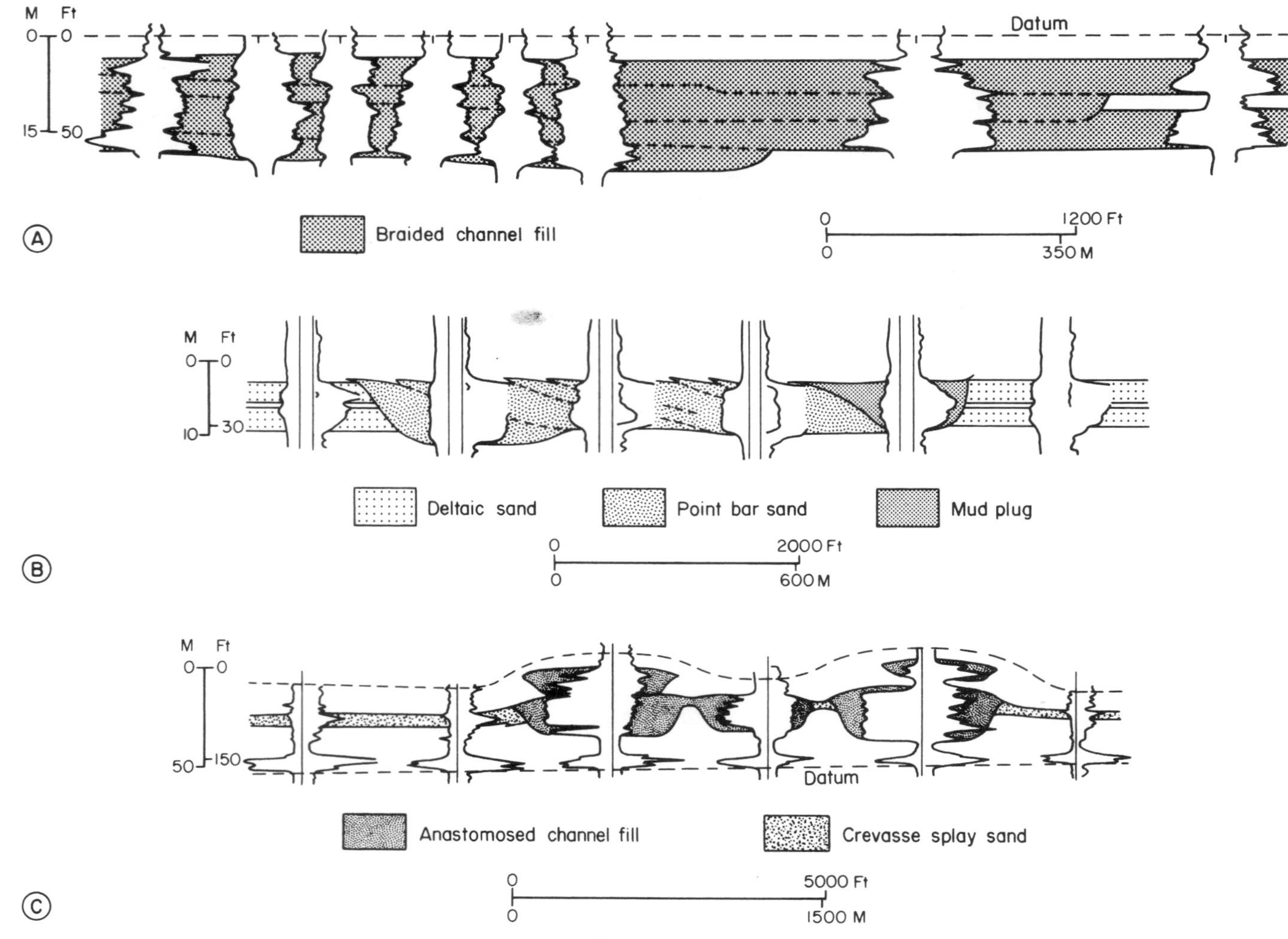

Figure 4-14. Channel-fill sand bodies of a braided bed-load channel (A), meandering mixed-load channel (B), and anastomosing, low-sinuosity suspended-load channel complex (C) as interpreted from electric logs. (A) Uranium-bearing Oakville Sandstone (Miocene), Texas Gulf Coast Basin, U.S.A. (B) Oil-bearing Fall River Sandstone (Cretaceous), Powder River Basin, U.S.A. (modified from Mettler, 1968) (C) Petroleum-bearing Cantuar formation of the Mannville Group (Cretaceous), Saskatchewan, Canada. (Modified from Christopher, 1974.)

4-14C), and forms observable sinuous or anastomosing patterns (Fig. 4-13). Channel-fill units are typically encased in fine-grained floodbasin deposits, and tend to stack vertically. Vertical sequences may fine upward or may show little vertical variation if the range of grain sizes is limited (Fig. 4-5).

The distributary channel model provides an example of a straight, suspended-load channel type. Point bars of the upper Mississippi delta plain record deposition by a sinuous suspended-load channel, and good examples of sinuous suspended-load channels are described by Taylor and Woodyer (1978) and Nanson (1980).

Interpretation of Ancient Fluvial Systems

A fluvial depositional system, or major segments of a fluvial system, can be categorized in terms of the dominant channel type (bed load, mixed load, or suspended load) and erosional or aggradational nature of the trunk stream(s). Thus, classification is based on geometric and compositional attributes of the framework sands, which, in turn, reflect important characteristics of the system as a potential conduit for flow of fluids. It is important to note that a wide range of channel types commonly coexists within a single fluvial system. For example, crevasse channels flanking the main channel will likely be enriched in suspended load. Tributaries can range from bed- to suspended-load type channels, depending on their provenance and gradient. The main channel itself may vary in type where it is locally affected by tributary input or incision into unusually sandy or muddy substrates. However, the dominant channel type of the trunk streams is a first-order basis for interpretation and categorization of the system.

An ideal fluvial system displays a systematic evolution of process, or morphology, and of depositional facies (Figure 4-15). This schematic fluvial system traverses two major depositional basins. The first is a montane basin opening into an aggrading riverine plain, such as the Murrumbidgee Riverine Plain of Australia, described by Schumm (1968a), or the Gran Chaco of South America's western Amazon Basin (Baker, 1978). The second depositional basin is a progradational coastal plain. The river is born as a network of alluvial fans and bed-load tributaries, undergoes considerable metamorphosis in response to downstream changes in gradient, sediment supply, and flow characteristics, and matures as a suspended-load distributary channel complex on a delta plain. The system illustrated may be considered "complete" in that the entire range of possible channel types is present. It is also possible that the fluvial system might reach its final depositional basin (be it lacustrine, inland sea, or oceanic) much earlier in its evolution, truncating development of mixed- or suspended-load channels. Alternatively, if the provenance of the system provides dominantly suspended load to the tributaries, bed-load channels will be absent. Generally, however, large integrated fluvial systems will display a broad spectrum of channel types along their course.

In addition to differing channel-fill facies, ancient aggrading bed-load, mixed-load, and suspended-load fluvial systems each display characteristic facies associations and early hydrologic history.

Bed-load fluvial systems consist dominantly of channel and channel-flank deposits; flood-basin facies are typically subordinate, and consist of floodplain sandy mud and silt. Associated eolian deposits may be common, even in subhumid climates, because broad expanses of unconsolidated sand are exposed to wind reworking during low water levels. The water table in interchannel areas may be quite deep, because such systems commonly occur in the topographically elevated upper reaches of the sedimentary basin, and the sandy floodplain and channel sediments are highly permeable. Well-developed crevasse splays contain significant amounts of bed-load sediment. Sand percentages for thick stratigraphic intervals along channel axes range from 50 to 90 percent. Modern examples include the Brahmaputra River (Coleman, 1969) and the Platte River (Smith, 1970).

Mixed-load fluvial systems typically preserve a high percentage of flood-basin deposits, including floodplain muds and silts or backswamp carbonaceous muds and clays. Sand percentages may be high along principal fluvial axes where channel-fill units are vertically stacked and amalgamated; however, overall sand percentage commonly averages 20 to 40 percent. Channel fills are flanked by crevasse splay and levee sands and silts. The water table is usually controlled by the water level in the main channels; thus levee and crevasse deposits are typically leached and oxidized. Much of the channel fill remains

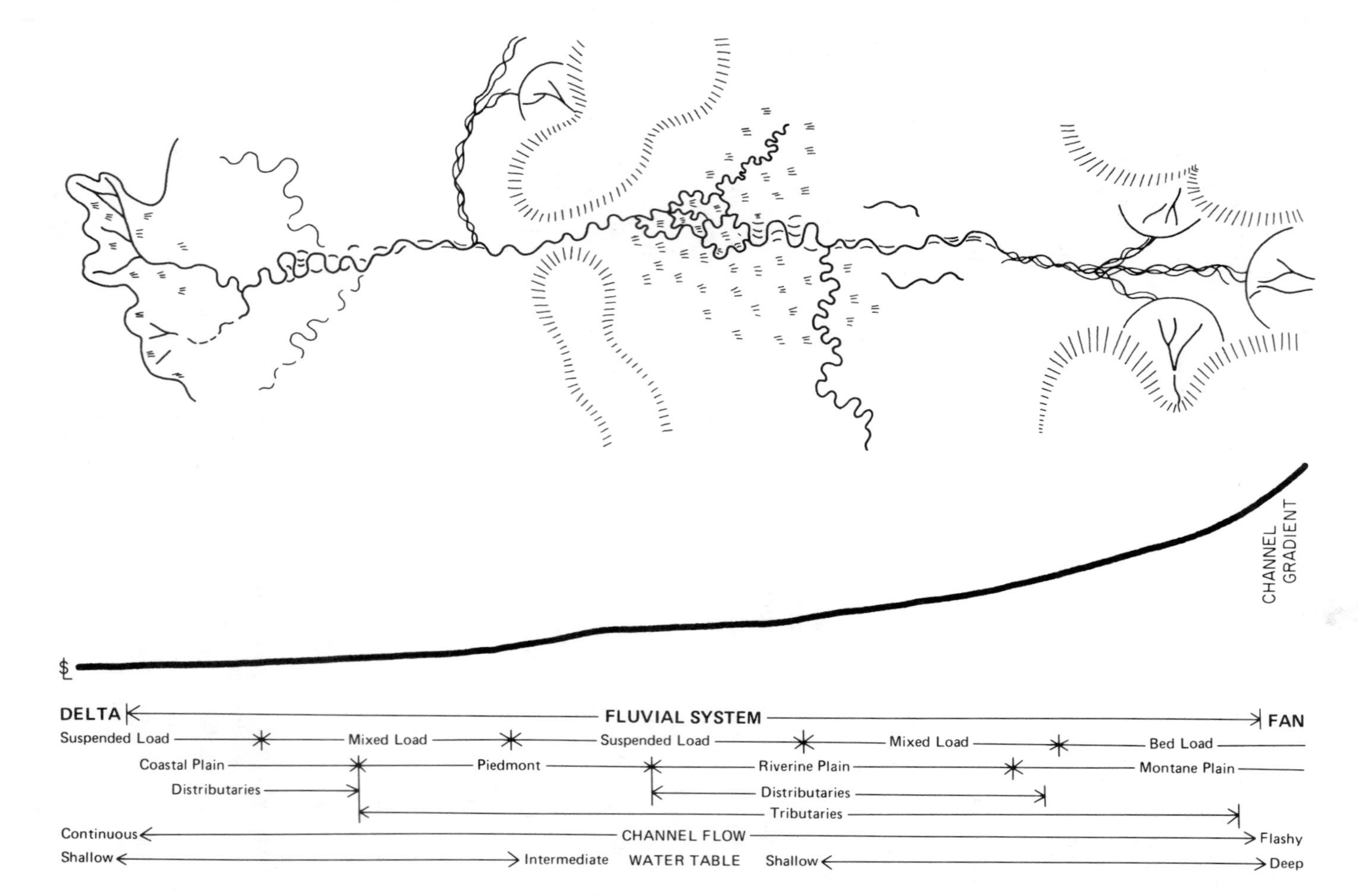

Figure 4-15. Hypothetical fluvial system traversing a montane basin and associated riverine plain, and then debouching onto a prograding, deltaic coastal plain. The trunk channel evolves through bed-load, mixed-load and suspended-load styles as sediment load, gradient, and discharge vary along the system.

saturated, and organic debris deposited within lower point bar and channel lag is commonly preserved. Broad, low-lying, backswamps also have shallow water tables, and organic debris, peat, and syngenetic sulfide minerals are commonly preserved within the mud and clay deposited in this environment. Modern examples are the lower Brazos River (Bernard et al., 1970) and the Mississippi River (Fisk, 1944).

Suspended-load fluvial systems have low sand percentages. Floodbasin deposits consist dominantly of backswamp and lacustrine muds and clays. Levees are well developed. Crevasse splays may be common but consist almost entirely of suspended-load sediment. Sand isolith maps show complex anastomosing or distributary patterns. The low topographic gradients and relatively impermeable sediments typical of suspended-load fluvial systems result in generally shallow to very shallow water tables that intersect the land surface along channels, and in backswamp and lacustrine basins. Considerable organic debris, bedded peat, and syngenetic sulfides are commonly preserved. Modern examples include the lower Mississippi and the Atchafalaya Rivers (Fisk, 1947, 1952; Frazier, 1967).

Eroding Systems–Valley Fills

Discussion thus far has emphasized facies development in unconfined aggrading systems. However, important segments of many ancient fluvial systems, as well as most modern examples of well-studied rivers, are confined within valleys. Channel bank and bed erosion are partially limited by comparatively consolidated bedrock. Even Pleistocene sand and mud form a much more durable substrate than do younger Holocene fluvial sediments. Such confined fluvial deposition is modified in several important ways.

Potential for channel avulsion is greatly reduced. Possible courses for the channel are restricted by the bounding bedrock valley walls or uplands. However, local chute and neck meander cutoffs may form and channel meandering continues within the confines of the valley.

The preservation of channel-fill deposits is accentuated at the expense of overbank and floodbasin facies. In relatively narrow valleys that approximate channel meanderbelt width, channel facies may essentially fill the entire valley, leaving only isolated remnants of floodplain and abandoned-channel mud plug sequences (Fig. 4-16). The resultant facies assemblage thus accentuates characteristics, such as high sand percentage, usually associated with bed-load channels.

The overall valley fill commonly consists of a composite, thick, fining-upward sand and gravel-rich unit (Schumm, 1977; Brown *et al.*, 1973). A valley fill can be distinguished from similarly fining-upward, meanderbelt sequences primarily by its size. Only the world's largest rivers have channel widths in excess of a mile, or depths of 65 ft (20 m) or more. However, valley dimensions can readily reach several miles in width and many tens of feet in depth. Many large, ancient channel deposits described in the literature may well be paleovalley fills instead (Schumm, 1977).

Recognition of eroding channels and their valley fills, though at first glance apparently complicating interpretation of ancient fluvial systems, is generally straightforward. Such sequences are deposited upon a regional unconformity. Delineation of the unconformable surface by various geologic methods is thus the critical first step to the interpretation of an eroding fluvial system. Scale and continuity of the erosional surface is important; all fluvial channels are erosional at the base. Regional stratigraphic analysis typically is necessary for proper classification of erosional fluvial systems. Delineation of the geometry of such systems is truly an exercise in paleogeomorphology.

Basin Analysis and Fluvial-System Evolution

Regional basin analysis necessitates both the delineation and interpretation of principal fluvial depositional systems and their component elements and description of the evolution of fluvial deposition through geologically significant intervals of time. Because parameters that influence fluvial depositional style, such as climate, nature and elevation of the source area, and rate of basin subsidence, may change, both the loci and nature of fluvial deposition may evolve through time. Thick sequences of fluvial deposits commonly display apparent cyclicity at several scales as well as abrupt and gradual changes in composition and sand-body geometry (Schumm, 1977; Miall, 1980).

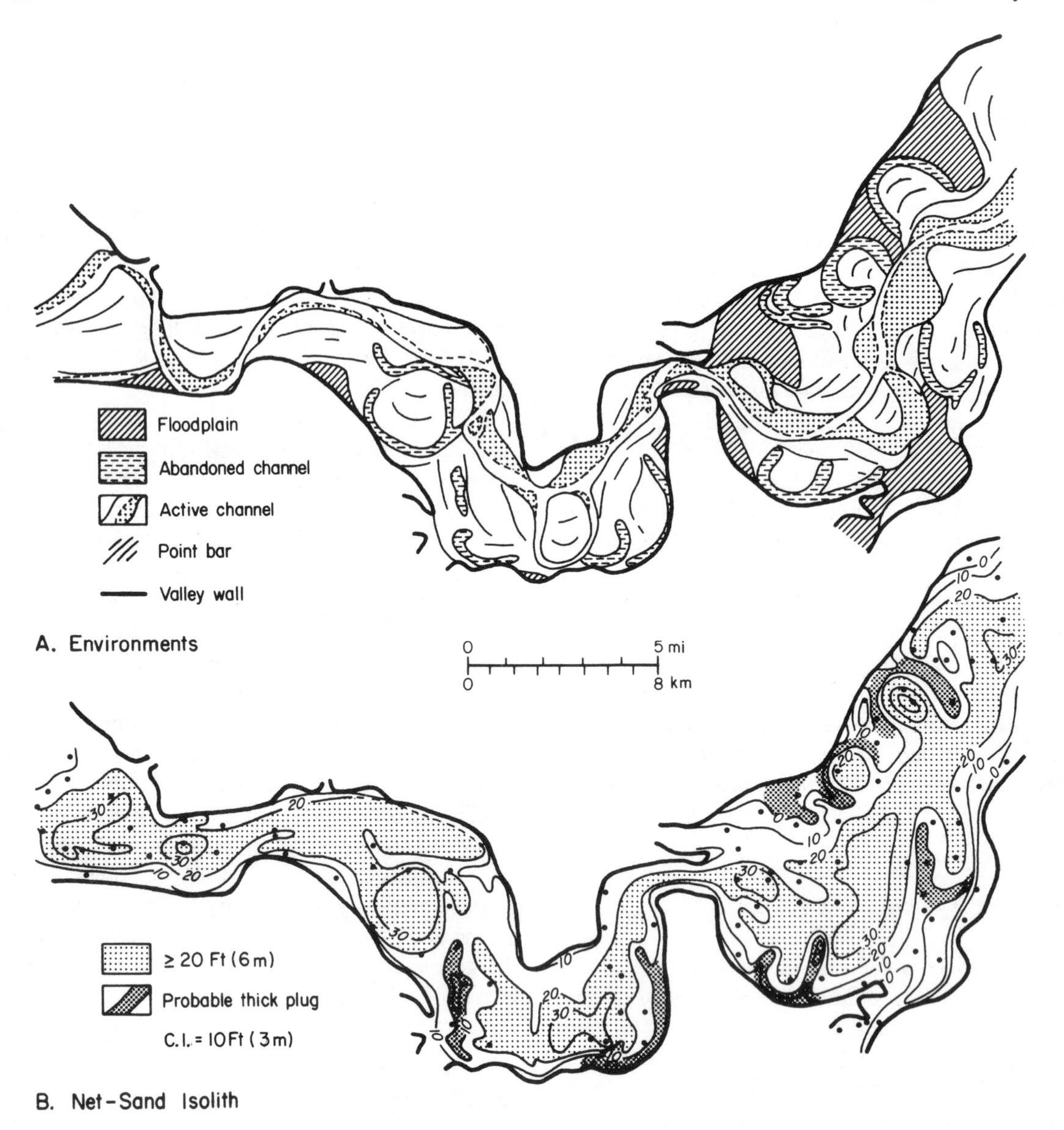

Figure 4-16. Surficial depositional environments (A) and interpreted sand distribution based on water well data (B) in the valley-confined Arkansas River (Oklahoma, U.S.A.) The valley is incised in Pennsylvanian sandstone and mudstone. (Data from Tanaka and Hollowell, 1966.)

The depositional architecture of an aggrading fluvial sequence includes both the three-dimensional interrelationships and geometries of component framework sand bodies and the distribution of bedding and internal structures within framework facies. Allen (1978) and Bridge and Leeder (1979) presented computer simulation studies of fluvial system aggradation and discussed implications of results for the interpretation of fluvial system stratigraphy. Combined with studies of thick fluvial sequences, such simulations illustrate several important generalizations about fluvial depositional style and its long-term evolution (Galloway, 1981).

1. First, aggradational fluvial sequences typically record deposition by multiple, contemporaneous drainage axes, ranging from large extrabasinal rivers to small streams that drain basin-fringe or intrabasinal terranes. A key initial step in basin analysis is the recognition and delineation of principal drainage axes and their deposits.

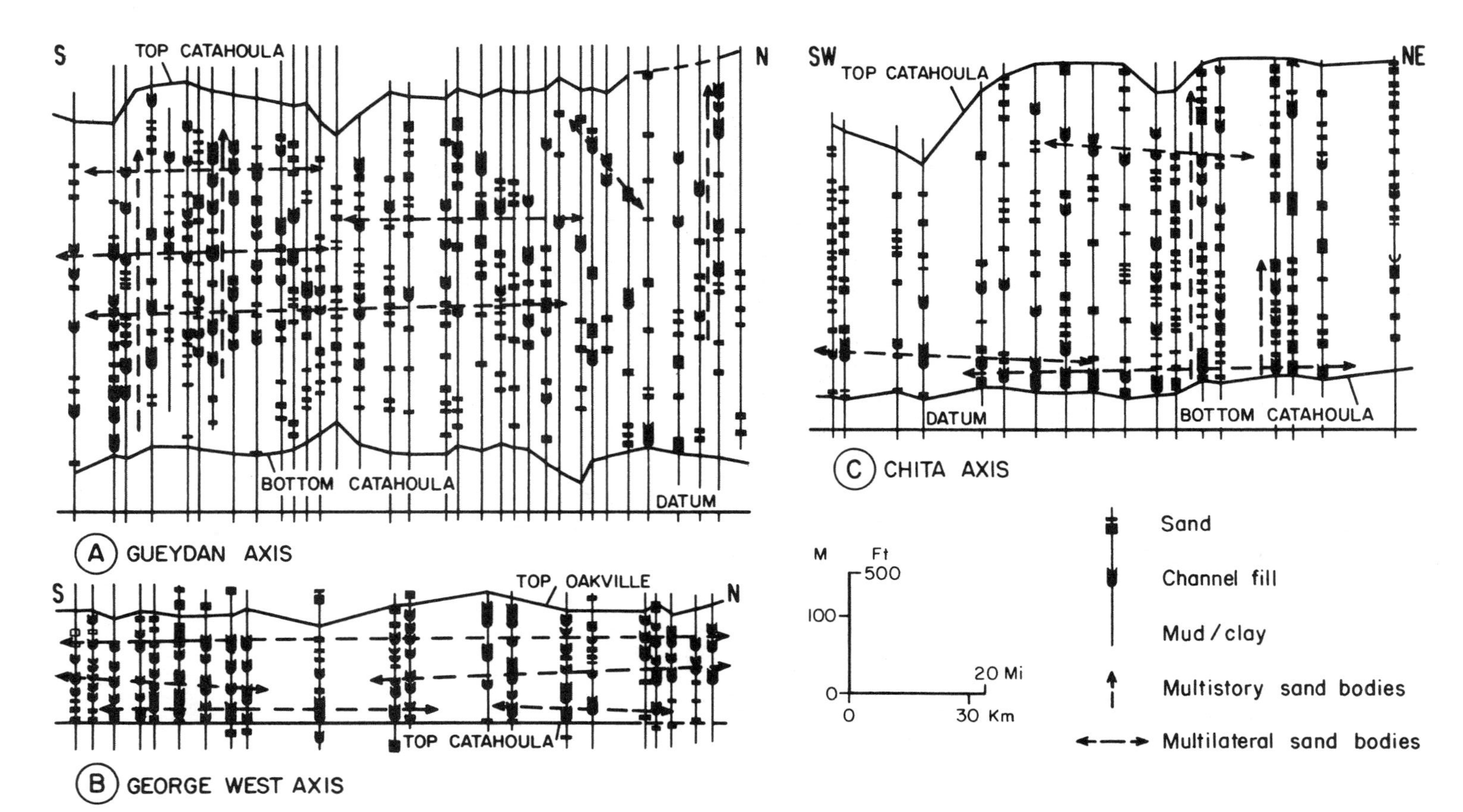

Figure 4-17. Regional subsurface strike sections through three fluvial depositional elements of Oligo-Miocene age, contrasting depositional architecture of different aggrading rivers (Texas Gulf Coast basin, U.S.A.). The Gueydan bed-load fluvial system (A) contains both multistory and multilateral channel-fill sand bodies. The George West axis of the Oakville bed-load fluvial system (B) is dominated by coalesced multilateral channel-fill sand bodies. The Chita axis of the Chita–Corrigan mixed-load fluvial system (C) is characterized by vertical stacking of channel fills within a thick sequence of floodplain deposits. Multilateral sand-bodies occur at the base of the aggradational sequence on a surface interpreted to be a low-relief, subregional unconformity. (Modified from Galloway, 1981.)

2. The contemporaneous rivers may possess greatly differing discharge characteristics, sediment load, and channel geometry. Consequently, each fluvial axis displays differing facies components and internal sedimentary features.

3. Large, extrabasinal rivers commonly enter the depositional basin at structurally or topographically stabilized points (Potter, 1978). Major fluvial axes generally parallel structural dip, and are concentrated along or adjacent to structural downwarps. Once on the unconfined alluvial plain of the basin, the river rhythmically switches its course so that the depositional surface is uniformly aggraded. Avulsion may occur at the point of entry into the basin (nodal avulsion), or may occur randomly at any point down channel. Consequently, facies maps, such as net sand isolith maps, showing deposits that accumulated during an extensive period of time, exhibit complex disributing and interweaving patterns. However, long-term stability of the points of sediment input into the depositional basin favors vertical persistence of fluvial systems and their major component drainage elements.

4. Within fluvial axes, the framework channel facies may stack vertically along relatively narrow, persistent trends or may form multilateral belts up to tens of miles or kilometers in width. Figures 4-17A and B illustrate two Oligo-Miocene fluvial systems of the Gulf Coastal Plain. Rapid subsidence along the axis of a continent-scale bed-load fluvial system, accentuated by deposition of large volumes of air-fall volcanic ash, produced multistory and multilateral sequences of channel-fill sands encased in finer-grained floodplain and splay facies (Fig. 4-17A). A younger version of the same drainage element, deposited during a period of decreasing rates of ash accumulation, produced a sand-rich stratigraphic unit composed dominantly of multilateral channel fills (Fig. 4-17B).

5. The evolution of river pattern in response to extrinsic or intrinsic changes in the drainage or depositional basin are recorded by metamorphosis of depositional style within individual fluvial axes. For example, the Chita axis, an Oligo-Miocene mixed-load fluvial depositional element of the Texas Gulf Coastal Plain, shows a basal sand-rich zone of multilateral sand bodies (Fig. 4-17C). Fining-upward sequences and internal structures show these sands to be typical meanderbelt deposits of mixed-load channels little different from channel-fills within the mud-rich middle and upper portions of the deposit. The rather abrupt evolution in depositional style within the same depositional element likely reflects changes in base level or rates of alluvial plain aggradation (Allen, 1978). In this example, the base of the aggradational sequence may be a subregional erosion surface. Such vertical evolution within a fluvial system may provide important information on source or basin tectonics, and illustrates the dependence of sand percentage upon tectonic and base-level changes as well as upon the type of fluvial system.

Fluvial Evolution Through Geologic Time

Schumm (1968b) discussed the profound impact of colonization of alluvial plains and upland source areas by land plants. Increasing plant cover has the effect of decreasing erosion rates over a wide range of climatic regimes, as well as modifying the texture of sediment entering the drainage network. In addition, vegetation plays an important role in the stabilization of channel margins and natural levees.

Pre-Devonian landscapes were essentially devoid of vegetation and closely resembled modern arid terranes. As pointed out by Cotter (1978), nearly all pre-Devonian fluvial systems are interpreted to have been characterized by braided or bed-load channels. With development of vegetated valleys and coastal plains through the later Paleozoic, channel stabilization was accentuated, and a more representative mix of straight and sinuous channels appears in the geologic record. As upland interfluves were colonized by relatively modern varieties of flowering plants during Mesozoic time, the effects of climate on the amount and character of sediment yield were accentuated, and the diversity of fluvial channel patterns increased. Development of grasses in Miocene time completed the evolution toward the spectrum of fluvial channel morphologies found today.

Ancient Fluvial Systems

Fluvial systems are important hosts of petroleum, uranium, and coal. Many examples will be described in subsequent chapters on these mineral fuels. The literature is replete with additional examples, though regional integration of large fluvial systems is less common. Braided, bed-load fluvial systems of the Proterozoic upper Witwatersrand sequence (Chapter 12) are a classic example of fluvial sedimentation in pre-Devonian landscapes. Young bed-load fluvial systems deposited in arid and wet climates are well represented by the Oakville Formation of the southwestern Gulf of Mexico coastal plain, and the upper Fort Union and Wasatch formations of the Powder River Basin, Wyoming. Mixed- and suspended-load systems are prominent in many late Paleozoic basins of the mid-continent of the U.S. and have been regionally mapped and described in both the outcrop and subsurface (Chapter 14). Additional, well-known units containing important fluvial components include the Old Red Sandstone of the British Isles, the Mauch Chunk and Pottsville–Dunkard wedges of the Appalachian foreland basin in the eastern U.S., the Permo–Triassic sequence of the Karoo Basin, South Africa, and Tertiary molasse of European Alps.

Chapter 5

Delta Systems

Introduction

A delta forms where a river transporting significant quantities of sediment enters a receiving basin. In one sense, few processes or environments are unique to the deltaic setting. However, the interaction of subaerial fluvial processes and subaqueous processes of marine or lake basins produces distinctive facies assemblages. In the following discussion, a delta is considered to be a river-fed depositional system that produces an irregular progradation of the shoreline into a body of water (Fisher *et al.*, 1969). As suggested by Moore and Asquith (1971), and consistent with the concept of a depositional system, the entire subaerial and subaqueous, contiguous sediment mass is included in a delta system.

Several important corollaries are implicit in this broad definition. First, the depositional architecture of a delta system is characteristically progradational. Secondly, sediments are derived from one or more definable point sources, although the apex may lie far inland. Thirdly, delta systems develop around the periphery of large basins, although deltaic progradation may ultimately fill the basin. Finally, because major fluvial systems are usually the principal sources of sediment entering a basin, delta systems commonly define loci of maximum deposition. Delta systems are thus readily outlined by isopach thicks within time-equivalent increments of basin fill, particularly in tectonically quiescent settings.

Delta systems are important hosts of petroleum, coal and lignite, and to a lesser extent, uranium. Because of their economic importance and geographic significance along many modern coastlines, deltas have been the subject of excellent three-dimensional stratigraphic and sedimentologic studies, beginning with the work of Russell and Russell (1939) and Fisk and coworkers (1952, 1954, 1955). A historical review of significant studies of deltas is provided by LeBlanc (1975).

Emerging from this body of study is the obvious and important observation that modern deltas are tremendously complex and variable in size, geometry, and composition. Recognition of this complexity led several authors, including Fisher *et al.* (1969), Fisher (1969), Coleman and Wright (1975), and Galloway (1975), to propose a series of end-member delta types and discuss their facies composition and processes.

Delta Process Framework

A delta is produced by an ongoing competition between deposition by constructional processes of the fluvial system and sediment reworking and redistribution by reservoir or destructional processes. By definition, in a delta system the river maintains at least a modest advantage in this interplay.

Constructional Processes

Flow of water from a channel, through an orifice, and into an unconfined, standing body of water is one of the few hydrodynamic processes unique to deltas. Overbank flooding and consequent floodplain and levee aggradation, as well as processes of channel incision, migration, and filling are all reflected in deltaic facies that resemble their counterparts within fluvial systems. However, other processes inherent in fluvial deposition, including channel avulsion and crevassing, assume greater importance in delta systems.

Channel-Mouth Deposition. As flow discharges from a confined channel into a reservoir, it spreads and mixes with waters of the basin. The rate and geometry of flow spreading and mixing is dependent on: (1) the momentum of the discharged waters, which is a function of discharge velocity and density; (2) the density contrast between the two water masses; and (3) bed friction, which is in part a function of reservoir depth at the channel mouth.

In marine basins, density of sediment-laden river discharge is typically less than that of

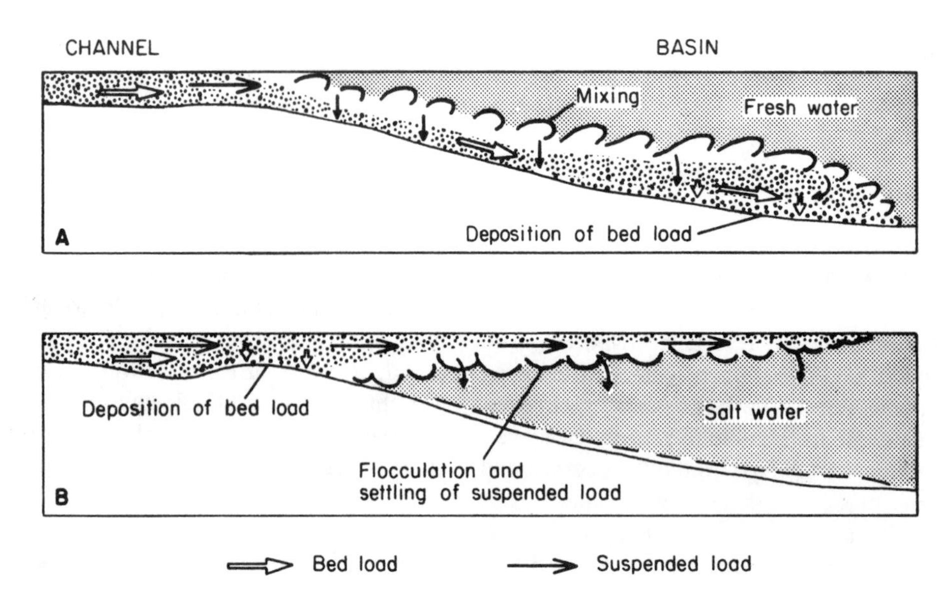

Figure 5-1. (A) Hyperpycnal flow from an orifice occurs when influent river water is denser than basin water, consequently flow remains attached to the bottom, sweeping bed load down the basin flank. (B) Hypopycnal flow occurs when influent water is less dense and rides out over basin water as a laterally expanding plume. (Modified from Bates, 1953.)

normal sea water. Consequently, river water and contained suspended sediment rides out over marine water as a laterally spreading, buoyant plume (hypopcynal flow; Fig. 5-1 B). Mixing occurs primarily at the base of the plume, and may result in flocculation and increased settling of suspended clay. Development of the salt-water wedge beneath the emerging river water, and consequent flow separation at the channel mouth, results in rapid and efficient deposition of bed-load sediment as a channel or distributary mouth bar (Fig. 5-2 A).

The discharge process efficiently sorts sediments introduced into the basin by the river. Sand is concentrated at the channel mouth. Very fine sand and coarse silt are swept onto the upper prodelta, where they progressively settle from suspension as velocity and turbulence wane. The finest suspended material is swept basinward, where it slowly settles, forming the prodelta slope, or is caught up in basin circulation patterns.

If the basinward slope away from the channel mouth is gentle, as is typical of many prograding shorefaces, the channel mouth bar is reworked and molded by bottom friction and turbulence. Marginal scours, which are separated by a middle-ground shoal or bar crest and flanked by subaqueous levees (Fig. 5-2), form in response to the symmetrical displacement of the threads of maximum turbulence along channel margins (Fig. 4-2). Resultant flow bifurcation, which may become increasingly stabilized by continued growth of the subaqueous levees, provides the seed for distributary channel branching. As would be expected, multiple distributaries characterize deltas prograding into shallow water, where the combination of discharge buoyancy and bottom friction play dominant roles in channel-mouth evolution.

In contrast, discharge of sediment-laden water into a fresh-water lake, or of very cold or highly sediment-laden water into a marine basin (hyperpycnal flow), causes the denser influent waters to remain on the bed, forming a density underflow (Fig. 5-1 A). If discharge rates are high, bed-load sediment can be swept far beyond the channel mouth. Lateral size grading occurs as current velocity and competence progressively decrease. In deep basins with steep sloping shorefaces, gravity may accelerate the water mass, producing a true density current than can spread far out into the basin.

Position of the salt wedge and consequent point of flow separation common to most marine deltas varies in response to river flow conditions. During low-flow stage the wedge may intrude far upstream. However, during major floods sediment-laden river water sweeps out of the channel

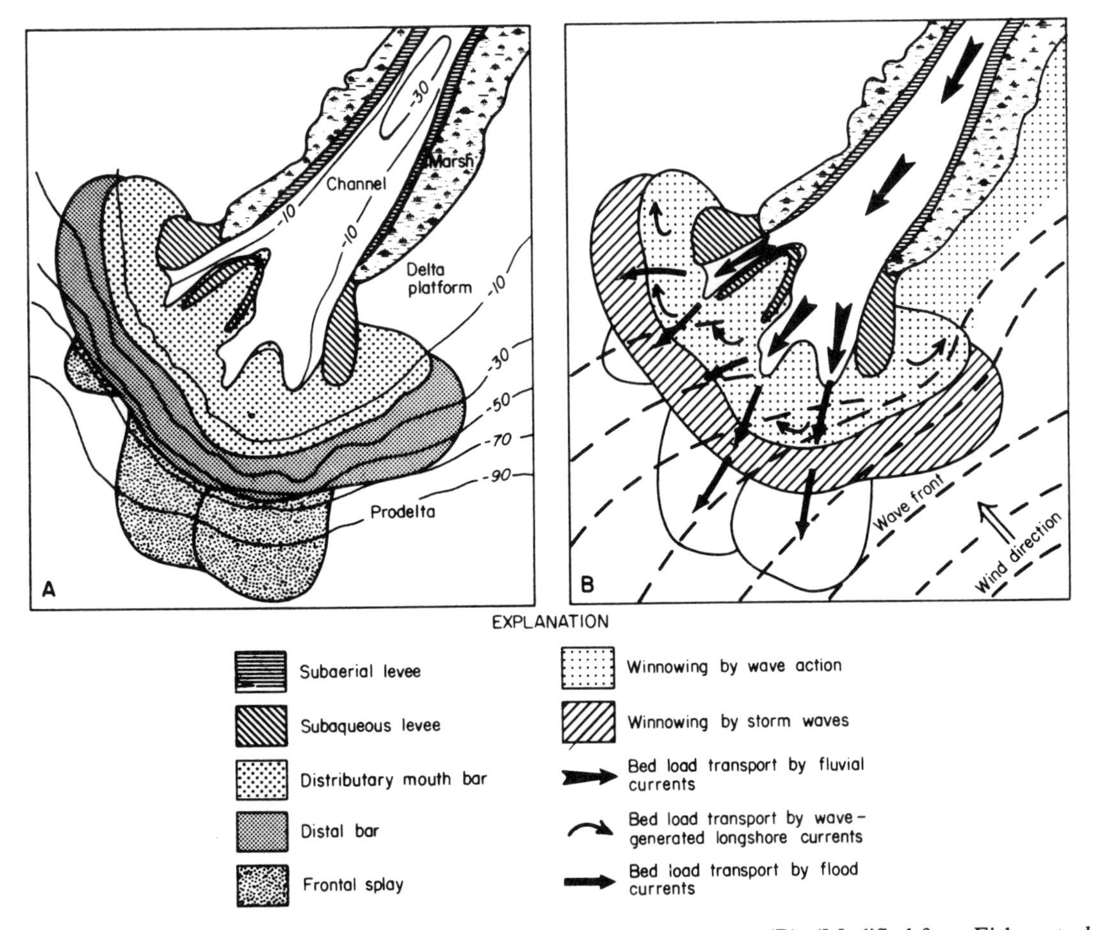

Figure 5-2. Channel-mouth depositional environments (A) and processes (B). (Modified from Fisher *et al.*, 1969; original figure from Coleman and Gagliano, 1964.)

mouth, scouring much of the mouth bar and pushing the salt wedge seaward down the prodelta slope. Bed-load sediment is swept by spreading currents onto the upper prodelta slope, forming aprons or lobes of sand called frontal splays (Fig. 5-2 A). The highly organized seaward-fining or stratigraphically upward-coarsening grain-size trend, which is produced at low stage, is thus punctuated by the flashy discharge characteristic of many fluvial systems.

Crevassing. Crevasse splays formed along delta distributaries are important in delta plain development. As in fluvial systems, crevasse splays form during flood events as water and sediment pour through breaches in the levees. However, formation of many delta-plain distributary splays is more complicated than in rivers. The splays may become, in effect, subdeltas that prograde into marginal interdeltaic embayments (Fig. 5-3). In the modern birdfoot lobe of the Mississippi delta system, a magnificent natural laboratory for study of delta processes, six major splays have developed within historic time (Coleman and Gagliano, 1964). The splays prograded onto the relatively shallow, subaqueous mud platform constructed by major distributaries into the deep Gulf of Mexico. The splays have, in effect, fleshed out the skeletal subaerial delta plain, which initially consisted of natural levees, by filling in interdistributary embayments such as that still separating Southwest and South Passes (Fig. 5-3). Welder (1959) documented the rapid growth of Cubits Gap, a major splay initiated during flooding in 1862. Cubits Gap has reached mature dimensions within about 35 years (Fig. 5-3), forming its own complex of distributaries and prograding channel mouth bars. However, splay progradation was highly sporadic, as new sediment was added primarily during brief floods.

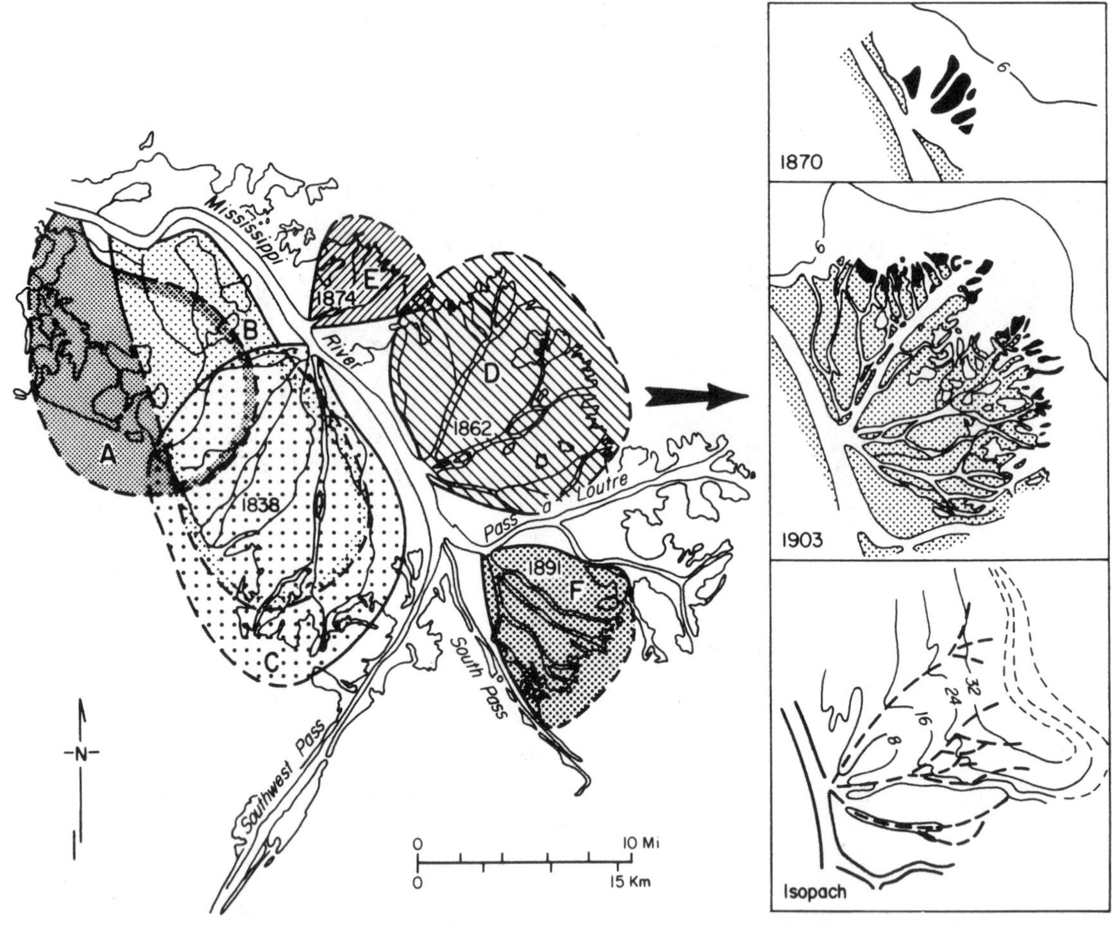

Figure 5-3. Major crevasse splays of the modern digitate lobe of the Mississippi Delta. Insets show the rapid growth of Cubits Gap splay and an isopach map of the splay unit. Contours are in feet. (Modified from Coleman and Gagliano, 1964, and Welder, 1959.)

Channel Avulsion and Lobe Formation. As the channel progrades basinward, effective hydraulic head is continually reduced. Instability increases with continued channel extension across a base level depositional surface. This leads to repeated channel avulsion and occupation of alternative courses across portions of the delta plain offering steeper gradients. Again, the highly constructional Mississippi delta system illustrates the importance of this process in delta progradation (Fig. 5-4). Detailed subsurface investigation and radiocarbon dating allowed Frazier (1967) to delineate 16 individual delta lobes constructed by the Mississippi River during the last 6,000 years of stable sea level. Together, the offsetting, partially imbricate lobes have produced a subaerial delta plain covering several thousand square miles. Frazier further suggested that many of the lobes represent more than a single progradational interval. Thus, the Mississippi River has altered its course by major channel diversion in the lower alluvial or upper delta plain more than 16 times during the last 6,000 years.

The delta lobe, comprising contiguous facies of a river and its distributaries deposited during occupation of a trunk channel course, is the fundamental building block of a highly constructional delta system. Significantly, in a large delta system such as the Mississippi, active deposition and delta construction occupy a relatively limited portion of the total system. The area of individual lobes of the Mississippi is on the order of one-tenth the area of the total delta system (Fig. 5-5). Through Holocene time the locus of deposition has repeatedly shifted as much as 150 miles (240 km) along the basin margin.

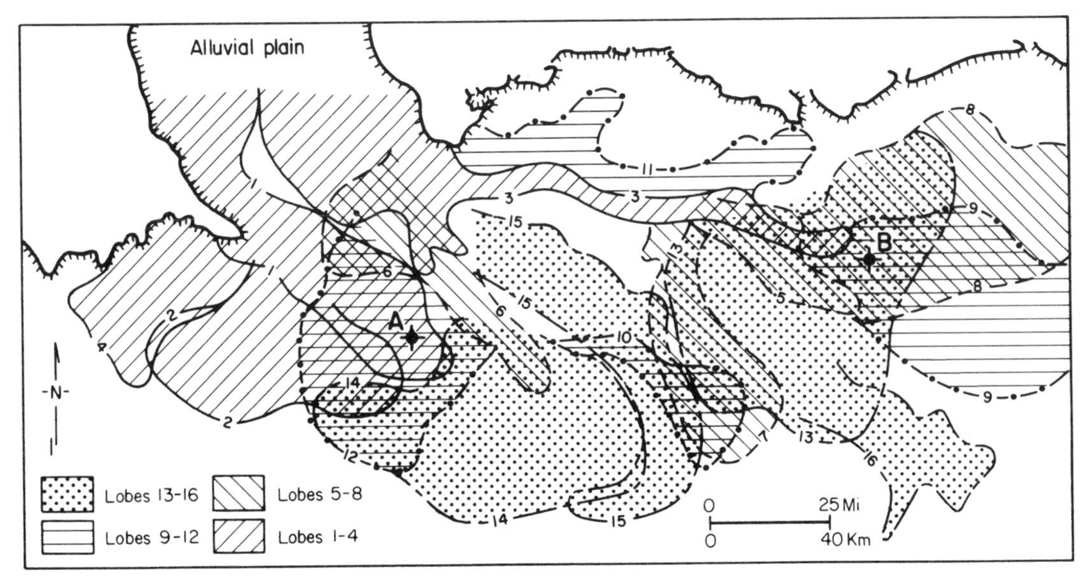

Figure 5-4. Principal lobes of the Holocene Mississippi delta system. At least 16 lobes have been identified by radiocarbon dating. Cyclic stratigraphic sequences would be penetrated in many areas of the modern delta plain, such as at hypothetical core locations A and B. (Modified from Frazier, 1967.)

Destructional Processes

A variety of reservoir processes that rework, modify, redistribute, or remove sediments deposited by constructional fluvial processes may be grouped as destructional processes. As emphasized by Fisher *et al.* (1969), a delta system can be viewed as the product of the competing effects of: (1) sediment input by a fluvial system, and (2) reservoir energy (Fig. 5-5). Sediment input and deposition by constructional processes are influenced by factors such as the ratio of bed load to suspended load, the total sediment volume, and the periodicity of sediment influx. The energy regime of the reservoir basin (Fig. 5-5) includes wave energy flux, tidal energy flux, intruding permanent basin currents, temporary wind drift, or wind-forced currents, and gravitational potential provided by elevation differences between the basin margin and floor (Swift, 1969; Galloway, 1975).

Wave and Current Redistribution. Wave energy exists in almost all reservoirs, whether marine or lacustrine. Significant tides and permanent currents are products of large, typically marine basins. Further discussions will focus primarily on marine basins.

Deposition of bed-load sediment in the channel mouth bars of delta or crevasse distributaries places it in optimum position for wave or tidal reworking. Breaking waves on the channel-mouth-bar crest create turbulence, remobilize sand, and transport it in the direction of longshore drift (Fig. 5-2 B). Sand deposited at the channel mouth is thus distributed laterally. If little sand is removed, the sand belt produced by channel progradation is simply broadened. If most of the mouth bar is reworked, the delta front may evolve as a series of coalescing arcuate beach ridges (sometimes called coastal barriers). In contrast, tidal currents move in and out of the channel mouth, alternately reinforcing and reducing or reversing the river currents. Remobilized sediment moves in a dip direction, forming elongate bars within the distributary mouth and spreading seaward as a broad, subaqueous delta-front platform. Intrusion of marine waters spreads, retards, and mixes flow, enhancing deposition of both bed and suspended load at the channel mouth, and widening the lower distributary segment to produce a funnel-shaped, or estuarine geometry. Both wave and tide reworking of mouth-bar and delta-front sediments further improves sorting.

Permanent marine currents may effectively redistribute sediment washed offshore into the prodelta environment. Thus, such currents pri-

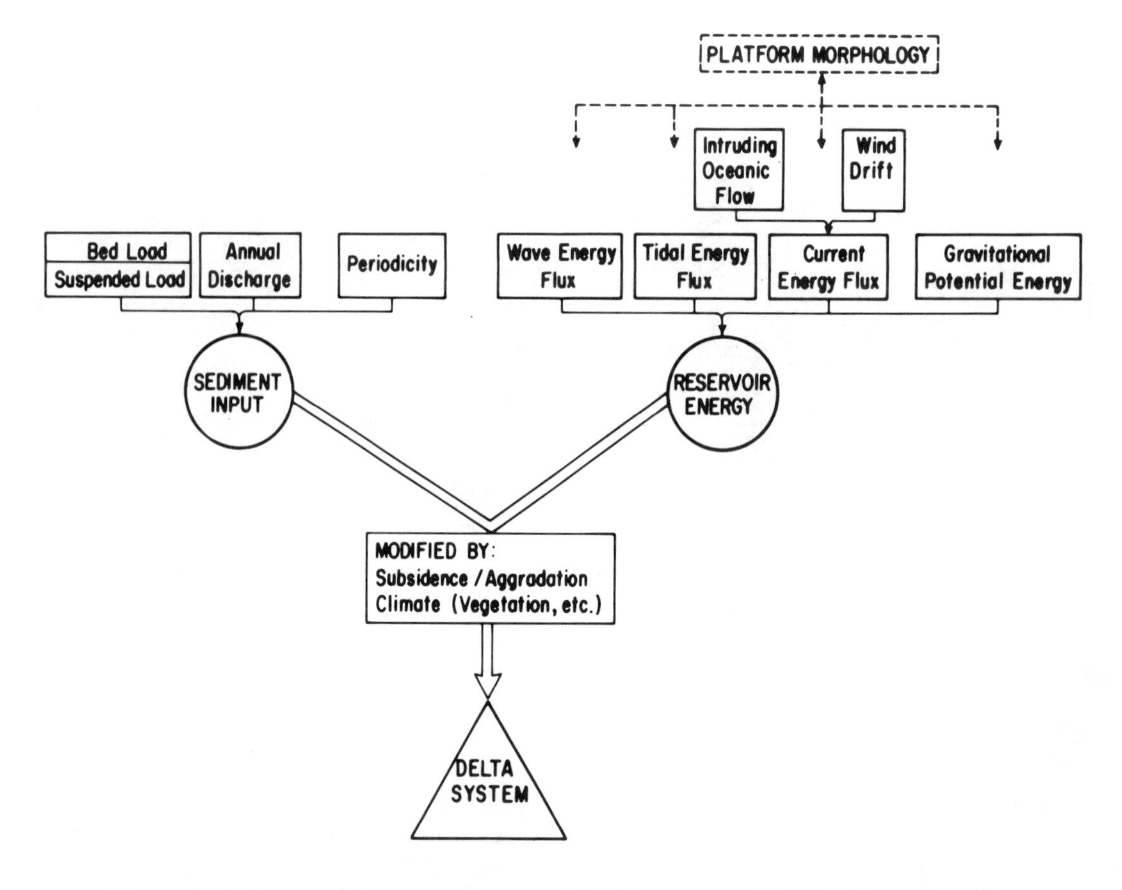

Figure 5-5. Diagrammatic process framework of delta systems. (From Galloway, 1975.)

marily affect suspended sediment. For example, most of the mud deposited by the Orinoco River is swept by the northwestward flowing North Equatorial Current into the Gulf of Paria between Trinidad and mainland South America (van Andel, 1967). The locus of mud deposition has been shifted about 95 mi (150 km) from the main body of deltaic sediment into a separate depositional basin.

Compaction and Mass-Gravity Transport. Channel-mouth sediments and their marine-reworked delta front equivalents are ideally situated for modification, disruption, or remobilization by gravitational potential energy. First, sand is deposited at the top of a commonly thick, rapidly deposited, water-saturated, and undercompacted prodelta mud platform. Secondly, the locus of deposition lies at the crest of a sloping prodelta apron. Although delta-front slopes rarely exceed a few degrees, such gradients are unstable in unconsolidated, subaqueous deposits. Thirdly, sediment input by large fluvial systems is measured in tens of millions of tons per year and is irregularly distributed in both time and space.

Rapid deposition of sand at the mouths of major distributaries loads underlying prodelta muds, creating a density inversion; the muds behave ductilely under stress because of high water content that may approach 80 percent. Instability of the undercompacted low-density, low-permeability prodelta sediment is further accentuated by *in situ* generation of gas from the bacterial alteration of organic debris (Coleman and Garrison, 1978). Sediment loading results in differential compaction and flowage, or fracturing of prodelta muds. Diapiric mud spines (Fig. 5-6), commonly called "mud lumps" in the Mississippi Delta, and ridges form by lateral and vertical flowage from beneath sites of most active loading (Morgan *et al.*, 1968; Coleman and Garrison, 1978). Underlying muds may extrude vertically as much as several hundred feet, causing corresponding rapid local subsidence of distributary

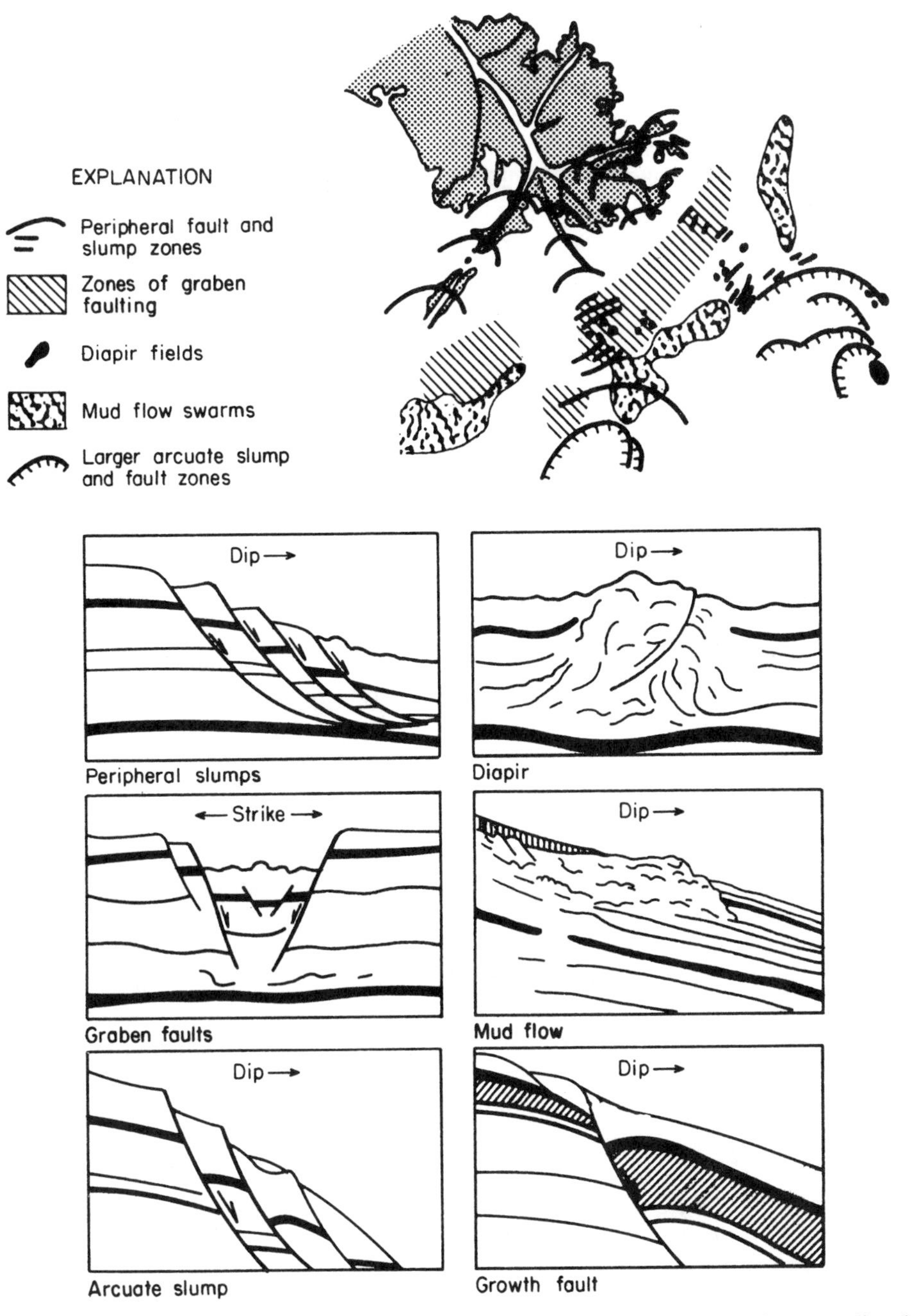

Figure 5-6. Styles of gravitational resedimentation and deformation along an actively prograding delta front. Map outlines the location and density of the various features around the periphery of the active lobe of the Mississippi Delta system. (Modified from Coleman and Garrison, 1977.)

mouth bar sands, which can thus accumulate to thicknesses in excess of 300 feet (100 m) in the Mississippi Delta (Coleman *et al.*, 1974). Rapid subsidence may effectively store most sand as mouth-bar facies. Shallow radial grabens on the prodelta slope reflect deeper mud intrusion.

Deformation on the prodelta slope also produces a variety of intraformational structural features (Fig. 5-6). Peripheral faults and slumps form near active channel mouths, offsetting mouth bar sands and associated facies. Large arcuate slump blocks, which may assume proportions of major, longlived growth faults penetrating underlying shelf and upper slope sequences, develop along the periphery of the delta platform and upper prodelta slope (Figs. 5-6 and 5-7). Large fault zones and associated mud flows initiated along the margin of prograding prodeltaic continental slopes may affect thousands of feet of section, and result in tremendous deformation and expansion of delta-front progradational sequences. The mechanics of growth fault evolution have been discussed by Bruce (1973), Dailly (1976), and Crans *et al.* (1980). The history of an individual growth fault zone is typified by three phases. (1) Rapid deposition at the channel mouth or delta periphery initiates uneven loading and consequent flowage of underlying muds. In shallower sediments deformation is accommodated by development of normal faults commonly showing *en echelon* patterns. (2) Continued loading results in rapid subsidence of the coastward, down-thrown side of the faults. A delicate balance is maintained with successive increments of sediment inducing equivalent subsidence of downthrown blocks. (3) Ultimately, deformation and compaction of the deep muds increases their bearing strength, and deformation ceases. The fault stabilizes when progradation of the delta front causes the zone of extension to advance into the basin, at which point a new zone of instability is activated and the process is renewed.

Growth-fault zones commonly form at consistent spacings along the advancing channel mouth or delta front (Fig. 5-7). Facies that show greatest expansion across the fault zone are those of the active delta front, including channel mouth bar sands and their marine-reworked equivalent delta-margin bars. Whether large or small, faults effectively outline the advancing periphery of the prograding delta platform and even of individual channel mouths (Fig. 5-6). Complexity of fault patterns reflects the complexity of the delta margin. Growth fault and associated deformational features constitute principal sites of hydrocarbon accumulation in thick deltaic sequences.

Significantly, mud flowage, diapiric intrusions, and growth faulting are characteristic of deltaic progradation at all scales. Tertiary deltaic sequences of the northern Gulf Coast Basin and the Niger Trough, for example, exhibit myriads of growth faults, shale ridges, and shale diapirs produced by differential loading of deltaic continental margins. Such features extend for miles horizontally and tens of thousands of feet vertically. In contrast, progradational deltaic sequences of Upper Cretaceous deltas of the Rocky Mountain foreland basin exhibit growth faults that cut and displace only several hundreds of feet of section (Weimer, 1973). Faulting on a similar scale (Fig. 5-7) is illustrated by Triassic deltaic sequences exposed in Svalbard (Edwards, 1976). Pennsylvanian strata of north-central Texas display delta-front growth faults that expand by several fold the delta front and channel mouth bar facies of small delta lobes that are less than 100 ft (30 m) thick (Brown *et al.*, 1973). Chaotic slump blocks and mud diapirs intruded into thickened channel mouth bar sands are also common in these abbreviated deltaic sequences, illustrating the ubiquitous nature of deltaic deformational processes, regardless of the scale of deltaic progradation.

In addition to intraformational deformation and differential compaction, sediment deposited at the crest of the delta platform is readily remobilized and transported down-slope. Surficial slump and mud-flow deposits are abundant on actively prograding delta fronts (Fig. 5-6). Rapid deposition of sediment at the crest of the prodelta slope may result in oversteepening and gravity sliding of shallow sediment layers (Hubert *et al.*, 1972). Slumping, as well as storms, periodically disturb and mobilize shallow delta-platform sediments. Once entrained, sediments may move down the fore-delta slope as mass flow or turbidity currents. Such mass gravity flow is inherent to subaqueous slope systems (see Chapter 8). Thus, delta-front sequences display a variety of sedimentary features, such as graded bedding, ball-and-pillow, and sole marks, which are typical of submarine fan systems. This is reasonable in that an actively prograding delta front is subject to the same suite of processes. However, from a paleogeographic as well as an economic perspective, the key feature is the development of an independent subaqueous sediment dispersal system capable of transporting large volumes of bed-load sediment from its initial site of deposition at the crest of the delta platform onto the basin floor. If such is the case, gravitational transport of sediment becomes a significant, perhaps dominant delta destructional process.

Contemporaneous delta-submarine fan couples appear to be common in many delta systems, including the Rio Balsas (Shepard and Reimnitz, 1981; Reimnitz and Gutierrez-Estrada, 1970) and Rhone (Bellaiche *et al.*, 1981), where marine

Figure 5-7. Growth-faulted delta-front progradational sequence exposed in the Triassic of Svalbard Cliff face approximately 1,000 ft (300 m) high. (Photographs courtesy of M. B. Edwards.)

reworking of the delta front is a dominant and ongoing process. Vertical stratigraphic transitions from shallow-water delta into prodelta or upper-slope sequences containing channeled sands and gravels, slump and mass-flow deposits, and well-bedded turbidite units displaying partial Bouma sequences have been documented in many basins (for examples, see Walker, 1966; Massari, 1978). As suggested by Burke (1972), the stratigraphic distribution of sand in such a couple should be bimodal, with greatest concentrations at the crest *and* base of the progradational sequence. Deltaic and submarine-fan sand facies are separated by sand-poor slope and prodelta muds laced with local lenticular submarine channel-fill units. Minimum bathymetric relief for development of contemporaneous slope and deltaic systems need only be several hundred feet, as illustrated by reconstructions of basin topography and depositional systems in the Pennsylvanian fills of the Midland and Central Pennine basins (Galloway and Brown, 1972; Reading, 1964; Walker, 1966; Collinson, 1969).

Lobe-Abandonment and Cyclic Destruction. Both marine and gravity processes continually modify and rework the active delta margin. However, in delta systems characterized by well-developed lobe growth and abandonment, destructional and constructional processes alternate, producing cyclic sequences that are inherent in deltaic deposition (Coleman and Gagliano, 1964).

Abandonment of a delta lobe reduces or removes the supply of fluvial sediment. However, marine processes continue to rework and modify the delta margin. Further, impact of the marine processes expands as continued compaction of the rapidly deposited, saturated prodelta mud substrate causes subsidence and transgression of the delta lobe. Basin subsidence may further accentuate foundering. Ultimately, much or all of the surficial sediment of the inactive lobe can be subjected to submergence and marine reworking, completing a sequence that is bounded above and below by marine shelf units.

Renewed progradation of a new lobe across a foundered lobe produces another upward-shoaling sedimentary cycle, which, in turn, will be abandoned and destructionally modified. Multiple lobe formation readily leads to repetitious sequences of similar facies assemblages. For example, hypothetical core sites A or B (Fig. 5-4) in the Mississippi Delta system would penetrate at least three whole or partial progradational cycles separated by destructional facies, all deposited within the last 6,000 years.

Process Classification of Delta Systems

Review of the process framework and associated morphology and sediment distribution patterns in modern deltas shows that three basic processes determine delta geometry and distribution of framework sand facies: (1) sediment input, (2) wave energy flux, and (3) tidal energy flux (Galloway, 1975). Intruding oceanic or barometric currents serve primarily to redistribute suspended sediment. By definition, significant removal of bed-load sediment by gravitational processes produces slope depositional systems that are independent of the delta system. Depositional stratigraphy and facies geometry are, of course, modified by the morphology of the sedimentary basin (deep versus shallow water) and rate of tectonic subsidence. Delta-plain aggradational facies are modified by climate, in part through its effect on the presence and nature of vegetation. However, the geometry, trend, and internal features of the progradational framework sand bodies are largely a product of the interplay between fluvial sediment input and wave and tidal components of reservoir energy flux (Fig. 5-5).

The process framework provides the basis for recognition of a tripartite classification of end-member delta types (Galloway, 1975). These end members—fluvial- , wave- , and tide-dominated deltas—form the points of a triangle into which modern or ancient delta systems can be placed (Fig. 5-8). Few modern deltas are ideal end members. Rather, most deltas, modern or ancient, reflect the combined impact of constructional fluvial and destructional wave or tidal processes. Fluvial-dominated deltas typically have elongate-to-irregular lobate areal geometries. Wave-dominated deltas record the overriding influence of strike reworking in their regular lobate to cuspate outlines. Tide-dominated deltas have irregular lobate to estuarine outlines. Most importantly, the morphology of each major type of delta is reproduced in the stratigraphy and geometry of

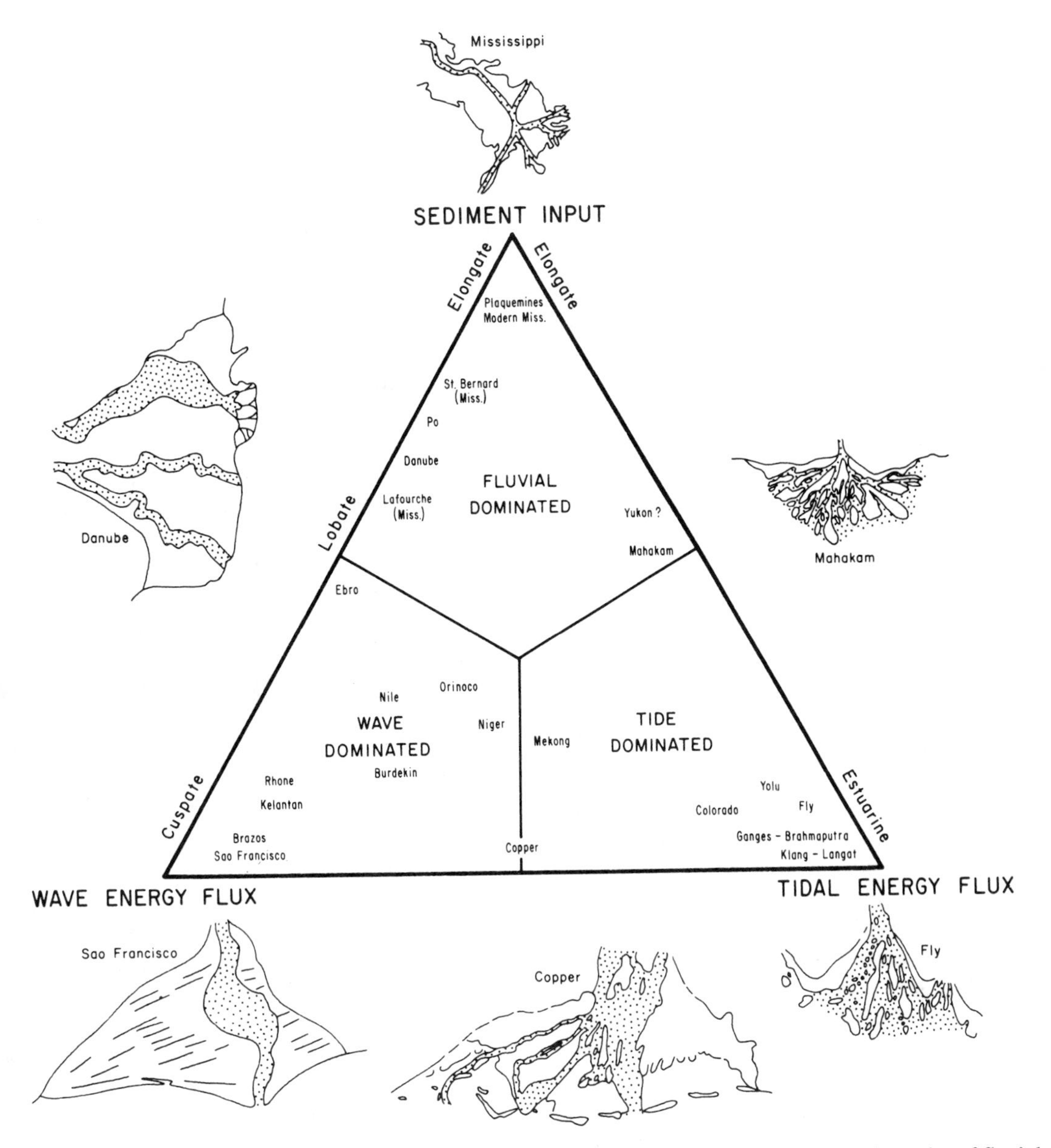

Figure 5-8. Morphologic and stratigraphic classification of delta systems based on relative intensity of fluvial and marine processes. (Modified from Galloway, 1975; courtesy of B. R. Weise.)

its framework sands, and can be recognized and interpreted on the basis of surface or subsurface stratigraphic information. Further, absolute sedimentation rates or quantitative coastal energy regimes (variables that are, at best, difficult to calculate for modern delta systems, and impossible to measure in ancient depositional systems) are not of great importance. Rather, the relative importance of each process determines delta stratigraphy. Thus, over a broad range of basin dimensions and coastal energy magnitudes, wave-, tide- , or fluvial-dominated deltas each retain basic similarities.

Fluvial-Dominated Deltas

In a fluvial-dominated delta system the rate and volume of sediment input exceeds the capability of reservoir energy flux to rework and substantially modify the active delta margin. Depending on the relative dominance of constructional fluvial processes, geometry of delta lobes ranges from highly elongate and digitate (birdfoot) to somewhat rounded or lobate. The Mississippi River delta system provides a well-studied example and, in its various Holocene lobes, illustrates a variety of depositional styles that may occur in fluvial-dominated systems. Important descriptions of the Mississippi system include those Russell and Russell (1939), Fisk (1952, 1955, 1961), Fisk *et al.* (1954), Scruton (1960), Coleman and Gagliano (1964), Coleman *et al.* (1964, 1974), Kolb and van Lopik (1966), and Frazier (1967).

Constructional Environments and Genetic Facies

The sand-rich environments and resultant framework facies of fluvial-dominated deltas consist of the distributary channel and channel mouth bar (Fig. 5-9) along with subsidiary laterally reworked delta-front sands and crevasse splays. Nonframework environments include the prodelta interdistributary and delta margin embayment and submerged delta platform, and delta-plain marsh, swamp, floodplain, levee, and lake basins. Volumetrically minor but prominent destructional bars and beach ridges fringe abandoned lobes.

The *distributary mouth bar* is the focus of sand deposition and storage in the fluvial-dominated delta system. In digitate delta lobes, such as the modern birdfoot lobe of the Mississippi system, rapid subsidence of the sands into thick prodelta muds, combined with limited wave energy flux, results in minimal reworking of mouth-bar deposits. In plan view, resultant sand bodies form a digitate complex of narrow, lenticular units that reflect the geometry of the delta lobe (Fisk, 1961; Coleman *et al.*, 1974). In cross section, mouth-bar sands thin laterally over short distances into bounding prodelta and delta margin muds (Fig. 5-10, section A'). Comparatively thin marginal bar sand may extend laterally and coalesce with similar reworked sands of adjacent mouth bars, forming a "webbing" between the bar fingers.

The mouth-bar sequence coarsens upward, as proximal mouth-bar crest, bar-front, distal-bar and upper prodelta deposits are superimposed during progradation. Upward coarsening may be reflected somewhat in average grain size but is more apparent in the increasing proportion of sand and silt relative to mud interbeds, thickness of sand beds, and scale of contained sedimentary structures (Fig. 5-11). Distal mouth-bar sequences consist of laminated mud, silt, and silty sand and display slump units, compaction and dewatering structures, slide features such as ball-and-pillow, and thin graded beds exhibiting sharp bases and incomplete Bouma sequences. Such graded units record the emplacement of both frontal splays and local turbidity current tongues. Middle mouth-bar deposits commonly consist of massive to thick bedded, clean, well-sorted sands containing trough and ripple stratification, dewatering structures, and climbing-ripple lamination. The uppermost deposits may display planar stratification and scour-and-fill structures of the bar crest, which are capped by finer ripple laminated sands, silts, and muds of the associated subaqueous levees. Thick but highly local slump masses of mouth-bar sand may occur within distal mouth bar or upper prodelta deposits, punctuating the upward-coarsening pattern.

Figure 5-11 is an idealized vertical sequence through the axis of a channel mouth-bar sand body illustrating features typical of the Mississippi and other fluvial-dominated deltas. The electric log profile is one of many upward-coarsening or funnel-shaped profiles produced in a strandline system. It is distinguished by the irregular, serrate basal transition zone into underlying prodelta muds, reflecting the abundance of slump, frontal splay, and turbidite beds common to many highly constructive channel mouth bars.

Mouth-bar sands and associated silts and muds contain abundant macerated plant debris, sometimes called "coffee grounds," as well as large chunks of woody material. Thin, discontinuous mud drapes and pockets of reworked mud clasts reflect the variability of fluvial scour and deposition. Burrows and trails may be present but are usually sparse because of high rates of deposition. Figure 5-10 shows a typical subsurface cross section of an Eocene bar finger sand. Of interest and potential use in detailed facies delineation is the development of the serrate basal transition zone only in the immediate position of the sand body axis (and inferred distributary mouth). Presence of abundant depositional features indicative of rapid, erratic loading of the upper shore face provides strong evidence of channel-mouth

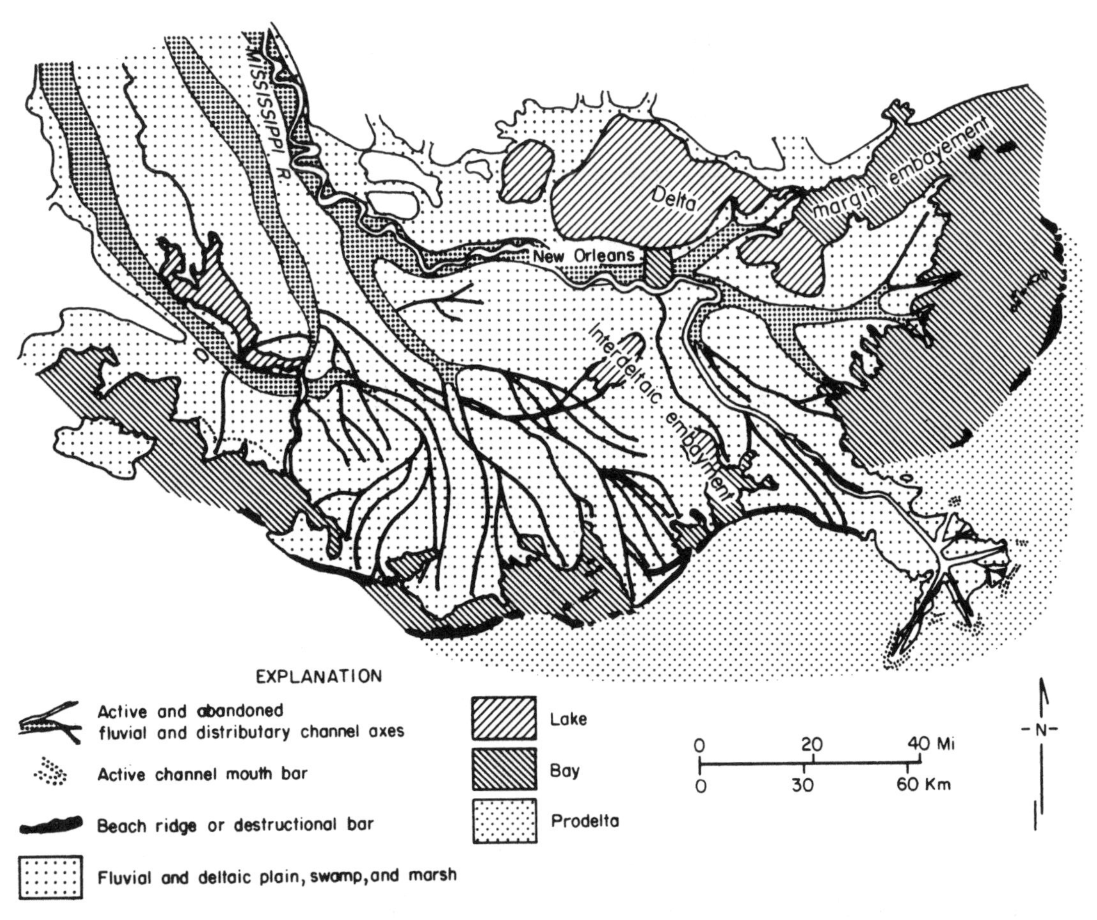

Figure 5-9. Generalized surficial depositional environments of the Holocene Mississippi fluvial-dominated delta system. (Compiled from Gould, 1970, and other sources.)

proximity and may be used to distinguish deltaic settings using both surface and subsurface data.

The channel mouth-bar sand unit forms an integral part of the overall progradational facies sequence. It is underlain by prodelta silts and muds and is overlain by delta platform facies, including levee and subaerial delta plain marsh or swamp. Laterally, mouth-bar sands grade into interdistributary sand, silt, and mud and may be overlain by crevasse-splay sequences. As the distributary advances, the mouth bar is partially channeled out (Fig. 5-10). Following distributary abandonment, the channel may fill wholly or partially with fine-grained sediment, forming a plug that runs along the axis of the sand body. In shoal-water basins where compactional subsidence is minimized, subsequent distributary incision may cut completely through the mouth-bar sand unit, leaving only remnants of the bar margin on either side of the distributary channel fill (Donaldson *et al.*, 1970; Morton and Donaldson, 1978). Such mouth-bar incision is characteristic of fluvial-dominated deltas in many intracratonic basins (Brown, 1979; Hobday, 1978a).

With increasing wave influence, delta geometry evolves from digitate to lobate, and sands are reworked along the upper shoreface and marginal beaches, flanking active distributary mouth bars to form a *delta margin sand sheet.* This delta front sand sheet (also commonly called the delta-fringe sand) connects the branching, dip-oriented distributary mouth-bar framework. Many older, shoal-water lobes of the Mississippi Delta system, such as the Lafourche complex (lobe 14, Fig. 5-4) contain extensive, thin, irregular sand and silty-sand sheets connecting and grading laterally into mouth bars of the myriad distributaries (Fisk, 1955; Frazier, 1967).

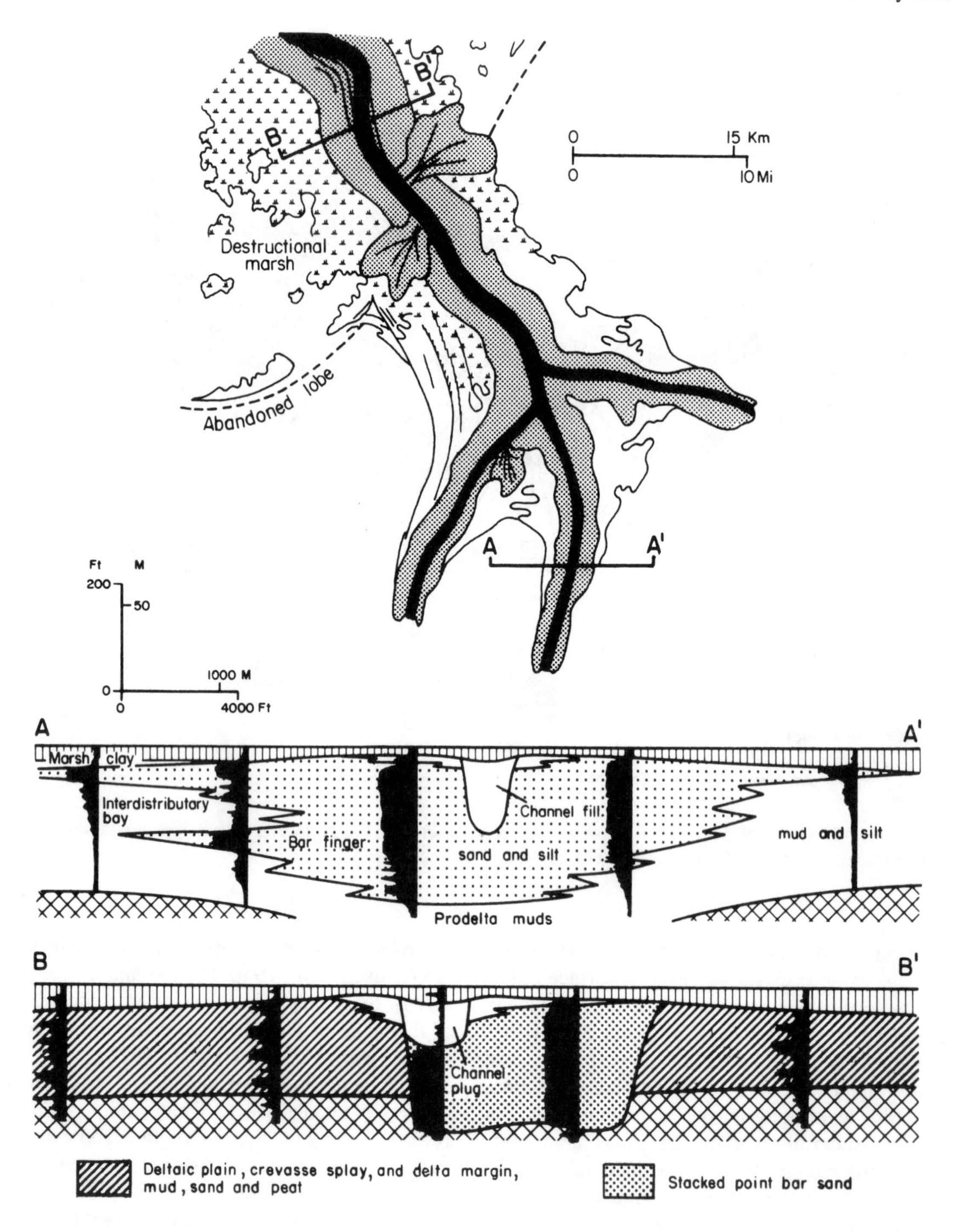

Figure 5-10. Interpretative facies cross-sections of (A) progradational channel mouth bar and (B) distributary channel-fill sand bodies of an elongate, fluvial-dominated delta lobe. Geophysical log profiles are based on bore hole data from the Wilcox (Eocene) Holly Springs delta system of the northern Gulf Coast basin. (Modified from Galloway, 1968.)

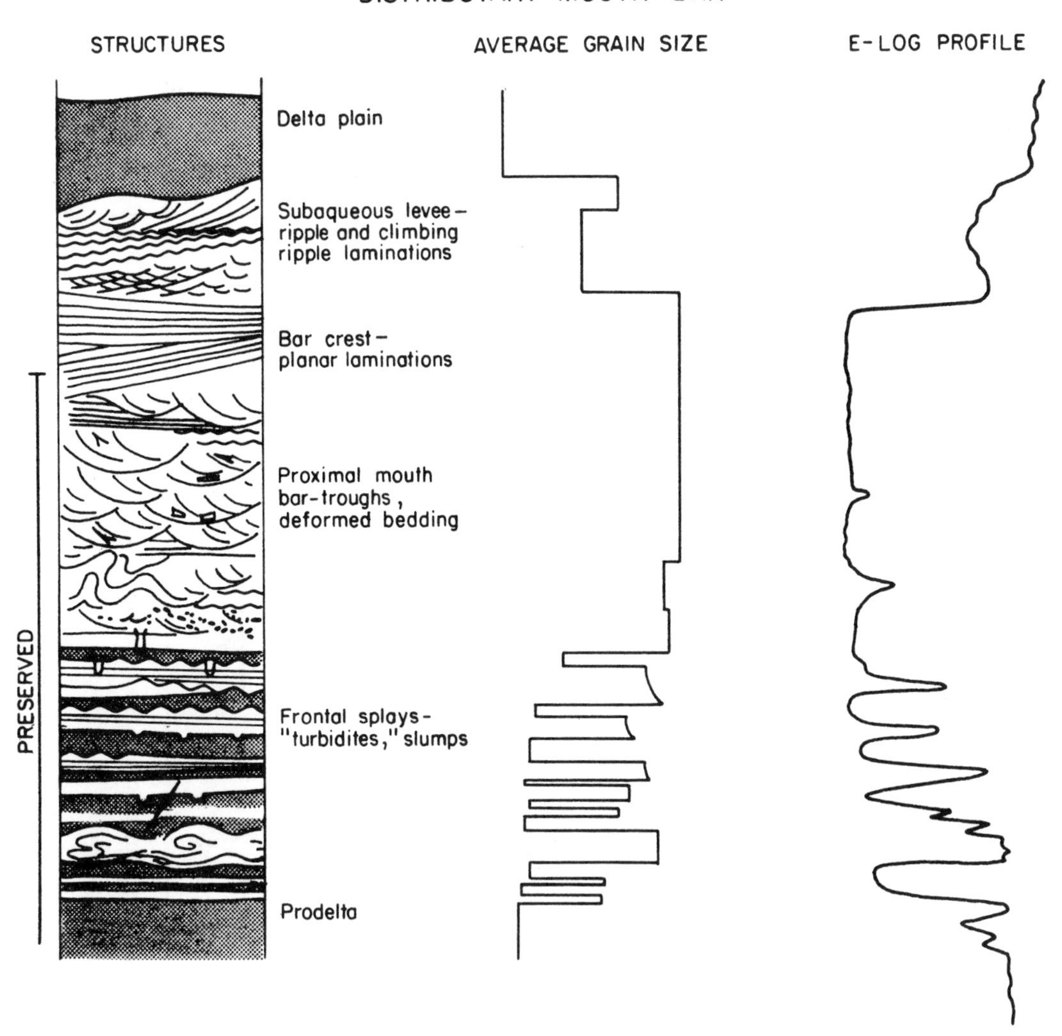

Figure 5-11. Generalized vertical profile through a channel mouth bar sand body. As in fluvial system profiles, sedimentary structures are illustrated schematically, grain size increases to the right on the average grain size plot, and the log profile is drawn to resemble either an S.P. or gamma-ray curve.

The delta front sand sheet fines away from associated mouth bars, commonly consisting of silty fine sand to sandy silt. In addition, delta fronts of fluvial-dominated deltas are relatively thin. Upward-coarsening textural sequences are well developed and reflect the dominance of longshore transport and sorting by wave or current action. Typical sedimentary structures of Mississippi delta-front sands and silts include abundant ripple lamination, low-angle planar (beach) lamination, and burrows and trails.

Delta-front sands overlie prodelta and delta platform muds, and are capped by aggradational delta plain or destructional-marine facies. Finely macerated plant debris is abundant; shell material is commonly present but is rarely preserved.

Distributary channel fills of the Mississippi Delta system are low-gradient suspended-load channels. Thousands of miles of anastomosing and branching distributary channel courses lace the Holocene Mississippi Delta plain (Fig. 5-12), reflecting the complex history of bifurcation and

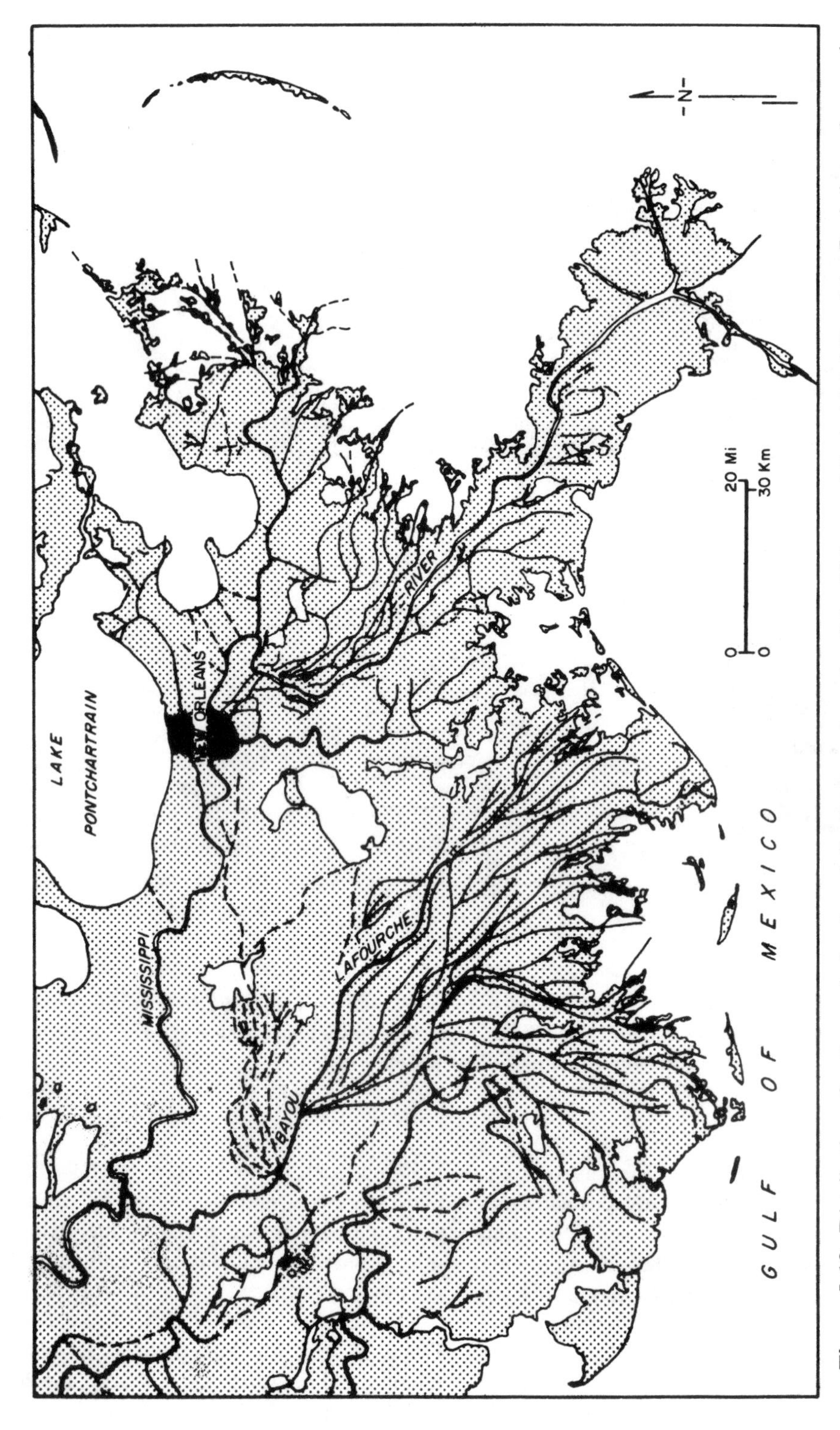

Figure 5-12. Distribution of active and abandoned distributary channels of the Holocene Mississippi delta plain. Buried or submerged distributary trends are not shown. (After Kolb and Van Lopik, 1966.)

avulsion typical of high constructive deltas. In addition to their anastomosing pattern, both active and abandoned distributaries display the features typical of suspended-load channels. In cross sections, distributary channels are narrow and deep (Fig. 5-10, section B′). Lateral migration is minimal and levees are prominent. Channel fill consists of interbedded fine sand, silt, and some mud, and resultant units are lenticular and nearly symmetrical. Abandoned channel plugs are abundant, particularly in the upper portions of the channel fill, and consist of mud and organic debris. Although other channel types might reasonably exist in high-constructional fluvial-dominated delta systems, large systems such as the Mississippi rapidly build a broad, low-gradient delta plain. Thus, in its lower reaches, the trunk stream likely has lost much of its bed load.

In addition to the suite of primary structures dominated by trough cross-stratification and ripple lamination (Fig. 4-5), delta distributaries have numerous features indicative of their rapid deposition on a water-saturated delta platform substrate. Dewatering, differential compaction, and liquifaction commonly destroy primary cross stratification. Slump structures, contorted bedding, and local mud diapirs and injections are typical, especially in the lower portion of the channel fill.

Vegetation is typically abundant on the delta plain and is readily incorporated and preserved in distributary-channel sediments. Shell debris scoured from underlying delta platform and embayment facies is common in the channel lag. Abandoned distributaries of the lower delta plain may be invaded by saline or brackish water, allowing colonization by burrowing organisms, such as the mud shrimp, *Callianassa*, normally associated with marine settings.

The geometry of distributary channel sand bodies is complex. Although average channel trend is in down depositional dip, progradation of elongate delta lobes and presence of large delta-margin embayments, such as Lake Ponchartrain, results in considerable dispersion in channel trend (Fig. 5-12). Distributary segments and even whole lobes may extend along strike or turn updip. Distributary and anastomosing channel patterns that characterize the delta plain are further complicated as avulsion produces new cross cutting or superimposed channel trends, which upon abandonment and burial become indistinguishable from earlier deposits.

Distributary channel fills constitute the sandy framework of the aggradational delta-plain facies assemblage. They cut down into progradational facies, and are laterally equivalent to levee, crevasse splay, marsh, swamp, and lake deposits of the active delta plain (Fig. 5-10, section B′). They are, in turn, overlain by younger aggradational delta or alluvial deposits, or may be capped by a veneer of transgressive destructional sediments.

Crevasse splays are common features of fluvial-dominated deltas, particularly on the lower delta plain where distributary levees are immature and poorly formed. Consequently, many splays become subdeltas, prograding into the shallow bays of the interdistributary delta platform. Splays reproduce, on a reduced scale, the environments and facies of the main delta lobe. Because wave energy is most likely low in the shallow, protected embayments, splays display minimal contemporaneous reworking. However, repetitive cycles of progradation and abandonment and destruction characterize splay sequences.

Splays form lobate wedges of sediment that spread from an apex located at the distributary margin (Fig. 5-3). Depositional architecture of large crevasse splays is well organized. Overall thickness of the splay unit thickens away from its apex, reflecting the depositional gradient away from the trunk distributary levee (Fig. 5-3). However, sand content and grain size are greatest near the apex and typically diminish toward the distal splay. Channeling and consequent sharp basal contacts characterize the proximal splay. Basinward, progradation of the splay deposits onto the submerged delta platform produces an upward-coarsening sequence that overlies embayment muds and silts. Stratigraphically, crevasse-splay deposits lie marginal to and largely above the distributary channel fill of the associated trunk channel. With the exception of the splay channel or channels that cut through the levee, the splay depositional unit and its contained sands pinch out against levee muds and silts. Even the headward portions of splay channels may fill with overbank suspension-load sediment, effectively isolating splay sands from framework channel-fill sands of the distributary system.

Splay deposits present a hodgepodge of sedimentary structures reflecting rapid but sporadic deposition and intervening reworking by physical or biogenic marine processes. Graded bedding, climbing-ripple lamination, and soft-sediment deformation characterize distal splay interbedded sand, silt, and mud. Proximal splay deposits exhibit complex scour-and-fill, channeling, and abundant cross stratification of a variety of types. Climbing-ripple lamination and organic debris are common. Progradational splay sequences may be capped by alternating beds of sand, silt, and clay displaying abundant root mottling. Early diagenetic iron-rich carbonate or pyrite nodules are plentiful. Carbonaceous mud or peat may cap the complete splay sequence.

Rapid deposition of suspension sediment seaward and marginal to active distributary mouths produces a thick sequence of laminated *prodelta and subaqueous delta platform muds* across the shelf and onto the upper slope. Prodelta muds may reflect their marine origin by dispersed burrows or shells. However, the high sedimentation rate commonly dilutes the effects of biogenic activity on proximal prodelta sediments. Contorted bedding, intraformational faults and fractures, and isolated lenses, beds, or blocks of sandy sediment reflect mud flow, fracture, and slumping on the prodelta slope. Prodelta muds are organic-rich, but here again, organic material dominates marine constituents. Resultant acidic pore waters commonly leach any shell material incorporated in the prodelta sequence.

The prodelta muds form one of the most homogeneous and laterally continuous units of fluvial-dominated delta systems such as the Mississippi. Thickness is dependent primarily on basin water depth. The muds grade basinward and downward into burrowed, fossiliferous shelf sediments. Landward, or stratigraphically upward, prodelta deposits grade either into distal channel mouth-bar deposits, delta-front sands, or interdeltaic embayment muds and silts deposited in a shoal-water delta platform setting.

Embayments represent environments temporarily bypassed by active distributaries. Consequently, their deposits form a veneer of burrowed, sporadically fossiliferous (restricted fauna) muds, silts, and some fine sands capping the progradational delta front and prodelta platform. Long-lived embayments can become isolated from the marine environment, evolving into fresh or slightly brackish lakes filling with mud and organic debris (Lake Ponchartrain, for example; Fig. 5-9). Crevasse-splay deposits typically fill embayments.

Aggradational *delta plain* facies of the lower Mississippi consist of natural levee deposits of massive to interbedded silt and mud, marsh and swamp peat and organic-rich clay and mud, and lacustrine mud and clay. Upper delta plain environments resemble comparable fluvial backswamp and floodplain settings. The Mississippi system is a major site of deposition and preservation of reasonably pure peat beds (Frazier and Osanik, 1969). It has been successfully utilized as a model of lignite and coal formation in numerous basins (see Chapter 12). With the exception of levee deposits, delta-plain facies are relatively thin and laterally discontinuous. They are segmented by contemporaneous distributary channel deposits and randomly cut by subsequent distributary channel fills.

Surficial environments of the delta system are, of course, the most sensitive to climate. However, even in subarid settings, vegetated interdistributary mud flats produce organic-rich sediments (Donaldson *et al.*, 1970).

Destructional Environments and Facies

Following lobe (or on a smaller scale, crevasse splay) abandonment, foundering, transgression, and marine reworking produces volumetrically minor, but distinctive destructional facies. In the Mississippi system, a typical facies tract begins with thin fossiliferous mud and sandy mud capping distal portions of the abandoned lobe as waves rework surficial mouth bar and delta front sands, producing a retreating series of thin, transgressive beach ridges and barrier bars (Fig. 5-13). Such reworked sands concentrate the coarsest material available, including abundant shell. Inland, portions of the delta plain are flooded, producing shallow bays and extensive, laterally continuous salt marsh. Thin veneers of mud and oyster shell accumulate in bays. Oyster reefs may grow on the more stable substrates produced by abandoned distributaries and associated levees (Coleman and Gagliano, 1964). Marsh peats blanket large areas and record the most inland impact of flooding.

As shown in Figure 5-9, arcuate, discontinuous belts of offshore barrier bars and beach ridges outline older, abandoned, and partially flooded lobes of the Mississippi delta system. Although

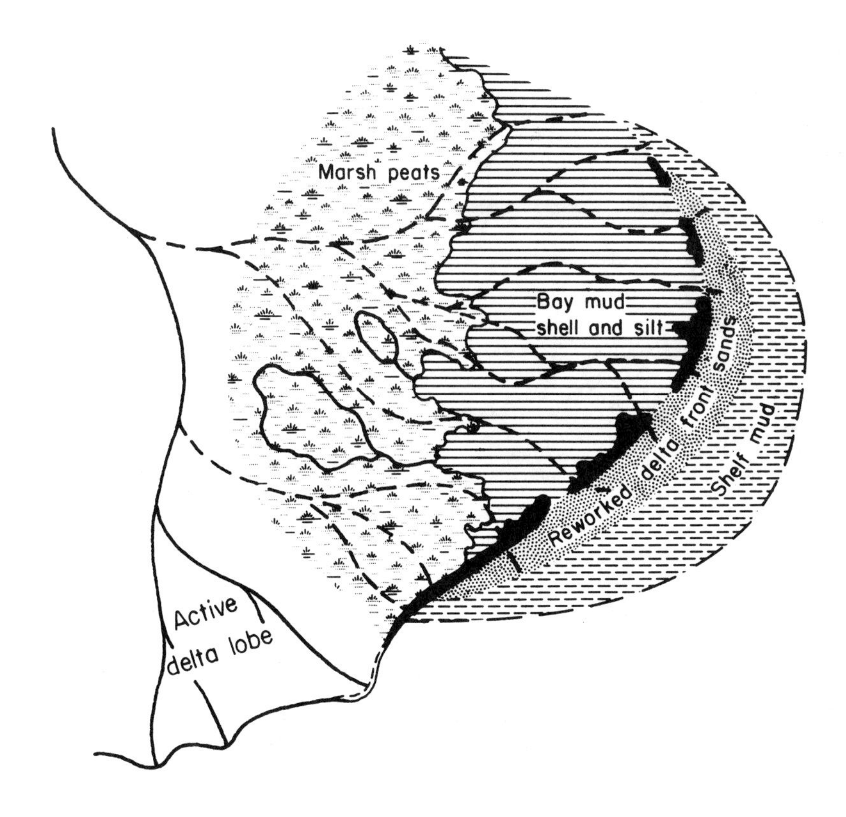

Figure 5-13. Idealized facies tract produced by abandonment, subsidence, and transgression of a fluvial-dominated delta lobe. (Based on Fisher *et al.*, 1969.)

sand bodies produces by destructional bars are thin, rarely exceeding 20 feet (6 m), barriers such as Grand Isle locally exceed 30 feet (9 m) in thickness (Fisk, 1955).

The stratigraphic importance of destructional units lies in their areal continuity and predictable lateral relationship, which favor their use for correlation and facies mapping within the deltaic complex. Careful analysis of thick deltaic sequences usually reveals thin shale, impure limestone, calcareous muddy sand, or coal beds that exhibit unusual continuity, and provide useful markers within the otherwise heterogeneous deltaic stratigraphy. In addition, although volumetrically minor, destructional sand bars are isolated sand bodies encased within impermeable muds. Thin shelf and bay mud blankets are effective seals for fluids trapped within constructional sand facies.

Facies Architecture

The facies architecture of fluvial-dominated deltas, although complex, is characterized by several important generalities.

1. Most vertical sequences through the delta plain reveal its dominantly progradational aspect by characteristic upward-coarsening textural sequences (Fig. 5-14).

2. The upper portion, or even all of the progradational section, is locally cut out and replaced by a complex network of distributary channel fills, the upper parts of which are laterally equivalent to the aggradational wedge that caps the progradational units (Fig. 5-14).

3. The progradational mouth-bar and distributary-channel sands together form a highly divergent but generally dip-oriented permeable sand framework of the component lobes of the delta system. Width of these skeletal depositional elements is increased by formation of laterally associated delta front sheet and crevasse splay sands.

4. Sediment accumulation is cyclic, reflecting alternating periods of construction and transgressive destruction (Fig. 5-14) inherent in distributary overextension and avulsion.

5. With the exception of destructional facies, lateral continuity of both framework sand and encasing mud facies is limited. Permeability pathways that do exist are circuitous at best, and

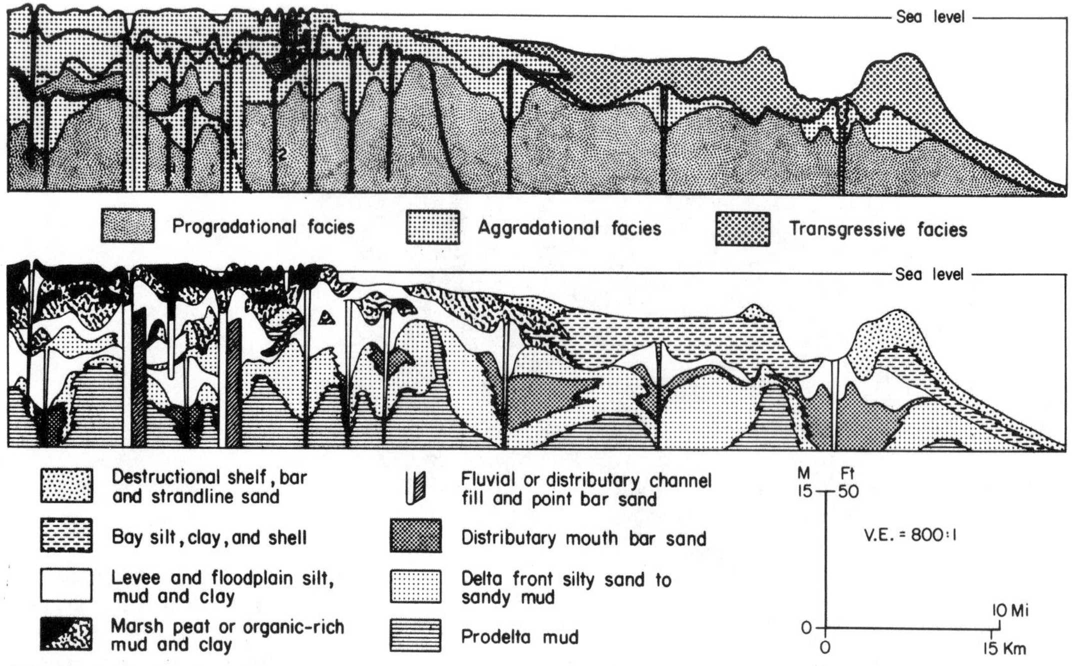

Figure 5-14. Depositional architecture and genetic facies relationships along a dip-oriented transect through the Holocene Mississippi Delta system. Section was based on 54 drill holes. (Modified from Frazier, 1967.)

sand bodies are commonly highly compartmentalized by cross-cutting muddy beds and lenses. Channel mouth-bar and delta-margin facies provide the best integrated pathways for fluid migration.

Depositional facies of the modern Mississippi Delta system are readily recognized in its Pleistocene precursor, which deposited as much as 15,000 feet (4500 m) of sediment in the northern Gulf Coast Basin (Caughey, 1975). Dip-elongate deltaic axes of the Pleistocene system contain an aggregate of several thousand feet of sand deposited in mouth-bar, delta-front, distributary-channel, and crevasse-splay environments.

Wave-Dominated Deltas

In wave-dominated delta systems, most bed load initially deposited at distributary mouths is reworked by waves and redistributed along the delta front by longshore drift. A smooth delta front, consisting of well-developed coalescent beach ridges, results. Delta-plain geometries are arcuate to cuspate. Numerous modern oceanic deltas, including the Rhone, Nile, Burdekin, Orinoco, Kelantan, and São Francisco are wave dominated. The Rhone Delta provides a well-studied Holocene example of a large, wave-dominated delta system and has been described by Kruit (1955), Van Straaten (1959), Lagaaij and Kopstein (1964), and Oomkens (1967, 1970).

Depositional Environments and Facies

Although not comparable in size to the Mississippi system, the Rhone has prograded nearly 10 mi (16 km) basinward, and deposited up to 230 ft (70 m) of sediment in Holocene time. Consequently, a suite of environments typical of wave-dominated deltaic systems is well displayed (Fig. 5-15). The dominant framework facies consists of coastal beach ridge and barrier sands deposited along the front and margins of active delta lobes; though not directly comparable to that of nondeltaic wave-dominated barrier coasts, this delta margin assemblage is commonly described as the "coastal barrier" facies. As shown by the generalized sand isolith of the delta system (Fig. 5-15), "coastal barrier" sands impart a first-order strike orientation to the delta framework. In addition, major distributary channel-fill sand bodies constitute secondary framework elements, and impart a dip orientation to isolith contours, which is especially evident on the landward side of the system.

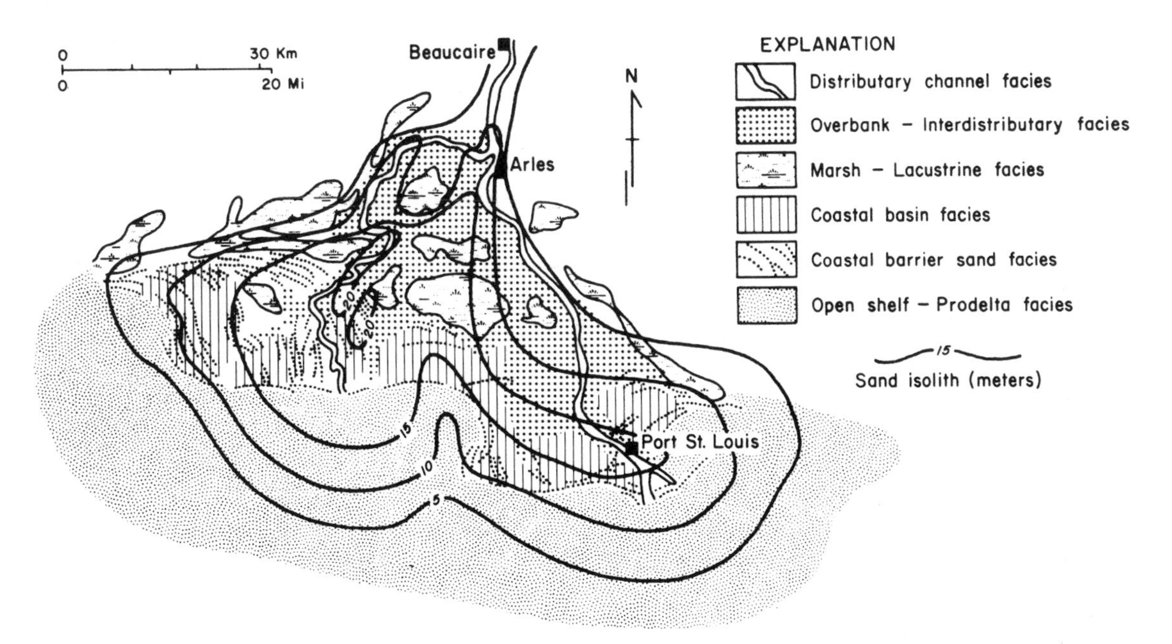

Figure 5-15. Generalized depositional environments of the Holocene Rhone wave-dominated delta system, Gulf of Lions. Contours outline sand distribution within the system. (Modified from Fisher *et al.*, 1969; courtesy of B. R. Weise.)

Nonframework facies include deposits of the prodelta and shelf, interbarrier coastal basins, interdistributary lakes and marshes, and channel margin levee and overbank settings. (Fig. 5-15).

The Rhone has occupied several discrete distributary channels during progradation of its delta. At any one time, however, flow was concentrated in a single trunk distributary. Unlike the Mississippi Delta, *distributary channels* are moderately sinuous, forming broad meanderbelts up to 3 mi (5 km) wide. Width/thickness ratios range from 100:1 to 1,000:1 (Oomkens, 1970). These broad belts branch and generally narrow basinward. Thus, typical mixed-load channels and even bed-load channels characterize the wave-dominated delta plain.

Internal sedimentary structures of the distributaries are also typical of mixed-load channel deposits. Rhone distributary channel fills fine upward from medium to coarse sand into fine sand and silt; they contain abundant trough cross-stratification that decreases in scale upward, and display common slump blocks, lenses, and beds rich in mud clasts. Lateral and basal contacts with surrounding delta-plain and coastal-barrier facies are abrupt and dominantly erosional. Detrital plant debris is a common accessory.

Progressive seaward accretion of Rhone *coastal-barrier bars* has produced a strike-elongate sand body up to 30 ft (10 m) thick and 30 mi (50 km) by 12 mi (20 km) in areal extent (Oomkens, 1970). Although marine energy is primarily responsible for deposition of sand in its ultimate site of storage along the barrier front, the influence of fluvial input is apparent locally near channel mouths, where distributary mouth bar sequences accumulate, and in the irregular outline of the deltaic coast and consequent variability in local trend of the coastal barrier sand bodies (Fig. 5-15). Thickness of the sand body reflects depth of wave base and marine shoreface development, as well as loading and sediment outwash from the channel mouths. Sites of mouth-bar deposition may be recognized by localized thicks in the progradational sequence, by the presence of a more erratic upward-coarsening sequence due to distal mouth-bar frontal splays, turbidite beds, and slumps, and by the presence of the capping distributary channel fill.

Coastal-barrier sands are better sorted and finer than associated distributary channel sands. Shell debris commonly accumulates as thin beds or lenses, and plant debris is abundantly dispersed through the sand body. Coastal-barrier sands coarsen upward, as is typical of sand bodies produced by progradation of a marine shoreface (Fig. 5-16). Distal-bar deposits consist of interbedded sand, silt, and mud. Bedding and con-

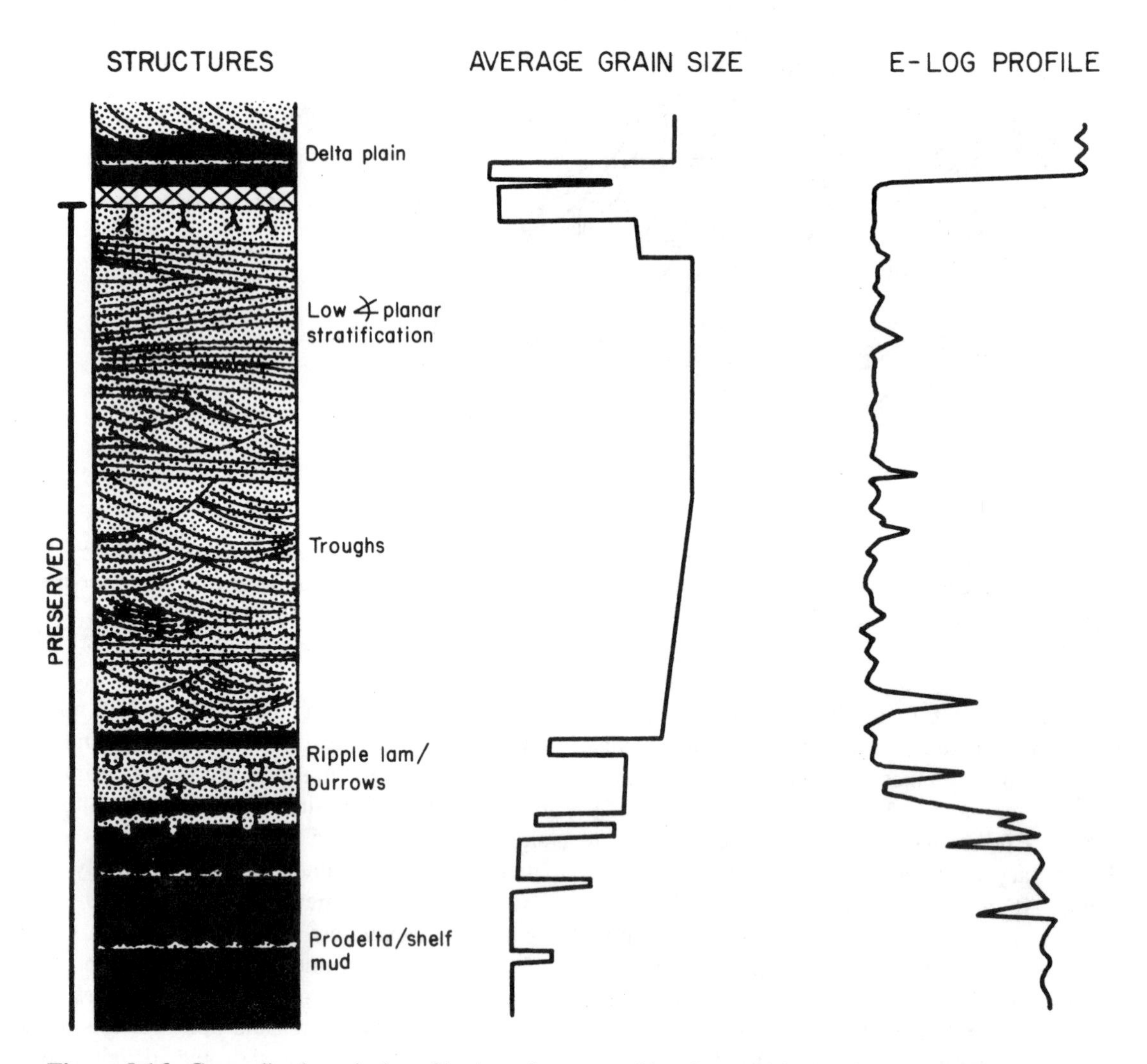

Figure 5-16. Generalized vertical profile through a coastal-barrier sand body of the wave-dominated delta margin.

tained primary structures are wholly to partially destroyed by burrowing. Upward, the proportion and thickness of sand and silt beds increases, burrows become less abundant, and primary structures, including planar and ripple lamination are preserved (Fig. 5-16). Slumped or contorted beds indicate sporadic gravity mass-flow of sediment down the sloping shoreface. The interbedded sequence grades upward into massive, planar or low-angle trough cross-stratified sand containing scattered burrows and discontinuous mud beds and lenses. The complete progradational sequence is capped by low-angle planar beach stratification, and the sand body is overlain by delta-plain marsh or coastal basin peat, carbonaceous mud, or mud (Fig. 5-16). Uppermost beds in the sand may show root disturbance, reflecting subaerial exposure.

Coastal-barrier sands grade basinward and stratigraphically downward into prodelta and normal marine shelf deposits, typically muds or silts. They are overlain by, and pass abruptly landward into, the suite of delta plain marsh, floodplain, lake, and coastal basin deposits. Ongoing subsidence combined with abundant, though locally intermittent, supply of new sedi-

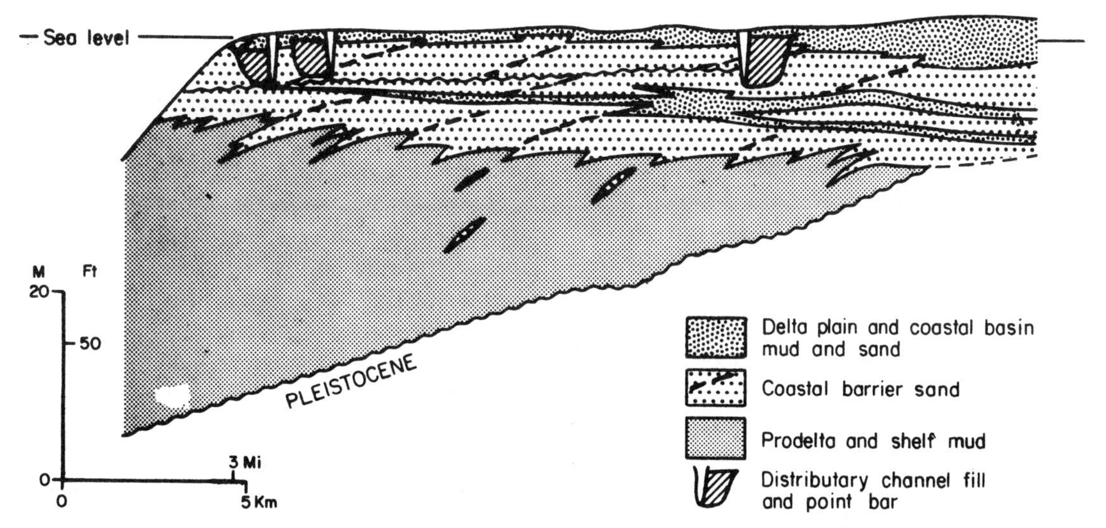

Figure 5-17. Dip-oriented facies cross section of the Rhone Delta illustrating the generalized depositional architecture produced by progradation of a wave-dominated delta system. Original section was based on nine drill holes. (Modified from Oomkens, 1970.)

ment may produce thick accumulations of amalgamated middle and upper coastal-barrier sands. For example, massive sand sequences interpreted to be coastal-barrier deposits ranging from several hundred to several thousand feet thick occur in the Upper Tertiary deposits of the ancestral Nile Delta system and in the Oligocene and Miocene Frio and Oakville Formations of the northwestern Gulf Coastal Basin (Galloway *et al.*, 1982). In addition, cyclic lobe progradation and abandonment may produce repetitive, stacked sequences of upward-coarsening coastal-barrier sands (Oomkens, 1970).

The greater intensity of wave reworking in a wave-dominated delta such as the Rhone results in widespread dispersal of suspended sediment. Consequently, a thick, localized *prodelta* facies may not be deposited. Rather, mud and silt slowly accumulate on large areas of shallow basin margin shelf and slope. These deposits differ little in composition, geometry, or accessory features from typical marine shelf facies. Prodelta deposits of the Rhone consist of burrowed, sparingly fossiliferous mud and silt. Species abundance and diversity, as well as intensity of burrowing, increase systematically basinward and laterally from the active delta front (Lagaaij and Kopstein, 1964), reflecting decreasing rates of sedimentation.

Nonframework *delta plain* facies of the subaerial Rhone Delta system form primarily by aggradation of overbank fluvial material washed into the floodplain, lake, and coastal basins. Deposits consist of discontinuous beds and lenses of clay, mud, muddy sand, and some peat. Root disturbance and pedogenic modification are common. Faunas, if preserved, indicate brackish- to fresh-water conditions. Crevasse splays may be common and resemble their fluvial counterparts. The abundant exposed sand of the coastal beach ridges provides a ready source of material for eolian reworking. Coastal dunefields are prominent features of many wave-dominated deltas, including the São Francisco (Brazil) and Burdekin (Australia) systems (Coleman and Wright, 1975).

Facies Architecture

The physical stratigraphic framework of wave-dominated deltas is grossly similar to that of fluvial-dominated deltas. Figure 5-17, which shows a generalized facies cross section of the Rhone delta system based on deep core holes, reveals the same basic vertical sequence recognized in Figure 5-14. Shelf and prodelta muds grade into an upward-coarsening sequence of delta front sands. In the wave-dominated system these sands form a broad, continuous sheet or apron, and reflect a greater impact of marine redistribution along strike than seen in the channel-mouth bar/delta-front sand sheet of fluvial-dominated systems.

Cyclic, stacked, progradational coastal-barrier sequences are typical (middle and landward portions of the Rhone section, Figure 5-17). However, contemporaneity of construction and destruction, as well as the less rapid and extensive progradation likely in the face of high marine energy flux, may induce stacking and amalgamation of coastal-barrier sand units (seaward margin of the section, Figure 5-17). Such stacking may greatly increase the thickness of the upper portion of the coastal-barrier sand. Single genetic sequences reasonably range from about 20 ft (6 m) to a maximum of about 100 ft (30 m), where wave heights are great or subsidence is rapid; composite sand bodies can be several times thicker. Growth faulting can expand upper prodelta and coastal-barrier sequences. Internally, coastal-barrier sand bodies display low to moderate angle (1–6 degrees), seaward-dipping bedding (Fig. 5-17), reflecting basinward offlap of the sloping upper prodelta and shoreface. Submarine erosion surfaces separate amalgamated coastal-barrier units. However, the well-sorted nature of coastal-barrier sands tends to minimize the impact of these internal bedding surfaces on lateral and vertical fluid movements.

Coastal-barrier sands are, in turn, overlain by a mosaic of finer, aggradational delta-plain facies, or perhaps by coastal dune sands. Upper portions of coastal barrier, as well as delta-plain facies, are cut by distributary-channel fills. Mixed- and bed-load channels typical of many wave-dominated deltas form broad upward-fining or uniform sand belts that locally segment the strike-oriented coastal-barrier sand bodies with dip-oriented facies elements.

Tide-Dominated Deltas

As tide range increases, tidal currents increasingly modify distributary-mouth geometry and redistribute bed-load sediment. In contrast to wave processes, sediment transport is primarily in the dip direction, out of the channel mouth and onto an extensive shoalwater prodelta platform constructed by rapid mixing and settling of suspended sediment. The channel mouth bar is reworked into a series of elongate bars that extend from well within the channel mouth out onto the subaqueous delta-front platform. Resulting delta-plain geometries are described as irregular or estuarine.

Tide-dominated deltas have not been the subject of detailed three-dimensional studies; their facies architecture is less well known than that of other delta types. Consequently, comparatively few ancient examples are described in the literature, although review of modern deltas shows that tide-dominated oceanic delta systems are quite common. Studies of the Mahakam Delta (Gerard and Oesterle, 1973, Magnier *et al.*, 1975), Niger Delta (Allen, 1965b; Weber, 1971; Oomkens, 1974), Klang-Langat Delta (Coleman *et al.*, 1970), and Colorado River Delta of the Gulf of California (Thompson, 1968; Meckel, 1975) provide insight into the sediments and facies framework of tide-dominated deltas. The Colorado Delta will be used as an example for discussion (Fig. 5-18).

Depositional Environments and Facies

Framework sand bodies of tide-dominated deltas are the product of deposition in estuarine distributary channels and tidal current sand ridge fields (Fig. 5-18). The two types of sand bodies merge as distributary mouths flare, and develop swarms of sand ridges. Subsidiary sand-bearing facies include crevasse splays, which are abundant on the lower delta plain in, or directly upstream of, the zone of tidal influence, and subsidiary tidal channels that do not connect updip to the fluvial system. Bounding facies include tidal flat and channel mud, silt, and fine sand, levee and floodplain mud and silt, tidal salt marsh and swamp peat and organic-rich mud, supratidal flat and pan muds and evaporites, and prodelta mud, silt, and minor sand.

Tide-dominated deltas display few to many *estuarine distributary channels* which are characterized by broad, open or funnel-shaped mouths, and narrow, sinuous upper reaches. In lower reaches, channels are symmetrical in cross section or exhibit multiple parallel thalwegs separated by elongate bars (Fig. 5-18, section A). Because the channel mouth is a locus of bed-load deposition, and tidal transport of sediment has a dominant up-channel component, estuarine dis-

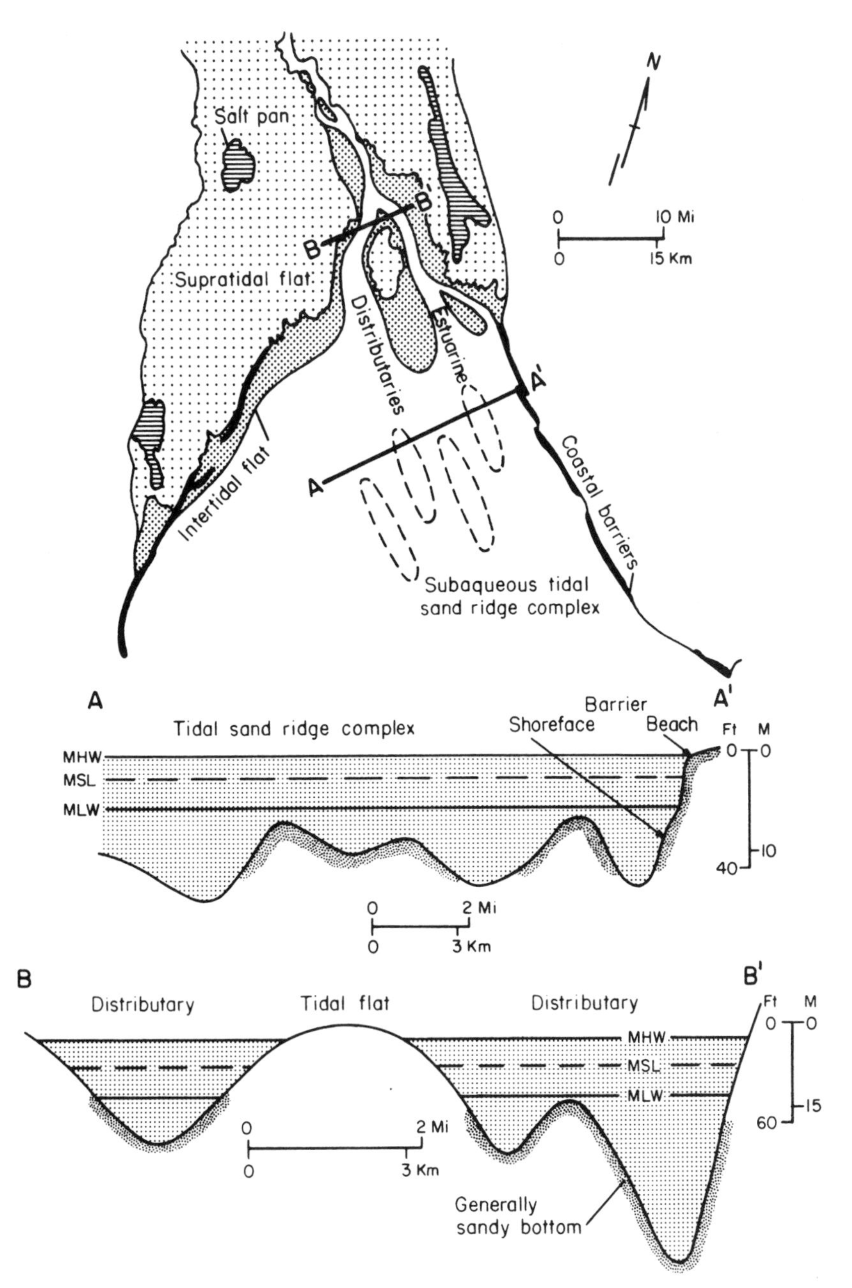

Figure 5-18. Surficial depositional environments of the Holocene Colorado tide-dominated delta system, northern Gulf of California. Cross sections A′ and B′ illustrate the subaqueous morphologies of the lower estuarine distributary and delta platform. (Modified from Meckel, 1975.)

tributaries tend to fill with sand, forming seaward-thickening and widening lenses. Sand-body geometry of estuarine distributary fill sequences is suggested by mapping of the Rhine delta system by Oomkens (1974). In the upper delta plain, channels are narrow, erosional and well-defined. Lateral accretion bedding, reflecting point bar deposition in a sinuous chanel, is likely present. Numerous dead-end tidal channels may extend from estuarine distributary margins into surrounding tidal flat and marsh facies. Both width and thickness of the sand body increase down channel, reflecting the estuarine geometry. Thickness decreases offshore where flow spreads from the channel mouth across the platform.

Sedimentary features and vertical sequence

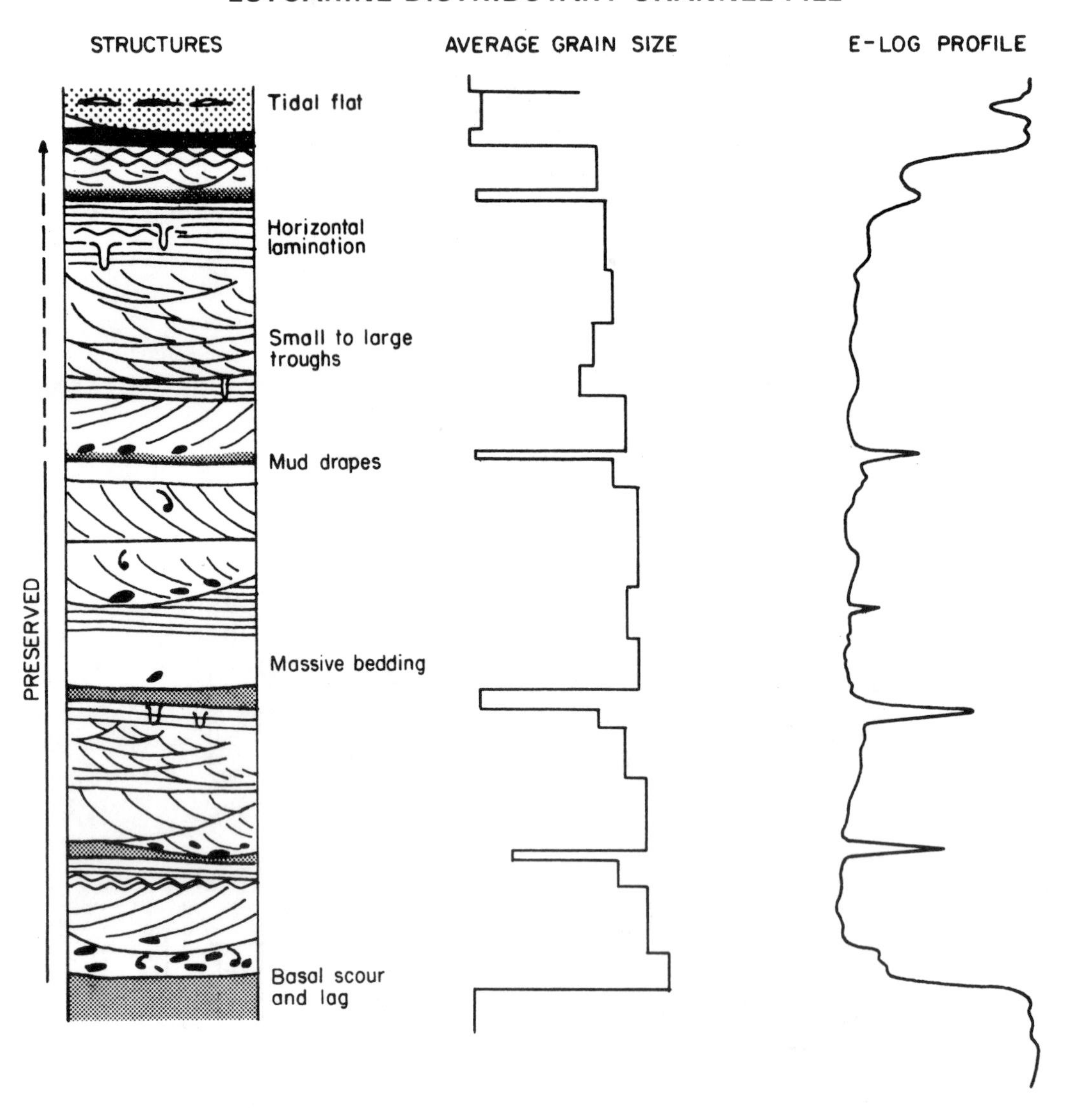

Figure 5-19. Generalized vertical profile through an estuarine distributary channel-fill sand body.

typical of estuarine distributary fills are shown in Figure 5-19. The channel-fill deposit is composed of multiple, superimposed, and variably preserved fining-up depositional units. Such units are interpreted to be the product of lateral migration of the thalwegs and bars present within lower portions of the distributary. Progradation results in superposition of deposits of the more confined, meandering upper distributary on the top of the sequence. In the Colorado system, internal depositional sequences range from a few feet to as much as 25 ft (8 m) thick; the composite estuarine channel-sand body may be 80–100 ft (25–30 m) thick (Meckel, 1975). Well-sorted sand constitutes most of the channel fill, and contains abundant discontinuous mud and silt drapes, laminae and clasts, laminae of macerated plant debris, and local shell. Primary sedimentary structures are diverse but show little regularity in vertical distribution (Fig. 5-19). Cross stratification, presumably of both trough and tabular types, is abundant. Well-developed herringbone cross-stratification is uncommon, but bidirectional dips are apparent if long cores or adjacent sections are compared. Parallel and ripple lamination are common throughout the sequence.

Scattered burrows may also occur, particularly in muddier intervals.

Estuarine distributary channels are flanked by diverse facies of the lower and upper delta plain. Inland, lateral contacts are abrupt and erosional, but in distal portions of the distributary, where surrounding flats and splays are flooded during high tides, channel-fill deposits grade in part into surrounding finer-grained sediments.

Elongate, usually dip-oriented, *tidal ridges* or bars occur on the marginal shelf or broad prodelta platform of tide-dominated deltas (Fig. 5-18, section B′), and are produced by complete remolding of channel-mouth deposits by tide-generated currents. The bars occur in fields or swarms as part of a widespread but irregularly distributed subaqueous sand sheet that breaks up into isolated bars in deep water. Colorado Delta sand ridges have a maximum relief of about 30 ft (10 m) and are regularly spaced at several mile intervals (Meckel, 1975). Dips of bar flanks are gentle, averaging only a few degrees. Bars are believed to migrate laterally in response to changing distributary mouth positions.

Internal features of a complete tidal ridge sequence are summarized in Figure 5-20. Interbar and distal-bar deposits consist of interbedded mud, silt, and sand. An upward-coarsening sequence is produced as well-sorted mid- and upper-bar sands are deposited. However, the common presence of discontinuous mud lenses, beds, and clast zones suggests that delta-associated tidal sand ridges contain an erratic vertical sequence visible as a serrate geophysical log response (Fig. 5-20). Sedimentary structures are dominated by parallel stratification and low- to high-angle cross stratification. Presence of large dunes and sand waves on submerged bars suggests that both trough and tabular cross stratification might be abundant. Thin sand and silt beds in the lower-bar sequence exhibit wavy and ripple lamination and are burrowed.

Tidal sand ridge deposits form a widespread, modified delta-front sheet-sand facies. Sands grade down-dip and basinward into prodelta and marine muds. They merge landward into intertidal flat, estuary, or wave-reworked beachridge muds and sands. Progradation of the delta, and consequent burial of tidal ridges, could superimpose tidel-flat mud and silt on the bar sequence (Fig. 5-20), or lead to incision and partial erosion by migrating distributary channels.

Subaerial environments of a high-tide range delta are dominated by *tidal flat and salt marsh*, and their associated tidal channels and gulleys. Tidal flat deposits generally resemble their counterparts on nondeltaic coasts (Chapter 6). They consist dominantly of mud with some interbedded thin sands and cross-cutting, mainly mud- and silt-filled tidal channels and gulleys. Periodic channel shifting and changing sediment supply along the delta margin results in periods of destruction and erosion of portions of the tidal flat. Thin, transgressive sand and shell berms or bars, which are interbedded with tidal-flat muds, reflect such alternation of constructional and destructional phases. In wet climates, extensive marsh or swamp deposits cover the supratidal flats, depositing beds of peat and organic-rich mud (Coleman *et al.*, 1970). In arid settings such as the Colorado delta plain, supratidal flats are the site of evaporite deposition.

Unlike fluvial- or wave-dominated delta systems, suspended sediment is preferentially trapped around the margin of tide-dominated delta as a *prodelta mud facies*. In deep basins, accumulation of prodelta mud constructs a broad seaward-extending subtidal platform onto which tidal-sand-ridge and tidal-flat deposits advance. In shallow basins, sediment is dispersed, and merges with normal marine facies. Sediments consist of mud, silt, and muddy sand. Minor constituents include plant and scattered shell debris. Though not documented by studies of modern tidal deltas, mass-gravity transport processes and growth faulting probably modify prodelta and overlying delta margin deposits.

Facies Architecture

The somewhat schematic cross sections of the Colorado Delta system (Fig. 5-21) published by Meckel (1975) and of the Klang–Langat system published by Coleman *et al.* (1970) describe a tide-dominated deltaic stratigraphy consisting of a progradational, upward-coarsening prodelta mud and tidal sand ridge sequence overlain by aggradational, delta-plain tidal flat, tidal channel, and marsh-swamp deposits. Aggradational and upper portions of the progradational facies are cut by dip-oriented estuarine distributary channel-fill deposits.

Both sand facies, though characterized by widespread external dimensions, are likely to be

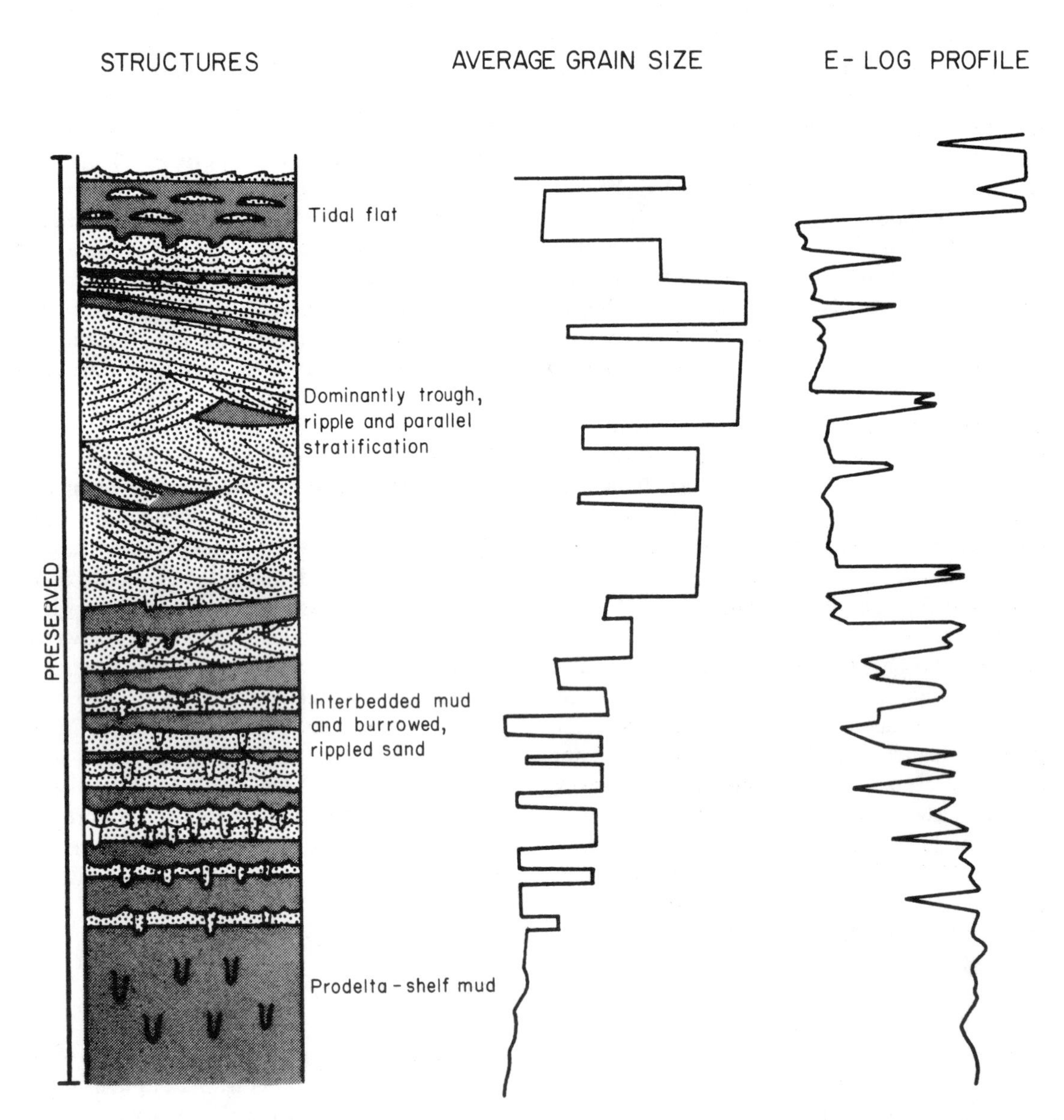

Figure 5-20. Generalized vertical profile through a tidal current sand ridge based on sedimentary features of the Colorado subaqueous delta-front platform.

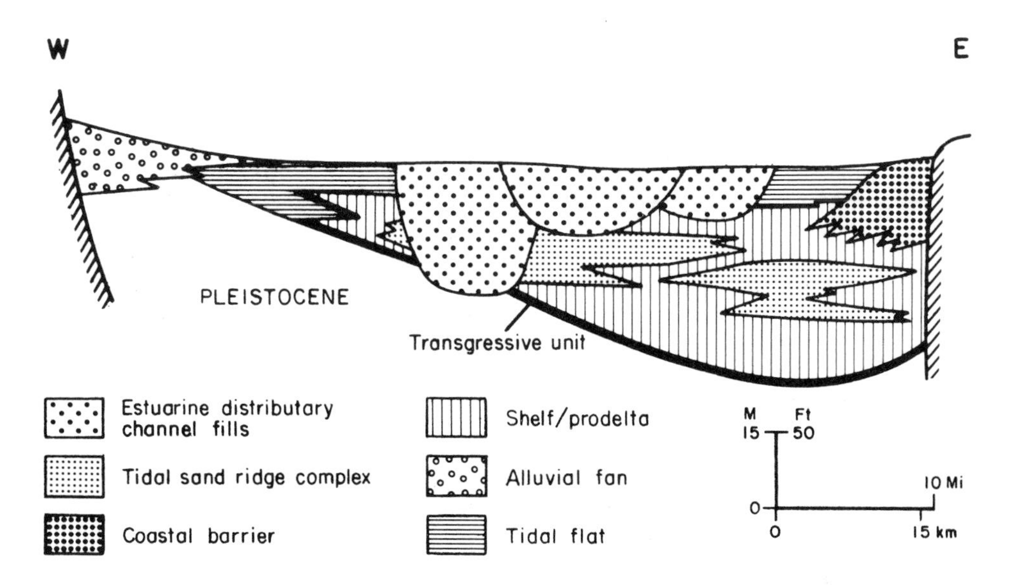

Figure 5-21. Schematic depositional architecture of a tide-dominated delta system as illustrated by a cross section through the Colorado Delta. Original section was based on five drill holes. (From Meckel, 1975.)

Table 5–1. Stratigraphic Characteristics of Deltaic Depositional Systems[a]

	Fluvial-Dominated	Wave-Dominated	Tide-Dominated
Lobe Geometry	Elongate to lobate	Arcuate	Estuarine to irregular
Bulk Composition	Muddy to mixed	Sandy	Muddy to sandy
Framework Facies	Distributary mouth bar and delta front sheet sand; distributary channel fill sand	Coastal barrier sand; distributary channel sand	Tidal sand ridge sand; estuarine distributary channel fill sand
Framework Orientation	Highly variable, average parallels depositional slope	Dominantly parallels depositional strike; subsidiary dip trends	Parallels depositional slope unless skewed by local basin geometry
Common Channel Type	Suspended-load to fine mixed-load	Mixed-load to bed-load	Variable, tidally modified geometry

[a]Modified from Galloway (1975).

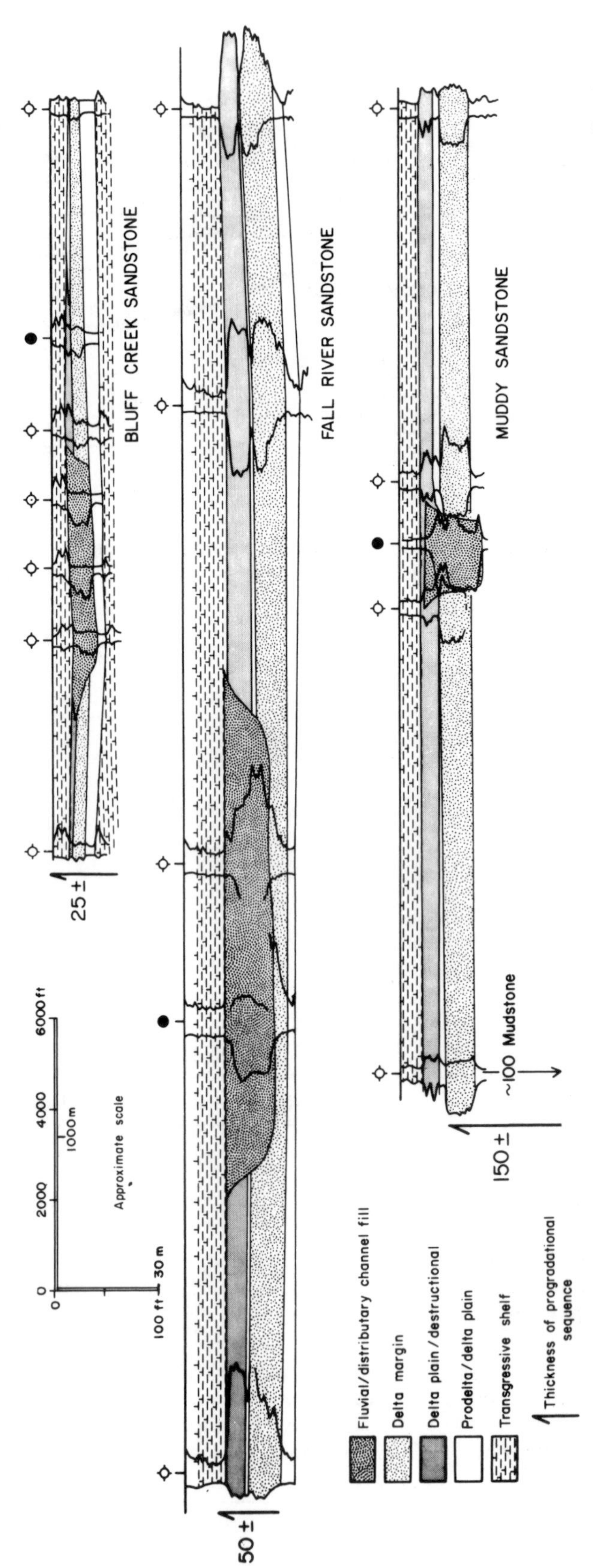

Figure 5-22. Comparative cross sections of distributary axes of three shallow-water, fluvial-dominated delta systems (Bluff Creek Sandstone, Pennsylvanian, Midland Basin; Fall River Sandstone, Cretaceous, Powder River Basin; and Muddy Sandstone, Cretaceous, Denver Basin). A basic facies pattern consisting of a lower, progradational sandstone body, overlying thinner aggradational sandstone body, and a cross-cutting lenticular, channel sandstone body characterize each section. (Muddy Sandstone section modified from Harms, 1966; Bluff Creek section from Galloway and Brown, 1971.)

quite heterogeneous internally. Although gross transmissivity of such sand bodies would be high, anisotropy and locally extensive permeability barriers might complicate fluid flux patterns. Cyclic progradation and destruction are poorly documented by limited studies of modern tide-dominated deltas, but some cores do yield repetitive sequences of similar facies.

Recognition and Interpretation of Ancient Delta Systems

Delta types can be differentiated primarily by recognition of the specific origin of the delta-margin sand facies. Key factors include the geometry and orientation of the progradational sand bodies, their spatial relationship to the distributary channel system and, to a lesser extent, the geometry of the distributary channel fills (Table 5-1).

Comparison of the generalized facies organization of each of the three end-member delta types (Figs. 5-14, 5-17, and 5-21) reveals basic similarities that mark delta system depositional style. In each, an upward-coarsening progradational sequence culminates in a marine-influenced delta-margin sand unit, which is overlain by a mosaic of dominantly aggradational facies, including cross cutting, erosionally based distributary channel and associated splay sand bodies. Figure 5-22 shows three subsurface cross-sections that illustrate variations of this recurrent stratigraphic theme. The examples are all elongate, fluvial-dominated deltas that prograded into shoal water, cratonic basins. Each section shows a simple succession:

1. A prodelta mud blanket underlies the sequence.

2. Above, a coarsening-upward sand body extends across the length of the cross-section segment. Regional cross sections show that these sands thin and pinch out within a few miles.

3. In most wells, the succession is capped by a second sandy unit that is typically thinner and more irregularly distributed, and is separated from the upward-coarsening sand body by a thin mud break.

4. A few wells penetrate a very different sequence. The upper sandy zone and all or part of the lower progradational sand are replaced by a thick sand body that exhibits a blocky or upward-fining profile on the electrical log. Boundaries of the facies are abrupt, suggesting erosion. In plan view, the sand body is comparatively narrow and elongate, and commonly occurs where laterally associated sand units are thickest.

5. The sequence in all wells is capped by open-marine shelf mud or limestone.

Although detailed facies mapping and core examination would be required for specific interpretation of all sand units, generalized facies interpretations are readily apparent. Laterally extensive prodelta muds are overlain by distributary mouth-bar and equivalent delta-front sheet sands. Together these units constitute the progradational platform. They are overlain by a thin sequence of interdistributary mud, which is capped by crevasse splay sands of the aggradational delta plain. Additional thin destructional sand units may also be included within the overlying sandy interval. Distributary channel sands cut through the more widespread delta platform and delta plain deposits, replacing the regular succession with massive to upward-fining sand bodies and local mud plugs. Such channels are likely sources of surrounding splay sands; consequently, the two-sand-body vertical sequence is best developed in proximity to the channel facies. The entire delta sequence is capped by transgressive marine-shelf deposits. Recognition of the genetic interrelationship of the three sand bodies illustrated on the sections enables early interpretation of their genesis, and extrapolation of their geometries.

Scale of Delta Systems

Because delta systems define the principal sites of sediment entry into many basins, they commonly produce major isopach thicks within time-equivalent sequences of basin fill. Maximum thickness of the deltaic sequences is a function of (1) water depth within the reservoir, and (2) rate of basin subsidence.

Depth of water affects deltaic sedimentation by determining the aggregate thickness of the prodelta platform. The comparatively rapid compactional subsidence of a thick muddy platform allows accumulation and preservation of the full suite of deltaic facies. Highly thickened sand bodies can result (Fig. 5-10). In extremely shallow water, thin progradational facies are cannibalized by the advancing distributary, which may actually cut a channel that is deeper than the receiving basin. In such settings vertical sequences are abbreviated, and facies show considerable lateral offset.

Regional basin subsidence (as opposed to differential compaction within the delta mass) has tectonic and isostatic components. Loading of the crust by a thick sequence of sediment is followed closely by isostatic adjustment. As pointed out by Weimer (1970), regional structural features such as broad arches or troughs may reflect patterns of basin filling by deltaic depocenters. Potential for isostatic subsidence is maximal in oceanic basins where several thousand feet of slope and deltaic sediment have prograded onto thin basaltic crust. Many of the great Tertiary deltaic depocenters of the world, such as the Niger and Gulf Coast basins, formed through this mechanism. Such "fat" deltaic sequences produce offlapping wedges characterized by extreme vertical stacking of deltaic facies, preserving tens or even hundreds of cycles of construction and abandonment. Large-scale development of growth faults and diapirs further expands the delta-front framework sand facies within syndepositional subbasins.

Whereas isostatic subsidence may be considered passive, subsidence induced by active compression or extension modifies basin geometry with little reflection of rates or preferred sites of sediment input. However, if a rough balance between subsidence and sedimentation is maintained, tectonic subsidence can accommodate the accumulation of large deltaic systems such as the Jurassic sequence in the North Sea basin and Atoka clastics (Pennsylvanian) of the Arkoma basin.

Tectonic stability, coupled with shallow water depths and cratonic crust, produces thin, elongate deltas that may traverse hundreds of miles and several depositional basins. Cratonic basin and platform deltas are characterized by lateral rather than vertical development of facies, extensive cannibalization of the deltaic facies by the equivalent fluvial systems, irregular digitate geometries, and laterally extensive, bounding marine beds that may produce cyclothemic sequences (Brown, 1979). Sand distribution maps commonly reflect mainly fluvial and distributary channel deposits, further exaggerating the digitate geometry. Dramatic lateral changes from thin elongate delta lobes to thick, vertically repetitive lobate deltas occurred in the Karoo Basin, where progradation crossed from shallow, stable platforms into locally subsiding intrabasinal troughs (Hobday, 1978a).

Evolution of Delta Systems

Delta systems respond to changes in the relative intensity of marine and fluvial processes. As a delta system progrades farther into a depositional basin, changing basin morphology can systematically modify marine processes. The nature and volume of fluvial input may also change with time as the tectonic framework, climate, or topography of the source terrane evolve. Succeeding delta lobes within a delta system will reflect such changes. Two examples of evolving delta depositional style described by Galloway (1975) illustrate possible trends.

Thick deltaic sequences of major depositional episodes along passive continental margins commonly display a recurrent evolutionary history. Initial deltaic lobes prograde onto thick, prodelta and continental slope mud platforms. Subsidence of mouth-bar sands into the underlying mud substrate results in deposition of digitate lobes with bar-finger sands similar to those of the modern Mississippi bird-foot lobe. Development of a skeletal sand framework across the platform, combined with the decreasing extent of undercompaction in prodelta muds, leads to progradation of deltas onto a more stable depositional platform. Marine processes have greater opportunity to modify distributary mouth-bar sands, producing lobate delta margins. As the source area is worn down, or as a relative rise in sea level occurs, volume and grain size of input sediment decreases. Delta lobes prograde onto foundered portions of the older delta platform, and marine reworking becomes increasingly dominant in determining facies distribution. Examples of this evolutionary sequence in Eocene through Miocene delta systems of the northern Gulf Coast basin have been documented by Fisher and McGowan (1967), Galloway (1968), and Curtis (1970).

Pennsylvanian deltaic systems of the Midland and Anadarko basins, Texas and Oklahoma, exhibit changes in deltaic depositional style resulting from progradation across a shallow stable platform, or shelf margin, into relatively deeper water of the basin center. On the platform, progradational deposits are thin and poorly preserved due to the shallow water depth (ranging from a few feet to tens of feet) and scouring by advancing fluvial and distributary channels. Deltas are inherently fluvial-dominated because wave energy was severely attenuated in such shallow water. However, as deltas prograded across the platform margin and into deeper water, several factors abruptly changed. Wave energy flux increased. Development of a thick prodelta platform allowed deposition and preservation of thickened, complete progradational delta-margin facies sequences. Further, presence of a gentle subaqueous slope with as much as 1,000 ft (300 m) of differential relief introduced the potential for sediment remobilization by mass-gravity transport. As a consequence of these changes, thick distributary mouth-bar and delta-front sand bodies assumed volumetric importance as lobate or even cuspate delta geometries evolved. Significant amounts of sediment, including sand, slumped down the prodelta slope or moved into submarine channels and down onto the floor of the basin, forming independent submarine fan systems.

Chapter 6

Clastic Shore-Zone Systems

Introduction

The shore zone (Fig. 6-1), excluding deltas, comprises the narrow, high-energy transitional environment that extends from wave base at a highly variable depth averaging 35 ft (10 m) to the seaward edge of the alluvial coastal plain, raised terrace, or cliffs. Although restricted in area, migration of shorelines through time has resulted in widespread shore-zone deposits in the rock record, with considerable bearing on the distribution of hydrocarbon, coal, and uranium resources.

Beaches, barriers, lagoons, and tidal flats may be component facies of other depositional systems, such as deltas, or they may in combination constitute independent shore-zone systems. In either case, their prime characteristics stem from marine inundation or reworking, as distinct from direct fluvial influx. Shore-zone systems are supplied by longshore transport of river-derived sediments, onshore transport of shelf sediments, erosion of local headlands, residual concentration, and by small coastal streams. Sands are concentrated in barrier-island complexes and low-tidal and subtidal sandbodies with finer sediments to landward, or they may accrete directly on the mainland as strandplain beaches (Fig. 6-2).

The characteristics of depositional coasts, as opposed to elevated coastlands undergoing erosion, vary in response to two fundamental energy factors: waves and tidal currents. Both of these factors are directly related to tidal range (Hayes and Kana, 1976, p. 35–37). Wave effectiveness is inversely related to tidal range because, with increasing range, wave energy is dispersed over a greater width of shore zone during each tidal cycle. Distribution of coastal features associated with the three tidal range categories (Fig. 6-3) shows the greatest proportion of barrier islands and related environments along microtidal coasts of 0–2 m (0–7 ft) tidal range, which tend to be wave dominated (Hayes, 1975). Tide-dominated or macrotidal coasts (4–6 m; 13–20 ft) show flaring estuaries with linear sand ridges. Mixed-energy or mesotidal coasts (2-4 m; 7-13 ft) have intermediate characteristics with stunted barrier islands and extensive tidal flats or marshes (Fig. 6-4).

Shoreline configuration and nearshore bathymetry can impart considerable local variation in the relative effectiveness of waves and tides. The brunt of the wave energy is expended on headlands and open coast, decreasing in embayments, where tidal effects are enhanced (Price, 1958). Broad shelves and shallow nearshore profiles similarly dissipate wave energy while increasing tidal range (Redfield, 1958).

Along modern coasts there is a complete range in variation among wave-dominated deltas (Chapter 5), barrier islands, and strandplains. Systematic coastwise changes from active or inactive deltaic protuberances through strandplains and transgressive barriers to regressive barriers reflect the interaction between sediment supply and nearshore energy. Inactive deltaic headlands are flanked by transgressive barriers followed by regressive barriers formed in interdeltaic bights (Morton, 1977, 1979). Similar lateral relationships are documented in the rock record (Fig. 14–17, for example). In ancient deposits it is generally not possible to distinguish among the deposits of wave-dominated deltas, strandplains, and regressive barriers on the basis of vertical sequence alone (Fisher and Brown, 1972); isolith or cross-sectional delineation of sandbody geometry, or environmental interpretation of landward bounding facies, is necessary, in addition.

Shore-zone sands are characteristically, but not invariably, quartzose and may be practically monomineralic. Depending upon provenance, other lithologies, such as volcanic rock fragments, may be the predominant grain types. Nonquartz material, with the exception of resistant heavy minerals such as zircon, tourmaline, and rutile, tends to be removed by breakdown of the less stable minerals and winnowing of fines. Increase in compositional and textural maturity depends on prolonged recycling by waves and currents,

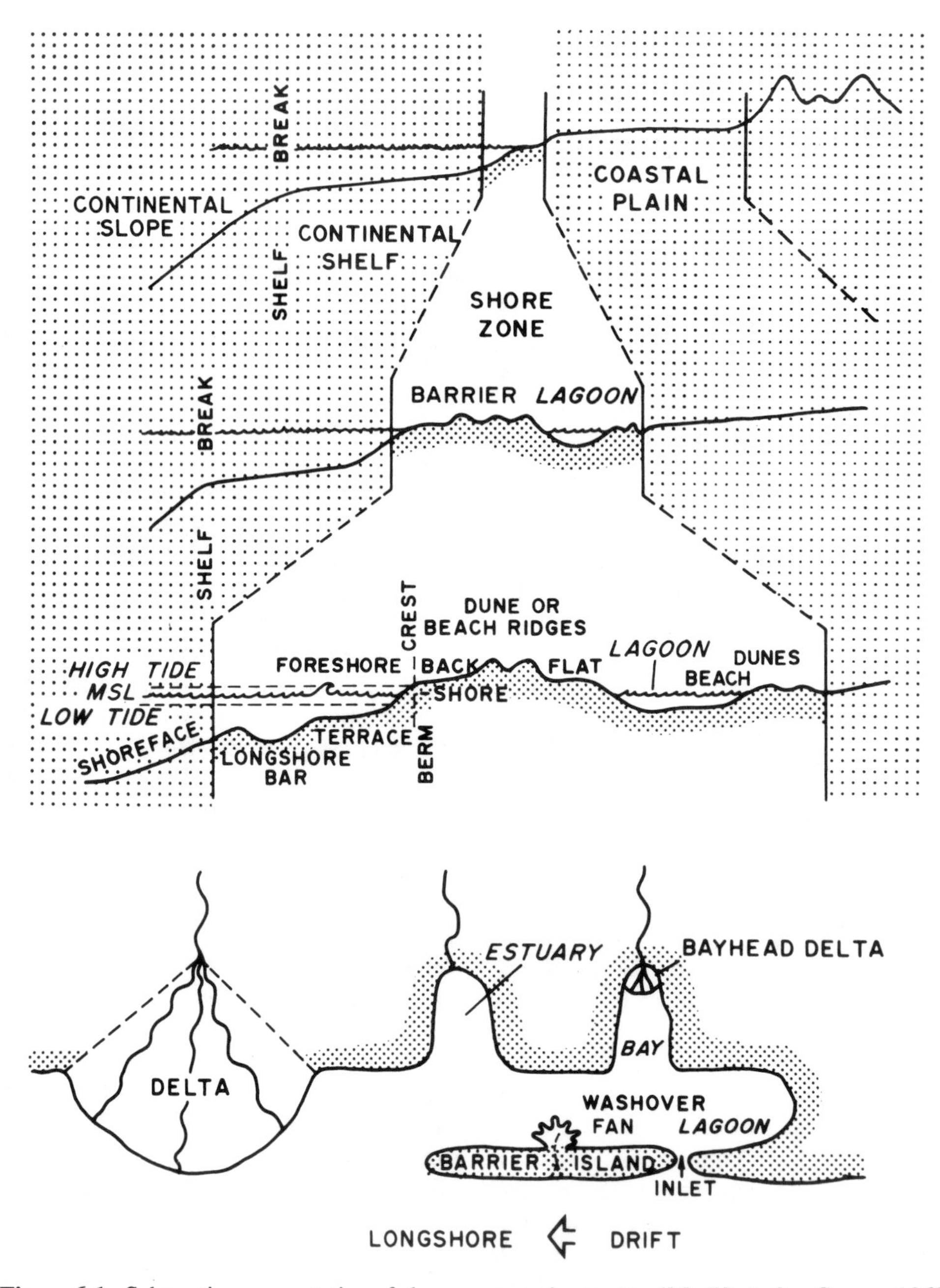

Figure 6-1. Schematic representation of shore-zone environments. (Modified after Curray, 1969.)

and continued grain abrasion during to-and-fro tidal transport can produce supermature rounding (Balazs and Klein, 1972). Interfingering of ancient marginal marine and coeval alluvial deposits is commonly accentuated by pronounced lithological contrasts between quartzarenites and the less mature sandstones updip (Ferm, 1974). Where fluvial systems supply quartzose sediment, this lithological distinction of ancient shore-zone is less clear (Vos, 1977). Some shore-zone deposits consist entirely of gravel-sized material, with fabric and bedding characteristics that distinguish it from alluvial gravel (Clifton, 1973).

Barrier-island and strandplain sand bodies are prime targets for petroleum exploration, with excellent primary porosity and high permeability. Landward and seaward interfingering with fine-

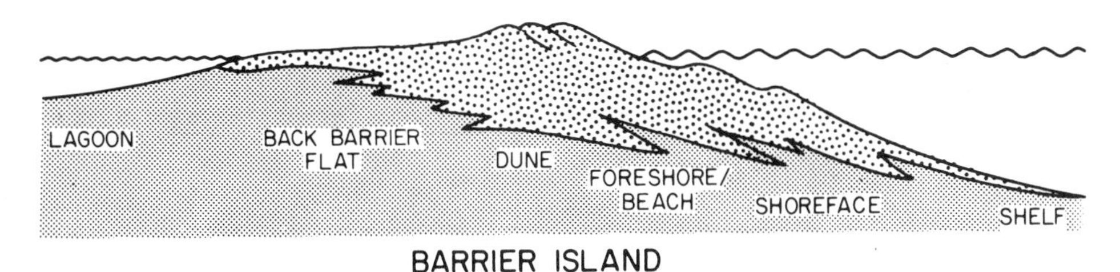

OLDER BEACH RIDGES
MARSH
BEACH RIDGE/ DUNE
FORESHORE/ BEACH
SHOREFACE
SHELF
STRANDPLAIN

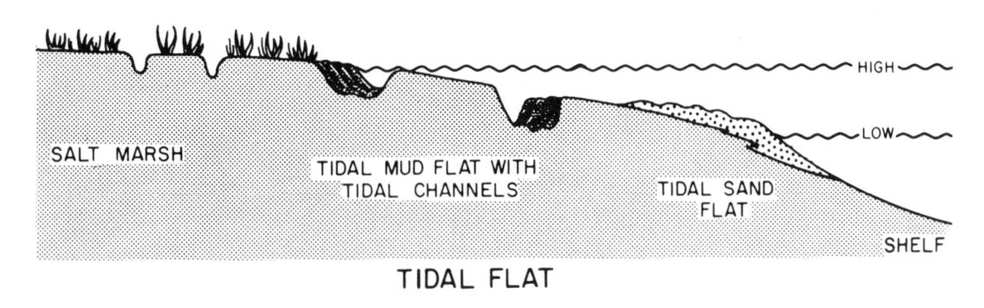

Figure 6-2. Contrasted morphologies of barrier/lagoon and strandplain coasts.

grained organic-rich sediments provide effective primary migration of hydrocarbons from source to reservoir. Furthermore, tidal delta and washover sands pinch out updip, so that barrier complexes commonly constitute excellent stratigraphic traps. Barrier sands are hosts to commercial deposits of epigenetic uranium, for example, in the Tertiary of southern Texas. Barriers are important in promoting marsh formation and peat accumulation, and back-barrier seams are mined in a number of coalfields. Diamondiferous beach placers are worked on a large scale on the arid southwestern coast of Africa and elsewhere, and a variety of metals, including gold, have been recovered from beach deposits. Quaternary beach sands are mined extensively for heavy minerals such as monazite, rutile, ilmenite, and tourmaline.

Depositional Processes

Short-term processes that modify the shore zone daily are tides, wind-driven currents, waves, and longshore currents. Winds not only affect the subaerial portions of barriers, but are important in coastal bays and lagoons, where they create waves and wind tides. Rivers contribute bed load and suspended load to lagoons, or directly to the sea coast. Longer-term processes of disproportionate significance are related to storms, whose effects on the shore zone are largely destructive.

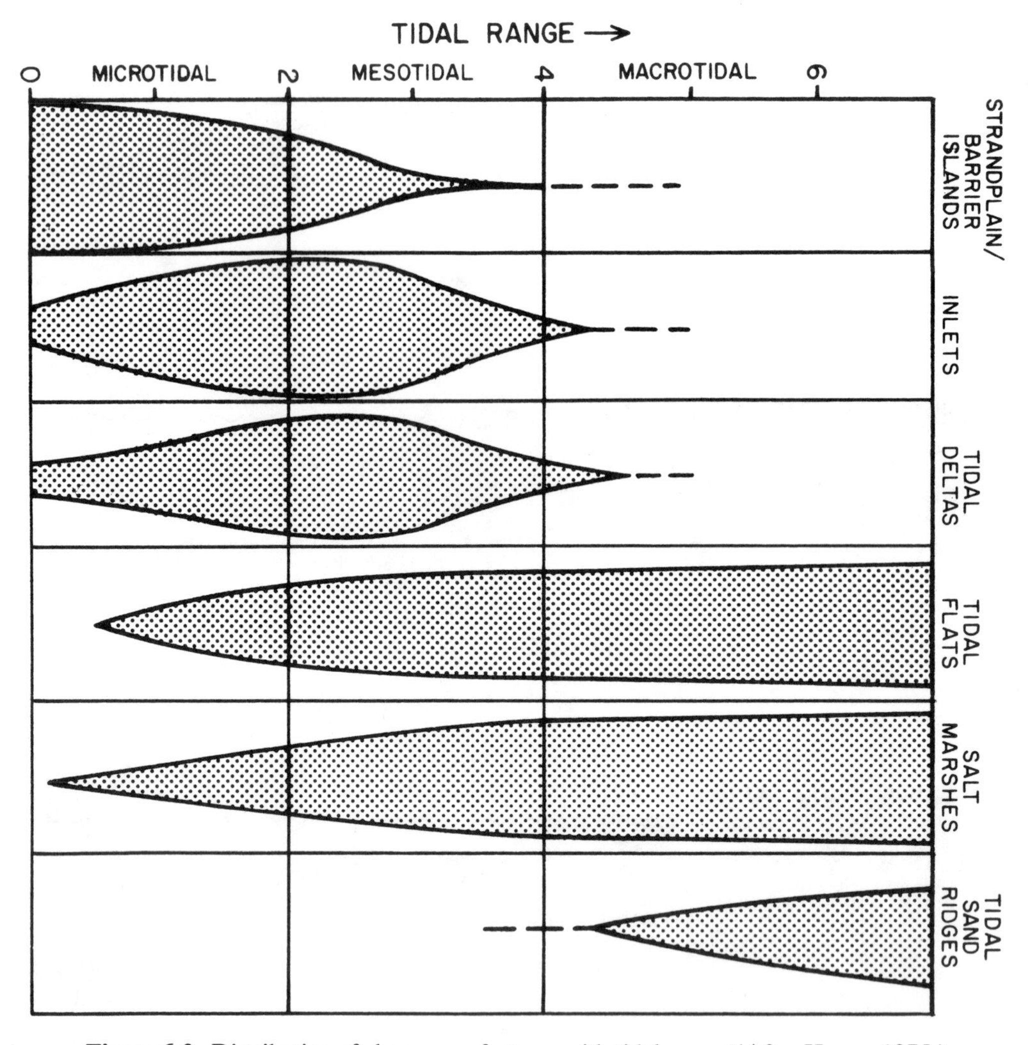

Figure 6-3. Distribution of shore-zone features with tidal range. (After Hayes, 1975.)

Relative changes in sea level, whether arising from eustatism, sedimentary compaction, or tectonism, are balanced against the effects of sediment influx to determine the fundamental style of sedimentation: progradational, aggradational, or transgressive.

Waves

The shoaling transformation from a trochoidal to a solitary form of waves approaching the shore occurs where the water depth is approximately five times the wave height. From this point, the wave becomes higher and steeper until it breaks where the water depth is about 1.3 times the wave height (Clifton *et al.*, 1971). Alternatively, breaker formation may correspond with the position of a longshore shoal. The orbital motion of waves approaching the breaker zone moves sand landward, and is opposed by longshore and rip currents. Ripples formed as a result of wave oscillation have long crestlines which branch or terminate in simple linear patterns. Profiles are symmetrical and rounded, rarely peaked, and internal lamination corresponds to the external form, dipping in two directions (Harms *et al.*, 1975, p. 58). With shoaling, the effects of directional flow become superposed on the wave motion, the crestlines become sinuous and less persistent, and external and internal asymmetry arises (Harms, 1969).

Breaking waves span a continuous spectrum (Fig. 6-5) from spilling breakers formed from

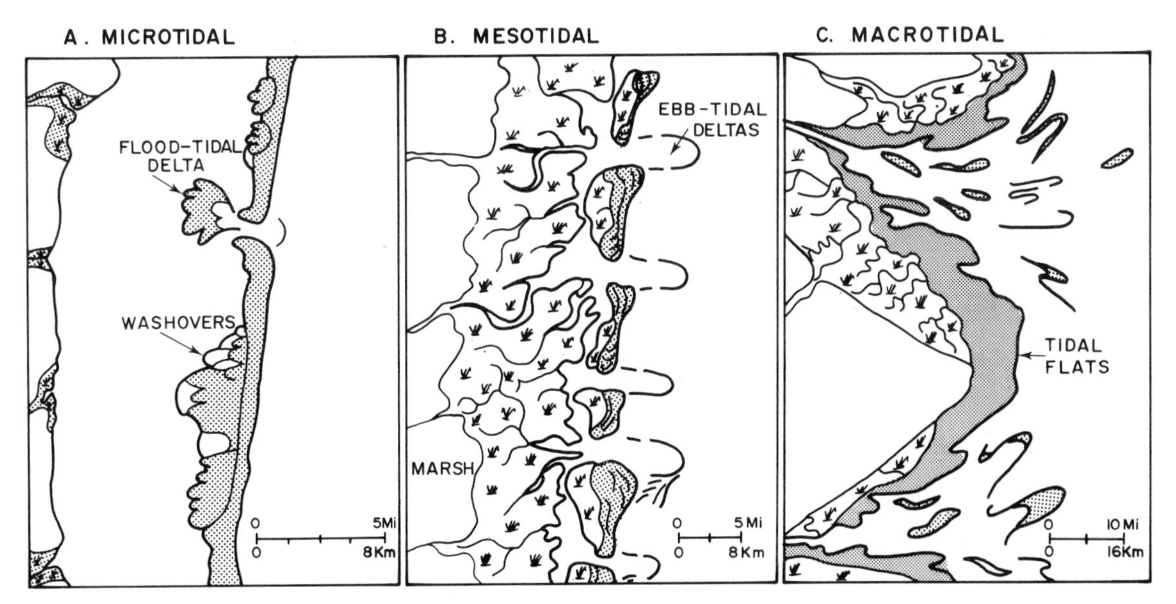

Figure 6-4. Variations in coastal sand-body geometry as a function of a tidal range: (A) long, narrow microtidal barriers, (B) short mesotidal barriers, (C) linear, shore-perpendicular tidal-current ridges of macrotidal estuaries. (After Barwis and Hayes, 1979.)

steep waves over a low-angle shoreface to plunging breakers from less steep waves over a high-angle shoreface (Galvin, 1968). Longshore break-point bars can be created by the large-scale vortices associated with plunging breakers (Fig. 6-6), but are obliterated by spilling breakers (Miller, 1976). Shoreward bar migration is favored by waves of intermediate form.

The incoming bore formed by wave collapse is compensated by backwash in the inner zone of the upper shoreface. Sand is thrown into suspension and is transported up the beach foreshore by wave swash. Bores of low initial energy are most effective in promoting foreshore accretion (Miller, 1976). Size fractionation in the swash zone decreases mean particle size, increases sorting,

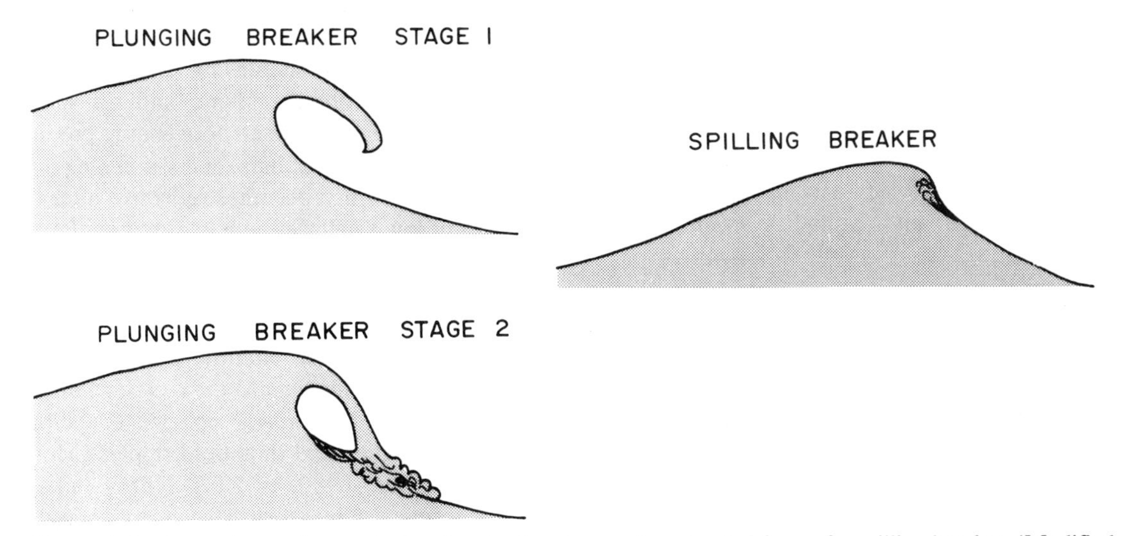

Figure 6-5. Stages in development of a plunging breaker, and contrasted form of a spilling breaker. (Modified after Miller, 1976.)

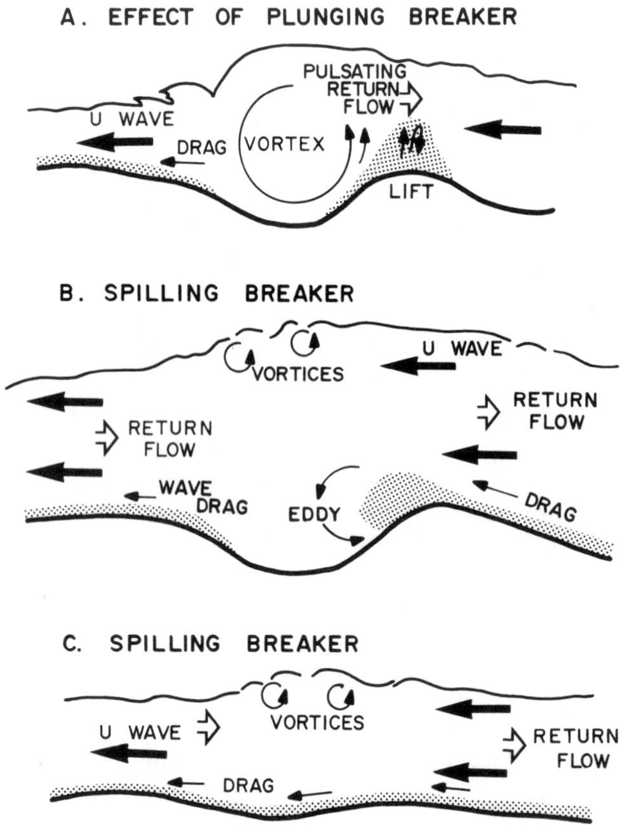

Figure 6-6. Fluid-sediment interactions: (A) creation of bar profile by a plunging breaker. U is the horizontal components of the wave,and is accompanied by creation of a vortex, pulsating return flow, and a low-pressure area of lift; (B and C) destruction of the bar by a spilling breaker. Note how the vortices are confined to the upper part of the wave. (After Miller, 1976.)

and reduces coarse skewness from low- to high-water mark (Greenwood, 1978).

Seasonal storms tend to produce a flat, truncated beach profile with scarps. The eroded sediment is transported seaward by storm-surge ebb (Hayes, 1967a), rip currents (Shepard *et al.*, 1941), or bottom-return flow (Morton, 1981); all of these processes are most effective during or immediately following storm-wave activity. The beaches are subsequently replenished and steepened during fair weather by landward migration of nearshore bars, which may emerge in the swash zone as ridge and runnel systems (Hayes, 1969).

These factors of wave climate and sediment supply lead to the recognition of two general categories of beach and nearshore profile: reflective and dissipative (Wright *et al.*, 1979). Reflective beaches are steep and linear with berms and beach cusps; they reflect much of the incident wave energy back to the ocean. Dissipative systems have a broad surf zone and a more complex nearshore topography with bars and rip-current cells. Shore-parallel interbar troughs contain longshore currents, which turn seaward at intervals as rip currents, persisting through channels cut across the adjacent bar. At times the bars are joined to the beach by transverse shoals. These basic elements may be arranged in different patterns which change in response to weather and wave climate (Wright *et al.*, 1979; Chappell and Eliot, 1979). With increasing energy, the bar moves farther from the shore, and rip spacing increases. Decrease in nearshore energy produces shoreward migration of bars, and steepening of the beach profile, which becomes reflective after long periods of low swell. Some bars remain relatively stationary due to the balance in sediment transport onshore and offshore (Greenwood and Mittler, 1979).

In the swash zone, shallow flow depths and high velocities contribute to the development of a plane-bed configuration with ephemeral bedforms such as antidunes and rhomboid ripples (Morton and McGowen, 1980, p. 128). Heavy minerals are concentrated by storm swash, by sheet-flow overtopping the berm and subsequent return flow from the back beach, and eolian winnowing.

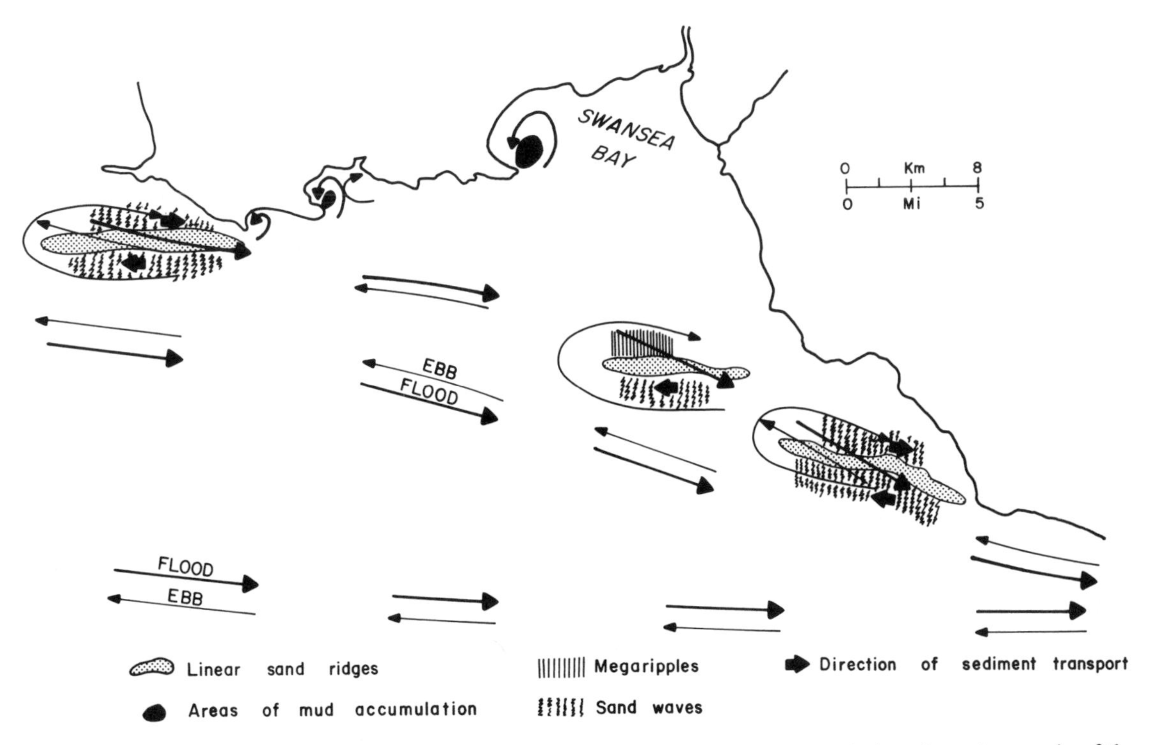

Figure 6-7. Tidal current circulation, sand ridge development, and mud accumulation along the margin of the Bristol Channel. (Modified after Ferentinos and Collins, 1980.)

Discoidal pebble shapes result from abrasion and shape sorting (Dobkins and Folk, 1970). Seaward-imbricate discoidal and blade-shaped clasts tend to be particularly abundant near the high-tide line, giving way to a greater proportion of subspherical clasts near low-water level (Bluck, 1967).

Tides

Astronomical tides may be semidiurnal or mixed. Tidal range varies from almost negligible in semienclosed seas to more than 33 ft (10 m) in embayments and on broad shelves bounding major ocean basins. Tidal currents tend to flow parallel to open coasts, reversing with the tidal phase (Fig. 6-7). Flow in and out of narrow embayments and drowned river mouths results in transport perpendicular to the regional shoreline trend. Effects of landward convergence and shoaling on currents entering such embayments may increase their velocities fourfold (Jago, 1980). Maximum velocities occur as the tidal bore enters the estuary early in the flood, and, during late stages of ebb withdrawal, as flow is restricted to a decreasing channel cross section. During neap tide, flow may be confined to channels throughout the entire tidal cycle, but during spring tide the adjoining marsh is extensively flooded (Oertel and Dunstan, 1981). Ebb-flood asymmetry of the shoaling tidal wave creates a sediment trap in the upper estuary. On the other hand, net seaward discharge during the phase of decreasing tidal range provides a mechanism for seaward escape of sediments (Allen *et al.*, 1980).

Nontidal gravitational circulation is also important (Officer, 1981). Suspended particles carried seaward in the upper water layers of a stratified estuary may subsequently be carried landward in the lower water layers, contributing to accumulation of fine-grained sediments in the upper estuary (Schubel, 1968).

Increased suspended-sediment concentration in summer is related to the incidence of thunderstorms and the higher level of biological activity (Ward, 1981). Turbid water issuing from estuaries and inlets is separated from the clear shelf waters by a transitional boundary layer (Oertel and Dunstan, 1981).

Coastal promontories significantly increase shore-parallel tidal velocities and generate powerful vortices that form offshore sand shoals (Pingree and Maddock, 1979). This increased tidal streaming results in mixing of a stratified water column. Offshore topography, too, has a profound effect on coastal development (Robinson, 1980). Sand bars and offshore shoals deflect tidal currents from their normal rectilinear pattern, increasing their onshore component. These currents transport sediment landward, where wave processes contribute to beach accretion.

Under macrotidal conditions, shore-perpendicular transport patterns prevail in estuaries and embayments, with ebb and flood currents commonly following mutually evasive paths. Macrotidal estuaries may change from a well-mixed state during spring tides to a partially mixed or stratified state during neap tides (Allen *et al.*, 1980).

Some macrotidal estuaries are characterized by differentially opposed sediment transport (Culver, 1980). In the Bristol Channel in the southwestern United Kingdom, sand is transported seaward as bed load, whereas finer particles are carried into the estuary as suspended load. Differential sediment transport may also be seasonal, particularly off continents with marked wet and dry seasons (Tucker, 1973). Fine material is carried seaward during the wet season, and coarse sediment moves up the estuaries during the dry season.

Astronomical tides in lagoons are considerably reduced in comparison with the open coast, and wind stress is important in generating currents and fluctuations in water level due to wind setup. These processes may be periodic, in response to diurnal wind patterns, with effects identical to those of astronomical tides.

The shallow subtidal and lower intertidal areas receive the highest physical energy for the longest period of time, concentrating the coarsest, winnowed sediment fraction. The duration of submergence decreases toward the high-tide mark, where progressively finer grained sediments accumulate in response to the lower physical energy. Whereas bed-load processes are dominant on the lower tidal flats, the proportion of suspension sedimentation increases landward. A broad intertidal zone is characterized by an alternation of traction processes and fallout of suspended sediments during slack water at high tide (Reineck, 1967; Klein, 1970). Variations in water depth and current intensity are typical, with intermittent subaerial exposure resulting in desiccation. During the late stages of ebb-tidal drainage, flow is deflected by minor bedforms, modifying these features and forming smaller ripples at right angles. Suspended sediment concentration increases up the high-tidal flat as a result of decreasing tidal current velocities, combined with settling and scour lag effects (Van Straaten and Keunen, 1957). Tidal channels are subject to high current velocities and may be ebb- or flood-dominated, or experience tidal reversals of equal intensity, but the majority are ebb-dominated. Sand is preferentially concentrated in these channels, which migrate laterally. Plants along channel margins and on the upper tidal flat effectively trap fine-grained sediments, while suspension-feeding organisms, too, can enhance accumulation of muds. On semiarid tropical tidal flats, biological activity reaches a maximum over the midtidal mangrove mudflats, declining above that level in response to increasing salinity (Semeniuk, 1981). In wetter environments high organic production continues across the flats leading to accumulation of salt-marsh peat and dark clays up to and above the spring-tide level. In arid areas, the salt marsh is replaced by extensive evaporitic mudflats containing stratiform evaporite deposits.

Shore-Zone Facies

Shore-zone facies show considerable intergradation one with another, and overlap with adjacent environments. Most characteristic is a facies trend parallel to depositional strike; even dip-oriented features, such as inlets, tend to migrate alongshore to produce sediment bodies which conform generally with this shore-parallel alignment.

Regardless of the energy mix, the primary facies inherent in all shore-zone systems is the shoreface. Additional facies, such as those associated with the foreshore, lagoon, and tidal inlets, are important locally.

Shoreface and Beach

Nearshore shallowing is accompanied by an increase in physical energy and a decrease in biological manifestations. Lower, middle, and upper shoreface subenvironments are each distinguished by a particular suite of textures, physical structures, and biogenic features. This shore-parallel zonation may become temporarily blurred or punctuated by the effects of major storms.

Lower shoreface deposits are located seaward of the break in slope at the base of the nearshore sediment prism (Reinson, 1979). Sand, silt, or mud layers a few centimeters thick may alternate irregularly, or may show cyclic arrangement indicating abrupt contribution of sediment followed by reduction in flow energy and prolonged reworking by organisms. Trace fossils typically include vertical tubular forms, such as *Asterosoma* (Howard, 1972), or a combination of suspension-feeder and deposit-feeder traces, such as *Thalassinoides* and *Teichichnus* (Cotter, 1975).

Each sedimentation unit is attributed to a storm event involving seaward transport by bottom-return flow (Morton, 1981) analogous to powerful rip currents. Indeed, offshore-directed cross bedding presumably reflecting this process is characteristic of some lower shoreface sandstones (Vos and Hobday, 1977). The basal shelly lag, too, may indicate offshore-current alignment of elongate fossils (Hobday and Morton, 1981). Parallel lamination, the dominant structure, is deposited under conditions of intense bottom shear (Kumar and Sanders, 1976) or by settling of clouds of sediment suspended by storm waves (Reineck and Singh, 1972). Wave and combined-flow ripples overlying the parallel-laminated interval mark the resumption of fair-weather conditions. Ripple crestlines are subparallel to the shoreline.

Middle shoreface environments are subject to more powerful waves and associated longshore and rip currents, leaving a complex depositional record. Longshore break-point bars move landward or occupy the same position over long periods of time. Bars studied by Davidson-Arnott and Greenwood (1976) contain low-angle lamination along the seaward slopes, horizontal lamination and trough cross beds of offshore and onshore orientation in the crestal portions, and onshore-dipping planar cross beds on the landward side. Comparable sequences of Miocene age in Spain change vertically from small-scale wave rippling through parallel lamination to longshore-directed dune cross bedding, commonly arranged in opposing directions; this sequence is ascribed to the effects of increasing wave energy with shallowing (Roep *et al.*, 1979).

Fair-weather deposits of a nonbarred, high-energy shoreface comprise medium-scale foresets dipping onshore or oblique to the shoreline, and subhorizontal lamination with rare straight-crested ripples (Clifton *et al.*, 1971). Ancient analogs with the same suite of structures indicate significant longshore currents with unidirectional or bimodal shore-parallel azimuths (Hobday, 1974).

Storm deposits of the middle shoreface are thicker and more lenticular than their lower shoreface equivalents. Individual units up to a meter thick contain a sporadic basal lag of gravel, shell, or mudclasts grading up into massive sand, parallel lamination, and ripple lamination with burrows. Some sandstone units are multistoried or overlap laterally, with zones of burrows separating the individual storm deposits. *Skolithos*, *Rosselia*, *Diplocraterion*, and *Ophiomorpha* are common trace fossils (Howard, 1972; Chamberlain, 1978). Broad, shallow channels approximately perpendicular to the shoreline are probably scoured by rip currents.

Upper shoreface environments corresponding to the upper part of the surf zone are dominated by powerful onshore, offshore, and longshore currents. The coarsest gravelly sediment fraction tends to be concentrated toward the stepped inner margin, with a large proportion of equant pebbles. Alongshore-directed trough cross beds are characteristic, with onshore-dipping planar cross beds deposited by bar migration, and irregular truncation surfaces reflecting storms.

Observations by Clifton *et al.* (1971) indicate that during fair-weather, seaward-dipping trough cross beds are deposited off gently sloping beaches, but with increasingly powerful longshore currents, the troughs dip parallel to the shoreline, as in a number of ancient examples (Hobday, 1974). Shore-parallel ripples form off steeper beaches, producing more complex stratification dipping either landward or seaward. Scattered trace fossils comprise long vertical burrows such as *Skolithos*.

Foreshore (beach face) environments correspond to the zone of wave swash. The dominant structure (Fig. 6-8) is planar lamination dipping gently seaward, with low-angle discordances representing adjustment of the beach to changes in wave regime or sediment supply. Heavy

Figure 6-8. Low-angle planar lamination with multiple truncation planes attributed to foreshore deposition overlying cross-bedded upper-shoreface quartzarenites at level of scale. Pennsylvanian Pottsville Formation, Cullman, Alabama.

minerals tend to be concentrated in discrete laminae, often alternating with quartzose layers. Inverse grading is common, with fine grains and heavy minerals merging upward into a coarser, quartzose layer (Clifton, 1969). Scattered pebbles are predominantly discoidal.

Upbeach migration of ridge and runnel systems leaves sequences of shore-parallel troughs overlain by large, convex foresets generated by the steep landward slipface, capped by parallel lamination and smaller structures formed on the seaward surface of the ridge (Hayes *et al.*, 1969). The dip of the ridge-accretion foresets flattens where they merge landward into a berm, and shell concentrations along the berm crest record the limits of wave uprush.

Backshore environments are separated from the foreshore by a berm or by a more subtle break in slope. The backbeach is flat, or slopes gently landward and aggrades by storm waves that top the berm crest and distribute their sediment load as a thin sheet, or by adhesion of windblown grains to the damp surface. Faint horizontal lamination is locally disrupted by burrowing, particularly by crustaceans.

Low coppice dunes and bare foredunes landward of maximum high-tide level can easily be breached by storm surge, and have zero preservation potential. However, the dunes play an important role in reducing foreshore erosion during storms. Truncation of the seaward dune margins supplies considerable volumes of sand; this accumulates offshore as a bar and is highly effective in dissipating wave energy (Leatherman, 1979). Higher vegetated foredunes may be partially preserved in the rock record (Chapter 10).

Barrier/Lagoon Facies

Barrier/lagoon environments contain three main facies elements: the shore-parallel barrier, the enclosed lagoon and bay, and inlets or channelways that facilitate exchange of water.

Barrier-island development is favored by relatively flat, low-gradient continental shelves, abundant sediment supply, and low to moderate tidal range (Glaeser, 1978). The origin of barriers has been attributed to at least three distinct mechanisms: (1) the vertical growth and emergence of offshore bars, (2) downdrift growth of spits, and (3) detachment of beach ridges from the mainland by a rise in sea level (Nummedal *et al.*, 1977). The third mechanism is most likely in coastal zones of low relief (Swift, 1975, p. 14), a morphology characteristic of passive-margin coasts (Glaeser, 1978). Barriers built by coastwise spit elongation may be more common along steeper, higher relief coastal zones of active margins, but shore-normal sand transport is responsible for barrier growth on some deeply embayed coasts (Roy *et al.*, 1980). There is some evidence of offshore bar emergence in modern barriers, for example along the Texas Gulf Coast (Fisk, 1959). Distinction among these various modes of origin is difficult, particularly as many barriers show evidence of composite development and modification (Schwartz, 1971).

Barrier morphology is largely determined by tidal range. Microtidal barriers tend to be long and narrow with conspicuous washover fans and few inlets. Mesotidal barriers are broader and are cut by numerous inlets with complex tidal sand shoals (Fig. 6-4). These stunted barriers typically have a wide, accretionary updrift end, a narrow midsection, and a recurved spit at the downdrift end (Hayes, 1975).

The degree of shoreline stability also exercises control over barrier morphology. High-profile barriers (Fig. 6-9) develop along stable or prograding coasts, whereas youthful, low-profile barriers are typical along transgressive coasts (Morton and McGowen, 1980). Distinct stratigraphic facies geometries are produced by barriers of transgressive, regressive, or aggradational phases (Fig. 6-10). Most modern barriers are eroding (Bird, 1976). Man's influence is locally a cause of coastal recession (Dolan, 1972), but many barriers have been retreating almost since their inception 7000 years ago (Kraft, 1971).

Transgressive Barriers

Transgressive barriers tend to be narrow sand strips except where extensive, thin washovers or tidal sand shoals are developed. Some have a

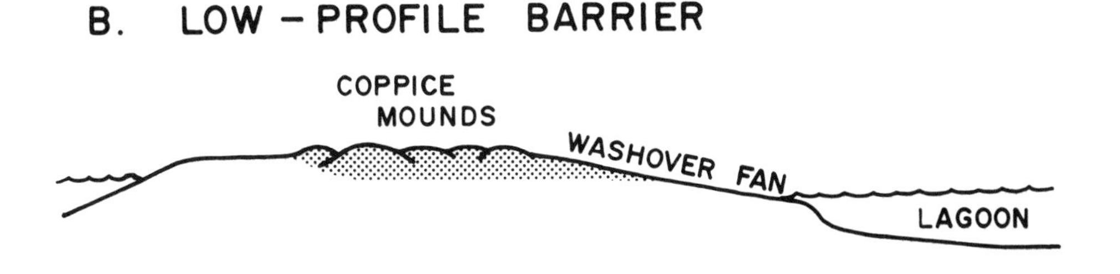

Figure 6-9. Contrasting high-profile and low-profile barriers and associated features. (After Morton and McGowen, 1980.)

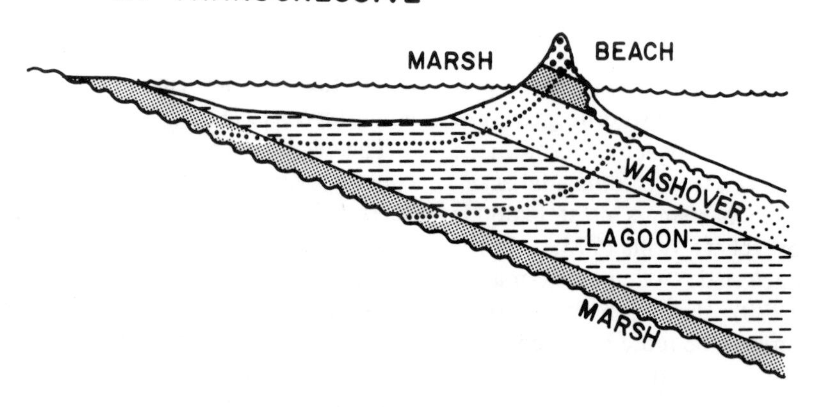

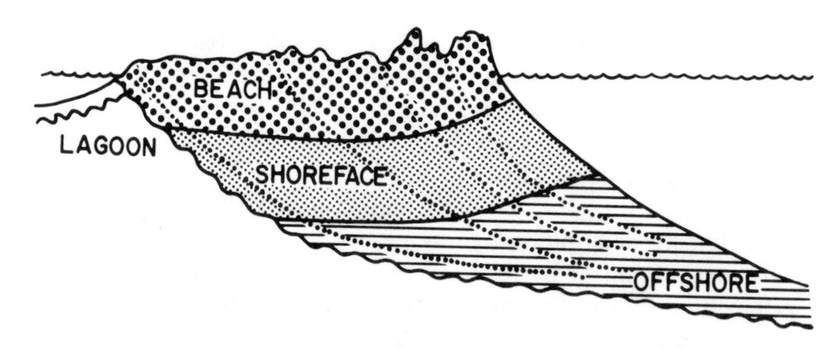

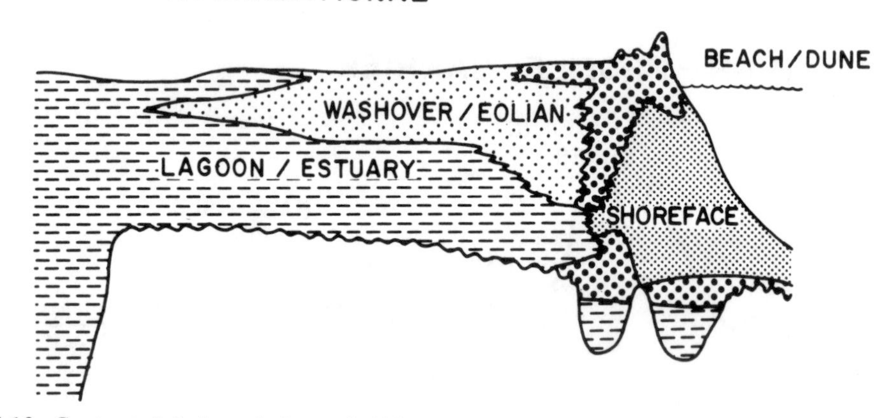

Figure 6-10. Contrasted facies relations of: (A) transgressive barriers (after Kraft and John, 1979); (B) regressive barriers (after Bernard *et al.*, 1970); and aggradational barriers (after Fisk, 1959 and Morton and McGowen, 1980).

broad, vegetated storm ramp which slopes gently lagoonward (Morton and McGowen, 1980).

Sand thickness across the entire barrier is limited, and tends to be further reduced or even removed entirely by passage through the surf zone. Back-barrier peat is exposed in many beaches as the barrier sand body moves landward (Fig. 6-10A). Thus the only record of a transgressive barrier may be a wave-eroded disconformity with a thin sediment veneer or ravinement lag (Swift, 1968). Much of the transgressive barrier sequence can nonetheless be retained if rapid sea-level rise is balanced by continuous sediment supply (Hobday and Jackson, 1979; Kraft and John, 1979). Preserved sand thicknesses of 1–10 m (3–33 ft) are typical. The barrier may be drowned in place, leaving multiple submarine sand bars parallel to the shoreline and encased in lagoonal and shelf muds (Sanders and Kumar, 1975); or, more commonly, sand is spread landward as a transgressive sheet. In all cases, the classical upward-coarsening shoreface sequence is lacking. Instead, barrier sands abruptly overlie lagoonal sediments, and are overlain in turn by shelf muds.

The main processes of barrier transgression are by washover and inlet-related deposition, with wind being important locally. Accelerated barrier retreat often corresponds to areas of temporary, storm-breached barrier inlets (Leatherman, 1979). Flood-tidal deltas form simultaneously, particularly along coasts of low tidal range, and are capable of spreading large volumes of sand landward during transgression. These sands build up a back-barrier platform for the development of washover fans and windblown accumulations. Severe or frequent overwash inhibits plant growth, and a very low profile is maintained. Where there is sufficient time for eolian dunes to form, this can afford the shelter necessary for establishment of back-barrier vegetation, which in turn acts as an effective sediment trap, promoting buildup of subsequent washover and windblown sediments (Leatherman, 1977).

Stable or Regressive Barriers

Stable or regressive barriers tend to have a broad beach and barrier flat showing parallel foredune ridges. Washovers are less common and tend to terminate subaerially.

The best-known accretionary barrier is Galveston Island (Bernard and LeBlanc, 1965; Bernard *et al.*, 1970), which has prograded locally almost 1¼ mi (2 km) in late Holocene time (Morton, 1974). At its eastern end are parallel accretion ridges underlain by dated profiles recording coastal offlap in the form of a complete lower shoreface through beach sequence (Fig. 6-10B). The typical regressive sequence coarsens upward from alternating sand and clay of the shelf and lower shoreface to shelly sand of the upper shoreface and beach (Fig. 6-11). Matagorda Island, another accreting Gulf Coast barrier, is a uniform, strike-oriented sand body which interfingers on either side with lagoon and shelf facies (Morton and McGowen, 1980, p. 133). Padre Island to the south extends almost 125 mi (200 km) with only one natural inlet, and has been subject to pronounced vertical aggradation (Fig. 6-10C) accompanied by growth of wind-tidal flats which in places attach to the mainland. Significant accretion can also be concentrated around cuspate spit headlands, a coastal topography that may reflect differences in wave approach (Nummedal *et al.*, 1977) or an earlier deltaic protuberance. Spit growth and progradation of the Cape Lookout headland, North Carolina, is related to a number of interacting processes including wave accretion of sand supplied by longshore drift, overwash, and dune-ridge growth (Moslow and Heron, 1981).

Vegetated barrier flats are produced by wind deflation to the level of the water table, and are underlain by homogenized, bioturbated sands. Back-island dunes represent beach-derived sand in transit toward the lagoon (Hunter, 1977), but are unlikely to be preserved.

Barrier Inlets and Associated Facies

Inlets serve to exchange water between back-barrier and open ocean environments during each tidal cycle, and thus are progressively closer spaced with increasing tidal range. Microtidal barrier inlets tend to be widely spaced and ephemeral, and migrate rapidly in the longshore drift direction by erosion of the downdrift margin accompanied by spit elongation on the updrift margin. Deeper mesotidal inlets are subject to less rapid longshore migration, but with increasing density may become a dominant sand facies.

The deepest parts of most inlets are dominated by ebb currents, with seaward-oriented sand waves leaving a record of large-scale planar cross beds, commonly showing flood-tidal modification in the form of reactivation surfaces (Kumar and Sanders, 1974). These overlie a deeply eroded surface covered irregularly by gravel or shells. Smaller scale, flood-oriented bedforms are present along the shallow inlet margins, whereas bidirectional bedform and cross-bed orientations are characteristic of intermediate depths. This inlet sequence (Fig. 6-12A) will vary somewhat in response to local inlet morphology and hydrodynamics (Hubbard and Barwis, 1976). Some inlets are entirely flood- or ebb-dominated, producing essentially unimodal cross bedding which decreases upward in scale (Oomkens, 1974).

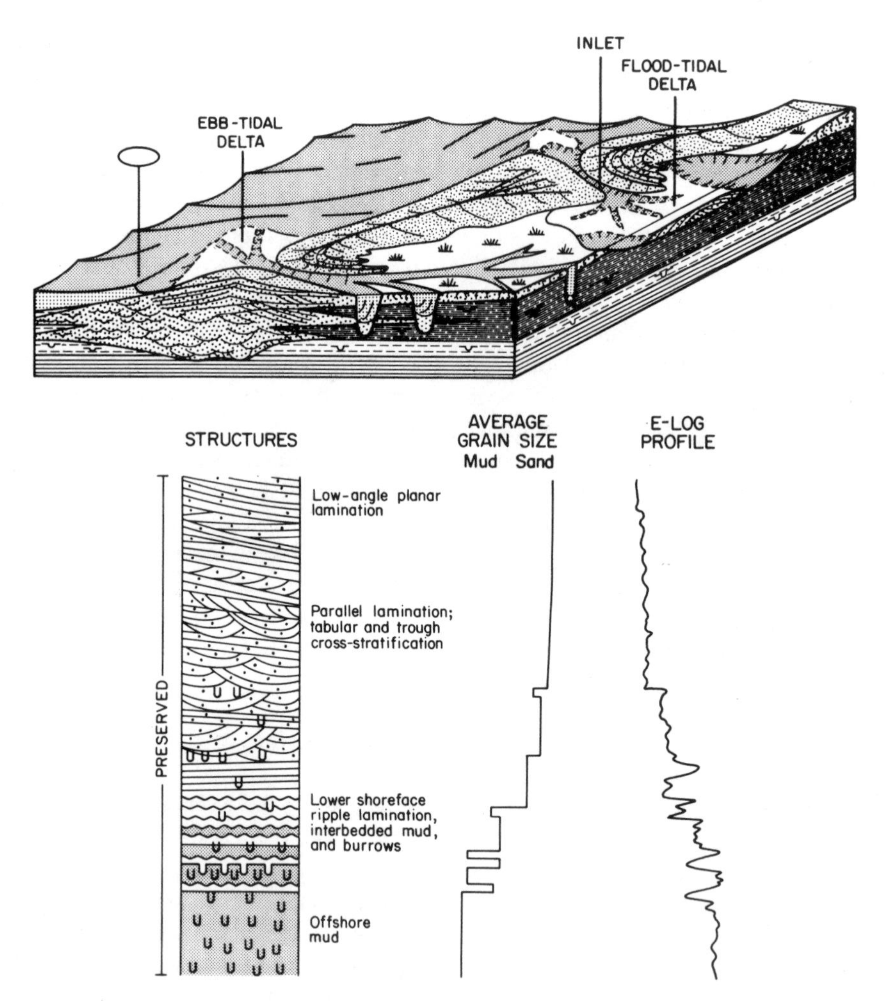

Figure 6-11. Barriers, inlets, and related facies along a mesotidal coast, and the vertical sequence produced by progradation. (In part modified after Ferm and Horne, 1979.)

Where tidal ebb is augmented by fluvial discharge, most of the cross bedding will be directed seaward (Van Beek and Koster, 1972). Such sequences might only be distinguished from fluvial deposits on the basis of associated facies, marine fauna, or trace fossils such as sparse callianassid burrows. Obliquity of inlets along some coasts results in cross-bed orientations almost parallel to the shoreline (Hubbard and Barwis, 1976).

Wave-generated structures, primarily low-angle swash lamination, cap most inlet sequences, reflecting reestablishment of the beach foreshore profile. Along a prograding coast, tidal flat, marsh, and lagoonal deposits will follow.

Longshore migration of inlets obliterates much of the typical sequence of barrier-beach and shoreface sediments, leaving in its place a record of inlet-fill deposits which may be as much as three times the thickness of the original barrier sands (Hoyt and Henry, 1967). Microtidal inlet deposits are typically 20–40 ft (6–12 m) thick, whereas mesotidal inlet deposits may attain 75 ft (20 m) but are less persistent along strike. This process of barrier-island modification is taking place along both prograding and retreating coasts, leaving two distinctive depositional patterns. Along prograding coasts, offshore and lower shoreface deposits are truncated by inlet-fill deposits, which are overlain by back-barrier and alluvial facies. The geometry of the inlet fill bears no relationship to the original inlet configuration, with the deposit generally elongate parallel to depositional strike and becoming more sheetlike with continued progradation.

A combination of inlet migration and barrier

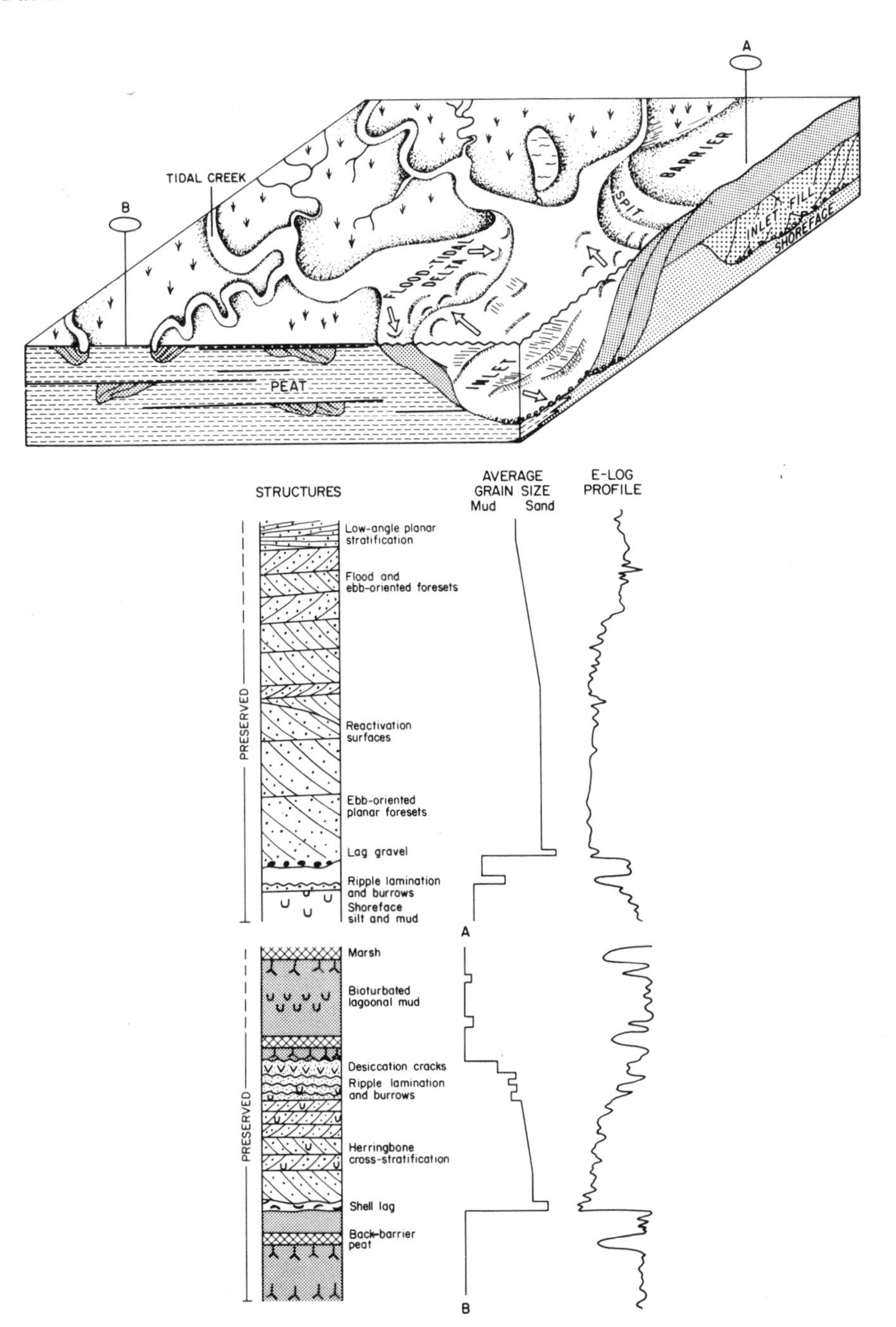

Figure 6-12. Migrating mesotidal barrier inlet and the inlet-fill sequence (A). The infilled lagoon is traversed by tidal creeks, which produce upward-fining sequences interspersed with peats and lagoonal muds (B). (In part modified after Barwis and Hayes, 1979.)

transgression leaves an even more complex record. As the surf zone moves landward across a shore zone that includes inlet fill as one of its thickest components, the subaerial to shallow subtidal portions are generally eroded. This leaves a sequence dominated by inlet fill which is bounded both below and above by pronounced erosion surfaces (Swift, 1968). Because barriers separate landward and seaward environments of grossly different faunal character, the shoreface erosion surface may be accentuated by a faunal jump which can easily be misconstrued as a significant stratigraphic hiatus (Barwis and Makurath, 1978). It is conceivable that because of preferential preservation, many of the extensive transgressive sandstones in the geologic record may be inlet-fill deposits (Kumar and Sanders, 1970). Fluctuating shoreline advance and retreat could leave a succession of stacked inlet-fill and associated deposits which might be difficult to decipher in detail.

Barrier-inlet deposits described in Carboniferous rocks by Hobday and Horne (1977) and in Silurian rocks by Barwis and Makurath (1978) are both very similar to modern examples, but differ from one another in patterns that exemplify the possible variations in bedform orientation. Both examples show the typical upward decrease in grain size and scale of sedimentary structures from deep axis to channel margin, but whereas the Silurian example includes a large proportion of bipolar cross bedding throughout, these reversals are restricted to the uppermost parts of the Carboniferous sequences. The latter contain other evidence of periodic fluctuations in current velocity attributed to tides, for example, rhythmically interlayered thick and thin cross-bed sets (Klein, 1970), high-angle and low-angle cross-bed sets, and plane beds interlayered with cross beds. Reactivation surfaces and clay drapes are common, along with rare marine fossils toward the top. These Carboniferous examples were part of a regressive complex with fluvial channels updip. The Silurian inlet-fill deposits, in contrast, are transgressive and formed along the margin of a shallow carbonate-rich basin isolated from direct fluvial influx.

Tidal Deltas. Constricted, high-velocity flow through barrier inlets is suddenly reduced as the currents disperse into open water at either end, depositing sand as flood-tidal and ebb-tidal deltas (Fig. 6-11). Like the adjacent inlet deposits, tidal deltas can grow laterally by inlet migration, and contain large volumes of sand as a result, with a good chance of preservation in the rock record. It is only in recent years that ancient tidal deltas have been positively documented, and most of these are flood-tidal deltas (Barwis and Horne, 1979), probably because more is known of the internal characteristics of their modern analogs. Furthermore, ebb-tidal deltas are more readily remolded or destroyed by waves as the inlet sand source migrates.

Mesotidal barriers commonly have a more or less equal development of flood- and ebb-tidal deltas, neither of which is particularly large, but which are continuously added to at the downdrift end as the inlet migrates. Hayes (1975) distinguishes among morphological subenvironments of both tidal delta types, with distinctive suites of bedforms and associated sedimentary structures giving rise to predictable sequences. On microtidal barrier coasts, flood-tidal deltas attain very large sizes in contrast to the poorly developed ebb-tidal deltas (Fig. 6-4). These flood-tidal deltas are broad, multilobate, or digitate sheets thinning landward (Barwis and Hubbard, 1976).

Flood-tidal delta deposits on mesotidal coasts are comparable to inlet sequences, but are thinner. Bidirectional or seaward-dipping planar cross beds at the base are deposited on the deep fringes of the sand body where the ebb shield deflects flow laterally along ebb channels (Hayes, 1975; Hubbard and Barwis, 1976). Above this level, thick cosets of landward-dipping planar cross beds reflect migration of successions of sand waves across the flood ramp. These sets become thinner upward and are overlain by mudflat and marsh deposits. As Reinson (1979) has commented, however, the detailed characteristics of the sequence will vary according to locality and inlet dynamics.

Flood-tidal currents are weaker in a microtidal situation, but much sand is transported through the inlet by storm wave surge. Extensive flat-topped bars (Fig. 6-13) build landward, resulting in cosets of planar cross beds with smaller reversed foresets reflecting ebb flow (Hubbard *et al.*, 1979). The idealized sequence is illustrated in Figure 6-13. Some large bars nucleate around shelly material, eventually becoming islands separated by channels. These islands develop small shell beaches and berms, burrowed tidal flats, and salt marsh (Morton and McGowen, 1980). Some tidal deltas have a broad sandy ebb shield, whereas others grade imperceptibly landward into successively finer sediments.

Ancient flood-tidal deltas of microtidal type have been recognized by Barwis and Horne (1979) in Carboniferous rocks in Kentucky, where they comprise erosively based quartz-arenite sheets thinning landward and containing extensive landward-dipping accretion surfaces of the ebb shield. Between these accretion surfaces are cosets of bimodal cross beds. A substantially thicker example in the Eocene of Texas probably owes its size to a contemporaneous regional rise in sea level along a coast of low tidal range (Hobday *et al.*, 1979). Landward-dipping planar

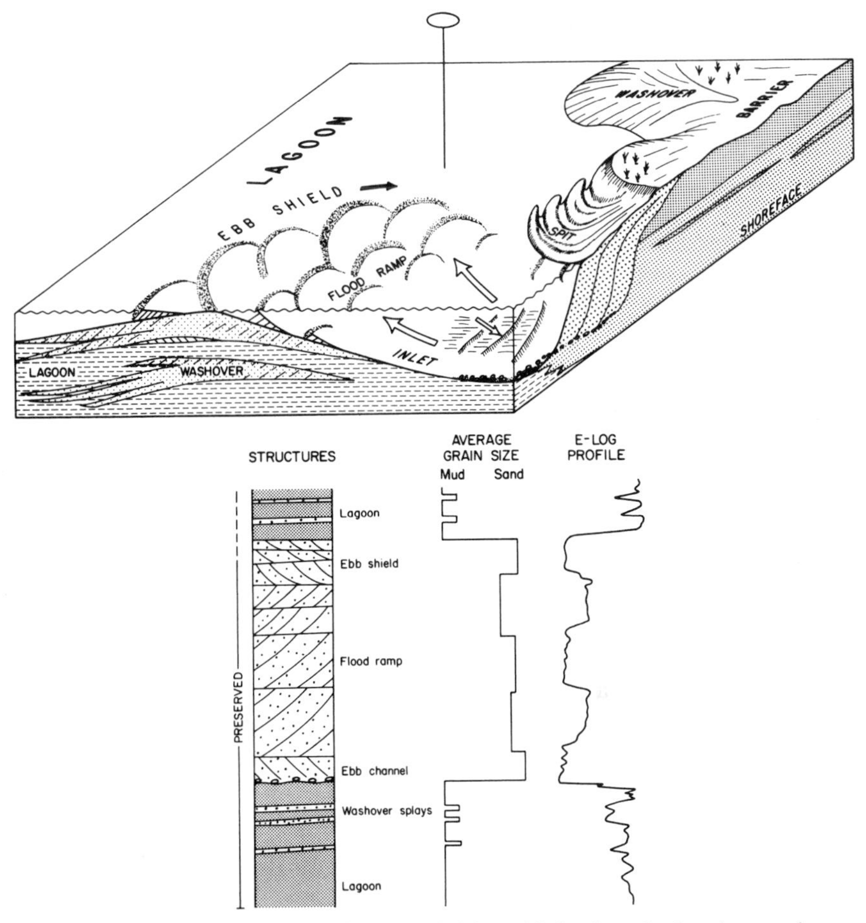

Figure 6-13. A microtidal inlet and large flood-tidal delta with landward-migrating sand waves and an ebb shield, together with a typical vertical sequence through a flood-tidal delta and lagoonal deposit. (In part modified after Barwis and Hayes, 1979.)

foresets show ubiquitous clay drapes deposited by settling of suspended fines during slack water, and layers of mudclasts, possibly derived from erosion of these clay layers and from nearby algal mudflats. Many of the foresets have superposed ripples, evidencing weak lateral and opposed flow related to lowering of water level and tidal ebb. These broad, gently dipping sand sheets interfinger with carbonaceous back-barrier mudstones and lignites.

Smaller, more complex Carboniferous flood-tidal deltas comprise wedges tapering landward from thick barrier-inlet deposits, which, in conjunction with substantial adjacent tidal-flat facies, suggest a mesotidal range. Bidirectional sand waves and troughs show transverse ripples, reactivation surfaces, and complex small-scale intrasets typical of tidal processes (Hobday and Horne, 1977).

Ebb-tidal deltas range from small, entirely subaqueous sand accumulations off microtidal inlets to more substantial, tidally emergent features off mesotidal inlets. Thicknesses depend largely on water depth and contemporaneous changes in relative sea level. For example, ebb-tidal delta deposits of the Texas Gulf Coast studied by Morton and McGowen (1980) are gradationally based and up to 40 ft (12 m) thick off accretionary barriers, but abruptly based and only about 20 ft (6 m) thick off transgressive segments.

Interaction among tidal currents, wave surge, and longshore drift, together with segregation of flood and ebb currents during different phases of the tidal cycle, control bedform distribution on ebb-tidal deltas (Oertel, 1974; Reinson, 1979). Ebb-dominated cross beds are generally deposited in the main channel, with a very large-scale, steep, solitary set produced by progradation of the terminal lobe slipface. Flood-oriented trough and planar cross beds are deposited in shallow marginal channels, with subhorizontal lamination characteristic of channel-margin linear bars.

Most ebb-tidal delta deposits are probably reworked by shoreface processes, making recognition difficult in ancient rocks. A hypothetical succession (Barwis and Hayes, 1979) shows a vertical decrease in set thickness accompanied by a change from ebb-oriented to more variable onshore, offshore, and alongshore azimuths, together with plane beds and subdued "washed out" bedforms. A comparable structural assemblage is present in suspected ebb-tidal delta facies along the seaward edge of barrier-island complexes in the Carboniferous of West Virginia (Hobday and Horne, 1977).

Back-Barrier and Estuarine Facies

A barrier-protected situation or an indented riverine embayment generally experiences a low level of wave activity, except during storms. The importance of tidal processes depends on the nature and configuration of the connection with the ocean. Tides in lagoons are generally reduced as they are propagated through narrow barrier inlets, but in funnel-shaped estuaries the reverse may be true, with significant increase in tidal range in comparison with open coasts.

Lagoons and Bays. Lagoonal environments of deposition are highly variable depending on climate, tidal range, storm frequency, and sediment supply. In addition to their shore-parallel alignment many lagoons have landward reentrants, or bays, representing the lower reaches of drowned river valleys; these tend to be filled longitudinally by small bayhead deltas (Fig. 6-1). Modern bays are commonly underlain by a transgressive succession from fluvial through deltaic to estuarine deposits, followed by upward-coarsening fill related to the present phase of bayhead delta progradation (Morton and McGowen, 1980, p. 83).

The tendency of lagoonal shores in some areas to subside and retreat as a result of fine-grained sediment compaction is augmented by localized shoreline erosion by waves, particularly where vegetation has temporarily been destroyed by fluctuations in water level or salinity. Inner-lagoon muds, either highly bioturbated or with graded lamination, accumulate by flocculation (Phleger, 1969) and biogenic pelletization of suspended clays (Pryor, 1975). The lagoonal muds are pyritic, glauconitic, or dolomitic, with a variety of shell fragments, mainly mollusks. Oyster reefs develop in this environment, and are commonly elongated perpendicular to the prevailing currents.

Wind-generated waves concentrate coarser sediments at the lagoon margins where they grade into tidal flat and marsh. Lagoon infilling is frequently accompanied by segmentation through spit or bar growth into smaller water bodies. Segmentation is most effective in tideless lagoons where variable winds generate longshore currents of alternating direction. Lagoon-margin sedimentation is greatly accelerated by vegetation, particularly mangroves, reed swamps, and aquatic plants which can trap and bind sediments at water depths of a meter or more (Orme, 1973).

Hypersaline lagoons in areas of excessive evaporation-potential tend to be shallow, with rare fresh-water inflow and limited marine exchange. Evaporitic algal mud, windblown sand, and shelly, burrowed muddy layers are typical, with shrinkage cracks and mudclasts along the margins. Very large lagoons commonly contain segregated water masses of contrasted salinity, with hypersaline conditions most characteristic of the landward portions (Phleger, 1969).

Progradation of lagoon margins generates a thin upward-coarsening sequence, with bayhead delta sands building from the landward side, and washover sands along the seaward edge (Fig. 6-14). Tidal-channel and tidal-flat deposits near the top of the sequence are overlain by marsh sediments. Similar sequences with a thin coal seam at the top are common in the rock record (Fig. 12-8) where they are situated landward of barrier-island sandstone (Hobday, 1974a). Subsurface isolith patterns reveal vast Cenozoic lagoonal tracts in Texas with tidal channel, tidal delta, and washover facies clearly represented (Fisher and McGowen, 1969).

Washover Fans. Washover fans form by storm surge over the berm crest, erosion of a washover channel across the backshore, and splaying out of lagoonward-convex lobes. The deposits are

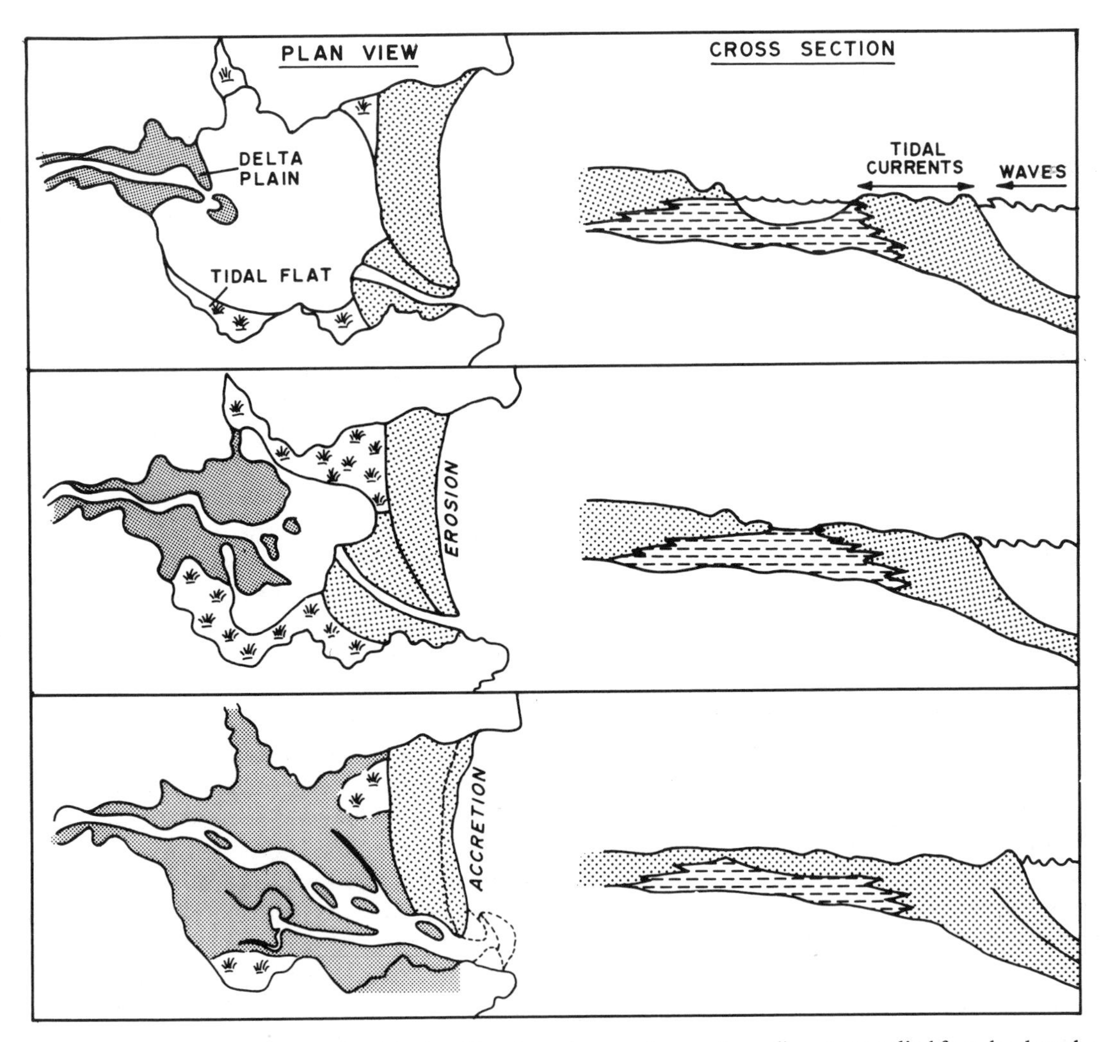

Figure 6-14. Stages in infilling of an idealized barrier estuary or lagoon by sediments supplied from landward as prograding bayhead deltas, and from seaward as washover sands and flood-tidal deltas, and on tidal flats. (After Roy *et al.*, 1980.)

wedge shaped in cross section, tapering gradually landward (Andrews, 1970). Over broad barriers, or where the overwash is small, much of the material is deposited subaerially on back-barrier flats and in interdune areas (Morton and McGowen, 1980, p. 129), contributing to aggradation.

Many of the largest washover fans are presently inactive except during the most severe storms. These fans were in some cases initiated during a phase of barrier accretion which accompanied the Holocene sea-level rise (Morton and McGowen, 1980, p. 129). Smaller fans are often flooded during spring tides, or even daily. Despite these differences in scale and level of activity, most washover fans tend to contain the same sedimentary structures (Deery and Howard, 1977).

Washover channels contain shell debris, driftwood, and heavy-mineral concentrations. Landward-dipping trough cross beds are deposited by bars in some channels. As the storms wane, a layer of suspended sediment accumulates. Eolian dunes that encroach on the washover channels constitute a secondary supply of sand for subsequent overwash (Deery and Howard, 1977). Quartzose sands of the washover fan proper display a characteristic subhorizontal lamination produced by upper flow regime transport. Inverse grading of laminae is common, but some monomineralic layers show normal grading (Leatherman and Williams, 1977). As the current velocity falls below the critical, ripples form, depositing a thin layer of ripple-laminated sand, usually destroyed by subsequent overwash.

During periods between storms, dry washover sands are subject to wind deflation and disturbance by plants and animals, so that successive storm increments tend to be separated by rootlets, heavy mineral and shell lags, and truncation surfaces (Leatherman and Williams, 1977), although some may be conformable. Crustaceans, in particular, cause intense bioturbation, or leave distinctive curved vertical burrows (Howard, 1978). Adhesion ripples develop on damp washover sands, and can be important diagnostic features in ancient deposits (Hobday and Jackson, 1979).

Washover-fan subfacies most likely to be preserved are the distal terminations which interfinger with tidal flat, marsh, or lagoonal sediments. Landward-dipping planar foresets form where overwash encounters standing water in the lagoon (Schwartz, 1975). These cross beds show reactivation surfaces and are commonly modified by tidal flow, with the development of trough cross bedding and ripple lamination. In some cases, the mud and organic content of these distal deposits is so high as to produce an almost imperceptible gradation into lagoon or marsh sediments.

Back-barrier Tidal Flats and Marshes. Broad tidal flats periodically inundated by astronomical or wind tides develop on the landward side of barriers over a platform of tidal-delta and washover deposits, and along some mainland shores where wave energy is reduced. The area of tidal flats generally increases in proportion to tidal range. Mesotidal lagoons tend to fill rapidly, resulting in extensive intertidal marshes traversed by tidal channel networks.

In humid climates, the tidal flats tend to be vegetated above the midtide level (Harrison, 1975) or near the high-tide level (Reineck, 1975) where muddy sediments become heavily disrupted. Salinity variations of more arid regions inhibit growth of rooted plants, but algal mats are common. These mats desiccate, crack, and peel on exposure, and are reworked as algal mudstone intraclasts, a very common component of ancient tidal-flat deposits, particularly in older Paleozoic and Precambrian sequences. Sands and biochemically precipitated carbonates commonly alternate with the algal mud layers, and lamination is disrupted by gases released during organic decay (Morton and McGowen, 1980, p. 130).

Migrating tidal channels produce lateral-accretion beds analogous to those of fluvial point bars, from which they are distinguished by textural alternations (Reineck, 1975), trace fossil types, and bimodal cross beds (Barwis, 1978). Barwis observed that vertical sequences generated by tidal channel migration (Fig. 6-12B) are very similar to the upward-fining sequence produced by progradation of tidal flats along open coasts.

Estuaries. Geological views of what constitutes an estuary are highly varied (Lauff, 1967; Cronin, 1975). In attempting to resolve this problem, Roy *et al.* (1980) distinguished between barrier estuaries, equivalent to lagoons, and drowned river valleys, or rias. The latter may be a component of extensive lagoonal systems, with which they are included; for example, the Texas "bays." Macrotidal estuaries are unique and are considered separately under macrotidal coastal systems.

Estuaries characterized by fresh-water and marine interaction in drowned river mouths and embayments are transient features, unless maintained by tidal scour. There is tremendous variation in process, physical characteristics, and biogenic features both within and between estuaries (Howard, 1975). These differences stem largely from the range in riverine discharge, tidal and nontidal circulation, and sediment supply. Sediments may be derived from landward, seaward, or both (Fig. 6-14).

The upstream parts of some estuaries are sandy as a result of fluvial influx, merging downstream into bioturbated muds. Other estuaries show the opposite trend, with upstream fining as a result of high rates of mud accumulation in the turbidity maximum (Postma, 1981). Such estuaries, with sand-rich seaward portions pinching out upstream into muddier sediments (Fig. 6-15) would provide attractive petroleum exploration targets in ancient deposits.

Three categories of Georgia estuaries recognized by Howard (1975) are also represented elsewhere. These are dominated by bioturbated muddy sand, alternating sand and mud, and sand, respectively. Sand is concentrated in estuarine channels and on shoals, with migrating sand waves and dunes generating large-scale cross bedding. Parallel lamination is produced by wave

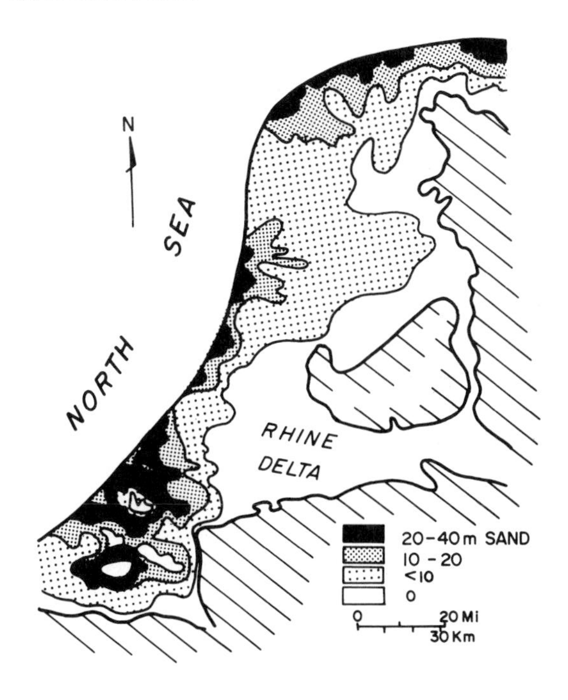

Figure 6-15. Sand-rich "blind estuary" deposits pinching out landward into muddy sediments of the Netherlands coastal plain. (Modified after Oomkens, 1974.)

swash, and by current flow over shallowly submerged bars. Mesotidal shoals and channels accrete rapidly to produce upward-fining sequences of flood- and ebb-oriented cross beds overlain by a thick succession of horizontal and low-angle laminated beds. In other estuaries with strong fluvial inflow, cross beds are unidirectional seaward, and might be difficult to distinguish from fluvial systems in the rock record.

Finer-grained estuarine deposits commonly comprise thin textural alternations reflecting periodic changes in current strength. Lenticular, wavy, and flaser bedding is characteristic. Although very few trace fossils are unique to estuarine environments, they are very useful in distinguishing among ancient estuarine facies (Howard and Frey, 1975). Polychaete burrows may dominate the inner estuary, but trace fossils generally increase seaward in abundance and diversity. Mollusk and echinoderm traces are locally abundant, and arthropods leave some distincitive burrows such as *Ophiomorpha*. Shell debris may be randomly distributed or concentrated as a lag.

As estuaries fill, the area affected by tidal channel migration increases. Consequently, some estuarine deposits are dominated by tidal-channel sequences (Howard and Frey, 1975; Barwis, 1978). These sequences (Fig. 6-12B) are upward fining, but differ from fluvial sequences in containing structures indicative of tidal current segregation and reversals, along with diagnostic biogenic features. Widespread sands resulting from prolonged tidal-channel migration will be capped by peat.

Remarkable exposures of ancient estuarine sequences studied by Horne (1979a) show evidence of several distinct evolutionary stages: fluvial abandonment, submergence, tidal scour, and infilling by tide-dominated processes.

Strandplain Systems

Sheetlike or strike-elongate sand bodies develop by successive seaward accretion of beach ridges. Narrow, marshy sloughs between ridges take the place of lagoons, and it follows that no inlets or associated environments exist. Concentric beach ridges are best developed near river mouths, where they represent extreme examples of wave-dominated depositional coastlines. As many as 280 parallel ridges are present across a strandplain width of 15 km (9 mi) on the Nayarit coastal plain of Mexico (Curray *et al.*, 1969). These ridges average 50 m (165 ft) in width, with heights typically ranging from 1 m (3 ft) on the crest to slightly below mean sea level in the intervening depressions. The Tabascan strandplain of Mexico attains 40 km (25 mi) width, with the largest beach ridges almost 4 m (13 ft) high (Psuty,

1966). Despite its broad extent, the strandplain sands are generally thinner than 10 m (33 ft).

According to Curray (1969), beach ridges accrete successively along the seaward margin by growth and emergence of longshore bars, mainly under conditions of low wave activity. Vertical growth of beach ridges is attributed by Psuty (1966) to storms that erode the lower beach and cause washover accretion on the berm. The resulting stratification dips consistently landward, comprising gently inclined topsets and bottomsets, and steep foresets. Beach progradation follows the resumption of fair-weather conditions, producing typical foreshore deposits and widening of the beach. Alternatively, aggradation of the strandplain may be accomplished mainly by eolian processes (Thom, 1964).

The vertical strandplain sequence is similar to that of a prograding beach/shoreface, grading from shoreface sand, silt, and mud into quartzose beach sands. Shore-parallel sand sheets characteristic of strandplains are present in a number of Cretaceous and Tertiary formations of the Gulf Coastal Plain (Caughey, 1977).

Cheniers. Chenier coasts represent a special case of mud-rich strandplains comprising isolated, shore-parallel bodies of shell and sand enclosed by prograding marsh and mudflat deposits. They grow in response to fluctuating supply of clastic sediment within a regressive facies framework. Coastal erosion and winnowing occur during episodes of reduced longshore sediment supply, with beach ridges marking the limit of marine transgression. These sandy deposits are subsequently stranded behind seaward-building mudflats corresponding to episodes of accelerated fine-grained sediment influx. Long-term variations in sediment supply are generally related to changes in the position of an adjacent delta system (Byrne *et al.*, 1959).

A chenier plain comprises several parallel beach ridges separated by prograded muds (Otvos and Price, 1979). The chenier plain of Surinam (Augustinus, 1980) comprises two types of sand accumulation (Fig. 6-16), produced, respectively, by onshore migration of nearshore sand bars and by longshore drift coupled with washover processes. The former variety is coarser grained and shows landward-dipping cross beds of both steep and gentle inclination. The second type comprises interlayered sand and mud with steep landward-dipping foresets and gentle seaward dips. The intervening muds are well laminated. An ancient example of lenticular chenier sandstones, with a maximum thickness of only a meter or two enclosed in organic-rich shales of coastal mudflat origin, is provided by Ferm and Cavaroc (1969).

Macrotidal Coastal Systems

Macrotidal coasts [range greater than 4 m (13 ft)] are normally characterized by extensive landward tidal flats, salt marshes, and estuaries, and shelfward tidal sand ridges and shoals (Fig. 6-17). Sand ridges tend to be concentrated toward

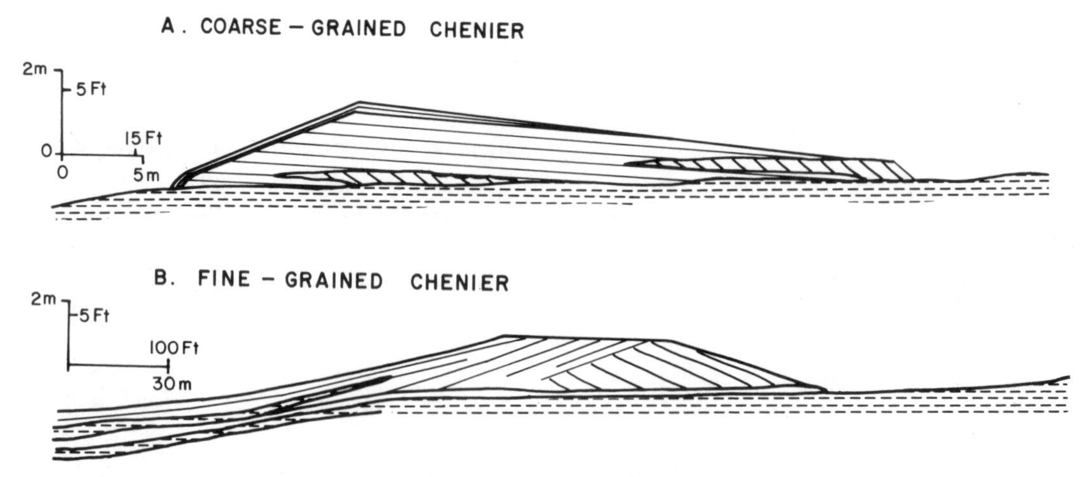

Figure 6-16. Schematic representation of cheniers of coastal Surinam and their internal structures. (After Augustinus, 1980.)

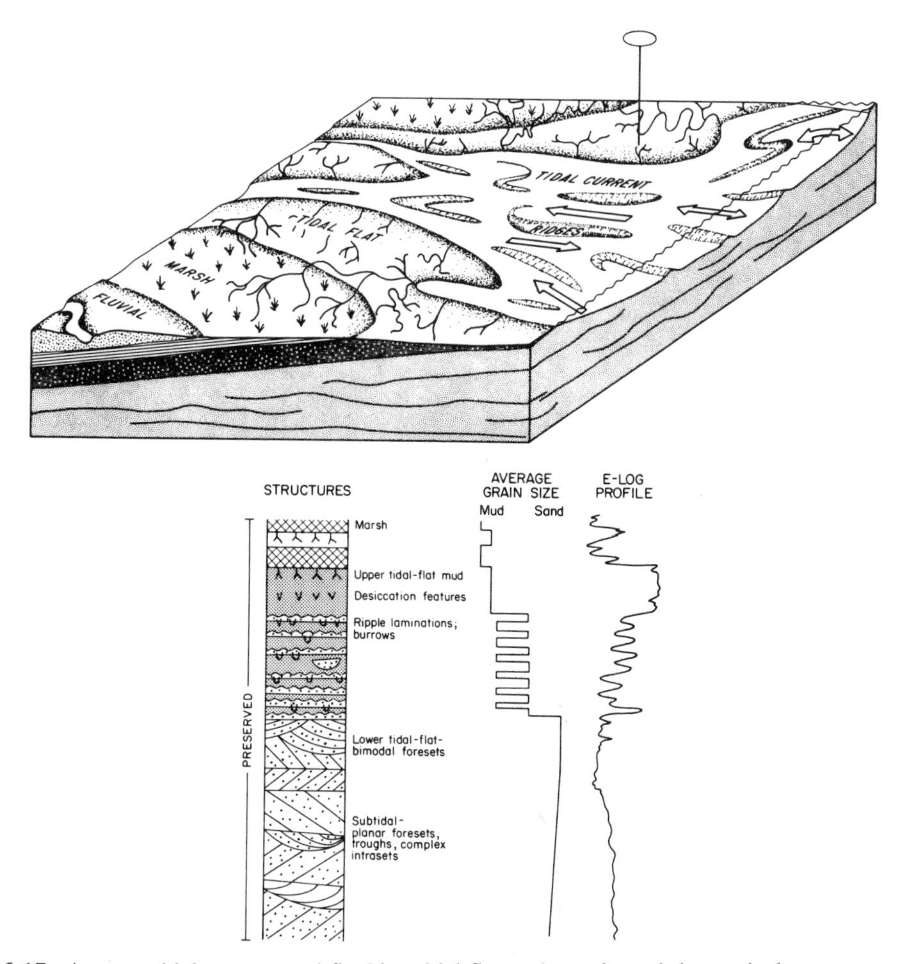

Figure 6-17. A macrotidal estuary and flanking tidal flat and marsh, and the vertical sequence produced by coastal progradation.

the axes of larger embayments or funnel-shaped estuaries, where they are aligned by tidal currents perpendicular to the mean shoreline trend. Off straight coasts, however, tidal currents and associated sand ridges are parallel to the shoreline, following a rectilinear or reversing pattern. Wave-produced structures are less important along tide-dominated coasts, being rapidly obliterated by tidal currents. Tidal range influences the amount of sediment in suspension, and maximum seaward escape of suspended sediments occurs when spring tides coincide with high river discharge (Castaing and Allen, 1981).

Landward reduction in grain size from subtidal to intertidal sands may be gradational, as in tropical Australia (Fig. 6-18) and The Wash of England (Evans, 1965), or it may be limited to a silty or muddy high-tidal flat, as in the Bay of Fundy (Knight and Dalrymple, 1975). Where little sand is available, as in the Gulf of California tidal flats (Thompson, 1975), silt and clay dominate even the lower intertidal and subtidal zone.

Tidal flats are generally traversed by an intricate network of migrating tidal channels, but where sand ridges and shoals are lacking, water may move across the flats as a broad, uniform sheet (Thompson, 1975). Tidal flats exposed to the sea are less channelled than are back-barrier tidal flats (De Jong, 1977).

Intertidal sands near low-tide level show complex bedforms which increase in size with increasing tidal range. At any particular location the bedforms are normally dominated by either the flood or ebb phase of the tidal cycle (Klein, 1970). Under the maximum tidal ranges, current velocities attain 2 m/s (7 ft/s) for a 1–4-hour period during the ebb to first half of the flood

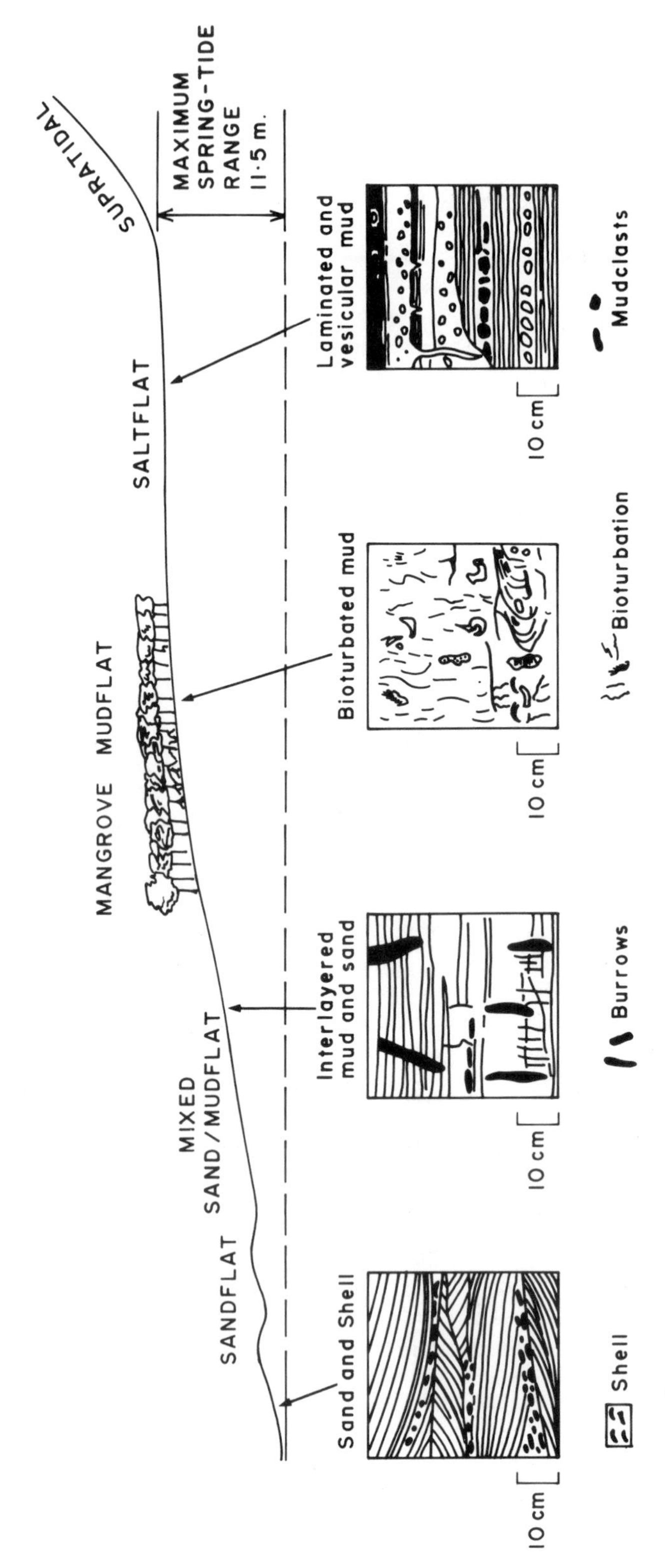

Figure 6-18. Topography, textures, and sedimentary structures of an Australian tropical tidal flat. (After Semeniuk, 1981.)

(Dalrymple *et al.*, 1978). Bedforms comprise sand waves, dunes, and ripples, commonly in complex patterns of superposition, with plane-bed surfaces best developed toward low-tide mark. The dominant internal structures are cross beds of simple or complex geometry, herringbone cross bedding, reactivation surfaces, and alternating cross-bed sets of different scale and foreset dip angle (Klein, 1970; Semeniuk, 1981). These structures have been recognized in a number of ancient quartzarenite units (Swett *et al.*, 1971; Eriksson, 1977). Ancient macrotidal deposits show an assortment of complex ripples, including double-crested, flat-topped, and ladderback forms; associated ripple lamination has rhythmic drapes of silt and clay.

The broad midtidal flats of most macrotidal coasts are underlain by interlayered sand and mud with abundant lenticular and flaser bedding (Reineck, 1972). A great variety of small wave-generated structures is characteristic (De Raaf *et al.*, 1977), and diverse ripples, including ladder-back, double-crested, and flat-topped forms, together with runzelmarks, rills, and desiccation cracks, reflect periodic shallowing and emergence (Klein, 1963; Reineck, 1967). Migrating channels produce lateral-accretion beds analogous to those of point bars, but are distinguished by textural alternations (Reineck, 1975), trace fossil types, and bimodal cross beds (Barwis, 1978).

Mudflats of the uppermost intertidal zone contain horizontal laminae which become progressively disrupted by roots, burrowing, shrinkage, cracking, salt crystallization, gas escape, and diagenesis. Sand and shell layers may be deposited during storms. Salt-marsh deposits which cap some intertidal sequences comprise a mixture of organic matter, silt, and clay, which becomes homogenized by roots (Evans, 1965). In hotter, arid environments the supratidal deposits are typically oxidized and deformed by desiccation and evaporite crystallization. Precipitation of gypsum and halite may elevate the surface to a level where it is no longer flooded even during extremes (Thompson, 1975).

Progradation of most macrotidal coasts generates an upward-fining sequence (Fig. 6-18), the thickness of which approximates to the tidal range (Evans, 1965); but paleotidal estimates may easily be exaggerated by the effects of contemporaneous subsidence and facies stacking. Finer grained, arid-region tidal flats will generate quite different sequences (Thompson, 1975), with suspension-deposited silts and clays dominant throughout, and no significant vertical change in texture.

Shore-Zone Systems in Basin Analysis

Shorelines are ephemeral and their deposits are highly variable. A large proportion is never entered in the rock record. Shoreline migration, which may be gradual and uniform or sporadic and reversing, is a response to tectonism, eustatism, or fluctuations in the rate or locus of fluvial sediment influx.

Elongate lenses of barrier quartzarenites may be isolated in shelf and lagoonal shales or may overlap. Such patterns probably result from abrupt changes in updrift delta systems, intermittent transgressive reworking, and overstepping. In contrast, a blanket-like barrier sand body, gradationally overlying finer-grained shelf sediments, is produced by persistent regression. Blanket sands may also be produced by continued transgression, and range from veneers over erosion surfaces to more substantial sands which grade upward into shelf muds. Finally, some barriers have undergone persistent vertical growth, producing a single thick, strike-oriented sand body with lagoonal and shelf muds on either side. Stacked barrier and strandplain sands in the Tertiary Frio Formation of the Texas Gulf Coast attain the exceptional thickness of 3500 ft (1000 m) and more (Galloway *et al.*, 1982).

Regressive sequences are normally thicker, and preserve a more complete range of shoreface subfacies than do transgressive shore-zone sequences. The vertical gamma or electric log pattern of a regressive shoreline may be indistinguishable from that of a prograding delta system. Furthermore, aggradational shore-zone sands may produce a geophysical log response identical to that of bed-load systems. Shore-zone systems are generally distinguished with confidence, however, by a shore-parallel alignment of sand isoliths, for example in the Frio Formation (Figs. 2–6 and 2–7A). This emphasizes the need for three-dimensional delineation in correctly identifying ancient depositional systems.

Regression and transgression of macrotidal coastlands produce upward-fining and upward-coarsening sequences respectively, although the

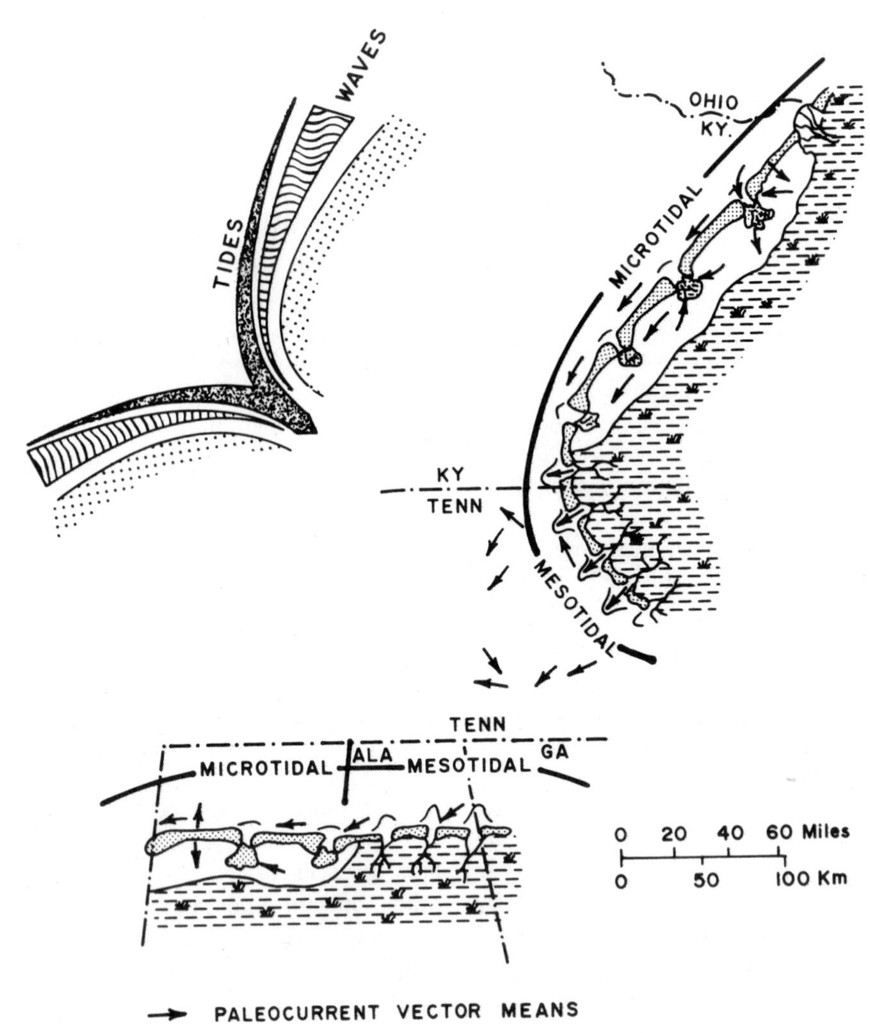

Figure 6-19. Changes in paleoshoreline configuration and corresponding changes in inferred barrier morphology resulting from differences in paleotidal range. (Modified after Horne, 1979c.)

latter may be very thin, or possibly represented only by a gravel-veneered erosion surface. Migration of tidal channels and shoals leaves a complex record of upward-fining, upward-coarsening, strike- and dip-oriented textural trends.

The effects of changing shoreline configuration on depositional patterns can be recognized in some basins, for example, in the Carboniferous of the southeastern United States (Fig. 6-19). Typical mesotidal conditions prevailed within the broad, open embayment subjected to amplified tides, merging onto microtidal barriers along the adjacent straight coasts (Horne, 1979b).

Shore-Zone Systems Through Geologic Time

Shore-zone systems were undoubtedly affected by changing dimensions and interrelationships of the evolving continents and ocean basins, by the evolution of invertebrate organisms, and by expansion of vegetation. The differences in shore-zone deposits of various ages are possibly less pronounced than is the case in fluvial and alluvial fan systems, and are certainly less-well documented. Some of the oldest known sedimentary rocks reflect wave and tidal conditions quite comparable to those of today (Eriksson, 1977).

The disproportionately large volume of shore-zone deposits in some very old successions (Rust, 1973; Von Brunn and Mason, 1977) may stem from the generally lower relief of continents as compared with the present, the prevalence of shallow-marine platforms and epeiric seas, and the instability of the unvegetated coasts. Anomalous tidal amplitudes have also been postulated as resulting from increased lunar gravitational attraction following early Precambrian capture of the moon (Singer, 1970). But earth–moon relationships appear to have remained fairly constant, with changes in structural setting, sediment texture, and degree of shoreline stabilization by biological agencies possibly accounting for most of the observed differences.

Archean shore-zone sedimentation was initially confined to margins of the small, evolving continental nuclei (Tankard *et al.*, 1982). The shore zone was very narrow, and separated coarse-grained fluvial systems from submarine fans. A continental shelf later formed by buildup of the underlying fluvial/turbidite wedge and flooding of the coastal plain (M.P.A. Jackson, personal communication, 1982). By late Archean and early–middle Proterozoic times, progressive cratonization provided broad, subdued landmasses that were intermittently transgressed by shallow seas, leaving thick sequences of shelf and tidal-flat deposits. Repetitive upward-fining regressive intertidal sequences of remarkably constant 12–20 m (40–65 ft) thicknesses are possibly a reflection of a prevailing macrotidal range (Von Brunn and Hobday, 1976). On the other hand, these sequences may have originated under a lower tidal range, with facies stacking and cyclicity produced by variations in subsidence rate. Such tectonically controlled sequences are common in other lower Proterozoic deposits, where they are separated by persistent transgressive discontinuities (Beukes, 1977).

Prior to the advent of terrestrial vegetation, larger volumes of sand would have been supplied to the coast by rivers, causing rapid progradation across the shallowly submerged platforms. A lower proportion of clay, and the absence of suspension-feeding organisms, which were later highly effective in speeding accumulation of fine-grained sediment, may account for the predominantly sandy intertidal sequences of some Proterozoic basins. But even in sediments as old as 3 billion years, algal binding was an important process (Mason and Von Brunn, 1977). Broad, unvegetated tidal flats were severely reworked by wind and storm-surge processes.

Unvegetated barrier coasts, too, would presumably have been rapidly modified. Storm-wave erosion carried large quantities of sand to the shoreface and inner shelf, and barrier overwash would have been common. Unstable barriers in a subsiding basin supplied by vigorous bed-load streams accumulated extremely thick quartzose sand successions which include shoreface, foreshore, inlet, and washover-fan facies (Hobday and Tankard, 1978) dispersed over a wide area. Preserved late Paleozoic and younger barrier systems are commonly linear features bounded landward and seaward by mudstones and shales, but transgressive and regressive sheetsands are also represented. Some of the thickest linear barrier and strandplain systems are preserved in the Tertiary Frio Formation of the Texas Gulf Coast (Galloway *et al.*, 1982). These extensive coastal systems received vast quantities of quartzose sand by convergent longshore drift from delta systems along strike. Sedimentation kept pace with continued subsidence, causing stacking of barrier sands to a thickness of thousands of feet. The vertically homogeneous barrier sands interfinger abruptly with the adjacent lagoonal and shelf muds, providing ideal hydrocarbon reservoir-seal relationships.

Many modern shorelines correspond within a few meters to former Pleistocene shorelines, thus possibly accounting for the extreme height of some stable barrier systems. Compound barriers, such as those along the Zululand–Mozambique coast of southern Africa, have a lithified or semilithified Pleistocene barrier/dune core, and with the capping Holocene dunes, are exceptionally long, narrow, and up to 200 m (650 ft) high (Hobday, 1976).

Chapter 7

Terrigenous Shelf Systems

Introduction

Terrigenous shelves include both epeiric (epicontinental) platforms and continental shelves, with a mantle of land-derived sediments, as opposed to biogenic and chemical precipitates. *Epeiric platforms* are broad, shallowly inundated continental areas. Modern examples such as the North Sea, Hudson Bay, and Gulf of Carpentaria are small by comparison with many of their ancient counterparts. *Continental shelves* are submerged continental margins, dipping very gently from the outer edge of the shore zone to a depth, generally between 300 and 800 ft (100 and 250 m), at which there is an abrupt increase in slope. If the shelf break is not well defined, the shelf is arbitrarily confined to depths shallower than 200 m (650 ft) (Bates and Jackson, 1980). Present-day shelves have a complex depositional and erosional evolution which commenced in the Mesozoic (Swift, 1969).

Modern shelves provide partial analogues for interpreting deposits of both epeiric platforms and shelf seas which, despite differences in origin and structure, were broadly similar in their fundamental processes and sedimentary response (R.G. Walker, 1979; Brenner, 1980). Because of the rapid Holocene rise in sea level, the deposits of modern shelves do not provide ideal models for direct stratigraphic application. Some shelf-sediment bodies originated in the shore zone, and are now at least partially out of equilibrium with shelf processes (Emery, 1952). However, modern studies recognize that most shelves are subject to dynamic processes at all depths (Field, 1982). An understanding of shelf processes is therefore more important than establishing detailed empirical rock models from present-day shelves, but as yet the variety and interaction of shelf processes are not fully understood. Storm events, of disproportionate geological significance, are naturally the most difficult to document. Oceanographic studies are nevertheless providing details of shelf circulation, sediment transport, and deposition, which are applicable to deposits of the vast terrigenous shelves that dominated large intracratonic areas during the Proterozoic, early Paleozoic, and late Mesozoic, and which can be applied to exploration of the hydrocarbon-rich shelf systems of Mesozoic and Cenozoic age.

Early Proterozoic shelf systems, such as the 12,000 m (39,000 ft) thick Transvaal Supergroup of South Africa, were a response to progressive cratonization of the earlier unstable Archean crust (Tankard *et al.*, 1982). Upper Proterozoic and lower Paleozoic shelf deposits locally attain prodigious thickness, as exemplified by the 5000 m (16,000 ft) succession in the Precambrian of Scotland (Anderton, 1976) and the 2000 m (6,500 ft) succession of shelf quartzarenites in the Cambro-Ordovician of southern Africa (Hobday and Tankard, 1978). The well-known Mesozoic terrigenous platform systems of western North America are relatively fine grained but contain hydrocarbon-rich sandstones (Brenner, 1980). Cenozoic shelf deposits tend to be thin but widespread, except on some rapidly subsiding margins.

Shelves vary according to their plate-tectonic setting (Shepard, 1973). Shelves along transform and rift-basin margins tend to be narrow, but some failed rifts and aulacogens, contain very thick shelf successions. Broad shelves and platforms are characteristic of:

1. Some convergent margins, where extensive shelf seas develop landward of island arcs in a backarc basin.
2. Divergent, trailing-edge, or passive continental margins. Passive continental margins commonly comprise several distinct elements (Fig. 8-1, p. 168) corresponding to stages in continental breakup (Falvey, 1974). The rift-valley stage is generally separated by an unconformity from the post-breakup progradational sequence.
3. Cratonic downwarps that open to the ocean.

The Holocene transgression inundated shore-zone deposits corresponding to lower sea levels, leaving relict sediments (Emery, 1952), some of which are now below the depth of effective wave and current reworking. Some Pleistocene dune ridges corresponding to lower sea levels were drowned and partially reworked, but still bear the imprint of their original environment.

The nearshore sand prism (Swift, 1969) extends from the edge of the shore zone across the shelf for a distance depending on shelf energy, gradient, and sediment supply. Its distal equivalent, a shelf mud blanket, is a youthful feature that on many shelves is still prograding seaward (Curray, 1965). Shelf-sediment distribution is strongly influenced by the climate of adjacent continental land masses (Hayes, 1967b), resulting in a broad climatic zonation of shelf sediments. Polar areas are characterized by gravelly, chloritic deposits; a sandy zone is deposited off temperate and arid tropical landmasses; and a muddy zone corresponds to hot, humid climates.

Terrigenous shelf sands have been a prime target in hydrocarbon exploration. They commonly have good primary porosity and permeability, and are in close proximity to potential source rocks of high organic content. Traps are represented by contemporaneous or early post-depositional structures and stratigraphic pinch-outs. Seals are provided by shelf or prodelta mudstones. Large volumes of hydrocarbons are produced from the Mesozoic Sussex and Shannon Sandstones and equivalent units, and from the Cardium Sand of Canada—all of which are regarded as good examples of ancient shelf deposits. Post-breakup progradation of passive margins (Fig. 8-1) superposes well-sorted shelf sediments on rift-stage deposits which include excellent lacustrine source rocks.

Processes on Shelves

The shelf processes most effective in transporting bed-load sediments are related to tides, storm waves, and wind-forced currents. Interacting with these are weaker currents related to temperature and salinity gradients, and Coriolis effects. Further complexity in shelf circulation patterns is introduced by seasonal convective sinking and overturn, upwelling of bottom waters, and downwelling. Tidal and storm processes are most influential on the inner shelf, whereas density stratification and oceanic circulation affect the outer shelf (Fig. 7-1).

Tidal Processes

Shelf tides are associated with semidiurnal, diurnal, fortnightly, and longer-term fluctuations in sea level resulting from gravitational attraction between the moon and the earth, and the sun and the earth. In major ocean basins these sea-level changes involve rotation of the tidal water bulge, or wave, around a central point of no tidal variation. These amphidromic systems may also exist in epeiric basins such as the North Sea. In the open ocean the tidal wave follows an elliptical path, but narrow rectilinear or reversed current patterns develop near basin margins. As the tidal wave is propagated from the open ocean onto the shelf it becomes asymmetric shoreward, and there is a tendency for flood velocities to exceed ebb

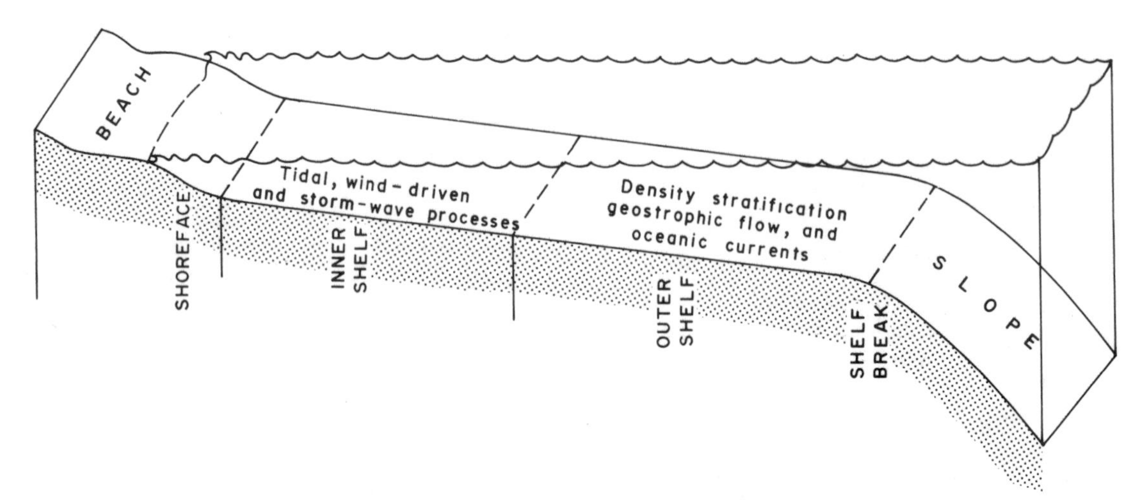

Figure 7-1. Shelf zonation and dominant processes. (Modified after Mooers, 1976 and Brenner, 1980).

velocities, resulting in net landward sediment transport; or it may move along the coast as a progressive tidal wave (Mofjeld, 1976). Tidal range and associated current velocities are highest fortnightly during spring tides, with additional reinforcement semiannually. Tidal-current velocity varies inversely with depth.

Resonant amplification of the tidal wave in semienclosed basins opening into a major ocean results in maximum tidal ranges of around 55 ft (17 m). These embayments are therefore dominated by high-velocity tidal currents. Enclosed seas such as the Mediterranean and Gulf of Mexico have very small tidal ranges, so tidal processes are negligible. But even these weak tidal currents may be significant when reinforced by wind-forced or wave-produced currents.

Broad shelves, too, have the effect of increasing tidal range as a result of transverse co-oscillation (Redfield, 1958). Many shelf seas, therefore, show progressively larger tidal range closer to the shore (Klein and Ryer, 1978). This shelf amplification is commonly superposed upon the effects of tidal-basin resonance.

Wave Processes

Waves affecting the shelf are mainly associated with storms, and are occasionally capable of disturbing the outer shelf to depths of more than 650 ft (200 m) (Komar *et al.*, 1972). Fair-weather waves have little effect on the shelf, except over the tops of shallowly submerged offshore bars.

Shallow-water waves involve circular orbital motion with a minor horizontal component. Wave surge over a shoaling bottom creates brief, high-velocity shoreward pulses as the wave crest passes, with longer but less intense seaward pulses beneath intervening troughs. Dominance of the shoreward pulses provides a wave-drift residual (Swift, 1969), causing net onshore sediment transport. The orbital wave motion also places bottom sediments in suspension where they can be transported by relatively weak unidirectional currents (Komar, 1976).

Threshold velocities for sediment movement on the Washington shelf were exceeded during five days of the year at the 550 ft (167 m) shelf break, and 53 days of the year at the 250 ft (75 m) mid shelf (Sternberg and Larsen, 1976). Sediments at the edge of the northwest Gulf of Mexico shelf are stirred by storm waves every five years or so (Curray, 1960). Wave oscillation ripples are long and branching, with sharp or rounded crests and symmetric profiles. Ripple troughs are flat or gently concave (Inman, 1957; Harms, 1969). Ripple spacing depends on wave characteristics and grain size, and indicates the near-bottom orbital diameter of waves (Inman, 1957).

Asymmetric, combined wave-current ripples with rounded, sinuous crests develop under the action of shoaling waves (Harms, 1969). These commonly preserve the effects of onshore wave-drift, so their crestlines tend to parallel the shore; but on some shelves they are markedly oblique to the paleoshoreline.

Storm waves affect shelf sedimentation out of all proportion to their infrequent occurrence. Onshore and offshore velocities associated with the orbital motion beneath steep storm waves are almost equal, so there is not the tendency toward net shoreward transport that characterizes fair-weather waves. On the contrary, storms tend to erode sediment from the beach and deposit it on the shelf and shoreface.

Storm Surge

Storm surge includes three components: forerunner, inverted barometric wave, and wind setup. Long-period swell, or forerunner, moves ahead of the storm, creating a slowly rising wave setup. Falling barometric pressure causes an additional rise in sea level accompanying the storm, and is compounded by wind-forcing to impound water against the shore, a phenomenon known as wind setup. Storm surge may be reinforced by the effects of resonance and bathymetry (Swift, 1969). The most significant storms are associated with tropical cyclones (hurricanes or typhoons), mid-latitude cyclonic low-pressure systems, and high-latitude meteorological disturbances.

Storm surge is capable of generating strong onshore currents, particularly where flow is concentrated over the tops of shelf sand ridges (Swift, 1970). Inundation of barrier coasts by storm surge, which attains a height of 13 ft (4 m) above normal sea level, is followed by gravitational outflow through washover channels and inlets. This storm-surge ebb carries shore-zone

Figure 7-2. Hummocky cross stratification in lower part showing upward convexity of laminae (arrowed), in inner-shelf deposits, Sydney Basin, Australia. The overlying pebbly sandstones show large-scale wave rippling, and are followed by storm-graded sandstones and siltstones with burrowed tops.

sediments seaward across the shelf, depositing them as sandy graded beds (Hayes, 1967a).

Transfer of storm-wave energy to shelf sediment becomes increasingly effective as water shallows. Storm-wave surge produces an irregularly undulating surface of low hummocks and swales, which becomes mantled with fine sand or silt settling from a sediment cloud suspended by passage of the wave (Harms, 1975). The resulting hummocky cross stratification (Fig. 7-2) extends between storm-wave base and fair-weather wave base (Walker, 1979).

Cyclic stresses produced by storm waves cause *in situ* liquefaction and resulting sand boils and mass flows (Field, 1982). Deformed and massive, poorly sorted shelf deposits in close association with large-scale hummocky cross stratification, for example, in the Sydney Basin of Australia, provide evidence of storm-induced liquefaction. Liquefaction of shelf sediments may also arise from earth tremors.

Density Differentiation

Density contrasts in water masses can be horizontal or vertical. Horizontal density differences arise from fresh-water inflow, insolation of shallow water layers, and contrasted evaporation rates. Lighter nearshore waters spread seaward, with landward flow of deeper shelf waters. During summer, the thermocline separating the two layers is around 100 ft (30 m) deep (Csanady, 1976). Wind stress affects only the upper layer directly. Excessive nearshore evaporation may cause seaward underflow of brines.

Suspended sediments are concentrated in near-surface and bottom (nepheloid) layers, and also within the thermocline (Drake, 1976). Estuaries trap a large proportion of fluvial suspended load, but where rivers discharge directly into the sea they effectively transport suspended particles (Emery, 1960; Drake, 1976). Mud accumulations below a depth of 65 to 175 ft (20–50 m) are

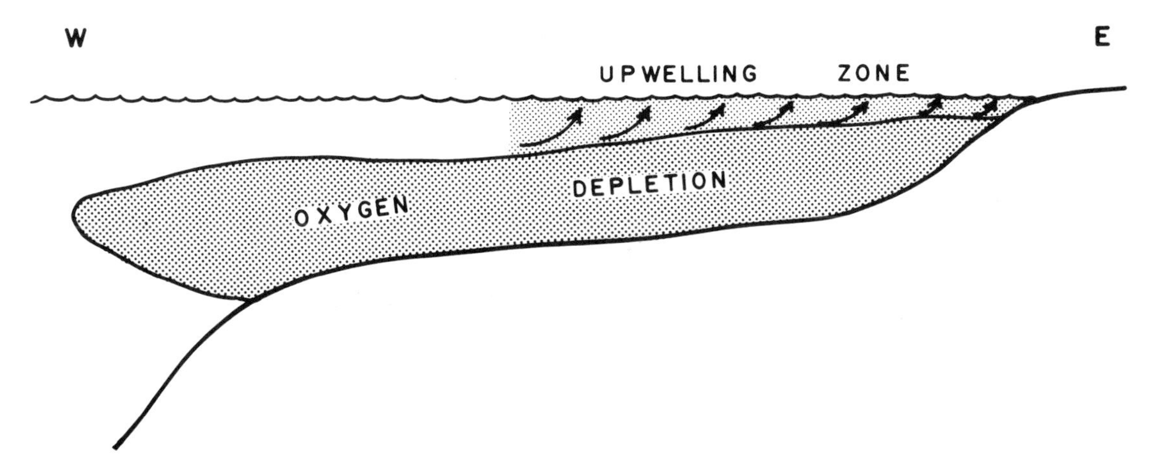

Figure 7-3. Oxygen depletion in shelf bottom waters in response to upwelling. (After Demaison and Moore, 1980.)

suspended during storms, and transported in part to the outer shelf and slope, countering the overall tendency for mud to be carried shoreward.

Density stratification involving seaward outflow of a tapering, warm, fresh-water plume, which overrides the denser, saline watermass, transports terrigenous clay across the shelf. These muddy plumes form after periods of high river discharge, and are generally asymmetric as a result of longshore drift. Flocculation speeds settling of clays. Weaker onshore flow of bottom water is capable of moving sediments entrained by wave oscillation.

Wind-Forced Currents

Unidirectional currents are generated by wind shear stress at the water surface, but deviate somewhat from the surface wind direction as a result of Coriolis forces. Wind-forced currents are particularly effective along storm-dominated coasts, where they act in conjunction with other storm-related processes. Weaker but more persistent wind-drift accompanies seasonal wind systems.

The most powerful currents accompany onshore or alongshore winds. Alongshore winds tend to produce a single-layer current system flowing approximately parallel to the shelf edge. Unidirectional currents of this type on the Washington–Oregon shelf exceed 80 cm/s (3 ft/s) and transport silt and sand (Sternberg and Larsen, 1976).

Onshore winds can produce a two-layer flow system, with the upper layer moving landward and the lower layer moving seaward (Forristall *et al.*, 1977). These currents operate in concert with wave processes, which assist in sediment entrainment, and may be reinforced or counteracted by tidal flow. Wind-forced currents are therefore important, not only on storm-dominated shelves such as the northern Gulf of Mexico (Murray, 1970) and western Atlantic (Swift *et al.*, 1969), but in tide-dominated basins such as the North Sea where they may augment or even briefly overwhelm the tidal component (Gienapp, 1973; Caston, 1976).

Upwelling is produced by displacement of surface water layers by offshore and shore-parallel winds. This upwelling is particularly marked on the east side of major oceans, or where associated with trade winds or monsoons (Ziegler *et al.*, 1979). Anoxic conditions commonly arise in the bottom waters (Fig. 7-3) beneath the nutrient-rich upwelled layer (Demaison and Moore, 1980). Shore-parallel winds can operate on a two-layer system (Fig. 7-4). Winds accelerate the upper layer, which may be deflected seaward by Coriolis force. Compensatory bottom-layer flow tilts the thermocline, which may eventually intersect the surface, causing upwelling (Csanady, 1976).

Bottom-return flow corresponds to the seaward-moving lower water masses which counter-

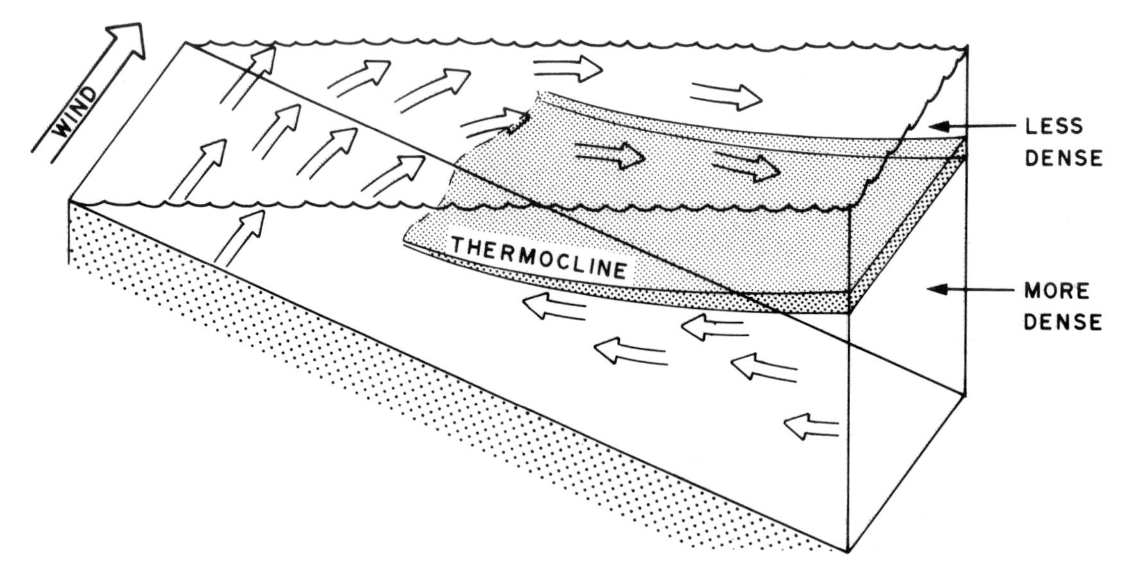

Figure 7-4. Stratification of shelf waters in response to temperature-related density differences and tilting of the thermocline by surface winds and compensatory lower-layer flow. (After Csanady, 1976.)

act the onshore movement of wind-forced waters during storms. This counter flow opposes the tendency of wave processes to transport sediment shoreward. Bottom-return flow during storms is probably the most effective mechanism for transporting sand from the shore zone across the inner shelf (Morton, 1981).

High current velocities, up to 200 cm/s (6.5 ft/s) occur at the height of a storm, or very shortly after the maximum wind stress is applied (Murray, 1970). Powerful offshore currents precede storm-surge ebb drainage from barrier coastlands. Bottom-return flow is most effective when relatively steep coastal landforms preclude overwash. It is therefore capable of transporting sand seaward off a variety of coastlands, including deltaic, strandplain, and barrier systems.

Semipermanent Ocean Currents

The outer portions of some shelves are affected by major oceanic circulation. The Agulhas Current of the western Indian Ocean produces persistent unidirectional flow southward along the outer shelf of southern Africa (Flemming, 1980). Velocities are sufficient to transport large volumes of bed-load sediments, some of which ultimately cascade down the heads of submarine canyons.

Oceanic currents of the northeastern Pacific shelf fluctuate seasonally, with reversals in direction. Although weak, these currents may be reinforced by storm waves and tidal or wind-forced currents during winter, thereby transporting suspended sediments (Johnson, 1978). The Panama and North Equatorial currents, too, are highly effective in transporting fine suspended sediments along the shelf.

Biological Activity

Biological processes on shelves are inversely related to the level of physical energy. Bioturbation is prevalent in areas of slow, fine-grained sedimentation, for example over much of the northern Gulf of Mexico shelf. On the other hand, in tide-dominated seas or on narrow, high-energy shelves, such as those of the southeast African margin where bed-load transport continues to the shelf edge, biological processes are unimportant. Coarse inner-shelf deposits off the Texas coast left by Hurricane Carla in 1961 still retain a graded character, but the primary structures have been obliterated by infaunal burrowing. On the higher energy California shelf, bioturbated sandy silt below a depth of 65 ft (20 m) contains isolated pods of shell debris (Howard and Reineck, 1981). On shelves receiving frequent pulses of coarse sediment, burrowing increases upward within each coarse unit. The trace fossils commonly reflect adjustment by the animals to alternating conditions of deposition and erosion (Goldring, 1964).

Faunal changes accompany an onshore textural gradation from mud to sand (Purdy, 1964). On many shelves, fine-grained seaward facies with a strong biogenic record grade through a transition in which physical and biogenic features are equally represented, to shoreface facies in which physical structures dominate.

The largest population of invertebrates is commonly found just below wave base (Howard, 1978), or at the inner-shelf/shore-zone transition. To landward there are suspension feeders in smaller numbers, while to seaward are few species and individuals of specialized deposit feeders. On the outer shelf generally, small horizontal grazing traces are dominant, becoming larger on the inner shelf where they are interspersed with sparse vertical burrows (Rhoads, 1975), the proportion of which increases toward the shore.

Shelf benthic fauna and their activities are strongly influenced by bottom water chemistry. Muddy shelf sediments beneath oxygenated bottom waters tend to be inhabited by bivalves, echinoderms, and polychaetes representing a combination of suspension feeders and deposit feeders. Suspension feeders leave distinctive trace fossils, while burrowing deposit feeders cause more intense disruption of the sediment. With lower oxygen content in the bottom waters, deposit feeders are represented by relatively inactive soft-bodied forms. With further reduction in oxygen content, only suspension feeders are present; these ultimately give way to anaerobic bacteria (Demaison and Moore, 1980).

The *Cruziana* ichnofacies of Seilacher (1967), which indicates a feeding pattern most characteristic of shelf depths, cannot be used as an unequivocal shelf bathymetric indicator because similar combinations of sediment type, energy, and water chemistry may be encountered at other depths. However, the gradual offshore change in the disposition of burrows such as *Rhizocorallium* (Ager and Wallace, 1970) from predominantly vertical in the nearshore, to oblique, and then horizontal, is a reliable indication of relative deepening from the shore zone across the shelf. Horizontal shelf burrows are typically lobate and branching with back-fill structures, and are accompanied by crawling and resting impressions. Other common shelf ichnogenera include *Thalassinoides, Teichichnus, Phycodes, Chondrites,* and *Pelecypodichnus.*

Marine organisms have an important effect on sedimentation rates on some muddy shelves. Very fine suspended sediments are concentrated as fecal pellets which settle more rapidly from surface water layers (Pryor, 1975). Although these pellets tend to disintegrate or become compacted beyond recognition, some remain intact and behave hydrodynamically like sand grains. Glauconite or chamosite are common diagenetic alteration products of fecal pellet precursors.

Shelf Facies

Shelf deposits vary in texture, geometry, and internal organization according to factors of sediment supply and dominant processes. Shelf sands are molded by tides and storms, whereas muddy shelf deposits represent slow settling of suspended sediments, which are reworked by organisms.

Sand Facies

Storm-dominated and tide-dominated sands are end members in the coarse-grained spectrum of shelf deposits (Brenner, 1980). Although tide and storm processes are readily distinguished on the basis of certain diagnostic criteria, for example, evidence of alternating flow directions as opposed to brief unidirectional pulses, their individual effects are often difficult to separate. This is because tidal currents, wave surge, and wind-forced currents frequently operate in combination (Swift, 1969). It is also possible for tidal currents at a particular locality to be essentially unidirectional, or for storm-related currents to be highly variable in direction.

Subtle distinctions can, however, be made between tidal and storm origins. For example, Precambrian sandstones in northern Norway ascribed by Hobday and Reading (1972) to an alternation of storm and fair-weather processes were subsequently shown to display a significant tidal component in addition (Johnson, 1977a).

The distinction between transverse shelf bedforms such as sand waves, and longitudinal bedforms such as ridges, may have genetic significance (Brenner, 1980). Longitudinal sand ridges are present on both tidal and storm-dominated shelves, but wind-forced currents, commonly superposed on tidal currents, are critical to their formation. Many shelf sand waves are of purely tidal origin, whereas others originate under semipermanent unidirectional currents.

Belderson and Stride (1966) documented a zonation of sediments and bedforms along tidal-current paths; these include sand ribbons, sand waves, and sheets and patches of gravel and sand. Farther out on shelves, reversing tidal currents give way to rotary paths, which favor larger longitudinal bedforms (Smith, 1969). But many of these sand ridges persist into shallower water off the mouths of estuaries.

Sand Ribbon Facies. Sand ribbons (Kenyon, 1970) are thin, longitudinal strips of sand that may be remarkably regular in width and spacing. Their origin is conceivably related to spiral vortices. Some sand ribbons are more complex, however, with superposed transverse bedforms (Flemming, 1980). Deposits of sand ribbons are unlikely to be volumetrically important in the rock record, but they undoubtedly contribute to coarse, basal transgressive layers.

Shelf Sand-Wave Facies. Shelf sand waves form transverse to the main tidal current paths. They may be symmetric or asymmetric, and their sense of asymmetry may be reversed as the tide changes (Granat and Ludwick, 1980). Opposed tidal flows of equal effectiveness produce symmetrical sand waves, whereas dominance of either the flood or ebb component produces a slipface between 3 and 35 ft (1 and 10 m) high inclined in the direction of net sediment transport. Whereas tidal currents dominate sand wave development in the North Sea, sand waves of the Virginia shelf migrate under conditions of storm-induced flow (Swift *et al.*, 1973, 1977).

Wave action prevents the formation of sand waves at very shallow shelf depths, the landward limit along the Dutch coast corresponding to the 18 m (60 ft) isobath (McCave, 1971a). The largest sand waves form just below this depth, and show the greatest complexity. Ripples and dunes are superposed on the upcurrent surface, and commonly on the slipface too, leaving a complex record of variable cross bedding (Langhorne, 1973). On the other hand, where dunes overlie the upcurrent surface only, the sand-wave deposits are likely to comprise large-scale solitary sets of planar cross beds overlain by smaller cosets with similar dip directions (Fig. 7-5). This sand-wave sequence overlies finer grained rippled and burrowed sediments (McCave, 1971a). Sand waves in deeper water are composed of finer-grained sand and are smaller and more regular in form. These, conceivably, leave a record of unidirectional planar cross beds produced by slipface accretion.

The maximum depth of sand-wave development corresponds to the limits of tidal-current effectiveness, and on some shelves they give way with depth to linear sand ridges. However, in the southern part of the North Sea the Dutch sand-wave field is downcurrent from major ridges (Fig. 7-6; Houbolt, 1968).

Diagnostic features of ancient sand-wave deposits (Nio, 1976; Johnson, 1978) include large-scale, locally unimodal foresets with variable, smaller-scale internal stratification, reactivation surfaces, clay drapes, and gravelly lenses on

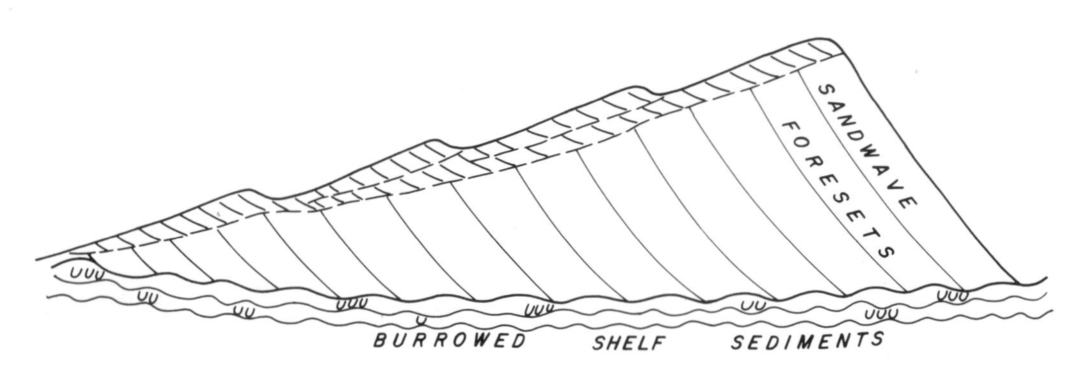

Figure 7-5. Sand-wave foresets overlain by smaller scale cosets produced by migration of overlying trains of dunes. (After McCave, 1971a.)

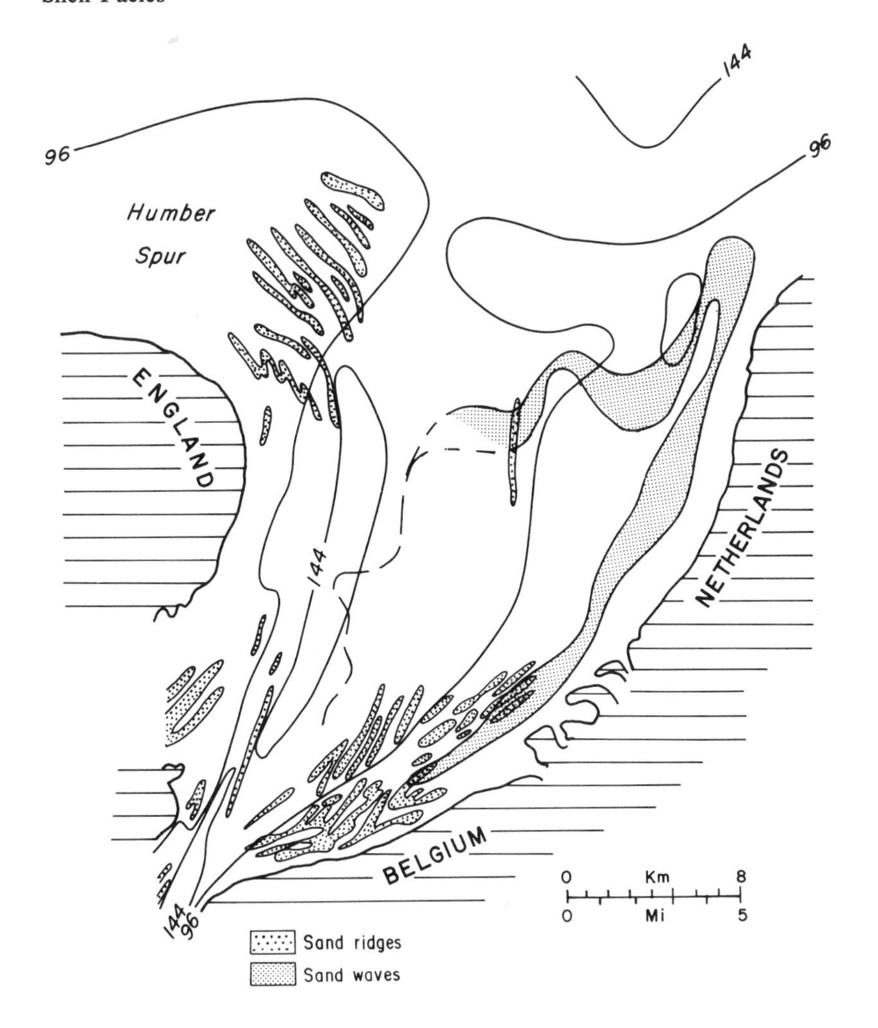

Figure 7-6. Numerous, shore-parallel sand ridges (dark stipple) of the North Sea, and the belt where sand waves (light stipple) are present. (After Houbolt, 1968 and McCave, 1979.)

foresets. However, dips on these major foresets are generally far steeper than the slipfaces of their modern counterparts (Walker, 1979). On a regional scale, coeval sand-wave deposits commonly show bimodal or polymodal patterns of major foreset azimuths reflecting the different areas traversed by ebb and flood currents. Alternatively, in some units such as the Jura Quartzite in the Dalradian of Scotland (Anderton, 1976), the major foresets are essentially unidirectional, but a tidal origin is suggested by minor foreset reversals reflecting weak opposed currents, and numerous mud-draped erosion surfaces.

Nio (1976) recognized distinct stages in the evolution of ancient sand-wave complexes (Fig. 7-7). Several small bedforms prograded one over the other as the water deepened to produce a thicker sand body; this body commenced to migrate by intermittent avalanching as a single sand wave as high as 20 m (65 ft). The sand-wave slipfaces were periodically modified by reversing currents, producing regularly spaced reactivation surfaces in the thick, solitary cross-bed sets. Next the sand waves were truncated and redistributed by more complex and variable currents, depositing planar cross beds with reversed dip directions.

Sand waves produced by semipermanent ocean currents are known only from the southeast African shelf (Flemming, 1978, 1980) where the Agulhas Current drives giant transverse bedforms up to 17 m (56 ft) high along the shelf. These sand waves give way seaward to sand ribbons and gravel. In contrast to the low-angle slipfaces of the North Sea sand waves, maximum dip angles on the Algulhas Current sand waves exceed 25 degrees. Similar bedforms related to intruding semipermanent ocean currents may have been responsible for large-scale, high-angle foresets in ancient shelf deposits (R.G. Walker, 1979). Very

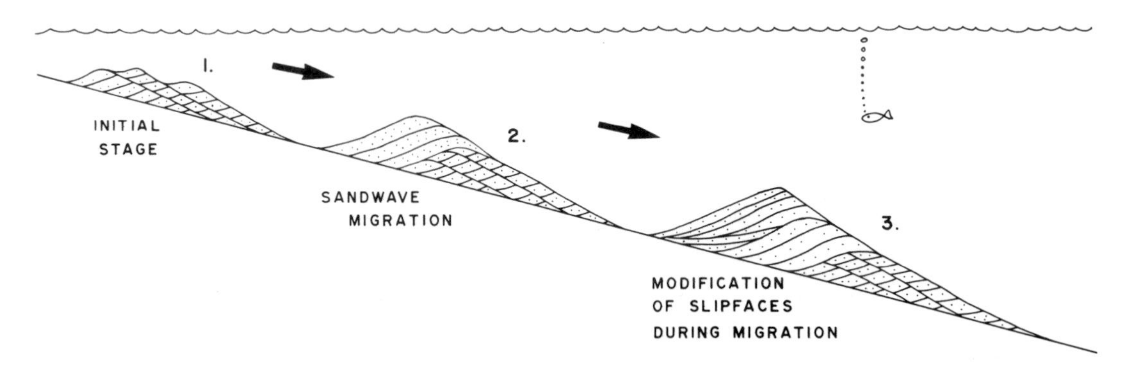

Figure 7-7. Stages in development of sand-wave complexes with progressive deepening during transgression. (Modified after Nio, 1976.)

large bedforms of this type are reflected in the Lower Paleozoic Cape sandstones of southernmost Africa (Hobday and Tankard, 1978); these indicate deposition by powerful currents flowing alongshore and obliquely offshore, with weak opposed flow attributed to tides.

Sand-Ridge Facies. Sand ridges form parallel to the dominant current flow, whether it be largely tidal in origin (Houbolt, 1968) or related solely to storm processes (Swift, 1976). Many of these features are thought to originate by inundation of shore-zone sands during transgression (Moody, 1964; Swift, 1976), producing a complex topography of ridges and swales parallel or somewhat oblique to the shoreline. Sand ridges in the tide-dominated North Sea (Fig. 7-6) are up to 40 m (130 ft) high (Houbolt, 1968), whereas the largest ridges off the storm-dominated coast of the eastern United States (Fig. 7-8) are less than 20 m (65 ft) (McKinney *et al.*, 1974). Currents flowing parallel to the ridges are thought to develop intense helical or two-dimensional crossflow patterns, which scour the swales and transport sand toward the ridge crests, contributing to ridge growth. Detailed studies by McCave (1979) show that some current directions do not conform with patterns predicted by helical flow, but do correspond with Eckman veering. The ridges are slightly oblique to tidal-current paths. Despite a likely origin in shallow water, many ridges are active under the existing shelf regime, representing a time-averaged response to a number of high-energy storms or tidal currents (Swift and Field, 1981). Ridges of the Atlantic shelf increase in spacing as the shelf deepens (Swift *et al.*, 1973, 1977).

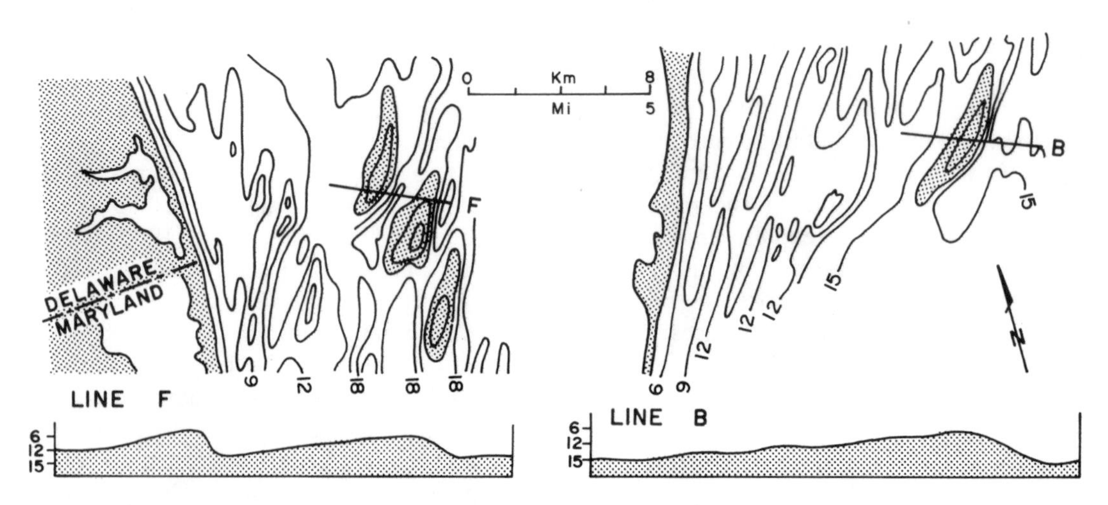

Figure 7-8. Sand ridges of the Delaware/Maryland inner shelf showing asymmetric transverse profiles. Depths in meters. (Modified after Swift and Field, 1981.)

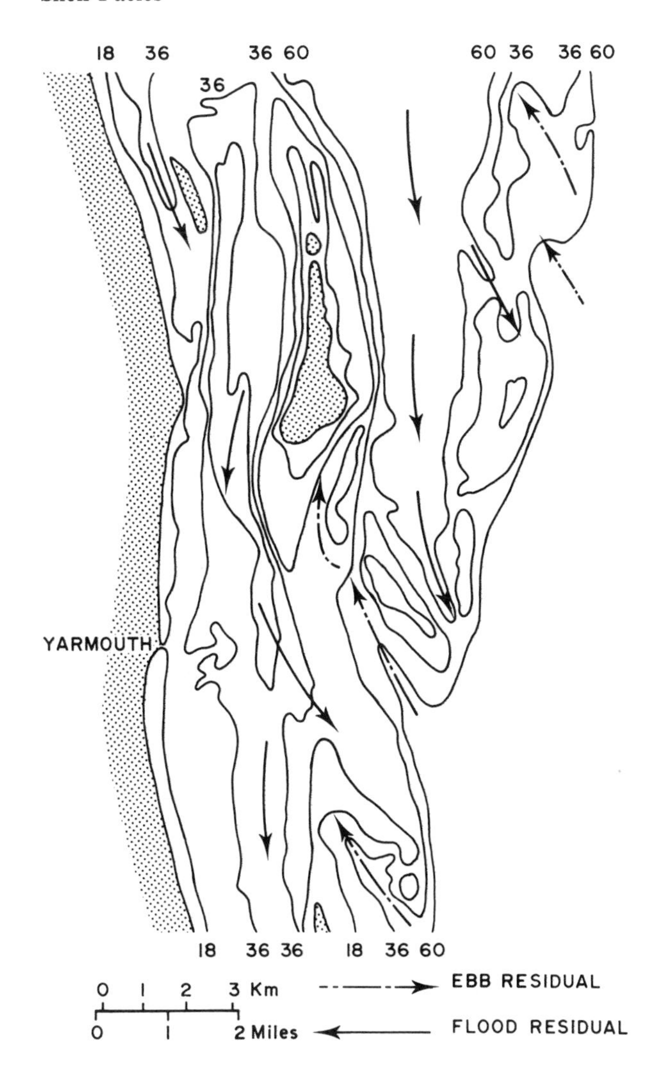

Figure 7-9. Ebb and flood channels off the East Anglian coast of Great Britain. Depths in feet. (Modified after Robinson, 1966.)

Current scouring leaves a gravelly lag in the swales, or may generate smaller transverse bedforms 2–5 m (7–16 ft) high which migrate parallel to the ridge axis. The tendency is for these bedforms to converge slightly toward the ridges and ascend obliquely toward the crests, corresponding to patterns of swale-to-ridge sediment transfer (Caston, 1972). Ridges generated purely by helical flow should be bilaterally symmetrical, with crestal sands somewhat finer than the lag deposit in the swales. Ridges generated by complex two-dimensional cross flow associated with rotary tidal currents will be symmetrical only if there is no ebb or flood current residual. Because of winnowing, crestal sands will be coarser than deposits in the swales (Swift, 1969). The two ridge types can develop in the same area. For example, on Georges Bank, rotary tidal-sand ridges (J.D. Smith, 1968) are superposed on larger, more gently rounded sand ridges, which, according to Swift (1969), may have been generated by helical flow.

In tidal seas, currents on either side of the ridges may flow in opposite directions. The ridge will migrate laterally in the direction of the weaker current. Offset flood- and ebb-dominated segments can produce a sinuous ridge, which may be subsequently breached at the bend to produce *en echelon* ridges (Caston, 1972; Brenner, 1980). In the Humber Spur region of the North Sea, linear ridges grade landward into sigmoidal ridges (Fig. 7-6) and the coastal ebb and flood channel systems (Fig. 7-9) described by Robinson (1966).

Internal stratification, although poorly documented in modern ridges, probably comprises trough cross bedding directed generally parallel to

Figure 7-10. Sand ridges with superposed smaller scale bedforms migrating obliquely up the gentle slope and at right angles across the steeper side. Complex internal stratification is inferred between the major sand-ridge foresets. (Modified after Houbolt, 1968.)

the ridge axis and slightly oblique to the shoreline. These sets might climb obliquely up low-angle (4–7 degrees) surfaces corresponding to the accreting ridge flank, and be directed perpendicular to the steeper foresets (Fig. 7-10). On asymmetric ridges, the deposits of transverse bedforms on the steeper ridge flank have higher preservation potential than those on the more gentle flank (Johnson, 1977a). Cross beds deposited by transverse bedform migration in swales may be approximately bipolar, although there is a tendency for only the stronger tidal component to be preserved. Deposits of fair-weather waves comprise wave ripples oriented parallel to the long axis of the ridges, but in most situations these will not be preserved.

The closest ancient counterpart to these modern linear sand ridges is documented by Johnson (1977a) from the Precambrian of northern Norway. Five upward-coarsening sequences of quartzose sandstone 2–10 m (7–35 ft) thick reflect alternating episodes of high- and low-energy sedimentation. Trough cross beds dipping parallel to the sand-body trend were deposited on the bar crest and flanks during storms, while sheets of sand accumulated contemporaneously in swales. Wave processes dominated during fair weather, when the bar surfaces were reworked and fine-grained sediments accumulated in the swales. One bar was dissected by mutually evasive ebb and flood currents.

A similar combination of features is in evidence in Mesozoic sand bodies of western North America (Davies and Brenner, 1973; Spearing, 1976; Campbell, 1973) and elsewhere, for example, in the Cretaceous of Nigeria (Banerjee, 1980). Paleocurrents tend to be parallel and perpendicular to the paleoshoreline (Fig. 7-11). The characteristic log patterns and sandstone geometries are illustrated in Figures 7-12 and 7-13. Many of the sand ridges were deposited over 100 km (60 mi) seaward of the shoreline and are surrounded by finer shelf sediments. Fair-weather processes played a role in shaping the relatively shallow bars of the inner shelf, but were insignificant on the outer shelf (Brenner, 1980). The sand ridges were supplied by shore-parallel sand transport related to powerful southward-moving storm or tidal systems, possibly augmented by semipermanent currents. Some ridges were severely channeled during storms and accumulated coquinoid channel-fill sands which coalesced laterally (Brenner and Davies, 1973). Others overlapped one another basinward, each forming during a temporary stillstand in the overall regression. Sand ridges on the outer shelf were breached by vigorous shelf-perpendicular flows which deposited lobate fringing sands analogous to subaqueous tidal deltas (Brenner, 1978, 1980).

Shelf Storm Facies. Shore-zone sediments are rapidly introduced to the inner shelf during storms. The best-known modern example is the graded gravelly sand bed of Hurricane Carla (Hayes, 1967a). The Carla deposit extends more than 15 mi (24 km) offshore, and is 8 to 10 in (20 to 25 cm) thick off Matagorda Island. It comprises a basal shelly lag gradationally overlain by sand in which the primary structures have been destroyed by burrowing in less than 20 years (Morton, 1981). These coarse sediments grade distally into interlaminated sand and mud, and finally into homogeneous shelf mud.

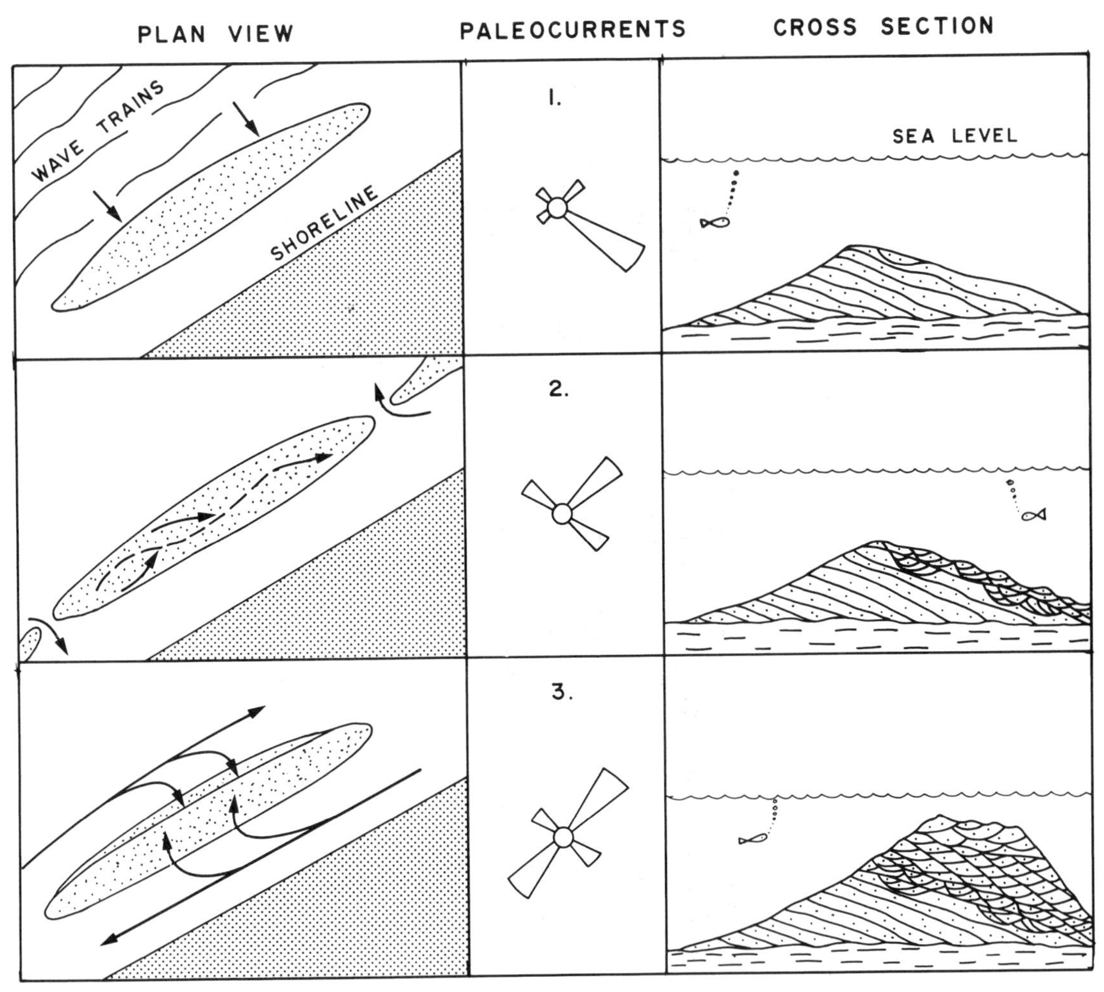

Figure 7-11. Inferred stages in growth of an ancient sand-ridge deposit under the influence of waves and shore-parallel currents. (Modified after Banerjee, 1980.)

A similar gradation is seen in considerable detail in exposures of the Cretaceous Washita Group of northeast Texas (Scott *et al.*, 1975; Hobday and Morton, 1983). The Early Cretaceous East Texas Basin was part of a storm-dominated protogulf. Sediments were supplied to the shore zone via small lobate deltas. A freshwater wedge may have contributed suspended clays to the shelf, but coarser sediments were carried seaward by bottom-return flow. Storm waves in the shoreface/shelf transition, in combination with seaward-directed currents, developed thick sequences of hummocky cross-stratified sands. Prolonged quiescent episodes are indicated by burrowed horizons consisting mainly of the spiral *Rosselia* separating successive sets of hummocky cross stratification. Graded units deposited below storm-wave base resemble those left by Hurricane Carla except that they are thicker and less bioturbated, possibly a consequence of more rapid sedimentation. A variety of shore-perpendicular sole marks and aligned *Turritella* in the basal lag confirm a seaward transport direction. The 0.2–1.3 ft (5–40 cm) sandy intervals are mainly parallel laminated, with subordinate very low-angle foresets merging up into ripple lamination, including climbing ripples. The upper parts of the sandstone are burrowed, with the intensity of bioturbation increasing into the gradationally overlying siltstones. Thin distal terminations of these sandstones are also heavily burrowed in the transition into shelf muds. Similar relationships have been observed elsewhere in ancient deposits (Howard, 1972; Goldring and Bridges, 1973; Cotter, 1975). Detailed process models interpreting ancient

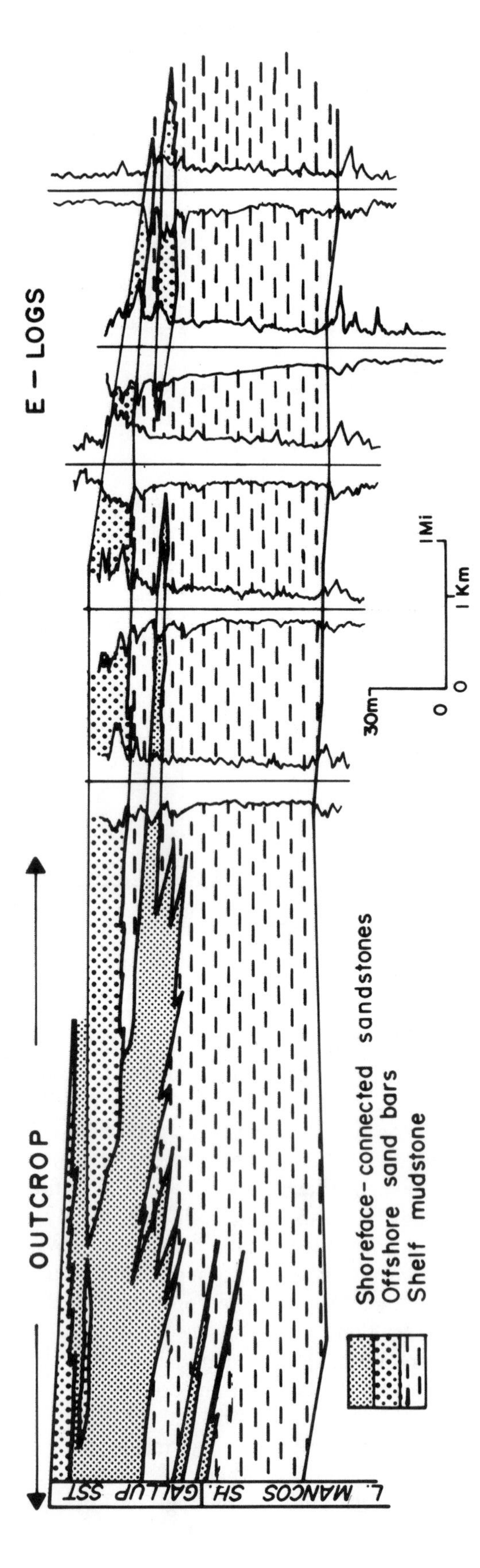

Figure 7-12. Stratigraphic cross section based on electric logs and outcrop through offshore bar deposits of the Cretaceous Gallup Sandstone, New Mexico. (After Campbell, 1973.)

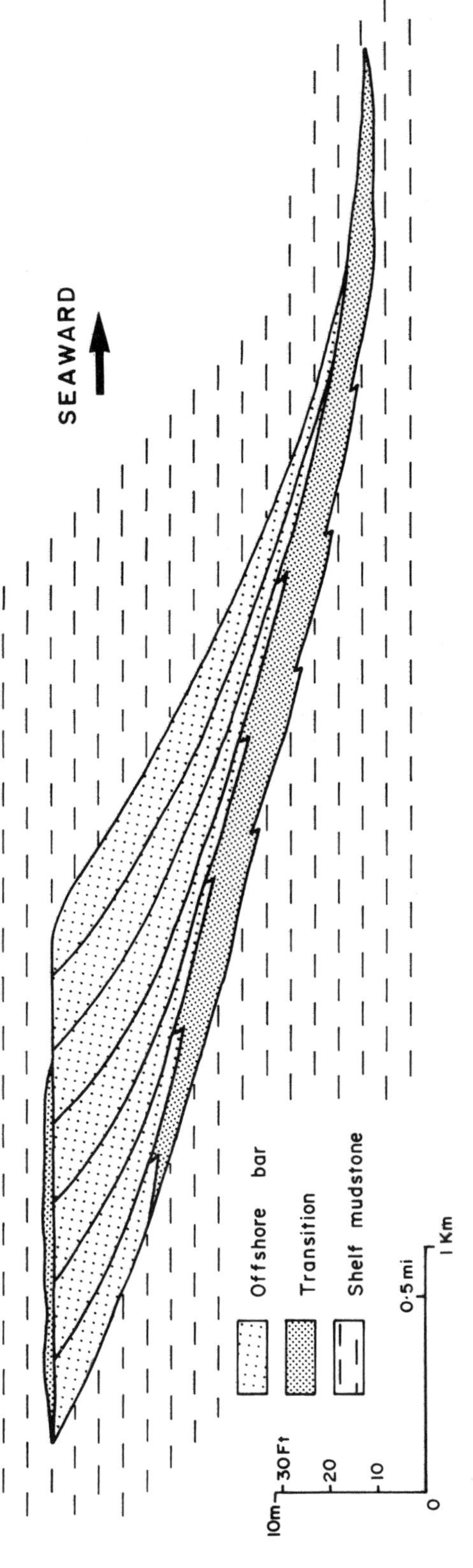

Figure 7-13. Diagrammatic cross section through an individual offshore bar sandstone body in the Gallup Sandstone. (After Campbell, 1973.)

sequences in terms of storm-wave processes and seaward-flowing currents penetrating the littoral energy fence have been proposed by Brenchley *et al.* (1979) and Walker (1979).

Mud Facies

Shelf mud, commonly with subordinate sand and shell, tends to be thoroughly bioturbated except where sedimentation rates are high or the bottom waters are severely depleted in oxygen. Modern shelf muds are most abundant in tropical seas (Hayes, 1967b), accumulating off river mouths and deltas where the particles are periodically resuspended and transported to lower energy environments alongshore and on the mid-to-outer shelf. Again, the modern situation is somewhat anomalous, with reworking of vast quantities of fine glacigene sediment over broad areas. On some shelves there is net onshore transport of mud (McCave, 1971). This may be in response to weak bottom currents compensating for fresh-water outflow (Bumpus, 1965), to onshore drift associated with upwelling, or to longshore transport from river mouths such as the Amazon. Wind and shoreline erosion contribute some fine sediment, but this amounts to less than 2 percent of the contribution from rivers (Drake, 1976).

McCave (1971b) has shown that mud can accumulate under relatively strong wave or current conditions provided the suspended sediment concentration is at least 100 mg/l or more. Furthermore, flocculation and biological agglomeration can accelerate the settling of clay particles by a factor of 10, and permit accumulation of clay that would not settle as single particles (Drake, 1976). At lower concentrations, waves, and to a lesser extent tidal currents, inhibit deposition of mud. On most shelves, the concentrations are normally 1 mg/l or less, so that mud will accumulate mainly in deeper or protected areas of low wave effectiveness. Off the Guiana coast, on the other hand, mud concentrations of up to 300 mg/l result in a shoreface and shelf mud belt as wide as 40 km (25 mi) (Van Andel and Postma, 1954). Concentrations of 260 mg/l off the Mississippi Delta decrease rapidly in the direction of nearshore current transport, with steady fallout of clay floccules (Scruton and Moore, 1953). Where shelf muds are disturbed by currents and waves, the highly turbid bottom nepheloid layer may flow as a density current (Komar *et al.*, 1974). However, where clay accumulates slowly in a stable environment, it compacts under its own weight, becoming far more resistant to erosion (Drake, 1976).

The outermost parts of many modern shelves are devoid of mud cover (Curray, 1965), possibly because of low suspended-sediment concentrations, incorporation into the substrate by burrowing organisms, or continued resuspension by minor turbulence. Organic disturbance too may cause resuspension (Stanley *et al.*, 1972).

Bioturbated, pelletiferous shelf muds beneath oxidized shelf waters contrast with the laminated organic-rich muds beneath anoxic bottom waters. Very high organic productivity in upwelled waters provides large quantities of organic material to the sea bed, causing oxygen depletion in the bottom water layer. The resulting high organic concentrations in the substrate, with carbon values of 20 percent or more, are ideal precursors to oil-prone kerogens (Demaison and Moore, 1980).

Some shelves and basins, for example, the vast Jurassic shelf sea of Europe, experienced periodic changes in bottom-water chemistry, producing alternations of strata with trace fossils reflecting:

(1) normal benthic assemblages,
(2) restricted deposit feeders,
(3) an inhospitable, oxygen-poor environment (Sellwood, 1971; Morris, 1979).

The Spectrum of Shelf Systems

A variety of rock-body geometries is possible in terrigenous shelf systems depending on shelf configuration, tectonic behavior, sediment supply, and basin hydrodynamics. The succession of shelf sediments is largely determined by the depositional phase: transgressive, regressive, or aggradational (Winker, 1980). But the sedimentary deposits within these three general categories may be complex and variable.

Modern shelves are in a state of disequilibrium, and there is commonly no straightforward relationship between shelf processes and shelf deposits (R.G. Walker, 1979). Nevertheless, comparing modern shelf sediments with the more numerous and complete ancient shelf successions permits recognition of a few general patterns.

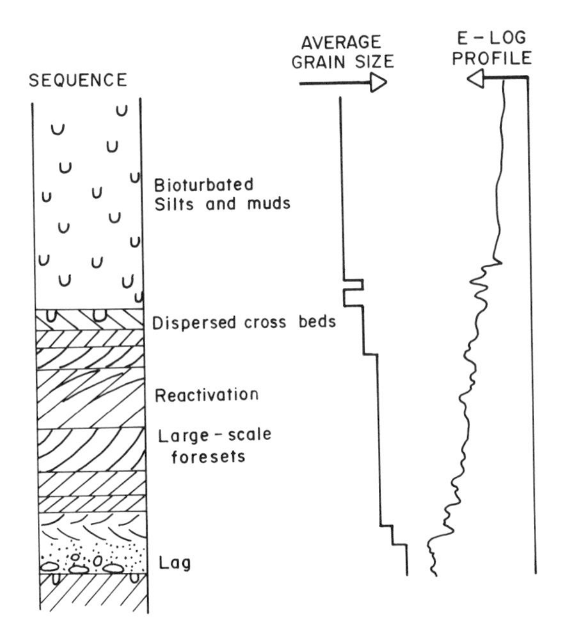

Figure 7-14. Schematic depositional sequence of a transgressive, tide-dominated shelf showing characteristic upward-fining pattern and corresponding SP or gamma log profile.

Transgressive Shelf Systems

Marine transgression generally reduces the contribution of sediment from rivers; tidal flats and estuaries act as efficient traps. Little sediment is provided by the retreating shoreline, but fluvial supply may continue despite accumulation in shore-zone traps. Transgression generally produces an upward-fining sequence from a basal, gravel-veneered surface of wave erosion, through sheetlike transgressive sands, to finer sediments representing successively deeper-water shelf accumulations. Many of these outer-shelf mudstones are phosphatic with abnormally high organic carbon content, and some are related to episodes of synchronous worldwide eustatic transgression (Parrish *et al.*, 1979; Vail and Mitchum, 1979; Demaison and Moore, 1980). Thick transgressive sequences attaining 650 ft (200 m) characterize tectonically active continental margins with adequate sediment supply, for example parts of the Cretaceous Circum-Pacific Belt (Bourgeois, 1980). In contrast, the sequences on stable shelves and epeiric platforms rarely exceed 300 ft (100 m) and may be as thin as 15 ft (5 m).

A distinction may be made between transgressive sequences of wave-dominated and tide-dominated shelves.

Transgressive tide-dominated shelf successions (Fig. 7-14) comprise basal gravels, massive sands, and cross-bedded sands overlain by finer grained sediments deposited by weaker currents in deeper water. The thickness and relative proportions of the various units depend largely on the rate of transgression and sediment supply. The deposits of sand ridges and tidal sand waves may be recognizable in the basal sands, for example the St. Peter Sandstone (Pryor and Amaral, 1971) and the Lower Greensand of the Isle of Wight (Nio, 1976). Large-scale, low-angle accretion surfaces along the inclined flanks contain cross beds inclined subparallel to the long axis of the sand bodies (Fig. 7-11), possibly ascending slightly toward the crest. The sand-wave deposits are expected to show an upward decrease in set thickness and internal complexity corresponding to increasing water depth.

There may also be evidence of sequential sand-wave growth, migration, and decay (Fig. 7-7). Very large cross beds corresponding to the phase of maximum sand-wave development are overlain by progressively thinner sets with a broad range in paleocurrent direction. The upper surface is moderately reworked by currents and by burrowing organisms, and is overlain by fossiliferous, bioturbated shelf silts and muds.

Transgressive storm-dominated shelf successions, too, are typically upward fining (Fig. 7-15); a basal erosion surface reflecting retreat of the shoreline is overlain by a sheetlike transgressive shore-zone deposit, which is commonly thin,

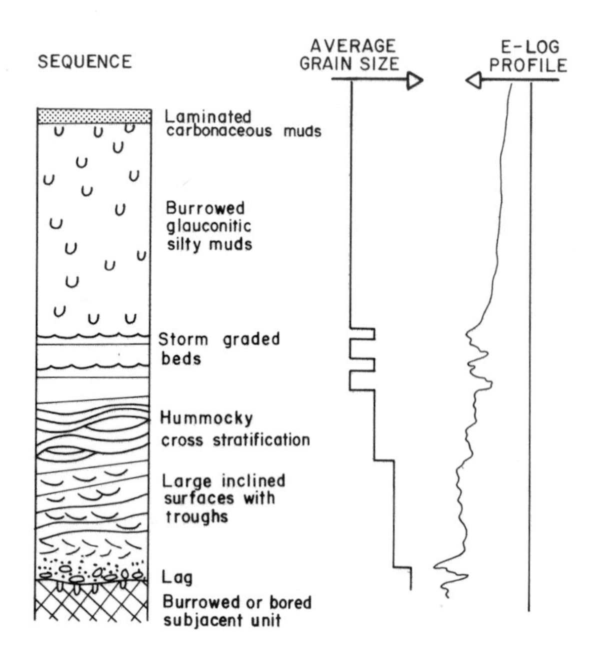

Figure 7-15. Schematic depositional sequence of a transgressive, storm-dominated shelf.

gravelly, and bioturbated. Long vertical burrows may extend into underlying units beneath the transgressive unconformity. Some relict shore-zone sand bodies, modified by passage through the surf zone, maintain a shore-parallel trend, but may be molded into sand ridges by shelf processes.

Other transgressive storm-dominated shelf sequences lack evidence of sand-ridge growth in their lower parts. The sequence above the basal transgressive lag merges upward through silts and muds containing thick lenticular sands with multistoried sets of hummocky cross stratification, into thinner, graded, parallel-laminated storm sands interspersed with shelf muds. Typically the graded sandy beds show an erosive, gravelly or shelly base, and finer, burrowed tops.

The hummocky cross stratification records storm reworking contemporaneous with offshore transport by bottom-return flow. Thinner sets are deposited as the water deepens, with graded, parallel-laminated units below storm-wave base. Biological reworking of the tops of these graded beds increases upward through the succession as the water deepens. Eventually, the very thin, graded silts, representing the distal extremities of the graded storm units, give way seaward to homogeneous shelf muds. The latter tend to be thoroughly reworked by burrowing, and contain abundant fecal material and shell debris, commonly bored during prolonged stable episodes below wave base.

The deepest shelf deposits at the top of the succession may comprise glauconitic or phosphatic muds, or laminated muds very rich in organic carbon and diatoms. Abnormal concentrations of uranium, nickel, and copper are present (Demaison and Moore, 1980).

On very broad, shallow shelves that are starved of clastic input during transgression, and on which wave energy is ineffective even during storms, the transgressive sediments are thin and composed mainly of pelletiferous silts and muds with little vertical pattern. Examples include the Weches and parts of the Reklaw Formations of the Tertiary Gulf Coast Basin.

Prograding Shelf Systems

Prograding shelves will in most cases generate an upward-coarsening sequence from burrowed shelf muds to cross-bedded or storm-graded sands. Without contemporaneous subsidence, shelf progradation is a self-limiting process incapable of building up great thicknesses of sediment (Winker, 1980). The total thickness of ancient shelf sediments from upper slope to shore zone provides some general indication of the original water depth, where effects of compaction and

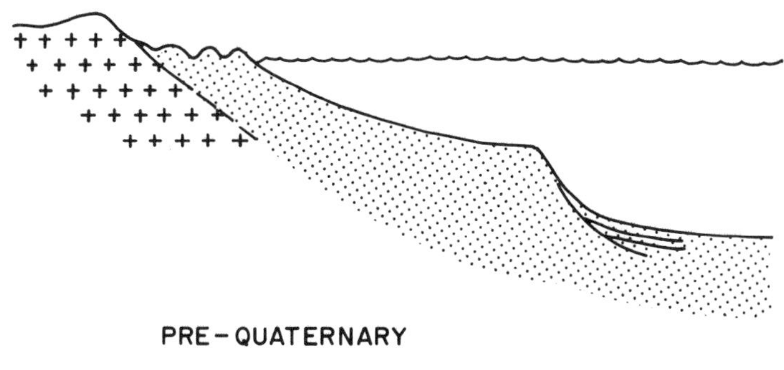

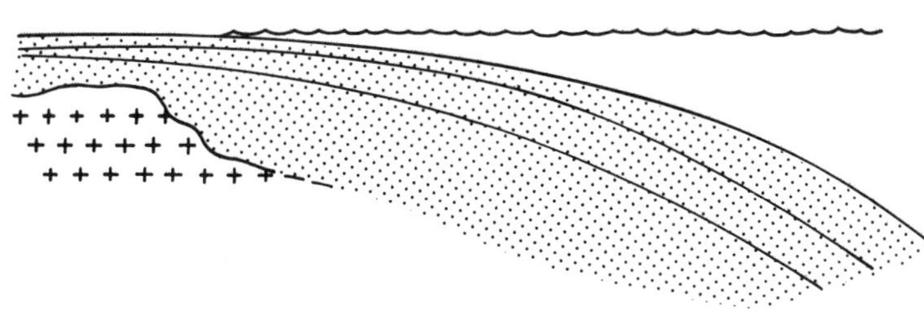

Figure 7-16. Comparison of Quaternary shelves with typical progradational pre-Quaternary shelves. (After Swift, 1969.)

syndepositional subsidence can be discounted. Because of their youthful stage of development, few useful observations can be made on the progradational sequences of modern shelves.

The climax shelf deposit (Swift, 1969), which is not represented in the Holocene, comprises a seaward-thickening prism (Fig. 7-16) in a state of dynamic equilibrium with factors of sediment supply and sea-level rise. A Cretaceous example from the Atlantic coastal plain grades laterally from clean, rippled and cross-bedded nearshore sands through finer, burrowed clayey inner-shelf sands to bioturbated mixtures of clay, silt and sand, and finally to mud of the outer shelf (Swift, 1969). Seaward fining was accomplished through progressive sorting by storm processes. This lateral gradation is characteristic of many other ancient shelf deposits, which, in accordance with Walther's Law, also show well-defined upward coarsening, and increase in bed thickness.

Two dominant vertical patterns are evident in ancient progradational shelves.

Prograding storm-dominated shelf successions (Fig. 7-17) are approximately the reverse of the transgressive succession, but tend to be thicker on average, depending on original shelf depth and rate of subsidence. Sands are restricted to inner-shelf deposits corresponding to the coarse-grained upper part of the succession. Much of the total thickness is made up of bioturbated, shelly shelf muds with graded silty laminae. Shelves with anoxic bottom waters accumulate laminated, carbonaceous muds. Upward passage into graded, parallel-laminated storm sands is very gradual and involves progressively thicker and more closely spaced sands. Parallel lamination gives way to hummocky cross stratification and ultimately to broad, low-amplitude concave lamination corresponding to the shore-zone transition (Walker, 1981). The classic example described by Hamblin and Walker (1979) from the Jurassic of the Canadian Rockies shows well-defined subfacies related to wave and associated processes in the shallowing shelf-to-shoreline transition (Fig. 7-18; R.G. Walker, 1979):

(1) Interbedded siltstones and graded sandstones deposited below storm-wave base by seaward-flowing bottom currents accompanying storms. This 50 m (165 ft) sequence becomes sandier upward.
(2) Sharp-based, thicker sandstones with hummocky cross stratification, emplaced by seaward-flowing currents, and simultaneously remolded by storm waves.
(3) Gently dipping, parallel-laminated beach sandstones.

Similar facies in the Permian of the Sydney Basin of Australia include a fourth unit between

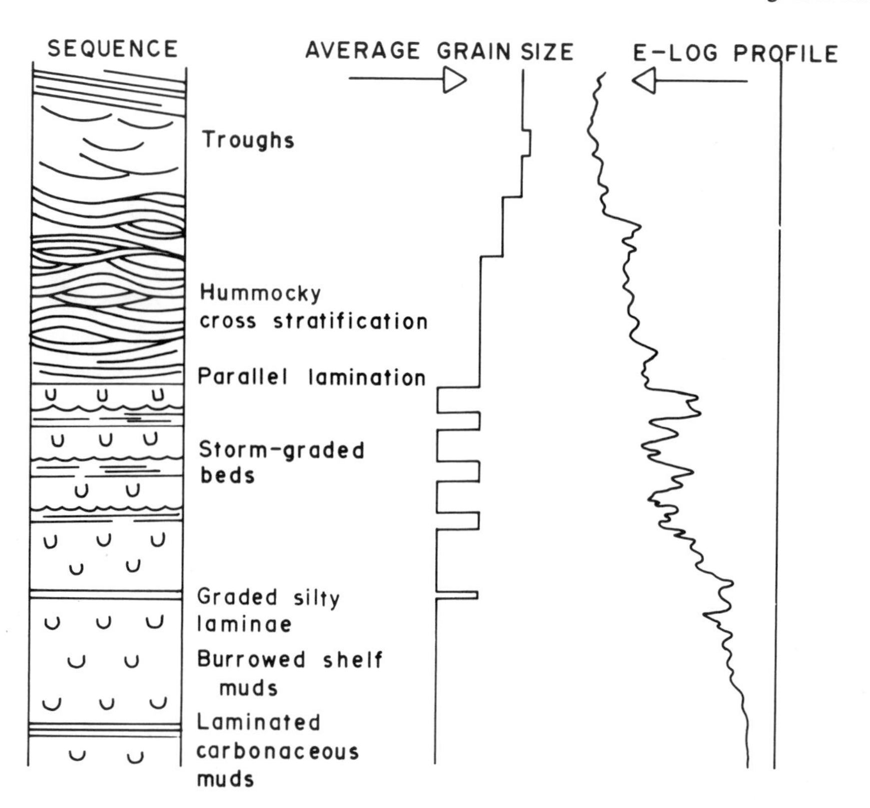

Figure 7-17. Idealized vertical sequence and log response of a prograding, storm-dominated shelf.

the hummocky cross-stratified and beach deposits. This comprises fine-grained sandstone with closely spaced, low-index wave ripples produced by weak wave activity above fair-weather wave base.

Prograding mixed-energy shelf successions (Fig. 7-19) reflect a combination of storms and tides, and possibly even oceanic currents, and include some of the best-known ancient shelf deposits (Brenner, 1980).

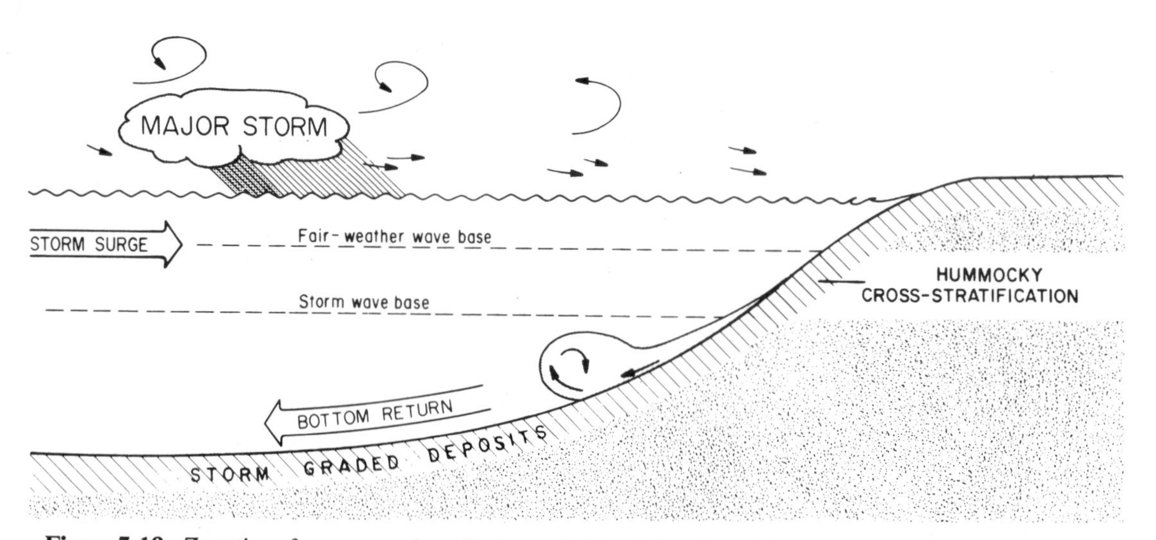

Figure 7-18. Zonation of processes along the inner shelf and shoreface, with progradation producing a vertical sequence of storm-graded sands and muds overlain by hummocky cross stratification and beach-foreshore sands. (After R.G. Walker 1979.)

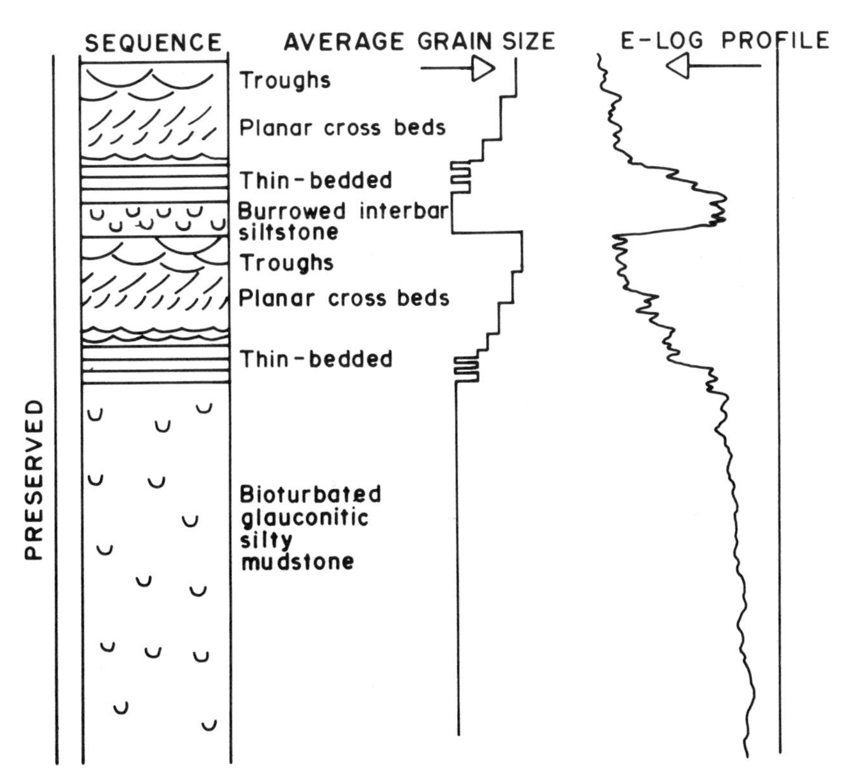

Figure 7-19. Idealized vertical sequence and log response of a prograding, mixed-energy shelf involving additive storm and tidal processes.

In tidally dominated seaways such as the North Sea, sediment transport rates are at times accelerated by a factor of twenty during storms (Johnson and Stride, 1969). Many ancient examples of shelf deposits are probably also a product of additive tidal and storm regimes. Mesozoic epeiric seaways of western Canada and the United States opened toward major ocean basins. For example, the Cretaceous Powder River Basin of Wyoming (Brenner, 1978) had a configuration resembling the modern Straits of Malacca (Keller and Richards, 1967). Superposed on the powerful and consistent tidal currents in the straits is a northwestward drift caused by monsoonal storms on neighboring seas (Coleman *et al.*, 1970). The elongate, embayed, or funnel shape of many ancient epeiric seaways would have been similarly conducive to interaction of tidal and storm processes. The relative importance of these processes varies even within a single basin, but the general sequence resulting from progradation (Fig. 7-19) is the same.

Deposits of the North American Mesozoic seaways typically comprise thick, basal shelf muds overlain by between one and five upward-coarsening sandstone units, each between 6 and 65 ft (2 and 20 m) thick. The sand bodies were aligned parallel to the active paleoshoreline and subparallel to the seaway axis. As in the Straits of Malacca some sand ridges grew far offshore. The thin-bedded lower parts of each unit consist of glauconitic sand alternating with clay layers. Some of the thicker sandstones are typical storm deposits with graded bedding and rippled, burrowed tops. The thinner sandstone intercalations contain combined wave-current ripples indicating intermittent longshore transport. The coarser, upper parts consist of moderately burrowed, cross-bedded glauconitic sand with clay drapes. Cross beds are mainly unidirectional and, less commonly, bipolar parallel to the sand-ridge axes in a response to storm-enhanced tidal current transport. Some units change upward from predominantly low-angle planar cross beds to troughs, consistent with shoaling and higher energy (Spearing, 1975, 1976). The vertical sequence can be interrupted by coquinoid sandstones deposited in transverse channels. Many of

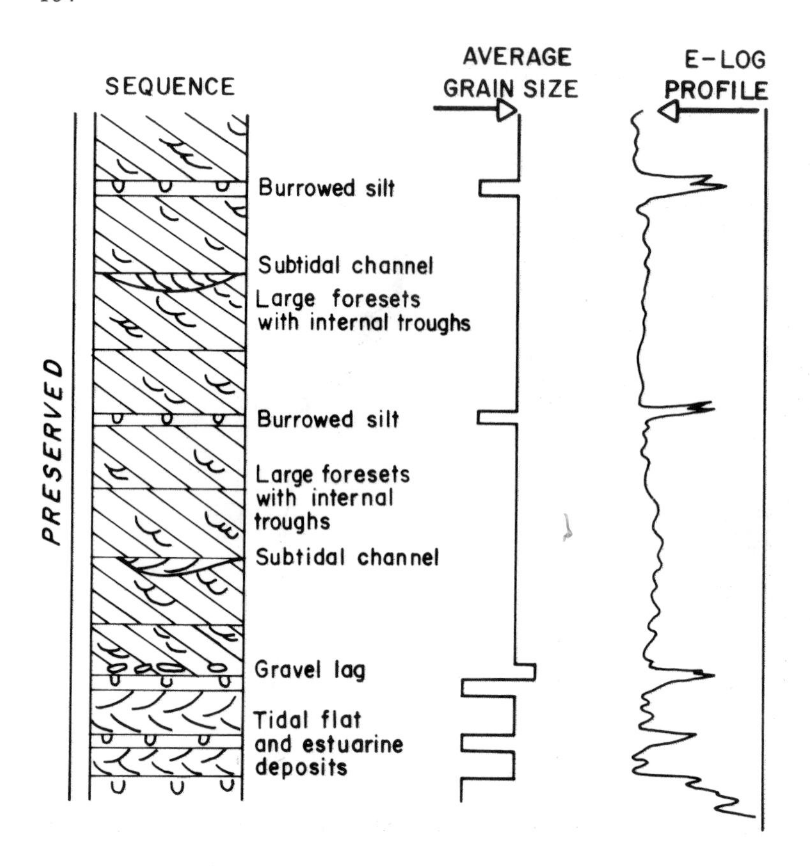

Figure 7-20. Idealized vertical sequence resulting from sand accumulation balanced by subsidence of an aggradational shelf dominated by storm and tidal processes. The log pattern is typically blocky.

these elongate, lenticular sandstone bodies are abruptly overlain by muddy interbar or shelf sediments.

Aggradational Shelves

Some thick shelf successions reflect a dominance of prolonged aggradation or facies stacking, with water depths fluctuating within shelf limits. Thick successions of starved-shelf mudstones reflect slow sedimentation during gradual sea-level rise, which may be global in extent, or related to local subsidence, commonly associated with growth faulting. Alternatively, aggradational shelf deposits may include large proportions of silt and sand resulting from continued sediment supply and strong traction currents. Some composite shelf sequences reflect intermittent progradation and transgression, with sandy inner-shelf facies alternating irregularly with thick deposits of more distal muds and silts. These composite sequences show an overall tendency toward shelf aggradation, with no evidence of emergence or drastic deepening. Log patterns include upward-coarsening, upward-fining and irregular, serrated sequences. Such successions are common in epeiric platforms. Other aggradational successions are sandy and of enormous thickness.

Aggradational mixed-energy sandy shelf successions commonly comprise thick, cross-bedded quartzarenites (Fig. 7-20) showing evidence of both storms and tidal processes. The thickest documented example is in the Jura Quartzite of Scotland (Anderton, 1976), where shelf sands deposited by storm events reinforced by reversing tidal currents aggraded to a thickness of 5000 m (16,500 ft). These deposits become thinner and finer grained in an offshore direction.

Aggradational shelf quartzarenites in the Cambro-Ordovician Cape Supergroup of southern Africa attain thicknesses of thousands of meters in two separate fault-controlled shelf basins in the south and east (Hobday and Tankard, 1978; Hobday and von Brunn, 1979). In the southern Cape they transgressively overlie tidal-flat and destructive-barrier facies; in the east they accumulated seaward of tide-reworked fan deltas. Tidal sediment transport permitted rapid shelf accumulation to keep pace with rift subsidence. There is no discernible change in the pattern of tidal sand ridges and channel-fill deposits through a thickness of 2000 m (6500 ft). Sediment transport directions are predominantly offshore, parallel to the axis of the rift basin, which opened into a major ocean.

Other examples of thick, laterally extensive shelf sandstones are also mainly of Proterozoic and early Paleozoic age (Banks, 1973; Johnson, 1977b). This may stem in part from the greater availability of sand introduced to marine basins by the sandy fluvial systems then prevalent. Furthermore, it is conceivable that meteorological disturbances may have been more severe in Precambrian times.

Many ancient shelf seas were very broad, and faced major ocean basins, factors that would have produced large tidal ranges and velocities (Klein, 1977). There has also been the suggestion that early Precambrian tides were significantly larger (Cloud, 1972), although work by Eriksson (1977) on Archean deposits suggests a range comparable to that of today.

Chapter 8

Terrigenous Slope and Basin Systems

Introduction

Slope and basin systems originate in relatively deep water beyond the shelf break. Although this environment encompasses a bathymetric range from upper slope at depths of several hundred feet to the abyssal plain at several thousand feet, many economically important ancient slope systems originated at more modest depths along the margins of shallow intracratonic seas (Galloway and Brown, 1972); others originated ahead of prograding oceanic delta systems and were carried to substantially greater depths by ongoing subsidence and growth faulting (Caughey, 1981; Galloway *et al.*, 1982).

The modern slope commences at the shelf break between 150 and 1000 ft (45–300 m) below sea level and is typically inclined at 1 to 3 degrees, locally approaching 10 degrees. Slope geometry varies according to tectonic setting, progradational history, and erosional modification. Prograding slopes may be dammed behind salt diapirs and roll-over anticlines, and displaced by slumping and contemporaneous faulting. Other slopes are nondepositional or predominantly erosional with bedrock exposed (Cleary and Conolly, 1974). Indeed, because of the rapid postglacial rise in sea level, the erosional character of many modern slopes is quite unrepresentative of most ancient slope systems. The upper slope today is typically an area of sedimentary bypassing with local progradation and canyon filling. Origins of submarine canyons are diverse (Shepard, 1981). Many submarine canyons, such as the Mississippi Canyon, which are located off the mouths of major rivers, originated during Pleistocene stages of lowered sea level, but others such as the famous Scripps Canyon, are totally unrelated to fluvial systems and apparently were scoured by gravity-flow abrasion, a process still operative in some.

On passive continental margins the sedimentary wedge of the upper slope is commonly separated from the thick accumulations of the lower slope and rise by a narrow zone of thinning over a basement high (Fig. 8-1). The lower slope and more gently inclined continental rise receive periodic terrigenous influx, which gives way to a dominance of grain-by-grain pelagic sedimentation of the flat abyssal plain. Some isolated abyssal basins are subject to gravity resedimentation from allochthonous and biogenic buildups, and may receive sporadic fine-grained terrigenous sediments.

Continental rises and slopes represent the new frontier of offshore oil exploration, yet they remain one of the least understood parts of the earth (Heezen, 1974). They constitute a vast area exceeding all of the onshore sedimentary basins. Ancient slope and basin systems commonly provide the ideal relationship between hydrocarbon-rich source rocks and sandy reservoir rocks. Most reservoir units are draped with muddy seals deposited during temporary cessation in aggradation. Updip pinchouts of slope sands into marine muds produce excellent stratigraphic traps, whereas growth faults, slumps, and diapirs generate structural traps.

Slope Processes

Subaqueous slope systems are characterized by the dominance of gravity mass-transport and density underflow processes and their depositional products. This is particularly true for sand-sized and coarser sediment within the slope system. Additional processes include pelagic settling and the flow of deep thermohaline and reversing tidal currents. Such permanent currents primarily affect suspended-load sediment but may be responsible for considerable morphologic modification of the slope and adjacent basin floor.

Because gravitational potential energy is the principal driving mechanism for down-slope mass-transport and density underflow, slope systems are distinguished from shelf, deltaic, and shore-zone systems by their inherent tendency to deposit coarsest sediment at the bottom of the

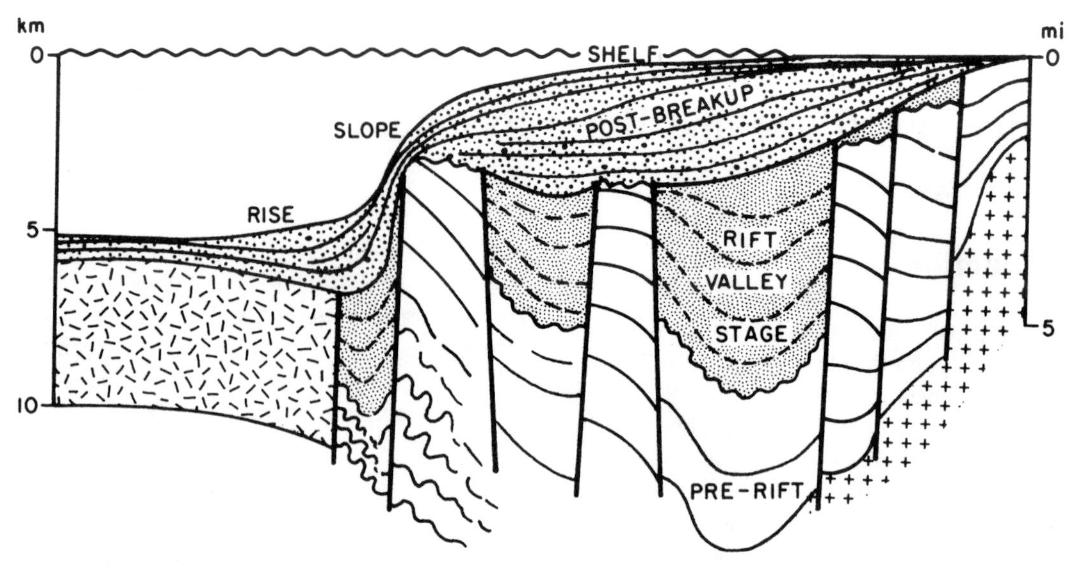

Figure 8-1. Generalized cross section through a passive continental margin, showing the accretionary shelf, continental slope, and continental rise. (After Falvey, 1974.)

depositional sequence and in topographically lowest areas of the sediment dispersal system. The upper slope is typically a zone of sand remobilization and bypass. In addition, the shelf edge and upper slope are a focus for erosional and channel-cutting processes, whereas the lower slope and adjacent basin floor are sites of sediment aggradation (Stanley and Unrug, 1972). In actively prograding basin margins, sedimentation obviously must predominate on the upper slope as well; nonetheless, channeling and bed-load bypass are prominent within the upper-slope deposits.

Other slope processes are increasingly significant where terrigenous influx is slow or lacking. Semipermanent traction currents rework sediment along the base of the slope. Airborne dust and muds from low-density riverine plumes settle along with organic material from the surface water layers as a continuous rain. Where the bottom waters are oxidizing, the fine-grained sediments are thoroughly reworked by organisms, but where anoxic conditions prevail, delicate lamination may be preserved in the organic-rich muds.

Fluid density contrasts, which induce downslope displacement of a discrete water mass, may originate from entrained sediment (sediment-density currents) or from temperature and salinity differences (fluid-density currents). Thus, there are three categories of gravitational resedimentation mechanisms:

(1) Gravity mass transport, where gravitational potential energy is converted to kinetic energy, and sediment is moved downslope, possibly setting the surrounding fluid in motion. Sediment may move as a coherent mass or be thoroughly dispersed in water.
(2) Salinity-related density currents causing overflow, interflow, or underflow.
(3) Thermal density currents, generally involving sinking and equatorward underflow of dense polar waters. This process is augmented by salinity differences.

Gravity Mass Transport

Gravity mass transport ranges from discrete free-falling rock masses, or olistoliths, to slow-moving, dilute, muddy suspensions or nepheloid layers (Fig. 8-2). Downslope movement occurs when the gravity shear stress exceeds the shear strength of the sediment, a function of intergranular friction and cohesion (Rupke, 1978). Thus slope failure may be brought about by a reduction in shear strength of the sediment, or by an increase in shear stress. The critical angle at which the shear strength of the sediment pile is exceeded

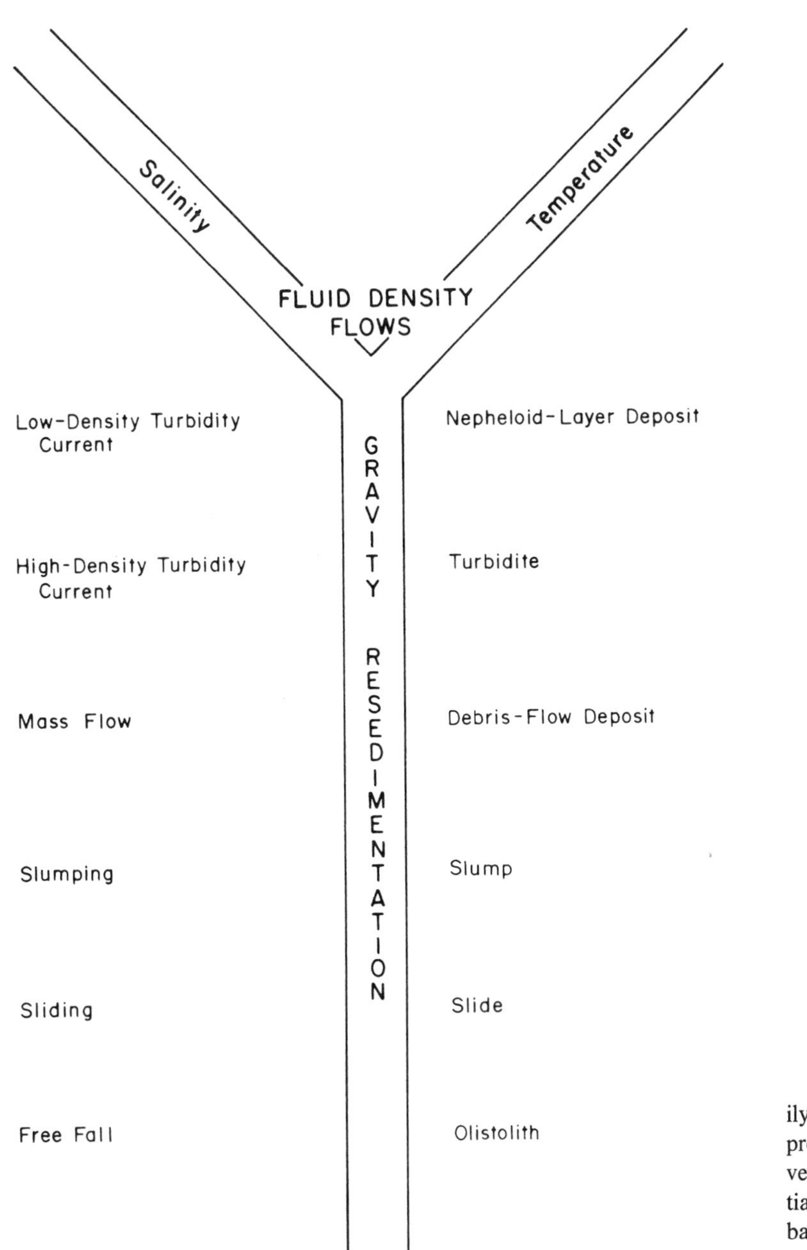

Figure 8-2. Schematic family tree of submarine slope processes produced by the conversion of gravitational potential to kinetic energy. From base to top, the transported material is less cohesive and increasingly fluid.

depends on sediment composition, texture, rate of accumulation, and conditions of pore-water escape. Fine-grained prodelta/slope deposits off the Mississippi are unstable at angles of less than one degree (Coleman and Pryor, 1980), whereas coarse sediments of the Gulf of California maintain a stable slope of up to 7 degrees (Weser, 1975, p. 8).

Gravity mass transport down the continental slope is preceded by sediment accumulation near the head of a canyon or at the shelf edge, or requires channelized bypassing of the shelf break. Such conditions would have been favored by fluvial discharge near the shelf edge during Pleistocene low sea-level stands, and at other periods of active continental shelf progradation.

Under existing sea levels, sediment is carried across submerged shelves into gravitationally unstable situations by oceanic currents, longshore drift, tidal currents, and storm-generated currents.

Ocean currents that sweep across the outer shelf (Flemming, 1978), and wave-induced longshore currents on the inner shelf, may contribute sediment to submarine canyons, whence it is funneled downslope. Powerful tidal currents are very effective in carrying sediment to the slope edge. Tidal sand ridges off the Bengal Delta converge toward a large submarine canyon which feeds the deep-sea Bengal cone (Weser, 1975, p. 11). On a smaller scale, tidal currents carry sediment from the mouth of the Columbia River to the head of a fan. Seaward-flowing bottom currents accompanying storms are probably an important process on narrow shelves. Rivers such as the Congo discharge directly into the heads of canyons. This may generate semicontinuous underflow or, more likely, sediment may accumulate in the upper canyon and move downslope during floods.

Olistoliths and Olistostromes

Large extrabasinal blocks, commonly of gigantic proportions, are produced by translation along narrow, tectonically active shelves. Rock falls may be accompanied by sliding and slumping. The coherent exotic rock bodies are enclosed in a matrix of younger slope sediments. The chaotic deposits of debris flows and related gravity resedimentation which incorporate the exotic clasts, or olistoliths, are referred to as olistostromes or sedimentary mélange.

Gravity Slumps

Subaqueous slumping involves shearing and rotation along an inclined glide plane, which may correspond to a bedding plane along much of its length (Lewis, 1971; Embley and Jacobi, 1977). Backward tilting is characteristic of the semiconsolidated slumped mass, and a series of small imbricate thrusts may be preserved at the toe. Alternatively, the slump toe may be a chaotic, jumbled mass. Where the displaced mass is thoroughly consolidated and maintains its form, it may be referred to as a slide. Slumps, on the other hand, show a variety of internal deformation features, although they retain some internal coherence (Fig. 8-3). Slump fold hinges tend to be strike-aligned, but friction may distort the axial trends (Rupke, 1976).

Figure 8-3. Deformed slump sheets in upper-slope deposits, Carboniferous of New South Wales, Australia. The slump sheets moved a distance of a kilometer or two from the vicinity of the shelf break. Scale bar approximately 3 ft (1 m). (Courtesy of C.G. Skilbeck.)

Slumping may occur on the lower or upper continental slope or at the shelf break ahead of prograding deltas (Uchupi, 1967; Lewis, 1971; Coleman and Pryor, 1980). Slumps range from large, discrete allochthons hundreds of square kilometers in extent (Roberts, 1972), to arcuate clusters of small slumps associated with prodelta growth faults (Coleman and Pryor, 1980).

A variety of factors operating singly or in combination can trigger slumping (Coleman and Pryor, 1980; Doyle *et al.*, 1979; Rupke, 1978; Carlson and Molnia, 1977; McGregor, 1977):

(1) Increasing pore-water pressures from rapid depositional loading of fine-grained sediments,
(2) Generation of gases by biochemical degradation of organic matter,
(3) Oversteepening by erosion or deposition,
(4) Differential sedimentary loading,
(5) Sea-level fluctuations,
(6) Cyclic wave-induced stresses associated with storms,
(7) Post-depositional tilting,
(8) Fluidization or thixotropy arising from earthquake shocks.

Thus, slumps may arise equally during the constructive and destructive phases of slope development.

On constructive delta-fed slopes, slumps typically form near the shelf edge and affect sediment packages as thick as 825 ft (250 m). They may be regarded as a component of prograding shelf-edge delta systems, so they occur at particular levels in the offlapping succession and show predictable electric-log responses (Fig. 8-4). The slump units tend to be sharp-based and blocky, with wide variation in grain size. The upper surfaces are intensely burrowed. Cores or dip-meter surveys show steeply tilted bedding planes, together with flow structures and closely spaced fractures. Many of the sandy slumped units are capped by marine clays (Coleman and Pryor, 1980).

Debris Flows

Debris flows are viscous laminar flows of sediment-water mixtures in which the larger clasts are supported by the yield strength of the clay-rich suspension (Middleton and Hampton, 1973, 1976) and by the buoyancy of the larger clasts (Hampton, 1979). The density of the clay–water suspension determines the competence of the debris flow. Hampton (1975) has shown that only 2–20 percent clay is necessary to support sand-sized material, and concluded that sandy debris flows might be indistinguishable from sands deposited by other processes if only textural criteria are considered.

Bouldery debris flows tend to be massive, with the larger clasts randomly oriented in the supporting fine-grained matrix. However, low-viscosity flows can show a vertical clast preferred orientation. Lamination and normal or inverse grading may be apparent in fine-grained debris flows (Hampton, 1975).

Modern submarine debris flows are encountered in a variety of deep-ocean environments (Klein, 1975), and may reflect considerable distance of transport. Debris flows are widely distributed on the continental rise off the eastern United States, where they were emplaced mainly during Pleistocene low sea-level stands (Embley, 1980). Some flows continued across the abyssal plain. Woodcock (1979) has suggested that comparable large-scale mass wasting events were rare prior to the Pleistocene, accounting for the limited occurrence of debris-flow deposits in many ancient slope and rise sequences. On the other hand, some sequences do contain a large proportion of pebbly mudstones (Crowell, 1957) of presumed debris-flow origin.

Turbidity Currents and Turbidites

Turbidity currents in oceans are usually catastrophic surges (Middleton and Hampton, 1976) resulting from the excess density imparted to sea water by the mass of suspended sediment. They may be preceded by debris, grain, or fluidized flows, or by slumps. Turbidity currents are initially channelized, but can spread out downslope, with gravitational acceleration offsetting frictional energy loss, and maintaining sufficient turbulence to keep most of the sediment in suspension (Middleton, 1970). Less commonly, sediment-laden river waters may generate steady, uniform turbidity currents off ocean margins as they are known to do in lakes (Middleton and Hampton, 1976).

The coarsest sediment fraction is concentrated in the head of the density current, where erosion

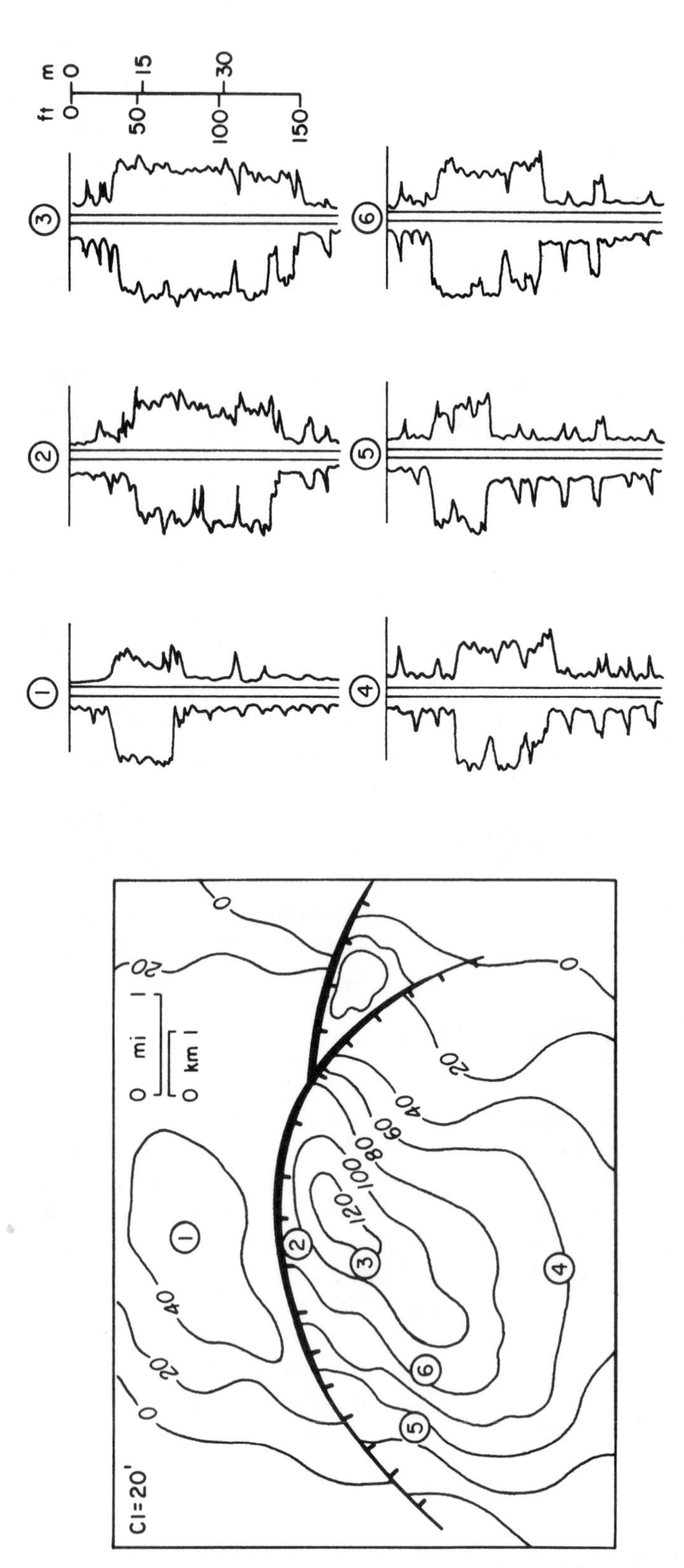

Figure 8-4. Sedimentary characteristics of large-scale slump deposits formed off-shore of the modern Mississippi delta. Diagrammatic geophysical log patterns reflect the abrupt upper and lower boundaries of the large mass of delta front and channel-mouth-bar sand emplaced within slope muds by the slump process. (Modified from Coleman and Pryor, 1980.)

may occur simultaneously with deposition from the more dilute body of the flow. Rapid deposition takes place proximally from sudden decrease in turbulence and distally from reduction in gradient. In the latter situation, scouring followed by slower deposition produces sole marks and small-scale traction structures (Middleton and Hampton, 1976).

The velocities, dimensions, and gradients of large turbidity currents are provided by reconstructions of the 1929 Grand Banks flow (Horn *et al.*, 1971; Uchupi and Austin, 1979). Flow was triggered by an earthquake, traveling down the Laurentian Cone, and continuing across the abyssal plain for 1000 mi (1700 km). Three hours after inception, it was still traveling at 20 m/s (66 ft/s) decreasing to 11.4 m/s (38 ft/s) after 13¼ hours, when, according to estimates by Walker (1980, p. 3), it was still capable of transporting particles of up to granule size in turbulent suspension. Over the final 800 miles (1200 km), the flow traversed the almost flat abyssal plain. Large flows such as this, which are triggered by earthquakes on the continental rise and slope, and which affect large areas of the deep basin plain, are infrequent in comparison with the turbidity flows generated by slumping off river mouths (Heezen, 1963; Walker, 1980, p. 5).

The waning flow sequence described by Bouma (1962) is widespread, but not all divisions are generally represented. The most common sole marks are grooves cut by the largest clasts in the base of the flow. Flutes are clear evidence of erosion by turbulent flow separation (Allen, 1971). The coarse-grained basal deposits tend to be graded, but Walker (1975) recognized inverse-to-normal grading and disorganized beds, in addition to normal size grading. Pebbly and massive sandstones show abrupt variations in thickness from a maximum of around 50 ft (15 m). Rapid deposition from turbidity currents, possibly supplemented by other gravity-flow processes, leads to sudden dewatering and the formation of dish and pipe structures (Walker and Mutti, 1973).

Plane-parallel lamination above the massive, graded interval probably records upper flow regime conditions, with superposed ripple cross lamination reflecting traction and rapid fallout (Middleton and Hampton, 1976). The upper parallel-laminated interval reflects gradual settling of the fine suspended fraction, and merges upward into hemipelagic shales.

Turbidites are volumetrically the dominant slope and basin facies by far, and alone may constitute as much as 45 percent of the total volume of all sedimentary rocks (Walker, 1980).

Low-Density Turbidity Currents

Highly diffuse bottom suspensions of mud, known as the nepheloid or turbid layer, represent low-density underflows, with velocities of the order of a few mi/h (Shepard *et al.*, 1977). The nepheloid layer may be constantly resuspended, or it may be continually replenished as a semipermanent feature. On the other hand it may be a shorter-lived response to stream inflow or infrequent storm-wave agitation. Finally, it may be the dilute tail of a high-density turbidity current (Rupke and Stanley, 1974).

Deposits of the nepheloid layer show subtle grading and delicate lamination, in contrast to thorough bioturbation or massive, structureless aspect of hemipelagic layers.

Fluid Density Underflows

Excess fluid density may be imparted by evaporative concentration or by cooling below the ambient temperature of the water mass.

Salinity Underflows. Underflows resulting from evaporative concentration of brines in broad shelf lagoons devoid of significant stream inflow may be an important process of continental-margin sediment transport. Although not documented in modern environments, the Permian Delaware Basin of Texas provides a convincing ancient example of salinity underflow deposits (Harms, 1974; Williamson, 1979). An extensive shelf-lagoon environment was enclosed by shelf-edge reefs. Hypersaline outflow through narrow breaches in the reef tract, continued down the continental slope as bottom-hugging fluid-density currents. These currents persisted across the deep-basin plain of the intermittently stagnant, density-stratified Delaware Basin where they were confined within broad, anastomosing channels (Fig. 8-5).

Traction-current structures dominate these salinity-underflow deposits. There are no well-defined proximal-to-distal variations in texture or structure, and there are no vertical sequences

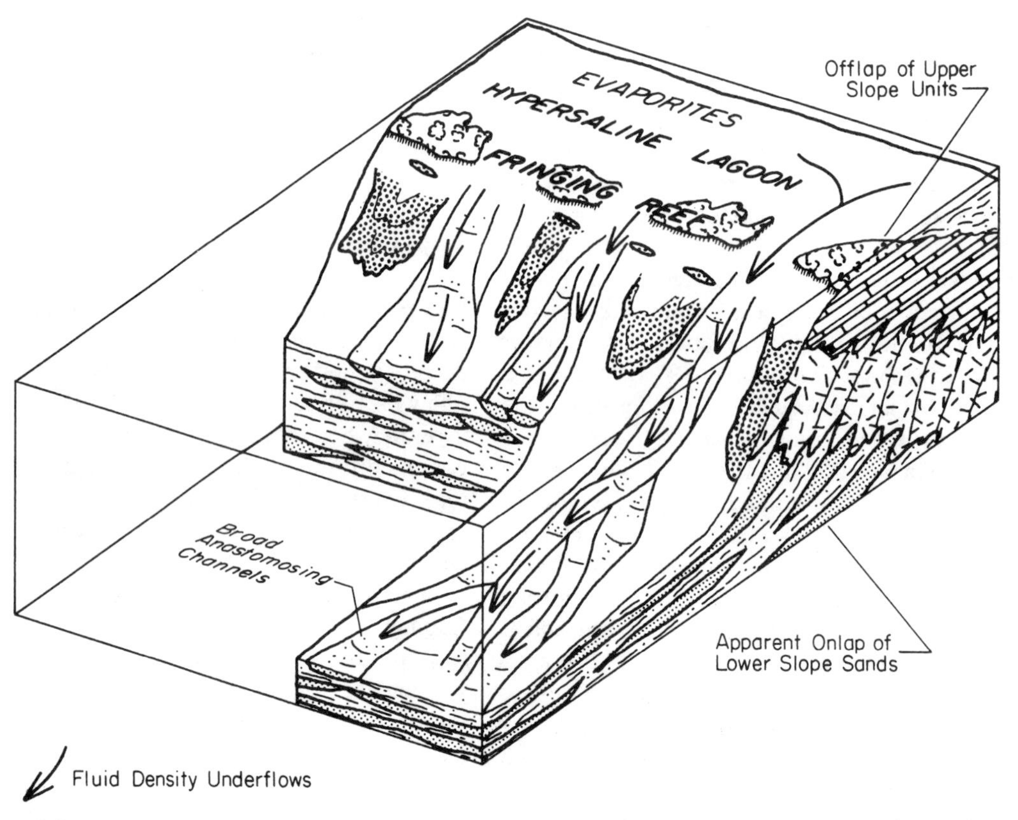

Figure 8-5. Depositional model for salinity-induced density underflow deposits. (Modified from Bozanich, 1978.)

comparable to the Bouma model. Between the cosets of cross-laminated silts are clays deposited from suspension during episodes of reduced flow. Normal basinal sedimentation is represented by dark, organic-rich shale.

Thermohaline Bottom Currents. Deep ocean waters are constantly moving, and complete stagnation is rare except in silled basins (Shepard, 1973, p. 60). Movement arises from surface wind stress, tidal forces, and, most importantly, regional thermohaline density differences produced by differential heating, evaporation, freezing, ice melting, and fresh-water inflow. Cold, descending polar waters converge toward the equator as bottom currents, which are deflected by Coriolis forces. These deep thermohaline or contour currents flow along isopycnic surfaces subparallel to the bathymetric contours, and are generally slow except in areas of convergence and reinforcement. Here they attain velocities of 16 ft/s (500 cm/s) (Betzer *et al.*, 1974), and mold streamlined drifts covered in places by complex fields of bedforms (Lonsdale, 1981). The western boundary undercurrent of the North Atlantic carries vast quantities of fine suspended terrigenous sediment, and has built the Blake–Bahama Outer Ridge. The South Pacific western boundary undercurrent, in contrast, is devoid of terrigenous sediment and for the most part sweeps over a scoured manganese pavement; reworked pelagic sediments accumulate where the current diverges and slows (Lonsdale, 1981).

The deposits of these thermohaline currents, known as contourites, tend to be well-sorted silts and fine sands showing ripple cross lamination in planar sets from a few mm to cm thick (Bouma and Hollister, 1973). Paleocurrent azimuths follow the slope contours (Rupke, 1978). Contour currents are important in shaping the continental rise, and may be clearly reflected in the rock record (Lovell and Stow, 1981). An ancient example that reflects interplay of gravity resedimentation and contour-current remolding is presented by Bein and Weiler (1976).

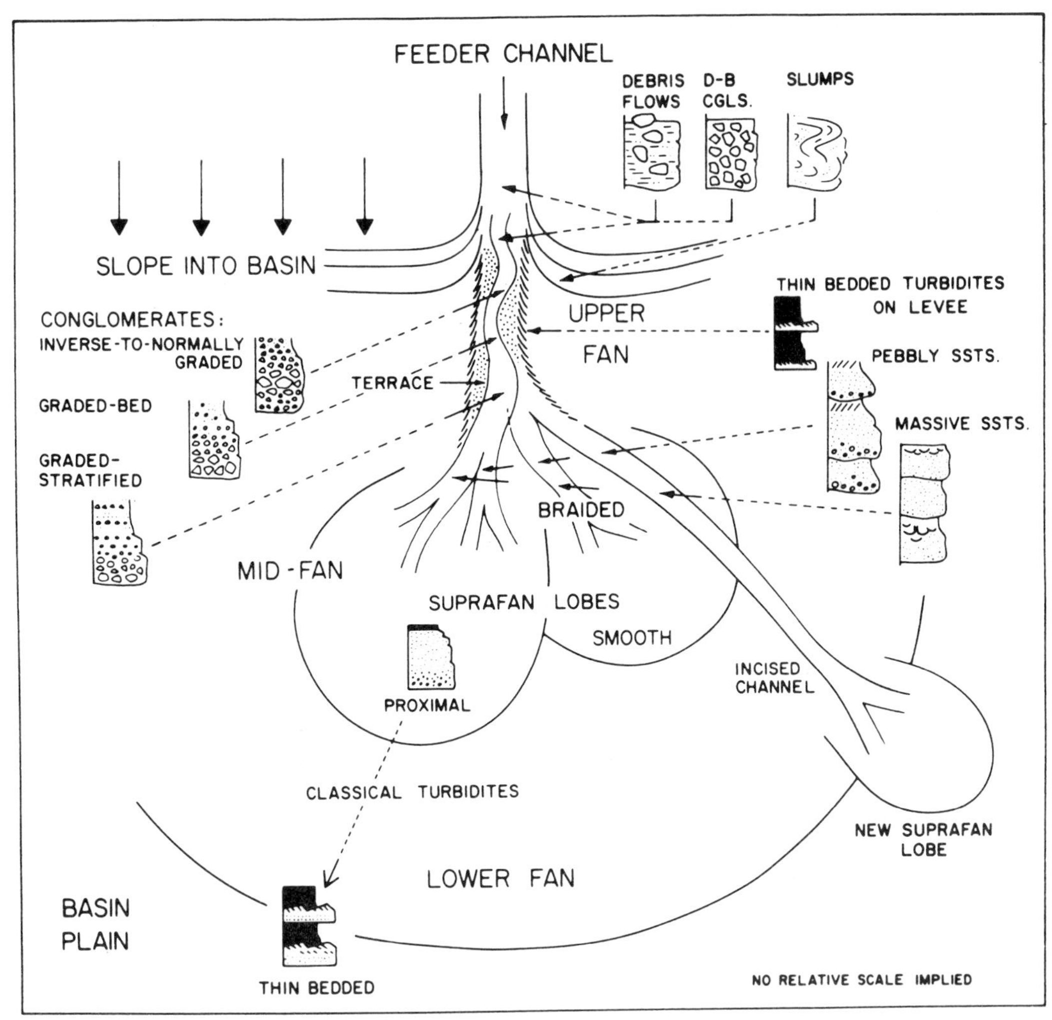

Figure 8-6. Depositional model of a sand-rich, unconfined submarine fan, relating morphologic features, depositional environments, and resultant facies. (From Walker, 1978.)

Deep-Ocean Tidal Currents

The geologic significance of deep, reversing tidal currents has been emphasized by Klein (1977, pp. 60–65). Tidal bottom currents at depths of up to 8000 ft (2400 m) in the Pacific Ocean rework biogenic carbonates into large and small bedforms (Lonsdale and Malfait, 1974). Deep Sea Drilling Project cores from Mesozoic and Tertiary sediments show a variety of small-scale sedimentary structures reflecting frequent alternation of traction and suspension deposition such as are commonly associated with shallow tidal environments (Klein, 1975). According to Klein, these ocean-basin deposits were originally supplied by turbidity currents, but were reworked by tidal currents.

Similar variations in current direction and strength are recorded in several ancient slope and basin systems. Devonian flysch basin deposits of New Zealand show typical tidal-current reversal patterns (Laird, 1972). Carboniferous slope facies in New South Wales, Australia, contain centimeter-thick sets of bipolar cross lamination and rhythmic clay drapes attributed to deep tidal processes (Skilbeck, 1982).

Pelagic Sedimentation

A steady rain of biogenic and airborne particles characterizes deep ocean basins well-removed from the direct influence of terrigenous sedimentation. Settling is accelerated by biological agglutination and flocculation.

Sedimentation rates are slow, possibly averaging around 2 in (5 cm) over 1000 years (Gorsline, 1980). Paradoxically, maximum biogenic sedimentation corresponds to episodes of maximum terrigenous influence, During stages of lowered sea level, rapid clastic influx was accompanied by stronger oceanic circulation and greater nutrient supply (Pitman, 1978; Berger, 1970). The composition of pelagic sediments is strongly influenced by solution of biogenic components, with oozes giving way to red clays at depths of 10,000 to 16,000 ft (3000 to 5000 m) (Griffin and Goldberg, 1963). Ancient pelagic sediments comprise fine-grained limestones, cherts, and marls, which are generally massive in appearance because of thorough bioturbation or recrystallization.

Major Types of Terrigenous Slope and Basin Systems

Synthesis of various slope models in a process-defined, three-dimensional classification remains tentative. Compared with other depositional settings, the slope remains relatively poorly known because of its inaccessibility to direct observation. Data are derived largely from three sources: (1) surficial sampling of modern slope deposits, (2) high-resolution marine seismic studies, and (3) field study of interpreted ancient slope systems. Because ancient slope systems have, of necessity, undergone significant uplift and consequently major tectonic modification, three-dimensional facies reconstruction is commonly interpretative. The effect of geologically recent eustatic transgression on modern processes also remains a point of controversy. The ongoing hydrocarbon exploration into relatively young, undeformed slope deposits may help resolve some of these ambiguities.

The "margin affected basins" (Gorsline, 1980) subject to terrigenous clastic influx may be divided into three distinct but interrelated morphogenetic types: submarine aprons, submarine-rise prisms, and submarine fans.

Submarine Aprons

Submarine aprons are products of mass wasting of the upper slope and shelf edge, and are supplied mainly by slumping and debris flows, which frequently terminate before reaching the base of the slope. Some large debris flows persist across the rise and onto the abyssal plain, producing lobate protuberances. Most of the slumps feeding the submarine apron are small, locally clustered, and intermittently active. Slump-generated turbidites are interbedded randomly with chaotic slump and debris-flow units. Matrix content is high and there is minimal reworking. As a result, submarine aprons are characterized by somewhat chaotic internal organization.

The Quaternary continental slope of the north-central Gulf of Mexico provides a well-described example of a slope apron. Seismic stratigraphic studies (Lehner, 1969; Sangree *et al.*, 1978) and deep coring (Woodbury *et al.*, 1978) provide information on component facies geometry and composition of this muddy progradational slope, which was constructed in front of advancing shelf-edge deltaic systems during Pleistocene low sea-level stands. The Quaternary slope is morphologically complex, containing multitudes of salt-cored bathymetric highs and interspersed, intraslope basins (Lehner, 1969). The shelf edge is further complicated by discontinuous belts of growth faults.

Three principal facies constitute the Quaternary slope system. *Pelagic mud drapes* are characterized by lateral continuity and uniform bedding, and generally parallel the underlying depositional surface. They consist almost entirely of mud, and were deposited by slow settling and accretion of hemipelagic sediment or by low-density turbidity currents (Sangree *et al.*, 1978). *Layered turbidite sequences* fill intraslope basins as well as local channels or scour depressions. Bedding is well displayed, and typically laps against adjacent bathymetric highs or channel margins. Sediments consist of mud and muddy silt, with scattered thin or rare thick beds of sand and silty sand. *Chaotically bedded tongues* and wedges are interbedded with layered turbidite sequences, and constitute a major portion of the slope sequence. The tongues consist of slump and debris-flow deposits transported down slope from the shelf edge or from crests of intraslope highs. In either case, sediments consist dominantly of mud and silt. The few sand units are discontinuous and commonly poorly sorted (Woodbury *et al.*, 1978). Bedding, if visible, is churned and deformed.

In summary, the Quaternary slope apron

system of the Gulf Coast is typified by low sand content (about one percent of the total core examined by Woodbury *et al.* consisted of sand), erratic interfingering of slump, turbidite, and hemipelagic deposits, and complex internal structure. Long-lived point-source sediment dispersal patterns that might funnel significant volumes of sand onto the lower slope and basin floor are absent in the north-central Gulf slope. An analogous slope depositional style appears to have persisted well back into Tertiary time (Caughey, 1981).

Current-Molded Submarine-Rise Prisms

Contour currents are instrumental in shaping the slope rise, and some rise prisms are built entirely by contour currents; for example, parts of the east coast of the United States (Rona, 1969). Silt and clay are dominant, with some well-sorted fine-grained sand in relatively thin beds. The rise prism off Cape Hatteras bears giant bedforms with heights of 35 to 350 ft (10 to 100 m) and wavelengths of up to 8 mi (12 km), with steeper south-westward slopes corresponding with the flow direction of the western boundary undercurrent. These may be relicts of episodes of accelerated thermohaline circulation accompanied by rapid sediment influx during Pleistocene glacial maxima (Rona, 1969). Similar large-scale features are known from continental-rise prisms in other ocean basins.

Submarine Fans

In contrast to aprons and rise prisms, submarine fans show distinct environmental subdivisions (Fig. 8-6). Fans are fed from point sources, either river mouths or submarine canyons, and receive the bulk of their sediments from turbidity currents. They are therefore located at the mouths of submarine canyons, off deltas, and in deep-basin plains.

Upper-fan feeder channels or canyons (Fig. 8-6) serve as sediment conduits. Many canyons, such as the Hatteras Canyon, are still erosional (Cleary and Conolly, 1974), whereas others are aggradational as a result of relative sea-level rise or continental-slope progradation. Coarsest sediment accumulates in the thalweg of the upper-fan channels (Fig. 8-7a), and occasionally spills out across the flanking levees and terraces, which are normally the site of fine sediment deposition in thin, graded units (Fig. 8-7b). These thin beds commonly display sharp, tool- or flame-marked bases and partial Bouma sequences consisting of parallel lamination overlain by convolute bedding or multiple thin sets of ripple cross lamination (Walker, 1978; personal communication, 1981). Where several channels form within the larger canyon, these fine-grained overbank deposits thicken toward interchannel depressions.

Upper-fan channels may be extremely large features. The single meandering channel on the upper part of the Rhone deep-sea fan is 1.3–3 mi (2–5 km) across, and is flanked by levees up to 250 ft (75 m) high (Bellaiche *et al.*, 1981). Each channel typically develops an upward-fining sequence (Fig. 8-8) between 30 and 160 ft (10 and 50 m) thick and possibly exceeding 300 ft (90 m) (Walker, 1978), comprising gravel, pebbly sand, or massive sand and graded fine-grained deposits. Some channel-fill units are uniformly coarse grained, producing blocky geophysical log patterns. The channels switch rapidly as the canyon fills, depositing interlensing series of dip-oriented gravels and sands separated by silts, muds, and subordinate sands.

The mid fan of sand-rich systems is characterized by shifting suprafan lobes (Normark, 1970) with smooth convex-upward surfaces (Fig. 8-6). Each lobe is supplied by bifurcating, distributary or braided channels which accumulate massive and pebbly sands showing lenticular bedding and shallow scour-fill structures (Walker, 1978). Interlobe sediments are partially to wholly reworked as the channels migrate. Finer, graded sediments are deposited in the upper parts of some channels and on flat surfaces as the channels die out downslope. Channel migration may develop multistoried, upward-fining sequences, whereas the distal suprafan lobe consists of a single upward-coarsening sequence, the upper parts of which are capped by the mud blanket of the abandonment phase (Fig. 8-8). Sands of the suprafan lobe range from 30 to 160 ft (10 to 50 m) thick (Walker, 1966, 1978; Hsu, 1977). A typical log pattern produced by an aggrading series of suprafan lobes and channels is shown in Figure 8-9.

Mud-rich submarine fans, such as the Mississippi fan of the eastern Gulf of Mexico basin, lack well-developed sandy mid-fan channels and suprafans. Rather, debris flow and slump

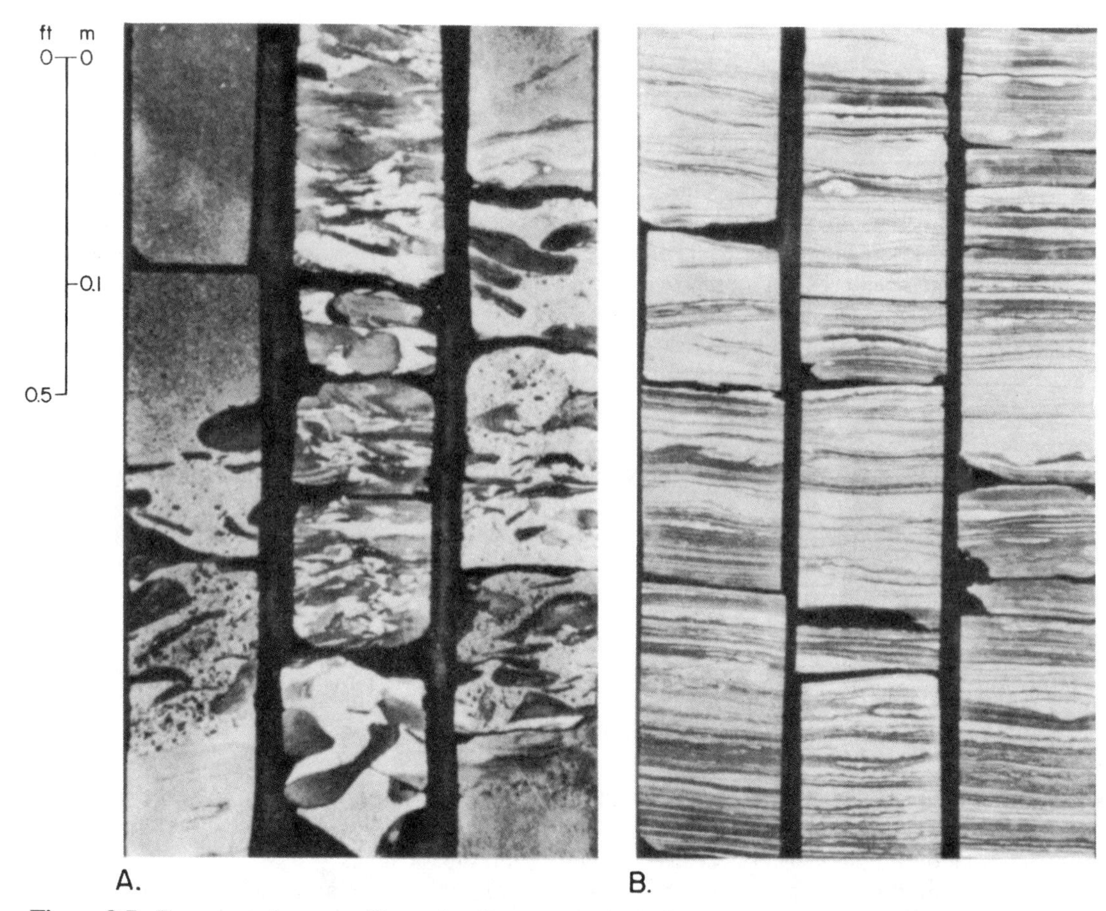

Figure 8-7. Representative cores illustrating features of middle-fan channel (A) and laterally equivalent levee deposits (B) of one Pennsylvanian submarine-fan system, Midland Basin, Texas. Channel deposits contain abundant mud-clast conglomerate, and display both normal and reverse graded bedding. Thinly bedded levee deposits exhibit micrograded bedding, ripple lamination, micro-faults, and abundant sole marks.

deposits are abundant and laterally extensive, and channelized sediment dispersal systems are poorly developed and largely filled with fine, muddy sediment (Moore *et al.*, 1978).

The lower fan has a smooth, gently sloping surface and receives slowly deposited increments of suspended sediment punctuated by pulses of fine-grained turbidites. The resulting graded beds are thin, laterally persistent, and monotonously superposed, commonly through considerable stratigraphic thickness. Log traces of distal fan sequences are highly serrate (Fig. 8-9).

Slope-Wedge Stratigraphy

Submarine-fan systems are subject to frequent switching of suprafan lobes fed by a single fan-head canyon and, on a larger scale, to switching point sources of sediment influx; for example, from distributaries of multilobate shelf-edge deltas. Together, these produce complex alternations of constructive and destructive phases, all recorded in a compound fan-cone corresponding to a major protuberance of slope sediments (Brown and Fisher, 1980). Progradation of one part of a fan cone is accompanied by destructive reworking of other parts, which, in turn, may be subsequently reactivated. Ultimately, the entire fan system is abandoned and blanketed with mud or, more rarely, carbonate debris derived from shelf-margin reef or bank complexes. Slope-wedge stratigraphy is commonly visible in reflection seismic records, and geophysical study of modern continental slopes has defined three genetically significant depositional styles (Brown and Fisher, 1980).

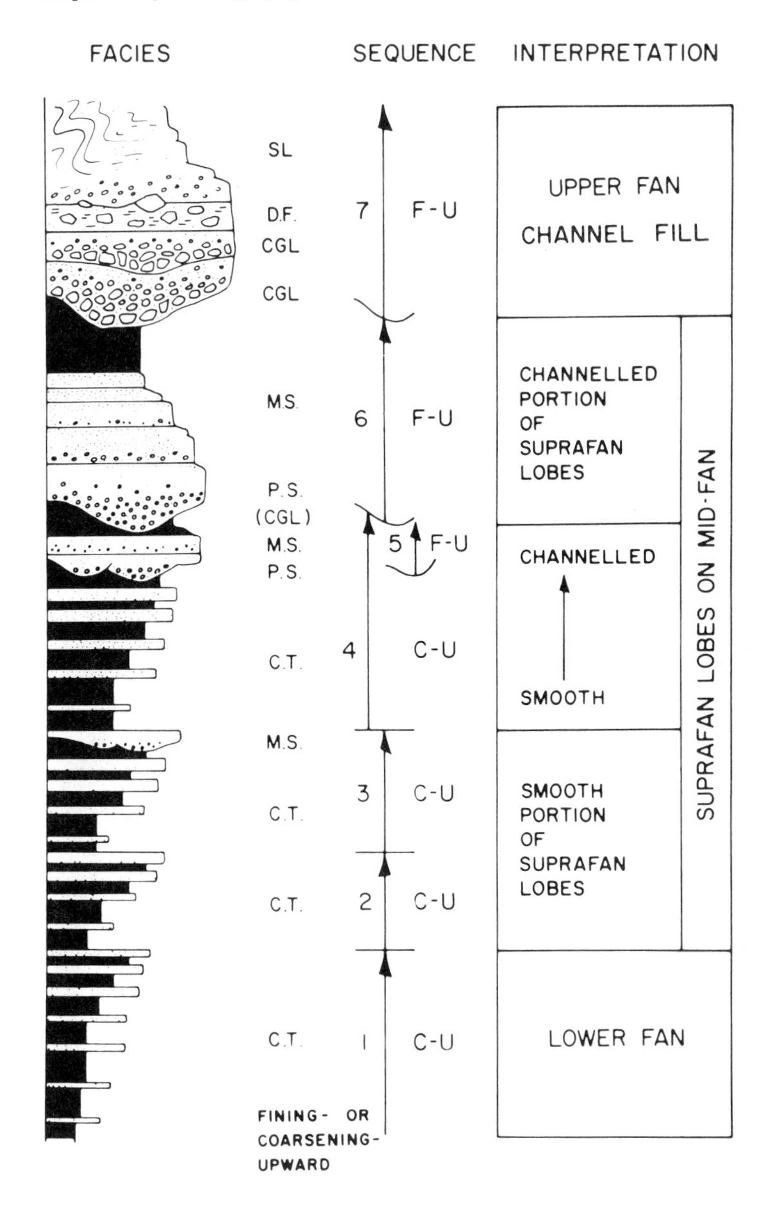

Figure 8-8. Hypothetical sedimentary sequence produced by progradation of a sand-rich submarine fan system. Rythmically interbedded distal fan deposits grade up into packets of suprafan sands consisting of upward-thickening sand beds. Superposition of the channeled suprafans and overlying, isolated upper fan channel deposits completes the sequence, with lenticular units characterized by upward fining. (From Walker, 1978.)

Offlapping Slope Wedge

Progradation of the slope system produces an offlapping depositional pattern (Fig. 8-10). The pattern of progradation is revealed by depositional or seismic surfaces inclined basinward at angles of 2–5 degrees, and is typically well displayed in basins subject to minimal contemporaneous down-warping, particularly along cratonic margins. Upper-fan, mid-fan, and distal-fan facies tracts tend to be well preserved in offlap sequences. Offlapping fan wedges are commonly associated with actively prograding deltaic coasts. Many large, modern deltaic systems, including the Rhone, Rio Balsas, Mississippi, Nile, Niger, and Irrawaddy, have slope-fan wedges that were actively accreting during Plestocene low sea-level stands.

A classic example of slope-fan progradation is provided by the Pennsylvanian Eastern Shelf of the Texas Midland Basin, a constructional platform capped by mixed carbonate and terrigenous sediments of fluvio-deltaic and carbonate-bank environments (Galloway and Brown, 1972). The offlap slope built intermittently westward more than 50 mi (80 km) by gravity resedimentation of delta-front sediments.

The preserved record shows massive shelf-edge

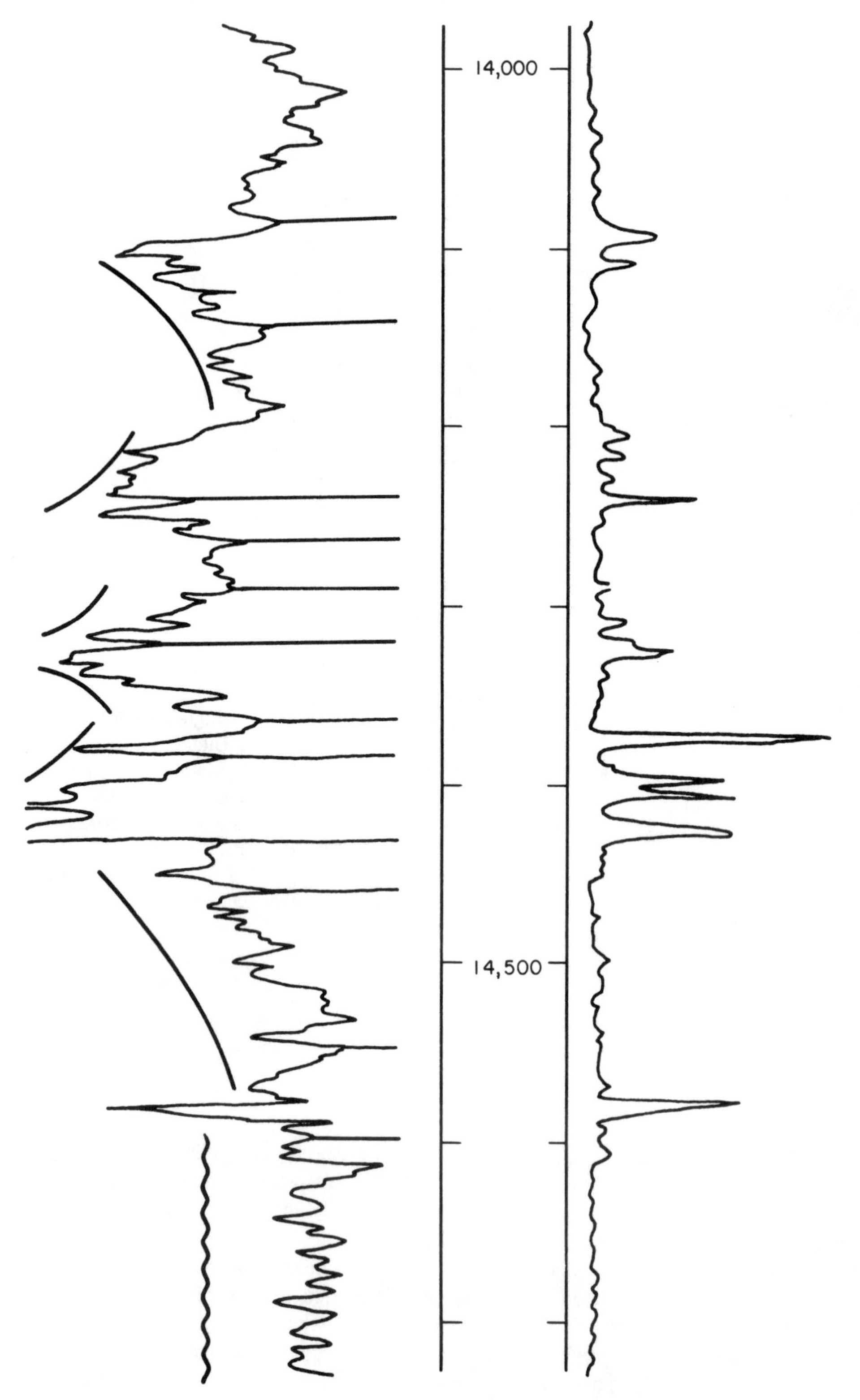

Figure 8-9. Electrical-log response of a submarine-fan sequence. Interbedded lower-fan deposits (14,600–14,700 ft) (4380–4410 m) produce a serrate log pattern. Upward-thickening suprafan packets [14,430–14,550 ft (4329–4365 m) and 14,100 to 14,200 ft (4230–4260 m)] bracket an interval characterized by upward-thinning mid-fan channel fills. Log is from the Hackberry Formation of the northern Gulf Coast Tertiary Basin. (Modified from Vormelker, 1979.)

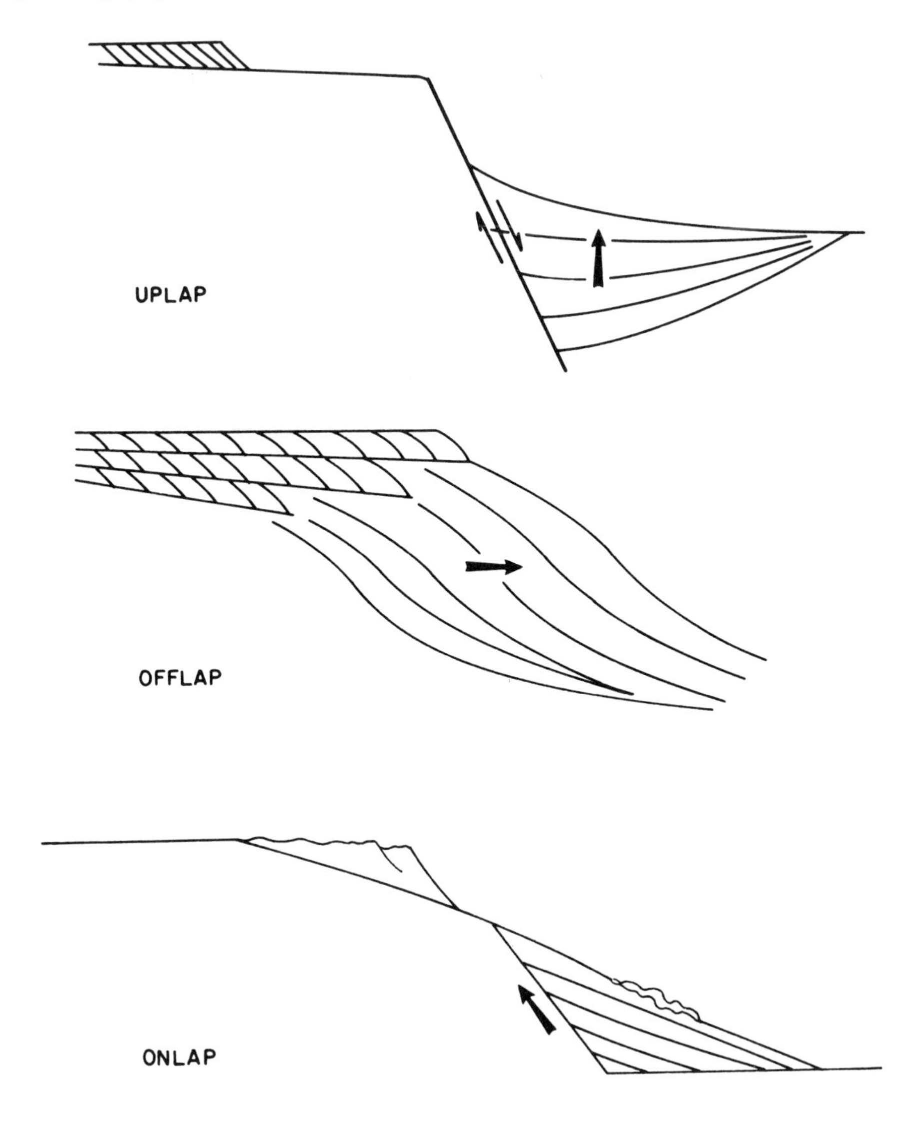

Figure 8-10. Classification of submarine slope wedges based on internal bedding architecture.

carbonate banks cut by dip-oriented sandstone stringers (Fig. 8-11). These sandstones are narrow where traversing the bank complex, but fan out into lobate-to-digitate delta lobes on the crest of the offlapping slope wedge. Sandstone is confined on the upper to middle slope to numerous dip-oriented channels, but splays out basinward to form irregular suprafan lobes (Fig. 8-11). Interlobe mudstones are generally massive, but include some thin graded, contorted, sandy beds deposited by overbank turbidity flows. Distal portions of inactive fans were mantled with siliceous micrite and basinal shales; middle to upper-slope deposits were encrusted by carbonate debris derived from mass-wasting of the bank margin.

The Eastern Shelf slope system and other comparable offlapping wedges of the intracratonic Midcontinent basins were characterized by large, low-relief terrestrial drainage basins, mixed-load fluvial systems deficient in sand and gravel, and rapidly switching delta systems providing sediment sporadically to the slope.

Onlapping Slope Systems

Episodes of marine transgression and high sea level cause significant destructional modification

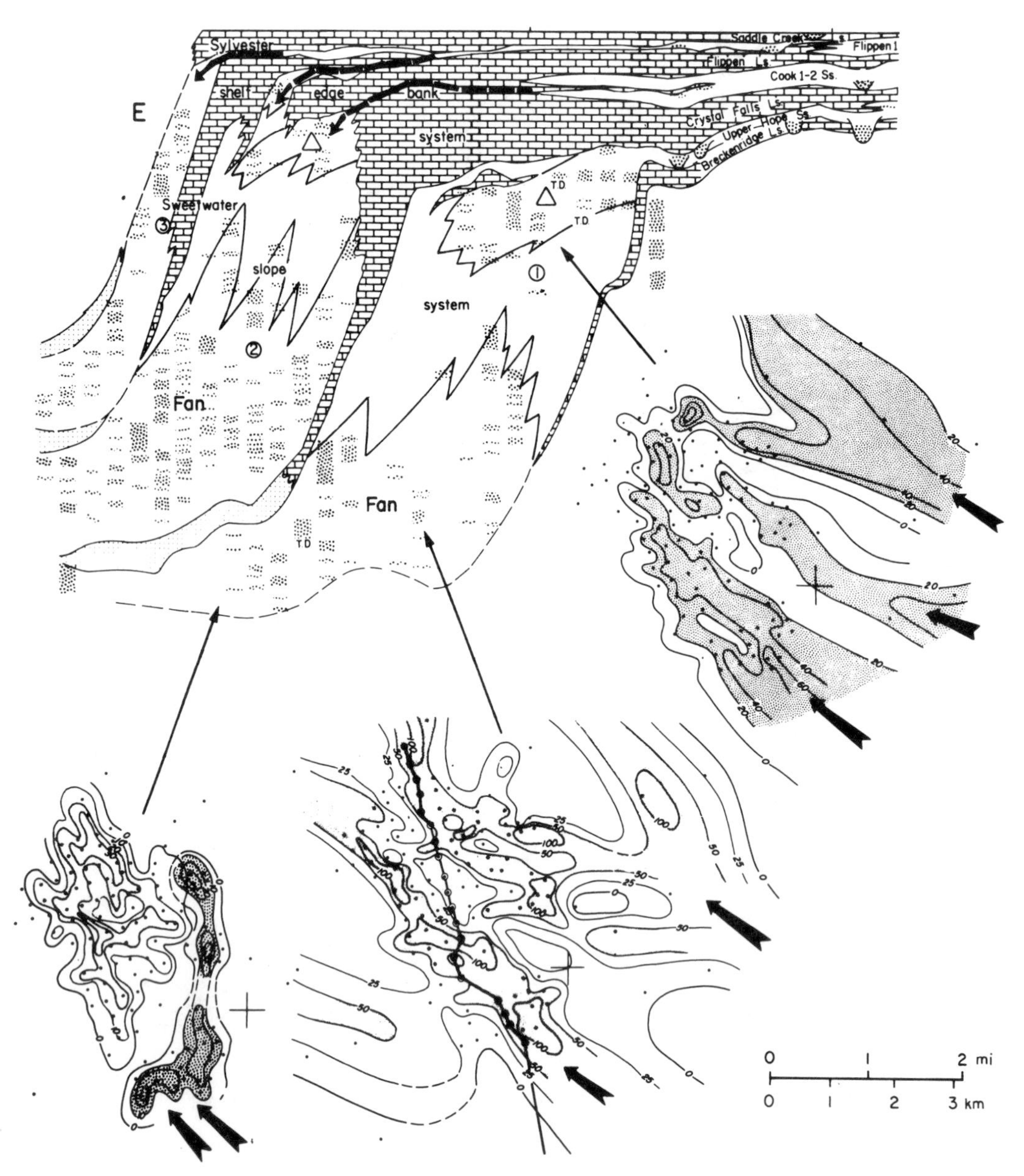

Figure 8-11. Sand-distribution isolith patterns in an offlapping slope wedge (Upper Paleozoic Midland Basin). Dip-oriented, digitate deltaic sand bodies cap the slope wedges, which are outlined by encrusting carbonate banks and debris aprons. As shown by the cross section, the upper slope is comparatively sand-poor. The middle portion of the wedge contains a complex network of interweaving submarine fan channel sandstones. The toe of the wedge contains thin, isolated, ameboid to elongate suprafan packages. Contours in feet. (Modified from Galloway and Brown, 1972.)

of fans, with redistribution of outer shelf and upper slope sediments. Progressive landward migration of submarine canyon and fan environments (Fig. 8-10) results in development of onlapping constructional fans or slope aprons which fill submarine canyons headward (Brown and Fisher, 1980). Subsequent onlap by continental-rise wedges can follow final abandonment of the fan complex. This inactive phase can also be marked by carbonate sedimentation on the upper slope and shelf (Fisher and Brown, 1972).

Examples of onlapping slope systems are provided by the deposits of the Yoakum Channel (Wilcox Group) and Hackberry Embayment (Frio Formation) of the Tertiary Gulf Coast

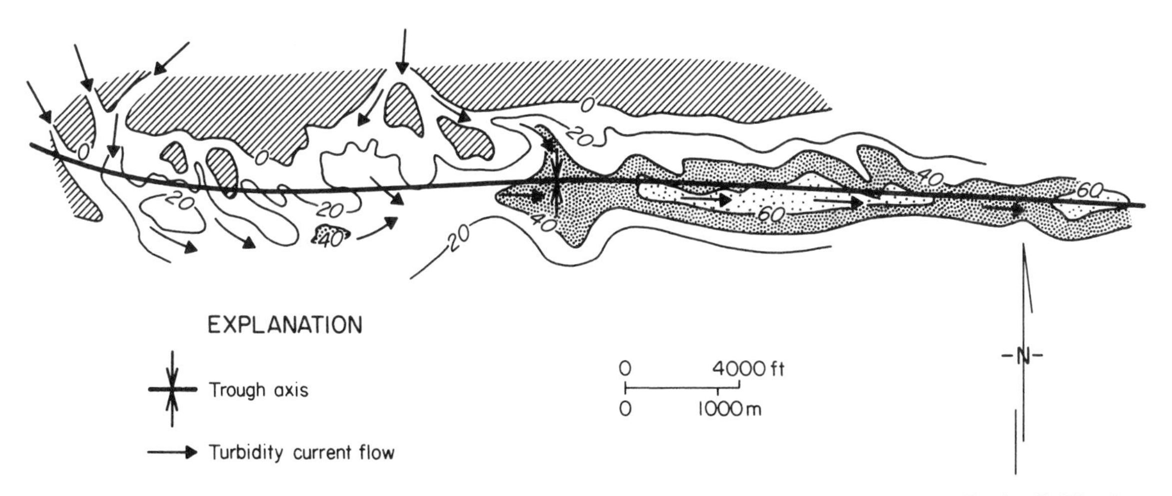

Figure 8-12. Sandstone isolith map of a single Lower Pliocene sandstone packet, Ventura Basin, California. The slope-fan sediment dispersal pattern was sharply directed from south to east along the axis of a rapidly subsiding structural trough. Abrupt facies changes and consequent sand pinchout to the north and south reflect the local paleobathymetry of the depositional basin. (Modified from Hsu, 1977.)

Basin. In both, large dissected canyons attaining depths of 600–3500 ft (180–1050 m) and widths of 4–10 miles (6–16 km) (Hoyt, 1959; Paine, 1971) were filled by onlapping turbidity-current channel and overbank deposits. Major submarine-fan sand bodies would be anticipated at great depth because sand-rich deltaic and shore-zone systems were cannibalized by canyon erosion (Paine, 1968; Brown and Fisher, 1980; Walker, 1980).

Within the Hackberry wedge, contemporaneous retreat of canyon-head gully systems by slumping tapped large volumes of shelf and nearshore sand, thereby stacking massive turbidite-channel sands along persistent dip-oriented axes (Paine, 1971; Galloway *et al.*, 1982).

Uplapping Fan Wedges

Uplap patterns are characteristic of fan wedges deposited in rapidly subsiding, fault-controlled basins such as those of the California Borderland (Hsu, 1977), mobile flysch troughs, and foreland basins (Fisher and Brown, 1972). Shelves tend to be narrow and flanked by a tectonically active hinterland. Braided streams and fan deltas supply sand and gravel, which are trapped in canyon heads and funneled downslope. These sediments are deflected and confined by the complex basinal topography. As a result, aggradational sedimentation patterns are typical. The basins accumulate vertically, and superposed fan wedges build to remarkable thicknesses if balanced by continued subsidence. Similar uplapping bedding characterizes interdiapiric basins in the Gulf of Mexico slope apron.

Because of this tendency toward vertical stacking, proximal–distal relations may be poorly defined, and the interplay between deformation and sedimentation commonly modifies simple, radial fan geometries (Fig. 8-12). Uplap slope systems or component facies wedges may be significantly displaced from equivalent shelf and shoreline sediments by regional tectonic or erosional discontinuities.

Chapter 9

Lacustrine Systems

Introduction

Terrigenous lacustrine systems are assuming new importance with the recognition of their role as hosts for a variety of minerals, including hydrocarbons, coal, and uranium. Some lake deposits are among the best known petroleum source rocks (Demaison and Moore, 1980), and locally contain good reservoirs. The energy resources of lacustrine oil shales are creating considerable interest, which is likely to increase in the coming decades. Commercial coal seams are widely developed in lacustrine and associated systems, and the permeability and geochemical gradients in fluvial-lacustrine transitions are ideal for epigenetic uranium mineralization. The association of bedded metalliferous deposits with ancient lake sediments has spurred additional economic interest, and lacustrine evaporites such as trona are fetching high prices on world markets.

The unique physico-chemical environments of lakes lead to the deposition of unusual mineral associations, and some species are diagnostic of lacustrine systems (Picard and High, 1972). But lakes whose primary role is as a precipitating basin for chemical sediments are not of present concern. Furthermore, many lacustrine water bodies such as floodbasin ponds, delta-plain lakes and bays, and playas, are constituent parts of other depositional systems, and are considered further in the appropriate chapters on fluvial, deltaic, and eolian systems.

Most large lakes are presently located in epeirogenic depressions. Lakes Tanganyika and Baikal are examples of rift-valley lakes, whereas Lakes Victoria and Eyre occupy major crustal sags. The depression covered by Lake Eyre was at one stage enlarged by wind deflation. A few nontectonic lakes such as lava-dammed Lake Kivu and the North American Great Lakes are large, but most modern lakes occupying deflated and glacially scoured hollows and meteorite-impact craters, or impounded behind alluvium, terminal moraine, lava, and landslide debris, are small and ephemeral. According to Picard and High (1981) the majority of existing lakes originated directly or indirectly from glacial processes, and provide poor analogues for interpreting ancient lacustrine systems related for the most part to tectonic controls. Some of these small lakes have been significantly reduced in size during the brief Holocene time span. Few modern lakes compare with the extent of some of their ancient counterparts, the largest fossil examples of which were the Triassic Popo Agie with a surface area of well over 50,000 mi^2 (130,000 km^2) (High and Picard, 1969) and Pleistocene Lake Dieri of Australia covering some 42,000 mi^2 (110,000 km^2) (Loffler and Sullivan, 1979).

Modern lakes vary markedly in depth relative to surface area. For example Lake Victoria with a surface area of 27,000 mi^2 (70,000 km^2), the world's second largest, has a maximum depth of only 260 ft (80 m), in contrast to nearby, smaller Lake Tanganyika which is 5000 ft (1500 m) deep. Lake Baikal, the world's deepest, attains 5750 ft (1742 m) with a surface area of only 12,000 mi^2 (31,500 km^2). The saline Caspian Sea is the largest inland body of water, covering 143,600 mi^2 (372,000 km^2), and has a maximum depth of almost 3300 ft (1000 m), although considerably shallower over most of its extent.

Similar broad variation is reflected in the preserved thicknesses of ancient lake systems (Feth, 1964; Picard and High, 1972), although many of these were subject to contemporaneous subsidence which caused vertical expansion, or underwent thinning due to compaction. Devonian lacustrine sediments of Scotland attain 13,000 ft (4000 m), but most of the well-studied examples in the western United States are thinner than 820 ft (250 m). Lakes Lillooet (British Columbia), Tahoe (western United States), and Geneva (Switzerland) all have around 650 ft (200 m) of Quaternary sediments, but large Lake Eyre and its Pleistocene predecessors together accumulated less than 65 ft (20 m) (Dulhunty, 1981).

Ancient lake deposits show a number of features in common with other clastic systems, for example epeiric basins and aggradational alluvial flood basins, with a large proportion of fine-grained sediment. These systems are commonly intergradational vertically and laterally. Furthermore, lacustrine systems include a variety of interrelated facies which in other situations might well be regarded as independent depositional systems, for instance deltaic, linear shore zone, shelf, slope, and deep basin. Some ancient lake systems have been neglected until recently because of undue emphasis on the uppermost, fluvially dominated facies.

Most modern lakes are surrounded by broad landsurfaces of subaerial denudation with soil development, and similar relationships are reflected in the Green River Formation where regressive lake deposits are correlated with paleosols and disconformities (Picard and High, 1972, 1981). On the other hand, many ancient lacustrine systems are gradationally surrounded by extensive eolian, fluvial, and deltaic systems (Glennie, 1972; Hobday, 1978a; McGowen *et al.*, 1979).

Lacustrine Processes

Factors influencing lake sedimentation may extend through the entire basin, for example those involving climate or lake level. Alternatively, they may be related to local sedimentation patterns affecting only a small portion of the lake.

Lakes of different size, depth, temperature, and water chemistry show a range of physical, chemical, and biological processes. Deep, sediment-starved lakes may be dominated by organic and chemical sedimentation associated with temperature and chemical layering of the watermass. Shallower lakes with terrigenous sediment influx develop shoreline features comparable to those of a marginal-marine environment, but on a smaller scale. Lakes, however, have rapid changes in water level and shoreline position. The resulting episodes of transgression, regression, and shoreline incision leave a complex stratigraphic record.

Temperature Stratification

Variation in surface water temperature with seasons produces density stratification of a warm, upper epilimnion and lower, colder, more dense hypolimnion, which are separated by a thermocline. Whereas the epilimnion is well oxygenated by continued circulation, the lower layer stagnates and becomes anoxic. Nutrients such as phosphates and nitrate are carried into lakes by streams, causing accelerated organic productivity in the upper water layers. Settling of this suspended organic material results in oxygen depletion in the lower water layers, a eutrophic condition inhospitable to most organisms. This situation is most pronounced in tropical areas where the initial dissolved oxygen content is lower because of high temperatures, and where there is a lack of seasonal overturn. Deep lakes of the low-latitude East African Rift system consequently tend to be eutrophic. Lake Tanganyika waters are anoxic below a depth of 150 m (500 ft) (Degens *et al.*, 1971). The shallower East African lakes are oxygenated, or oligotrophic, as are all of the large temperate lakes of the Northern Hemisphere which undergo annual mixing (Demaison and Moore, 1980). Even Lake Baikal, the world's deepest, is well oxygenated (Swain, 1970). Aeration of bottom waters by seasonal overturn may be augmented by underflow of cold, oxygenated stream waters, and by turbulence in shallow lakes.

Chemical Stratification

Temperature stratification may be reinforced by salinity layering. Increase in salinity due to evaporation and inflow of sea water produces density differences, with the more saline waters sinking to the bottom. A halocline separates surface waters of low salinity from more saline bottom waters, which are commonly charged with hydrogen sulfide. Direct discharge of ground waters by saline springs on the floor of a lake can produce stratification of the water column, as in Lake Kivu (Stoffers and Hecky, 1978), or Lake Eyre (Dulhunty, 1981).

Stream Inflow

Inflowing stream waters disrupt pre-existing stratification of the watermass to produce complex circulation patterns (Fig. 9-1). These processes can vary seasonally, with overflows, interflows, and underflows occurring within the same lake at different times of the year in response to density changes. Overflow of a less dense, warm, fresh-

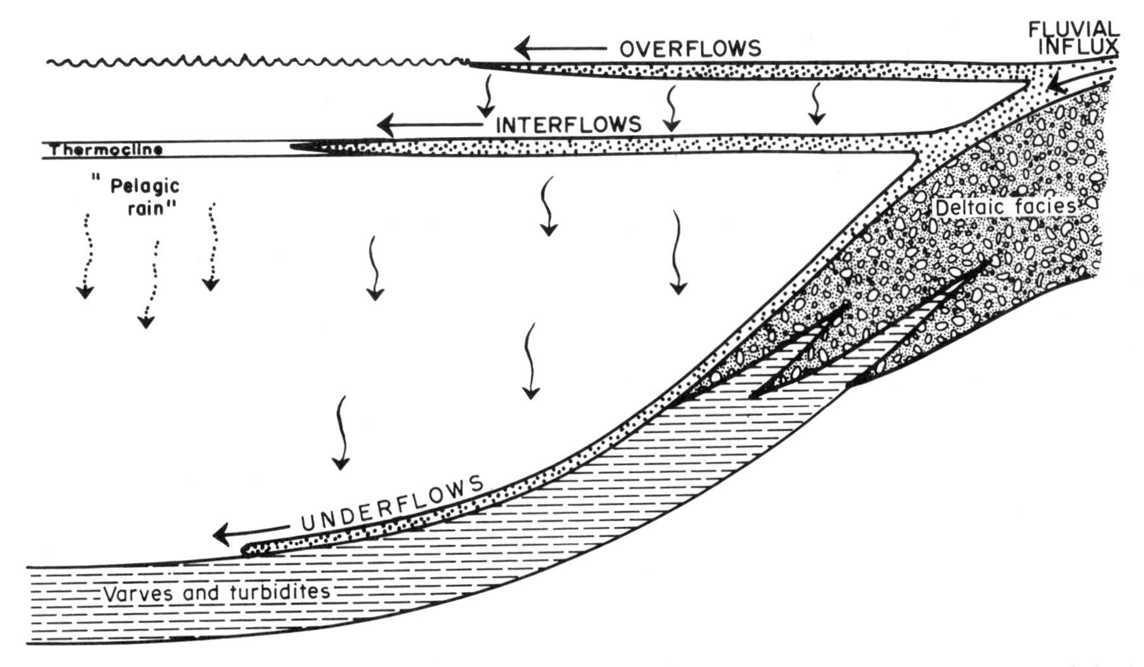

Figure 9-1. Flow mechanisms and sediment types in stratified lakes receiving significant volumes of clastic sediment. (Modified after Sturm and Matter, 1978.)

water plume disperses sediment basinward in a progressively fining pattern. The earth's rotation causes these inertial currents to be deflected, resulting in rotating circulation patterns. In the case of interflows, the stream waters are intermediate in density between the hypolimnion and epilimnion, so that flow occurs across the top of the thermocline. These currents, too, follow a rotating path, with velocities of up to 5 cm/s at 25 m (0.2 ft/s at 82 ft) (Nydegger, 1976). Fine sediments are dispersed over a broad area of the lake floor, and the finest fraction remains trapped within the thermocline, to be released during the seasonal overturn. This rapid release of sediments from the thermocline provides a winter lamina, which, in combination with the distinctive lamina provided by the continuous rain of suspended sediment during the summer months, constitutes a varve couplet (Sturm and Matter, 1978).

Introduction of cold, saline, or sediment-laden stream waters leads to underflow beneath the less dense lake waters. This process often accompanies the inflow of turbid glacial meltwaters during the spring thaw. The stream sediment load and caliber depend on climate and drainage-basin characteristics. High relief and a semiarid climate provide flashy discharge and a dominance of bed-load transport. Streams draining a hinterland of low relief, or evenly distributed rainfall, tend to contain a higher suspended and dissolved load. Underflow transports a mixed array of sediments to the deeper parts of the lake basin as semicontinuous turbidity currents. Above the level of the thermocline, interflows behave as density underflows.

Where the inflowing and lake waters are of equal density, rapid three-dimensional mixing causes rapid deposition of bed load, with the suspended material settling a short distance offshore. This localized sedimentation favors delta development.

Ground-Water Discharge

Discharge of ground water from springs is particularly important where the lakes are fringed by alluvial fans. These springs, which may constitute the only perennial water supply to closed lacustrine basins in arid regions, are concentrated along the contact between coarse, permeable alluvial-fan deposits and the underlying, relatively impermeable mud-flat or lake-floor sediments (Hardie *et al.*, 1978). Elsewhere, the zone of discharge is more broadly dispersed through the distal fan, the surface of which can support a dense stand of vegetation.

Outflowing spring waters feed small streams and rills, form ponds, percolate back into the substrate, or enter the lake directly. The discharging ground waters may be highly saline (Eugster, 1970) or relatively dilute. Where evaporation exceeds recharge, evaporite minerals are precipitated, the most diagnostic of which are travertine and tufa, which form coatings, mounds, and pinnacles (Scholl, 1960; Hardie *et al.*, 1978). Other chemical deposits related to springs include sinter, gypsum, halite, and glauberite. Plants may speed precipitation by photosynthetic uptake of carbon dioxide.

Lake systems supplied exclusively by groundwater inflow are readily identified in the rock record on the basis of extensive teepees, crusts, and caliche (Marjorie Muir, personal communication, 1982). Primary current-produced structures can be sparse or absent, in contrast to lakes supplied by rivers.

Waves

Shoreline wave processes resemble those along marine coasts, but are reduced in intensity and scale. Where the fetch is sufficient, waves rework the lake-shore sediments, particularly in areas of reduced clastic sediment supply. Storm-wave erosion may be dramatic because of ready removal of large portions of the narrow beach. Broad transgressions and regressions corresponding to changes in lake level spread a thin veneer of shore-zone sediments over a wide area.

In very shallow lakes, wave surge may be the dominant mechanism in sediment transport, planing off highs and redistributing the sediment into the intervening depressions. This process accounts for the regular depths of some coastal lakes subject to extreme wave activity (Orme, 1973). Hummocky cross stratification has a distinctly shorter wavelength (16 to 32 in; 40 to 80 cm) than in open marine shelf environments. Storm-wave mixing in deep, stratified lakes may cause seasonal lowering of the thermocline.

Wind Forcing

Impoundment of lake water against a shore by wind stress and its subsequent release are important in creating wind-tidal flats, currents, and seiches. These processes may be periodic, for example in response to diurnal wind changes, or they may be intensely seasonal. Alternating conditions of wind-forced submergence and emergence of a narrow lake margin simulate the effects of lunar tides, particularly those of microtidal coasts. Multidirectional, intermittent, low-energy wave activity, deposition of suspended mud, emergence runoff, and desiccation accompany these meteorologically produced changes.

Wind-forced currents on many lakes are likely to resemble the one-layer or two-layer systems of marine shelves (Chapter 7). These currents are equally capable of carrying large volumes of sediment both alongshore and offshore. Single-layer flow along the shallow inner lake results in circular water motion and the development of convergence patterns, which may ultimately lead to segmentation of an elongate lake (Bird, 1965; Hobday, 1976).

Seiches, produced by wind-forcing of a watermass against the shore and its sudden release, are solitary wave forms up to several meters in height. The wave is reflected from opposite shores, at a period determined by the lake dimensions. Seiches may disrupt stratification of the watermass, or lower the thermocline.

Climatic Changes

Physical and chemical processes in lakes change significantly in response to climatic fluctuations. Lowering of water level in response to aridification causes stream entrenchment, coarse bedload deposition at the stream mouths, and increased flocculation and chemical precipitation (McGowen *et al.*, 1979). Subaerial exposure of the marginal lake floor results in desiccation, while stranded shoreline deposits become dissected and modified by wind.

Rise in lake level with the onset of more humid conditions leads to flooding of incised valleys and adjacent flats, the reduction of stream gradient, and consequent reduction in sediment grade. This, coupled with stabilization of stream banks by vegetation, leads to a change in fluvial style to a generally meandering pattern.

Shallow lacustrine depressions may undergo profound fluctuation in surface area. For example, Lake Chad which has changed from 25,000 to 10,000 km^2 (9650 to 3900 mi^2) during the present century had attained 300,000 km^2 (116,000 mi^2) during the Pleistocene (Grove, 1970; Collinson, 1978b). Late Pleistocene Lake Dieri, which was the vast predecessor of Lake Eyre, Australia, developed during a wet climatic phase, but dried up with a return to aridity around 14,000 years ago. Slight climatic amelioration and a rise in the water table established the ephemeral Holocene Lake Eyre (Dulhunty, 1981).

Periodic changes in climate lead to alternating conditions of chemical and clastic sedimentation (Van Houten, 1965). Seasonal effects may be superposed on these longer term cycles.

Lacustrine Facies

Lacustrine facies reflect two general categories of sedimentation: lake margin plus shallow lake floor, and deeper water. Lake-margin sediments may be contributed by shoreline erosion and reworking, or may be concentrated near stream mouths. Deeper water (basinal) sedimentation is strongly influenced by density stratification.

Deep, Oxygenated Lake-Basin Facies

The deposits of most deep lakes in cold or temperate climates reflect seasonal overturn, which causes mixing of oxygen-depleted bottom waters and highly oxygenated surface waters, coupled with density underflows of colder, sediment-laden stream waters. Overturn results in characteristic nonglacial varving. Underflows may be more or less continuous during the rainy season. Larger-scale gravity resedimentation involves slumping and avalanching, with associated turbidity currents and debris flows. This combination of processes produces a characteristic alternation of coarse graded units and lake-floor muds.

In Lake Brienz, Switzerland, for example, catastrophic floods and landslides once or twice a century severely erode the delta-front and slope environments, depositing turbidites up to 1.5 m (5 ft) thick at the base of the slope, extending across the basin plain (Fig. 9-2). These massive sands merge distally into graded silts, and finally into discrete silt and clay laminae. The more frequent low-density turbidity underflows related to seasonal floods deposit thin graded sands that merge over a short distance into basin-plain silt laminae. These underflow laminae are practically indistinguishable from associated summer varves related to overturn (Sturm and Matter, 1978).

Underflow of the clear waters of Lake Geneva by cold, turbid waters of the Rhone River has produced a complex subaqueous topography of channel, levees, and fan (Houbolt and Jonker, 1968). Ripples and sand waves up to 50 cm (1.7 ft) high form in shallow water under currents averaging 5 cm/s (0.17 ft/s). Finer suspended sediments spread out across the thermocline while sand continues to move downslope as ripples (Shepard and Dill, 1966). Graded, horizontally laminated sands with subordinate cross bedding characterize the deep channel floor, with upward-fining wavy-laminated silts in the flanking levees. Cross bedding and convolute lamination increase downslope toward the fan, which consists of somewhat finer sand flanked by graded and interlaminated sand, silt, and clay of the fan margins. The thickness of the Lake Geneva fan sequence is approximately 50 m (165 ft) (Houbolt and Jonker, 1968).

Depending upon the relative effectiveness of

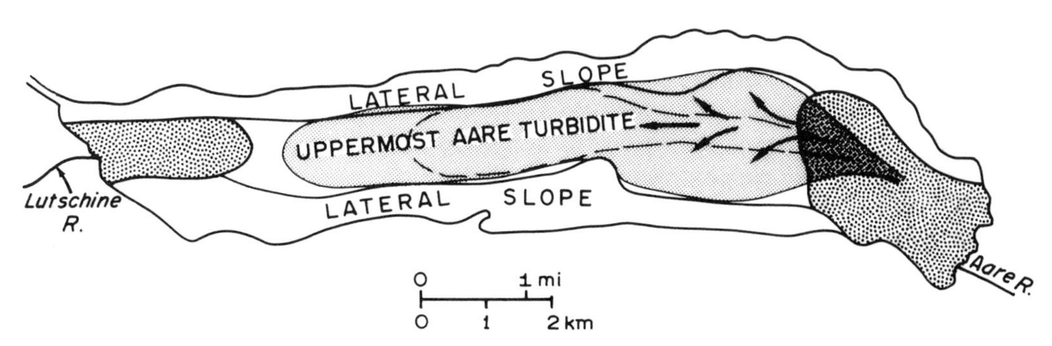

Figure 9-2. Lake-head deltas and the extent of one turbidite bed in Lake Brienz, Switzerland. (Modified after Sturm and Matter, 1978.)

carbonate productivity and dilution by terrigenous influx, carbonate content can increase or decrease toward the lake center (Picard and High, 1981). Offshore decrease in clastic grain size is generally accompanied by an increase in dispersed organic matter (Fig. 9-3), which settles permanently only below storm-wave base. Proximity to the lake shore in ancient lake systems may be marked by concentrations of vertebrate footprints, and remains of insects, conchostraca, ostracods, and charophytes (Bradley, 1925; Van Dijk *et al.*, 1978; Burne *et al.*, 1980).

The ideal vertical succession developed by infilling of a deep, oxygenated lake subject to active fluvial influx and gravity resedimentation would be upward-fining (subaqueous fan to slope with channel-fill), and then upward-coarsening (prodelta slope to delta front and delta plain). A simpler, and possibly far more widely represented sequence (Fig. 9-4), arises from fluvial infilling of lakes dominated offshore by slow suspension sedimentation. The thickness of the sequence would provide some indication of the original water depth. Upward gradation from storm-graded laminae in a predominantly shale sequence, to hummocky cross stratification produced above storm-wave base, to rippled sands reflecting fair-weather wave processes, to a dominance of current ripples and channel-fill, permits a general reconstruction of nearshore bathymetry and processes (see Chapter 7). Sequences of this type are well represented in the eastern Karoo Basin (Hobday, 1978b), but as pointed out by Picard and High (1981) the majority of carefully documented ancient lake systems show somewhat unpredictable vertical patterns.

Deep Anoxic Lake Basin Facies

This facies is distinguished by organic-rich shales and marls that accumulate in deep-water environments hostile to most organisms. The shallow nearshore zone is aerated by inflowing waters, wave agitation, and circulation above the thermocline. Below the thermocline a reduced substrate is maintained beneath stagnant, anoxic bottom waters. Organic oozes produced by settling of microorganisms from the aerated surface-water layers alternate seasonally with silt, clay, and authigenic sediments to produce extensive varvites. These bottom sediments typically have a very high organic carbon content, for example reaching 15 percent in Lake Kivu sediments. These high values are ascribed by Demaison and Moore (1980) to lacustrine processes of fresh-water fermentation, which contrast with the more efficient sulfate-reduction mechanism operating in anoxic marine environments.

Several ancient deposits of large anoxic lakes are excellent oil source rocks or oil shales. The Mahogany oil shales of the Eocene Green River Formation accumulated on a permanently anoxic lake bottom, as evidenced by absence of bioturbation in minute varves of the brown dolomite marlstone (Smith and Robb, 1973). The lower part of the Green River Formation originated in a fresh-water lake that subsequently became alkaline.

Vast oil-shale resources are present in the Permian Irati Formation of Brazil, which with correlative carbonaceous shale of the Karoo Basin of southern Africa probably originated in permanently stratified lakes (Anderson *et al.*, 1977). These bituminous and pyritic black shales with carbonate bands grade into sandier facies toward the lake margins. Triassic rift-basin lake deposits of the Connecticut Valley also show a shore-parallel zonation, with black pyritic-mudstone facies of the deep central basin bounded by shallower-water gray mudstones containing couplets of dolomite and black shale. These thin couplets are thought to mark seasonal changes (Hubert *et al.*, 1976).

Perennial Alkaline Lake-Basin Facies

These deposits reflect precipitation from a dense, saline lower water layer as material rains down from the fresh-water upper layer. Progressive enrichment of bottom brines deposits a sequence that typically comprises a fresh-water, organic-rich unit overlain by alkali-earth carbonates, followed by gypsum and halite (Hardie *et al.*, 1978). The composition and proportion of saline minerals varies according to the brine chemistry. In large saline lakes, evaporite precipitation would be expected to increase with shallowing, particularly along embayed or otherwise restricted shores. But precipitation is generally quite evenly distributed across lake floors (Picard and High, 1981). This pattern is confirmed in several ancient deposits of saline lakes, for example by

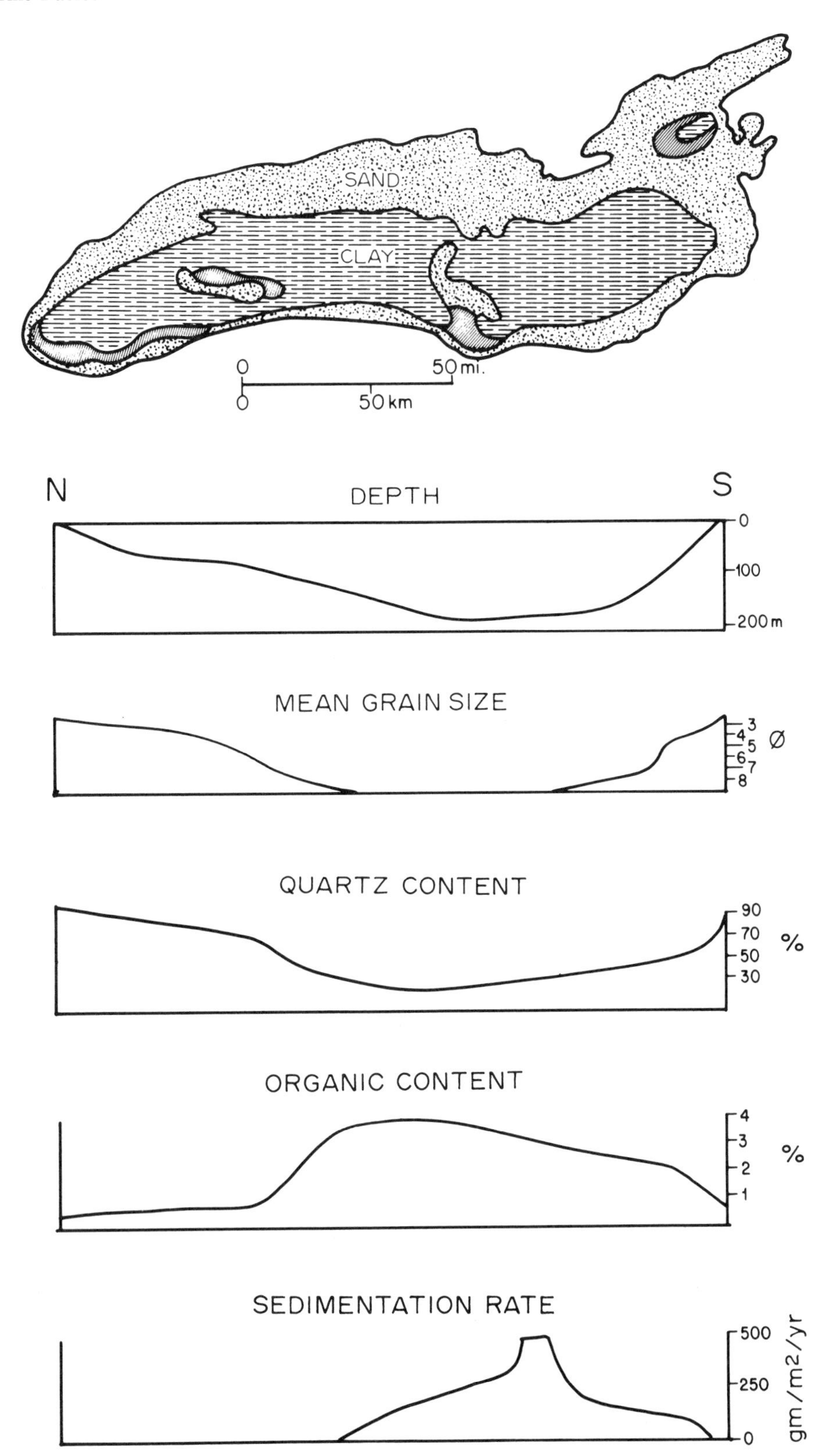

Figure 9-3. Distribution of clastic sediments in Lake Ontario and asymmetric north–south distribution of depth, composition, and sedimentation rates. (After Picard and High, 1981.)

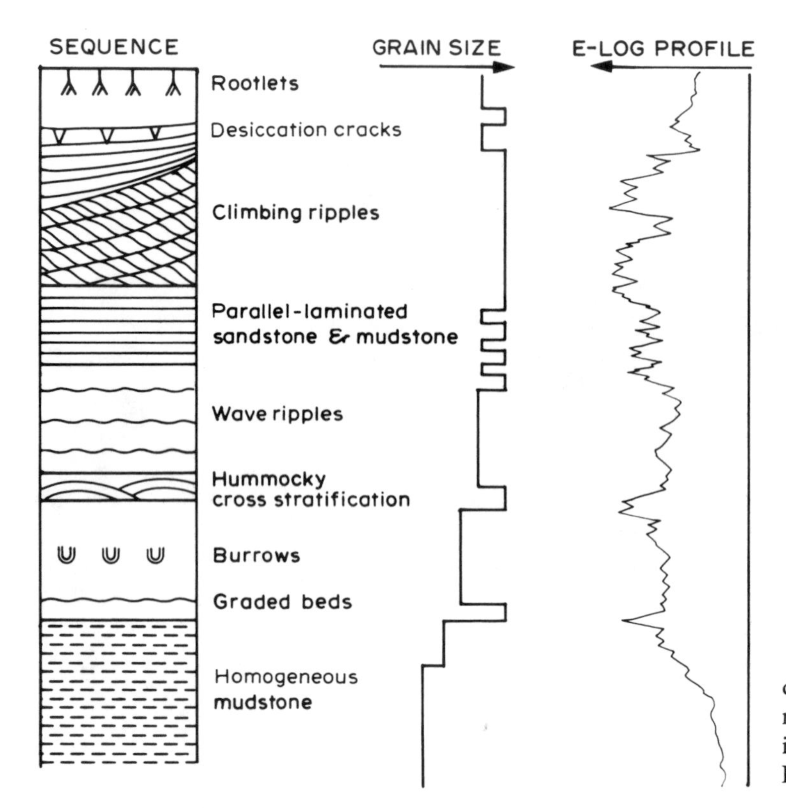

Figure 9-4. Idealized vertical sequence arising from simple lake-margin regression as a result of fluvial influx. Based on the eastern Karoo Basin (Van Dijk *et al.*, 1978.)

the lateral persistence of glauberite crystals in the Lockatong Formation (Van Houten, 1965).

Thin mud partings tend to be interlayered with the saline minerals. These clastic-chemical couplets normally represent irregular timespans, and are therefore not true varves. Continuous chemical precipitation is punctuated by clastic influx. Most of the fine clastic sediment layers are introduced by slow settling, but during rare floods, unconfined sheetfloods may continue beneath the lake as a dense underflow that deposits a graded clastic unit (Hardie *et al.*, 1978).

Oolite shoals and algal mats, oncoliths, and bioherms grow in nearshore areas and on shallow offshore platforms. These features are documented in the Great Salt Lake, Eocene Green River Formation (Picard and High, 1981), and Pliocene Glenns Ferry Formation (Swirydczuk *et al.*, 1980). Oolitic carbonates built large terraces (Fig. 9-5) along the margins of a Pliocene rift-valley lake in the Snake River Plain, and responded to cyclic changes in water depth. The terraces prograded during episodes of stable lake level. Ooids grew on the bench surface prior to avalanching down the steep foresets inclined at angles of 26 degrees. Grain-flow deposition on the upper foresets produced inverse grading, whereas fluidized flow on the lower foresets produced normal grading with dish structures. The foreset-bedded unit attains 60 ft (18 m), and is overlain by low-angle topsets and massive, burrowed, transgressive units a meter thick (Swirydczuk *et al.*, 1980). Falling lake level led to subaqueous erosion of the terrace. The algal-capped oolitic shoal sequences described by Picard and High (1972) were also related to high lake levels (Fig. 9-6), and were overwhelmed with terrigenous sediments as lake level fell.

Borates and trona are unique to saline lake deposits, and certain assemblages of authigenic minerals produced by reaction of saline lake waters with clays, feldspars, or volcaniclastic sediments are also diagnostic (High and Picard, 1972). These include diverse zeolites, particularly analcime, and other silicates, carbonates, sulfates, and phosphates.

Postglacial, temperate hard-water lakes of North America typically develop a sedimentary sequence that reflects progressive shrinking in size and encroachment of marsh and swamp (Fig. 9-7). Basal clastic sediments of glacial-drift origin are overlain by marly and organic-rich clays as organic productivity of the lake increases.

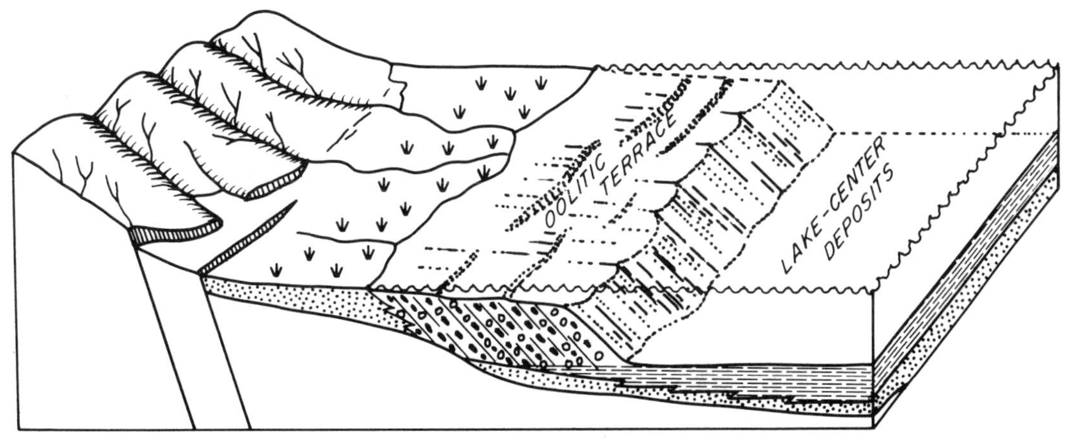

Figure 9-5. Schematic reconstruction of oolite terraces along a lake margin. (After Swirydczuk *et al.*, 1980.)

Floating sedge peats extend basinward as the lake shallows (Dean, 1981).

Sodium-rich bottom waters are indicated by the shale geochemistry of some oil shales, for example in the Green River Formation, implying a salinity-enhanced permanent stratification of the watermass, with bottom conditions lethal to macrolife (Demaison and Moore, 1980).

Lake-Shore Facies

Large lakes are fringed by a variety of shoreline features including beaches, spits, bars, deltas, and broad wind-tidal flats periodically inundated and exposed in response to changes in wind velocity. Wind-tidal flat and delta facies are volumetrically most important.

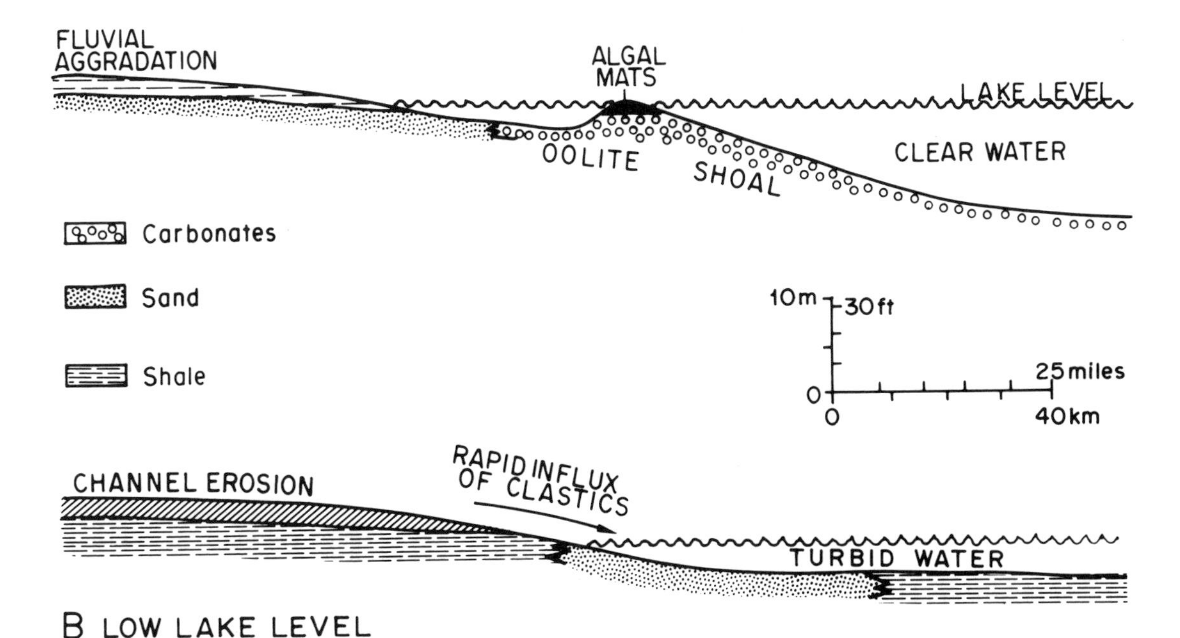

Figure 9-6. Changes in sedimentation related to fluctuating lake levels. (After Picard and High, 1968.)

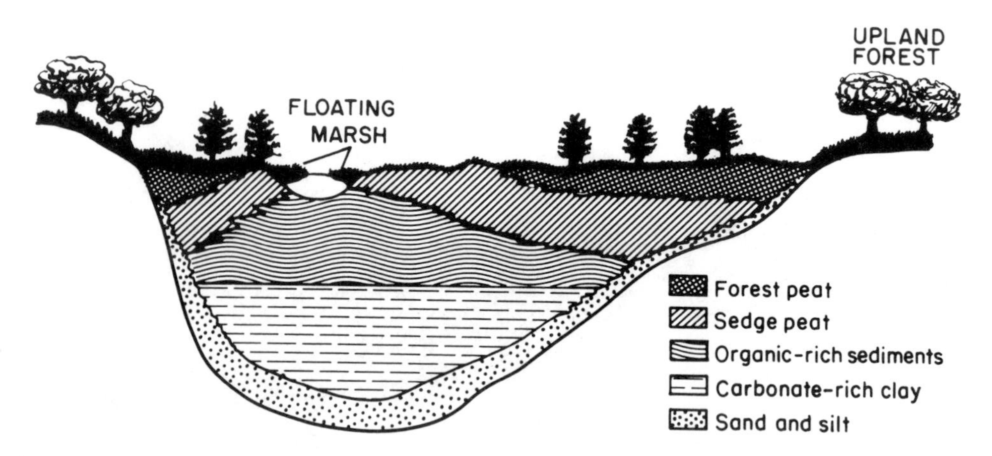

Figure 9-7. Sediment-starved lake basin subject to peat encroachment over a substrate of clay, silt, and sand. (After Dean, 1981.)

Lacustrine delta sub-facies vary according to factors of stream discharge, sediment caliber, wave fetch, and water depth. They have variable preservation potential. For example, areally extensive but thin deltaic deposits produced during flooding of Lake Eyre are rapidly remolded into eolian dunes as the lake falls to its normal low level. In more stable lake systems with significant fluvial influx, deltaic facies will be an important component of the sedimentary record. Textural patterns and internal geometry of the deltaic facies are strongly influenced by relative water densities. Where river and lake waters are of approximately equal density, homopycnal flow leads to rapid mixing and the development of classical Gilbertian foresets with a height determined by water depth, and inclinations up to the angle of repose. The finer sand and silt fractions contain abundant climbing ripple lamination, reflecting rapid fallout from suspension. Foreset avalanching is normally discontinuous, with intervening periods of wave, current, and biological reworking. An ancient example comprising conglomeratic sandstone foresets as much as 100 ft (30 m) high (Fig. 9-8) shows evidence of intermittent migration of smaller bedforms down, along, and occasionally directly up the slope of the major foresets. Muddy partings and rippled siltstones thicken and merge into the fine-grained bottomsets.

Where cold, turbid stream waters are sufficiently dense for hypopycnal underflow to occur, the lacustrine delta-front geometry will be more complex as a result of partial erosion and redistribution of the sediments. Foresets of deltas entering Lake Brienz are traversed by channels which funnel coarse sediments to the deep basin plain (Sturm and Matter, 1978). These channels switch position, scouring the delta front irregularly and replacing it with well-sorted massive sands with pebbly bases, representing the coarsest reworked fraction. Distal deposits which mantle the lower delta front and prodelta slope are alternating sand, silt, and clay layers showing grading and lenticular bedding. Similar deposits are also present more proximally on levees and in interchannel areas.

Deltas of elongate, lobate, and arcuate types are all represented along lacustrine shores, but river-dominated elongate or lobate forms are prevalent because of the low efficiency of shoreline processes relative to the high rate of sediment supply by major streams. Wave reworking characterizes very small deltas of small or intermittent streams, particularly those with a sandy load. The resulting wave-dominated arcuate deltas grade laterally into spits, bars, and subdued strandplains.

Permian Karoo Basin examples of lake-delta progradation fed by mixed-load streams comprise wave-rippled siltstones grading upward into silty sandstone with wave and current ripples (Fig. 9-4). The overlying mouth-bar sandstones are erosively based near major distributary axes, but are gradationally based over most of their extent. Delta-plain facies such as levee, splay, pond, and marsh tend to be thin. Rootlet beds, concretion layers, and coal laminae are characteristic. Some

Figure 9-8. Large foresets produced by lacustrine delta progradation, northern Sydney Basin, Australia.

of these layers are strongly oxidized, with large polygonal desiccation cracks, and may record seasonal lowering of ground-water level. The remains of swamp foragers such as the mammal-like reptile *Lystrosaurus*, which occupied a habitat comparable to the modern hippopotamus, are common (Van Dijk *et al.*, 1978). Smaller aquatic reptiles are typical of the contiguous delta-front and prodelta facies.

Wind-tidal flat subfacies of lake shores contain diverse small-scale structures comparable to those of tidal flats, the only distinction being an absence of well-defined textural zonation. This is because wind-tidal flats vary markedly in position with change in lake level. Features of wind-tidal flat deposits which bounded major lakes include lenticular, wavy, and flaser bedding, complex ripple forms with mud drapes and desiccation cracks, runzel marks, adhesion ripples, rills, and etch marks (Van Dijk *et al.*, 1978). Associated fresh-water fossils are typically fish, conchostraca, mollusks, insects, and plant debris.

Shallow, Oxygenated Terrigenous Lake Facies

Deposits of very shallow but fairly persistent waterbodies resemble wind-tidal flat deposits in their diversity of shallow-water structures and widespread oxidation. But there is no evidence of subaerial exposure except along the margins, or of evaporative precipitation of minerals, which sets these deposits apart from typical wind-tidal flat and playa-lake facies. Typical deposits are massive or laminated clays, rippled sands and silts, and graded parallel laminae. Mudstone and siltstone facies are generally light gray, greenish, or reddish. Organic matter tends to be sparse; only impressions of stems and leaves are preserved, but some thin beds contain carbonized plant material and abundant pyrite.

Preservation of these nonoxidized beds is attributed to reducing conditions within the shallow substrate (cf. Sellwood, 1971), with particularly organic-rich layers persisting beneath the constantly oxygenated lake bed. But turbulence generated by wind stress, together with infaunal burrowing, generally lead to pervasive oxidation.

The graded parallel laminae result from sheet floods, differential settling of storm-suspended sediments, and the effects of intense wind shear on shallow water. Muddy intercalations are burrowed, and locally preserve delicate arthropod tracks, and trails left by fish brushing against the lake floor. Ripples and associated cross lamination record a combination of wave oscillation and current processes, with linear ripple trends, in places, providing an indication of shoreline orientation. Commonly, however, these ripple trains intersect at large angles, reflecting only the shifting wind patterns. Complex shoreline associations of desiccation-cracked, concretionary, and root-penetrated units interfingering with lake sediments may indicate seasonal lowering of water level. Stable episodes during the retreat of Triassic lakes in southern Africa are marked by well-defined terraces with emergence features; these terraces were subsequently preserved beneath a covering of muddy sediments corresponding to renewed expansion of the lakes (Stear, 1978). Drying out of marginal depressions produced calichified, diagenetically reddened surfaces scattered with vertebrate remains reflecting the activities of terrestrial carnivores (Smith, 1978). Deeper portions which persisted unchanged during episodes of maximum contraction of the lake system are represented by gray and purple mudstones containing large numbers of fresh-water fish and semiaquatic herbivore remains.

Ephemeral Saline-Lake Facies

The deposits of ephemeral saline lakes, or playas, reflect broad fluctuations in surface area with repeated desiccation leaving layers of evaporites (Reeves, 1968, p. 87). Some, such as Lake Eyre, Australia, are vast, whereas others form in small interdune hollows less than a kilometer across. Significantly larger lake systems of this type formed during Quaternary stages of maximum aridity. For example, Central Australia was covered by a shallow lake system some six times larger than at present (Loffler and Sullivan, 1979).

The Lake Eyre Basin extends over an area of 6200 mi^2 (16,000 km^2), with a catchment basin covering one-sixth of the Australian continent (Dulhunty, 1981). The lake is situated in the southwestern Great Artesian Basin, the locus of maximum ground-water discharge. Because of prolonged residence time in the enormous tract of Mesozoic strata, these ground waters discharging into Lake Eyre are saline, and concentrations of dissolved salts are rapidly increased by evaporation. Infrequent but violent surface flooding lowers the overall salinity dramatically, but the saline lower layer may be maintained by density stratification. As the fresh-water plume spreads across the top and mixes, saline ground-water discharge replenishes the dense bottom layers. Fresh-water fauna and flora become widely dispersed across the lake system. These flow patterns, in conjunction with a southerly lake-floor gradient, have led to the development of three distinct facies tracts (Dulhunty, 1981): a northern well-drained playa facies of red and yellow clay with thin salt crusts; a central zone of gelatinous flocculated clays and organic material moulded into giant transverse bedforms during floods; and a southern zone of maximum subsidence where a thick salt crust overlies gypsiferous slush charged with hydrogen sulfide.

Clastic sediments are introduced to playa lakes by sheet floods, channelized flow, and eolian processes. Deltas at the mouths of ephemeral streams tend to be destroyed as lake level falls. Sheet-flood deposits, comprising graded, parallel-laminated sand and silt with sole marks, have higher preservation potential, and are frequently identified in the rock record (Tucker and Burchette, 1977). Farther out in the lake, graded silt and clay laminae settle from suspension. Very fine sediments tend to be sparse, possibly because of the general deficiency of clay-size weathering products in an arid environment. These thin storm-generated clastic layers are commonly overlain by a lake-floor algal mat produced by a major bloom of blue-green algae from spores carried in by the flood waters (Hardie *et al.*, 1978). Where the water table drops substantially and wind deflation ensues, the dried-out algal mat and underlying clastic and gypsiferous layers are reworked into transverse dunes, or lunettes (Stephens and Crocker, 1946), on the downwind lake margin.

Prolonged evaporation increases salinities until halite, trona, or other salts precipitate as a continuous layer. Salt crystals grow simultaneously in the underlying clastic layers, which become black and anaerobic as a result of bacterial reduction of sulfates (Baas-Becking and Kaplan, 1956). Comparable deposits in the rock record are pyritic black shales interlayered with evaporates. Precipitation of evaporites may be concentrated in a zone of ground-water discharge around the lake margin; alternatively, it may be greatest near the center of the lake (Amiel and Friedman, 1971). Concentric zonation of facies surrounding lake-center halite deposits was noted

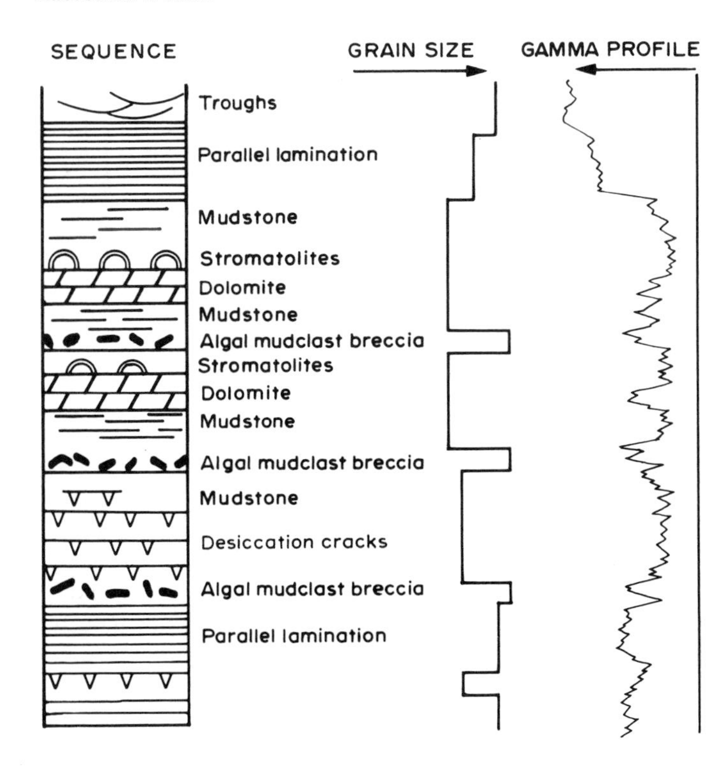

Figure 9-9. Schematic alternation of playa-lake and lake-margin clastic sediments; showing generalized grain-size trends and gamma-log profile. Gamma log indicates increasing sand or limestone with a deflection to the left.

by Handford (1982); sheet floods and rill flow deposit sand to mud-sized peripheral tracts grading into saline mud flat with displacive halite crystals, followed by chaotic mud-halite saltpan beds. Very similar facies arrangements are present in Wilkins Peak Member of the Green River Formation and in the Permian evaporites of Texas and Oklahoma (Smoot, 1978; Handford, 1980).

Some sedimentary deposits previously ascribed to deep eutrophic lakes are now interpreted in terms of a shallow-lake model (Eugster and Surdam, 1973), and display cycles of transgression and regression (Eugster and Hardie, 1975). At the base are desiccation-cracked mudstones overlain by a transgressive lag of dolomitic mudclasts. Oil-shale facies comprising laminated organic-rich dolomites and breccias accumulated in shallow water that periodically dried out. Delicate fungal and insect remains are preserved in the oil shales. Stromatolites grew in hypersaline conditions along the lake margins, whereas trona and halite precipitated from alkaline brines.

Thin cycles are also recognized in several hundred meters of playa-lake deposits in the Cambrian of Australia (White and Youngs, 1980). These cycles comprise sapropelic mudstone containing evaporite pseudomorphs, overlain by silty dolomitic mudstone with chert pebbles in its upper part, capped by algal boundstone and mudclast breccia. The depositional environment is visualized as playa lakes that underwent annual flooding and drying out, as is the case in some modern Australian playas (Von der Borch and Lock, 1979). Where the water table did not fall below the sediment surface during the dry seasons, organic matter was preserved under reducing conditions. These deposits are excellent potential sources of petroleum.

Ancient playa-lake systems may attain astonishing thicknesses of 1000 m (3300 ft) and more (Reeves, 1977). The proportion of evaporites to terrigenous sediments in the successions varies from almost entirely evaporitic to largely clastic. Apart from the above-mentioned small cycles, vertical sequences in playa-lake deposits tend to be somewhat irregular, but infilling by terrigenous lake-margin sediments produces a characteristic upward-coarsening sequence (Fig. 9-9). Predominantly clastic sequences of this type in the Triassic Clarens Formation of South Africa preserve a variety of lake-margin structures, including the abrupt termination of raindrop impressions coincident with minute lake-edge terraces, and footprints produced by hordes of dinosaurs (Van Dijk *et al.*, 1978).

Geometry, Distinguishing Characteristics, and Evolutionary Patterns in Lake Systems

The diverse range in size, shape, depth, tectonic behavior, climatic setting, and water chemistry of lake basins precludes the development of anything but the most general models for lacustrine systems. Facies architecture and interrelationships of lacustrine facies may not be unique to lakes. For example, some large, ancient lake systems invite comparison with epeiric marine systems, and in Precambrian rocks the distinction between these two depositional systems may be very difficult to discern. Most Phanerozoic lake systems have been identified on a paleontological or mineralogical basis (Picard and High, 1972). On a smaller scale, lake facies are commonly regarded as components of regional fluvial or deltaic systems.

Being essentially closed basins, lakes trap a very large proportion of the sediment that is transported into them. As a result, terrigenous sedimentation rates in lakes may be at least ten times more rapid than in oceanic environments (Sly, 1978). This rapid infilling, coupled with limited wave fetch and no astronomical tides, will generally be reflected in a lower degree of physical and biological reworking compared with marine basins.

The most diagnostic sedimentological, geochemical, and mineralogical characteristics of lake systems arise from their sensitivity to climatic factors which affect water temperature and composition and bring about rapid changes in lake level.

The Vertical Succession

Progressive infilling of lake basins with clastic sediments should produce an upward-coarsening sequence (Fig. 9-4) with a thickness reflecting approximately the original lake depth. The ideal autocyclic lacustrine sequence (Visher, 1965) is confirmed in small lakes where coarse peripheral clastics encroach on lake-center silts, muds, and chemical sediments, but it generally does not hold for the deposits of larger lakes (Picard and High, 1981). Unpredictable vertical textural patterns may arise from an overprint of climatic or tectonic effects, an irregular areal distribution of facies tracts, or a prevalence of gravity resedimentation. The latter may produce a thick, coarse-grained complex of lake-basin fan sediments at the base of the succession. But because lakes are transient and eventually fill with sediments of prograding peripheral environments, the gross pattern will tend to be upward-coarsening.

Cyclicity in Lake Systems

Cycles of the order of tens or hundreds of years to 21,000, 100,000, 500,000 years and longer have been attributed to solar causes or perturbation in the earth's orbit, but their precise duration is open to question (Picard and High, 1981). The associated climate changes and their effects on lake sedimentation are well documented (Bradley, 1929; Van Houten, 1964). Comparable cycles could conceivably result also from tectonic controls.

The effects of lake-level fluctuations on integrated facies patterns are illustrated by Picard and High (1972), McGowen *et al.* (1979), and Swirydczuk *et al.* (1980). Figure 9-6 is a reconstruction of fluvial aggradation and the development of oolite shoals and algal mats in the clastic-starved nearshore area during episodes of high lake level. A drop in lake level caused subaerial dissection and soil formation, rapid deposition of coarse clastic sediments in the nearshore zone, and accumulation of mud at the center of the lake. The Triassic Dockum lake system (Figs. 9-10 and 9-11) was surrounded by fan deltas along its more rugged margins, with elongate or lobate deltas along its more subdued margins which were prevalent during high lake stands. Falling lake levels during arid climatic episodes caused headward erosion of lake-margin ravines, which cut into fluvial and deltaic facies deposited during the preceding progradational episode. The eroded sediments were deposited as smaller fan deltas composed largely of reworked mud clasts and other fine-grained sediments. The ravines were filled with lacustrine and deltaic sediments as water level rose once again.

Pronounced cyclicity is also apparent in the deposits of oolitic lake-margin terraces (Fig. 9-5) which prograded during high lake level and were subject to subaqueous erosion as the lake level fell. Massive oolitic deposits accumulated on the terrace during renewed rise in water level.

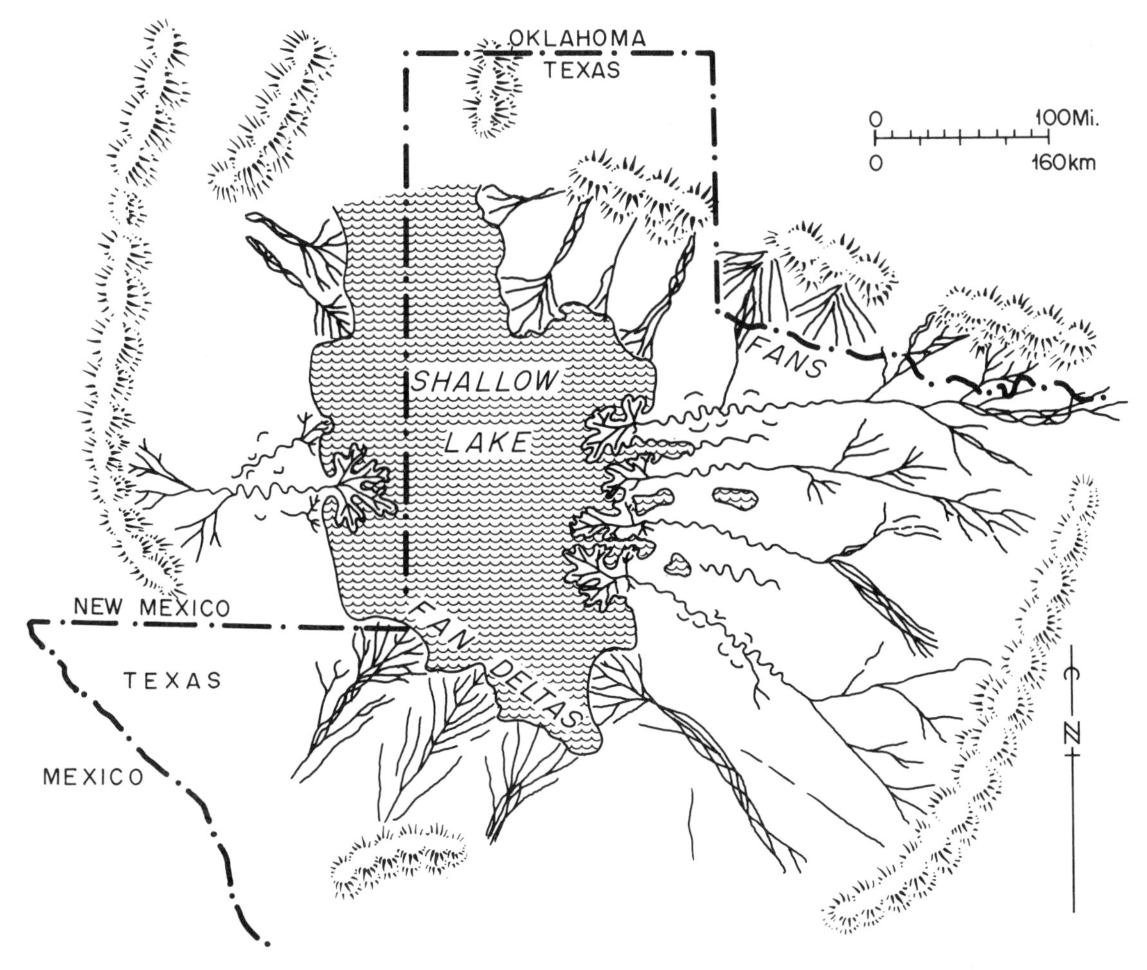

Figure 9-10. Convergent fluvial and deltaic systems of contrasted styles surrounding an extensive shallow lake of fluctuating outline, Triassic Dockum Group. (After McGowen *et al.*, 1979.)

Evolution of Lacustrine Basins

Lacustrine systems are related to a variety of structural settings, the most important being rift and foreland basins, pull-apart basins associated with strike-slip faulting, and intracratonic downwarps. Rifts and accompanying lake systems form during the initial stages of continental breakup (Falvey, 1974; Veevers, 1981), and rift-basin fill, including lake deposits, is an important part of the tectono-sedimentary record of passive continental margins (Fig. 8-1). This record typically consists of pre-rift basement, pre-rift unconformity, rift-basin deposits, post-breakup unconformity, and offlapping post-breakup clastics and carbonates. Some of these rift-basin deposits comprise thick fluvial arkoses and lacustrine evaporites, but others include organic-rich shales and marls of deep, stratified lakes. These shales may be so impermeable as to prevent migration of the abundant hydrocarbons that they generate, thus constituting oil-shale reservoirs. Some hydrocarbons may accumulate in lake-margin or deeper-water clastic reservoirs, with large trapping structures generated by contemporaneous basement tectonism.

The thickness of lacustrine deposits largely depends upon tectonic setting. Thus, in the western United States, abnormally thick Tertiary successions amounting to thousands of meters (Feth, 1964) resulted from intense subsidence associated with plate movements (Picard and High, 1981). Stable cratonic lake deposits tend to be thin, although some cover very large areas.

Many larger lakes evolved from epeiric basins, and some fresh-water lakes converted into silled

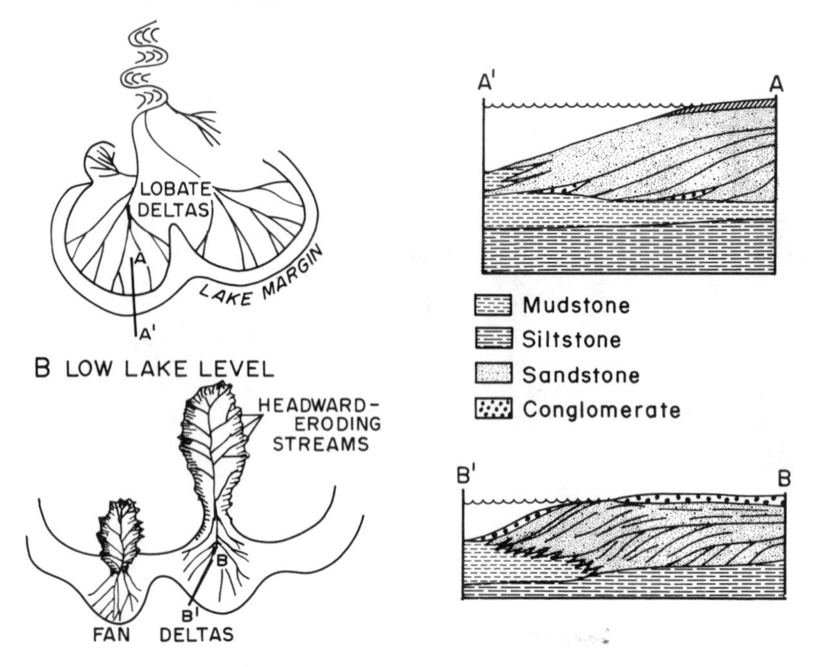

Figure 9-11. Changes in deltaic sedimentation brought about by changes in climate and lake level. During high-stand humid phases, meandering streams fed extensive lobate deltas. These deposits were eroded as the climate became more arid and lake level fell, with fan deltas developing along the lake margin. (Modified after McGowen *et al.*, 1979.)

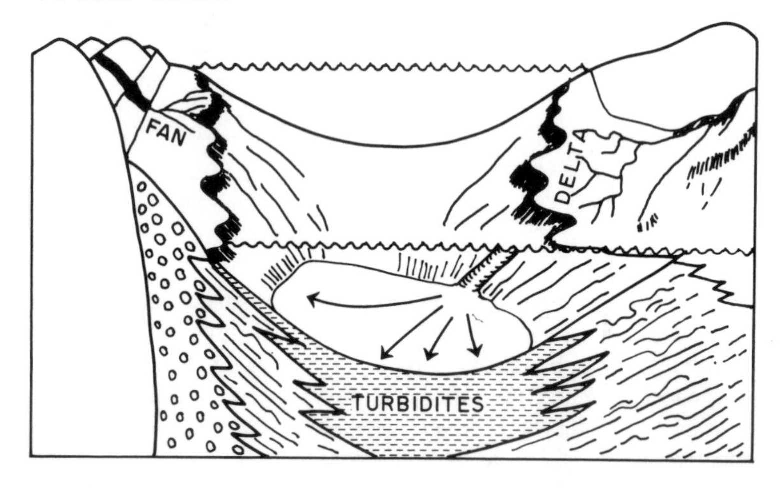

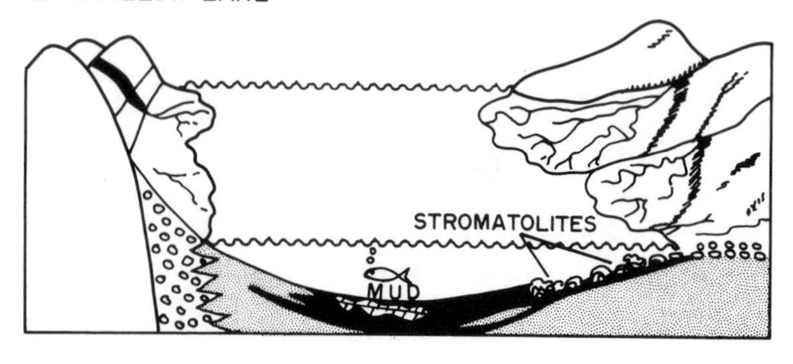

Figure 9-12. Transformation of a narrow, tectonically active basin to a shallow lacustrine basin, Ridge Basin, California. (After Link and Osborne, 1978.)

marine basins or shallow seas. The Black Sea was a fresh-water lake 22,000 years ago and was subsequently invaded by the rising waters of the Mediterranean. Indeed, the Black Sea was a deep fresh-water lake comparable to Lake Zurich during much of late Neogene time (Hsu and Kelts, 1978). As a Miocene remnant of the Tethys, the Black Sea became gradually isolated, with black shales of a late-Miocene brackish environment giving way to predominantly chemical sedimentation. These deposits varied with changes in water level and chemistry. Calcite chalk accumulated under deep, stratified, fresh-water lacustrine conditions, but was replaced by aragonite, magnesian calcite, and dolomite as it changed to a brackish environment and then into a shallow salt lake (Hsu and Kelts, 1978).

The Pliocene Ridge Basin of California (Fig. 9-12) changed from a deep, semi-restricted marine embayment into a closed lacustrine system in response to faulting. During its subsequent history as a shallow fresh-water lake, the Ridge Basin accumulated muds and carbonates, some of which are dark and organic rich (Link and Osborne, 1978). Typically, the lacustrine system gave way to entirely subaerial environments as alluvial-fan and fluvial sediments filled the basin.

Chapter 10

Eolian Systems

Introduction

Modern environments in which wind is the dominant agent of deposition range from tropical to polar, the sole requirements being the availability of uncohesive sediment and an incomplete cover of vegetation. Hot, arid regions, the most favored locale for eolian accumulation, are typical of tropical to subtropical high-pressure regions, of mid-latitude rain shadows and landlocked continental interiors, and of the western margins of continents swept by cold ocean currents or subject to upwelling. Humid regions dominated by eolian deposition generally involve copious sand supply, as along beaches and broad fluvial tracts, or very cold, unstable conditions that inhibit plant growth. High-latitude deserts result from the aridity associated with divergent air flow. Pleistocene circulation of this type redistributed glacial sand into substantial dune systems in North America and northern Europe (Glennie, 1970, p. 6). Tropical and subtropical deserts, too, expanded dramatically during the most arid phases of the Pleistocene, with the "Kalahari Sands" extending into parts of the Congo and Zambezi drainage basins which are densely wooded today.

Deserts cover approximately one-fifth of the earth's landsurface, but the spectacular sand seas, or ergs, constitute only one-fifth of the total desert area (Cooke and Warren, 1973), the remainder comprising alluvial fans, playas, eroding highlands, and vast stony plains or regs. Linear coastal dunes are areally far less significant, but coastal regression can produce an extensive eolian blanket sand.

Studies of eolian systems have been spurred by their association with hydrocarbons, particularly in Pennsylvanian to Jurassic formations of the United States Western Interior and the Permian Rotliegendes of the North Sea, which contain chemically cemented interdune facies that act as an excellent seal. Elsewhere, eolian sandstones with outstanding porosity and permeability are singularly unproductive because of distance from source beds and lack of seals, but recent recognition of intraformational seals (Lupe and Ahlbrandt, 1979) has significantly enhanced their prospectivity. Eolian systems are locally important as hosts for epigenetic uranium. Although they are excellent aquifers, eolian sands tend to be barren geochemically, and epigenetic uranium mineralization, as in the Poison River Basin (Galloway *et al.*, 1979b), therefore requires the introduction of a reducing agent. Finally, early-cemented beach-dune ridges of Quaternary age locally enclose extensive peatswamps, for example along the southeastern coast of Africa (Hobday, 1976). It is conceivable that some mined coals have a similar origin, although the amounts are likely to be small.

Texture

Wind is by far the most effective sorting agent. Eolian deposits are typically fine to medium-grained, quartzose sand, although there are examples of gypsum (McKee, 1966) and clay dunes (Huffman and Price, 1949). Eolian transport involves a dominance of saltation (Bagnold, 1941) with some creep of larger particles, suspension of fines, and gravitational transport on slopes. Continued grain impact provides effective rounding, pitting, and frosting. The smooth eolian grain surfaces observed under electron microscopy were attributed by Krinsley and Doornkamp (1973) to prolonged attrition and temperature-dependent solution and reprecipitation.

Sands of interior deserts tend to be medium-grained and moderately to well sorted (Ahlbrandt, 1979). Interdune deposits, in contrast, are poorly sorted and often polymodal. Coastal dunes inherit a degree of textural maturity from their beach-sand source, and tend to be fine grained and well to very well sorted. Skewness and kurtosis of dune sands are variable. Graphical methods have proved somewhat equivocal in distinguishing among eolian, fluvial, and littoral sands, but discriminant analysis has been quite effective (Moiola and Spencer, 1979).

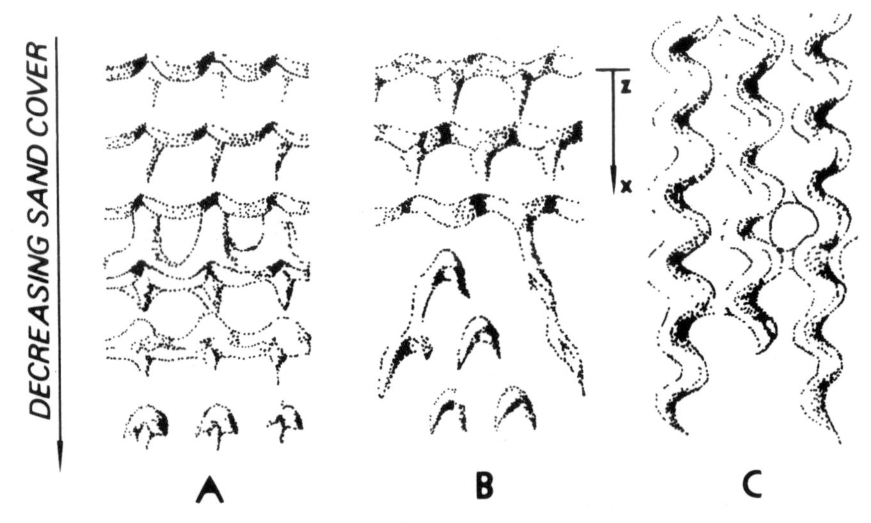

Figure 10-1. Eolian bedform patterns resulting from a combination of longitudinal and transverse elements, and changes that result from a decrease in sand cover. Transverse dunes evolve into barchanoid forms in A and B, with prominent longitudinal arrangements in B and C. (After Wilson, 1972.)

Silt particles are not readily entrained by wind, but once in suspension can be transported considerable distances before accumulating as loess, which is characteristically massive, soft but cohesive, and highly porous. Most of the widespread Pleistocene loess deposits were probably derived from areas of glacial outwash. Another source is arid terrain undergoing deflation, as exemplified by the Dust Bowl clays of the western United States (Swineford and Frye, 1945).

The red iron-oxide pigmentation of many desert sands derives from decomposition of ferromagnesian silicates such as hornblende and augite, and from adhesion to particle surfaces of iron-bearing clay components of airborne dust (T.R. Walker, 1979). There is a general tendency toward increasing reddening with age.

Eolian Bedforms and Structures

Eolian bedforms are present in a variety of sizes and shapes, of which Wilson (1972) recognized three major groupings of different wavelengths: ripples 0.01 to 10 m (0.4 in to 33 ft), dunes 10 to 500 m (33 to 1650 ft), and complex stellate dunes, or draas, 500 to 5000 m (1650 to 16,500 ft). Each group includes transverse and longitudinal components (Fig. 10-1); for example, the crescent-shaped barchan with horns pointing downwind consists of ridges intersecting at right angles. Wilson showed that under steady-flow conditions the bedforms, and the airflow pattern with which they interact, reach equilibrium, so the shapes remain fairly constant even though they may migrate. But ripples, dunes, and draas frequently coexist, with no intermediate forms. Since transverse bedform elements tend to migrate at a rate about inversely proportional to their height, ripples often overtake the larger bedforms on which they are superposed, or lose their identity when they avalanche down the slipface of the major dunes. In contrast to the sudden initiation and rapid movement of ripples and small dunes, the giant Algerian draas migrate at rates as slow as 1.6 cm (0.6 in) a year, and require a timespan of some 10,000 years to develop fully (Wilson, 1972).

Ripples

Impact ripples, the most common variety, are asymmetric transverse bedforms a few centimeters high produced by the impact of saltating grains, their wavelength being proportional to the distance of saltation (Bagnold, 1941, p. 34). Slipfaces are poorly defined or absent, and ripple index is high, exceeding 15. These ripples are commonly fashioned of the coarsest sand or fine gravel fraction (Folk, 1971), and the largest grains and heavy minerals are concentrated along ripple crests. Many wind ripples are remarkably straight crested. Sinuosity in others probably

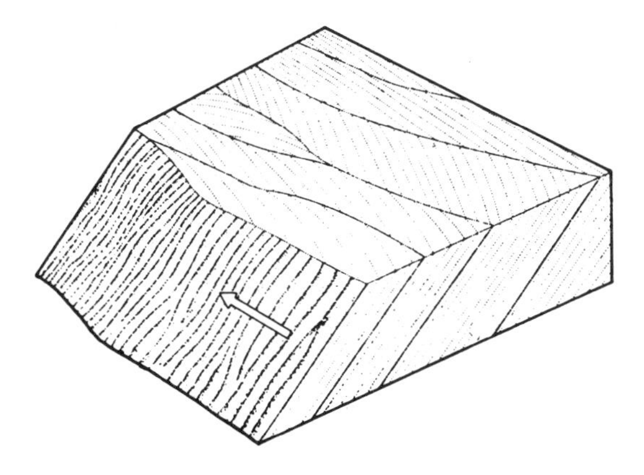

Figure 10-2. Eolian foresets with transposed ripples directed at right angles to the foreset dip. (After Walker and Harms, 1972.)

arises from secondary spiral vortices in ripple troughs.

Impact-ripple deposition produces three forms of cross lamination, depending on rate of sedimentation and ripple-form preservation: obliquely climbing cross laminae with truncated foresets, similar laminae with complete foresets, and superposed ripple forms with both stoss side and foresets preserved.

Climbing-ripple cross lamination with a low angle of ascent is estimated by Hunter (1981) to make up between 10 and 40 percent of many of the Pennsylvanian to Jurassic eolian sandstones of the United States Western Interior. This small-scale structure is particularly prevalent in the gently inclined lower foresets of large cross beds. According to Hunter, some entire outcrops of the Casper Formation and Entrada Sandstone are dominated by climbing ripples such as these, defining low-angle lee slopes of dunes, the upper parts of which are generally not preserved. The minute ripple-foreset cross lamination may be difficult to detect, but the associated larger scale inclined surfaces up which the ripples migrated are more obvious. These large, gently inclined surfaces define *climbing translatent stratification* in Hunter's terminology, and are probably equivalent to the eolian accretion bedding of Bagnold (1941). Inverse grading is characteristic of these thin, even, low-angle beds. Because the direction of ripple migration often departs significantly from true wind direction, ripple-foreset azimuths show wide dispersion.

Ripples form on the slipfaces of major bedforms, particularly barchans and transverse dunes (McKee, 1945; Walker and Harms, 1972). The crestlines are oriented at a variety of angles relative to foreset dip, and are commonly transverse (Fig. 10-2). Ripples formed during a brief change in wind direction (Glennie, 1972) are preserved by renewed avalanching or grainfall, particularly where the sand surface is moistened. Traces of climbing eolian ripples of this type may be observed on foresets even where the actual ripple form is not preserved (Hunter, 1981).

Aerodynamic ripples (Wilson, 1972), with a pronounced longitudinal component and larger size and more continuous crestlines than associated transverse ripples, are relatively uncommon. In combination with impact ripples, they produce a diagnostic diamond-shaped or fish-scale pattern.

Adhesion ripples form when saltating grains adhere to the damp sediment surface. Capillary rise of moisture causes continued accretion of layers of wind-blown grains. Individual adhesion ripples build upwind as the depositional surface aggrades, producing an irregular warty surface that is recognizable in the rock record.

A distinctive variety of eolian cross lamination produced by climbing adhesion ripples has been documented in interdune areas by Hunter (1973). Downwind-inclined layering represents the bounding surfaces between successive climbing adhesion ripples, whereas irregular small-scale cross lamination dips upwind and is commonly convex-up.

Dunes

The formation of large bedforms is attributed by Wilson (1972) to localization of pre-existing secondary flows by an obstacle or irregularity in the sand surface. Of the two general dune categories with respect to prevailing wind direction, longitudinal and transverse, longitudinal (also known as linear or seif) dunes are the more common in modern sand seas. In the Northern Hemisphere deserts, crescentic dune complexes of barchanoid, and transverse dunes are locally dominant, whereas in the Southern Hemisphere Australian, Kalahari, and central Namib Deserts, longitudinal dunes are ubiquitous (Breed and Grow, 1979); true barchans are almost entirely absent from some deserts, for example in Australia (Madigan, 1946).

Barchans, barchanoid ridges, and transverse dunes are a response to essentially unidirectional winds, and commonly represent a gradational series corresponding to an increase in sand supply (McKee, 1979b). Barchans are best defined where the sand supply is limited, for example toward the edges of sand seas and along the vegetated margins of beaches.

Barchans show a remarkable variety of sizes and spacings depending on the volume and texture of sediment, the nature of the substrate, and wind regime. Apart from the laterally linked patterns of barchanoid ridges, barchans may also be arranged *en echelon* in an asymmetric chain oblique to the mean wind direction (Wilson, 1972). These oblique chains comprise alternating longitudinal and transverse elements. Longitudinal groupings of barchan dunes are observed, and barchanoid ridges are also transitional into straight-crested transverse dunes. Avalanche slipfaces are only developed on barchanoid and transverse dunes more than about 30 cm (1 ft) high (Glennie, 1972).

Barchans and barchanoid ridges investigated by McKee (1957, 1966) show large-scale planar cross beds in the axial portions, dipping consistently at angles greater than 30 degrees, but some have complex internal truncation surfaces that steepen in the downwind direction. Sporadic troughs represent small blowouts. Cross-bed sets become thinner, and the foresets less steep, toward the barchan horns, with an asymmetric, lobate distribution of azimuths. The lower windward barchan slope may be overlain by low-angle lamination. Steep planar-tabular cross-bed sets of transverse dunes show remarkable lateral continuity. Set boundaries have a tendency to dip downwind, but flatten as the successive sets become thinner upward.

Several basic genetic types of small-scale eolian structures (Fig. 10-3) have been recognized in barchan and other transverse dunes (Hunter, 1977, 1981), and these may apply equally to many of the larger scale bedforms. Successive increments of climbing-ripple cross lamination contribute to the development of translatent stratification, which may be volumetrically very important. With stronger winds, traction operates on a planar surface to produce plane-bed lamination, most commonly in the crestal area of the dune where preservation potential is low. Net accretion of plane-bed lamination characterizes the gentle upwind surfaces of some dunes.

Downwind of the dune crests, sand settles through the zone of flow separation to blanket the lee slope with parallel laminae at angles of up to 28 degrees. For this grainfall lamination to be preserved, the slope must remain insufficiently steep to avalanche. Like plane-bed laminae, grainfall laminae are thin, even, and indistinct; indeed, there is a complete gradation between these two bedding types which are arbitrarily distinguished on the basis of dip angle. The proportion of grainfall deposits in modern and ancient dunes is variable, but locally constitutes some 80 percent of the Lyons Sandstone (Hunter, 1981). Steeper slopes generate avalanching sandflows with resulting cross beds inclined up to the 34 degree angle of repose. These steep avalanche foresets are relatively thick and lenticular along strike, and show a concentration of the coarsest grains near their sharp, angular basal contacts. Bigarella *et al.* (1969) observed avalanche foresets inclined at angles steeper than 36 degrees, possibly due to slight increase in cohesiveness resulting from high humidity in the coastal environment.

Climbing-ripple cross lamination tends to be preserved in those parts of the dune that are convex downwind in plan view, whereas high-angle avalanche cross bedding is preserved where the dune is concave downwind. Grainflow lamination occupies intermediate areas (Fig. 10-4; Hunter, 1977).

Dune migration might thus be expected to produce a vertical succession of structures commencing with adhesion-ripple lamination of interdunal origin at the base, overlain by large-scale foresets of avalanche or grainfall origin, followed by climbing-ripple cross lamination and plane-bed lamination of the upwind portions of the dune. However, this complete dune sequence is rarely preserved in the stratigraphic record because a migrating dune truncates its own deposits (Brookfield, 1977), with the thickness of net

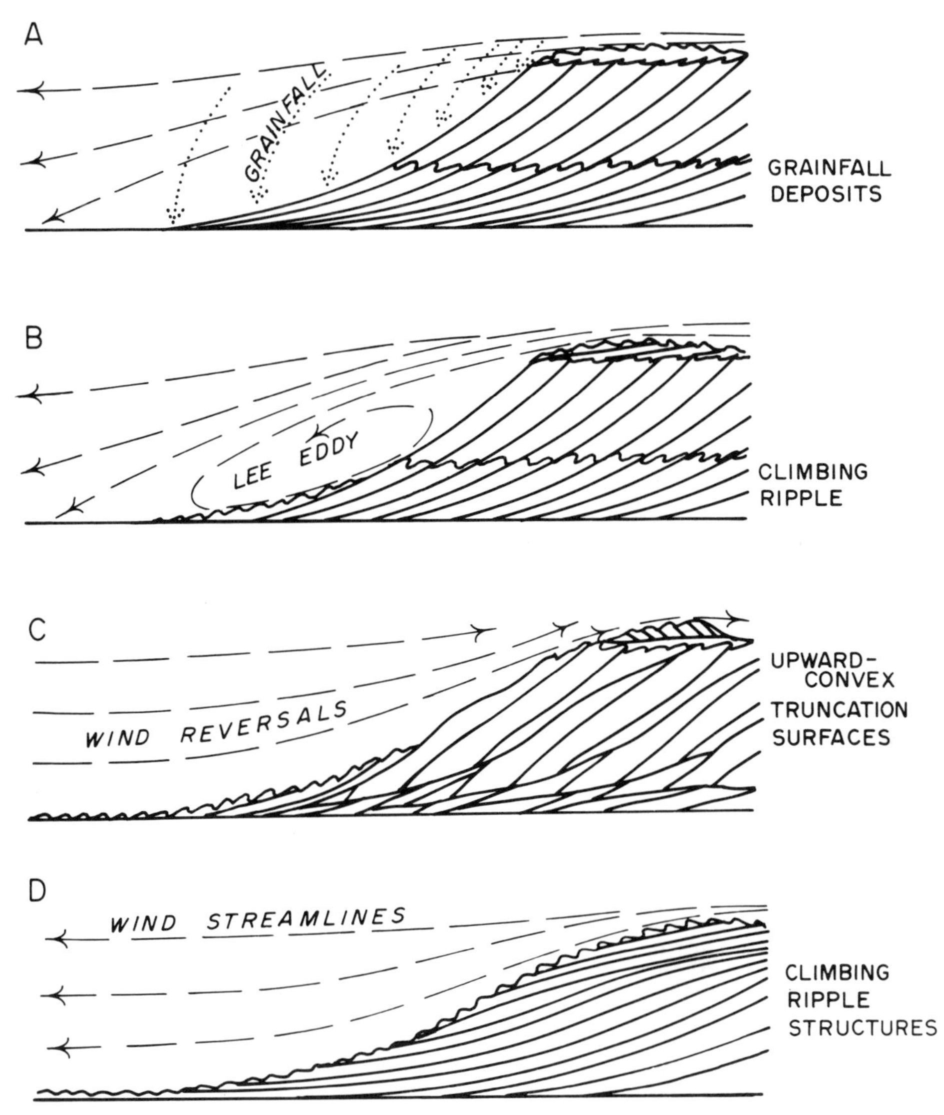

Figure 10-3. Schematic cross sections through eolian dunes showing: A. grainfall deposition on and ahead of the slipface of a small dune; B. development of a lee eddy on slopes oblique to the wind; C. dune affected by wind reversals; and D. a gentle lee slope without flow separation. (After Hunter, 1981.)

accretion commonly amounting to a small fraction of the dune height (Walker and Middleton, 1979). Some large migrating dunes in Namibia leave no record (V. von Brunn, personal communication, 1981). Furthermore, large areas of dunes may be deflated to the level of the water table, producing widespread horizontal surfaces (Stokes, 1968). Changes in wind direction commonly produce inclined truncation planes and planar-wedge sets (McKee 1979b).

Stereographic projection of cross-bed azimuth and dip (Reiche, 1938) may be of value not only for determining paleowind direction, but also for indicating the type of dune (Glennie, 1970, p. 101). For example, in the case of barchan foresets, low dips might be expected to converge toward the direction of steeper foresets of the axial slipface, which is inclined in the downwind direction (Shotten, 1956); this results in a conical girdle of stereographic poles. Cross beds of transverse dunes would show narrower dispersion, forming a point maximum on stereographic projection (Fig. 10-5).

Dome dunes are circular in plan with no distinctive slipfaces, and are thought to originate by modification of barchanoid dunes as a result of

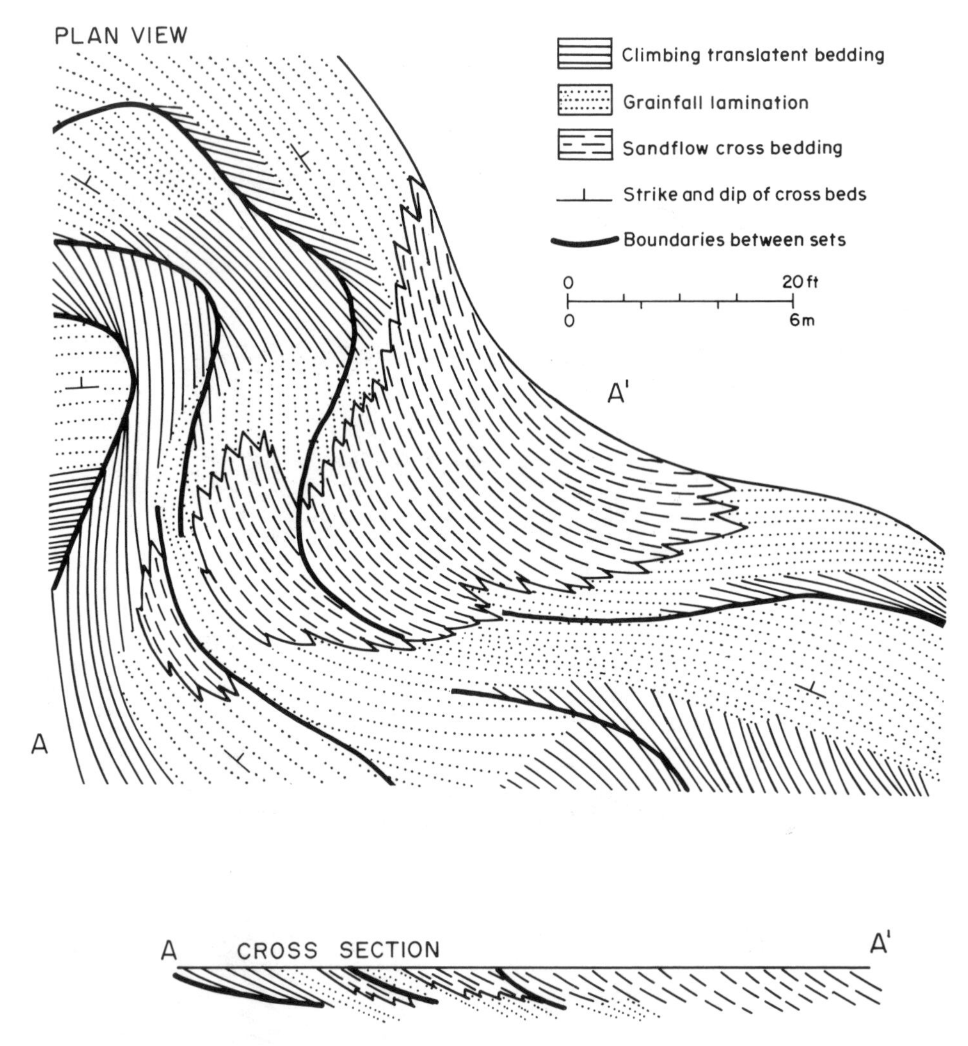

Figure 10-4. Plan view and cross section of a planed-off, sinuous transverse dune showing relationships of bedding types. (After Hunter, 1977.)

strong winds (McKee, 1966). Although common in coastal regions where moisture and vegetation play a role in their stabilization, they are also developed in some interior sand seas. The deposits of dome dunes are distinguished by the gradation of flat topset laminae into foresets, which can dip in every direction (Bigarella, 1972). Numerous cut-and-fill structures are also characteristic.

Reversing dunes are intermediate between transverse and star dunes, resulting from a close balance between opposed winds (McKee, 1979a). Slipfaces form on opposite sides, and these would presumably be far more simple than in the case of longitudinal dunes, which should lead to ready recognition in the rock record.

Longitudinal (linear, sand-ridge, or seif) dunes apparently arise by a variety of mechanisms under wind distributions varying from unimodal to complex (Fryberger and Dean, 1979). Broad areas of uniform sand accumulation and a low water table are most conducive to their development. Sand-drift patterns are more variable than in the case of barchanoid and transverse dunes, and wind velocities are generally higher—the stronger the winds, the higher the dunes and the wider their spacing (Glennie, 1970, p. 95). McKee (1979a) distinguishes among several varieties of longitudinal dunes, including the zigzag type of Libya, feather type of Saudi Arabia, and the converging type of Australia. In all cases the dune axes are parallel or subparallel

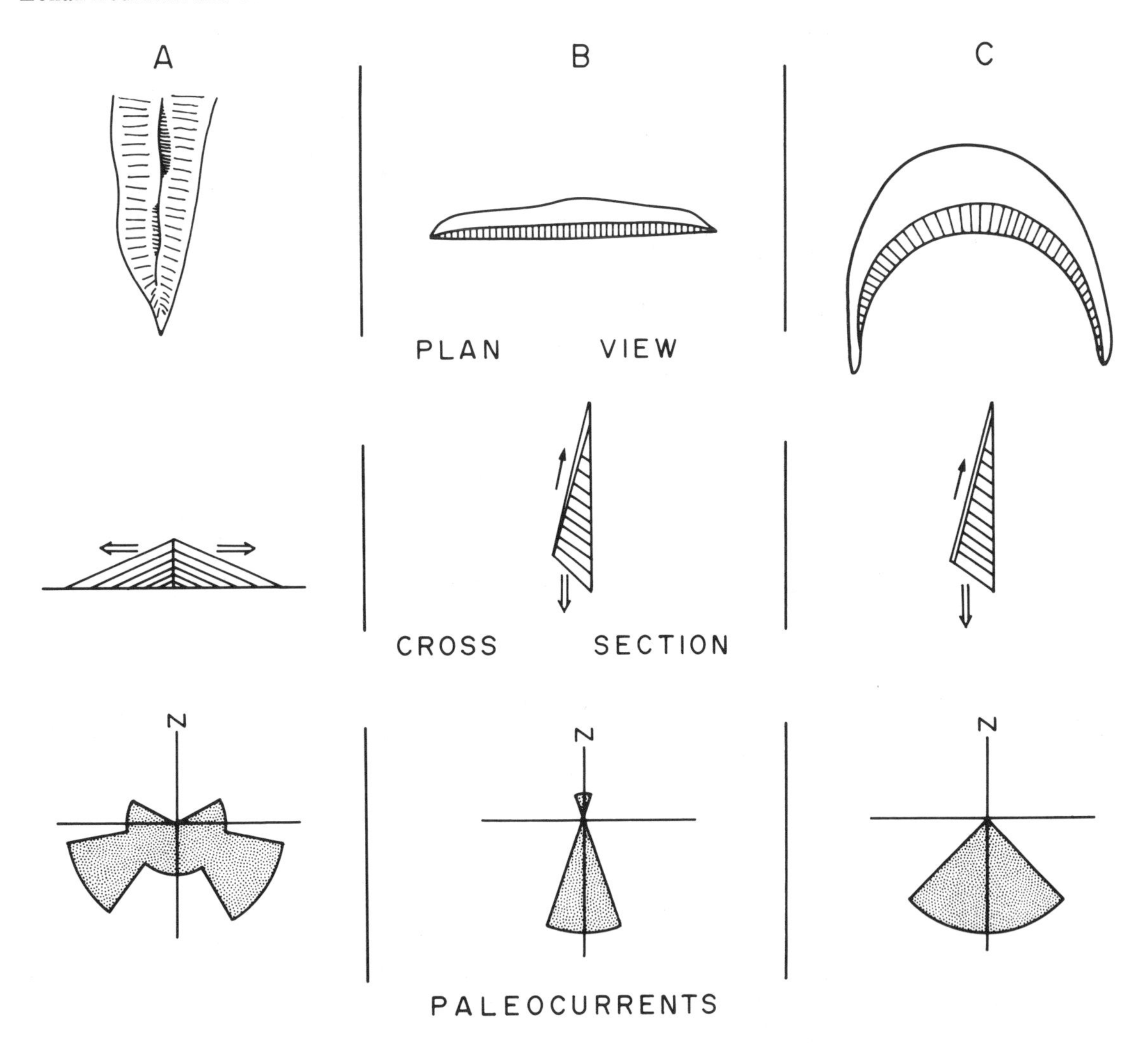

Figure 10-5. Plan view, cross section, and paleocurrents of: A. longitudinal; B. transverse; and C. barchan dunes. (Modified after Glennie, 1970.)

to the dominant wind (Fig. 10-6), although reversed, convergent, or side winds may contribute to their formation. Helicoidal processes may be primarily responsible for the wind-parallel alignment (Bagnold, 1953). Sand is transported obliquely from the interdune areas to accumulate at the convergence of adjacent spiral vortex systems. Thus, the dunes tend to be equally spaced and self perpetuating because the heated dunes become the sites of convective uprising (Folk, 1971); upwind branching of dunes at a low angle is regarded as further indication of helicoidal flow processes. McKee and Tibbits (1964) provided evidence of longitudinal dune development controlled by two winds at right angles, but Glennie (1970, p. 89) suggests that the wind blowing parallel to the dune axis is dominant. Side winds thus appear to contribute mainly toward modifying the form of dunes aligned parallel to the dominant wind. Some longitudinal dunes show slipfaces along alternate sides (Wilson, 1972); others have slipfaces along both margins and at their downwind ends (Glennie, 1970).

The longitudinal-dune studies by McKee and Tibbits (1964) showed large-scale bimodal, almost bipolar, planar cross beds dipping laterally from the dune crest. High-angle foresets are typical of the upper part, with low-angle accretion bedding on the lower flanks. This bimodal cross bedding is approximately perpendicular to the dominant wind (Glennie, 1970, p. 100), but one mode may predominate. A third cross-bed mode can be directed downwind.

Longitudinal dunes are relatively stationary

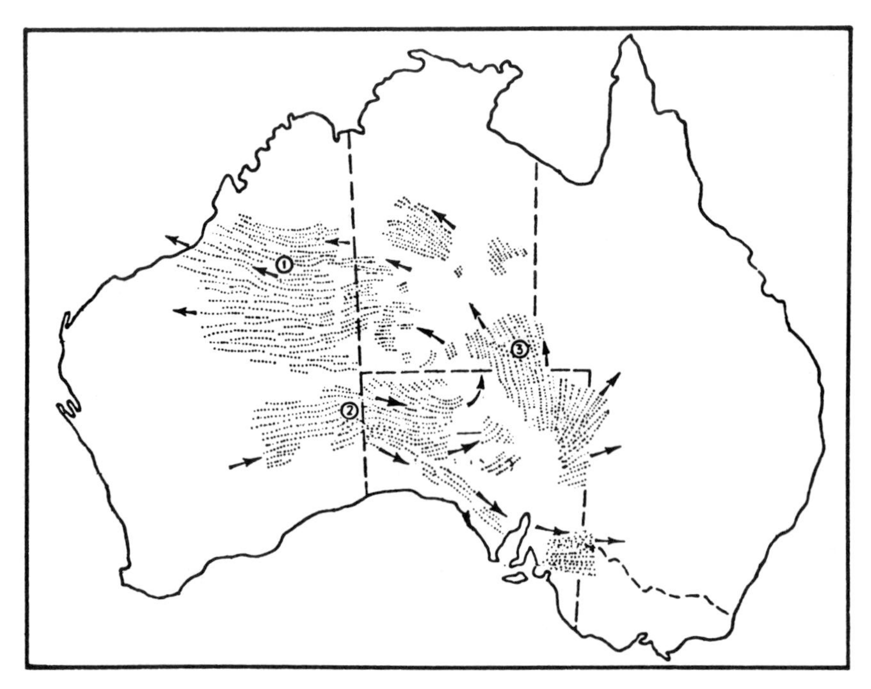

Figure 10-6. Relationship of longitudinal dune systems to causal winds in Australia. Arrows show the direction of dune convergence coincident with the wind resultant. 1. Great Sandy Desert; 2. Great Victorian Desert; 3. Simpson Desert. (After King, 1960.)

forms that commonly involve significant erosion, and are therefore less likely to be preserved in the geologic record. Some have a core of older sediments, which attests to their stability (Folk, 1971). Because they grow downwind, the subordinate, third cross-bed mode may have the greatest chance of preservation (Walker and Middleton, 1979).

Blowouts and parabolic dunes result from breaching of partially stabilized dune ridges, commonly in coastal regions. They are generally a response to a single wind direction, but in some areas reflect complex opposed patterns. The arms extending from the flanks of the blowout remain anchored while the lobate dune front migrates (McKee, 1979a). The curvature of the dune front produces broad dispersion of foresets. Distinguishing internal structures of parabolic dunes are upward convexity of foresets, abundant organic-rich partings, and numerous penecontemporaneous deformation structures (Bigarella *et al.*, 1969).

Shadow dunes are wind-parallel accumulations in the lee of obstacles, most commonly a clump of vegetation, forming in the zone of horizontal flow separation characterized by reversing vortices (Hesp, 1981). Dune height is determined by the width of the obstacle and the texture of the sediment. Although these features generally have low preservation potential, they may be important in nucleating larger eolian bedforms.

Draas

Where fully developed, draas comprise large stellate rosettes with a high central peak, radiating arms, and superposed smaller dunes of various forms and orientations (Wilson, 1972; McKee, 1979a). Their formation is thought to be related to intense, multidirectional wind systems in areas of high sand-drift potential (Fryberger and Dean, 1979), different grain sizes responding to different winds. More rarely, two or three regional trends that are normally apparent in draa complexes may be aligned parallel, normal, and oblique to prevailing winds; oblique segments can also arise from alternating longitudinal and transverse components. Draa heights range from a few meters to over 400 m (1300 ft), with spacings of 0.5 to 5 km (0.3 to 3 mi), increasing considerably in areas of sporadic sand cover (Wilson, 1973). Large draas are commonly related to proximity of coarse alluvium, draa height decreasing with reduction in grain size, although some 200 m (650 ft) high draas in the Namib are composed of very fine sand (M.P.A. Jackson, personal communication, 1982).

Because of the long timespan required for their formation, many draas display haphazard, hummocky, and sinuous forms representing incomplete stages of development. Draas are regarded by some as stationary because wells and oases have persisted for centuries in hollows between named draas; some draas may be subject only to vertical aggradation (McKee, 1979a). Wilson (1972) points out, however, that although they are stable in human terms, some Algerian draas have migrated between 20 and 500 km (12 and 300 mi) in the past million years. Transverse draa elements do indeed show slipfaces, attesting to migration. The location of some radially symmetrical draas was attributed to the positioning of nodes of stationary waves (Clos-Arceduc, 1966), and it has been suggested that some may be positioned in response to thermal turbulence in the troposphere (Bagnold, 1953) brought about by rapid heating of the lower air layers; but Wilson (1972) has shown that even these very large draas show transverse components that migrate.

Some ancient eolian cross-bed sets 100 ft (30 m) or more thick could only have been produced by giant sand accumulations on the scale of draas (Walker and Middleton, 1979).

Sand Sheets

Flat to gently undulating sand sheets are ubiquitous features of all interior sand seas (Breed and Grow, 1979), and are particularly common around the margins of dune fields (Fig. 10-7). The example documented by Fryberger *et al.* (1979) which occupies an area of 280 mi^2 (720 km^2) lies upwind of a dune field, and is the first site of sand accumulation along the transport route. Internal bedding is typically low angle, ranging from horizontal to 20 degrees. The foresets, in contrast to the typical asymptotic eolian foresets of dunes, have angular basal contacts. Complex alternations of coarse and fine layers have high-index ripples in the coarser fraction, whereas laminae deposited during sandstorms tend to be graded and convex-upward. Burrows and plant rootlets are common. Because of the low topographic elevation of the sand sheets, streams deposit gravel, and silt accumulates from ponded floodwaters.

Fryberger *et al.* (1979) point out that sheet-sand facies, being transitional between high-angle dune deposits and noneolian or extradune deposits, are useful indicators of ancient dunefield margins (Fig. 10-7).

Biogenic Structures

Despite statements to the contrary, "bioturbation is a very common, if not ubiquitous, feature of eolian deposits" (Ahlbrandt *et al.*, 1978), particularly in strata of Carboniferous age and younger. Many organisms specifically adapted to eolian environments leave distinctive traces. Dunefield arthropod burrows, such as those constructed by sand wasps, wolf spiders, or crickets, are reinforced by cementation, agglutination, or web collars. Unwalled burrows can be preserved in damp, cohesive sand, and some contain well-defined backfill structures. Traces in slipface and interdune deposits have the highest chance of preservation because of rapid burial.

Footprints of four-footed animals are preserved in exceptional detail in eolian deposits (McKee, 1979b, p. 193). McKee proved experimentally that these tracks originated in dry sand near the angle of repose. Hyena tracks are not uncommon in Pleistocene "eolianite" of southern Africa.

Plant roots leave molds or "dikaka" (Glennie and Evamy, 1968) which are commonly accentuated by mineralized lining. These range from horizontal runners to tap roots 5 m (16 ft) deep. Intense root growth results in complete obliteration of primary structures. The oldest known dune plants are preserved in Lower Cretaceous eolian deposits in Germany. By Tertiary time, a number of rooted plants had adapted to an arid dune environment (Glennie, 1970, p. 115). Ahlbrandt *et al.* (1978) note that contrasted colors in the central and peripheral parts of the sediment fill distinguish root traces from faunal burrows.

Penecontemporaneous Deformation

Eolian deposits show many small-scale deformation features resulting from slumping, grainflow, loading, and drag (McKee *et al.*, 1971; McKee and Bigarella, 1972). Generally these

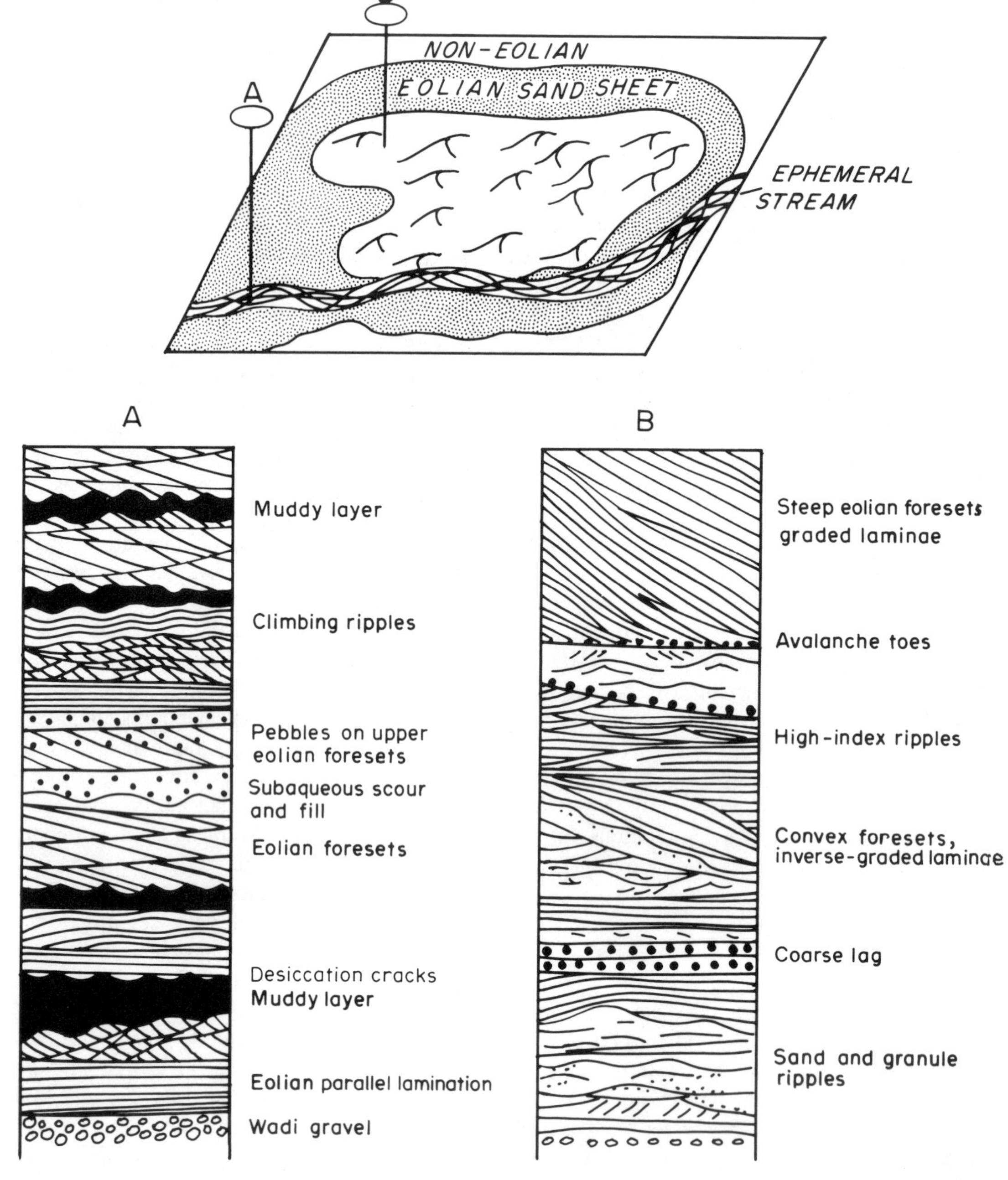

Figure 10-7. Schematic areal distribution and stratigraphic relationships of: A. eolian sand sheet and ephemeral stream deposits; and B. eolian dune deposits with large-scale foresets. (After Fryberger *et al.*, 1979.)

structures are of tensional origin on the upper parts of dune slipfaces, and of compressional origin toward the base of slopes. The type of deformation is largely dependent upon moisture content. Dry sand is subject to gentle folding and blurring of laminae, whereas damp sand with greater cohesion shows steeper asymmetric folds, rotation of bedding planes, fractures, and brecciation. Arising mainly in damp sand, structures such as those in Figure 10-8, could be particularly helpful in paleoenvironmental determinations.

Figure 10-8. Small-scale faulting in damp coastal dune sands, South Africa.

Deformed layers are bounded above and below by totally undeformed strata, and may reflect sliding of rain-saturated surface layers over dry sand (Glennie, 1970, p. 109). Load deformation is most common in alternating muddy and sandy deposits of interdune areas.

Interdune Facies

In an excellent discussion of interdune facies, Ahlbrandt and Fryberger (1981) recognize two process-related end members—deflationary and depositional, both of which include dry, wet, and evaporitic varieties. Interdune facies vary further according to dune type; they are thickest and most continuous in association with seifs and draas that form under multidirectional winds, and are thinner than 2 m (7 ft) and lenticular in association with the more rapidly migrating transverse, barchan, and parabolic dunes formed under unidirectional winds. Where dune migration is particularly fast, interdune deposits may be very thin or lacking.

Deflationary interdune areas vary from broad, featureless plains to narrow, windswept corridors, and irregular blowouts. Lag gravels may cover wide areas known as gibber plains, serirs, or regs. Multifaceted ventifacts, commonly aligned with the strongest winds, tend to be regularly spaced. Pebble lags develop a shiny desert varnish resulting from leaching and precipitation of iron and manganese by diurnal fluctuations in moisture. Granule ripples produce thin, poorly sorted, inversely graded layers (Ahlbrandt and Fryberger, 1981). Some deflationary interdune areas are virtually devoid of unconsolidated sediments, and biological or pedogenic processes dominate. These are preserved in the rock record as a disconformity overlain by a discontinuous winnowed lag.

Depositional interdune areas range from permanently saturated, through episodically wetted, to completely dry. Enhanced precipitation over desert highlands promotes wadi sedimentation in adjacent interdune areas, and deposits of ephemeral streams and small alluvial fans dominate some deserts. Intermittent flooding can erode parts of the dune system, redistributing sand to the intervening depressions, but wadi discharge is commonly impounded by dunes, creating a series of temporary lakes (Glennie, 1970, p. 39). Elsewhere the dunes are breached, or flow persists longitudinally between dune chains, developing an interdunal braided floodplain. As the floodwaters recede, the ephemeral stream deposits are veneered with mud which later dries, cracks, and is redistributed by wind.

Interdune depressions may also accumulate ground-water discharge, which is normally saline. Where the water table undergoes a gradual rise, growth of adhesion ripples produces layers a meter or more thick (Ahlbrandt and Fryberger, 1981). Some interdune lakes are semipermanent, with coarse sedimentation along their margins, and differential settling of silt and clay giving rise to graded laminae (Glennie, 1970, p. 58), which are commonly disrupted by burrowing, slumping, and crystallization. These clays are characterized by a low boron content, and can be rich in organic material. Fresh-water invertebrates, vertebrate remains, and footprints are common. Interdune deposits described by McKee and Moiola (1975) comprise thin, flat or irregular silts and sands 1 to 4 ft (0.3 to 1.3 m) thick with root contortion. In humid coastal areas, interdune deposits locally include black clays and peat, and commonly comprise an alternation of colluvium and mud with paleosols (Bigarella, 1979). In salt-encrusted depressions with playas or inland sabkhas (Chapter 9), the fluctuating water table generally lies just below the surface. Although many playas involve concentration of salts through surface evaporation, some result from the rise of already-saline artesian water. Playa sediments are highly variable, depending on the clastic and dissolved materials available and the evaporite and later-stage authigenic minerals which develop. Typical minerals include carbonates, sulfates, halides, borates, and nitrates, with uranium concentrations in some areas. Gypsiferous sandy layers originate by the adhesion of sand to a wet, salty surface (Glennie, 1972). Varicolored clays reflect significant oxidation, and display evidence in the form of runzelmarks, etch marks, rills, and adhesion warts of a shallowly ponded or damp surface that subsequently dried to leave desiccation cracks, raindrop impressions, and evaporite layers and pseudomorphs.

Scale and Associations of Eolian Systems

By comparison with most other depositional systems, eolian sediments are generally thin. Despite their enormous areal extent, dune sands of the Sahara are generally thinner than a few hundred meters at most. Average sand thickness of the large Algerian sand seas is only 26 m (85 ft) (Wilson, 1973). The sands of the Australian deserts are even thinner, amounting to a broad veneer over older alluvial deposits and bedrock. Among ancient examples, the Mesozoic Navajo Sandstone and correlatives locally attain 3000 ft (900 m), which is regarded as exceptional. In contrast, the Triassic Botucatu Sandstone of Brazil, which covers a preserved area over three times larger than the Navajo, is generally only 10 to 20 m (33 to 65 ft) thick (Bigarella, 1979; McKee, 1979c).

Eolian systems commonly show close association with fluvial, lacustrine, or marine sediments. The Permian Rotliegendes of Europe, comparable in area to the modern Sahara, comprises an association of eolian dune, playa, and wadi deposits. But the thickest part of the Rotliegendes, approximately 1500 m (5000 ft), corresponds to the position of a large saline lake. Adjacent dune sands were derived from deflation of alluvial-fan sediments (Glennie, 1972), and attain only 200 m (650 ft).

Sand Seas (Ergs)

Sand accumulates preferentially on surfaces that are already sand covered (Glennie, 1970, p. 23), a feature most characteristic of low-lying desert basins with less than 15 cm (6 in) annual precipitation (Wilson 1973). Wilson noted that sand seas are present at practically all latitudes, and were probably even more widespread prior to Paleozoic colonization of dry land by plants, and their adaptation to desert environments by Permian time (Glennie and Evamy, 1968). Widespread eolian sand deposition was associated with Pleistocene development of broad, poorly vegetated periglacial areas in high latitudes. Today the largest ergs cover an area of around 1×10^6 km^2 (4×10^5 mi^2), but the average is 10^3 km^2 (400 mi^2) (Wilson, 1973).

Sand seas form in topographic basins downwind of older deposits, generally alluvial or lacustrine, that are subject to rapid deflation. These eolian sands either interfinger with their fluvio-lacustrine precursors, or may be far removed from their source.

Persistent onshore and alongshore winds in Namibia redistribute large quantities of beach and river sand, whereas in Australia the sand seas are supplied in part by deflation of playa depressions and lake-margin beach ridges (Stephens and Crocker, 1946), in addition to reworking of older fluvial sands (Folk, 1971). Other sand seas, for example, in Algeria, are interdigitated with elevated, eroding source areas, but all of the Saharan ergs contain a large proportion of sand in transit from one desert basin to another, with transport rates of up to 800 km (500 mi) per century.

Most sand seas are characterized by discontinuous sand cover. A complete cover develops only where the wind has been saturated with sand as a result of decelerating flow (Wilson, 1973). The great variation in sediment distribution and thickness gives rise to differences in major bedforms.

The sand is sufficiently thick, and sufficient time has elapsed, for draas to develop in the Saharan, Namibian, Arabian, and Asian sand seas; but the thinner sands of the Australian deserts preclude draa formation. Dunes tend to become more complex as they enlarge (Breed and Grow, 1979), which possibly accounts for the persistence of longitudinal dunes in deserts with a thin sand cover, such as the Simpson and Kalahari. Some deserts, such as the Namib, show a regular downwind or inland change from barchanoid and transverse through longitudinal dunes to draas. Others change from scattered barchans through barchanoid ridges to higher transverse dunes (McKee, 1979a). Elsewhere the progression is from dome to transverse, barchan, and parabolic dunes (Fisher and Brown, 1972). Anomalous patterns are not uncommon; for example some draas grow atop coalesced barchan dunes, while others are superposed on longitudinal dunes (Fryberger and Dean, 1979). The Thar Desert of India and Pakistan is dominated by parabolic dunes, whereas dome dunes are present in the northern deserts of Saudi Arabia (Breed and Grow, 1979). No regular and predictable progression of dune types is therefore apparent in most sand seas.

In spite of the limited information concerning the internal character of large eolian bedforms, a combination of paleoenvironmental indicators can provide convincing reconstruction of the deposits of ancient sand seas (e.g., Walker and Harms, 1972; McKee, 1979b). Large-scale cross beds inclined at high angles (as much as 34 degrees and even steeper) are regarded as typical. Foresets tend to be truncated in their steepest upper parts and show long, sweeping, tangential bases (Fig. 10-9). Individual cross-bed sets attain enormous thicknesses of 100 ft (30 m) or more, but more commonly are in the range of 3.5 ft (110 cm), and are monotonously superposed. Cross-bed directions may be unimodal, bimodal, or polymodal, but examples of unidirectional azimuths are surprisingly prevalent in the stratigraphic record. McKee (1979b) recognized two dominant varieties of cross bedding. Tabular-planar sets show flat lower bounding surfaces of erosional or interdune origin, whereas wedge-planar sets are bounded by inclined truncation surfaces representing a response to changing wind directions.

There is no vertical sequence or model involving both textures and sedimentary structures that can be applied to the deposits of sand seas. McKee (1979b) did determine that planar-tabular cross beds tend to become thinner and less steeply inclined upward through a single dune, and the contacts between successive sets may flatten with increasing elevation, but the textural changes that are so useful in both outcrop and subsurface recognition of other depositional systems are absent. Successions such as the Permian Coconino Sandstone, deposited by rapidly migrating barchanoid and transverse dunes (McKee, 1979c), consist of monotonously stacked sets of steep, large-scale cross beds. On the other hand, successions typified by the Mesozoic Navajo Sandstone and correlatives contain thin interdune lenses of siltstone, mudstone, or limestone between cosets of eolian cross beds (McKee and Moiola, 1975). Initial porosity contrasts arising from finer grain size and relatively poor sorting of interdune deposits, as compared to dune deposits, are accentuated during burial. These porosity and permeability contrasts (Fig. 10-10) are clearly apparent in resistivity logs, and supplement evidence from dipmeter logs that distinguishes between the steeply dipping eolian foresets and the relatively flat bedding in interdune and

Figure 10-9. Typical large, sweeping tangential foresets of the eolian Navajo Sandstone, Zion National Park.

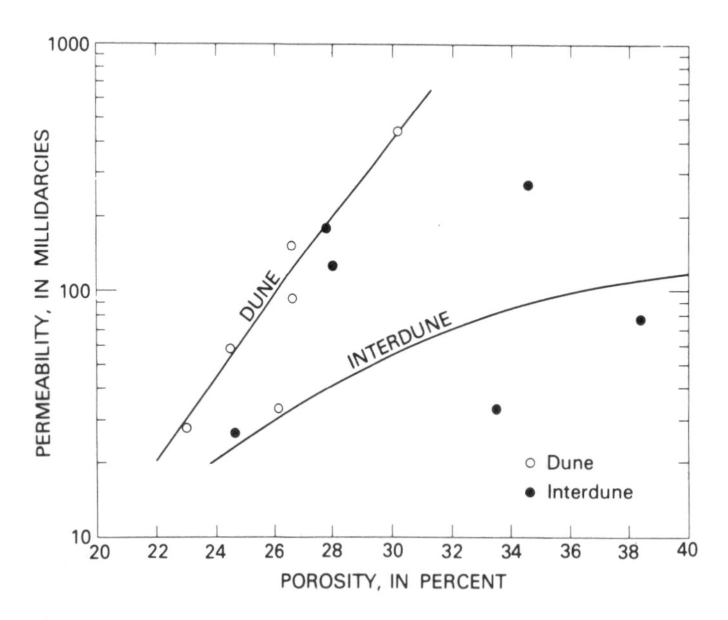

Figure 10-10. Porosity and permeability contrasts between dune and interdune deposits. (After Lupe and Ahlbrandt, 1979.)

associated noneolian facies (Fig. 10-11). The detailed studies by Lupe and Ahlbrandt (1979) relate the effective fluid migration and seal properties of the Nugget Sandstone, equivalent to the Navajo, to these textural contrasts provided by interdune deposits. Elsewhere, too, these thin permeability barriers provide excellent seals in otherwise highly porous eolian sandstones.

On the other hand, where the eolian systems are totally enveloped by extradune deposits of lakes, rivers, tidal flats, and beaches, Lupe and Ahlbrandt (1979) show that the texturally in-

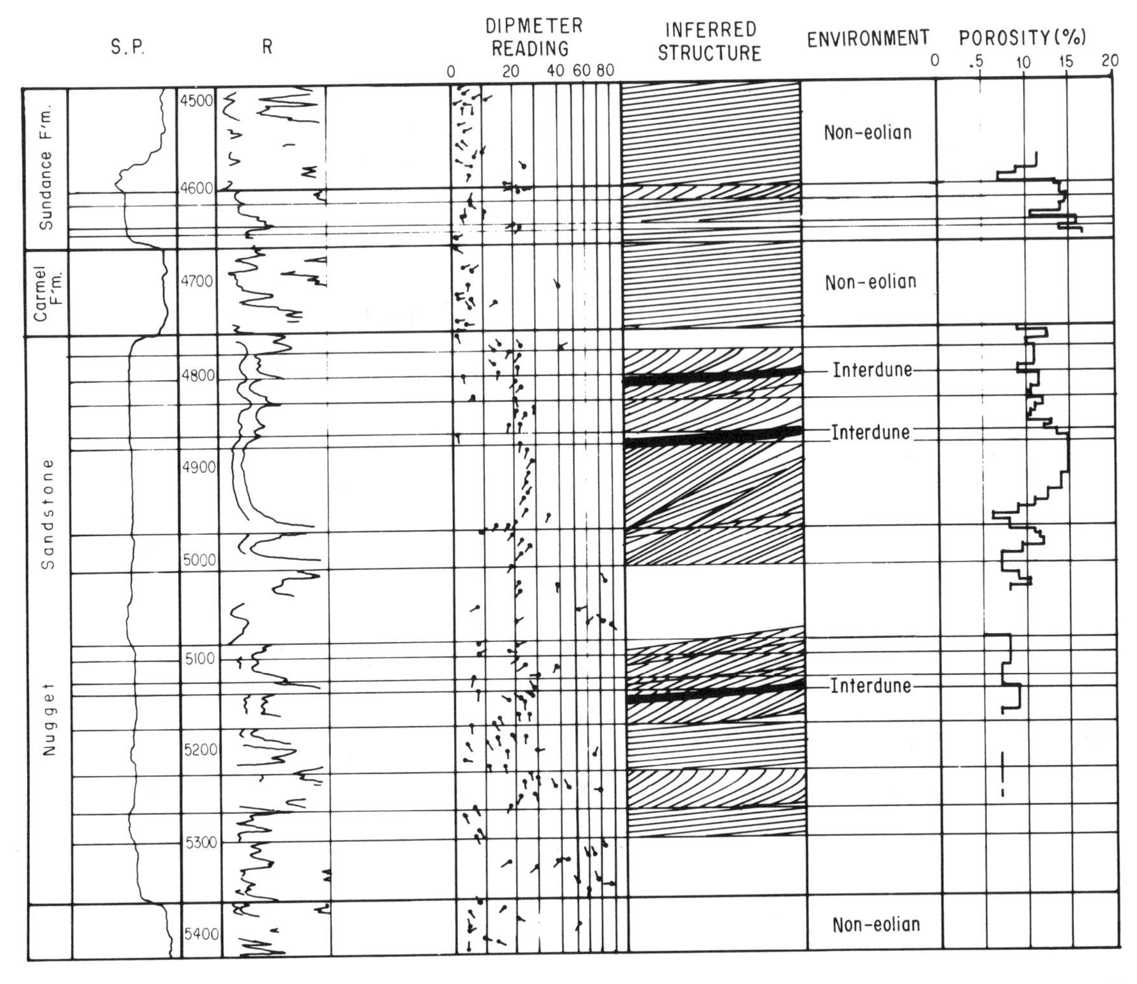

Figure 10-11. Resistivity (R) contrasts and associated porosity changes in dune and interdune deposits, with lower angle dipmeter readings of the latter. (After Lupe and Ahlbrandt, 1979.)

homogeneous reservoirs have poor fluid migration properties. This situation exists in the Weber and Tensleep Sandstones. The effectiveness of dipmeter patterns, used in conjunction with sonic and gamma logs, in distinguishing between eolian and fluvial deposits is apparent in Figure 10-12.

In summary, some of the features that might be regarded as characteristic of ancient interior sand seas are:

(1) Vast areal extent.
(2) Peripheral gradation into fluvial and alluvial-fan systems.
(3) Incorporation of the deposits of sabkhas and broad deflation surfaces with ventifacts.
(4) Dominance of planar-tabular and planar-wedge cross beds.
(5) Monotonous superposition of very large-scale, steep cross-bed sets, or alternation of large-scale cross beds and thin, flat-bedded interdune deposits (reflected in dipmeter patterns).
(6) Vertical contrasts in texture arising from dune-interdune environments (reflected in gamma or resistivity logs).
(7) Recognition of specific dune types through cross bed dispersion—bimodal or trimodal azimuths of longitudinal dunes, narrow dispersion of barchanoid and transverse dune foresets, with low-angle stoss laminae, and giant, broadly dispersed sets of draas.
(8) Delicate tracks and trails produced by typical desert fauna.

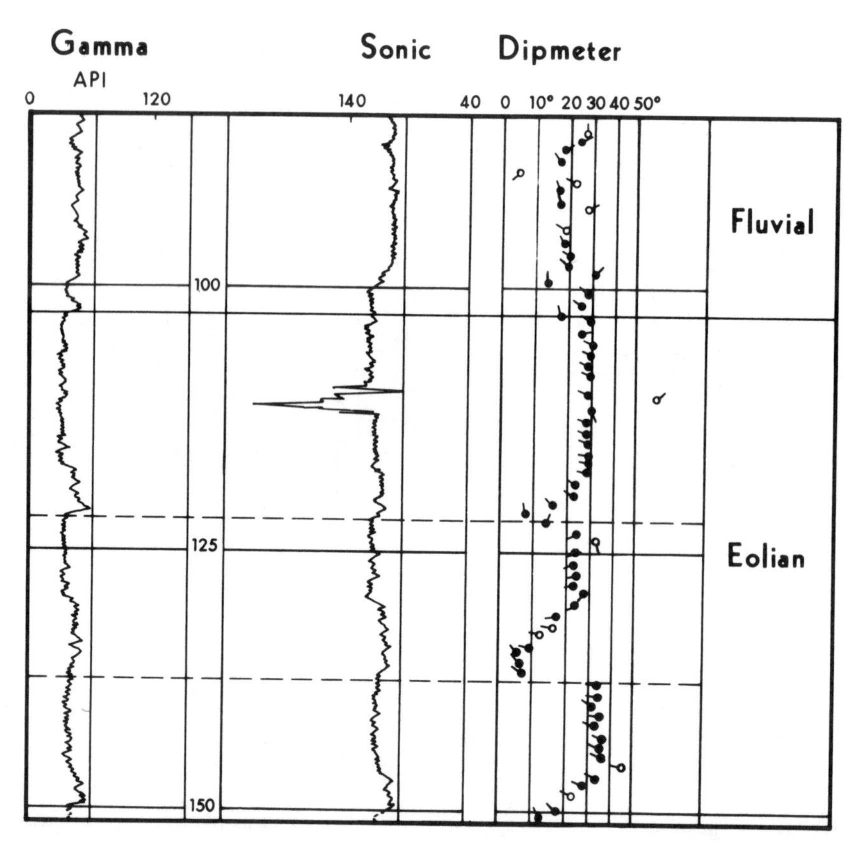

Figure 10-12. Eolian deposits of the Permian Rotliegendes showing steep dips of 30 degrees or more, with lower angle interdune deposits and variable dips of fluvial facies. (Modified after Selley, 1978b.)

Coastal Dunes

Coastal dunes form along both arid and humid coastlines, but are broadest and best defined on low-rainfall coasts with onshore trade winds, for example the Namib coastal tract of southern Africa. Ancient coastal dune systems such as the Casper and Lyons Sandstones of the United States Western Interior are characterized by a broadly linear trend and lateral interfingering with open marine facies (McKee, 1979c). Coastal dune sands are on average finer grained and better sorted than those of interior deserts.

Vegetation and high atmospheric moisture content play an important role in coastal dune development. As a result, forms such as coppice mounds, shadow dunes, fore-island dunes, retention ridges, dome dunes, blowouts, and parabolic dunes are all characteristic of coastal areas. Fore-island dunes and retention ridges are vegetation-stabilized, linear ridges of the back-beach area (Bigarella *et al.*, 1969; McGowen, 1979). Although much of the primary stratification is disrupted by roots, planar-wedge cross beds are preserved, with dip modes reflecting seasonal winds (McBride and Hayes, 1962). Longitudinal, barchanoid, and transverse dunes are all represented on back-barrier sand flats (Brown *et al.*, 1977; Hunter, 1977).

With sand supply constantly being replenished from the littoral source, the inland extent of sand transport is limited only by vegetation. Along the humid, densely vegetated coast of southeastern Africa the coastal dune fringe is narrow and high, attaining 200 m (650 ft) peaks and continuing almost unbroken for hundreds of kilometers. This cordon was initiated during the last glacial lowering of sea level, with modification continuing to the present day (Hobday, 1976). Blowouts are developing, with parabolic dunes extending obliquely inland. Internal stratification is complex, including giant troughs and multidirectional planar-wedge sets. Older, more subdued cordons mark stillstands during Quaternary regression, and show darker shades of reddening with increasing age. On many sandy coasts, beach foreshore

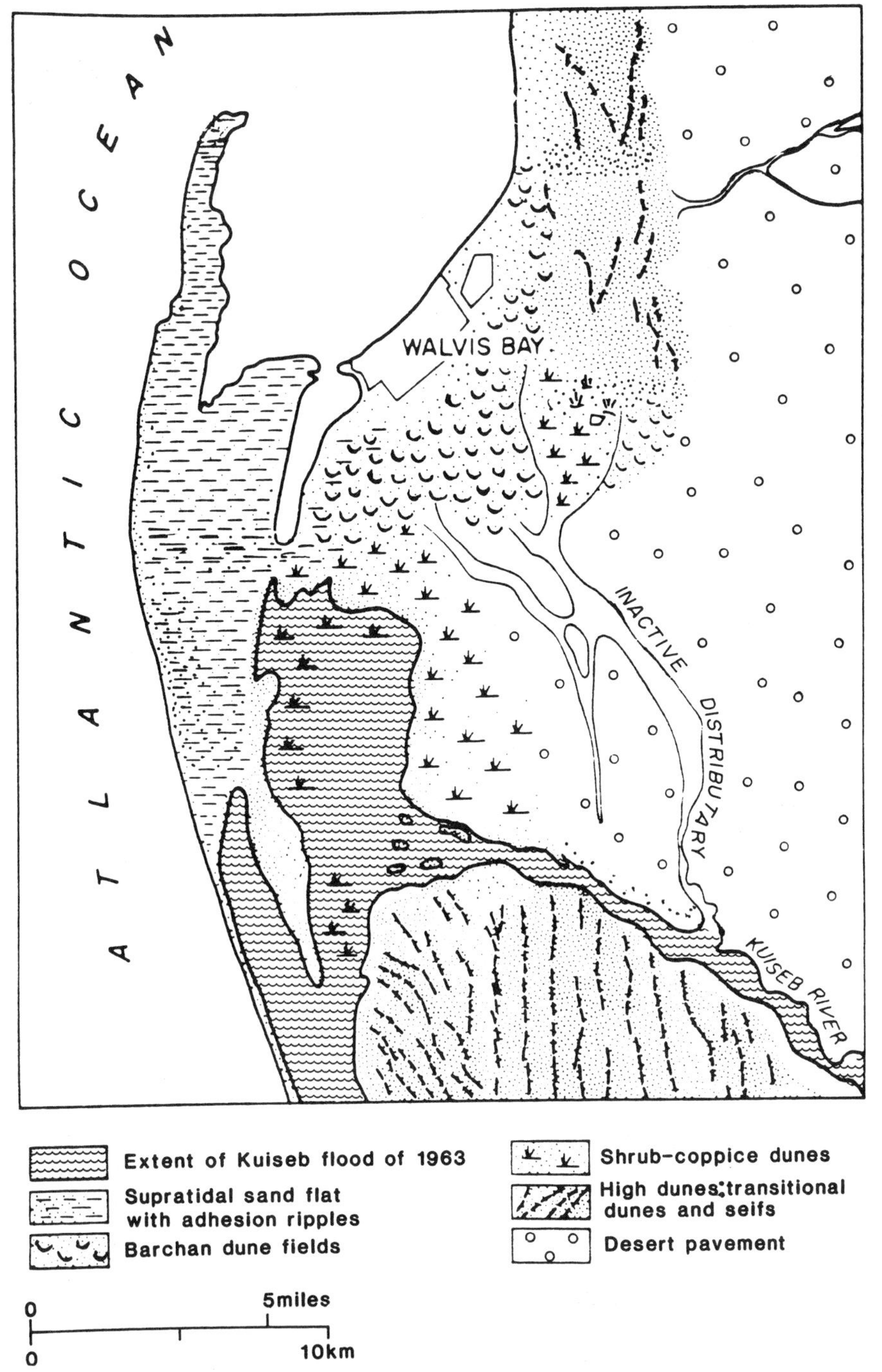

Figure 10-13. Coastal seif, barchan, and coppice dune fields and associated ephemeral stream and supratidal sand-flat environments, Walvis Bay area of the Namib Desert. (After Nagtegaal, 1973.)

concentrations of heavy minerals, particularly ilmenite, monazite, rutile, and zircon, were subject to even more effective winnowing and enrichment by wind action, permitting large-scale commercial exploitation. In less vegetated areas such as South Texas the dunes move slowly inland to occupy a broad belt where, depending on climate, eolian processes may alternate seasonally with aqueous reworking.

Along desert coasts such as Namibia, dune migration is unimpeded by vegetation. Inundation by high tides and storms leads to the development

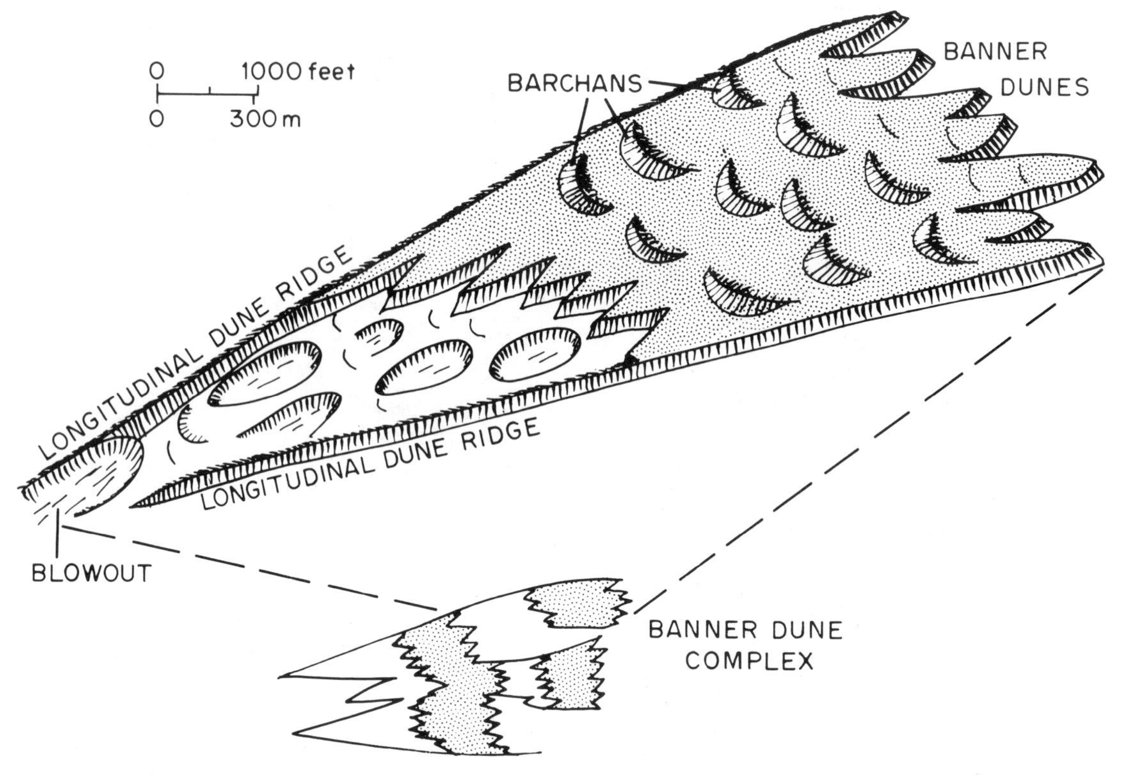

Figure 10-14. Banner dune complexes of coastal Texas comprising longitudinal dune ridges, barchans, and parabolic downwind terminations. (After Brown *et al.*, 1977.)

of sabkhas with their characteristic mud-cracked algal mats, halite crusts, and gypsum crystals. The Walvis Bay area of the coastal Namib Desert (Fig. 10-13) is also subject to extensive river flooding once a decade or so. Evaporite precipitation proceeds rapidly in the vadose zone, while surface cementation by halite plays an important role in preservation of barchan foresets and adhesion ripples (Nagtegaal, 1973).

Blowouts form during episodes of temporarily reduced sand supply or by breaching of the dune cordon by local destruction of vegetation. Parabolic dunes generally encroach landward from the blowout, and may eventually become detached and migrate. Complex examples of parabolic dune complexes in South Texas, referred to as banner dunes (Brown, *et al.*, 1977) (Fig. 10-14), have smaller barchanoid and transverse dunes superposed. These smaller dunes convey sand from upwind deflation areas to the downwind parabolic slipfaces (Price, 1958), but banner dunes eventually become immobilized by vegetation. The features typical of parabolic dunes and associated forms are scour-fill structures, large trough sets, broad dispersion of planar foresets that commonly show upward convexity, organic partings, extensive penecontemporaneous deformation, and abundant plant roots (Bigarella *et al.*, 1969).

The Upper Paleozoic Casper Sandstone with its enormous festoons (Knight, 1929), or troughs, and high-index eolian ripples with crestlines parallel to foreset dip, may have originated as parabolic dunes and blowouts (McKee, 1979c).

Dome dunes in a coastal setting result from modification and coalescence of barchanoid dunes under the influence of strong winds. Gradations of topset laminae into foresets and rippled silty bottomsets, with major erosional discontinuities, are sufficiently distinctive to permit detailed recognition of individual dune forms in the rock record, for example, in the Triassic Keuper Sandstone of England (Thompson, 1969).

The Permian Lyons Sandstone (Walker and Harms, 1972), possibly the best ancient example of a coastal dune belt, is elongated north–south, was bounded to the east by a seaway, and was supplied with arkosic sediments from the ancestral Front Range to the west (McKee, 1979c). Reworking by wind fashioned complex dunes, probably parabolic, with diverse slipface azimuths, and eroded, broad, scour troughs resembling blowouts. Preservation of small eolian ripples and distinctive eolian slumps was facilitated by moisture.

Coastal dunes are particularly susceptible to changes in climate and sea level. Rainstorms modify the dune morphology, and produce broad sandy ramps. Clays, hydrous iron oxides, and humic compounds are carried into the dune sands by infiltrating waters, accumulating as irregular, wispy dissipation layers obscuring primary stratification (Bigarella, 1979). These dissipation layers may form cyclically, alternating with dune cross bedding. Soils form rapidly in the coastal dune environment, and advanced podsolization with indurated ferruginous zones characterizes Quaternary dunes of Zululand (Hobday, 1975). Diagenetic breakdown of ferromagnesian minerals in a moist, alkaline environment can cause reddening shortly after deposition. Some Holocene dune sands have already undergone partial cementation by redistribution of carbonate from wind-blown forams and molluscan fragments, whereas some late-Pleistocene coastal dunes are thoroughly indurated.

Salient features of coastal dune deposits are, in summary:

(1) Elongate, coast-parallel sand-body geometry.
(2) Lateral interfingering with shore-zone and marine-shelf systems.
(3) Broad spread of cross-bed dip azimuths.
(4) Development of large troughs and other multiple erosion surfaces. Abundant trough cross beds may be diagnostic.
(5) Common distortion, brecciation, and microfaulting of foresets favored by high moisture content.
(6) Preservation of high-index ripples transverse to foresets.
(7) Unique topset–foreset–bottomset continuity of laminae.
(8) Cyclic development of colloidal dissipation layers and paleosols.
(9) Abundant root penetration and thin organic layers.
(10) Characteristic heavy mineral associations.

Although coastal dunes tend to be obliterated during marine transgression, there is evidence that in some situations they may survive passage through the surf zone and be preserved in the geologic record. Prominent offshore ridges of Pleistocene "eolianite" on the Australian, southern African, and other continental shelves accumulated as coastal dunes during Quaternary low sea-level stands (McCarthy, 1967; Hobday, 1975; B.G. Thom, personal communication 1982). Rapid cementation preserved these dune systems.

Chapter 11

Depositional Systems and Basin Hydrology

Introduction

Although a discussion of hydrogeologic processes and basinal flow systems may seem to be a departure from this examination of genetic stratigraphic analysis and energy mineral occurrence, it is included for several compelling reasons.

First, the distribution or presence of certain nonframework facies of many systems is determined mainly by the configuration of syndepositional ground-water flow, rather than by depositional setting. A prime example, and one of direct economic interest, is the formation of peat. Plant detritus can accumulate in any fluvial backswamp setting. However, the high organic productivity and subsequent preservation necessary for economically significant deposits also requires specific conditions of regional ground-water discharge. Thus, a combination of both depositional environment and hydrogeologic setting defines exploration targets for coal and many other syngenetic deposits.

Secondly, epigenetic deposits, such as sandstone uranium ore bodies, are produced by the concentration of dissolved material from moving ground water. Ground-water flow patterns, which reflect the geometry and distribution of permeable framework facies, determine sites of accumulation. Thus, localization of deposits is indirectly related to facies distribution, but is neither controlled by, nor uniquely associated with, the depositional environment of the host unit. In addition, the geochemical preconditioning or "host preparation" necessary for concentration can be controlled by syndepositional hydrogeology. For example, preservation of dispersed organic material within fluvial channel-sand facies is likely determined by the depth of the water table during deposition. Organic debris and early diagenetic iron disulfide provide the matrix reducing capacity necessary to precipitate dissolved uranium from meteoric ground waters that may subsequently intrude the sands. Obviously, understanding and predicting distribution of uranium and other epigenetic sedimentary mineral deposits necessitates reconstruction of both depositional environment *and* ground-water flow history.

Thirdly, moving ground waters are believed by many researchers to play an important part in the primary and secondary migration of hydrocarbons, in the localization of hydrocarbon pools, and in the diagenetic modification of reservoir sands. Geochemical modification of oils at depths of up to several thousands of feet by circulating ground water has been well documented. Although each of these topics is a focus for active research and controversy, an understanding of the potential role of basin circulation systems is a necessary prerequisite for testing and application of new ideas.

Distribution of the mineral fuels—coal, uranium, and petroleum—within the fill of a sedimentary basin thus reflects the combined influence of both depositional and ground-water flow systems. Attempts to understand their distribution will be successful only to the extent that both systems are considered and accurately reconstructed.

Fundamentals of Ground-Water Flow

Although many texts review the principals of hydrology, few specifically emphasize natural flow in large hydrologic basins. An excellent text for expanded discussion of the fundamentals briefly reviewed here is Freeze and Cherry (1979).

With the exception of a thin veneer capping its subaerial portions, sedimentary basin-fill is saturated by fluids. By far the greatest volume of fluid is water. Although movement of such ground water may be slow, even by geologic standards, long-term stagnation is unusual. More commonly, ground water-systems are dynamic. Measurement of movement may be expressed as flux: the volume of water crossing a unit cross-sectional area per unit of time. Quantification of flux is given by the well-known formulation of Darcy's Law:

$$Q = KAI \text{ or } Q/A = KI,$$

where Q is the volumetric flow rate per unit time (L^3/T), A is the measured or assigned cross-sectional area (L^2), I is the hydraulic gradient, or difference in potential energy, expressed as the change in elevation of the water level per unit distance ($\triangle L/L$), and K is a proportionality constant known as the hydraulic conductivity. Examination of the equation shows that K has dimensions of L/T. Its value is a function both of properties of the rock matrix and of the fluid. Water level may be thought of as the elevation to which water in the aquifer would rise in an open well bore. Darcy's Law is an empirical expression that has been validated over a wide range of matrix types and gradients typical of subsurface conditions.

A singularly important implication of Darcy's Law is that ground water moves in response to gradients within a potential field. Consequently, the physics and mathematics of ground-water flow are directly comparable to those of other potential fields such as magnetic and gravitational fields. Placement of a water particle within such a potential field defines its energy level, and determines the direction in which it will migrate in order to convert potential energy to work and move to a lower energy level. A mathematical treatment of fluid potential was developed by Hubbert (1940) in a classic paper. The magnitude of the hydraulic potential is related to the hydraulic head by the gravitational constant g. At any point within the flow system, hydraulic potential is determined by the sum of two components: elevation relative to the base level (usually sea level), and pressure. A water particle at a higher elevation has a greater gravitational potential energy level, or head, than all particles at a lower elevation. Similarly, water within a regime of high pressure has a higher potential energy level, expressed as a capacity to do work, than water in a lower-pressure regime. In natural ground-water flow systems, either component of the total hydraulic head may dominate and determine the direction and magnitude of fluid flow.

The basic device for measurement of fluid potential is the piezometer, which in simple form is a tube open at one end to a point within the saturated portion of the flow system, and to the air at the other. The level to which water rises within the tube, which is measured as an elevation, defines the potential energy or hydraulic head of the ground water at that point in the flow system. The water table is a special case of such a head surface. A series of head measurements, providing a three-dimensional array of points, describes the potential field. Using the array, equipotential contours can be drawn in horizontal or vertical planes to describe the succession of equipotential surfaces that define the field. Along such a contour or surface, measured head is contant. Maximum head gradient is obtained by flow perpendicular to equipotentials. Thus, direction of ground-water flow within isotropic media is uniquely defined by the field; ground-water flow lines are always normal to equipotentials (Fig. 11-1). The combination of equipotential contours and flow lines within a plane of section is a two-dimensional solution to the general equations of flow within a potential field, and describes both magnitude (flux) and direction of fluid migration.

Ground-water flow through saturated media may be either steady state or transient. Steady-state flow occurs when the magnitude and direction of flow are constant with time, and transient flow occurs when either magnitude or direction changes with time. In conditions of shallow ground-water flow, recognition of the two flow types is relatively straightforward. Under steady-state conditions, conservation of mass (as defined by the equation of continuity), combined with the incompressibility of water, requires that the amount of water entering or recharging a segment of the flow system is balanced by an equivalent volume of water discharging from the flow segment. Thus, a simple flow system consists of at least three components (Fig. 11-1): a recharge area, a zone of lateral flow, and a discharge area. Along flow, potential energy is converted to work and decreases, but the volume of water entering the system is the same as the volume discharged from the system.

Differentiation of steady-state and transient flow is more difficult when the long-term hydrologic evolution of large sedimentary basins is considered. On a geologic time scale, significant volumes of water are added to the basinal flow systems by compaction and mineralogic transformations, and both gravitational and pressure head play roles in flow. In general, transient flow characterizes periods when pressure head determines the direction and magnitude of ground water flux within significant portions of the basinal flow system.

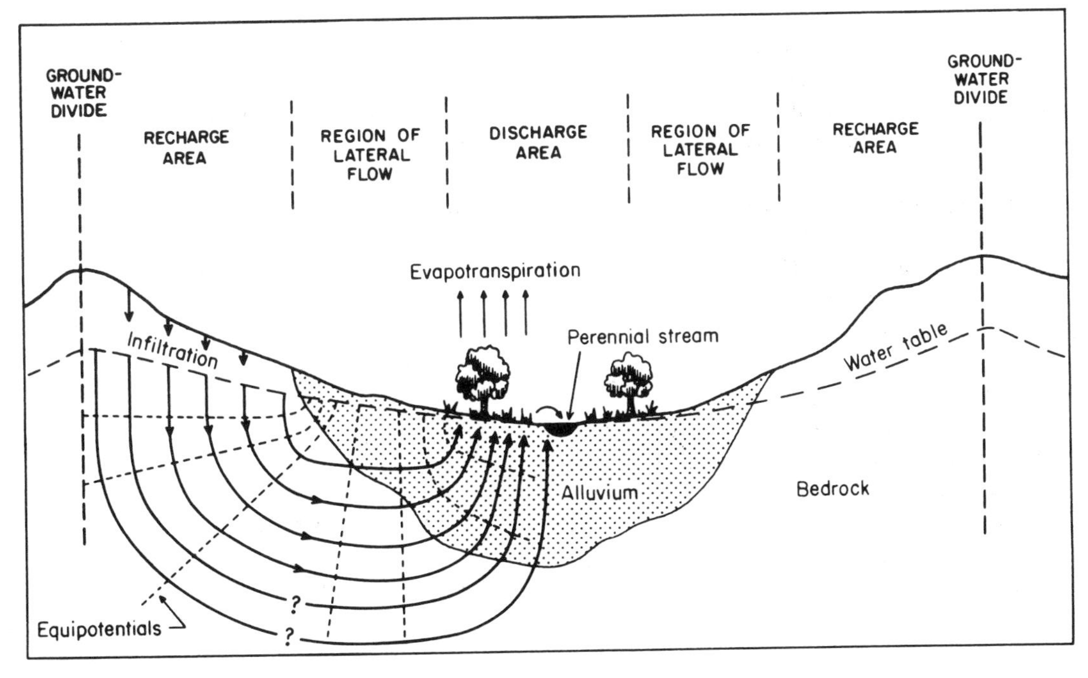

Figure 11-1. Simple ground water flow cell within the shallow meteoric regime. Flow lines cross equipotentials at right angles, indicating isotropic permeability. (Modified from Galloway *et al.*, 1979b; original from Williams, 1968.)

Limitations to Darcy's Law

Darcy's Law summarizes an empirical relationship between fluid flux, a potential field, and matrix and fluid properties over a wide range of conditions. Mathematical as well as experimental derivations of the relationship do indicate possible limitations.

Inherent in Darcy's Law is the assumption that flow is laminar, responding to viscous and frictional forces rather than to inertial forces. This requirement may be violated in media where large pores, such as fractures, and high head gradients exist. However, under normal head conditions in granular media, intergranular flow is laminar and not turbulent. Thus, analogies to open channel flow that incorporate turbulence or eddy formation are inappropriate for description of ground-water flow processes.

At conditions of low head differential in materials characterized by extremely small pore diameters approaching molecular dimensions, non-Darcian behavior is suggested by some authors (Swartzendruber, 1962). Causes and significance of this phenomenon are subject to debate. The principal implication is that there exists a threshold hydraulic conductivity that must be exceeded for flow to occur under a specific gradient within fine materials such as clay. Conversely, a threshold gradient must be exceeded for flow to occur in media characterized by low hydraulic conductivity. In other words, existence of a head differential does not guarantee fluid flux, but may persist indefinitely in a totally stagnant system.

Mass and Energy Transfer

Moving fluids can transport both mass in the form of dissolved ions, complexes or collodial suspensions, and energy in the form of heat or electrically charged particles. The primary direction and magnitude of transfer (or flux) of mass or energy are directly proportional to the volume and direction of ground water flow. Thus, stagnation or reduced-flow conditions are diametrically opposed to large-scale accumulation of soluble materials because the flux of mass is proportionally reduced.

Solutes can also move through ground water by molecular diffusion. Where a concentration gradient exists, dissolved species move under the influence of their molecular kinetic energy from areas of high concentration to areas of low concentration. Diffusion may thus bring new material to a site of precipitation because the precipitation process removes the constituent from the water, reducing concentration and regenerating the gradient. However, in saturated granular media, diffusion is an extremely slow process. Calculations using measured diffusion coefficients show that diffusion flux is typically overwhelmed by bulk transfer of the constituent by moving ground water (Freeze and Cherry, 1979). However, in materials characterized by very low hydraulic conductivities and extremely low flux, such as deeply buried shales and mudstones, diffusion may be an important process of mass transfer over significant periods of geologic time.

Transport of mass or energy by moving ground water is characterized by gradual spreading and dilution. This phenomenon is called hydrodynamic dispersion and occurs at both microscopic and macroscopic levels. Microscopic diffusion primarily results from molecular diffusion, and variability of pore dimensions and geometries. The larger-scale mechanical dispersion has been shown to be a cumulative product of flow variability induced by nonuniform stratification characteristics, irregular bedding and reservoir geometries, and textural inhomogeneities (Schwartz, 1977). Thus, degree of dispersion and mixing in a ground-water flow system is determined in part by depositional or facies characteristics of the host depositional system.

Mass transfer of solutes, as well as thermal energy, by ground water provides natural "tracers" that may be useful in determining ambient ground-water flux (Wallick and Toth, 1976; Jones, 1975). If dissolved constituents react with the rock matrix, a record of cumulative flux geometries may be preserved and provide insight into the paleoflow history of the basin. The discussion of common geochemical interactions between circulating ground waters and the rock or sediment matrix presented in a subsequent section indicates many such geochemical tracers that reflect flow history.

Properties of the Aquifer Matrix

An aquifer is the medium in which ground water flows. In most contexts, the term aquifer is used only for units within the stratigraphic succession that have the greatest capacity to transmit ground water. In a general sense, an aquifer is any stratigraphic unit, be it a specific sand body, genetic facies, depositional system, or composite of any of these, that is capable of transmitting significant quantities of ground water under the conditions and within the time frame pertinent to the discussion. Thus, almost any sedimentary unit can serve as an aquifer. In the following discussion, the term is used in its most general sense as any sediment or sedimentary rock capable of transmitting ground water.

Physical Properties

The fundamental properties of the aquifer matrix are its *porosity* and *permeability*. Porosity (ϕ) is defined as the ratio of pore volume to the total volume of a sample of material. Pore spaces include primary intergranular pores, open fractures, and secondary leached pores. Permeability (k) is a measure of the ease with which a fluid can pass through a porous material. If the fluid density and viscosity are specified, then measured permeability is an intrinsic property of the sample. Permeability has dimensions of L^2, and is commonly expressed as darcies or millidarcies in petroleum literature, and as square centimeters in ground-water literature. Permeability is related by hydraulic conductivity (K) in the Darcy equation. Hydraulic conductivity expresses the combined intrinsic property of the media, its permeability, and properties of a specific fluid, water, at ambient or standardized physical conditions. Conductivity dimensions are L/T and it is commonly expressed in units such as centimeters per second or gallons per day per square foot.

The range of permeability and hydraulic conductivity in natural sedimentary materials is given in Table 11-1. The range is immense, covering nearly 14 orders of magnitude. Even sediments within a single textural class, such as sand, may have a spread exceeding four orders of magnitude.

Like many physical parameters, permeability and hydraulic conductivity commonly are not uniformly or homogeneously distributed within an aquifer. In such a case, the aquifer is described as heterogeneous. Further, the measured value of either parameter may vary depending on the orientation of flow relative to bedding. Such an

Table 11–1. Range of Values of Hydraulic Conductivity and Permeability (from Freeze and Cherry, 1979)

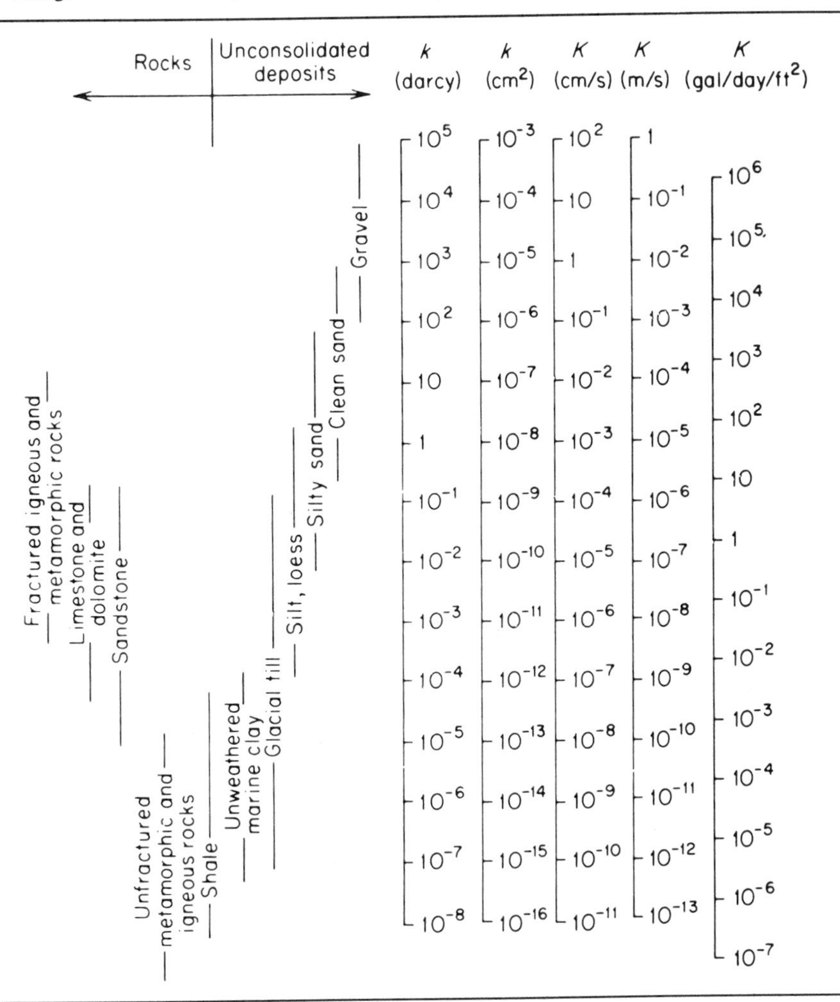

aquifer is said to be anisotropic. Both heterogeneity and anisotropy can vary randomly or can show a definite pattern or orderly distribution within an aquifer. In sedimentary strata, maximum values of permeability are commonly parallel to bedding and minimum values are vertical to bedding. The ratio can be as much as 10 to 1 in clays and shales (Freeze and Cherry, 1979) and even some sandstones (Davis, 1969, Table 4).

Porosity and permeability are interrelated. Most commonly, permeability shows a positive correlation with porosity. However, the exact relationship varies from unit to unit because pores may have varying sizes, degree of interconnection, and origins. For example, very small pores in mud are abundant, but tortuous fluid circulation through the microscopic pore throats entails more surface contact, and consequent viscous frictional drag, than less tortuous flow through large, open pore throats in a sand possessing equivalent porosity. Isolated secondary leached pores may increase porosity but have little effect on permeability.

Primary intergranular porosity is largely a function of sorting, compaction, and cementation. Permeability varies with these factors as well as with grain size. Well-sorted sediments have higher permeabilities than do poorly sorted sediments of equivalent average grain size.

Transmissivity. Permeability measures the ease with which fluid moves through a unit cross-sectional area of the aquifer matrix. Transmissivity is the product of permeability and aquifer thickness, and is the cumulative water-bearing capacity of an aquifer. Total transmissivity of heterogeneous aquifers may be calculated by summing the transmissivities of individual permeability layers. Together, permeability and transmissivity describe the potential for fluid flow through beds or facies within a basin fill. However, the actual volume, velocity, and direction of flow are also determined by regional hydraulic head. Thus, measurement or calculation of transmissivity or permeability distributions within a sedimentary sequence does not, *per se*, determine which portions of the sequence have been flushed by circulating ground waters. Rather, they define optimum pathways or "plumbing" for groundwater flow should an appropriate hydraulic head exist.

Because transmissivity is a product of aquifer thickness and of permeability, which is in turn controlled by texture and bedding, the three-dimensional water-bearing capacity of a sedimentary sequence is closely related to facies distribution. Framework facies define the highly transmissive "plumbing." Bounding lithologies form a confining, but nonetheless leaky matrix around permeable, transmissive elements. Consequently, transmissivity distribution may show both vertical and lateral gradients as well as abrupt boundaries reflecting transitional, abrupt, or erosional facies boundaries. In most depositional systems, transmissivity is a highly directional property, reflecting the trend of framework sand bodies. The importance of transmissivity anisotropy is illustrated in Figure 11-2, which shows two equally permeable and thick sand bodies typical of coastal-plain depositional systems. Although the fluvial channel and barrier sand body have comparable permeabilities and thicknesses, only the fluvial sand body produces a highly transmissive element oriented in the direction of the probable basinward hydraulic gradient. For water to move down gradient through a succession of barrier sand units, it must cross the intervening low-transmissivity lagoonal facies. Flow is across the depositional grain and is much less efficient.

Regional and local transmissivity maps can be compiled from quantitative facies maps if sufficient permeability or pump-test data exist to calibrate the most permeable facies. Because of the orders-of-magnitude differences between permeability of coarse and fine end-member sediments, only the coarsest, most permeable units are likely to contribute significantly to total transmissivity of the system. Bounding facies and even finer-grained framework facies can be ignored.

In a series of regional studies of major Tertiary aquifer systems of the northern Gulf Coast, Payne (1970, 1975) demonstrated straightforward correlations between average permeability, aggregate aquifer transmissivity, regional sand trend and thickness, and depositional facies. Axes of maximum transmissivity and observed flushing by ground water coincide with belts of fluvial channel-sand deposits. Similar results for uranium-bearing aquifers are reviewed later in this chapter.

Hydrostratigraphy. Recognition of hydrostratigraphic units provides a parallel to conventional lithostratigraphy and is based on the examination of basin fills from the perspective of the water-bearing properties of component units. Hydrogeologic units function as either water-bearing or water-retarding beds relative to adjacent strata. Tóth (1978) distinguished a hierarchy of such aquifer or confining units based on their bulk permeability, regional extent, and lithostratigraphy. The *hydrogeologic* formation closely conforms to conventional lithostratigraphic units. Two or more formations that function as a relatively conducting or retarding complex are combined to form a *hydrogeologic group*. The third-order and largest units are the *hydrogeologic systems* which are distinguished by "the coherencies of their flow and pressure systems instead of rock parameters" (Tóth, 1978, p. 808).

Recognition and delineation of basin hydrostratigraphy, combined with three-dimensional description and mapping of aquifer transmissivities and possible flow boundaries, provide the logical starting point for interpretation of extant ground-water flow and reconstruction of paleoflow systems. Correlations between aquifer properties and specific facies, framework isolith maps, and observed head distributions and ground-water compositions provide the basis for such a detailed hydrostratigraphic synthesis. Though such syntheses have rarely been attempted, a quantitative three-dimensional hydrostratigraphic framework would seem to offer a powerful and necessary tool for interpretation and extrapolation of epigenetic, diagenetic, and migration processes that require significant fluid flux. An example of such a synthesis will be reviewed in Chapter 13.

In summary, physical hydrologic properties of sedimentary units are related to physical and genetic stratigraphic features at a variety of levels (Galloway *et al.*, 1979).

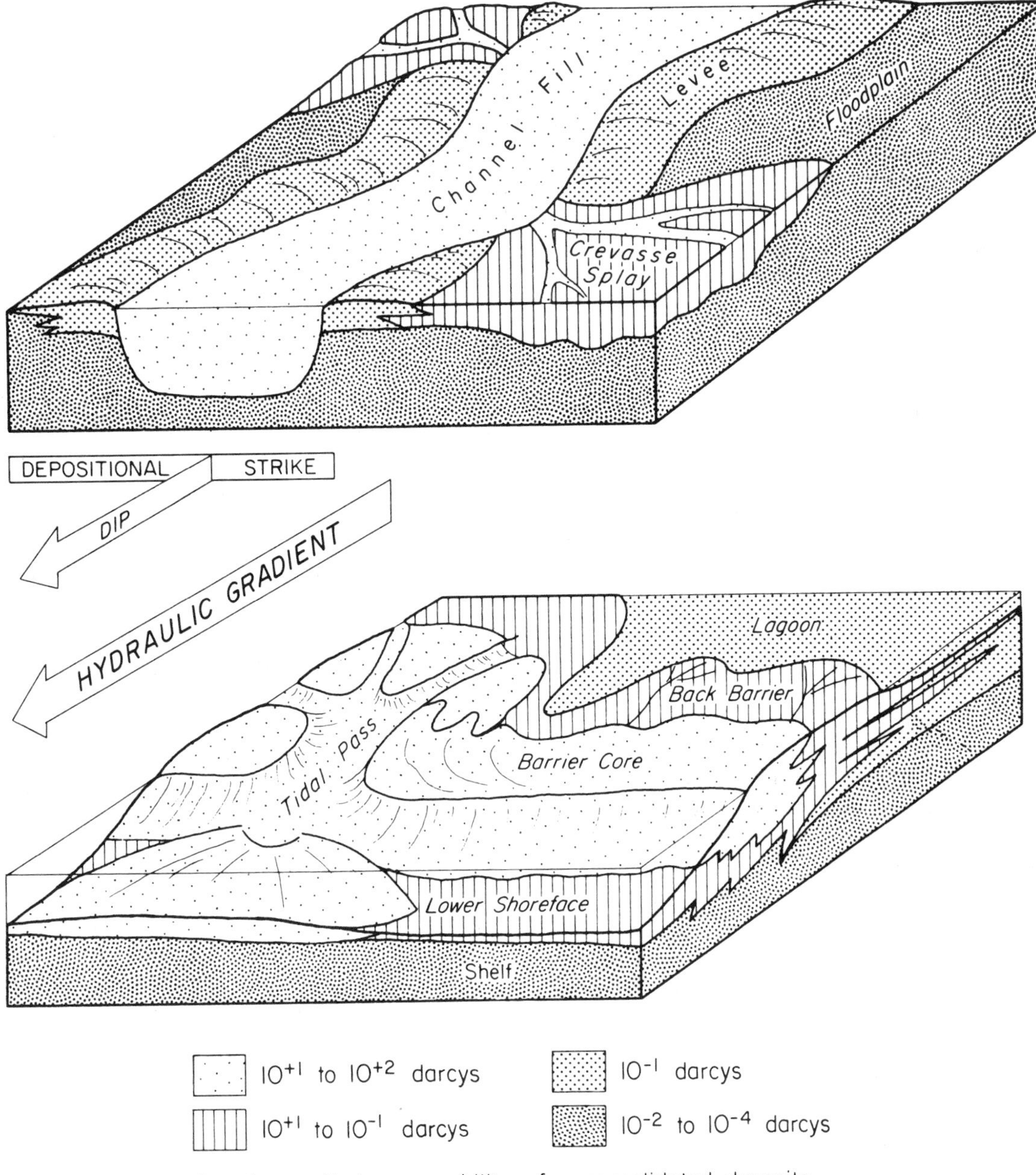

Figure 11-2. Contrasting geometry, lateral relationships, and trends of equivalently transmissive framework sand facies of a fluvial and barrier-lagoon system. Both produce highly anisotropic transmissivity axes. The dip-parallel fluvial channel fill is the preferred aquifer for the regional coastward hydraulic gradient. Permeability ranges for unconsolidated sediments based on data in Davis (1969). (From Galloway *et al.*, 1979b.)

(1) Texture, fabric, and degree of cementation determine the porosity and permeability of the aquifer matrix.
(2) Within depositional units, sedimentary structures, bedding architecture, and irregular or systematic vertical and lateral textural changes introduce both random and ordered heterogeneity and anisotropy to three-dimensional permeability distribution.
(3) Dimensions and orientation of depositional units, particularly framework facies, determine aquifer transmissivity and commonly

impart major anisotropy to water-bearing capacity.

(4) The abundance and hydraulic continuity of framework facies determines the overall transmissivity of the depositional system or of its component facies assemblages and effects its hydrostratigraphic characterization as an aquifer or confining unit.

(5) The vertical and lateral relationships of various depositional systems within the total basin fill commonly define the limits and dimensions of the hydrogeologic systems.

Geochemical Properties

Sediments and sedimentary rocks contain various detrital and diagenetic phases that may react with circulating fluids. Such reactions change the chemical character of both the fluid and the aquifer matrix. Specific reactions depend upon the chemistry of the invading ground waters, but matrix reactivity may be defined in four basic ways.

Soluble constituents that may be selectively leached from the matrix are present in most sediments. Although solubility reactions may be quite complicated and poorly documented at the low to moderate temperatures and pressures prevailing in sedimentary basins, they are primarily controlled by the pore-fluid chemistry and availability and grain size of solid phases within the rock matrix. Reactions may be as simple as solution of sodium chloride by fresh ground water, or as complicated as the dissolution and replacement of plagioclase by potassium feldspar during burial diagenesis. Thus, mineralogic composition is a fundamental property of sediments that predetermined potential dissolution/precipitation reactions.

Reducing and buffering capacity are measurements of the matrix Eh and pH. Matrix Eh is a function of the content of reactive elemental oxygen as well as elements such as iron that exist in two or more valence states. Iron disulfide and organic material are two common constituents in many sediments that impart reducing capacity to the rock matrix. Buffering capacity, or pH of the matrix, is a function of the abundance of constituents capable of releasing or consuming protons upon reaction. Oxidation of iron disulfide, for example, releases four moles of hydrogen ions (protons) for each mole of oxygen consumed. Obviously, pH and Eh are interdependent electrochemical properties of the sediment.

Exchange capacity of a sample measures the abundance and reactivity of colloidal-sized constituents, primarily clay minerals, as well as certain mineral phases such as zeolites. Electrochemical and physiochemical properties allow these phases to adsorb or take up ions from solution, commonly releasing other ions in their place. Exchange capacity and adsorptive capacity are intrinsic properties of a sediment, but vary according to the ionic strength and composition of the species in solution.

Hydration of the aquifer matrix is a sum of chemically and physiochemically bound water contained within hydrous mineral phases, such as zeolite and structured water in clay minerals. Such water is subject to release upon application of heat and pressure.

As in the case of physical properties, geochemical properties of the aquifer matrix reflect the potential for water–rock interaction. Specific reactions depend upon the degree of flushing, and chemistry of the migrating waters. Alteration of the aquifer matrix by reactive fluids, in turn, leaves a record within the aquifer that may be defined by traditional paragenetic and diagenetic studies.

Basin Geohydrology

The geohydrologic framework of a large, actively compacting sedimentary basin consists of ground water circulating within several different regimes (Fig. 11-3). Although the boundaries of the regimes and their local terminology are sometimes difficult to reconcile (Bogomolov *et al.*, 1978; Kissin, 1978; Jones, 1975), they nonetheless form important end members characterized by their relative positions within the geography and history of basin development.

The *meteoric regime* typically occupies the shallow periphery of the basin. Waters recharged by infiltration of meteoric precipitation move toward the topographic basin center under the influence of gravitational head. On a geological time scale, circulation is rapid. Discharge and evaporation complete the traditional hydrologic cycle (Fig. 11-3).

The *compactional*, or elisian, regime is characterized by upward and outward expulsion of pore

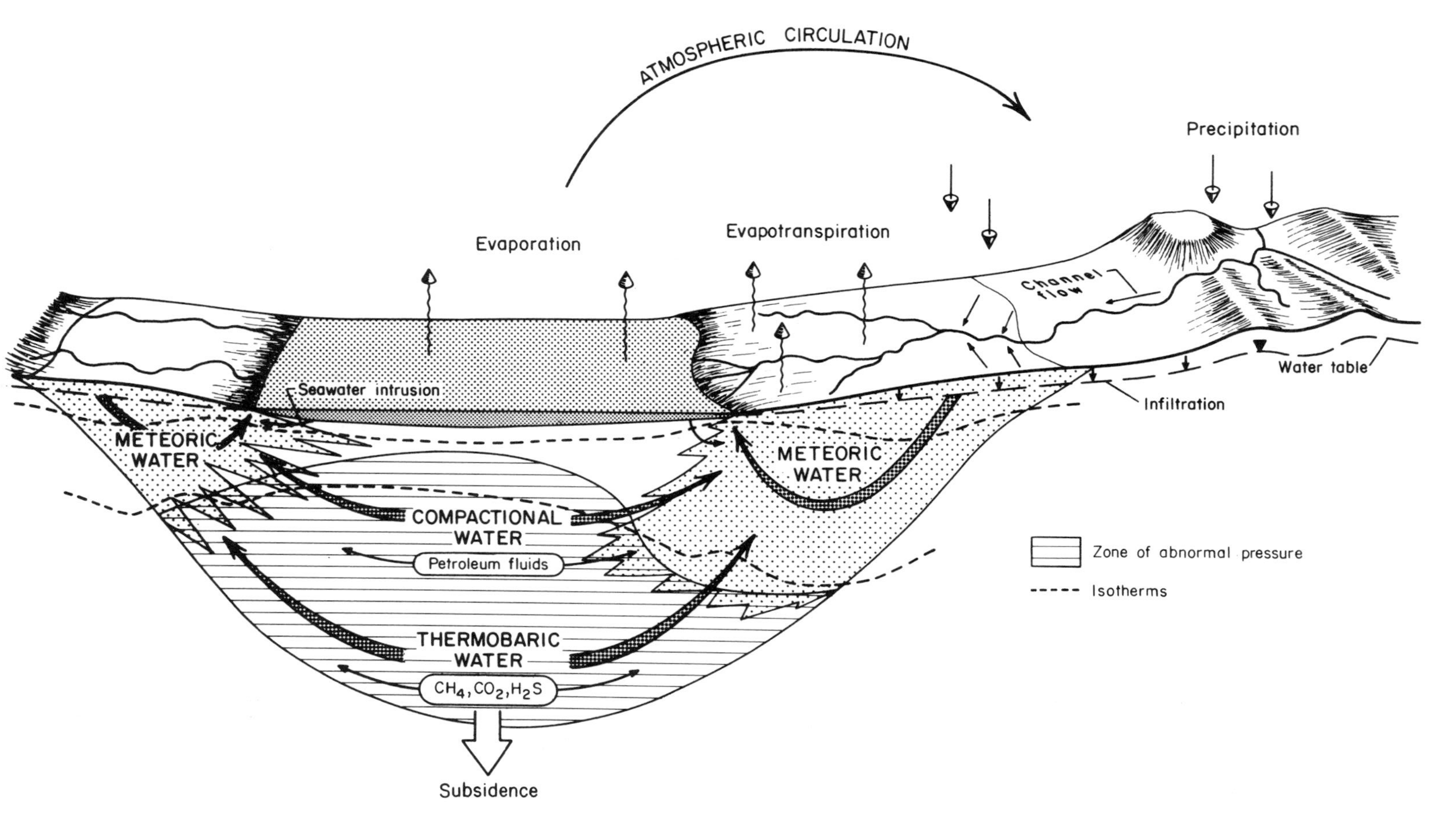

Figure 11-3. The geohydrologic cycles of a large, depositionally active sedimentary basin.

waters contained within the compacting sediment pile. Such waters may be evolved connate water, that is, water deposited with the sediment and subsequently modified by water–rock interactions, or may be meteoric water that has been buried below the zone of active meteoric circulation. Pressure head, generated by lithostatic loading or compressive tectonic stress, is the driving mechanism. If circulation is impeded by low vertical transmissivity, waters in this regime may develop pressure heads substantially greater than that at the base of a free-standing column of water of the same height.

The *thermobaric*, or abyssal, regime lies in the deepest portions of the basin fill where temperature and pressure are high. Water is released by dehydration reactions of clays and other mineral phases. Fluids move in response to pressure head created both by phase changes, such as generation of gases, and by lithostatic loading. However, the extremely low permeabilities produced by compaction and cementation of confining units commonly restrict water circulation, and geochemically modify expelled waters.

Recognition of the thermobaric and compactional regimes as important components of the hydrologic basin requires larger view of the hydrologic cycle as it evolves during filling of major sedimentary basins. Large volumes of water are diverted from the meteoric and surface flow systems and entombed within the sediment fill as pore or mineralogically bound water. These waters discharge into and mix with the meteoric cycle slowly, if at all. Unlike the meteoric system, which continually recirculates unlimited volumes of water, the supply of compactional and thermobaric waters, though large, is finite. Interpreted processes or products ascribed to circulation within these systems must recognize and accommodate this limitation.

Evolution of the Hydrologic Basin

The compactional and thermobaric regimes persist as long as active basin subsidence and filling continue. However, with cessation of basin subsidence and possible uplift of the basin margin or more extensive tectonic segmentation, the flux of compactional waters and thermobaric fluids decreases and ultimately ceases. Strata of such a "mature" basin are increasingly flushed by meteoric waters recharged along the uplifted margins (Tóth, 1980; Coustau *et al.*, 1975). Regional flow is centripetal toward the topographic floor of the hydrologic basin. Well-described basins in which ambient flow is dominated by regional meteoric circulation driven by gravitational head include the Paris Basin (Korotchansky and Mitchell, 1972), the Great Lowland Artesian Basin of Hungary (Erdelyi, 1972), and the Western Canada Mesozoic foreland basin of Alberta (Hitchon, 1969a and 1969b). Long-term flushing by regional meteoric circulation replaces residual connate waters with geochemically-evolved meteoric waters (Clayton *et al.*, 1966). With tectonic stability and erosional leveling of basin margin relief, circulation of meteoric water slows, and the hydrologic basin may become effectively stagnant. The vertical pressure gradient throughout the basin fill is hydrostatic (Coustau *et al.*, 1975), or may become less than hydrostatic.

Dynamics of Ground-Water Flow

The geometry of ground-water flow is determined by (1) the slope and configuration of the potentiometric surface (as defined by the water table in interconnected hydrostratigraphic systems or by pressure head within isolated systems) and (2) the three-dimensional permeability distribution within the saturated basin fill. Water-table configuration is a subdued image of the land surface, rising in areas of greater surface elevation. Thus, in the meteoric regime, ground-water flow is a function of topography and geology (Freeze and Witherspoon, 1968). Figure 11-4A and B illustrates two simple situations. In both models, regional flow of water is to the left, down the regional hydraulic gradient established by the slope of the water table. In both examples, flow is concentrated or focused within either a dipping (model A) or horizontal (model B) permeable layer that offers a path of least resistance appropriately oriented to move ground water down the regional gradient. Note that recharge can be direct where the aquifer intersects the water table (model A) or can occur by cross-stratal flow through an overlying confining unit (model B).

Model C (Fig. 11-4) illustrates the effect of an irregular water table surface, such as might occur in the dissected or hilly topography of a basin margin. Flow is divided into local and regional cells and multiple zones of recharge and discharge. Tóth (1962) emphasized the importance

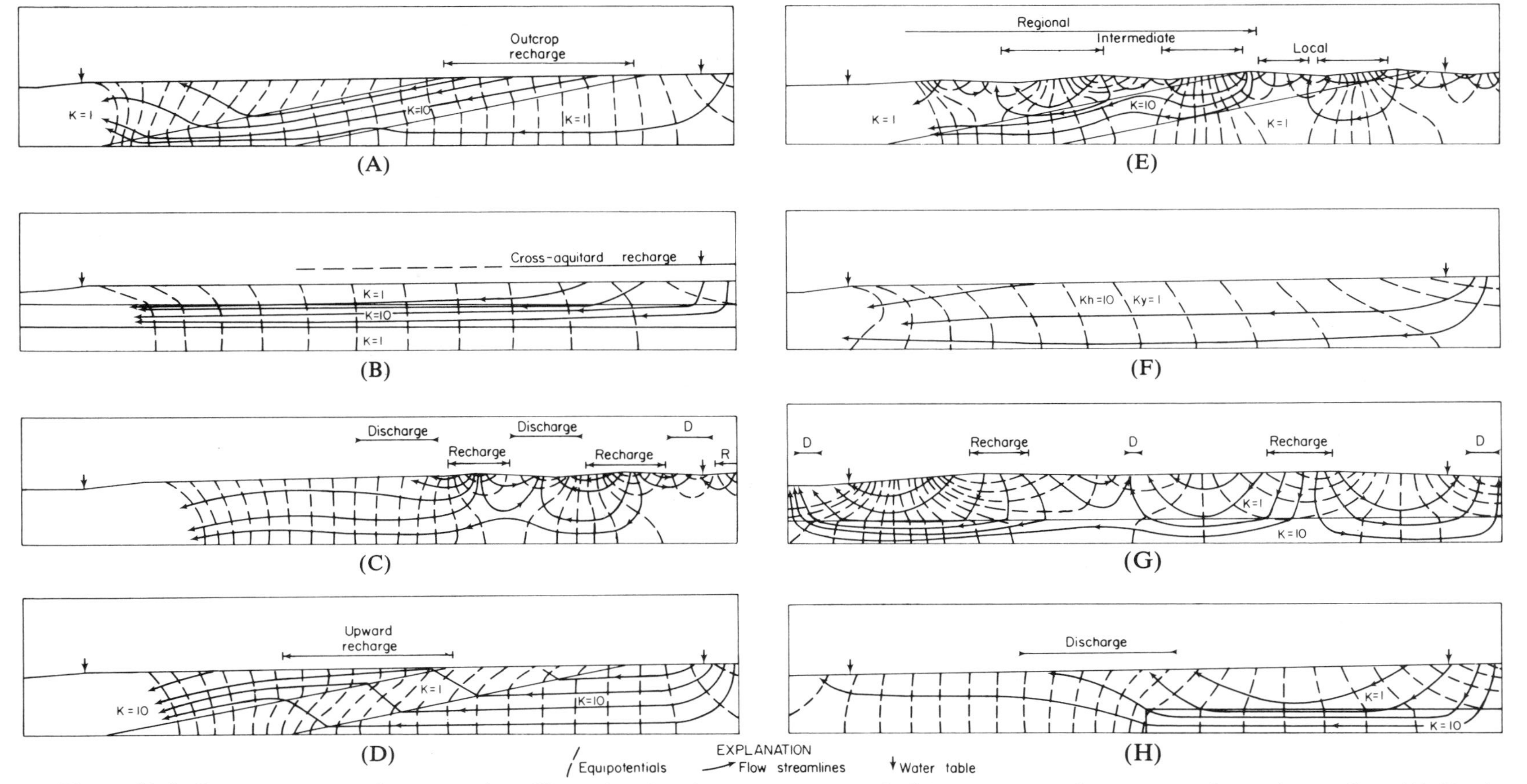

Figure 11-4. Computer-generated cross sections illustrating the effects of various geologic parameters on the geometry of ground-water flow. (A) Simple outcrop recharge and down-gradient (to the left) flow within a permeable layer. (B) Cross-aquitard recharge of a buried permeable layer. Flow moves downward into the permeable conduit, then basinward down hydrodynamic gradient. (C) Complex local and intermediate flow cells generated within an homogeneous aquifer by irregular water-table topography. (D) Upward flow of water across a dipping confining layer and into a shallower permeable layer. (E) Concentration of regional basinward flow within a permeable layer. Irregular topography results in complex local and intermediate flow cells, particularly within less permeable units. (F) Accentuation of lateral flow within a layer displaying strong horizontal permeability anisotropy. Flow lines are not normal to equipotentials in anisotropic media. (G) Collection of regional flow in a buried permeable layer. Flow lines originate from several local cells produced by topographic irregularities. (H) Flow dispersion and surface discharge produced by down-gradient pinch out of a permeable layer. In all models hydraulic conductivity (K) of layers is expressed as a relative value. Absolute values would determine the volume but not the geometry of flow. (Modified from Freeze, (1969).

of a hierarchy of flow cells within a regional ground-water basin. An analogous hierarchy could form within the compactional system as complex pressure differentials develop between low permeability muds and irregular, interspersed permeable sands. Figure 11-4E illustrates the effect of a permeable layer on a hierarchical flow system. Flow threads tend to collect within the aquifer unit, accentuating development of the regional flow system at the expense of the local cells.

Flow intersecting a confining layer (Fig. 11-4B) tends to refract across the less permeable zone, thus minimizing flow distance within it. This model, combined with models A, B, and E reinforce the important concept that fluid flow preferentially utilizes the most permeable avenues available to move down regional gradient. Differentially conductive materials distort the potential field that controls flow direction, just as an iron bar within a magnetic field distorts the field lines. Relative permeability rather than absolute permeability of the transmissive and confining zones determines flow geometry. Volume of flux is, of course, directly proportional to absolute transmissivities of the system.

In anisotropic layers, flow along the direction of maximum permeability is emphasized (Fig. 11-4F). Thus flow typically parallels bedding within sedimentary sequences, as long as boundary conditions and regional head slope permit. Note that in anisotropic media, flow lines do not cross equipotential lines at right angles.

The final two models (Fig. 11-4G and H) depict effects of continuous and discontinuous buried, or "blind" aquifers. Buried aquifers serve to efficiently collect recharge through overlying confining units and move it in the direction of regional gradient. Pinchout of such an aquifer results in dispersion of the collected flow as local increases in hydraulic gradient (indicated by closer spacing of equipotentials) and increase of the effective cross-sectional area of flow (indicated by divergence of flow lines) compensate for the tenfold decrease in permeability of the total section. Flow dispersion or defocusing results in discharge of fluids to the surface, or into adjacent permeable layers.

Structural features may produce permeable or impermeable boundaries that cross-cut stratigraphic bedding. Fault and fracture zones, in particular, may form significant vertical permeability conduits that connect stacked aquifers. However, it is important to emphasize that ground-water flow is directed down regional hydraulic gradient, and is not controlled by structural dip. Because the directions of bed dip and slope of the land surface often coincide in subsiding basins, flow commonly parallels dip. However, if dip reversals do occur, flow direction remains basinward across bedding.

The Meteoric Flow Regime

The meteoric regime may extend to depths approaching 10,000 ft (3,000 m) around the margins of active basins, and to greater depths in stable or uplifted mature basins. Residence times in the subsurface may be as great as several millions of years (Tóth, 1978). However, all waters originate as recharge by downward percolation across the water table and ultimately return to the surface hydrosphere as discharge. Development of hierarchical flow cells is common. Meteoric circulation is likely responsible for formation of epigenetic uranium deposits, commonly determines the distribution of swamp and marsh environments, and affects the secondary migration and surficial alteration of hydrocarbons.

Geochemical Alteration

As it enters and moves through the basin fill, meteoric ground water is modified by contact with the aquifer matrix. Extent of the modification depends on several factors, including the nature of the recharge zone, mineralogy of the sedimentary fill, residence time in the subsurface, and cumulative degree of flushing already experienced by the aquifers. In addition, the relatively simple downflow evolution may be abruptly or gradually modified by mixing with basinal waters discharged from the compactional or thermobaric regimes.

Initial recharge of the meteoric system is by downward percolation through the vadose zone, a veneer of soil or sediment in which pores are filled with air (Fig. 11-5). Fresh waters are unsaturated with respect to soluble minerals, contain dissolved oxygen and CO_2, and are capable of mechanically transporting very fine clay or colloidal particles. Pedogenic features, including leached zones and horizons of accumulation of carbonate minerals, metal oxides, and clay

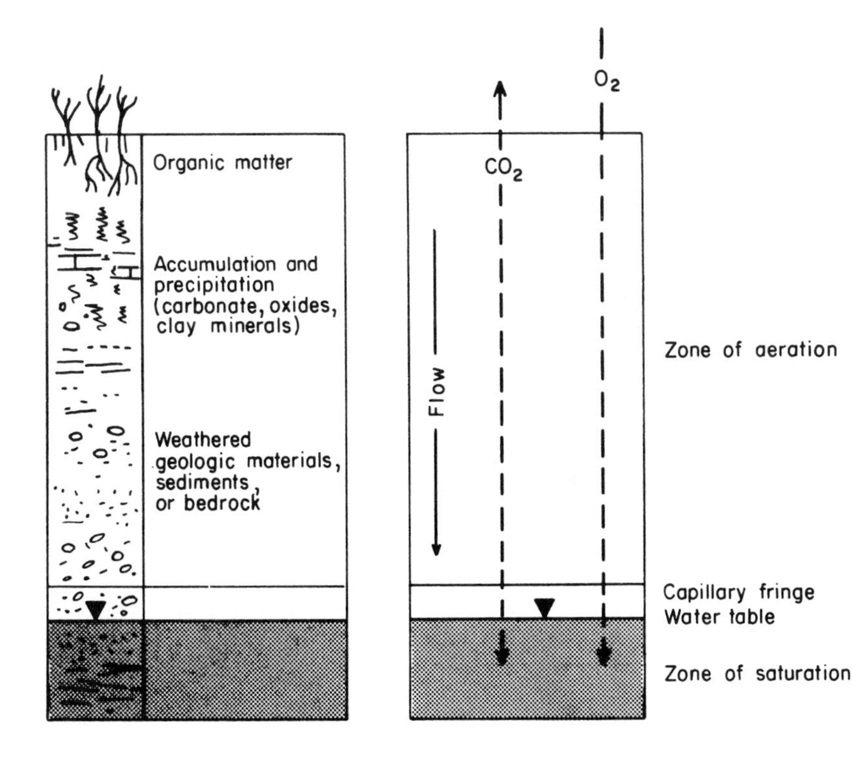

Figure 11-5. Geochemical and saturation environments of the recharge zone. (Modified from Freeze and Cherry, 1979, p. 240.)

minerals, are direct products of the downward percolation of infiltrated waters through the zone of aeration. Bacterial degradation of plant debris consumes oxygen and generates CO_2, which is partly dissolved in the water and partly discharged into the atmosphere. Oxidation is pronounced in well-aerated soils with a deep water table, and may be incomplete in poorly drained soils supporting abundant vegetation. Addition of CO_2 and depletion of O_2 decrease Eh and increase pH as precipitation enters the surface hydrologic regime and then infiltrates into shallow aquifers (Fig. 11-6). The typical resultant shallow ground water is slightly oxidizing and mildly acidic to mildly basic.

Once below the water table and in the zone of lateral flow, ground water begins a geochemical evolution that, at its completion, commonly produces a moderately reducing brine. Regional down-flow geochemical modification of ground water and development of a succession of hydrochemical facies was first described by Chebotarev (1955). Although the specific succession of hydrochemical facies he recognized does not characterize all sedimentary basins, general geochemical trends (Fig. 11-7) are typical of most meteoric flow systems contained within clastic aquifers (Galloway *et al.*, 1979; Freeze and Cherry, 1979, p. 241).

Oxidants, including dissolved O_2, NO_3^-, SO_4^{2-}, and CO_2, are consumed down flow by reactions with reducing organic and mineral components of the aquifer matrix, thus decreasing Eh (Champ *et al.*, 1979). Reactions with silicate phases tend to increase pH as well, producing a mildly basic water containing increasing amounts of dissolved HCO_3^-. Other major dissolved ionic species include Na^+, Ca^{2+}, Cl^-, and sometimes SO_4^{2-}. Shallow, fresh meteoric ground water is dominated by Ca cations. Down-flow, the Ca^{2+} replaces Na^+ 1 for 2 in exchange sites on clays. Concomitantly, Cl^- is dissolved from dispersed sites within the sediment matrix (leaching experiments show that even fluvial muds have significant quantities of leachable Cl^-). The resultant ground water is dominated by Na^+ and Cl^-. Content of total dissolved solids (TDS) increases as additional salts are leached, resulting in a moderately saline brine.

Distribution and abundance of sulfur species is more complex. Moderate amounts of SO_4^{2-} are generated in shallow portions of the flow system by oxidation of sulfide minerals or organic sulfur within the aquifer matrix, and by leaching of sulfate minerals. Down flow, anaerobic reduction of sulfate may produce H_2S or HS^- (Stumm and Morgan, 1970; Champ *et al.*, 1979). In the absence of bacterial activity, SO_4^{2-} persists as a

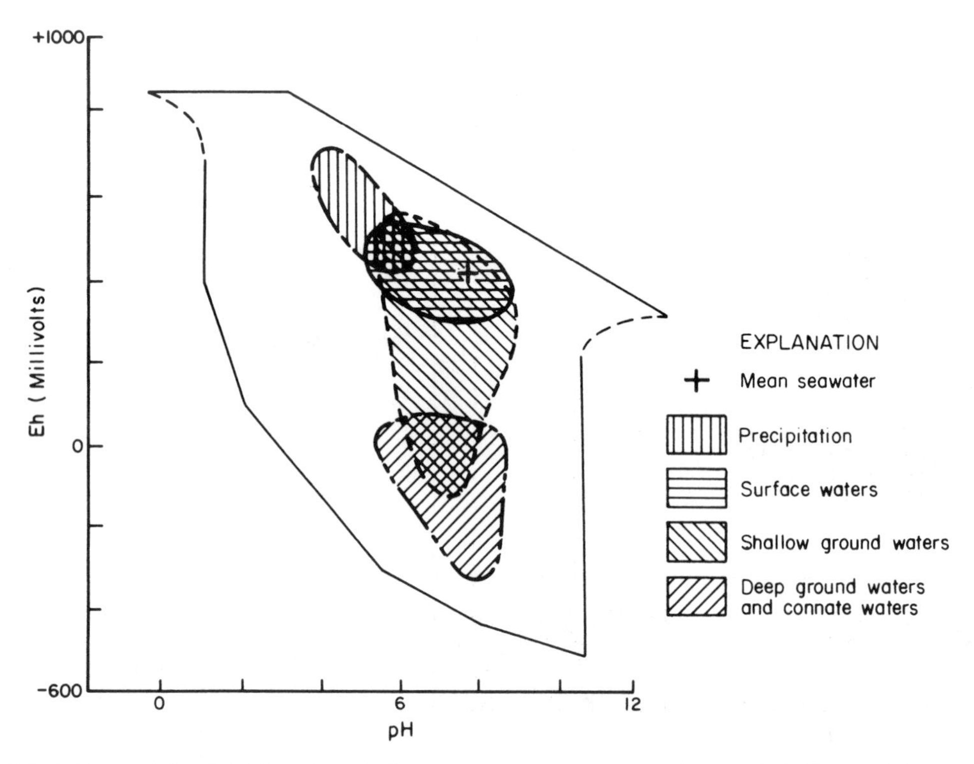

Figure 11-6. Typical Eh–pH fields of precipitation, sea water, surface waters, and shallow and deep ground waters showing the systematic electrochemical evolution through the hydrological cycle. (Modified from Baas-Becking *et al.*, 1960.)

metastable phase at low temperatures, and can increase if additional sulfate minerals are leached.

In addition to geochemical changes, meteoric ground water is heated as it moves to greater depths basinward. Both the trends in geochemical evolution and perturbations in isothermal surfaces or contours (Fig. 11-3) provide traces for interpretation of flow dynamics of regional meteoric flow systems (Tóth, 1972).

Water–Rock Interactions. As meteoric ground waters evolve by interaction with the rock matrix, physiochemical and chemical reactions alter the composition not only of the water but of the host rock as well, leaving a partial record of the passage of the water in the diagenetic mineralogy and geochemistry. Variations in flux of waters through various portions of the aquifer determine the sites at which the reactive waters contact the rock matrix. Ground-water flux is, in turn, a product of the physical aspects of the flow system (permeability distribution, flow boundaries, and hydrologic gradient).

Examination of alteration patterns suggests that their geometry is significantly controlled by physical aspects of aquifer systems. One economically important example of this interplay between the physical stratigraphy and ground-water flux is the formation of an oxidation tongue. Because meteoric ground waters of many flow systems are mildly oxidizing at the recharge area, flow results in oxidation of reductants in the permeable rock matrix. With time and accumulated transport of oxidant into the system, portions of the aquifer are thoroughly oxidized. If there are reasonably uniform abundances of reductants (including organic material, reduced iron, and sulfide minerals) in the rock and oxidant within the ground waters, the volume of rock oxidized along a flow line will be directly proportional to the ground-water flux along that flow line. Areas of high flux will be characterized by development of down-gradient projections, or tongues, of oxidized rock matrix.

Figure 11-8 shows oxidizing water recharging a fluvial aquifer sand unit and moving down the regional hydraulic gradient. Flow is focused by

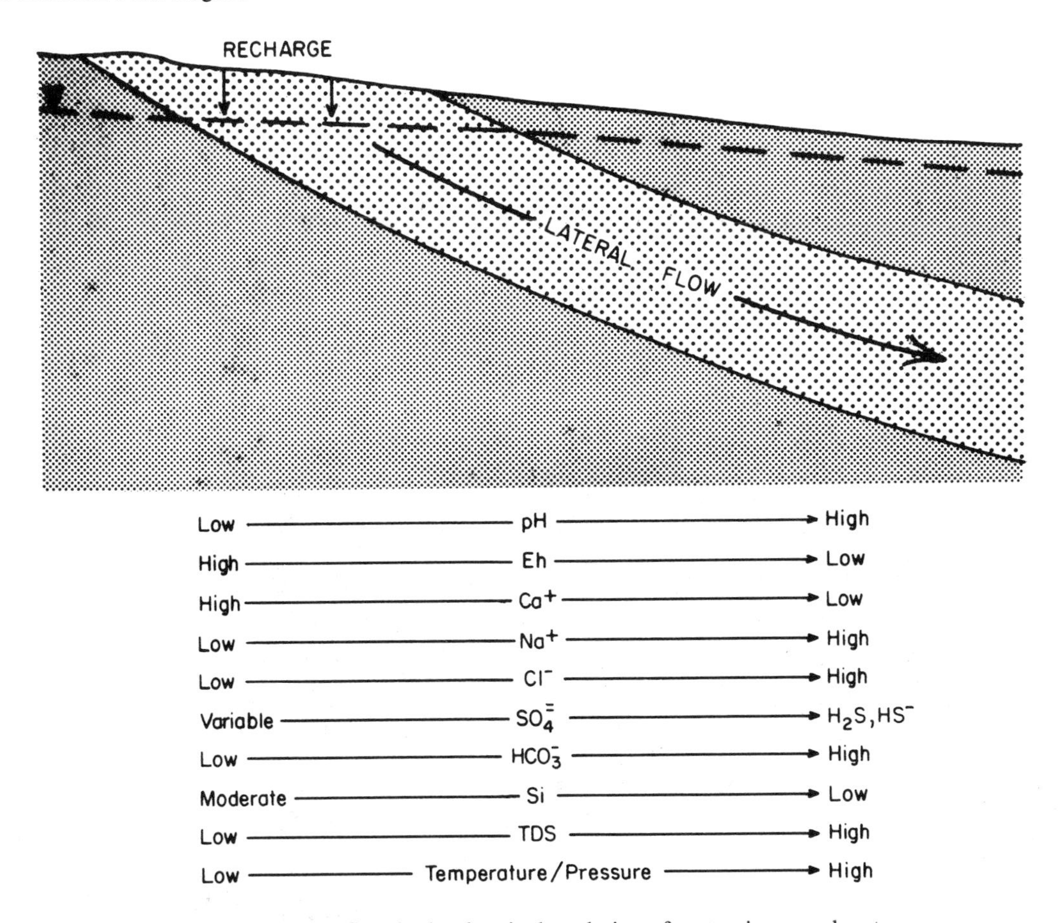

Figure 11-7. Downflow hydrochemical evolution of meteoric ground water.

converging sand trends into area A. Thus, the flux of oxidant into this relatively limited portion of the aquifer is increased, and the oxidation front migrates most rapidly downflow, producing a salient. Flanks of the salient, or oxidation tongue, are characterized by little crossflow and appear as poorly defined boundaries between oxidized and reduced ground. Other areas of active crossflow, such as B, also form a well-defined geochemical interface, but less oxidant is imported into these areas than into areas of flow concentration, so the front migrates less rapidly. Where flow lines diverge, as at C, front migration is minimal, resulting in remnant-reduced islands or embayments in the oxidation tongue. Lastly, large volumes of highly permeable, transmissive sand may be bypassed altogether or may lie beyond the reach of the alteration front (Fig. 11-8, area D).

Discharge Phenomena. Sites of active discharge, particularly of regional meteoric systems, are of particular significance. Discharging waters maintain large areas of shallow or even emergent water table, allowing rapid growth of plants and preventing oxidation of detrital vegetal matter in the soil environment. Bogomolov and Kats (1972) estimated that over 60% of the marshlands of Russia are fed by ground-water discharge; they further pointed out that the chemical constituents added to the marsh deposits by ground-water make them unusually fertile soils. Development of marsh and associated bedded organic debris along the topographic axis of the Great Valley in California, in the face of an arid climate, attests to the geologic importance of meteoric ground-water discharge in the formation and preservation of marsh and swamp deposits. Coastal marshes, though in part related to the availability of surface water, occur along what is inherently a zone of regional ground-water discharge for the coastal aquifer system.

Highly mineralized ground waters may of course prove inimical to vegetation. In arid settings, evaporative concentration may lead to

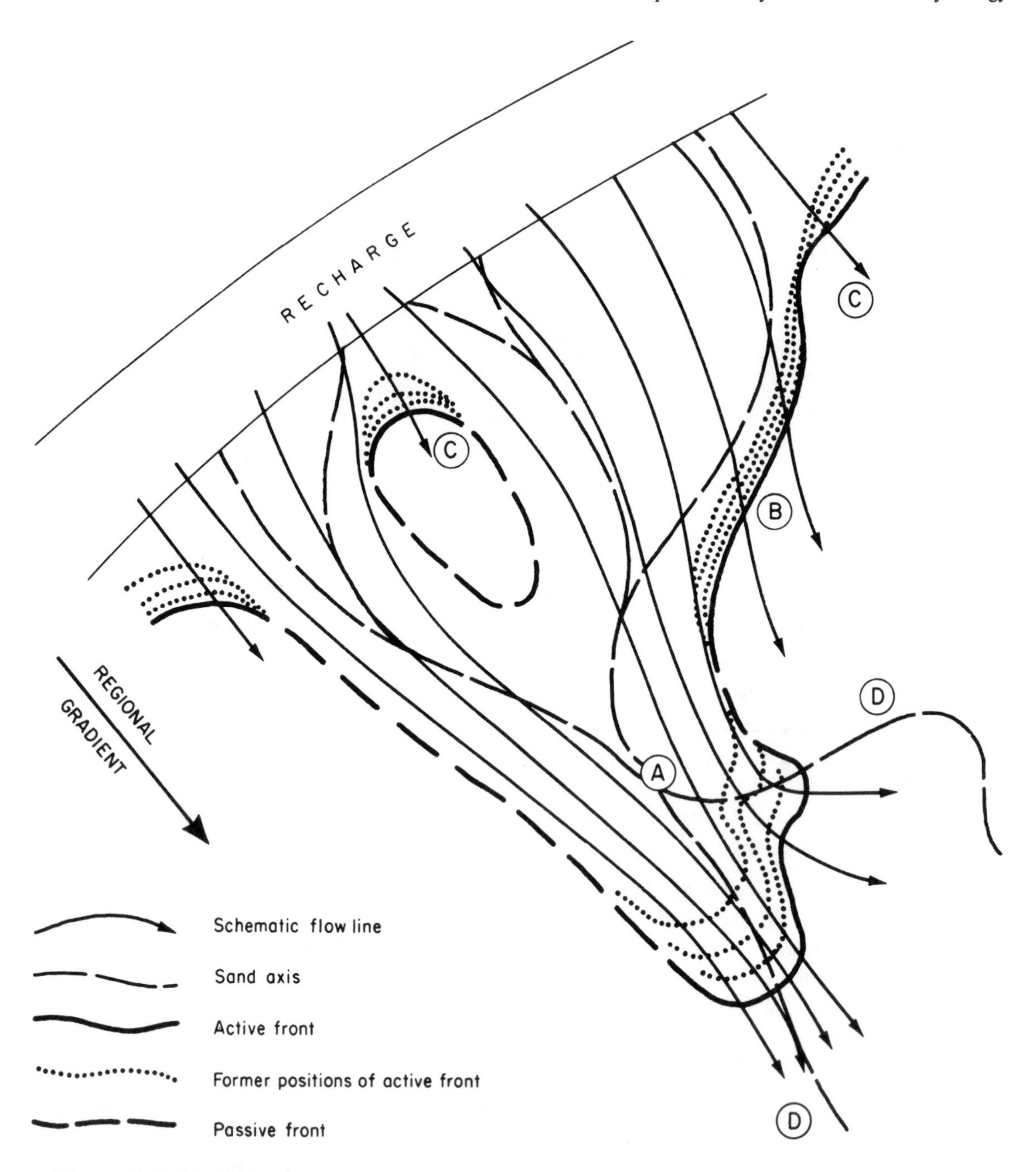

Figure 11-8. Idealized oxidation tongue produced by the flux of reactive, oxidizing ground water through a uniformly reduced aquifer. (A) Convergent flow and oxidation tongue. (B) Deflected flow. (C) Areas of divergent flow. (D) Bypassed permeable zone. For further discussion of lettered phenomena A through D, see text. (From Galloway *et al.*, 1979b.)

intraformational precipitation of evaporative minerals such as gypsum or halite. Discharge may provide the principal source of water for playa lakes in closed basins. In contrast, recharge areas are characterized by a comparatively deep water table, deficit of soil moisture, oxidation, and leaching of soluble solids (Tóth, 1972).

Aquifer Evolution

Beginning with its deposition within the meteoric regime, an aquifer may experience up to four evolutionary stages during a simple cycle of basin filling and uplift (Galloway, 1977; Galloway *et al.*, 1979). The stages are illustrated schemati-

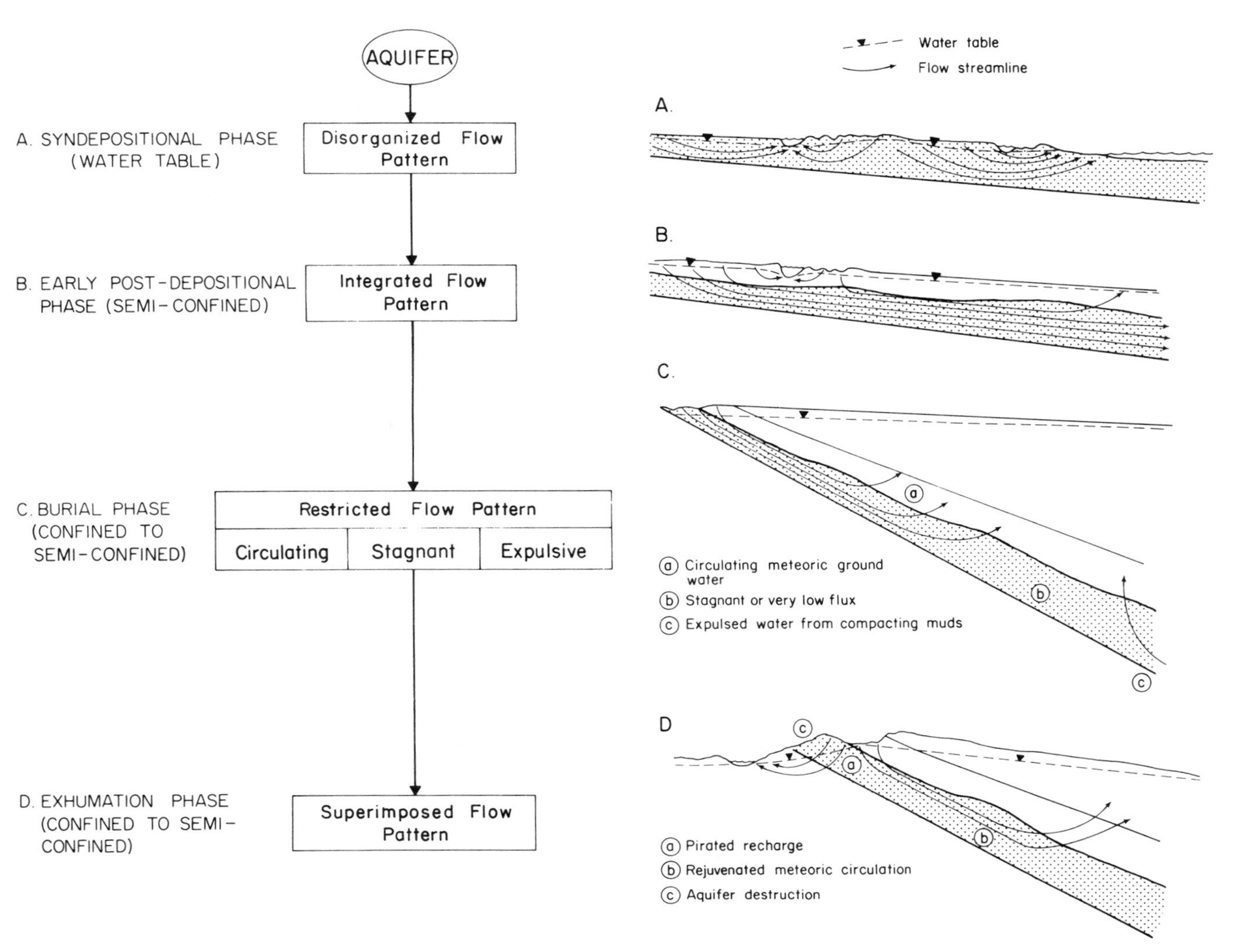

Figure 11-9. Evolutionary phases of flow in a terrigenous aquifer system. (From Galloway, 1977.)

cally in Figure 11-9, and consist of syndepositional, early post-depositional, burial, and exhumation phases.

Syndepositional Phase. Ground-water circulation is directly controlled by the water table, which fluctuates considerably in response to seasonal or erratic climatic variation, and defines a complex surface related to the topography of the depositional surface. Consequently, flow cells are typically localized, variable, and, in a geological time frame, ephemeral (Fig. 11-9A). However, permeability of framework and nonframework facies is at a maximum.

Post Depositional Phase. Ground-water circulation becomes semiconfined by deposition of low-permeability units, such as floodplain, lagoonal, or lacustrine facies, above permeable framework sands (Fig. 11-9B). Depths of burial remain shallow (tens or a few hundreds of feet or meters), permeabilities of both framework and nonframework facies are still quite high, and recharge is primarily by downward flow of water through overlying less permeable beds. Concentration of flow into buried high-permeability units produces, for the first time, a well-integrated flow pattern responding to the regional hydraulic gradient (Freeze, 1969).

Burial Phase. With greater depths of burial of the aquifer, flow is increasingly confined by overlying retarding units, recharge becomes more closely linked to existing outcrop areas of the aquifer, and permeability of the aquifer is decreased by compaction and diagenesis (Fig. 11-9C). Faulting affects flow patterns, and assumes increasing importance in the cross-stratal discharge of water from the aquifer across strata. Flow patterns are regionally coherent, but increasing basinward confinement and decreasing permeability of the aquifer tend to decrease total flux; increased basin margin elevation and consequent higher potentiometric gradient may counteract this decrease, however. Three idealized subdivisions of the flow system are possible within the aquifer.

(1) Actively circulating meteoric ground water continues to dominate shallow portions of the aquifer.
(2) Areas of stagnant conditions or of very low flux develop in portions of the aquifer that lie downdip from zones of vertical discharge or that are otherwise isolated by faulting or diagenetically induced permeability barriers.
(3) At greater depths, expelled pore waters discharging toward the surface may enter shallow aquifers and mix with slowly circulating or nearly stagnant meteoric waters.

Exhumation Phase. Erosion of the updip margins of aquifers may begin during the burial phase. Regional uplift or tectonic segmentation of the basin resulting in exhumation and drainage incision climax this destructional aspect of aquifer history. The results include local pirating of regional recharge, rejuvenation of shallow meteoric circulation due to increased head, and possible exposure of the aquifer in discharge zones. Such a reorganization of flow dominates the later history of tectonically active depositional basins, and may recur in areas of complex geologic history.

Example of a Coastal Plain Aquifer System

Uranium and petroleum-bearing Oligocene and Early Miocene units of the northwestern Gulf Coastal Plain provide examples of the complex flow dynamics and hydrochemical evolution typical of aquifers within the meteoric regime of a depositionally active basin (Galloway, 1977; Smith *et al.*, 1980; Galloway, 1982). Major aquifers in this sequence include the Oakville and portions of the Catahoula sands at outcrop and in the shallow subsurface; both formations consist of the deposits of major mixed-load and bed-load fluvial systems. High-quality pump-test data show that in areas not influenced by faulting, average permeability of bed-load channel fill sequences is about twice that of mixed-load fill sand bodies. Both are much more permeable than bounding mudstone and splay facies (Galloway *et al.*, 1982b).

Tertiary aquifers of the Gulf Coastal Plain have experienced moderate erosional exhumation along their outcrop belt due to basin margin uplift and intervals of low standing sea level during the Quaternary. Greatest topographic relief occurs where valleys of major rivers cut across the strike-oriented outcrop belt. The aquifers dip gulfward beneath increasing thicknesses of younger sediments, and fluvial systems grade into major deltaic systems in the deep subsurface. Thus, deeper circulation remains little affected by exhumation, and is more characteristic of the burial phase.

Interpretation and synthesis of vertical and lateral head, temperature, resistivity, and hydrochemical distributions for ground waters of the Oakville aquifer reveal flow patterns characteristic of a coastal plain setting (Fig. 11-10). Factors shown to influence flow directions and volumes include: (1) outcrop distribution and topography; (2) distribution and orientation of the permeable channel axes (as delineated by net-sand isolith maps); (3) presence and location of small-displacement, strike-parallel fault zones, some of which appear to serve as loci for discharge of deeper, confined aquifers; and (4) the increasing coastward burial and confinement of the Oakville sands (Galloway, 1982; Smith *et al.*, 1980).

Local, intermediate, and regional flow cells are present, though most local cells are beyond the resolution of the data base and are not shown in Figure 11-10. Recharge occurs by infiltration along the outcrop and through the thin, updip edge of overlying confining units. Regional flow extends from the recharge zone basinward along sand axes toward a broad discharge belt that lies as deep as several thousand feet below sea level (Fig. 11-10). There, waters percolate upward across a large cross-sectional area of the overlying confining unit and into shallower aquifers. Waters typically evolve down flow from a Ca^{2+} and HCO_3^- dominated composition to a Na^+ and Cl^- dominated composition. Hydrochemical anomalies characterized by unusual abundances of Cl^- or SO_4 suggest active leakage of water from underlying aquifers and typically are encountered along fault zones. General geochemistry of waters of the meteoric system is summarized in Table 11-2.

Intermediate flow cells, characterized by flow along strike and discharge to the surface or into alluvium of major river valleys, interrupt regional flow for distances of as much 20 mi (30 km) along strike (Fig. 11-10). Topographic relief between interfluve highs and valley flood-plains of approximately 100 ft (30 m) or less affects the slope of the potentiometric surface, and hence the groundwater flow to depths of as much as 700 ft (210 m). Waters are typically of the Ca^{2+} HCO_3^- hydrochemical facies unless modified by leakage along faults.

Together with the younger Quaternary sediments of the northwestern Coastal Plain, the Oakville and Catahoula aquifers indicate the complex physical and chemical hydrology of the meteoric regime in an active depositional basin. Hydrology of terrestrial basins, such as the San Joaquin Valley in California, illustrates many of the same principles and is equally complex (Miller *et al.*, 1971; Davis *et al.*, 1959).

Compactional and Thermobaric Systems

Thick piles of subsiding sediment provide two sources for internally derived ground water. First, interstitial water trapped between sediment particles may be squeezed out by mechanical compaction. Second, chemically bound water associated with hydrous mineral phases may be released by reactions brought on by increasing temperature and confining pressure. Occupying a somewhat gray area between these two end members is the loosely bound inter-layer water in smectite and smectite-illite mixed layer clays. Release of this bound water requires an increase of temperature and pressure to overcome binding energy. However, such water is released at much lower temperatures and pressures than are generally associated with the thermobaric system.

Because muddy marine sediments are typically deposited with high porosities (commonly exceeding 50 percent), physical compaction offers a source of water equal to the volume of sediment. However, compaction and pore space reduction occur rapidly during incipient burial of a few tens to hundreds of feet or meters. Representative volumes of water released by successive 1,000 meter (3,300 ft) increments of homogeneous mud displaying an average consolidation gradient were calculated by Bjørlykke (personal communication) and are compared graphically in Figure 11-11. Deposition of the first 1,000 meters of mud releases over 21,000 cm^3 (0.74 ft^3) of interstitial water per sq. cm. of surface area. At a depth of 1,000 m average porosity is reduced by physical compaction to slightly over 20 percent. Remaining compaction to near zero porosity at

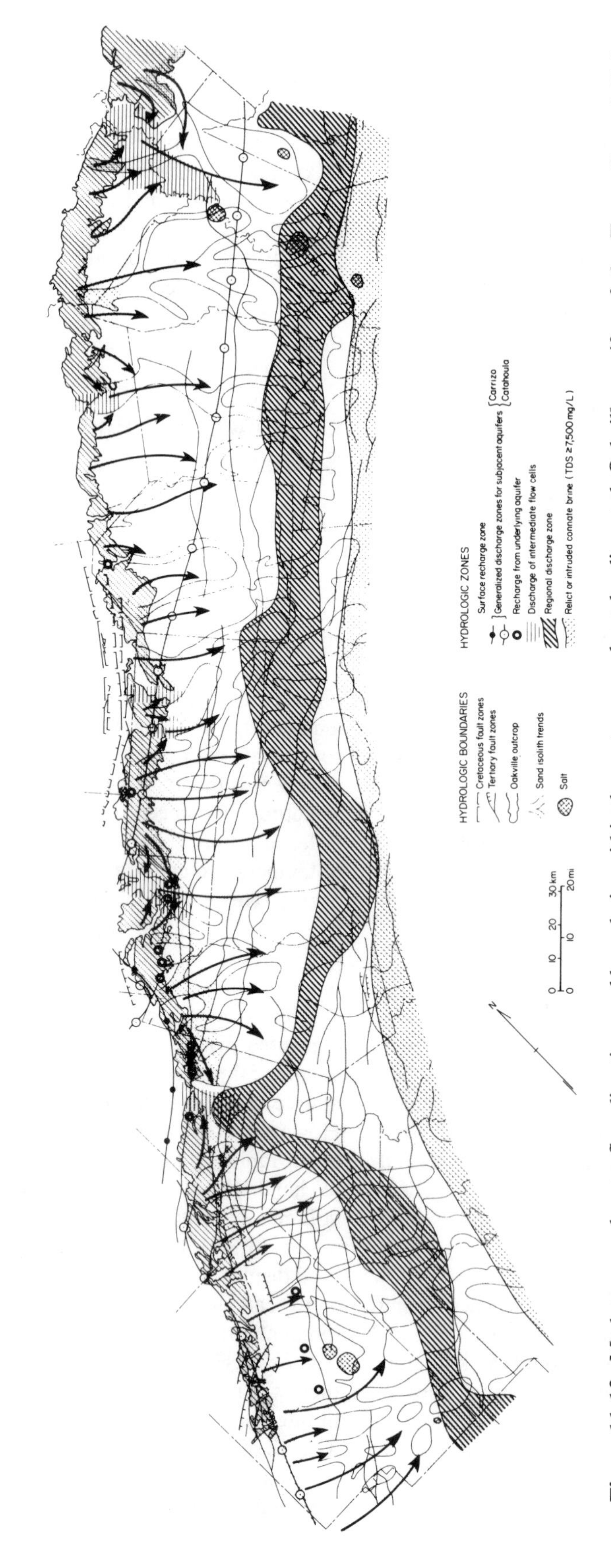

Figure 11-10. Modern ground-water flow directions and boundaries within the mature, moderately dissected Oakville aquifer of the Texas Coastal Plain. Regional hydraulic gradient is gulfward, to the bottom of the map. Flow lines are interpreted from both head and geochemical data. (From Galloway *et al.*, 1982b.)

Table 11–2. Geochemical Characterization of Principal Ground-Water Regimes of the Northwestern Gulf Coastal Basin (mg/l unless otherwise noted)[a]

			Thermobaric	
Depth Range (Generalized)	Meteoric 0–2,000 ft; 5,000 + max 0–600 m; 1,500 + max	Compactional 2,000–12,000 ft ± 600–3,600 m ±	Terrigenous <12,000 ft ± <3,600 m ±	Carbonate >20,000 ft <6,000 m
TDS	10^2–10^4	10^4–10^5	10^4–10^5	10^5
Ca	10–10^3	10^2–10^4	10^2–10^4	10^4
HCO_3	10^2–10^3	10^2	10^2–10^3	n.d.
SiO_2	10^0–10^2	10–10^2	10–10^2	n.d.
Fe	10^{-1} to $<10^0$	≤ 10	10^{-1}–10	10^3–10^4
Heavy metals *(Cu, Zn, Pb, Ni)*	$<10^{-1}$	$<10^0$	$<10^{-1}$–10	10–10^2 + (Zn; Pb)
Se, Mo, As	$<10^{-2}$	$\leq 10^{-2}$	$<10^{-2}$	n.d.
U_3O_8	$\leq 10^{-1}$	n.d.	$\leq 10^{-5}$	n.d.
SO_4	10^{-1}–10^3	$<10^0$–10^2	10^0–10^2	10–10^2
H_2S, HS^-	≤ 10	<10	$<10^{-2}$ mol % in gases; $\leq 10^0$	10^{-2} to 10^{-1} mol % + in gases
Eh	+500 to −200 mV	Reducing	Reducing	Reducing
pH	6.8–8.2	5.5–7.5	4–6.5	n.d.

[a] From Galloway (1982); original data compiled from various sources.

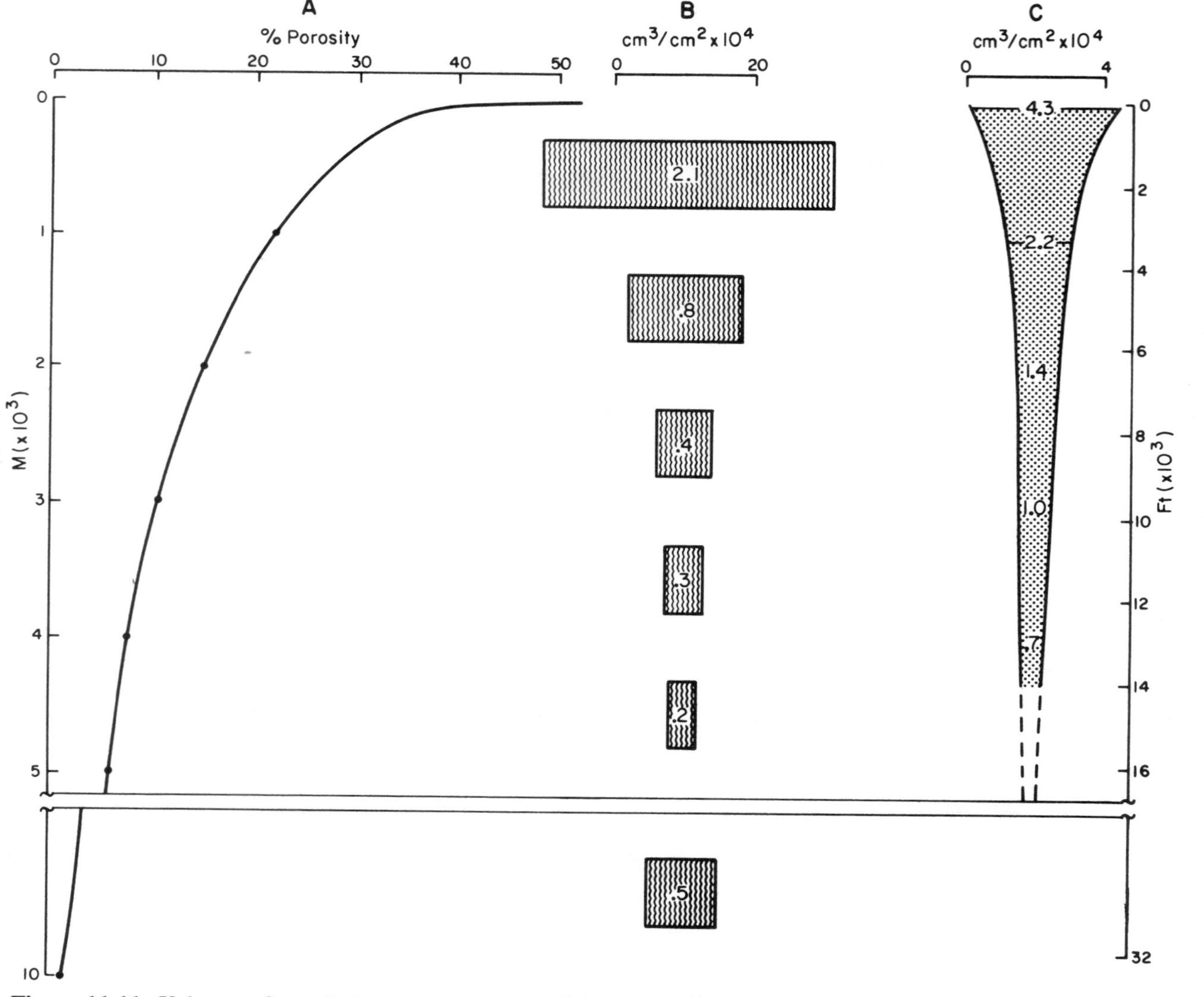

Figure 11-11. Volumes of expelled pore water generated by burial of a typical mud. (A) Generalized consolidation profile expressed as decreasing porosity vs. depth. (B) Comparative volumes of water generated by consolidation of a 1000-m-thick column of sediment for each 1,000 m of burial per square centimeter of surface area. (C) Cumulative volume of fluid remaining to be expelled by compaction below successive 1,000-m depth levels. (Data provided by Knut Bjørlykke.)

depths of 10,000 m (33,000 ft) releases a comparable volume of water in decreasing proportions with depth (Fig. 11-11, column B). Potential upward flux of water across an arbitrary datum (Fig. 11-11, column C) is no greater than the cumulative release of water by underlying sediment volumes. The illustration reemphasizes the important fact that, unlike meteoric water, availability of connate pore water is finite. However, total volumes of pore water released during basin compaction are tremendous. For example, Hitchon (1968) calculated that average porosity of mudstones in fill of the Western Canada sedimentary basin decreased from over 50 percent to a present value of slightly less than 14 percent. Total volume decrease of mudstones approximately equals 175,000 mi^3 (730,000 km^3); comparable volumes of pore water were expelled during consolidation.

Volume release of thermobaric waters is poorly documented. Clay dewatering is the best described process within large volumes of smectite-bearing argillaceous sediments of many young basins. Powers (1967) suggested that smectite dewatering might release as much as 20 percent by volume (of typical Gulf Coast mud) of additional free water into intergranular pores, and thus into the flow system. However, Burst (1969) refined the quantitative description of the dehydration process for a smectite-rich Gulf Coast shale, and suggested an increment of about 15 percent by volume of additional water. Magara (1978, p. 100) argued reasonably that even Burst's figure is likely to be too high. Regardless of the exact figure, clay dewatering offers a mechanism for "recharge" of the compactional regime by substantial volumes of fresh water (Jones, 1969).

Clay dewatering occurs in discrete steps at depths which are a function of temperature, pore water composition, and confining pressure (Burst, 1969; Perry and Hower, 1970). Smectite is converted to illite by incorporation of potassium and concomitant expulsion of interlayer structured or bound water. In the Gulf Coast basin, Perry and Hower found that about 65 percent of the smectite layers converted within the depth range of 7,000 to 10,000 ft (2100 to 3000 m), depending on geothermal gradient. The second pulse of dehydration occurred at depths ranging from 10,000 to 18,000 ft (3,000 to 5,400 m), again depending on temperature and pressure gradients. Approximately 80 percent of the smectite layers were converted by the end of this dehydration, leaving about 5 percent by volume water remaining within unaltered smectite layers. This water would be released by deep burial metamorphism.

Flow Dynamics

Confining pressure, supplied by the sediment pile, is the principal mechanism forcing water up and out of the basin fill. Volume increase, associated with release of bound or mineral-phase waters, as well as phase transformations or decomposition reactions producing fluids, become increasingly important within the deep, hot core of the fill. Although the migration pathways and flow history of both connate and thermobaric waters remain poorly known (in comparison with the meteoric regime), several key aspects are generally accepted.

1. In thick, homogeneous sections, expelled waters rise toward the surface or toward the base of the meteoric regime. However, as pointed out by Bonham (1980), water movement is upward relative to stratigraphic horizons only. In a subsiding basin with a normal compaction profile, the uniformly advancing fluid front never reaches the surface; in fact, it is buried under increasing sediment cover. Thus, in a homogeneous basin fill, connate or expelled pore fluids would not be discharged either directly to the surface or into the meteoric regime.

2. If comparatively permeable beds, such as sandstones, are interstratified with the compacting fine-grained sediments, permeability becomes highly anisotropic, and flow patterns become more complicated. Fluids may drain more readily through the permeable conduits towards their basin margin outcrop or shallow subcrop (Magara, 1976). Locally, flow from the compacting interbeds may drain both up and down towards the closest permeable bed. Suggested evidence for downward drainage of compaction water is contained in the consolidation profiles of mud rocks overlying permeable units (Magara, 1968, 1978). Once in the permeable strata, waters could migrate updip, thus accomplishing the necessary vertical component of flow out of the basin. Such flow channeling by both permeable strata, or by cross cutting permeable structural zones, results in localization of dis-

charge of compactional and thermobaric fluids around the basin margin.

3. In a subsiding progradational sequence, early compaction and dewatering of subaqueous muds (be they marine or lacustrine) expels connate waters into interbedded strandline or delta-margin sands. Thus, a dynamic interface forms between the meteoric regime, moving waters down slope through subaerial deposits, and the early compactional regime, moving waters updip and landward within elements of the same basin margin plumbing (Jones, 1969). Syndepositional structures, such as growth faults, may provide conduits that localize the upward flow of compaction waters. Mixing of the chemically different waters has several implications for early diagenesis or alteration of flushed sediments.

Development of Abnormal Pressure. Clays and shales are characterized by low permeability (Table 11-1). Thus, the assumption of equilibrium between physical loading and concomitant compaction and drainage, and application of overburden stress is readily violated in thick, rapidly deposited argillaceous units. When drainage is unable to keep pace with loading, the pore fluid begins to bear some of the weight of the overlying sediment pile. Observed fluid pressure gradient exceeds the normal hydrostatic gradient of about .45 psi/ft (10 kPa/m). The interval of abnormally high fluid pressure, which may comprise a large volume of the basin fill, is commonly called the overpressure or geopressure zone.

Although numerous hypotheses to explain development of overpressure have been suggested, all rely on at least partial hydraulic isolation of the affected sections. Such isolation commonly reflects the paucity and discontinuity of permeable strata within the section. Under such conditions, fluid drainage is slow, even in a geologic time frame and in the face of large cross-sectional areas. As Q (flow volume) is restricted, pressure head builds.

The simplest explanation for development of overpressure (and one that seems to apply well in many Tertiary basins characterized by rapid deposition of deltaic, slope, and coastal plain systems) is that rate of sediment loading exceeds the potential for fluid discharge under a modest pressure head (Dickinson, 1953; Bredehoeft and Hanshaw, 1968). Consequently, pressure head builds as interstitial waters assume a greater share of the weight of the overburden. Jones and Wallace (1974) estimated that in the Gulf Coast basin, formation waters of the overpressure zone support 50–75% of the overburden weight. Maximum pressure gradients attainable by this process would be less than the usual lithostatic gradient of approximately 1.0 psi/ft (22.5 kPa/m). With time, fluids are slowly expelled, leading to a long-term degeneration of the abnormal pressure head. Various physical models combined with the variability of hydrologic properties of confining units suggest that isolation in a thick interval may begin very early in the burial history, or may require some minimal burial and compaction before fluid discharge is sufficiently restricted to initiate pressure buildup (Magara, 1978). Bredehoeft and Hanshaw (1968) demonstrated the pressure gradients approaching lithostatic could be established and maintained in the Gulf Coast basin, assuming average sedimentation rates of 1500 ft per million years (500 m/10^6 yr) and hydraulic conductivities of 10^{-8} cm/s (10^{-2}md). Obviously, the much higher local rates of sedimentation characterizing deltaic depocenters could induce significant overpressure in relatively conductive unconsolidated muddy sediments. Buried, isolated permeable lenses, beds, or structural salients will assume the high pressures generated by the surrounding mud blanket.

If one assumes a perfectly closed system, allowing no fluid leakage, the continued burial and consequent heating of water would lead to aquithermal pressuring (Barker, 1972). Such pressurization would result because density of water decreases (and consequently volume increases) with increasing burial temperature. The condition of perfect sealing is, however, difficult to envision in terrigenous clastic systems. The high pressures that could result do provide an appealing mechanism for autofracturing and permeability development within massive mudrock units believed to be major hydrocarbon sources.

The pulses of bound water released during clay dewatering have also been proposed as causes of overpressure development at depth (Burst, 1969). Again, highly restricted circulation is required if the high differential pressures are maintained for geologically significant lengths of time. Most authors now discount this mechanism as a principal cause of overpressure. Magara (1978) convincingly argues that the mechanism does not explain other phenomena associated with the overpressure zone.

Thermal generation of less dense phases, principally gases, has been recently recognized as a dominant mechanism for producing abnormally high fluid pressures in low permeability sediments of older Paleozoic and Mesozoic basins (Hedberg, 1974). Such a mechanism is little dependent on initial burial conditions or depositional facies. An abundance of organic materials suitable for generation of natural gas is required. However, postdepositional tectonic or diagenetic isolation of sand-rich sequences must precede pressure generation. Natural flow dynamics of such a two-phase fluid systems is complex. Wells that penetrate such "tight gas" reservoirs commonly produce large volumes of gas, even though water saturations are high. If indicative of natural conditions, large-scale flow of gas and water might characterize such systems. Unlike other mechanisms of progressive overpressure generation and fluid expulsion, this mechanism produces a pressure pulse late in the history of basin formation.

Abnormal pressure gradients have several effects on the physical properties of the sedimentary sequence. Physical consolidation, which affects primarily the mudrocks, is retarded within the zone of overpressure. Thus, parameters that measure porosity, including *bulk density* and *interval travel time*, deviate from normal compaction trends. The amount of deviation provides a guide to the extent of overpressure. The inverse of travel time, the *interval velocity*, can be calculated from modern common depth point (CDP) seismic data. Declines or reversals in the velocity/depth gradient thus provide indirect indicators and measurements of overpressure. Although the causes are more complex than a simple correlation with porosity, *shale resistivity* commonly reflects deviations from the hydrostatic pressure gradient by abrupt or gradational appearance of abnormally low values with depth. Finally, thermal conductivity of undercompacted muds is lower than that of normally consolidated sequences. Thus, assuming uniform heat flow, *thermal gradients* increase within the zone of overpressure. All of these phenomena are commonly observed and utilized in basins such as the Gulf of Mexico, Jurassic of the North Sea, and McKenzie Delta.

Hydrochemistry of Compactional and Thermobaric Waters

The initial chemistry of connate pore waters is variable, but two end members are common. Trapped sea water, modified to some extent by shallow diagenetic processes, fills pore spaces of sediments deposited in shoreface and marine environments. Sediments deposited at or above sea level were probably flushed by circulating meteoric waters. Thus, initial waters range from fresh to moderately saline NaCl solutions. Complexities are introduced in continental basins filled by fluvial and lacustrine systems. Water released by thermobaric processes from clays or other solid phases is free of solutes.

Long residence times, combined with increasing temperature and pressure, decreasing pore throat size, and increasing importance of membrane effects resulting from burial initiate a variety of water–rock interactions that grossly change the hydrochemistry of connate waters. Burial diagenesis of permeable sand facies includes grain and cement leaching, precipitation of pore-filling cement, grain replacement or alteration, and reduction. Increasing thermal exposure leads to successive generation of petroleum liquids and gases, CO_2, and H_2S, which act as separate fluid phases or dissolve in associated pore waters.

Data on composition and electrochemical properties of compactional and thermobaric waters are scattered, and sometimes questionably representative of *in situ* conditions. Extensive data on deep basin waters are reviewed by White (1965), Collins (1975), and Rieke and Chilingarian (1974). However, differentiating waters of deep meteoric regimes from those of the compactional and thermobaric regimes is often arbitrary. Galloway (1982) summarized the hydrochemistry of the three regimes in the northern Gulf Coast basin, which appears similar to many other depositionally active basins filled by mixed fluvial, deltaic, strandline, and slope depositional systems.

Thermobaric fluids, including waters, solution gases, and free gases, occur within two significantly different types of basin fill. Cretaceous forereef slope and basinal carbonate and shale units form the stratigraphic foundation of the thick clastic Tertiary section. Lying at depths exceeding 20,000 ft (6,700 m), these faulted forereef deposits have been penetrated by the drill only along their updip margins. Thick continental slope, prodelta, and delta-margin Paleocene and lower Eocene mudstone and minor sandstone occur at depths below about 10,000 ft (3,300 m), and form the bulk of overlying hot, overpressure zone.

Thermobaric waters are moderately to highly saline, sodium-chloride brines. Total dissolved solids content ranges from near 10,000 to more than 200,000 mg/l (Gustavson and Kreitler, 1977; Prezbindowski, 1981; Fisher, 1982). Predictably, greatest values occur in parts of the Coastal Plain characterized by salt diapirism; lower values typify deep brines of the South Texas Coastal Plain where bedded salt is sparse. Very few systematic data on the trace element content of thermobaric brines are available. Data, though scanty, are summarized in Table 11-2. Dissolved uranium values from analyses of six downdip geopressured brines from Oligocene and Miocene reservoirs ranged from 0.003 to 0.03 μg/l. Although the Eh of these very deep waters has not been measured, they can be confidently inferred to be reducing relative to iron because of the presence of sulfide and common occurrence of pyrite in the host sediments. A few careful measurements of pH suggest that deep, overpressured waters may be strongly acidic (Kharaka *et al.*, 1980). This observation is consistent with the common occurrence of kaolinite, which is favored by acidic conditions, as a product of deep burial diagenesis in many Tertiary reservoir sandstones.

The thermobaric regime is a major reservoir of gases. Light hydrocarbons (principally methane), hydrogen sulfide, and carbon dioxide are generated by thermal degradation of organic debris at the temperatures characteristic of this zone (Hunt, 1979). High concentrations of hydrogen sulfide, which is extremely soluble in subsurface waters, are typically restricted to carbonate reservoirs. In sand–shale sequences, abundantly available iron reacts with reduced sulfur species to form disulfide. Hydrogen sulfide is a common constituent in gases produced from forereef Edwards Limestone reservoirs, but is rarely present in measurable quantities in gases produced from Gulf Coast sand reservoirs, and typically measures less than 5 mg/l in produced waters (Kharaka *et al.*, 1980). Carbon dioxide content of gases from both limestone and sandstone reservoirs of the thermobaric regime averages several mole percent; methane is the dominant constituent. Both gases are highly soluble in brine.

Fluids of the compactional regime include sodium-chloride brines, hydrocarbon liquids, and gases. Waters do not vary greatly from their deeper Tertiary counterparts in overall composition (Table 11-2) and are of the evolved marine connate type (Kharaka *et al.*, 1977).

Concentration of hydrogen sulfide is low; both waters and gases of the compactional regime are sweet. Further, gas solubility decreases at the lower pressures. Produced gases commonly contain 0.1 to 1 mole percent carbon dioxide. The hydrocarbon fraction includes abundant methane plus heavier fractions, distillate, and liquid petroleum. Most produced hydrocarbons of the Tertiary Gulf Coast Basin, and essentially all oil, are contained within the compactional zone. Measured values of fluid pH commonly range between 6 and 8 (White, 1965; Kharaka *et al.*, 1977). Eh is negative (Fig. 11-6). Host sediments are dominantly reduced, containing iron disulfide or dispersed organic material, or both.

Hydrology of Depositional Systems

Although depositional environment is only one of several parameters that determine the early hydrology of sedimentary sequences, paleogeographic reconstruction provides a logical starting point for reconstruction of paleohydrologic history. A primary distinction can be drawn between systems deposited largely subaerially, and thus above the regional hydrologic base level established by sea or lake level, and subaqueous systems deposited below regional hydrologic base level. Transitional systems, such as deltaic and interdeltaic strandline systems, exhibit predictably complex syndepositional hydrologic characteristics.

Subaerial Depositional Systems

Most environments of subaerial systems, including alluvial fan, fluvial, and eolian systems, lie above the regional hydrologic base level established by the adjoining subaqueous depositional and topographic basin. Thus, water tables lie below the depositional interface, and ground water moves in generally the same direction as regional surface flow. However, topographic irregularities produced by depositional or tectonic processes may induce large second-order flow systems within the subaerial basin fill. Consequently, the shallow, regional meteoric flow regime may be segmented into several local to subregional flow cells controlled by major topographic features. Alluvial fan systems provide good examples of such segmentation.

The inherent morphology of an alluvial fan results in a predictable distribution of recharge, lateral flow, and discharge zones in the shallow, unconfined to moderately confined fan aquifer sands (Fig. 11-12). The topographically high proximal fan is an area of deep water table and active ground-water recharge. Conversely, the distal fan, which is an area of abruptly decreasing topographic and hydraulic gradient, is a major discharge area. Here, the water table is shallow, or even emergent. Thus distal and inter-fan sequences interfinger with marsh, swamp, mudflat, or evaporative playa facies. In open hydrologic basins, such as the San Luis Valley, Colorado (Fig. 11-12), discharge constitutes a major source of base flow into the through-flowing fluvial system. Even in subarid environments, fan toes are commonly sites of relatively luxuriant plant growth and preservation of organic debris.

In addition to their inherent hydraulic gradient and boundary zones, alluvial fan systems, particularly wet or fluvial-dominated fans of perennial streams, produce highly transmissive aquifers. Framework sands are coarse, abundant, well interconnected, and oriented parallel to the existing hydraulic gradient. The great agricultural areas of the desert southwestern United States, such as the San Joaquin and San Luis Valleys, rely on the tremendous volumes of ground water contained in alluvial fan and associated depositional systems.

The great flux of freshly recharged ground water commonly leads to extensive oxidation of upper- and midfan deposits, particularly in arid settings where the water table is comparatively deep (Davis *et al.*, 1959). Oxidation of detrital ferromagnesian minerals and mechanical dispersal of colloidal products favors geologically rapid formation of red beds (Walker, 1967). Conversely, shallow water table and regional discharge insulate the distal fan and associated facies from oxidation in all but arid settings. In contrast, highly vegetated fans of wet climates with well-developed humic soils and abundant shallow ground water may be little affected by early post-depositional oxidation.

The syndepositional hydrology of fluvial systems is much less predictable. Regional components of lateral flow tend to parallel the flow within the trunk stream. However, areas of recharge and discharge are determined by details of basin physiography and tectonic setting of the alluvial plain. For example, the San Joaquin River (California) flows down the topographic axis of an open-ended intermontane basin bounded on both sides by coalesced alluvial fans. Consequently, the axis of the San Joaquin fluvial system lies dominantly within an area of regional and abundant ground-water discharge. In contrast, fluvial systems flowing across a featureless aggrading coastal plain will traverse both regional updip recharge and downdip discharge zones. Valley fills lying below the regional land surface are inherently foci for local flow cells. Consequently, framework facies of fluvial systems may range from highly leached and oxidized to carbonaceous and reduced.

The transmissivity of various fluvial systems ranges from high to low, depending on the type and abundance of channel-fill facies preserved. Transmissivity of bed-load systems is very high; that of suspended-load systems lowest. Permeability is extremely heterogeneous and anisotropic because of the internal variability and lenticular geometry of fluvial sand bodies. Orientation of permeable framework elements is, however, inherently parallel to a component of regional flow as both surface and ground water respond to the same topographic gradients.

Eolian sands form extensive, highly permeable aquifers. Permeability tends to be homogeneous and isotropic, though interbedded interdune flat and playa deposits may severely reduce vertical permeability where they occur in abundance. The necessity of an abundant source of dry sand and the topographic relief of large dunes imply deposition well above the water table. Consequently, eolian sands are subject to considerable oxidation and leaching by downward percolating meteoric waters. Larger interdune flats, draas, and sabkhas may constitute areas of local ground water discharge, evaporative concentration, and precipitation of soluble salts.

In summary, terrestrial systems are characterized by intrusion and flow of waters of the

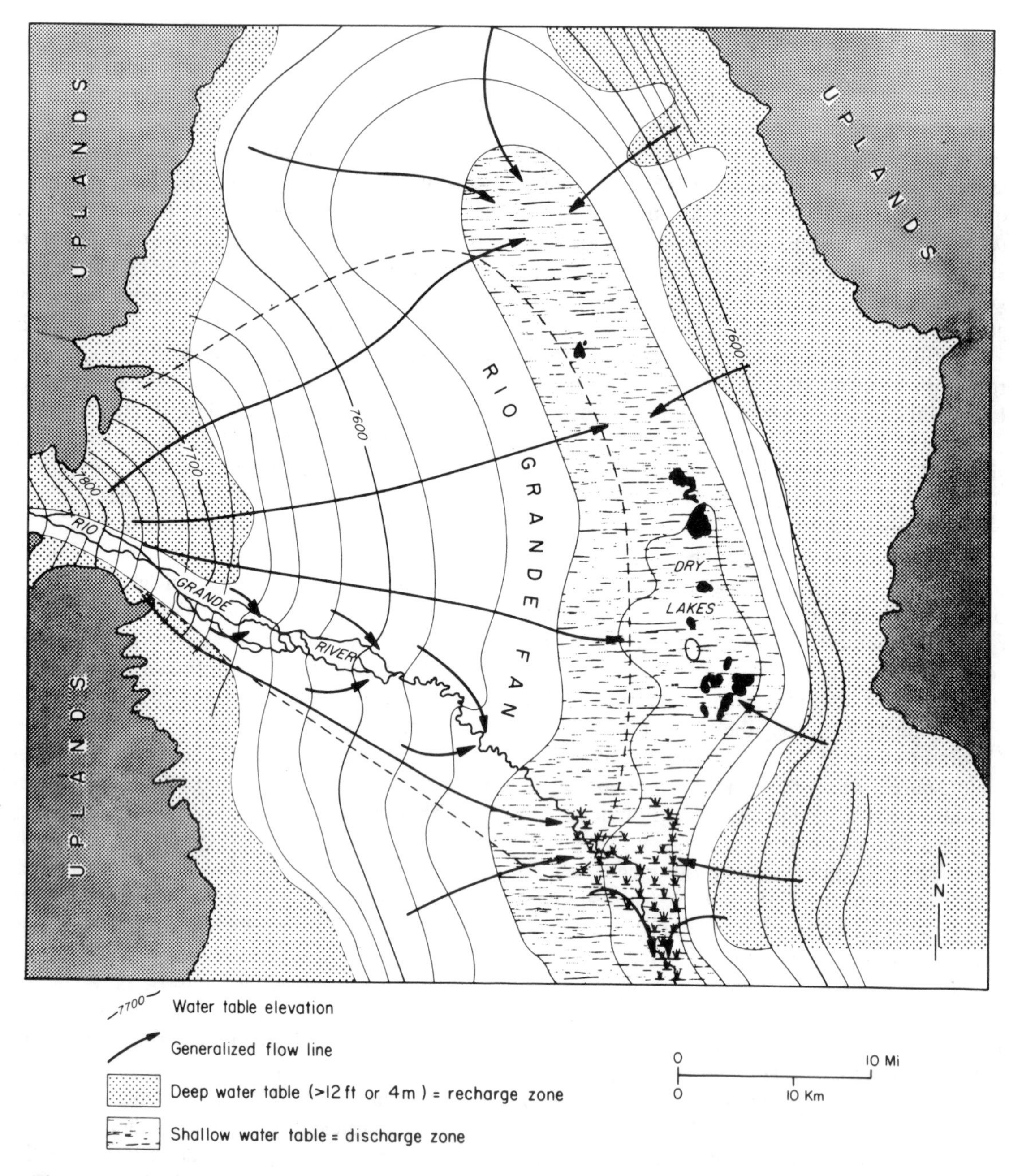

Figure 11-12. Physical hydrogeology of the large alluvial fan of the Rio Grande River, San Luis Valley, Colorado. The modern channel is slightly incised across the fan surface; thus, it is a focus for local discharge. Regional discharge characterizes the distal fan and basin center. (Modified from Emery, 1971.)

meteoric regime. Even pore waters later expelled by burial and compaction are meteoric rather than connate in origin. Contained pore waters are generally fresh to only slightly saline, but a variety of salts can concentrate in discharge zones within climatically dry, closed basins. Prediction of recharge, regional flow direction, and discharge sites is possible if paleotopography can be reasonably interpreted. Depositional or diagenetic features associated with recharge and discharge zones, such as pervasive oxidation, accumulation of organic debris, or occurrence of evaporative precipitates, also provide important clues to the regional paleohydrology.

Transitional Depositional Systems

The strandline defines a zone of regional meteoric ground-water discharge. Positioning of the zone is determined by the flexure between the sloping water table of the subaerial basin margin and the horizontal base level established by average lake or sea level. Although confined aquifers may discharge far basinward of the strandline, shallow semiconfined or unconfined aquifers are shown both by analytical models (Hubbert, 1940) and by field observation to focus discharge along the shore zone. Common manifestations of regional discharge include coastal swamp and marsh environments, shallow lakes, and saline mud-flats.

Within the regional coastal discharge zone, complex local meteoric flow cells are produced by minor elevation variations associated with depositional features such as natural levees and barrier-island or beach-ridge dune fields. Depth of penetration of local fresh water cells is a function of maximum water-table topography, thickness of shallow permeable facies, and the density contrast between fresh and saline basin waters. Head differentials of only a foot can induce local fresh water penetration to depths of several tens of feet in thick, permeable strandline sand bodies.

Both regional and local circulation result in mixing of meteoric and basin waters. Width of the mixing zone is a function of many factors, including rates of ground water flow, aquifer heterogeneity, and tidal range. Mixing of two geochemically diverse water masses provides the opportunity for a variety of reactions involving both organic and inorganic species. For example, Swanson and Palacas (1965) documented the flocculation and concentration of soluble humate by mixing of fresh and saline waters in coastal sands of northwest Florida. Deposits of inorganic precipitates may result from mixing of ground water and lake waters (Hardie *et al.*, 1978). The tufa mounds of Mono Lake, California are a striking example.

Framework facies of coastal depositional system make good to poor conduits for circulation of regional flow. Deltaic systems contain generally dip-oriented channel-fill sand units. However, presence of distributaries with their variable trends, and isolation of lenticular channel facies within delta-plain and prodelta muds reduces the overall transmissivity of the facies assemblage. Delta-margin facies, including coastal barrier and delta-front sheet sands, are highly permeable, but sand bodies are oriented at various angles to the regional flow direction. Barrier island and sand-poor strandplain sand bodies constitute highly permeable conduits, but are poor regional aquifers because of their strike-parallel orientation and common isolation within bounding mud facies. Thus, heterogeneous lithostratigraphic variations commonly reinforce the complex flow patterns produced by interaction of two water masses of contrasting densities. Sand-rich strandplain and strongly wave-dominated delta systems produce the most highly transmissive, laterally uniform aquifer facies.

Large volumes of mud and sand deposited subaqueously within deltaic and strandline systems are little affected by meteoric circulation. Pore waters are trapped lacustrine or marine water. With initial burial, connate waters are flushed from compacting muds into adjacent permeable zones, where they may mix with relict or actively circulating meteoric waters. Thick prodelta muds tend to drain poorly, resulting in early undercompaction and abnormal fluid pressure. With time and increasing burial, the prodelta sequence and contained discontinuous sand bodies become increasingly isolated and develop substantial overpressure.

Sediments of strandline and deltaic systems are commonly reduced soon after deposition. Organic material is abundant in both subaerial and subaqueous environments. The water table is shallow, and circulation is dominated by discharge of regionally evolved meteoric ground water, thus inhibiting shallow oxidation of detrital organic debris. Anaerobic bacterial activity reduces sulfate, which is abundant in sea water that may invade periodically submerged sediments, producing early iron-disulfide minerals (Berner, 1970). Diagenetic blackening of barrier sands by dispersed iron sulfide may occur in a matter of weeks following deposition. Coastal red beds appear to be limited to extremely arid settings where even regional meteoric circulation systems remain oxidizing, and hypersalinity retards both subaerial and subaqueous organic productivity.

Shelf and Slope Systems

Subaqueous, offshore deposits of terrigenous shelf and slope systems generally lie beyond the limits of the meteoric regime. Trapped connate waters fill pores and reflect initial geochemical conditions of the overlying water column. However, a variety of early diagenetic reactions modify the original composition of the pore waters. Ionic composition is modified by exchange reactions, carbonate precipitation or dissolution, and sulfate reduction by anaerobic bacteria. Sediments may be oxidized or reduced, depending on abundance of organic material, degree of physical reworking and exposure to oxidizing bottom waters, abundance of benthonic infauna, and bottom water chemistry. Important early bacterial diagenetic processes include the conversion of organic debris into kerogen (Tissot and Welte, 1978).

As shown by Bonham (1980), expelled pore waters from underlying, compacting muds do not uniformly flush newly deposited bottom sediments. However, conduits could focus expelled waters, and discharge them to the bottom or into shelf sand facies, flushing and mixing them with less evolved connate waters.

In general, both the early chemistry and dynamics of ground waters within terrigenous shelf, slope, and basinal depositional systems are dominated by processes of the compactional regime acting upon connate waters. Meteoric intrusion requires major tectonic or eustatic disruption of original depositional patterns. Stratigraphic onlap or offlap in a marine basin will result in analogous onlap or offlap of hydrologic flow systems, and may superimpose early meteoric or compactional regimes on facies not normally associated with that regime.

Conclusion: Paleohydrology

Interpretation and mapping of depositional systems provide a description of three-dimensional permeability distribution-trend, geometry, and spatial relationships of transmissive conduits, as well as insight into flow boundaries and gradients extant during the syndepositional flow phase. Postdepositional structural and topographic evolution of the basin determine the changing boundary and head conditions that must be inferred to interpret later flow phases. The distribution of uranium, petroleum, or any other potentially mobile constituent, ultimately reflects this historical interaction between the physical geology and the evolving, dynamic subsurface fluid flow systems. An immutable consequence of this interaction is the fact that description of the physical system alone is inadequate to uniquely explain or predict the distribution of epigenetic mineral deposits or petroleum fields. Syngenetic concentration shows direct relationships to depositional facies patterns, but here also, early hydrologic factors can play the decisive role in determining the location of the most favorable facies.

The following chapters on major sedimentary fuel resources—petroleum, coal, and uranium—will repeatedly illustrate this interplay between depositional and hydrologic processes, reinforcing the need for reconstruction of both paleodepositional as well as paleohydrological systems.

Chapter 12

Coal

Introduction

Coal-forming environments have existed since mid-Paleozoic time in areas where substantial quantities of vegetal matter are accumulating and being preserved. Widespread peat accumulation only became possible with the appearance of psilophyte vegetation during the Silurian, which accounts for the first Devonian coals and then the extensive Carboniferous seams. Evolving Mesozoic and Tertiary vegetation occupied an ever expanding range of environments. Today tropical swamps and marshes are undergoing the most rapid accumulation of peat, with rates in Borneo of 17 m (55 ft) in 4000 years (Stach *et al.*, 1975, p. 15); but peats are also forming in temperate and even polar regions, and at high altitudes. The same geographic diversity is apparent in the origin of coal (Fig. 12-1). The majority of early seams were laid down in humid tropical regions, but in Permian and later times coal formation shifted to temperate and polar regions (Diessel, 1970). The lower proportion of coals between the 15- and 30-degree latitudes results from the warm, arid conditions that prevail.

Many ancient swamps and marshes were very large by comparison with those of today and represented prolonged adjustment to equilibrium conditions. Modern swamps, in contrast, are still adjusting to the unstable conditions that characterized much of the Quaternary, and are therefore rather anomalous. Furthermore, vigorous clastic influx from the unusually elevated terrain of today, in comparison with the subdued landmasses of much of the geologic past, has the effect of limiting the areas of swamplands. Drainage systems in many ancient coal basins were substantially smaller, and the deltas would have been more intensely reworked by waves, creating widespread barriers conducive to development and preservation of broad peat blankets. Most modern peat-swamps therefore provide incomplete analogues for coal seams. However, studies of sedimentary processes operating in these swamps and in adjacent clastic environments allow interpretation of most coals in terms of their depositional setting.

Most coals formed from predominantly *in situ* organic precursors with subordinate transported components, but the proportions vary widely even within individual seams. An *in situ*, or autochthonous, origin is accepted for most Northern Hemisphere coals, but some authors consider the Permian Gondwana coals to have a larger proportion of transported, or allochthonous, material. Certainly there are pronounced differences in petrography and substrate characteristics of Gondwana and Upper Paleozoic Northern Hemisphere coals, but factors of climate and vegetation were also important along with possible differences in emplacement of the dead vegetation.

The striking similarities in coal-bearing deposits of the various continents and of various ages demonstrate that many paleoenvironmental concepts and models have broad application in coal exploration, extraction, and reserve estimates. Regional differences may very often be regarded as variants of the general models. Application of the depositional systems concept tends further to reduce the confusion that may be introduced by a multiplicity of minor facies viewed as disparate elements, and eliminates the need for elaborate stratigraphic nomenclature.

Of the total volume of sediments in the Earth's crust, only about 4 percent is potentially coal bearing (Fettweis, 1979, p. 40). About 40 percent of the world's coal resources are in Upper Carboniferous and Permian rocks, with about 50 percent in Upper Cretaceous and Tertiary rocks; in the United States the figures are about 37 and 54 percent, respectively (E.T. Hayes, 1979). Coal is therefore concentrated in rocks that represent a restricted range of depositional environments with specific conditions favoring organic preservation, and is confined to a narrow geologic timespan.

Coal seams range in thickness from centimeters to over 300 ft (100 m) in exceptional cases, attaining a maximum of 765 ft (233 m) in

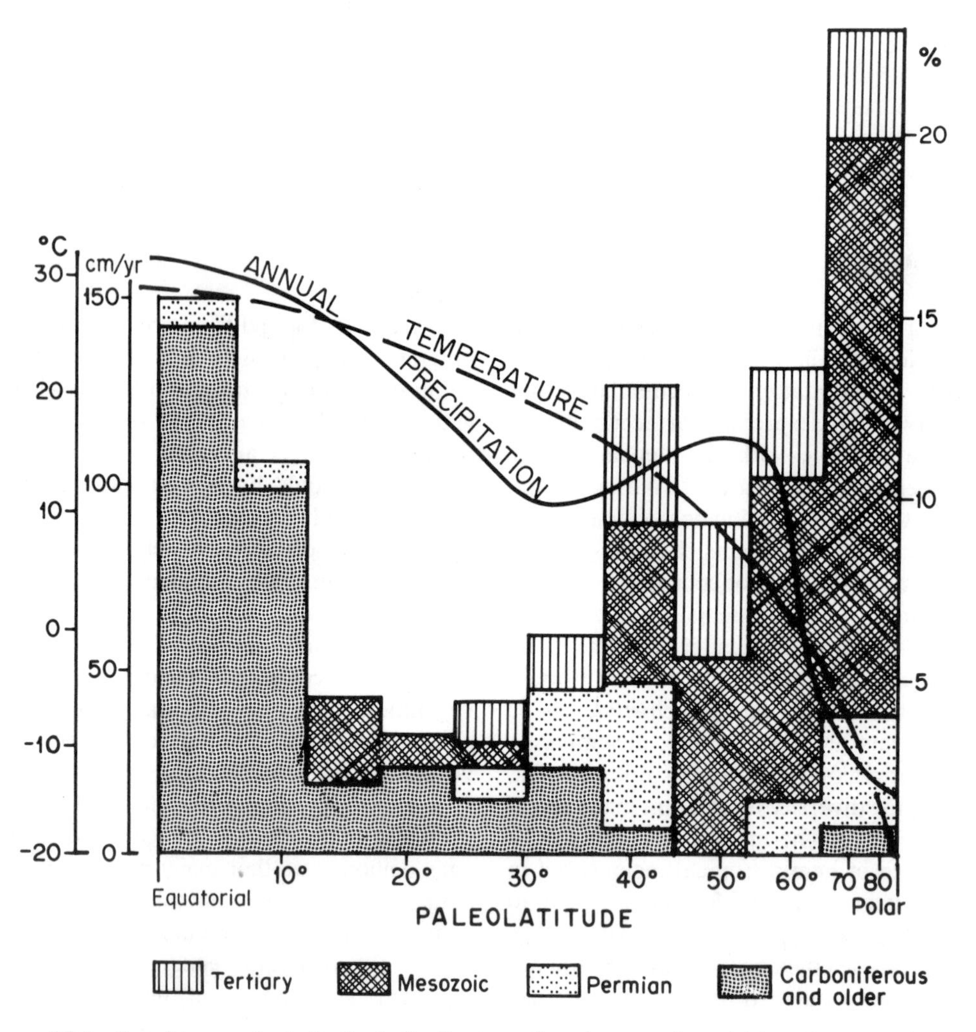

Figure 12-1. Equal-area paleolatitudinal distribution of coal seams for which paleomagnetic control was available. Temperature and precipitation curves based on modern data. (After Irving, 1964 and Diessel, 1970.)

Tertiary lignites of Victoria, Australia. Generally, a thickness of 2–3 ft (70–100 cm) is the lower limit for both open-cut and underground mining. In most coalfields, open-cut mining is limited to depths of less than 300 ft (100 m) because of potential ground-water, slope-stability, and environmental problems. The maximum ratio of overburden to coal in these mines usually ranges between 8 to 1 and 25 to 1. With increased coal prices and improving technology, several open-cut mines are extending to greater depths, with 1000 ft (300 m) planned for some Australian mines, and 1650 ft (500 m) being considered in Europe. Open-cut mining removes some 90 percent of a coal seam, with a small proportion frequently left in the floor. Only in the United States is auger mining of any consequence, providing a very small proportion of the total coal mined.

Underground mining accounts for around 60 percent of world coal production. In many older mines, bord and pillar (room and pillar) methods are employed. This method leaves the pillars to support the roof, but today usually involves two stages in extraction, first from the stalls or tunnels and later from the supporting pillars. Depending on the coal thickness, roof and floor strength, and mining procedure, between 10 and 90 percent of the coal is left in the ground in bord and pillar mining. The more modern technique of longwall mining utilizes hydraulic jacks which support the roof and are moved forward as coal is removed from the face. Longwall mining has been increasingly mechanized but still results in coal losses in some mines of as much as 50 percent (Fettweis, 1979, pp. 68–93). The world's deepest coal mines are around 5000 ft (1500 m), but few exceed 3300 ft (1000 m) because of problems associated with high temperature and rock stress.

Coal at even greater depths is potentially recoverable using remote-mining methods. In hydromechanical extraction, coal is loosened by high-pressure jets and brought to the surface by water flow. In underground coal gasification the seams are partially burned and the gaseous products piped to the surface.

Exploitation of coal is being restricted in places by environmental considerations such as land disfiguration, acid mine drainage and other forms of water contamination, air pollution, and waste disposal. Remote mining poses the threat of ground-water contamination and land subsidence (Kaiser, 1974). Labor unrest and the voluminous and often conflicting regulations of governmental agencies continually burden the coal industry in several countries despite calls for increased production.

The largest coal resources are in the U.S.S.R., United States, People's Republic of China, and Australia. Most coal is used as metallurgical coke in blast furnaces and as steam coal in electricity generation. Coal's importance as a fuel is being further extended by its use in the production of synthetic liquid fuels. Production of gasoline and diesel fuel using Fischer–Tropsch synthesis has been a commercial operation in South Africa since the 1950s. Higher-efficiency second-generation coal liquefaction and gasification have not yet reached the commercial stage.

The organic properties of coal that govern its use fall into two categories: type and rank. Coal type depends upon the basic plant constituents, or macerals, which are comparable to the mineral constituents in other rocks. Rank expresses the level of maturity or degree of coalification. Inorganic constituents, or ash, together with chemical precipitates, may severely restrict coal use. But improved beneficiation methods have permitted the mining of increasingly dirty coal. For example, in the Ruhr Field of West Germany an ash content of up to 13 percent was acceptable in 1935, but this had risen to 40.3 percent in 1973 (Fettweis, 1979, p. 81). Furthermore, coals with a high pyrite content, which may be unsuited to combustion because of emission problems, may be excellent for liquefaction purposes since the conversion of pyrite to pyrrhotite early in the process is thought to catalyze the reaction (Renton, 1979).

Coal-Forming Environments

For a mat of dead vegetation to be preserved as peat, which is ultimately transformed into coal, the area must be protected from detrital influx, and the water table must remain at or near the ground surface, with no pronounced seasonal lowering. Ideally, the water table should rise slowly and continuously as the area subsides to keep pace with vertical growth of the peat mat. Vegetable matter is rapidly oxidized on exposure, whereas prolonged submergence terminates peat accumulation. A low pH of the peatswamp waters is critical; total degradation results at values higher than 5.0 (Renton and Cecil, 1979). Even a temporary decrease in acidity may degrade the peat surface to the extent that inorganic constituents of the cellular structure of swamp vegetation are preferentially concentrated on the surface, subsequently to be preserved in coal seams as ash bands. Similar authigenic partings in seams may also arise from burning or emergence oxidation of the peat surface. Prolonged marine incursion is another factor that significantly reduces the preservation potential of a peat (Cecil *et al.*, 1979b).

Conditions favorable for peat accumulation prevail in paralic coal-forming environments of humid coastal plains as southern New Guinea, parts of West Africa, and the Gulf and Atlantic coastal plains of the United States. According to Diessel (1970) over 90 percent of all coals formed in a paralic setting. The other broad category of coal-forming environments, limnic, is represented by alluviated interior basins such as the upper Nile and Amazon, and infilled lake systems such as the Okavango of Botswana.

The bulk of the peat mat tends to be provided by collapse and decay of the plant cover, but this is augmented in some areas by driftwood and floating leaves. Rate of peat accumulation largely depends upon climate, being some four times more rapid in tropical forests than in swamps with lower rainfall. Open marshes tend to be dominated by reeds and other herbaceous plants, and

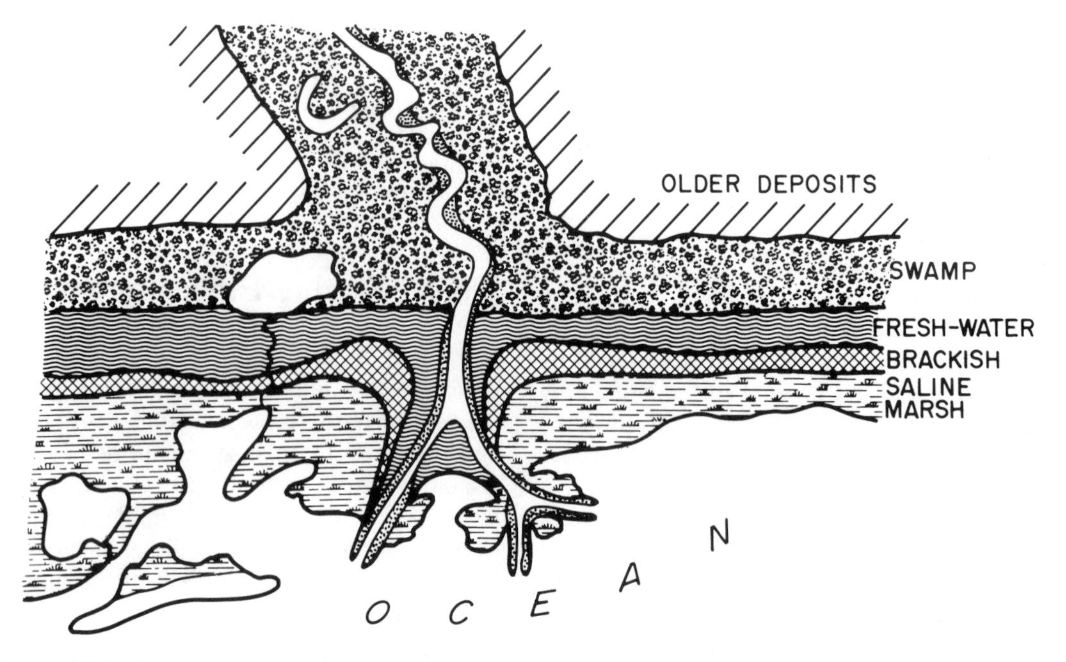

Figure 12-2. Schematic zonation of saline, brackish, and fresh-water marsh and swamp. (Modified after Coates *et al.*, 1980.)

accumulate large quantities of fungal spores. Marshes, by definition, are devoid of trees, so the resulting peat has no woody material. Swamps contain a variable amount of woody vegetation which increases in proportion in forest swamps and drier forests with dense stands of large trees. Aquatic and semiaquatic plants contribute material to abandoned river channels, ponds, and lake floors, which commonly also receive transported debris. These lake-floor accumulations include the degradation products of pre-existing peat beds, to which are added algae and spores (Moore, 1968).

All of these different plant communities give rise to distinct maceral types in coal. For example, stunted herbaceous plants and grasses produce inertinite-rich coal, whereas an abundance of large trees gives rise to a high vitrinite content (Smith, 1968; Smyth, 1980). But maceral content is also influenced by factors of hydrochemistry and level of the water table.

Most vegetated coastal lowlands show a parallel zonation of plant communities, reflecting differences in maturity, substrate consistency, salinity, and ground-water level. The ground-water level affects not only the living community, but also controls the biochemical environment in which the dead vegetal matter accumulates and is preserved (Diessel, 1970). In modern tropical areas, mangroves extend to the limits of aeration corresponding approximately to neap high water. Mangroves commonly show a well-defined zonation depending upon frequency of tidal inundation, and merge landward into dense fresh-water swamps; marshes are conspicuously absent (Coleman *et al.*, 1970). Away from the tropics, mangroves gradually give way to a belt of marsh in the most youthful, seaward portions of a prograding lowland; this merges inland into tree-bearing swamps. Where riverine discharge is large and unconfined, fresh-water marsh may extend to the water's edge. More commonly, however, a belt of saline marsh forms along the zone affected by tidal incursion, and is succeeded inland by a transitional zone of brackish marsh followed by fresh-water marsh and swamp (Fig. 12-2). Brackish and salt marshes are generally not areas of significant peat accumulation. Where present, these thin, marine-influenced peats tend to have a high H_2S content, giving rise to pyritic, high-sulfur coals.

Fresh-water marsh is a more important coal-forming environment. Large areas can be dominated by vegetation floating in shallow, stagnant pools. These floating-marsh peats thus accumulate on a root-free underclay (Frazier and Osanik, 1969), and could be mistaken for allochthonous coals in the rock record. In other fresh-water

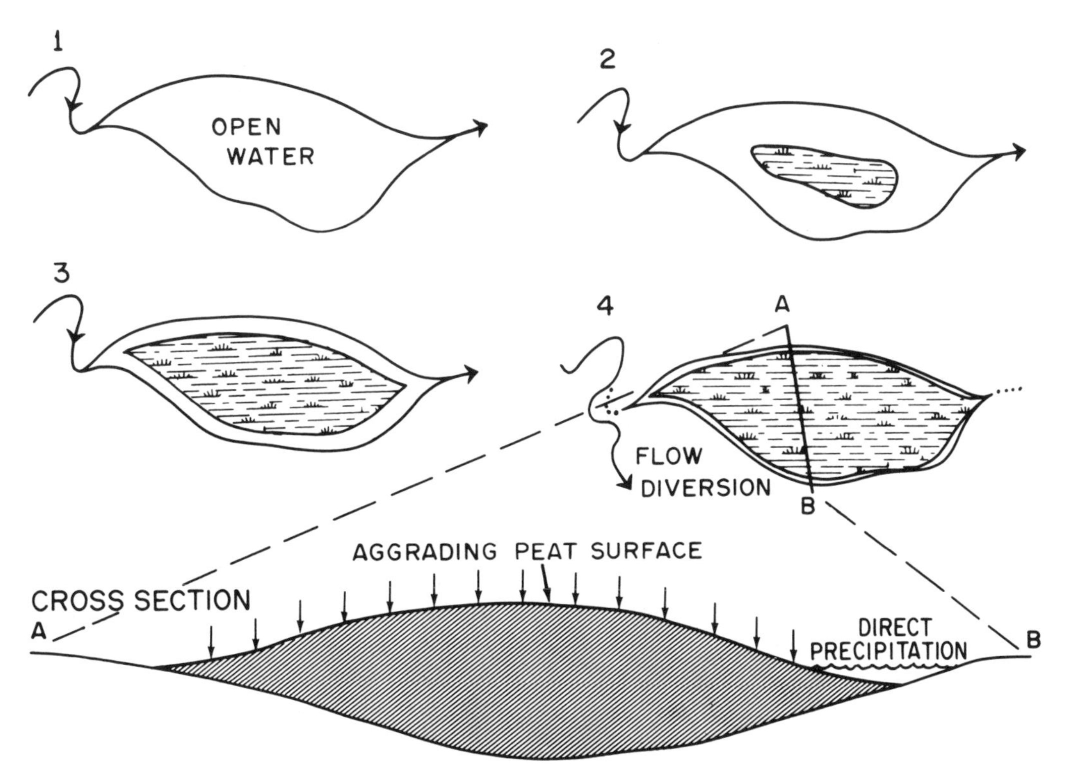

Figure 12-3. Stages in evolution of a peat bog with eventual diversion of inflowing waters and vertical growth of peat, with direct precipitation on the peat surface. (Modified after Moore and Bellamy, 1974.)

marshes the rootmat of living plants is very thick, with roots commonly penetrating to depths of a meter or more. The marshes grade landward into more stable swamps, which may extend vast distances up alluvial lowlands toward the continental interior. Many of the thickest peats accumulate in these fresh-water swamps, and most economically important coal seams were probably of similar swamp origin.

Raised bogs may be locally important as a peat-forming environment, and account for some exceptionally thick seams (Smyth, 1980). These bogs are related to perched water tables, and vary from moss swamps in humid temperate regions to forests comprising a limited number of tree species in the tropics. Raised bogs of Borneo carry dense forests with one or two tree species dominant. The peat surface is typically convex upward, with steep slopes near the margins of the peat domes. Toward the rivers and coast the number of tree and shrub species increases before passing into coastal mangroves (Stach *et al.*, 1975, p. 15).

Peat growth may evolve through several stages before becoming a peat bog (Moore and Bellamy, 1974; Fig. 12-3). Initial peat accumulation from allochthonous material, or as a partially floating mat, forms an island that diverts flow across progressively narrower areas of the basin. The peat becomes less frequently inundated as a consequence, with additional water being supplied by direct precipitation or by local drainage into the depression. The rising peat surface develops its own water table and can therefore accumulate to considerable thickness. In the case of a 33-m (110-ft) seam in Queensland, Australia, Smyth (1980) suggests that the final stage of peat-bog formation was rapidly attained, and that most of the seam accumulated in a peat-bog environment. This mode of accumulation also accounts for the very low detrital mineral content and the high-inertinite composition.

Although often difficult to reconstruct because of poor preservation and mixing, the paleoecological relationships of some ancient coal-forming plant communities have been established. These reflect a zonation into open water, marsh, and swamp environments (Table 12-1) not unlike those of today, although spanning a narrower range of adaptations. The fernlike floras of the

Table 12–1. Plant Types of Various Coal-Forming Environments[a]

Modern Environment	Tertiary of Germany	Northern Hemisphere Carboniferous	Southern Hemisphere Permo Carboniferous
Open water (aquatic plants)	*Marcoduria*	Algae	*Botryococcus*
Marsh	Reed marsh, sedge, *Salix, Myrica*	*Calamites,* lycopods	*Phyllotheca,* lycopods, *Schizoneura, Paracalamites*
Wet swamp	Nyssa–Taxodium swamp	*Sigillaria, Lepidodendron* ?	*Glossopteris, Sphenophyllum, Cyclodendron*
Drier swamp	Myricaceae–Cyrillaceae moor, *Sequoia* forests	*Cordaites, Lepidodendron*	*Dadoxylon, Glossopteris*

[a]After Diessel (1970) and Stach *et al.* (1975).

Northern Hemisphere Carboniferous are poorly represented in Gondwana deposits, which characteristically show evidence of autumnal leaf accumulations (Plumstead, 1969, p. 41).

Coalification

Once a layer of peat is covered by detrital sediments and buried to progressively greater depths, it is compressed, thermally insulated, and converted through a series of increasing coal ranks from lignite (brown coal) through subbituminous and bituminous coal, to semianthracite and anthracite. Biological, physical, and chemical changes are involved in this coalification sequence. Conversion to lignite rank is accomplished mainly by biological activity, whereas upgrading to bituminous coal and anthracite involves a combination of physical and chemical processes (Teichmuller and Teichmuller, 1968; Stach *et al.*, 1975).

Gradual increase in geothermal energy with burial is the most important factor in coalification, which can be accelerated by an anomalously high heat flow associated with tectonism or igneous intrusion. The increase in coal rank with depth was shown by Hilt in 1873 to result from increase in carbon and decrease in volatile content. Geothermal gradient varies markedly with the structural setting. The relationship between coal rank and depth is strongest in young continental-margin basins, whereas intracratonic basins of lower heat flow may show only a minor increase in coal rank with depth (Wilson, 1976).

Lithostatic pressure is relatively unimportant during conversion of peat to subbituminous coal, apart from its role in expelling moisture by compaction. At higher rank, pressure may even retard coalification, and may account for unusually low coal ranks below certain thrust sheets (Hubbert and Rubey, 1959). Frictional heating associated with tectonic deformation does produce locally elevated coal rank (Middleton, 1981), but is generally of little significance. Variable pressure conditions can, however, influence the partial pressures of volatile products of coalification, thereby providing anomalies in coal rank and organic maturation indicators (Cecil *et al.*, 1979a).

Time is an important factor in coalification only if the temperature is above a critical minimum. For example, Carboniferous coals in the Moscow Basin have only reached lignite rank. This is because of a very low geothermal gradient and a temperature at the depth of the coals of only 50 to 60°C (120 to 140°F) (Karweil, 1956). Upper Miocene coals of Louisiana have reached the rank of high-volatile bituminous coal, whereas some Carboniferous coals at comparable depths in Europe have less than half the volatile matter of their Louisiana counterparts (Teichmuller and Teichmuller, 1968b).

Increase in rank is generally accompanied by progressive darkening and increased opacity, calorific value, and vitrinite reflectance. Solubility, together with oxygen, volatile, and moisture content decrease concomitantly. Reversals in some of these trends are apparent at the coalification break at the level when the volatile content of vitrinite falls below 30 percent. At this stage, there is an abrupt change in various parameters of coalification. Both calorific value and hydrogen content decrease, and the volume of methane suddenly increases before expulsion as coalification progresses (Wilson, 1976). As a consequence, this threshold is an important consideration in the assessment of underground mining hazards.

Peat stage is achieved by surficial biochemical diagenesis which extends to a depth of around 30 ft (10 m), where purely chemical processes become dominant. Decomposition of dead plant material is accomplished largely by microorganisms. Cellulose and protein are decomposed, progressively concentrating lignin and humic acids, and increasing the carbon content to between 50 and 60 percent. Plant material is transformed into peat by massive removal of both organic and inorganic components of the plant debris. Of the latter, oxalates and alkali and alkali-earth elements are almost totally removed. Clay minerals form rapidly *in situ*, presumably from amorphous inorganic material in woody plant tissue (Renton *et al.*, 1979). The activities of aerobic bacteria, fungi, and actinomycetes in the surface layers are replaced downward by anaerobic bacteria.

Lignite (brown coal) stage is normally attained at depths of 650–1300 ft (200–400 m) as a result of compaction and associated reduction in moisture content. Some of the giant brown coal seams of Victoria, Australia, still contain 67 percent moisture. Most lignites are Cenozoic in age, but a few are as old as Carboniferous. Plant remains tend to be still recognizable, thus providing the opportunity to compare them with modern plant communities. For example, in the Tertiary lignites of Germany, coals of coniferous and angiosperm forests can be distinguished (Teichmuller and Teichmuller, 1968a).

Increase in rank through the lignite stage is reflected in progressively higher carbon content and calorific value, and continued reduction in moisture content. During this process the humic acids condense into larger molecules forming humines, and the coal gradually becomes black and lustrous as it is transformed to bituminous rank by vitrinization.

Bituminous stage generally requires burial to depths of 1500–4000 m (5000–13000 ft) depending on geothermal gradient. By this stage humic acids have been totally transformed into humines. In contrast to the very gradual transition from lignite to bituminous coal, there is a very abrupt coalification break at the boundary between high-volatile and medium-volatile bituminous coal. This break corresponds to a carbon content of about 87 percent and a volatile content of around 29.5 percent (Stach, 1958; Teichmuller and Teichmuller, 1968b). Bituminous rank increases by removal of alicyclic and aliphatic groups and aromatization of humic complexes. Vitrinite reflectance rises accordingly to a level of around 1.0.

Anthracite rank is marked by a strong increase in reflectance and optical anisotropy. Hydrogen content is dramatically diluted by release of methane and water. Many coal basins of predominantly bituminous rank, for example, the Karoo coalfields of South Africa, contain anthracites in the contact aureole of igneous intrusions. Larger plutons and laccoliths convert greater volumes of coal to anthracite rank (Wilson, 1976).

Coal rank is strongly influenced by the tectonic behavior of a basin. Because coal is carried down to deep crustal levels of high geothermal energy, coal rank increases laterally in the direction of maximum deformation and heat flow. For example, Upper Permian coals of Queensland, Australia, originated in a foredeep plain west of the main orogenic axis. The east-central part of the basin was severely deformed during Late Triassic orogeny, and this area of maximum heat flow now contains semianthracites. These high-rank coals grade into bituminous and subbituminous coals toward the stable craton (Wilson, 1976). The same changes are evident between the anthracite and bituminous coal regions of the eastern United States. Local hot spots in orogenic terranes may be sensitively recorded by coals of higher rank, as may be uranium-rich granitoid platforms.

Coal Type

Maceral Groups

Microscopic study of coal recognizes basic morphological constituents, or macerals (Stopes, 1935; Stach *et al.*, 1975), distinguishable by their optical properties. Some macerals can have several possible origins, whereas others reflect a certain botanical precursor or burial history. The proportions of the various maceral groups have a profound effect on the commercial uses of coal. The components of the three maceral groups and their forms and possible origins are outlined in Table 12-2.

Table 12–2. Typical Form and Origin of the Three Maceral Groups[a]

Macerals	Form	Origin
Vitrinite Group		
Telinite	Well-preserved cell walls	Cell walls of trunks, branches, stems, leaves, roots
Collinite	Well-preserved cell fills	Cell-fill material
Telocollinite	Homogeneous	Tissues which have become structureless
Gelocollinite (rare)	Homogeneous	Formed from a humic gel
Desmocollinite	Fragmentary groundmass	Fragments of humic material
Corpocollinite	Homogeneous, massive circular, or oval bodies	Circular or oval bodies are probably plant cells
Vitrodetrinite	Fragmentary	Detrital fragments of vitrinite
Exinite Group		
Sporinite	Spore exines, well-preserved botanical forms, usually compressed except in very low rank coals	Skins of spores and pollens
Cutinite	Well-preserved botanical forms, sometimes cellular (often characterized by saw-tooth edge)	Outer layers of leaves, needles, shoots, stalks, and thin stems
Resinite	Pod-shaped, globular, or irregular bodies	Resins, oily secretions
Alginite	Circular or pod-shaped, well-preserved botanical forms	Algae
Liptodetrinite	Fragmentary	Unidentifiable fragments of exinite
Intertinite Group		
Semifusinite	Well-preserved cell walls } may show collapse structure (Bogen-structure)	Oxidized cell walls of trunks, branches, stems, leaves, roots, as above, frequently charcoaled
Fusinite	Well-preserved cell walls } may show collapse structure (Bogen-structure)	
Sclerotinite	Rounded or oval bodies with well-preserved cell structures inside	Fungal remains
Macrinite	Amorphous groundmass, structureless rounded fragments	Oxidized gelified plant material
Inertodetrinite	Fragmentary (less than one complete cell)	Cell-wall fragments, cell fillings
Micrinite	Specks, 1–2 μm	Controversial

[a] Compiled by Michelle Smyth.

Vitrinite is the most important maceral group. It is produced from woody tissue and bark in a generally reducing environment which, because of rapid subsidence and a high water table, was abruptly sealed from the atmosphere. As a result, there is no evidence of subaerial degradation. Vitrinite is further subdivided on a basis of botanical attributes and postdepositional alteration. Huminite is a soft precursor to vitrinite in lignite.

Inertinite may have the same botanical origin as vitrinite, but suffered oxidative changes during prolonged subaerial exposure prior to its eventual burial. This maceral group, therefore, indicates areas of slow subsidence and a low or fluctuating water table. It may also represent charcoal left by swamp fires or the oxidized remains of nonwoody plants. Inertinite is nonreactive during coalification.

Exinite, or liptinite, originates from specific botanical components such as spores, cuticles, and resins, and shows a great diversity of forms. It has the highest hydrogen content of any maceral group.

Vitrinite is the dominant maceral group in bright coal, whereas inertinite content tends to be high in dull coal. A coal composed largely of vitrinite with a small proportion of inertinite may constitute a better coking coal than one composed of vitrinite alone, and is significantly better in this regard than a vitrinite with a small amount of exinite.

The three maceral groups change independently during coalification. Vitrinite changes at a more or less constant rate, and for this reason measurements on vitrinite, usually vitrinite reflectance, are used to express coal rank. Inertinite is rich in carbon, and apart from changes during the early stages, it remains relatively stable during coalification. It does, however, blend with vitrinite at anthracite grade.

Microlithotypes

Microlithotypes, composed of macerals, are determined from very thin coal layers, down to 50 μm. They reflect the variety of combinations of maceral groups. For example, vitrite and fusite are equivalent to monomacerals composed of vitrinite and fusinite, respectively. Clarite and durite are bimacerals composed of vitrinite plus exinite, and exinite plus inertinite, respectively. The bimaceral microlithotype vitrinertite is composed of vitrinite plus inertinite, and the trimacerals clarodurite and duroclarite are intermediate in composition between clarite and durite (Stach, 1968).

Lithotypes

Fine banding develops during the transition from lignite to subbituminous coal. These bands represent an alternation of bright and dull laminae which consist of four lithotypes: vitrain, clarain, durain, and fusain. They are characteristic of bituminous coals, and disappear at anthracite rank.

Vitrain, composed of the microlithotypes vitrite and clarite, is glassy black with conchoidal fracture. Clarain, made of of these alternating layers of vitrite, clarite, durite and fusite, is black with a silky luster and is finely striated. Durain (durite and trimacerite) is dull grayish-black and massive, whereas fusain (largely fusite) is dull black and sooty (Stach *et al.*, 1975, pp. 113–134). Modern macroscopic coal logging procedures record the various combinations of bright (vitrain plus clarain) to dull (durain plus fusain) bands.

Petrologic Cycles in Coal Seams

Regular vertical changes in the petrographic characteristics of coal seams commonly involve upward decrease in vitrinite content as reflected in gradation from bright to dull coal. These changes stem from factors related to the type of parent vegetation and degree of exposure to the atmosphere before burial. Shrubs and grasses are the primary source material for inertinite, and trees for vitrinite, so that the vertical changes in coal seams are commonly a response to changes in the original vegetation cover (Smyth, 1980). However, a change from vitrinite-rich to vitrinite-poor coal can also result from slight lowering of the water table (Shibaoka, 1972). In many seams these two factors probably operated in concert.

As the peat surface is raised and becomes progressively better drained, the increase in peat thickness is accompanied by a replacement of the dominant tree cover by shrubs and grasses (Smith, 1968). Any slow increase in the rate of subsidence, or rise in the water table, encourages reoccupation of the peat surface by trees. Rapid inundation of the swamp surface destroys the vegetation, and may be recorded in a coal seam as a prominent clastic parting. Prolonged exposure causes severe oxidative degradation of the peat surface, and is preserved as a band of inertinite and ash. Fires sparked by lightning or spontaneous combustion leave a charcoal layer subsequently recorded as a thin fusinite band.

Because rapidity of burial is an important requirement for vitrinite preservation, coals in unstable foredeep basins tend to be vitrinitic. Declining rates of tectonic subsidence produce a higher inertinite content in the succeeding seams. Seams in a stable intracratonic setting tend to be inertinite-rich throughout (Smyth, 1970; 1980). Seams with "brassy tops," or pyritic upper parts, originate from marine incursion across peat-swamps. In back-barrier and lower delta-plain coals this pattern can be repeated in successive seams.

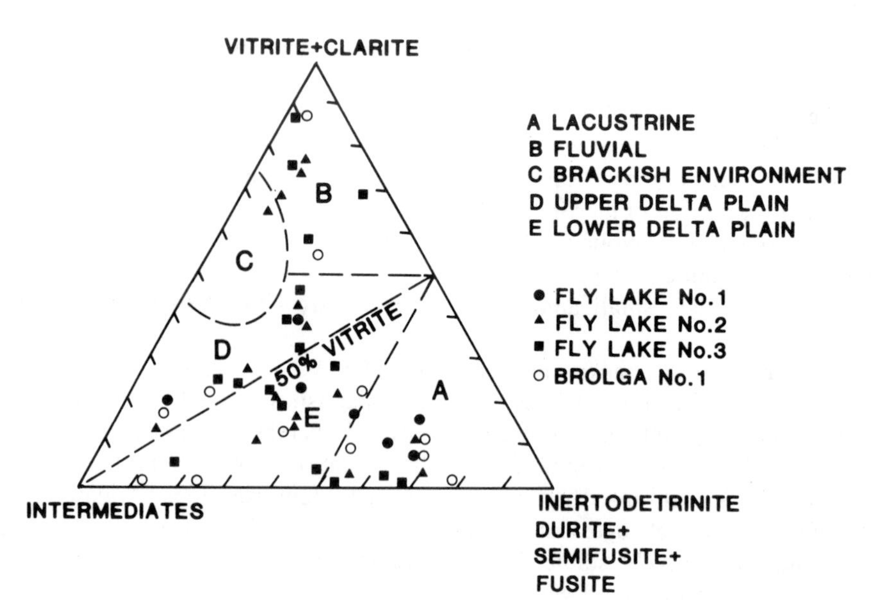

Figure 12-4. Microlithotype compositions of coals from the Fly Lake and Brolga exploration wells of the Cooper Basin, Australia. (After Smyth, 1979.)

Coal Petrography and Paleoenvironment

The petrographic characteristics of coal seams have been linked to depositional environment (Britten *et al.*, 1973; Smyth, 1979). Figure 12-4, with microlithotype compositions of coals from the Permian Cooper Basin of Australia, shows the petrographic differences among lacustrine, fluvial, "brackish," upper delta plain, and lower delta accumulations. These distinctions are supported by sedimentological investigations in the Cooper and other Permian basins. Lacustrine coals have a vitrite-plus-clarite content of less than 50 percent, but are typically rich in inertodetrinite and semifusite (Fig. 12-4, area A). Coals rich in vitrite-plus-clarite (area B) characterize fluvial systems with a large sediment load. Area C represents coals known to have accumulated in a saline or brackish shore-zone environment. Deltaic coals (D and E) tend to be rich in intermediates (duroclarite-plus-clarodurite-plus-vitrinertite).

Coal studies such as these have further application in hydrocarbon source evaluation. Dispersed organic matter of plant origin yields hydrocarbons when buried to sufficient depth. Alginitic and exinitic matter generate oil, whereas vitrinitic matter is gas-prone (Smyth, 1979) but can also produce oil.

The Late Paleozoic flora was highly susceptible to desiccation and oxidation during peat formation, and the originally low exinite content was further depleted. Jurassic and younger coal-forming flora were better able to resist decomposition during early diagenesis. This high survival rate stemmed from abundance of cuticle, spores, and resins, so that the original exinite-rich character was maintained. Coal-bearing sequences of Jurassic to Tertiary age consequently have the potential for significant liquid hydrocarbon generation, as evidenced by the Gippsland Basin Oil Fields of Australia (Thomas, 1982).

Tectonic Setting of Coal Basins

Tectonic setting influences the number, thickness, continuity, and quality of seams, and determines their attitudes, degree of disruption, and present depth. This in turn determines their mineability. Coal basins that have been subjected to tectonic deformation, with folding and faulting over a broad area, are the most important economically. Over 90 percent of the world's coal production comes from regions affected by late Paleozoic and end Mesozoic–Tertiary tectonism, with only a small amount from the vast, stable intracontinental basins such as the Moscow Basin (Wilson, 1976).

Coal Basins Related to Subduction

Orogenic foredeeps and back-arc basins undergo locally tight folding, thrusting, and, where oblique collision is involved, major strike-slip faulting. Rapid subsidence through high geothermal gradients produces coals of subbituminous to anthracite rank. In areas of maximum subsidence, coals may be thick and numerous but contain large detrital splits. Further, the seams commonly show a very complex three-dimensional geometry (Britten, 1972).

A modern example of a subduction-related coal basin is the southwestern foredeep of New Guinea, which is underlain by as much as 13,000 m (42,000 ft) of Cenozoic sediments containing abundant coal seams (Stach *et al.*, 1975, p. 17). The Permian Sydney Basin of Australia is remarkably similar to the New Guinea foredeep basin in geometry and inferred origin (Jones and McDonnell, 1981).

Coals in these orogenic basins are rapidly covered with sediment, and therefore tend to be bright. Although deeply buried during tectonic subsidence, subsequent basin-edge tilting and erosion, or the onset of later tectonic uplift, elevates many seams to levels at which they can be mined without difficulty.

Coal Basins Related to Epeirogeny

Intracontinental rift grabens and strike-slip pull-apart basins can involve large vertical displacement, but deformation is less intense than is the case in orogenic basins. However, except in areas of persistent block faulting, even the most tectonically active intracontinental basins are generally subject to only moderate tectonic subsidence. Thus, the total coal-bearing succession tends to be thinner. The number of seams is limited, but their geometry and thickness are highly variable. Patterns of splitting are less complex than in the deposits of subsiding continental margins, and the seams contain a greater proportion of dull bands because of slow burial. Many of the more rapidly subsiding epeirogenic basins tend to fill with conglomeratic sediments devoid of coal. Significant exceptions include the Cantabrian coal basins of Spain (Heward, 1978a and b), where peats accumulated during quiescent episodes. Lignites of Victoria, Australia, occupy a Cenozoic aulacogen which opened to the sea; these seams are among the thickest known.

Stable Cratonic Coal Basins

Coal-bearing successions of cratonic basins are normally thin and comparatively uniform, with few seams. Individual seams may, however, attain considerable thickness, and some cover vast areas of 4000 mi^2 (10,000 km^2) and more; other seams are restricted to paleotopographic depressions, terminating abruptly against basement highs (Le Blanc-Smith, 1980).

Smyth (1980) has shown that most Australian coals thicker than 15 m (50 ft) accumulated in small, stable intracratonic basins subject to very gradual subsidence.

Coal-bearing successions in the intracratonic Karoo Basin change in response to elongate downwarps in the regionally stable basement platform of Precambrian rocks (Mason and Tavener-Smith, 1978). Coals are more numerous in the downwarps, within which regressive deltaic sequences are more abundant than they are on the adjacent stable platform. Individual seams commonly split in the direction of downwarped areas. This simple pattern of interfingering contrasts with the complex splits in orogenic coal basins (Wilson, 1976).

Coals of cratonic downwarps are predominantly of low to medium rank because of low geothermal gradients and limited depths of burial. The Cooper Basin of Australia, with a radioactive crystalline basement in parts, provides an exception. High temperatures associated with igneous intrusion produce pockets of high-rank coal unrelated to depth. Burned coal, or natural coke, forms at the intrusive contact. The generally dull nature of epeirogenic basin coals reflects prolonged exposure, responsible for the high inertinite basin and ash content.

Passive-Margin Coal Basins

Terrigenous clastic wedges that prograde over a progressively deepening shelf, for example the northern Gulf Coast Basin, commonly contain major coal deposits in the updip facies (Fisher and McGowen, 1967; Kaiser *et al.*, 1978). The coal-bearing succession can show evidence of listric growth faulting, but the coal-bearing strata do not as a rule display the dramatic expansion

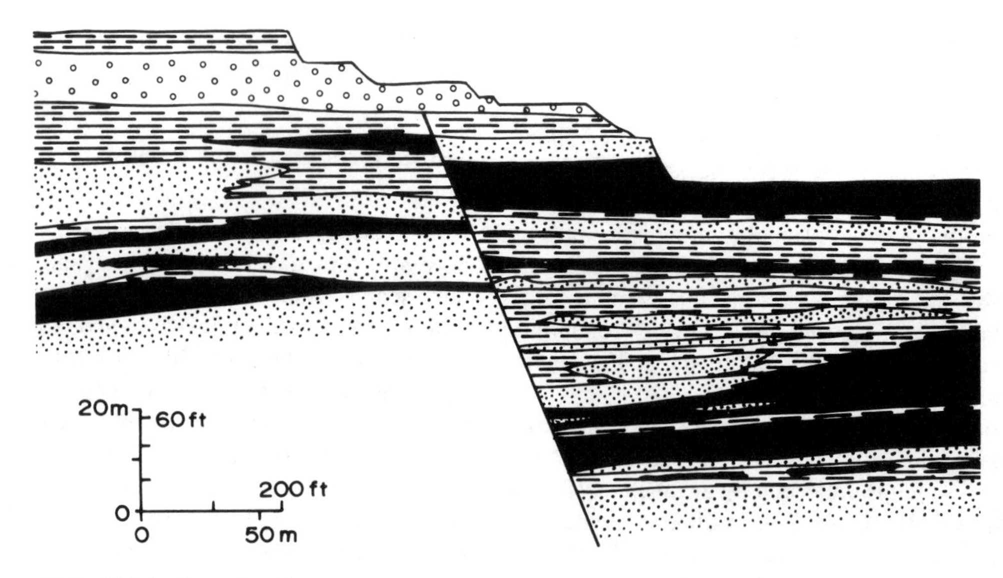

Figure 12-5. Thickening of coal seams across contemporaneous basement faults. Modified after Junghans (1958) in Teichmuller and Teichmuller (1968a).

characteristic of the hydrocarbon-bearing facies farther basinward (Galloway *et al.*, 1982).

Effects of Contemporaneous Tectonism on Coal

Contemporaneous tectonism produces dramatic changes in coal-seam thickness and petrography. Displacement may be related to synclinal warping (Vetter, 1980), graben development (Teichmuller and Teichmuller, 1968a), or growth faulting (Horne *et al.*, 1976). The effect of contemporaneous block faulting on Tertiary coal sequences in Germany is illustrated in Figure 12-5. Coals of the downthrown block are not only considerably thicker, but are of higher rank. In parts of the Appalachian Basin subsidence appears to have limited peat accumulation on the downthrown block, whereas thick coals are located along the edge of the upthrown block, which provided a platform for peat accumulation. Elsewhere in the Appalachian Basin the coals may thin or merge onto the upthrown block, and some coals are preserved only on the downthrown block, where they survived the destructive effects of oxidation at higher levels (A.J. Tankard, personal communication, 1982). Subtle differences in elevation of peatswamps subject to minor tectonism might be reflected in the microlithotypes, chemistry, or ash content of the coal.

Changes in Tectonic Setting of Coal Basins

Coal-forming paralic environments, in which the bulk of the world's coal resources originated, commonly precede limnic environments during coal-basin evolution. For example, the Late Carboniferous foredeep north of the Variscan highlands of Europe first contained large paralic coal basins along the prograding fluvio-deltaic margin, followed by the limnic coals of intermontane basins (Mackowsky, 1968). Similarly, in the classical thick and relatively undeformed clastic wedges that result from asymmetric basin filling (Ferm, 1974), there may be a gradual vertical and updip progression from distal paralic to proximal limnic coals.

Coal-Bearing Cycles and Paleoenvironments

The degree of ordering, or cyclicity, that is evident in most coal-bearing rocks has been variously attributed to a number of factors including:

(1) glacially induced climatic changes,
(2) periodic or continuous changes in relative sea level due to tectonism or eustatism, or

(3) lateral migration and switching in position of channels and delta lobes accompanied by compactional subsidence.

Eustatic sea-level changes undoubtedly occurred during deposition of Carboniferous coal-bearing sequences, and tectonic deformation, too, is known to have produced cyclic sedimentation patterns in places. Nevertheless, many three-dimensional rock geometries in coal basins are satisfactorily explained in terms of processes that are readily observable in modern depositional environments (Ferm and Horne, 1979). Within this context, however, units of unusual thickness or lateral persistence can reflect external controls. In some coal basins, particularly those that are fault-controlled or contain a large proportion of uncompactable coarse sediments, tectonic controls were probably dominant. The resulting cycles may be of basinwide extent. Similar patterns would result from water-level changes produced by eustatism or varying rates of inflow and evaporation in a land-locked sea or lake. It is conceivable that eustatic sea-level changes, whether glacio-eustatic or resulting from changes in rate of sea-floor spreading, could permit broad correlation among coal-bearing successions originating in different parts of the world. But even in these situations, paleoenvironment exercises an important influence on lateral changes in coal thickness and quality.

Knowledge of coal-seam geometry relies on close stratigraphic control. In the absence of this control, a common error is to overestimate the continuity of seams. Major road excavations in the Appalachian region, for example, have shown that most seams are far less continuous than previously thought (Ferm, 1974).

Coals originate in a wide range of paleoenvironments, but most economic seams formed in a restricted number of depositional settings, of which river-dominated delta plains and mixed-load alluvial plains are of paramount importance. Fluvio-deltaic progradation results in a coal-bearing continuum ranging from lower delta plain to alluvial plain. The most landward components of such regressive successions may include coal-bearing deposits of bed-load streamplains, and even coarse alluvial fans. Other economic seams have originated in periglacial environments, along barrier-lagoon shorelines and prograding tidal flats, and by processes of swamp encroachment on shallow interior lakes. But greatest emphasis will be placed on fluvio-deltaic systems because of their vast coal resources. Earlier representation of these coal-bearing deposits as one-dimensional cyclothems has been supplanted by more sophisticated three-dimensional models (Ferm and Williams, 1963; Horne *et al.*, 1978; Ferm, 1979). These models are flexible, and have widespread, albeit not universal, application.

Differences in alluvial and deltaic coals of various regions stem largely from the variety of river and delta types and their subenvironments, which provide a range in settings for peat accumulation. These variations occur on two scales, firstly among the major river systems, for example, bed-load versus mixed-load systems (Chapter 4), and wave-dominated versus river-dominated systems (Chapter 5), and secondly on a more local facies or subfacies scale within each of these major depositional systems.

General Factors Affecting Coal Seam Distribution

The most important factor influencing the areal distribution of peat is the nature of the platform upon which it accumulates (Ferm *et al.*, 1979). Most peat-forming platforms are at or slightly above the contemporaneous sea level. The platforms may be flat over a wide area, or they may rise gently inland at a rate determined largely by the water table. Vegetable matter deposited above the ground-water level has little chance of preservation except in consistently wet, cool climates.

Along the seaward margin of a prograding coast, peat growth commences only when the effects of brackish or marine intrusion are excluded. Inland peat accumulation is dependent upon a delicate balance among subsidence, aggradation, and maintenance of a high water table. Inundated depressions and active channels preclude organic accumulation, and may be recorded in coalfields as "wants," or local areas lacking coal. Peat encroaches across the surfaces of depressions and swales that are subject to shallowing but which lack vigorous sediment incursion. Abandoned channels and swales become filled with mud or organic material, the latter giving rise to elongate coals of considerable thickness. Surface irregularities in the peat-forming platform are thus reflected in the base of overlying seams.

Flow character and sediment load of adjacent streams are additional factors influencing peat accumulation. Channels with a highly peaked discharge and a large sediment load deposit extensive sheets and tongues of overbank sediment. This disrupts peat accumulation. Shelf storms can have the same effect by washing sediment over coastal peatswamps. Rivers with a more evenly distributed discharge tend to contribute only fine suspended sediments during occasional overbank flooding. This generally has little effect on peat accumulation other than producing a thin, muddy detrital band in the resulting coal seam.

Peat-forming platforms are in places contemporaneously deformed by differential compaction of underlying sediments. Areas where the substrate includes significant proportions of clay and organic material subside most rapidly. Provided that the supply of organic material keeps pace with subsidence, thicker peats will accumulate over these compacting areas than over adjacent areas underlain by sand. Where subsidence is more rapid than the rate of organic supply, the peat surface becomes ponded. Even though there may be no detrital influx, degradational release of intrinsic inorganic material will result in an increase in ash content (Donaldson *et al.*, 1979).

The effects of differential compaction on coal seam thickness is even more pronounced where the sediments accumulate over a high-relief unconformity. In the Karoo Basin, the lowermost seams are most drastically affected by Precambrian highs, pinching out against their flanks and thinning over their tops, where the peat was degraded as a result of its higher elevation. The influence of paleotopography is propagated upward through several coal-bearing sequences, but these effects of differential compaction decline with increasing stratigraphic elevation (Le Blanc-Smith, 1980). In other areas it appears that peat accumulated preferentially on the flanks of topographic highs, with major channels following the depressions.

Relocation of channels across the top of peatswamps may significantly reduce the peat thickness, or even erode completely through it. In most basins, new channels tend to select areas of thickest peat where compactional subsidence offers the steepest stream gradient (Ferm *et al.*, 1979). But in some basins a high proportion of mud in the subsurface causes rapid subsidence of channel-fill sands. This produces a vertical stacking of fluvial channel sands along a persistent path (Brown, 1969). Although channeling has drastically reduced the volume of some seams such as the Pittsburg Seam of West Virginia and Pennsylvania (Donaldson, 1979), the erosive effects of other major channel-fill sandstones resting directly on coal seams has been negligible.

Coals in Relation to Delta Type

Chapter 5 distinguishes among the spectrum of delta types molded by interplay between fluvial and marine processes. Associated coals display significant differences.

Fluvial-Dominated Deltas

Substantial coal seams trending approximately parallel to depositional dip are associated with birdfoot-type or elongate river-dominated deltas, accumulating along levees and infilled bays. Coals capping the thick, upward-coarsening delta-front and interdistributary bay sequences are extensive but seldom attain significant thicknesses. These seams are relatively free of washouts because of the limited number of distributaries, which remain stable while extending their courses basinward. Detrital splits are common, however, because of the ineffectiveness of the levees in containing floodwaters.

Coals of lobate deltas accumulate during both the constructive and destructive phases of deltaic sedimentation. During the constructive phase, the delta plain is traversed by numerous distributary channels. Peat formation is restricted to the interdistributary tracts, which in many cases are of limited extent. The coals are therefore characterized by "wants" corresponding to the positions of the active distributary channels. Ash content can be high because low levees allow frequent inundation of the marsh. In contrast, blanket peatswamps originating during abandonment of lobate deltas can be thick and remarkably widespread. The seaward edge of the abandoned delta lobe is attacked by waves, creating transgressive barriers backed by shallow, brackish bays. Peat accumulates on the more slowly subsiding delta platform landward of the maximum marine influence. Blanket peats of the Mississippi delta plain cover areas of up to 190 mi^2 (500 km^2) (Coleman and Smith, 1964). As transgression

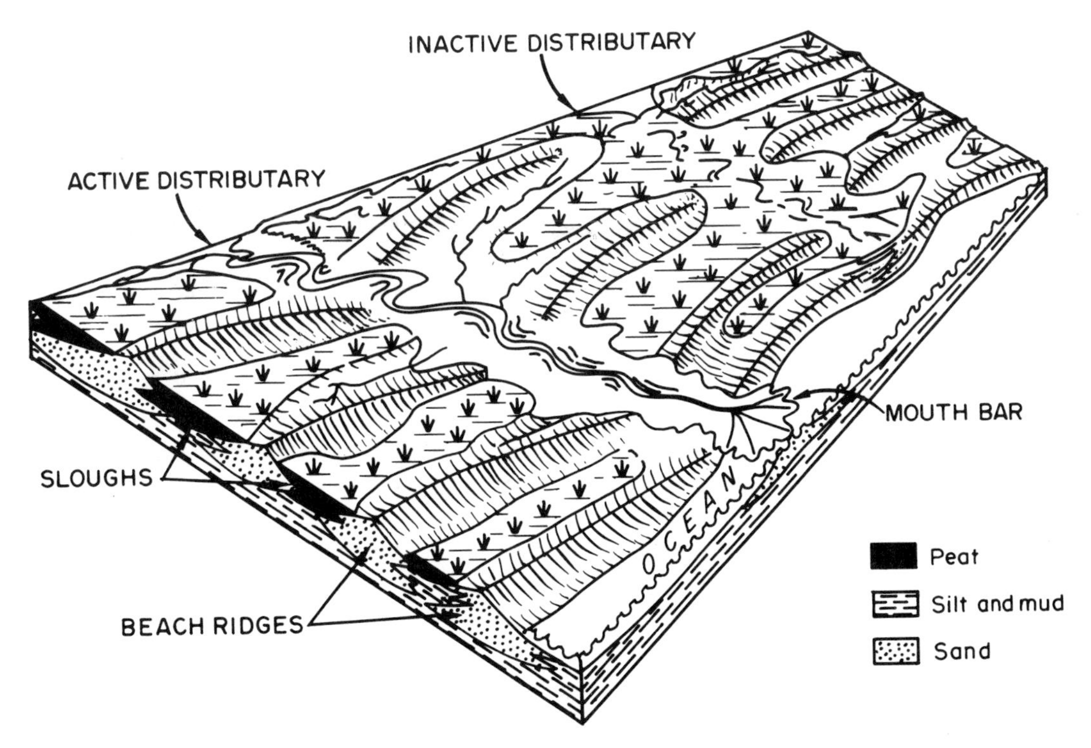

Figure 12-6. Peat accumulation between prograding beach ridges of an idealized wave-dominated delta.

progresses, the peats are likely to be subject to marine influence, which impairs the chemical quality of the coal, at least in its upper parts.

Tertiary lignites of the Gulf Coastal Plain originated partly as thick blanket peats formed in a variety of inactive delta-plain environments. The lateral extent of these peats was determined by the size of the underlying platform, which in places may have comprised several adjacent, inactive delta lobes (Fisher and McGowen, 1967; Kaiser *et al.*, 1978). In other coal basins, too, the most persistent delta-plain seams probably originated as similar blanket peats. These seams are readily distinguished from the far more numerous but laterally more restricted seams representing peat accumulation during the active delta phase.

Fluvial-Dominated Deltas

Wave-dominated deltas display consecutive shore-parallel beach ridges with intervening sloughs and embayments that become peat-swamps as infilling proceeds (Fig. 12-6). Associated coals thus tend to follow belts parallel to the shoreline, but extend perpendicularly through abandoned channels and inlets. The seams are interrupted in places by washover fans, and contain wind-blown detrital grains. Abandonment-phase seams are likely to be moderately strike-elongate.

Tide-Dominated Deltas

Coals related to tide-dominated delta systems are rare, with the notable exception of those in the Cretaceous Horseshoe Canyon Formation near Calgary, Alberta (Rahmani, 1981). This is surprising, in view of the luxuriant swamps that characterize a number of tropical tide-dominated deltas. The compound delta of the Klang and Langat Rivers of tropical Malaysia provides a good example (Coleman *et al.*, 1970). Mangrove swamps are extending seaward as the coast progrades at rates of over 20 ft (6 m) per year at some localities. A tidal range of some 13 ft (4 m), augmented by a general northwesterly flow, fashions sand waves up to 40 ft (12 m) high in the Malacca Strait. Delta-front sands grade upward into tidal-channel and tidal-flat deposits (Fig. 12-7), representing a thick, upward-fining sequence. Mangrove clays are organic-rich, but are

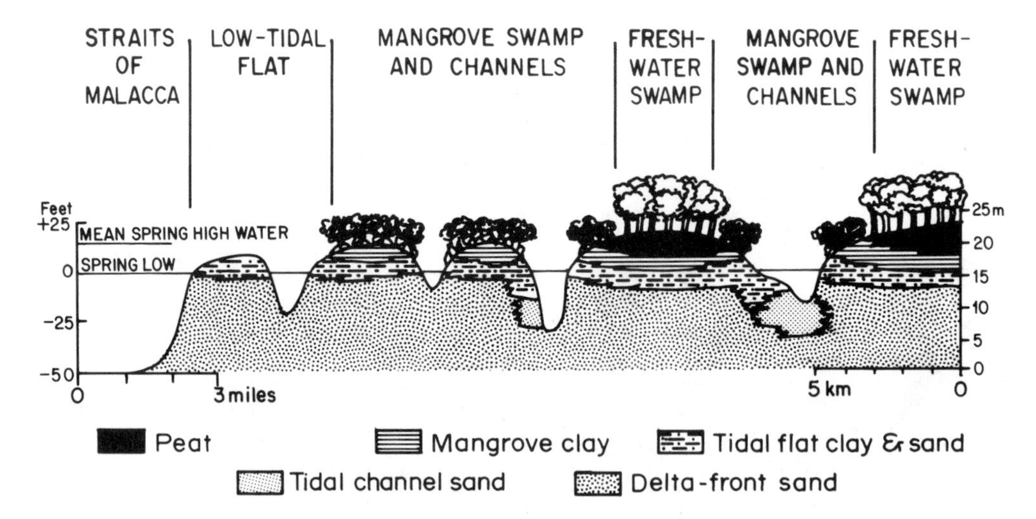

Figure 12-7. Peat in a tide-dominated deltaic environment, Malaysia. Progradation results in an upward-fining sequence capped by peat. (After Coleman *et al.*, 1970.)

light colored in their upper parts, where they resemble seat earths. The peat forms in freshwater swamps, and consists of up to 16 ft (5 m) of decomposing wood fragments in a fine organic matrix. Coleman *et al.* (1970) estimated an accretion rate for the peat of some 4 in (10 cm) a century. They attribute this rapid rate to a combination of high organic productivity, the continuously high humidity beneath the dense tree canopy, and conditions toxic to plant-decomposing bacteria. The paleoenvironmental reconstruction by Rahmani (1981) for part of the Cretaceous coal-bearing succession of Western Canada shows a macrotidal estuary bordered by tidal flats and extensive peat swamps (cf. Fig. 6-12). However, on many other macrotidal coasts the high rate of sedimentation, coupled with physical reworking and oxidation resulting from frequent high-energy marine inundation, may account for the rarity of thick peats. The carbonaceous mudstones that cap some upward-fining intertidal sequences associated with ancient delta systems are possibly derived from advanced degradation of peat.

Coal-Bearing Deltaic Facies

Variations in seam thickness, lateral continuity, ash, sulfur, and trace elements are all determined in large part by the fluvio-deltaic subenvironment (Horne *et al.*, 1978). These relationships have been most rigorously documented in Carboniferous rocks of the Appalachian region where a vast amount of outcrop, subsurface, and mine data are available (for example, Baganz *et al.*, 1975; Ferm, 1974, 1976; Donaldson *et al.*, 1979). But many of these relationships are equally valid in post-Devonian strata elsewhere, such as the Carboniferous "coal measures" of Europe (Elliott, 1975; Steel *et al.*, 1977b), Permian deposits of the Gondwana continents (Casshyap, 1970; Hobday and Mathew, 1975), and Cretaceous coal-bearing strata of the western United States (Marley *et al.*, 1979; Ryer *et al.*, 1980; Flores, 1981). These repetitive, predictable patterns are explained by comparison with modern coastal and alluvial environments and processes.

Back-Barrier Coals

Barriers that form during delta progradation or a subsequent phase of delta destruction are backed by shallow bays and lagoons, which may fill with sediments supplied from land and sea and evolve into back-barrier marsh. The resulting peats generally show a pronounced shore-parallel trend except where they extend up infilled reentrant bays, former inlets, and abandoned channel tracts. Many back-barrier coals are thin, discontinuous, and have a prohibitively high sulfur content, but some such as the Beckley Seam of

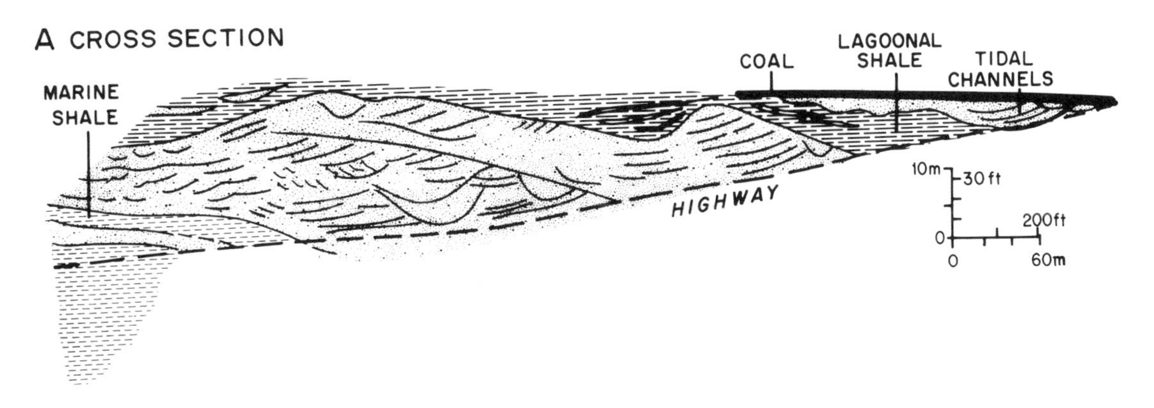

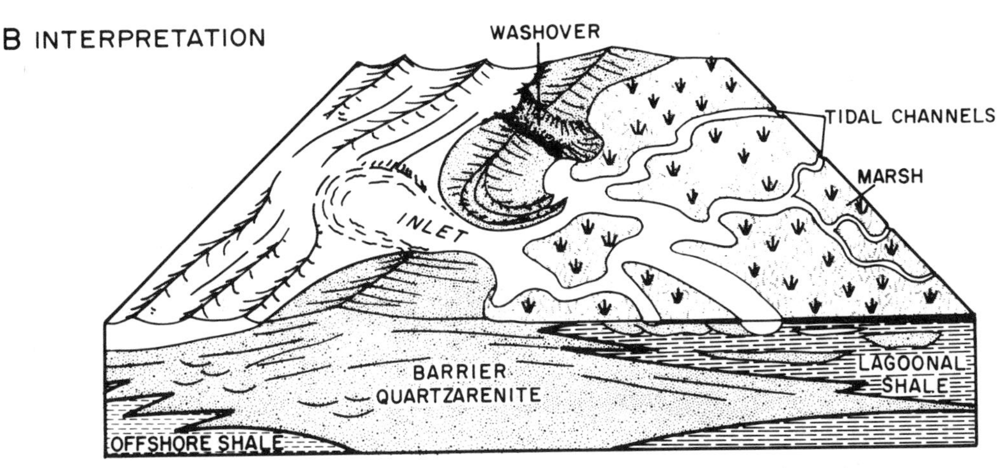

Figure 12-8. Relationships among Carboniferous back-barrier coal, other back-barrier facies, and back-barrier quartzarenites, Rockledge, Alabama: (A) observed facies geometries, (B) interpreted origin. (Modified after Hobday, 1974a and Ferm and Horne, 1979.)

West Virginia and others in western North America are commercially important. In some cases, only the upper parts are pyritic.

Back-barrier coals (Fig. 12-8A) tend to overlie upward-coarsening sequences of carbonaceous lagoonal shales and siltstones with rootlets and burrowed, sideritic clay-ironstones. Restricted, brackish faunas and finely divided plant debris are characteristic. These lagoonal sequences intertongue landward with deltaic facies such as bayhead deltas and bay-fill with levees and crevasse splays. They interfinger seaward with quartzose sandstones of barrier origin, including washover, flood-tidal delta, and tidal channel-fill deposits. Back-barrier peats also cap tidal-flat sequences. The generalized back-barrier peat-swamp setting is illustrated in Figure 12-8B. Some seams lie directly on barrier sandstones, and are overlain by washover sandstones (Fig. 12-9).

The Snuggedy Swamp of South Carolina provides a modern example of back-barrier peat accumulation (Staub and Cohen, 1979), albeit behind an abandoned barrier ridge. Upward-coarsening sequences of lagoonal and tidal-flat sediments are separated by root-penetrated horizons or salt-marsh peats. The latter would be preserved as thin, discontinuous, highly sulfurous coals, or simply as a black shale or "bone coal" (Renton and Cecil, 1979). Thick marsh peats are commonly situated adjacent to the barrier-island sands which provide an additional source of fresh water, or create islands above subsurface sand highs subject to slower rates of subsidence. Back-barrier peat underclays are devoid of lamination and have a high kaolinite content because of leaching of montmorillonite by acid waters draining from the overlying peat (Staub and Cohen, 1978).

When the rate of peat accumulation in the

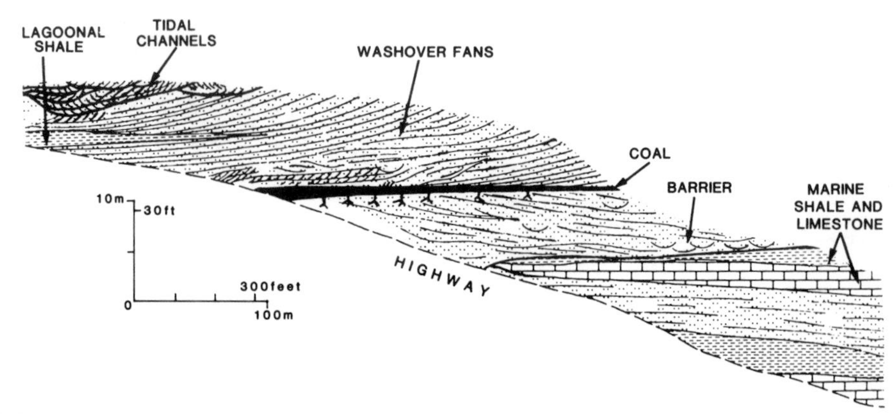

Figure 12-9. Landward-thickening Carboniferous coal overlying barrier quartzarenites, and overlain in turn by landward-dipping washover-fan sandstones cut by tidal-channel deposits. (Modified after Hobday, 1974a.)

Snuggedy Swamp exceeded the rate of Holocene sea-level rise, the islands of fresh-water peat grew laterally and coalesced into a more extensive peat layer (Fig. 12-10). At times these rates were in balance, and the peats built vertically, but did not enlarge in area. When peat growth could not keep pace, the invading sea water converted the area into a shallow salt marsh and lagoon (Staub and Cohen, 1979). These back-barrier peats and their relationship to transgressive, strongly aggradational sandy facies are illustrated in Figure 12-11.

Similar relationships are displayed in raised Pleistocene deposits of Zululand (Hobday and Jackson, 1979). There the peat overlies lagoonal clays containing a typical lagoonal fauna of oysters, crocodile, and hippopotamus. The lagoonal clays are traversed in places by sandy tidal-channel fill. A few large upright trunks anchored in the peat protrude into the overlying transgressive washover sands. Rafted vegetation is conspicuous, however, with poorly aligned, compressed logs of swamp trees still common on the modern coastal plain. The peat has a very high sulfur content, abundant dispersed wind-blown grains, and thin washover sand partings. Severe contemporaneous deformation is a feature of these Pleistocene peats (Fig. 12-12). Buckling, thrusting, and normal faulting, including growth faulting, resulted from a combination of clay diapirism and gravity gliding. Comparable structures in a coalfield would make mining extremely difficult. Some southern Appalachian coals that were subject to transgressive overlap by eroding barrier sands show small-scale glide deformation and differential compaction of lagoonal clays.

The Beckley Seam of West Virginia originated behind a compound barrier system, and persists along a coast-parallel trend interrupted by tidal channels. The coal rests on lenticular, wavy, and flaser-bedded tidal-flat facies, which in turn overlie lagoonal shales. As peat growth expanded across this platform, the smaller tidal creeks became clogged with plant material, and only the major tidal channels remained open. As a result, the coal is thickest over the sites of the inactive channels, and thins over the highs. Even the major tidal channels were eventually abandoned, but only after they had been partially filled with clastic sediment, so that their courses are marked by belts of thin coal. The thickness of the Beckley Seam varies markedly as a result. Additional complexity arises from extensive splay and barrier-washover deposits. The back-barrier Sewell Seam, in contrast, reflects only minor splay activity, probably because of the lower energy of the associated channels (Horne *et al.*, 1978).

Back-barrier coals occupy distinctive positions within geophysical log sequences (Fig. 12-13A). Tertiary lignites of the lower part of the Jackson Group, Texas, are conspicuous components of the fine-grained back-barrier facies that overlie upward-coarsening regressive barrier-strandplain sands (Kaiser *et al.*, 1980). Sand isopachs of this stratigraphic interval show a shore-parallel trend (Fig. 12-14A), with a similar alignment evident in the lignite seams (Fig. 12-14B) which formed behind the beach ridges and barriers.

Less common Appalachian coal-bearing facies occupy major estuarine scours in a lower delta-

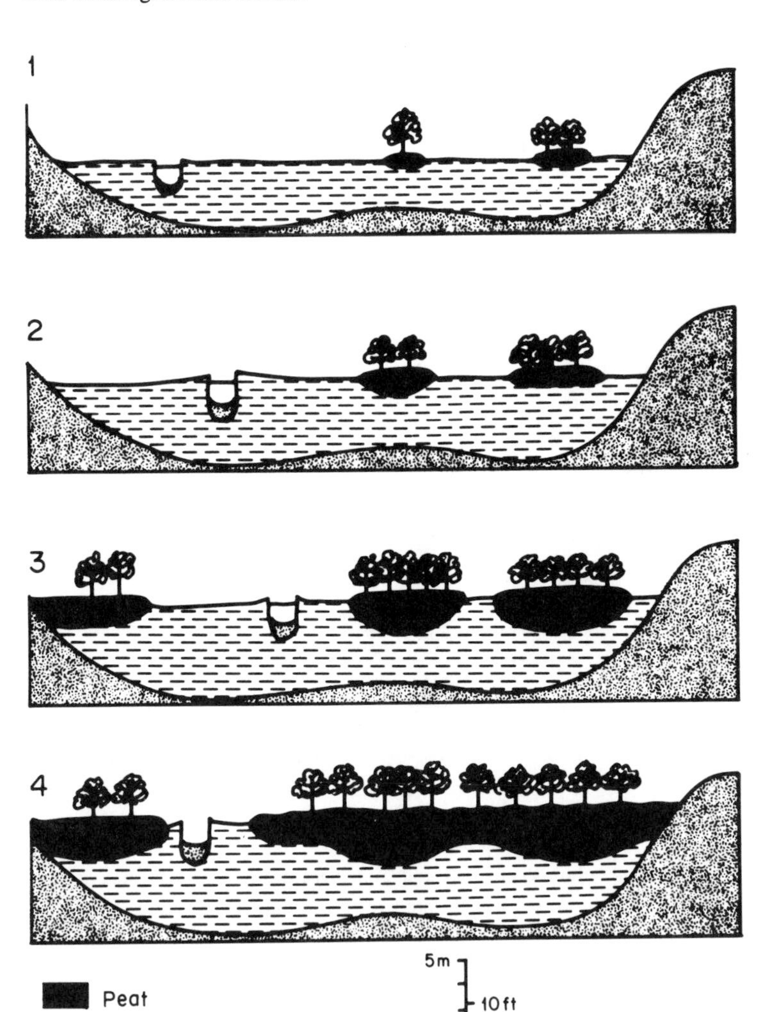

Figure 12-10. Stages in development and coalescence of back-barrier peat islands: Stage 1, colonization by fresh-water vegetation; 2, initial development of peat islands with deposition of silt and clay in intervening areas accompanying sea-level rise; 3, expansion of peat islands; 4, development of extensive peat mat with an irregular basal contact. (After Staub and Cohen, 1979.)

plain to back-barrier setting. These estuaries originated by abandonment of very large distributary channels, and were subject to gradual marine transgression. Tidal scour contributed to temporary enlargement, but the main processes involved infilling by slow subtidal accretion of bed-load and suspension-deposited layers. Alternating with these episodes of vigorous tidal exchange were phases of temporary isolation due to construction of a littoral sand bar across the mouth. These abandonment phases are marked by black, organic-rich shales. The estuarine sequences are capped in places by thin seat earths and coals (Hobday and Horne, 1977; Horne, 1979a).

Ferm (1976) has observed that barriers can play another critical role in sealing off the delta plain from the oxidizing effects of sea water. This is clearly demonstrated in Tertiary lignites of the Lower Rhine Bay, Germany, which was protected by an extensive chain of spits and barriers (Teichmuller and Teichmuller, 1968a), allowing the alluvial-plain environment of maximum peat accumulation to extend to within a few kilometers of the coastline. Prolonged peat growth, represented by 50 m (160 ft) of lignite, must have been accompanied by continued aggradation of the shoreline sandbodies. The Karoo Basin also provides evidence of barrier influence on coal distribution. Coals that originated behind a barrier shoreline typically extend farther basinward than do coals in delta systems that lacked barriers.

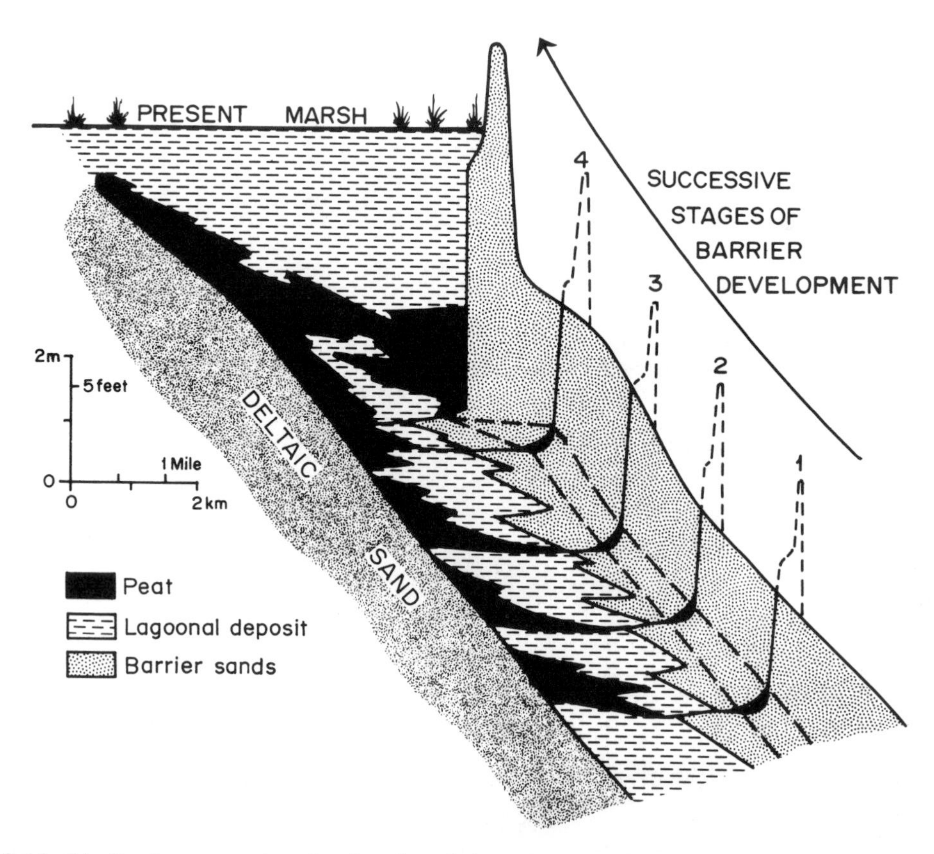

Figure 12-11. Idealized cross section showing the relationships of peat and lagoonal deposits to successive positions of the transgressing barrier. The entire system was characterized by pronounced aggradation. (After Woollen, 1976.)

Marginal Delta-Plain Coals

Alongshore from active delta lobes, peripheral areas receive sediments transported largely by longshore drift. Waves rework the coarser sediment fraction into beach ridges and cheniers that subsequently become isolated from the shoreline by intermittent progradation of fine-grained sediments. Extensive peatswamps may form between the successive beach ridges as the strandplain enlarges (Coleman, 1966). The geophysical log pattern of a prograding marginal delta front is shown in Figure 12-15C.

Where unidirectional longshore currents prevail, areas updrift of the delta lobes are characterized by organic accumulation in a marshy environment. The area may remain starved of clastic sediment over a long period, while active lobes switch position along strike. The geologic record consists of numerous thin, chemically impure coals separated vertically by a few meters of homogeneous, sideritic dark shale with rare thin, lenticular chenier sandstones.

Lower Delta-Plain Coals

On river-dominated deltas, peats of the lower delta plain accumulate on a platform of upward-coarsening sediments which include prodelta and bay clays and silts overlain by, and laterally gradational into, mouth-bar and splay sands (Fig. 12-15A). Delta-front sands of a prograding distributary mouth bar are gradationally based, except in the immediate vicinity of distributary channels. The overall sand-body shape in strike section (Fig. 12-16) has a broad, flat base and tapers gradually upward. Lower delta-plain coals tend to be thin but persistent, particularly in the paleoslope direction, but seam continuity along strike is limited by distributary sand axes. Sulfur

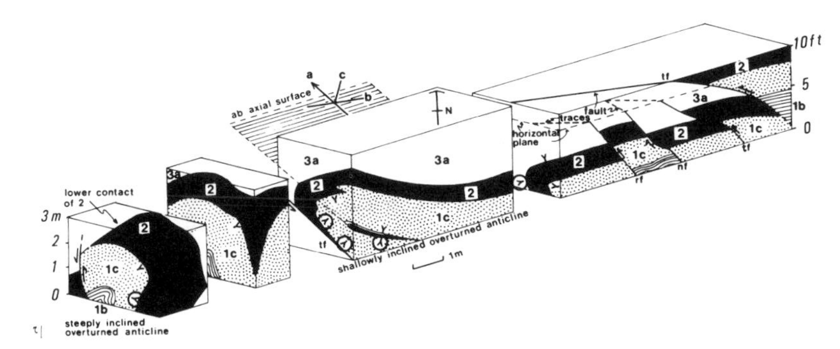

Figure 12-12. Intense deformation of an Upper Pleistocene peat bed as a result of gravity gliding and diapirism of lagoonal clays, Zululand, South Africa. (After Jackson and Hobday, 1980.)

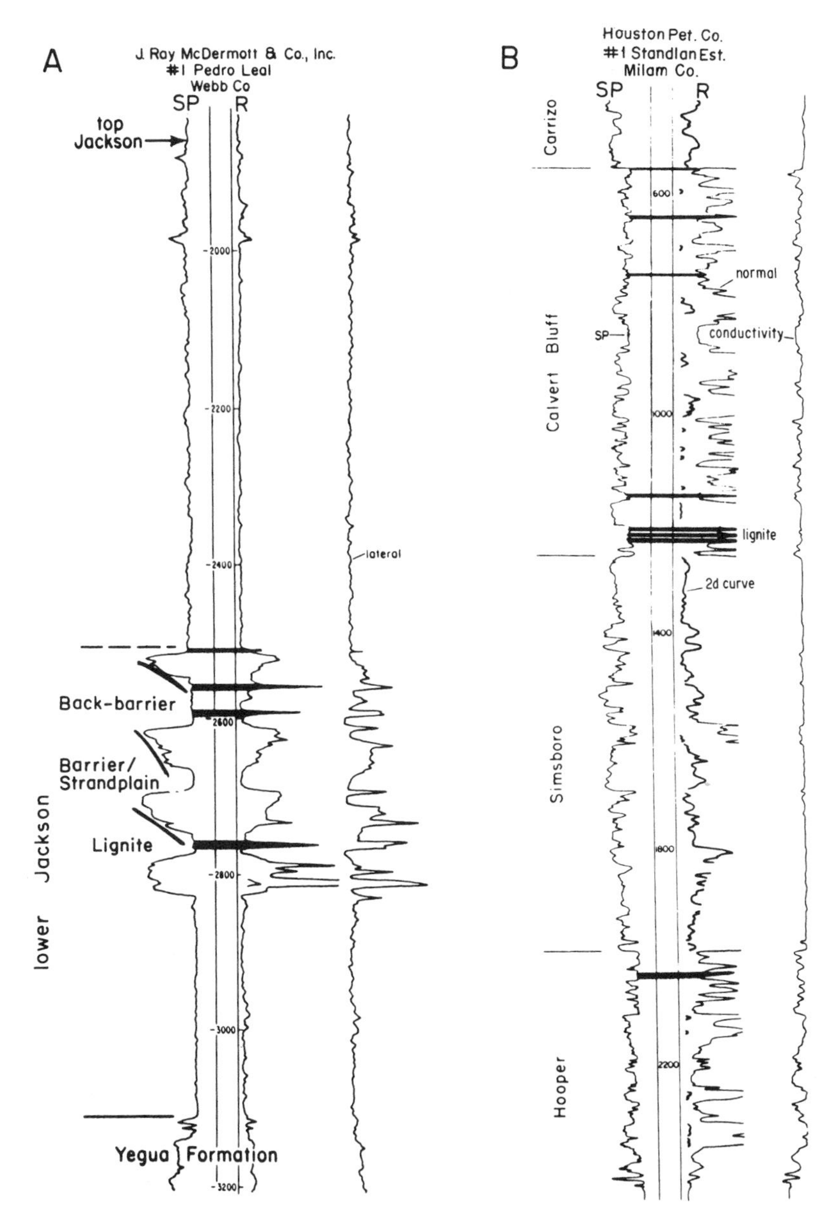

Figure 12-13. Log response of Tertiary lignite-bearing intervals of the Texas Gulf Coast. Note the flat SP and the sharp resistivity spikes of the lignites: (A) the lower part of the Eocene Jackson Group of prograding barrier/strandplain origin; (B) the Eocene Wilcox Group with lignites concentrated in the Calvert Bluff Formation. (After Kaiser *et al.*, 1980.)

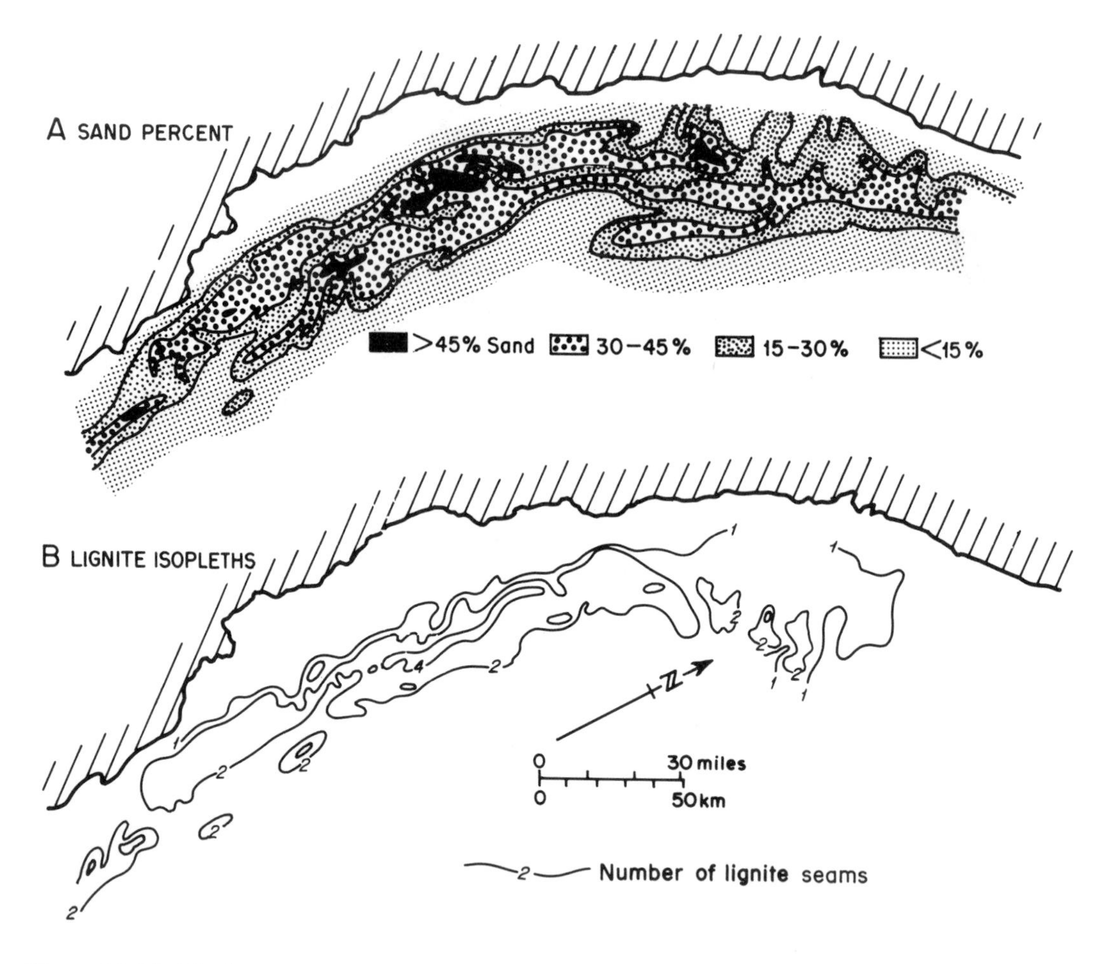

Figure 12-14. The lower part of the Jackson Group of Texas showing: (A) sand percent map; (B) lignite isopleths of characteristic barrier/strandplain pattern. (After Kaiser *et al.*, 1980.)

and trace-element content of these coals tends to be variable.

Swamps, marshes, tidal flats, and bays are sheltered from vigorous currents and are dominated by chemical and biological processes in a generally reduced setting conducive to organic preservation. Peat growth commences in narrow belts along the levees, gradually extending across the surfaces of adjacent bays as they shallow and fill. The main processes of shallowing involve fine-grained overbank flooding, accumulation of coarse-grained crevasse splays, and onshore transport of sediment by wave drift and tidal currents. Crevassing is the most effective process in expanding the subaerial extent of the lower delta plain. The shallow splay surfaces soon become vegetated and promote rapid expansion of the peat area (Fig. 12-17). Marsh surfaces effectively trap subsequent splay sediments and inhibit scouring by splay feeder channels, so that splays projected over the marsh become inactive as the floodwaters recede. These sandy splay lobes subside by compaction of the underlying spongy peat and clay, and soon become incorporated in the growing peat mat. The detrital splits that result from this process are particularly abundant in lower delta-plain coals because of the low levees (Fig. 12-18).

Mississippi Delta salt marsh (Fig. 12-19) receives substantial volumes of clastic sediment from the sea during storms, a process that builds the surface a meter or so higher than the fresher water marshes inland (Saxena, 1976). Consequently, the salt marsh is well drained and has a

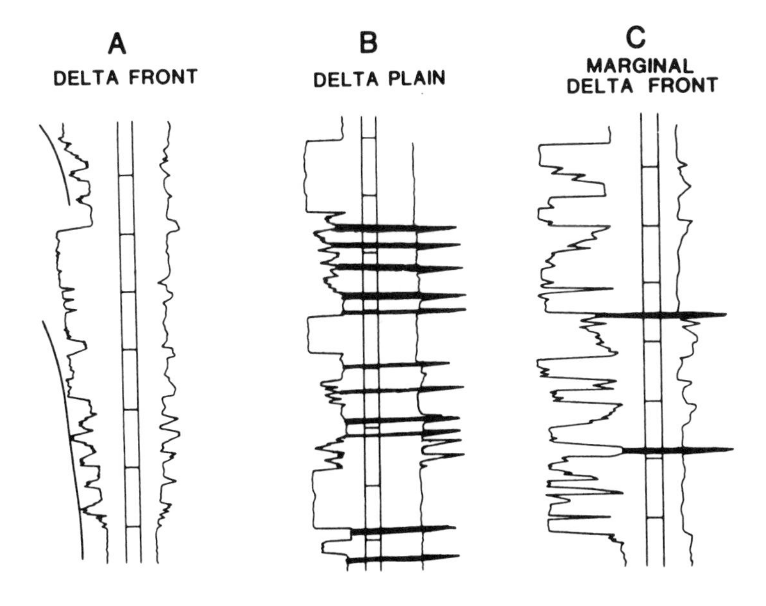

Figure 12-15. Typical log patterns showing: (A) upward coarsening progradational delta-front facies overlain by; (B) lignite-bearing delta-plain facies—the blocky patterns represent distributary channel sands; (C) the development of lignites in marginal delta-front facies. (After Fisher, 1969.)

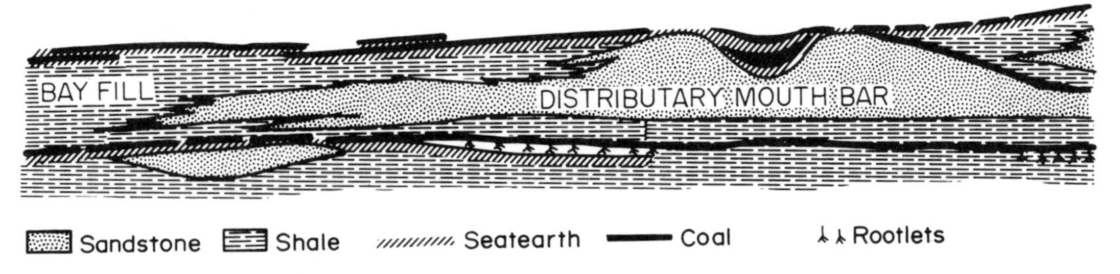

Figure 12-16. Cross section through a flat-based distributary mouth bar sandstone with the typical upward-tapering profile and grading laterally into upward-coarsening bay-fill. Coals are developed along the top of the bay-fill sequence and thicken dramatically across a notch representing an abandoned distributary channel. Carboniferous of the Ohio Valley. (Modified after Ferm and Cavaroc, 1969.)

firm substrate of laminated clay, silt, and sand. Disseminated organic matter is abundant, but peats are thin or absent. Nodules of iron oxide and carbonate tend to be abundant, and are represented in the rock record by root-penetrated, ferruginous pedogenic horizons; many of these are laterally continuous, seaward extensions of lower delta-plain coals.

The brackish marsh tract is normally dominated by a limited number of plant species. It is a poorly drained area of the Mississippi Delta traversed by a few tidal creeks. Root-disrupted and burrowed nodular clays are typical, and peats are subordinate.

Fresh-water marsh tends to be dense, with thick peat mats riddled with roots. Compactional subsidence produces small ponds, many of which contain floating vegetation. Underclays beneath these progressively thickening peat rafts consequently show little sign of root penetration. Fresh-water marsh peats are generally massive, apart from bedding related to clay bands produced by fires or overbank flooding. Banding of the coal results from compaction and coalification, and does not reflect original sedimentary layering (Renton and Cecil, 1979). Swamps with a significant proportion of woody vegetation develop initially on a firm levee substrate flanking the main distributary channels. In tropical areas, however, mangrove swamps can dominate the entire delta plain.

Evolution from open bay to marsh and swamp

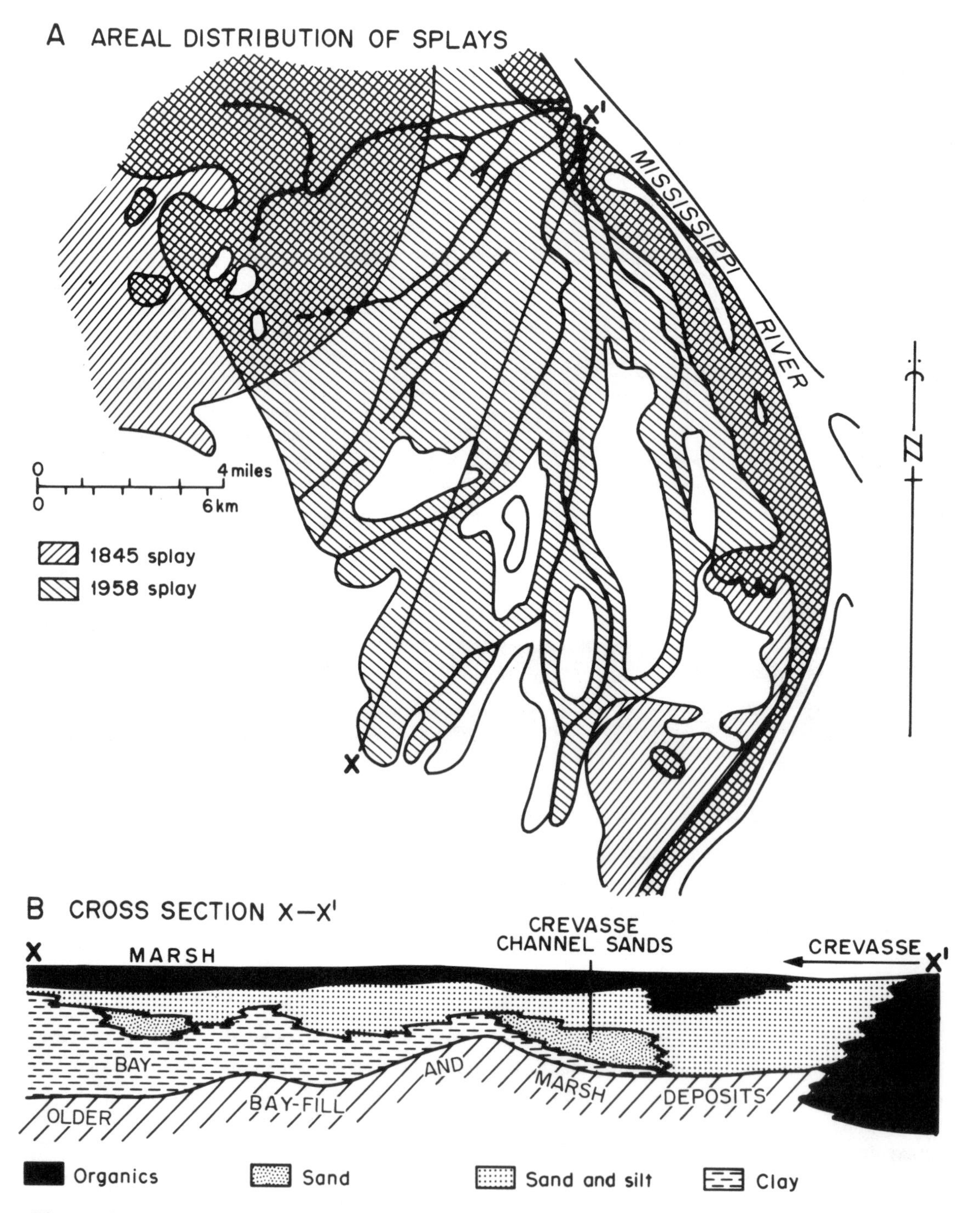

Figure 12-17. Infilling of a bay of the Mississippi lower-delta plain by two major crevasse splays. (A) Marshy surfaces of the 1845 and 1958 splays. (B) Typical proximal-to-distal change in coal-capped crevasse splay and bay sequence. Note the characteristic upward-coarsening bay-fill sequence except near the splay channel. (After Coleman and Gagliano, 1964.)

can be reversed after sediment cutoff and subsidence (Saxena, 1976). Marsh surfaces gradually become firmer and more elevated as organic and detrital layers are accreted. This permits swamp trees to become established. Corresponding changes are recorded in many ancient delta-plain sequences. Carbonaceous streaks, rootlet bands, and chemically precipitated ironstones mark the

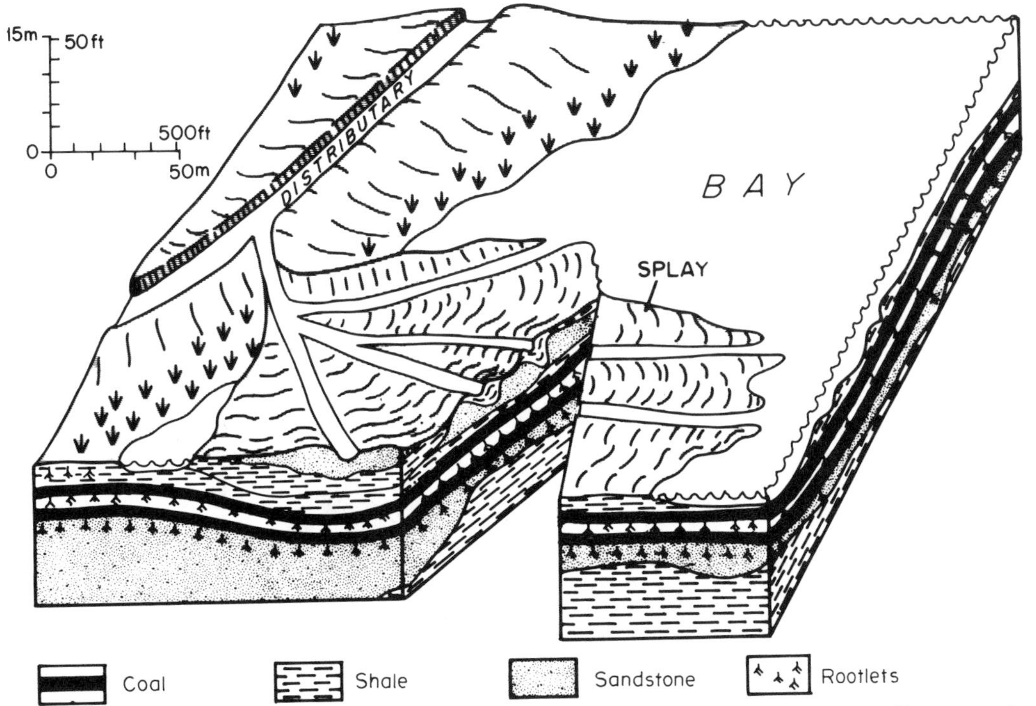

Figure 12-18. Reconstruction based on almost complete exposure of crevasse-splay sediments projecting laterally over bay shales and coals. Carboniferous of eastern Kentucky. (After Baganz *et al.*, 1975.)

initial stages of marsh formation at the tops of bay-fill sequences. The overlying coals of fresh-water marshes and swamps overlie a platform of overbank, splay, and bay-margin deposits.

Subsidence of the peat platform causes gradual reversion to open-bay conditions. Silts and clays accumulate over the peat surface prior to re-occupation of the site by crevasse-splay lobes. Peat growth is rapidly renewed on the platform provided by the splay. This process generates thin, coal-capped, bay-fill sequences so characteristic of the lower delta plain. Intraseam splits are produced by frequent overbank flooding and splaying across the narrow, low levees.

Distributary channels of the lower delta plain tend to be relatively straight and fixed in position, particularly if the banks contain a high proportion of clay. Rapid upstream diversion and abandonment of a distributary leads to filling with suspended clays or rafted organic material, represented by channel-filling impure coals (Fig. 12-16).

The Lower/Upper Delta-Plain Transition

The transition between the lower and upper delta plains is optimal for the development of thick, laterally extensive coals that are low in sulfur, with generally stable roof conditions (Horne *et al.*, 1978). Seams in this environment tend to be slightly elongated in the depositional strike direction. Although best documented in Carboniferous rocks of the Appalachian region (Baganz *et al.*, 1975; Horne and Ferm, 1976), this pattern is confirmed in Upper Cretaceous deposits of Utah (Figure 12-20), where the thickest coal is in a 6-mi (10-km) wide belt parallel to the paleoshoreline (Ryer *et al.*, 1980). This broad belt is crossed at right angles by the deposits of contemporaneous distributaries.

Detrital rocks of this important coal-bearing environment have distinctive geometries. The basal contacts of the major sandstone bodies are generally flat, although channelized in places, suggesting an origin as a sandy splay sheet or minor mouth bar. The lowermost coals of the distributary system onlap the margins of these splay sheets, but are thickest near the axes of interdistributary depressions. Narrowing of the sandstones above this level indicates progressive confinement of flow by levee growth as the distributary channel evolved. Lateral reduction in grain size away from the sandstone axes reflects a gradation from distributary channel-fill through

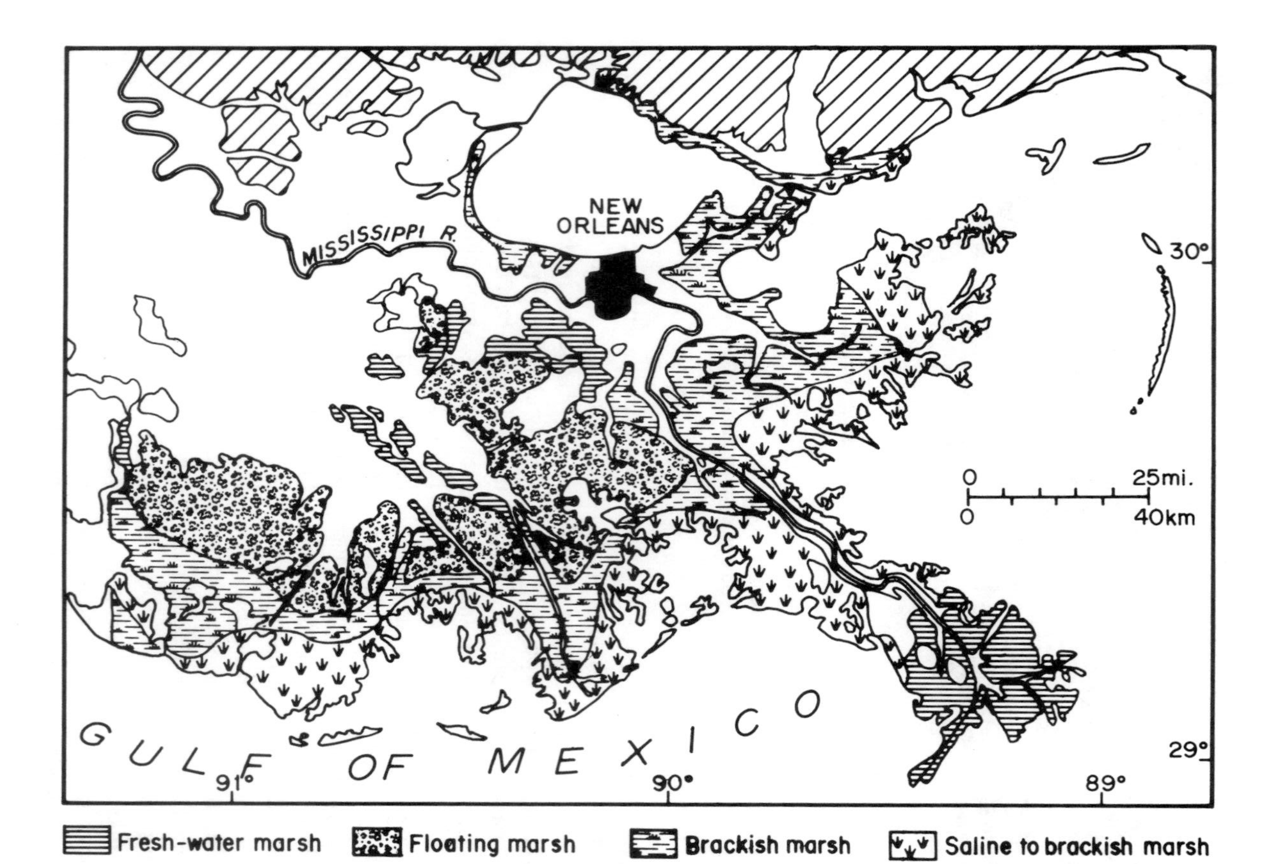

Figure 12-19. Marsh zonation of the Mississippi Delta plain. Note the roughly parallel succession from the coast inland except on the modern birdfoot delta where fresh-water marsh extends to the sea. (After Frazier and Osanik, 1969 and Saxena, 1976.)

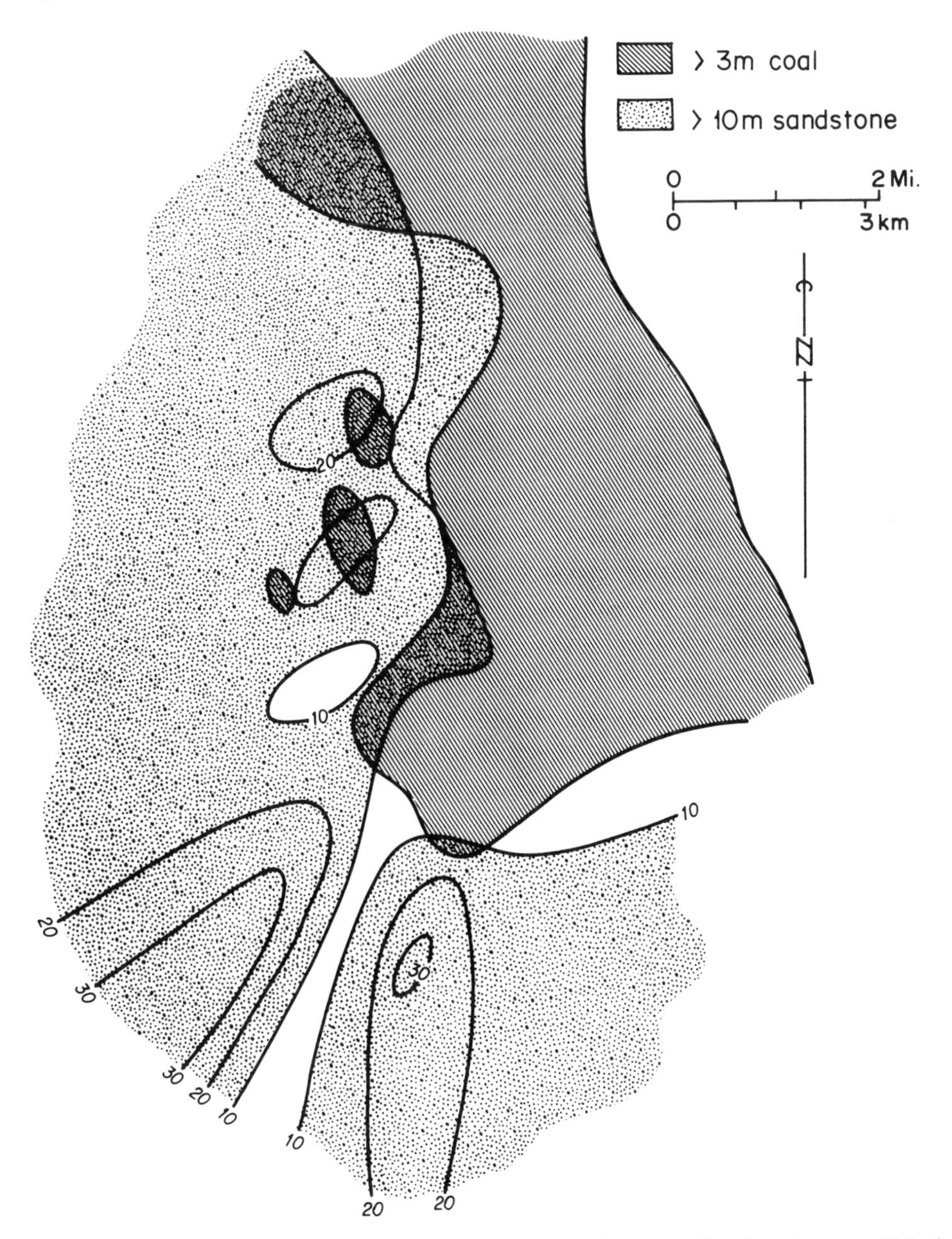

Figure 12-20. Sandstone isopachs and area of thick coal development showing shore-parallel elongation, Cretaceous Ferron Sandstone of the U.S. Western Interior. (Modified after Ryer *et al.*, 1980.)

levee and bay to backswamp peats. Continued vertical aggradation of channel-fill sands is limited by the tendency for discharge to be deflected by the steeper flow gradient toward the interdistributary lows. As a result, multiple splays characterize the upward-widening top of the distributary sandstone complex. These splays interrupt or inhibit peat accumulation, but their effects diminish rapidly away from the channel. Areas of thickest coal are therefore in the central backswamp areas, where they also tend to have the lowest ash content. Major coal seams thin toward the distributary channels, with increasing numbers of detrital splits.

Thicknesses of bay-fill sequences in the transition zone, typically 5–26 ft (1.5–8 m), are intermediate between the thick, shaly bay-fill sequence of the lower delta plain and the very thin interchannel sequences of the upper delta plain (Horne, 1979b). As in the lower delta plain, they

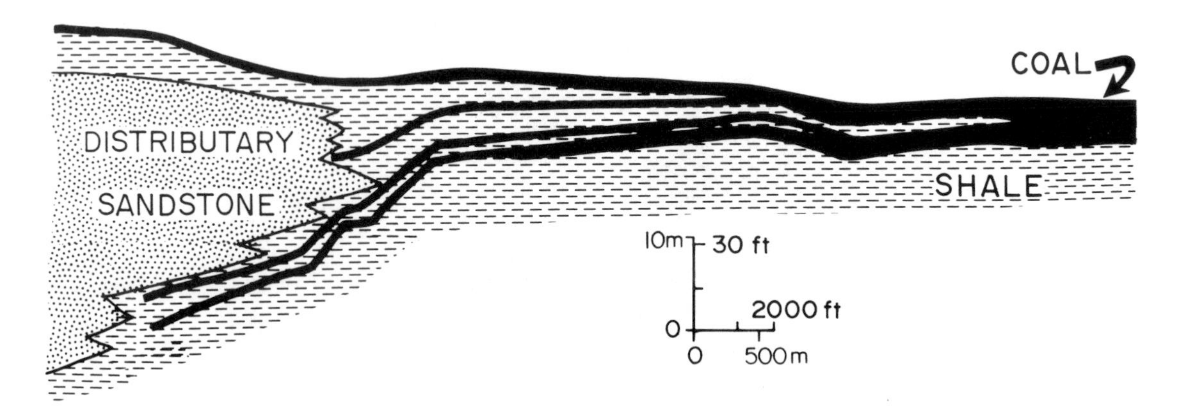

Figure 12-21. Deterioration of a coal seam by splitting in the direction of distributary channel sandstone; Carboniferous of West Virginia. (Modified after Howell and Ferm, 1980.)

commonly contain marine or brackish faunas in their fine-grained lower parts and are heavily burrowed in places.

Upper Delta-Plain Coals

The area of peat accumulation is less extensive in the upper delta plain, but the environment is relatively stable. As a result, the coals are thick and lenticular with abrupt lateral changes in seam characteristics. The coals are situated between scour-based channel-fill sandstones which splay out in their upper parts, interfingering with freshwater pond and backswamp facies. The coals are thickest along the interchannel axes and therefore show a pronounced downdip alignment. Splay tongues extending from the channels on either side cause extensive "fishtail" splitting of the seams. Typically, coal "benches" split off the base of the seams (Fig. 12-21). These are separated from the main seam by thin shales that coarsen channelward into poorly sorted sand of the proximal splay (Howell and Ferm, 1980). Viewed along the splay axis the splits are lenticular, with thin ash partings continuing as "tails" for considerable distances on either side of the sandy lens.

As a distributary channel becomes inactive, the flanking swamplands encroach over the subsiding clastic lens, depositing a layer of peat over the top. Some coals reflecting this process thicken dramatically over steep-sided depressions corresponding to the abandoned channel.

New channel courses created by avulsion tend to correspond with the area of maximum thickness of underlying peat. This zone is not only the lowest topographically, but it suffers the greatest compactional subsidence. The resulting lateral offsetting of successive distributary-sand trends (Fig. 12-22) produces a predictable pattern of clastic framework sands, fresh-water shales, seat earth, and coal (Ferm and Cavaroc, 1968).

Fluvial Coals

Coal is commonly at the top of upward-fining sequences deposited by mixed-load rivers, accumulating in backswamps adjacent to migrating channel tracts (Fig. 12-23). These fine-grained meanderbelt deposits may be laterally extensive or even blanket-like (Visher, 1972) as a result of persistent channel migration and point-bar accretion. On rapidly aggrading alluvial plains, channels sweeping across in one direction and then the other may produce an intricate zig-zag geometry in cross section (Britten, 1972). More characteristically, however, channels change position abruptly toward low-lying interfluves, resulting in dip-oriented sands of lenticular cross section enclosed by carbonaceous shales and coals (Ferm and Cavaroc, 1968). The sand bodies tend to be multistoried in contrast to the solitary channel-fill sequences of the upper delta plain. These thick, composite fluvial sands commonly show thicknesses in multiples of the 15–50 ft (5–15 m) most typical of individual channel-fill sequences, and may attain 200 ft (60 m) or more.

Peatswamps form on the inclined surface of the well-defined levees, and thicken toward the back-

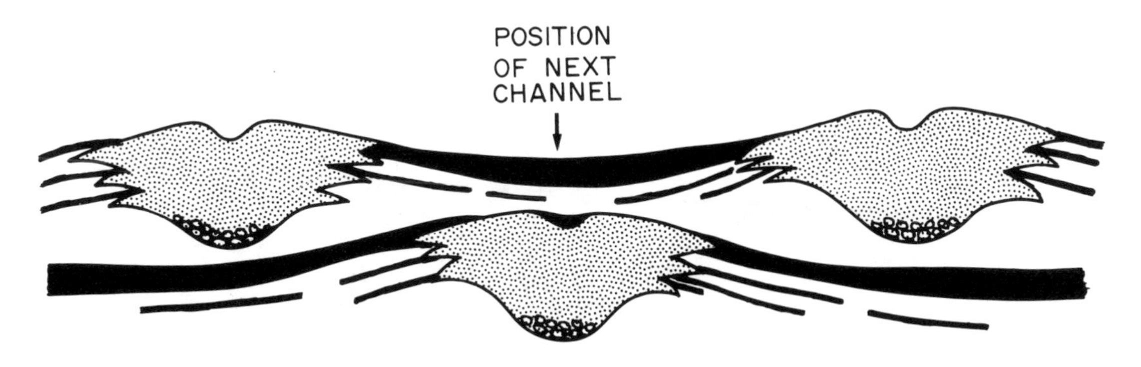

Figure 12-22. Pattern of lateral offsetting of successive channels.

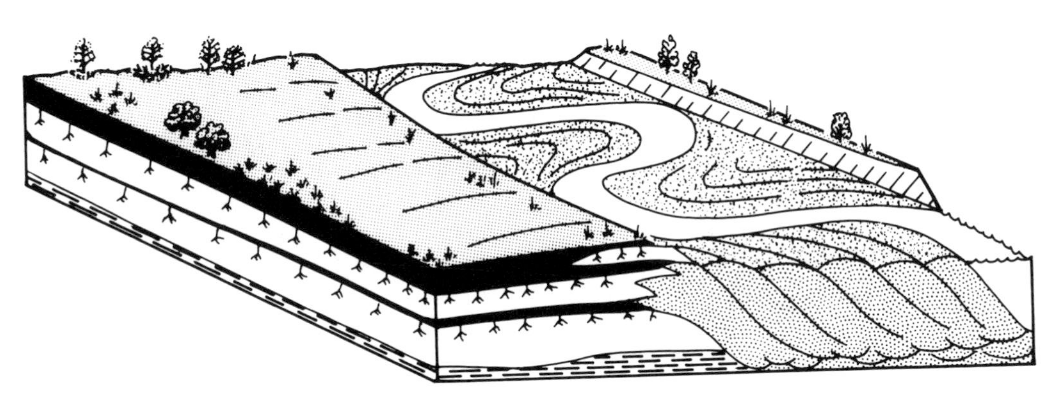

Figure 12-23. Pattern of peat accumulation in backswamps adjacent to meandering fluvial channels. (After Ferm and Horne, 1979.)

swamps separating the major fluvial axes (Fig. 12-23). Subsequent compaction further steepens the dip of the coals along the channel flanks, contributing to additional potential mining problems in an area already plagued by splits. The lateral gradation from point-bar sands through levee into backswamp is also represented vertically as compactional subsidence superposes levee silts and backswamp peats directly upon the channel-fill sands. These simple vertical cycles are common in coal basins, but may not contain the commercially most important alluvial coal seams. The thickest coals tend to be between major sandstone lenses, splitting and thinning as they ascend over the sandstones, but thickening over the channel notch. The Calvert Bluff lignites of the Eocene Wilcox Group of Texas are typical (Figs. 12-13B, 12-24). Coals overlie stacked, thick, upward-fining fluvial sequences, but commercial lignites are confined to broad, subsiding interchannel areas where they are associated with finer overbank deposits (Kaiser *et al.*, 1980). These stable backswamps were between widely separated meandering channels analogous to those of the Holocene Mississippi alluvial plain. Hardwood swamps of a tropical environment persisted landward between dendritic stream courses. The channels remained relatively fixed in position, permitting the gradual accumulation of hardwood peat, which accounts for the high proportion of woody material and low ash content of the lignite. As with other Gulf Coast lignites, the seams thicken down dip-oriented belts toward the fluvial-delta plain transition (Kaiser, 1974; Kaiser *et al.*, 1980).

Coals associated with Paleocene and Eocene fluvial systems of the Powder River Basin of Wyoming and Montana are commonly more persistent, with some seams up to 30 ft (10 m) thick extending along outcrop for distances of 12 mi (20 km) with only moderate merging and splitting. This lateral persistence has been attrib-

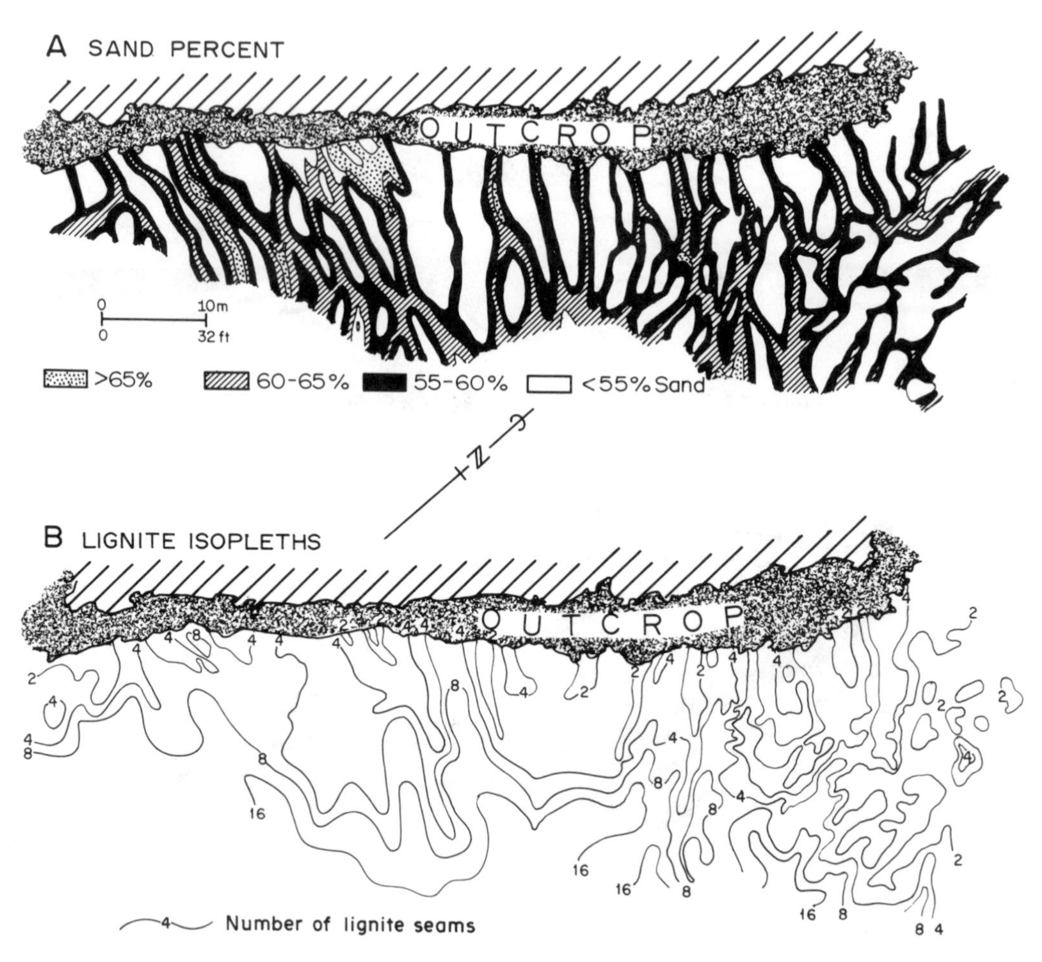

Figure 12-24. Calvert Bluff Formation (see Fig. 12-13B) of Texas: (A) sand-percent map showing well-defined dip alignment of channel-fill sand axes; (B) distribution of lignite, reflected in the number of seams, illustrating the relationship to channel sand framework, particularly in updip areas. (After Kaiser *et al.*, 1980.)

uted partially to fault-controlled subsidence of the intermontane basin (Obernyer, 1978). Another factor was the stability of channel positions (Beaumont, 1979), with the thickest, laterally most extensive coals arranged in belts peripheral to major fluvial channel axes, where ground-water discharge provided an elevated water table and abundant nutrients (Ethridge *et al.*, 1981). These peats extended over a platform of abandoned channel and splay deposits, attaining great continuity as depressions were filled and vegetated. The growing peat bogs assumed a domal form with a rising water table; these peat domes were later transformed by compaction into broad lenses (Flores, 1981). The elevated peat surface was immune to flooding, thus accounting for the low ash content of the coals. Major fluvial tracts were flanked by extensive fresh-water lakes, which Flores compares to the modern Atchafalaya Basin of the Mississippi complex. Lacustrine deposits of the Powder River Basin consist of carbonaceous shales, thin coals, and fresh-water limestones, with upward-coarsening cycles representing lacustrine delta progradation (Flores, 1981; Ethridge *et al.*, 1981).

Washouts are common in alluvial floodbasin coals, originating from rapid channel diversion because of the gradient advantage through the backswamps depressions. Sand washouts (Fig. 12-25) are produced by rare major floods which cause the river to straighten and deepen its course, resulting in a "low sinuosity" washout pattern (Donaldson, 1979). Broader, shallower washouts which do not cut entirely throught the seam are created by entrenchment at meander bends during less severe floods. Some relatively

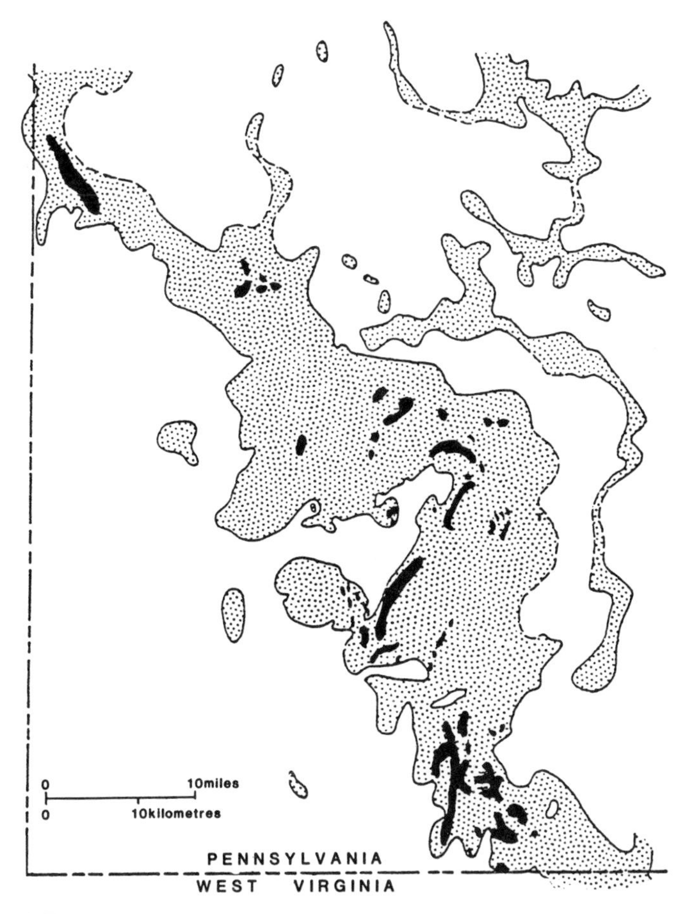

Figure 12-25. Linear distribution of the fluvial Pittsburgh Sandstone and the pattern of reported cutouts of the underlying Pittsburgh Seam. (After Donaldson, 1979.)

small stream courses initiated during flooding are unable to scour into the resistant peat mat, and the sands of these shortlived channels may later be incorporated in the peat as an elongate split (Cairncross, 1980).

Abandoned channel courses, such as meander loops subjected to neck cutoff, accumulate rafted and aquatic vegetation together with variable amounts of inorganic matter. Some abandoned channels are not filled entirely with coal, whereas others contain coal alternating with detrital sediments deposited during episodes of temporary reoccupation by the river.

Roof conditions are generally good, and chemical impurities seldom pose a problem in coals associated with mixed-load alluvial systems. Exceptional coal thicknesses of 65 ft (20 m) or more are not uncommon, but these coals tend to deteriorate laterally to subeconomic thickness over a relatively short distance.

Fluvial systems in which the banks are completely stabilized by vegetation may follow a single course or they may assume a multiple-channel anastomosing pattern. In either case, aggradation of channel-fill sediments concomitant with peat growth produces narrow, thick, dip-oriented sand bodies bounded by levee, splay, and peat facies (Smith and Smith, 1980). Comparable anastomosing stream systems within thick coal sequences have been identified in Tertiary strata of the Canadian Cordillera (Long, 1981) and in the Permian Karoo Basin (Le Blanc-Smith and Eriksson, 1979; Cairncross, 1980). The Canadian seams attain thickness of over 20 m (65 ft) but have a high ash content ascribed to frequent overbank flooding. Karoo examples show distinct high-ash trends paralleling the paleochannels (Fig. 12-26).

In some areas of very thick fluvial peat accumulation, the effects of compaction and

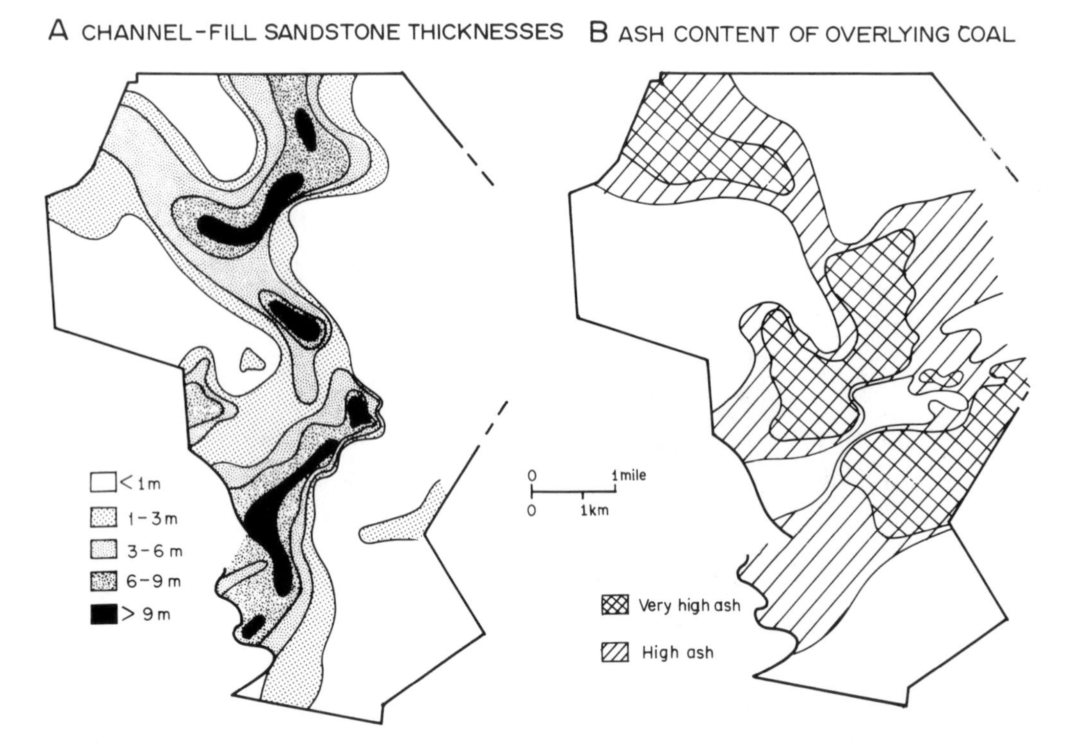

Figure 12-26. (A) Isopachs of a fluvial sandstone unit in the Van Dyks Drift Colliery area of the northern Karoo Basin, (B) Ash content of the overlying coal. (Modified after Cairncross, 1980.)

downwarping of the peat mat along the margins of the channel-fill sands cause the channel or associated splays to move laterally into the flanking depression (cf. levee-flank depressions associated with thick Mississippi River channel sands; Russell, 1936). This process of automigration is clearly reflected in some Australian coalfields (Mallett *et al.*, 1983). The seams are bent steeply under the flanks of the thick channel and splay lens which built sideways over the depressed peat surface (Fig. 12-27). Upright trees rooted in levee siltstones are preserved in cross-bedded channel-fill sandstones as much as 50 m (160 ft) thick, indicating that automigration was extremely rapid. Other spectacular examples of seam steepening, sandstone deformation, and the generation of large-scale, inclined pseudobedding by contemporaneous compaction of peat are provided by Britten *et al.* (1973).

In drainage networks transporting sediment of sand size and coarser, dense vegetation is probably a prerequisite for the development of coarse-grained meanderbelt systems (McGowen and Garner, 1970) in preference to the braided stream systems that prevail in the absence of an effective cover of vegetation (Schumm, 1972; Miall, 1978). Coal-bearing, coarse-grained meanderbelt systems in the Simsboro Sand of Texas (Fig. 12-13B) comprise thick multilateral sands along linear, dip-oriented trends. Dendritic tributary patterns are apparent at the upstream ends. As in the mixed-load Calvert Bluff Formation, lignite is concentrated in belts between the main sand axes (Kaiser *et al.*, 1980). Elsewhere the coals are interlayered with vertically homogeneous bed-load channel-fill sandstones. Coals in sandy fluvial systems may be more tabular than their mixed-load system counterparts, but because of ineffectual confinement by the very subdued levees they are commonly interrupted by sheetlike sandy splits. Irregular, thin coals follow abandoned channel courses and occupy small chutes in the sandy point-bar surfaces. Both roof and floor conditions are excellent in these sandstone-dominant successions, and coal quality is good, apart from locally very high ash content.

In linear or confined braided-stream systems, oxidizing conditions prevail, the water table fluctuates markedly, and pH tends to be high. Vegetable matter has a very low preservation potential as a result. This is true of both sandy braided streams and their more proximal gravelly

Figure 12-27. A Permian seam in Queensland, Australia, showing the effects of differential compaction of peat, clay, and the less compactable crevasse-splay lens in the middle foreground. This was in response to lateral migration, toward the left, of major fluvial channels. Moura Mine.

equivalents. An important exception is where well-integrated, highly transmissive aquifers allow continuous discharge of ground waters. These conditions favorable for peat accumulation are prevalent on wet alluvial fans, which include a large proportion of braided-stream facies. Because of the unique geometric relationships of these braided-stream facies, they are more appropriately discussed under the category of alluvial-fan coals below. Another factor militating against peat accumulation in a braided-stream environment is the frequent erosion of finer grained, organic-rich overbank sediments. It is conceivable, however, that the peaty topstratum could be preserved in upward-fining braided stream sequences of the type deposited in the Donjek River (Williams and Rust, 1969), which in some respects resemble the deposits of meandering channels (Miall, 1978). Even more conglomeratic coal-bearing channel-fill sequences resembling the Scott-type of Miall have been described by Long (1981). These fluvial conglomerates, 2–25 m (6–80 ft) thick, are followed by plane-bedded and ripple-laminated sandstone, laminated and root-disrupted mudstone, and thin coals. Preservation of peats in a braided-stream environment is more likely in a cold climate because of slower evaporation rates.

Many coal-bearing fluvial successions merge up the paleoslope into progressively coarser grained sediments with decreasing proportions of siltstones, mudstones, and coals. Overbank clastic sediments tend to be reddish, light gray, or green, in contrast to the darker colorations in the lower alluvial plain. There is commonly little trace of organic material in the most proximal alluvial deposits. Updip equivalents of coals are root-disturbed beds and other fossil soils.

Fluvio-Deltaic Progradational Succession

A typical idealized succession resulting from fluvio-deltaic progradation is illustrated in Figure 12-28. Although derived from Carboniferous rocks of the Appalachian region (Ferm *et al.*, 1971), this generalized pattern is observed, with some differences, in many other coal basins.

The basal limestones include shallow-water bioclastic, oolitic, and micritic deposits. The overlying shelf shales are gray, green, or reddish, with a typical open-marine fauna, and grade into shoreface facies. Although generally conformable, the overlying barrier sandstones are erosively based near tidal-inlet facies. Because of wave and tide reworking, the barrier deposits are lithologically more mature than associated deltaic sandstones. Back-barrier shales are dark gray with brackish-water faunas, and coarsen upward into washover and tidal channel sandstones overlain by thin, sulfurous coals.

The succeeding lower delta-plain deposits overlie upward-coarsening bay-fill or prodelta sequences 50–200 ft (15–60 m) thick. Distributary mouth bar sandstones are broad-based and generally conformable, becoming narrower and coarser grained upward. Distributary channels, which locally cut through the mouth bar into the underlying shales, show two distinctive types of fill: sandy active fill and fine-grained or coaly abandonment fill. Levee facies are thin, and widespread crevasse-splay sandstones are present near the tops of bay sequences; these are overlain by persistent lower delta-plain coals, some of which extend over the tops of adjacent channel and mouth-bar sandstones.

Sandstones of the upper delta plain thicken at the expense of finer bay-fill deposits, attaining 80 ft (25 m). They are typically scour-based and upward-fining; but in the transition from lower to upper delta plain the sandstones show a characteristic "hourglass" shape, with a sheetlike base and top, narrowing midway. Levees of the upper delta plain are typically thick and steeply dipping, and merge into upward-coarsening pond or lake sequences with thick coals over the top. Gradation into the alluvial plain is indicated by evidence of lateral channel migration, thinning of overbank deposits, local thickening of lenticular coals, and pinchout of other seams into root-penetrated, pedogenic zones.

An unexplained feature of Figure 12-28 is the presence of barrier quartzarenites stratigraphically below the delta-front sandstones, which according to the classification in Chapter 5, should be at approximately the same stratigraphic level, or reversed. According to A.J. Tankard (personal communication, 1982), the quartzose barrier sandstones of some of the Appalachian coal basins originated on shoal-water tectonic highs well seaward of the deltas. As such, they were not classic barriers directly related to deltaic systems. Despite problems in interpretation, the vertical succession illustrated is sufficiently widespread to represent a norm for comparison with other coal-bearing successions.

In many coal basins, for example Permo-Carboniferous Gondwana basins, the basal carbonates are thin or absent, and the barrier facies may not be present. Vertical transitions illustrated in Figure 12-28 may be less regular or predictable because of sporadic tectonism or eustatic fluctuations. Further departures from the general model result from differences in delta morphology and fluvial style. Some coal-bearing deposits comparable in geometry to upper delta-plain systems originated in an entirely terrestrial, fluvially dominated environment with broad floodplain lakes (Gersib and McCabe, 1981), but can be distinguished on the basis of their structural, paleogeographic, and stratigraphic setting.

Alluvial-Fan Coals

Coals are associated with alluvial-fan and fan-delta systems in a variety of intermontane and foredeep plain settings. Pull-apart basins and other downfaulted segments of strike-slip fault complexes are particularly favorable locations for coal-bearing alluvial-fan systems. These have been documented in northern Spain (Heward, 1978a and b), the Canadian Cordillera (Long, 1981), Indonesia (Koesmanidata, 1978), and eastern Australia (Packham, 1969). Coal-bearing alluvial-fan systems are also present in rift basins, for example, in southeastern Africa (Du Toit, 1954, p. 505) and the Latrobe Valley of Victoria, Australia (James and Evans, 1971).

Basins associated with both strike-slip and rift displacement tend to undergo sporadic tectonic movement during sedimentation, with complicated patterns of fan diversion and local base-level development. This leads to complex juxtaposition of coals, conglomerates, sandstones, and shales. Contemporaneous downwarping in the absence of detrital influx may result in the formation of extraordinarily thick coals, in some cases exceeding 100 m (330 ft) (Vetter, 1980). In

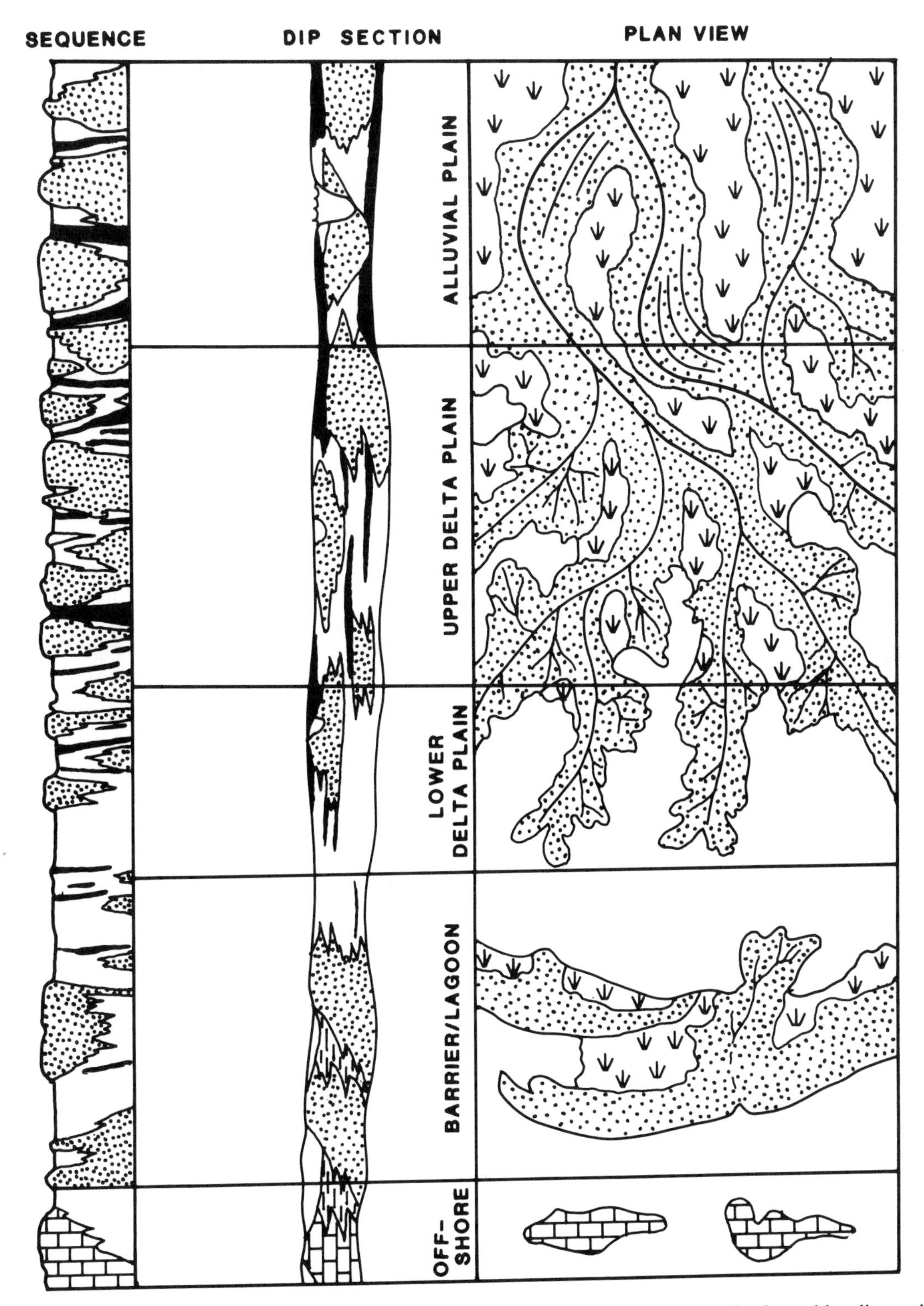

Figure 12-28. Idealized vertical sequence characteristic of several Appalachian coal basins, with a dip section showing changes from the alluvial plain through the delta plain to the barrier fringe, and its interpretation according to the delta model for that area. (Modified after Ferm *et al.*, 1971.)

addition, periodically renewed subsidence of these fault-bounded depressions can result in stacking of coal-bearing sequences to substantial thicknesses (Long, 1981).

Carboniferous coalfields of northern Spain are present in a number of semi-isolated, east–west striking synclinal basins (Heward, 1978a and b). The basal deposits consist largely of coarse fanhead and canyon-fill conglomerates occupying paleovalleys up to 300 m (1000 ft) deep. Deposition varied locally from angular scree and colluvium to matrix-supported debris-flow conglomerates arranged in irregular upward-fining sequences. The finer-grained coal-bearing sediments overlie diagenetically reddened conglomerates representing a stable episode of tropical weathering. Accumulation of peats in the sediments directly above these proximal-fan deposits implies subsidence of the conglomerates below the water table. Coals are also present in sequences of siltstones and fine-grained, rippled sandstones which accumulated in depressions between mid-fan lobes of conglomerate; the peatswamps expanded markedly during prolonged episodes of nondeposition resulting from fan diversion and abandonment. Distal fan sandstones are interlayered with coals and root-disturbed beds. Sheet-floods transported sand into a predominantly muddy lacustrine environment, with preservation of numerous trees in upright position and a fresh-water fauna. Some of the thickest coals attaining 20 m (65 ft) and more record peatswamp development and slow subsidence along the distal fan-delta lake margin (Heward, 1978a and b).

The other important settings for alluvial-fan coals are the alluvial margin of orogenic foredeeps, as represented by the Sydney and southern Karoo Basins of Gondwana, and paraglacial outwash plains as in the northern Karoo Basin. Both environments are conducive to the formation of wet alluvial fans, with the preservation of organic material.

Coal-bearing Upper Permian alluvial-fan systems of the Sydney Basin were deposited in an arcuate foredeep bounded to the north and east by tectonically active crystalline highlands. Fifteen or more seams are present in the 400 m (1300 ft) thick succession. Large gravel fans, probably resembling those of southeast Alaska, accumulated in a high-latitude situation. These merged downslope into sandy fan deposits dominated by braided-stream processes. Coastal marshes formed along the muddy distal fringes of the fans as they prograded into low-energy lacustrine basins (Fig. 12-29). These marshes resembled the lake-margin fan-delta coals in northern Spain and the Canadian Cordillera but were generally more extensive and thinner. Most seams are 1–3 m (3–10 ft) thick, with a maximum of 9 m (30 ft), and cover areas of up to 400 km^2 (160 mi^2) (Davis, 1974). Peat accumulated initially on the muddy, compactionally subsiding distal fan and in interlobe depressions, but encroached up the slope of inactive fans so that coals are preserved directly upon coarse con-

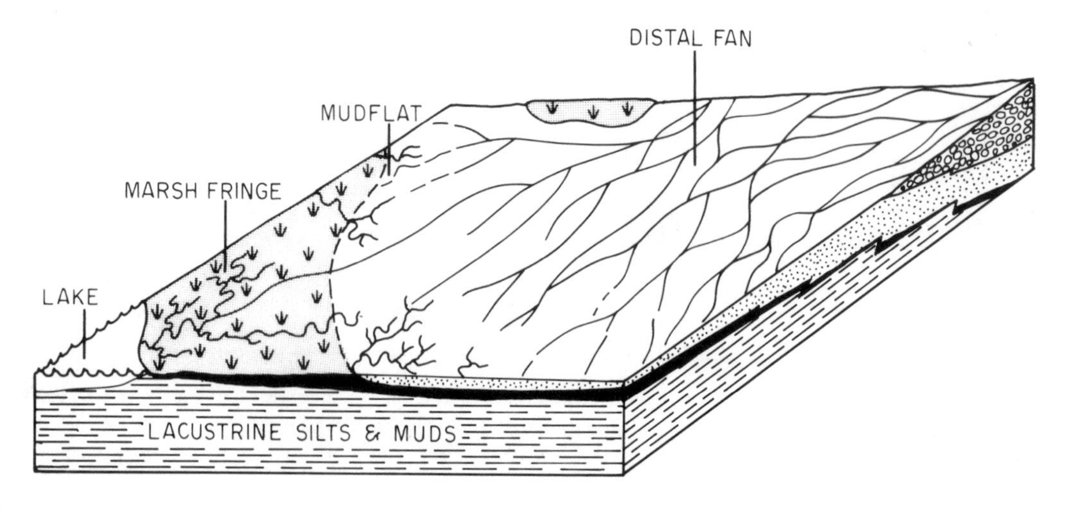

Figure 12-29. Schematic representation of peat development along the distal edges of lacustrine fan deltas. Progradation results in a vertical sequence of lacustrine silts and muds overlain by coal, sandstone, and conglomerate.

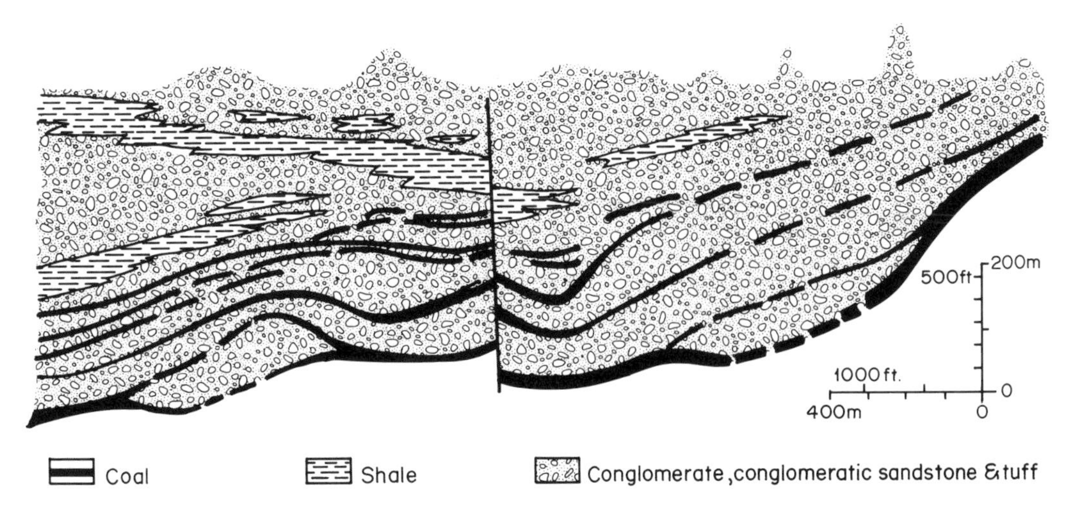

Figure 12-30. Dip section through alluvial fan systems of the northern Sydney Basin, Australia, showing the relationship of thick coal seams to conglomerates. (Modified after Packham, 1969.)

glomerates (Fig. 12-30). Many of these seams are also succeeded by conglomerates. These relationships suggest that ground-water discharge was instrumental in promoting peat accumulation and preservation (Fig. 12-31). Ground-water recharge by heavy rainfall in the adjacent highlands was probably augmented by snow melt; the gradient and permeability of the fan conglomerates would have permitted rapid ground-water flow. Reintroduction of gravels above the peats followed headwater flow capture, diversion, and reoccupation of the area by active fans. The upper parts of the seams are only slightly eroded, and most of the updip reduction in seam thickness is attributed to slower subsidence and a gradually lower water table up the paleoslope. Upward decrease in the vitrite-plus-clarite content of the seams, manifest in duller coals, was ascribed by Smyth (1970) to a progressively drier peat surface as it was raised by organic accumulation. This caused the stagnant swamp-forest vegetation to gradually give way to more aerobic, open

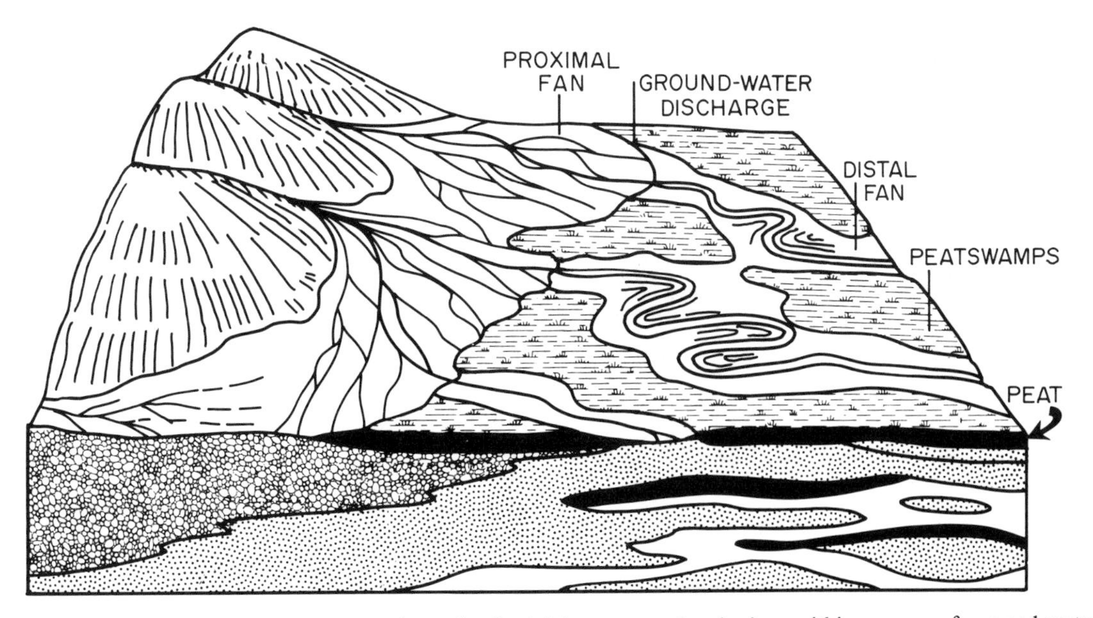

Figure 12-31. Idealized reconstruction of alluvial-fan peats developing within zones of ground-water discharge.

vegetation. Alternatively the duller coals may have originated on topographically higher or more proximal gravels. The alluvial-fan seams in general contain fewer bright bands than their downdip fluvial facies equivalents.

Basal coals of Permo-Carboniferous age in the northern Karoo Basin overlie the deposits of fluvioglacial outwash fans produced by the debris from the waning Gondwana icesheets (Le Blanc-Smith and Eriksson, 1979). Valleys eroded during the preceding stage of glacial advance were occupied by lakes in which density underflows deposited graded sandstones; these beds are preserved within finer, varved lake-floor deposits containing dropstone erratics. The fan-delta foresets are 5 m (16 ft) high and were supplied by braided streams of the broad, gravelly fan. Peats formed from shallowly rooted Arctic-type vegetation covering the fan-delta plain. Three or more prominent and areally extensive coal seams resulted, the lowermost of which in places onlaps tillite ridges (Fig. 12-32). These coals display alternating bright and dull bands attributed to fluctuations in ground-water level, which may have been seasonally controlled.

In addition, some large fan-delta foresets directly overlie major coal seams, for example in the Sydney Basin, a relationship that suggests accumulation of hydroponic vegetable matter (Conaghan, 1982). According to this theory, some Gondwana seams originated at water depths of 10–50 m (33–160 ft), as indicated by the magnitude of the overlying fan-delta foresets. Waterlogged organic material is visualized as constantly settling to the bottom from a hydroponic blanket that covered much of the lake surface. Contamination by clastic sediment was minimized by the shielding effect of the floating vegetation mat, and by homopycnal mixing of inflowing and lake waters which caused rapid dumping of the entire sediment load in the immediate vicinity of the delta front. The resulting coals are consequently low in ash and rich in vitrinite. Conaghan attributes the alternation of dull and bright plies in some seams to cyclic disruption in density stratification of the lake waters. Foundering of abandoned fan-delta lobes caused the hydroponic blanket to encroach on the delta platform and to coalesce with fan-delta peats.

Paleoenvironmental Control Over Mineral Matter in Coal

Mineral matter in coal comprises elements other than organic carbon, hydrogen, nitrogen, oxygen, and sulfur. It may consist of detrital grains such as quartz that are washed or blown into the peat-

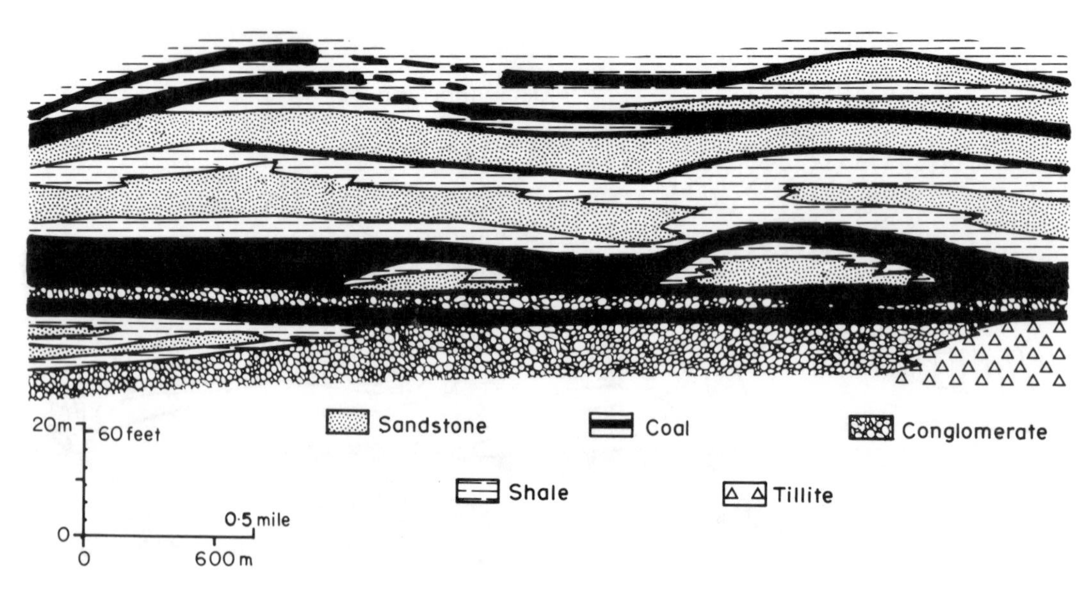

Figure 12-32. Basal Permo-Carboniferous coals of the northern Karoo Basin resting on the tillite and proglacial alluvial fan deposits. The overlying coals originated in the lobate fluvial-dominated delta environment. (After Cairncross, 1979.)

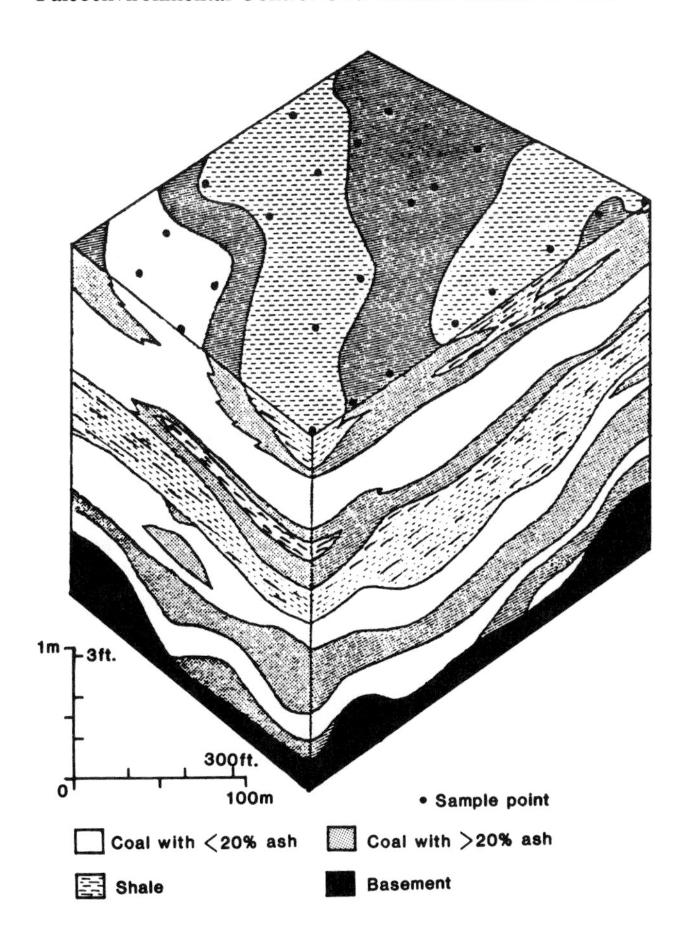

Figure 12-33. Alternation of high-ash and low-ash coal layers in mined portions of the Waynesburg Seam of West Virginia and Pennsylvania. (After Donaldson *et al.*, 1979.)

forming environment; it may be introduced by ionic exchange from interstitial waters; it may comprise elements incorporated in the vegetal precursors and released during peatification; or it may be chemically precipitated (Cecil *et al.*, 1979b). According to these authors, the pH during peat formation is critical in determining ash and sulfur content. In highly acidic environments mineral matter in peat may be removed by leaching. Peat degradation associated with less acidic conditions concentrates mineral matter in the peat. Fluctuating conditions may bring about a more or less regular alternation of low-ash and high-ash coal (Fig. 12-33). Dissolved calcium carbonate promotes sulfur reduction and organic-decomposition by bacteria, so that iron is fixed in sulfide or carbonate phases.

Although most original inorganic components of plant tissue are drastically depleted during peat formation, clays are deposited by the transformation of amorphous aluminosilicates of plant origin. Kaolinite forms in the more acid swamps, whereas mixed-layer clays dominate in less acidic back-barrier and deltaic swamps (Renton *et al.*, 1979). The montmorillonite interlayers of mixed-layer clays convert to illite upon burial, and the ions released thereby contribute to the formation of carbonate minerals during early coalification. The low-temperature ash content of woody tissue in modern swamps is around 2.4 percent on a dry weight basis (Renton and Cecil, 1979).

Clastic sediment partings and splits increase in thickness and abundance toward major drainage courses and barriers. Clay-rich fluvial splays tend to affect only the margins of the swamp. This is because of the trapping efficiency of dense swamp vegetation, and rapid flocculation of clays in brackish swamps. These argillaceous splays become compacted and preserved in coal seams as thin, laterally restricted splits. Fires affect a far larger area and leave a layer of black fusinite grading up into clayey peat. Increase in clay content is attributed to concentration of the incombustible organic content of plants by fires, and the tendency of overbank floodwaters to occupy these burned depressions subsequently.

The effects of swamp fires are preserved in coal seams as carbonaceous shale partings with a thin basal inertinite layer.

The iron disulfides marcasite and pyrite are present in coal as euhedral crystals, coarse-grained replacements of plant material, joint-filling material, and minute disseminated spheres and framboids. Framboidal pyrite poses the greatest environmental problems. It is finely dispersed and decomposes rapidly on subaerial exposure, so that in the absence of associated carbonates it can produce acid drainage problems (Horne *et al.*, 1978).

Framboidal pyrite is thought to be produced by the action of sulfur-reducing bacteria in marine or brackish water where anaerobic conditions and high pH promote sulfur fixation (Cecil *et al.*, 1979b). Some framboidal pyrite forms directly during the early stages of plant degradation (Renton *et al.*, 1979). Studies on modern peats by Cohen (1973), and on Carboniferous coals by Mansfield and Spackman (1965), confirm the association between high sulfur content and marine influence. The environment succeeding peat deposition is most important in determining sulfur content (Williams and Keith, 1963). Thus modern peats with abundant framboidal pyrite are forming in marshes undergoing marine transgression. A shield of overbank or washover sediments may protect the peat from bacterial effects, thus accounting for the acceptable sulfur content of some seams in a generally transgressive environment. Less acid conditions favoring pyrite formation may also arise without marine incursion, simply as a result of evaporation or influx of carbonate-rich waters (Renton and Cecil, 1979). Some pyritic coals originated in flooded depressions where abundant ferric iron was available (Reidenouer *et al.*, 1967); this was subsequently reduced to ferrous iron in the substrate.

Dramatic evidence of the effect of marine incursions on coal composition is provided by Karoo coals of South Africa. These seams are noted for their low sulfur content, the exceptions being lower delta-plain coals which are directly overlain by persistent beds of marine-reworked, glauconitic sediments (Cadle and Hobday, 1977). In the Permian of eastern Australia, widespread marine sandstone sheets transgress a variety of coeval fluvial and deltaic facies (Britten, 1972). These transgressions, which affected several basins, were probably a response to eustatic rise in sea level. The coals beneath the marine units are characterized by an anomalously high sulfur content. Trace-element content shows a similar increase where peats were exposed to marine influence. Boron, titanium, vanadium, and zinc, for example, affect industrial use. Some sulfides and other minerals are introduced by solutions or form by alteration during the coalification process (Stach *et al.*, 1975).

Paleoenvironmental Control Over Roof Conditions in Mines

Roof conditions in underground coal mines are determined by a number of factors, including rock type, degree of lithification, bedding characteristics, relationships among adjacent lithofacies, and a variety of structures resulting from biogenic activities, penecontemporaneous deformation, differential compaction, and later tectonism. In addition, the stresses at depth and the mining techniques employed are important in influencing roof stability.

According to Horne *et al.* (1978) the best roof conditions are encountered in upper delta-plain and alluvial-plain sequences, where the upward-fining arrangement of rock types superposes resistant, massive sandstone on the coal seams. Local roof falls may be promoted by channel-lag concentrations of clay pebbles and coal spar or compressed, coalified driftwood. The most dangerous conditions are associated with rotational slump blocks which form along the cutbank margins of paleochannels.

Back-barrier sandstones originating as washover and tidal-channel deposits are generally quartzose and resistant, providing excellent roof conditions except where they are closely jointed. Conditions of roof stability caused by jointing are also encountered in the fractured homogeneous black mudstones that accumulate on delta plains subject to gradual flooding during delta decay.

In the horizontally interbedded sandstone and siltstone or mudstone of many lower delta-plain deposits, roof quality is largely dependent upon bedding thickness (Horne *et al.*, 1978). Beds averaging less than 2 ft (60 cm) require bolting. Optimum conditions are provided by beds between 2–10 ft (60–300 cm), but thicker beds may be dangerous because they commonly exhibit compactional deformation which promotes roof failure.

Widespread upward-coarsening sequences of lagoonal and lower delta-plain origin overlie coal seams directly in many places. Compactional rolls and associated microfaults provide an irregular, unstable roof (Krausse *et al.*, 1979). Intense bioturbation is another factor in reducing roof strength substantially, and in some mines even bolting proves inadequate. Concretions up to several feet in diameter pose an additional threat to miners (Damberger *et al.*, 1980).

Seat earths are inherently weak, so that where they are present directly above a coal in a multiseam succession they generally have to be removed. Thin overlying or rider coals, too, may cause roof failure. Coalified upright tree trunks or "kettles" are common in roofs of seams, and frequently collapse into the mine without warning.

Geometry and Evolution of Coal Basins

Different categories of coal basins may be distinguished by a combination of features such as the tectonic setting, geometry, and mode of infill. Plate-tectonic theory recognizes a great variety of basin types, but as far as coal basins are concerned, a few general categories stand out (Fig. 12-34). These coal basin types are by no means of equal importance, with the bulk of the world's coal resources contained in the first two types.

Asymmetric Foredeep Basins

Some of the most important coal-bearing successions are located in elongate basins filled transversely by vigorous sediment influx from a nearby tectonically active source. The active basin margin is subject to marked downwarping, producing a stacking of sedimentary facies (Fig. 12-35). Bordering an orogenically active terrane, commonly with low-angle thrusts, coals originate on the expanding alluvial and deltaic foredeep plain. The shallower, more stable cratonic margin receives limited volumes of sediment from a subdued hinterland, and later in larger amount from the opposite side. As the depocenter migrates across the basin, sediment loading promotes crustal downwarping accompanied by faulting; some are true growth faults, whereas other faults extend into the basement.

The Pocahontas and Warrior Basins of the Appalachian Plateau (Ferm, 1974), the Sydney Basin of Australia (Conolly and Ferm, 1971), northern England (Johnson, 1960), and the Rocky Mountain coal-bearing molasse basins all show an asymmetric cross-section geometry in which the main coal seams are present in a major fluvio-deltaic wedge. This is generally built out from the steeper basin flank, but in some basins the directions are reversed. Marine carbonates, muds, and quartzose sands were deposited on the shallow, stable basin platform. Progradation yields an upward-coarsening succession of the type illustrated in Figure 12-28, grading from marine to continental facies.

The coal-bearing succession commences at the level of the quartzarenites, where peats accumulated in a back-barrier environment (Fig. 12-35). In the absence of barrier-island facies, the lowermost coals are related to a delta-plain environment. These basal coals tend to be thin and sulfurous. Coals in the offlapping deltaic and fluvial wedge above become thicker upward at the expense of lateral continuity.

Fault-Bounded Basins

Important coal seams are present in some rift basins, block-faulted grabens and halfgrabens, and pull-apart basins associated with strike-slip faults (Fig. 12-34B). The nature of the sedimentary fill is largely a response to direct tectonic controls. Rock geometries and coal seam distribution may be difficult to predict as a result. Many of these basins are located over zones of persistent crustal weakness and are intermittently active over a long period of time. In Africa, for example, some rift basins have been sporadically reactivated since their Precambrian inception (Green *et al.*, 1980). Coal-bearing strata are present in the upper parts of several reactivated basins. Pull-apart basins develop along sinuous portions of major strike-slip faults such as those of the California borderland (Crowell, 1974). Complex block-faulting provides a local sediment source. Differentially subsiding grabens may be arranged in subparallel belts.

In many fault-bounded basins, whether related to strike-slip fault systems or diverging plates, a vigorous phase of vertical tectonism, commonly accompanied by volcanism, causes sporadic influx of talus and debris flows followed by growth

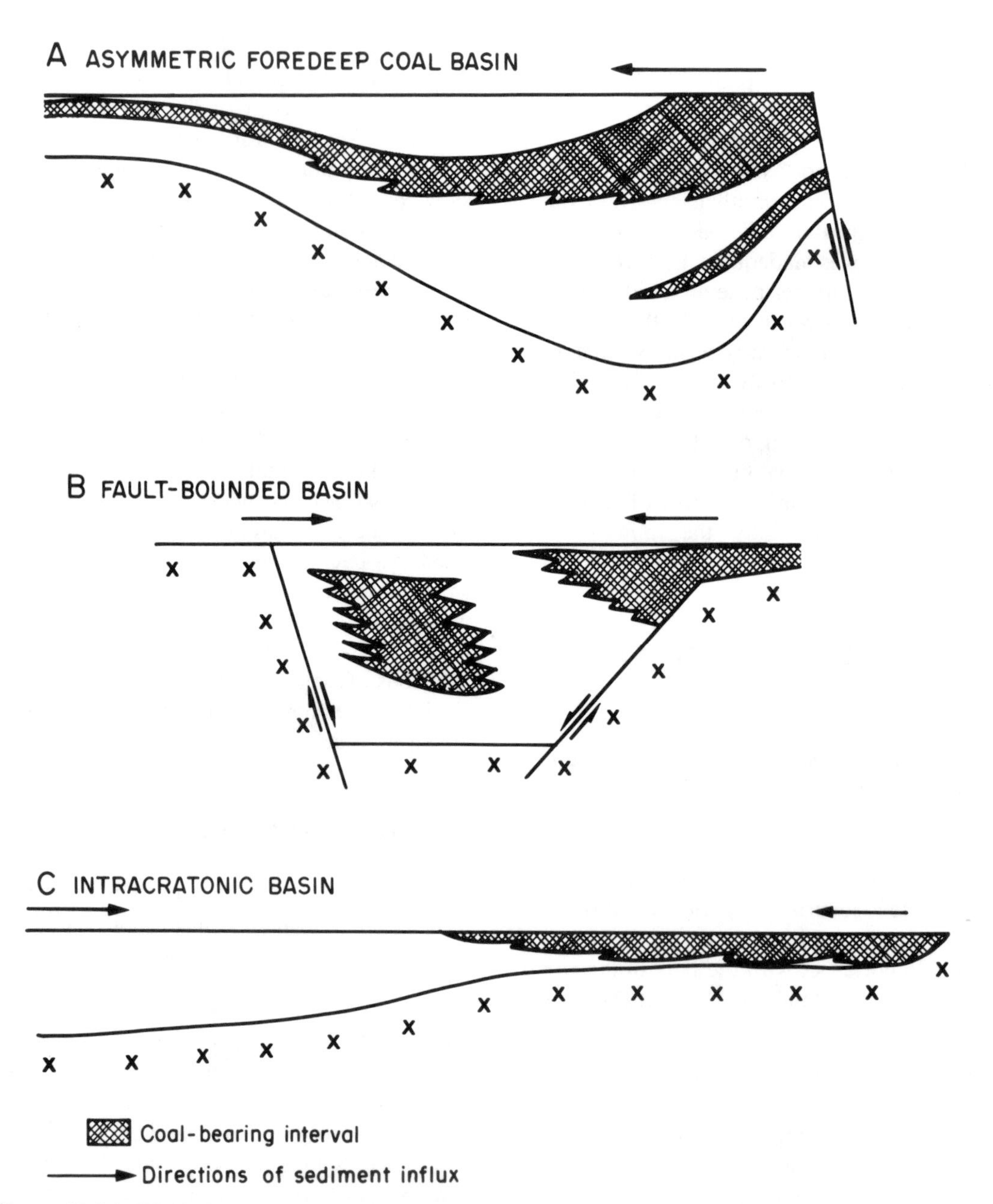

Figure 12-34. Highly idealized structural configurations of some coal basin types. (A) Asymmetric foredeep coal basins as exemplified by the Sydney Basin (Australia), Warrior, and Pocahontas Basins. (B) Fault-bounded basins such as those associated with rifts (Powder River Basin) and pull-aparts (Cantabrian basins of northern Spain). (C) Intracratonic basins such as the northern Karoo and Cooper Basins.

of alluvial fans. Coarse canyon-fill and proximal-fan sediments are onlapped by progressively more sandy mid-fan and distal-fan facies (Heward, 1978a). In humid climates, peatswamps form in interfan depressions, on surfaces of inactive fans, and along the lacustrine margins of fan deltas. Individual seams may transgress a variety of facies to an extent determined largely by the degree of ground-water discharge. Subsequent sedimentary cycles may be upward-coarsening, upward-fining, or poorly ordered, depending on factors of tectonism, fan piracy, and relief lowering. In many basins, mixed-load fluvial systems tend to gradationally overlie the alluvial-fan systems. Some of these fluvial systems include major coal seams. Sporadic tectonism

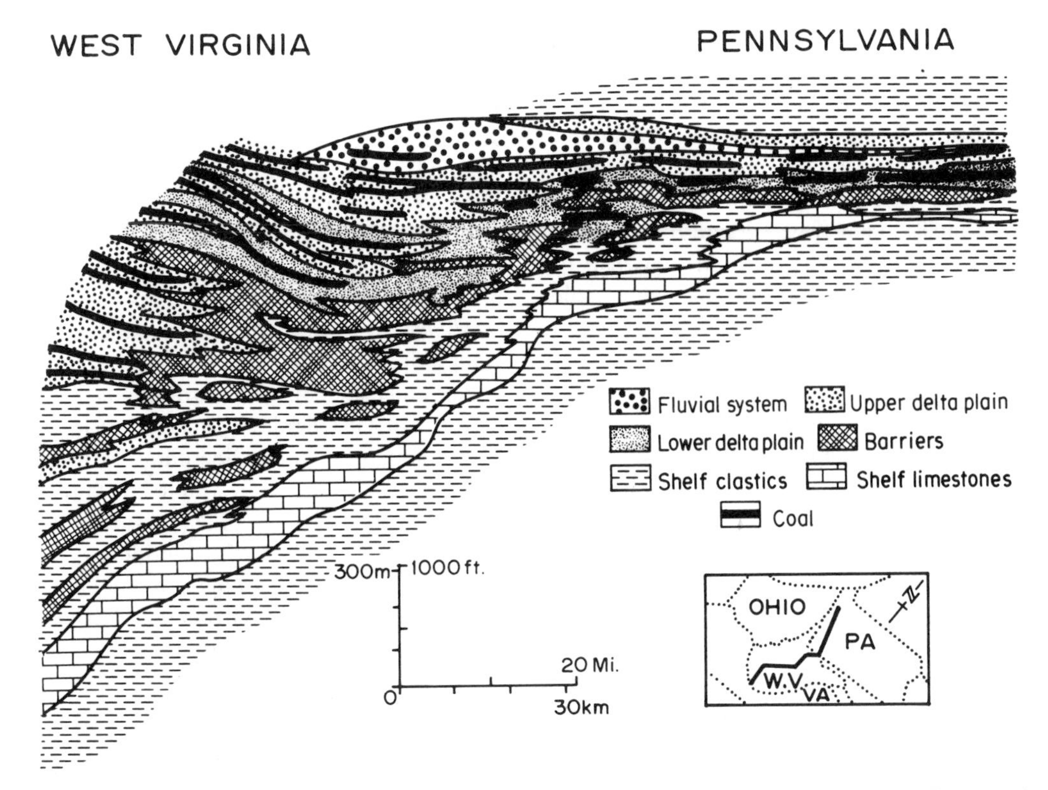

Figure 12-35. Generalized stratigraphic section through the Pocahontas and Dunkard Basins of the Appalachian Plateau. Sediment influx was from the south. (Modified after Ferm and Cavaroc, 1969.) Similar patterns are present in the Warrior Basin to the south.

generates renewed phases of alluvial-fan sedimentation, in some cases grading laterally through alluvial floodplain to elongate rift-lake environments.

Less active fault-bounded basins, including some large aulacogens, contain more regular, vertically repetitive sequences of prodelta or bay-fill overlain by delta-front, and coal-bearing delta-plain and alluvial-plain facies (Whateley, 1980; Green *et al.*, 1980). The uppermost sequences overlap the faulted basin margins onto the bounding craton.

Anorogenic Intracratonic Basins

Broad, intracratonic basins characterized by mild epeirogenic subsidence over vast areas may give a misleading impression of their coal endowment, which is generally restricted to a relatively thin stratigraphic interval. The origin of such basins remains speculative, with crustal attenuation, thermal tumescence, or passage over an irregular asthenosphere (Dickinson, 1974) providing possible mechanisms. Some intracratonic coal basins, such as the Cooper of Australia, were entirely fresh water; the Karoo Basin was briefly connected to a major ocean basin; others such as the Chesterian Illinois Basin and Cretaceous Mississippi Embayment opened seaward at one end for most of their history. The two latter examples exemplify mild downwarp of symmetrical troughs (Potter and Pettijohn, 1977, pp. 235–240), with the thickness of the sedimentary fill increasing down the basin axis but nowhere exceeding 1600 ft (500 m) or so. Longitudinal transport patterns prevailed, in which mixed-load streams fed elongate, fluvially dominated deltas. Coals are present in the delta-plain facies but are commonly thin.

Broad intracratonic basins such as the Cooper and Karoo are dominated by bed-load and mixed-load fluvial systems and lobate or elongate delta systems which prograded across a broad, shallow platform. The Cooper Basin was filled progressively from the south and west. The Karoo Basin, in contrast, shows a centripetal paleodrainage pattern, but was later bounded on the south by a major fold belt, whence it received a great volume of its sedimentary fill. Commercial coals are virtually confined to the stable northern platform (Fig. 12-34C). The total coal-bearing succession is rarely thicker than 150 m (500 ft) and contains up to five mineable seams. Near the basin margin the lowermost seam rests locally on glacial moraine and gravelly fluvio-glacial outwash fan systems, but farther basinward prodelta shales overlie the glaciogenic deposits gradationally. Lobate deltas built rapidly across the shallow platform, coalescing laterally. Transgressive glauconitic sandstones represent barriers formed after delta abandonment; these destructive deposits in places overlie extensive delta-plain coal seams, the upper parts of which originated as a blanket peat over the foundering lobe (Cadle and Hobday, 1977).

All of these coal-bearing Karoo systems were related to moderate but persistent influx from the stable cratonic flank, but this was eventually overwhelmed by fluvial sediments from the southern orogenic mountainland, which had filled the intervening foredeep and onlapped the cratonic platform. These conglomeratic sandstone wedges show radiating paleocurrent patterns typical of vast alluvial fans. Despite the abundance of vegetation, very little was preserved as coal, possibly because of a fluctuating water table in a warmer, seasonal climate, or high pH. Coals are, however, preserved across the top of the most regressive Indwe alluvial fan wedge, and these have been exploited locally.

Coal Resource Estimates

Resources comprise the mineable, or potentially mineable, portion of the coal endowment. Reserves, in contrast, comprise the small fraction of the total resources that can be extracted at a profit under existing economic, technological, and legal conditions. These two terms are commonly misused, resulting in a confusion that is compounded by the different criteria employed by various nations in defining resources and reserves.

Adequate appraisal of coal resources requires some knowledge of basin geometry and depositional systems, which allows comparison with better-known, analogous deposits elsewhere. More precise estimates require core drilling in selected areas. Regional and national resource appraisals are based on aggregation of numerous detailed local appraisals (Blondel and Lasky, 1956; Van Rensburg, 1980).

Resources are classified according to degree of certainty. Ideally each category involves standard procedures in calculation.

Measured resources are calculated in areas where the geology is well understood, and involves close spacing of control points. In the case of the Tertiary lignites of Texas, for example, where abundant subsurface data are available, a spacing of around 0.25 mi (0.4 km) is used (Kaiser *et al.*, 1980). This permits individual seams to be correlated with confidence. Optimum spacing in other coal basins is largely determined by the depositional environment. Discontinuous fluvial and upper delta-plain coals obviously require closer-spaced control points than do the laterally more continuous lower delta-plain coals. A spacing of around 0.3 mi (0.5 km) was found to be necessary in upper delta-plain coals of West Virginia (Howell and Ferm, 1980). Drilling at closer intervals would involve wasted exploration effort, and at wider intervals would provide insufficient information for meaningful correlation.

Kaiser *et al.* (1980) employed a method of calculating measured resources that uses driller's logs and geophysical logs to construct a strike cross section and several dip sections. Seams are projected updip to within 20 ft (6 m) of the surface, and downdip to a depth of 200 ft (61 m). The area underlain by coal is planimetered, and near-surface resources are calculated by multiplying this area by total lignite thickness and a factor representing coal tonnes per hectare-meter.

Indicated resources require less dense borehole control, with spacing in the case of the Texas lignites (Kaiser *et al.*, 1980) of up to 6 mi (10 km). Productive area underlain by lignite is determined from a variety of sources. One of the most useful methods is the geologic extrapolation of known coal-bearing facies. For example, isolith mapping of channel sandstone axes outlines dip-oriented coal-forming environments adjacent to or between them.

Demonstrated resources are the aggregate of the measured plus indicated resources. *Inferred resource* estimates are more tenuous and involve subjective assessment on the basis of geologic environments favorable for coal occurrences. This category commonly applies to coal in unexplored areas adjacent to demonstrated resources.

Chapter 13

Sedimentary Uranium

Introduction

Though not a fossil fuel, uranium shares with petroleum and coal many similarities in both geologic occurrence and utilization. The uranium fuel cycle includes a series of extraction and enrichment processes beginning with discovery and mining of a uranium deposit and including: (1) milling and enrichment of the uranium ore, which has a typical initial concentration (ore grade) of only a few hundredths to a few tenths of a percent of U_3O_8, to produce nearly pure U_3O_8 (yellowcake); (2) concentration of the U^{235} isotope in gaseous UF_6 (uranium hexafluoride); (3) fabrication of solid fuel pellets; and (4) fission of the uranium in a reactor, generating heat to drive turbine generators. Existing fission reactors "burn" U^{235} in that they consume it, producing waste by-products of the induced fission process. It is noteworthy that less than one percent of naturally occurring uranium is the desired U^{235} isotope. The remainder consists of the major uranium isotope, U^{238}, and traces of U^{234}. Currently, nuclear reactors produce approximately 10 percent of electric power consumed in the United States, and nuclear power accounts for substantial portions of the generating capacity of several European countries, as well as of Japan.

Most estimates of world uranium resources place over 90 percent of known United States uranium reserves within sedimentary host rocks. Approximately 65 to 75 percent of the total world reserve of uranium occurs in sedimentary strata. Much of the remainder occurs in epigenetic vein-like ore bodies within metasediments and are of problematic origin. Thus, an estimated 90 percent of known uranium deposits are the product of mobilization and concentration by surface or ground water (Robertson *et al.*, 1978).

Uranium Geochemistry

Uranium is a metal and is the heaviest naturally occurring element. It is highly reactive and readily combines or complexes with many other metals, nonmetallic elements, and anions.[1] Uranium can also substitute for calcium in the crystal lattice of a few common sedimentary minerals, particularly apatite.

The single most important geochemical characteristic of uranium is its occurrence at earth surface conditions, in two valence states. Uranium is highly insoluble in naturally occurring reduced ground waters. In this environment it forms U^{4+} minerals, including uraninite (UO_2) and coffinite ($USiO_4$). Solubility increases at low pH and at higher temperatures. In contrast, the U^{6+} form is highly soluble in oxidizing surface and ground water. Solubility of U^{6+} is further enhanced by the ready formation of stable complexes of U^{6+} and common anions, including CO_3^{2-}, HPO_4^{2-}, and SO_4^{2-}, over a broad range of pH conditions (Langmuir, 1978). Although complexes of U^{4+} can also form, they are important only at extreme values of pH rarely encountered in sedimentary environments.

The valence state of dissolved uranium is determined by its Eh and pH environment. The uranium reduction "fence" approximates the position of the important sulfate/sulfide and ferric/ferrous iron reduction fences. Thus, the probable valence state of uranium in a sediment can readily be inferred from the presence of more abundant mineral phases such as pyrite, hematite, or limonite.

In addition to its geochemical attributes, uranium is, of course, one of the family of elements characterized by natural isotopes that spontaneously disintegrate. Such atomic disintegration produces both energy, in the form of gamma radiation, and particles, including alpha and beta particles and daughter elements. Decay of uranium isotopes produces radioactive daughter products which, in turn, decay through a series of daughters until stable lead isotopes are pro-

[1] Langmuir (1978) provides a thorough summary of uranium hydrochemistry that expands on the now classic work of Hostetler and Garrels (1962).

duced. Several steps in the decay sequence produce high-energy gamma radiation, which can be measured by surface and subsurface counters. A gamma-ray log, similar to that used for lithology determination but set to a much less sensitive calibration, is the principal tool for downhole uranium exploration and detection. Scintillometers or other radiation detection devices have long been used for surface or airborne exploration surveys.

Uranium is a lithophile element. Consequently, it is most abundant in highly differentiated products of magmatism and is concentrated in the crust. Although uranium is a ubiquitous element, average concentrations in crustal materials are low. Uranium content averages 2.7 ppm in sedimentary rocks, 2.7 ppm in siliceous igneous rocks, and 1.7 ppm in mafic igneous rocks (Gabelman, 1977). Highly siliceous rocks are enriched in uranium. Granites average 3.6 ppm and rhyolite averages 5.0 ppm; such comparatively enriched rocks are commonly inferred to be primary sources of uranium concentrated within sedimentary strata. Obviously, processes capable of mobilizing uranium and concentrating it by orders of magnitude are necessary to ores containing several hundreds to thousands of parts per million of the metal.

Uranium Ore Deposits

The largest world reserves of commercial uranium occur in three broad classes of deposits: quartz-pebble conglomerates, paleosurface-related vein-like masses, and sandstone ores. In addition, large subcommercial resources occur in black shales.

Uraniferous quartz-pebble conglomerates occur as basal units within thick sequences of compositionally mature quartzite and siltstone that unconformably overlie crystalline Archean rocks. The conglomerates contain abundant pyrite, as well as uranium and, commonly, gold. Host strata are drab and reduced. Younger Precambrian strata covering the uraniferous conglomerates include mineralogically diverse red beds and hematitic Superior-type banded iron formation.

Paleosurface or unconformity-related vein-like deposits are characterized by their localization within fault, breccia, or fracture zones immediately beneath a regional unconformity. Deposits rarely extend up into the overlying structurally simple strata that blanket the paleosurface. Ores are commonly rich, dominated by uranium, and contain abundant gangue consisting of hematite, carbonate minerals, and chalcedonic quartz. Geologic associations, disappearance of the deposits with depth, and low-temperature nature of alteration all suggest a supergene origin contemporaneous with or postdating the development of the erosion surface.

Sandstone ore deposits consist of elongate mineralization fronts or clusters of isolated pods, tabs, or lenses containing uranium dispersed within interstices between detrital grains. Deposits may occur at boundaries between drab, reduced portions and red-brown, oxidized portions of the host unit, or as localized concentrations in either oxidized or reduced hosts. Associated metals include molybdenum, selenium, and, in some districts, vanadium.

Although not commercial because of the typically low concentrations of uranium (averaging a few tens to hundreds of ppm U_3O_8), thin, sulfidic organic shales of restricted cratonic-basin origin constitute a major uranium resource. Areal extent of these units may exceed several hundred square miles.

The four major types of uranium concentrations exhibit definite patterns of occurrence through geologic time (Fig. 13-1) (Robertson *et al.*, 1978). Quartz-pebble conglomerate ores were deposited during the Proterozoic from 2,200 to 2,800 m.y.b.p. The primary uranium minerals include uraninite and pitchblende, which occur as detrital placers deposited during a period predating evolution of a highly oxygenated atmosphere. Under such conditions, reduced uranium minerals would not be oxidized and destroyed by surficial weathering and transport. As a result, quartz pebble-type deposits are restricted to hosts deposited more than 2,200 m.y. ago, before partial pressure of oxygen in the atmosphere became adequate to actively oxidize exposed sulfide minerals as well as uranous mineral phases (Robertson *et al.*, 1978).

Paleosurface-related vein-type ore bodies formed between 2,000 and 1,500 m.y.b.p. Age of the unconformity and deposits is bracketed by the age of the underlying, eroded strata and the unmineralized capping strata. These deposits seem to require the high uranium mobility in ground and surface waters that has existed since the evolution of an oxygen-enriched atmosphere

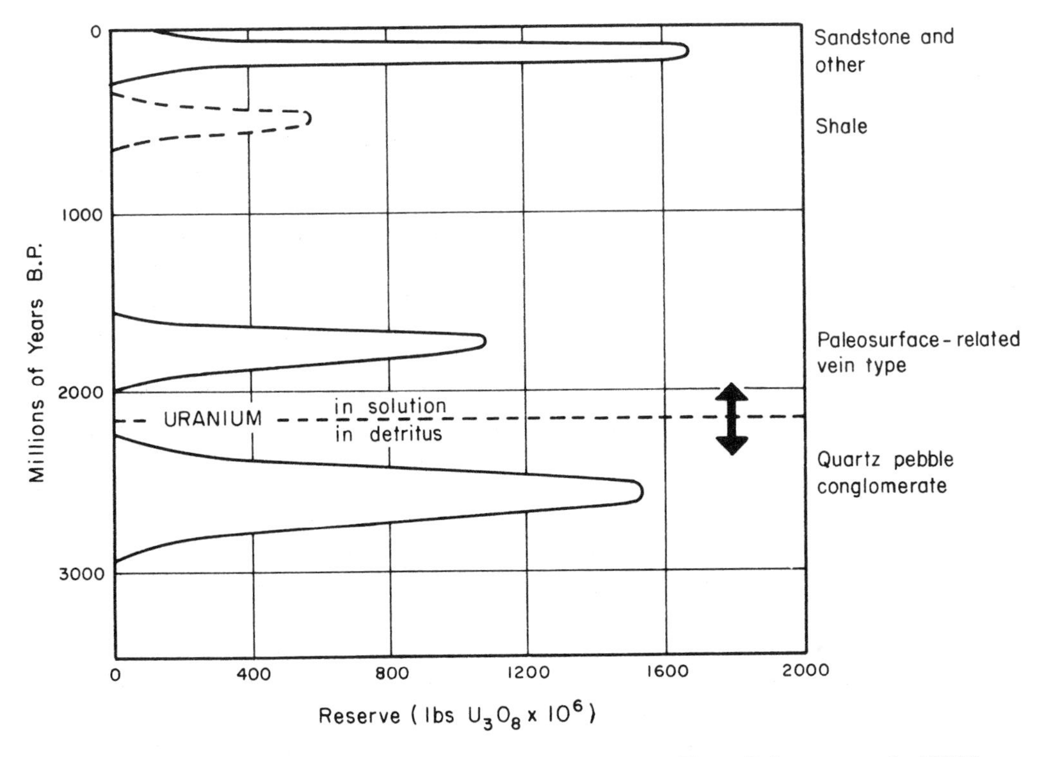

Figure 13-1. Time-bound distribution of uranium reserves. (From Robertson *et al.*, 1978.)

at about 1,700 m.y.b.p., and their tremendous age-restricted enrichment poses important questions about the exploration favorability of younger but geologically analogous terranes. The age restriction may reflect an unique enrichment of surface and shallow ground waters as uranium in older strata deposited under the oxygen-deficient conditions was exposed and rapidly dissolved in an increasingly oxygen-enriched environment (Robertson *et al.*, 1978). Because the emphasis in this book is on sediment-hosted fuel minerals, this extremely important class of deposits will not be further discussed.

Sandstone uranium ore deposits are concentrated dominantly in rocks of Permian, Mesozoic, or Cenozoic age. Small amounts of sandstone ore also occur within Paleozoic rocks. Major peaks of uranium occurrence lie within strata of Jurassic and early to middle Tertiary age, suggesting that this younger style of mineralization is characterized by distinct metallogenic epochs (Malan, 1968). Nearly all such deposits are interpreted to reflect solution transport of uranium in the oxidized form. In many, age dating indicates that some remobilization and concentration of uranium has continued into the Quaternary. Extensive, uranium-enriched black shales are dominantly early Paleozoic in age. Together, these two classes of young sedimentary deposits are largely the product of recycling of the crustal endowment of uranium, with additional uranium liberated from exposed igneous masses.

The Uranium Cycle and Its Products

Uranium continually recycles through sediments and waters during basin filling. Under certain geochemical or mechanical conditions, such recycling may produce localized enrichment of uranium within a portion of the basin fill. Under optimum conditions, sufficient enriched material is produced to constitute a potential resource.

The basic sedimentary uranium enrichment cycle consists of three parts.

1. Uranium is leached from low-grade source rocks. Although some authors argue that average crustal abundances are adequate, the spatial as well as temporal localization of uranium deposits of all types suggests that unusually rich or available sources are necessary for large-scale mineralization (Malan, 1968; Grutt, 1972;

Galloway *et al.*, 1979b). This view is consistent with the observation that large uranium-producing districts are commonly associated with uranium-enriched source rocks, particularly granites or siliceous volcanic debris. Furthermore, the geochemical and hydrologic processes responsible for transport and concentration of uranium are neither unique nor unusual.

2. Released uranium is then transported to a site of accumulation by flowing surface or ground water. In environments with low oxidizing potential, detrital uranium minerals, such as uraninite (UO_2), can be transported mechanically (Grandstaff, 1976). In oxidizing conditions typical of present surface waters and shallow ground waters (Fig. 11-6), uranium is rapidly dissolved and transported in solution. Thus, uranium transport paths parallel the general directions of both sediment and meteoric waters through depositional systems.

3. Uranium is concentrated by mechanical, geochemical, or physiochemical processes. The specific process depends on the transport mechanism, the chemistry of the transporting ground or surface water, and the composition of the sediment or rock matrix which the uranium contacts. In all, five mechanisms of uranium concentration in sedimentary deposits have been demonstrated in laboratory or field investigations.

Hydraulic sorting concentrates grains with anomalous density or grain shape on the channel bed. Uranium minerals have a high specific gravity (9–9.7), and are readily concentrated as a lag along with the heavy mineral fraction of the sediment. Efficiency of the process is accentuated where fine, dense grains can settle into interstices between larger, less dense grains such as quartz.

Precipitation of highly insoluble U^{4+} solid phases, such as the minerals uraninite (UO_2) and coffinite ($USiO_4$), is controlled by Eh and pH (Hostetler and Garrels, 1962; Langmuir, 1978). As indicated by Eh–pH diagrams such as Figure 13-2, reducing conditions (low Eh) and low pH both favor precipitation.

The Eh–pH stability fields of coffinite and uraninite are believed to be nearly identical; which mineral forms is determined by the concentration of dissolved SiO_2. A SiO_2 concentration of 60 mg/l ($H_4SiO_4^{o}$ activity of 10^{-3}) is commonly assumed for the UO_2–$USiO_4$ equilibrium concentration. This value is above the average in ground water (17 mg/l) and below saturation with amorphous SiO_2 (120 mg/l) (Langmuir, 1978); but approximates the equilibrium saturation with silica glass.

Coffinite (or uraninite) has a large stability field (Fig. 13-2), suggesting it can precipitate from very dilute ground water containing only a few μg/l uranium. However, experimental work (Kochenov *et al.*, 1977) indicates that uranium does not precipitate from solutions containing less than 100 μg/l uranium in the absence of an adsorbent. Initial precipitates are commonly amorphous phases; thus, the amorphous $USiO_4$ field (Fig. 13-2) best characterizes the Eh–pH conditions favoring uranium precipitation from ground waters enriched in uranium and carbonate species. High pH and low uranium activity in solution require more powerful reducing conditions for precipitation to occur. Likewise, the presence of strong complexing agents lowers the Eh threshold necessary to initiate precipitation.

Solution Eh and pH changes resulting in precipitation of uranium may occur in a variety of ways. Simplest and best demonstrated is the reduction of oxidizing meteoric ground water in a reduced aquifer. Normal redox reactions establish an Eh gradient, which may be characterized by one or more rather sharp boundaries (Chapter 11). Precipitation occurs at the appropriate boundary. Alternatively, diffusion mixing of oxidized, uraniferous waters with reduced waters or gases may initiate precipitation along a fluid–fluid interface.

Uranyl (U^{6+}) minerals may form under special conditions. Economically important species are carnotite ($K_2(UO_2)_2(VO_4)_2 \pm H_2O$), autunite ($Ca(UO_2)_2(PO_4)_2 \pm H_2O$), uranophane ($Ca(UO_2)_2(SiO_3)_2(OH)_2 \cdot 5H_2O$), and schroeckingerite (a hydrous uranyl salt). These phases are common in arid settings where evaporation produces shallow, oxidizing ground waters enriched in uranium as well as vanadate or phosphate, and depleted in bicarbonate due to $CaCO_3$ precipitation or low CO_2 partial pressure inherent in organic-poor soil horizons. Phase diagrams and solubility calculations show the uranyl minerals are consistently most stable in the intermediate pH range, and that carnotite is the most stable of the various phases. Under reasonable conditions of vanadate content, shallow oxidizing waters in equilibrium with carnotite could contain as little as 1 ppb U (Langmuir, 1978).

Physiochemical concentration by adsorption,

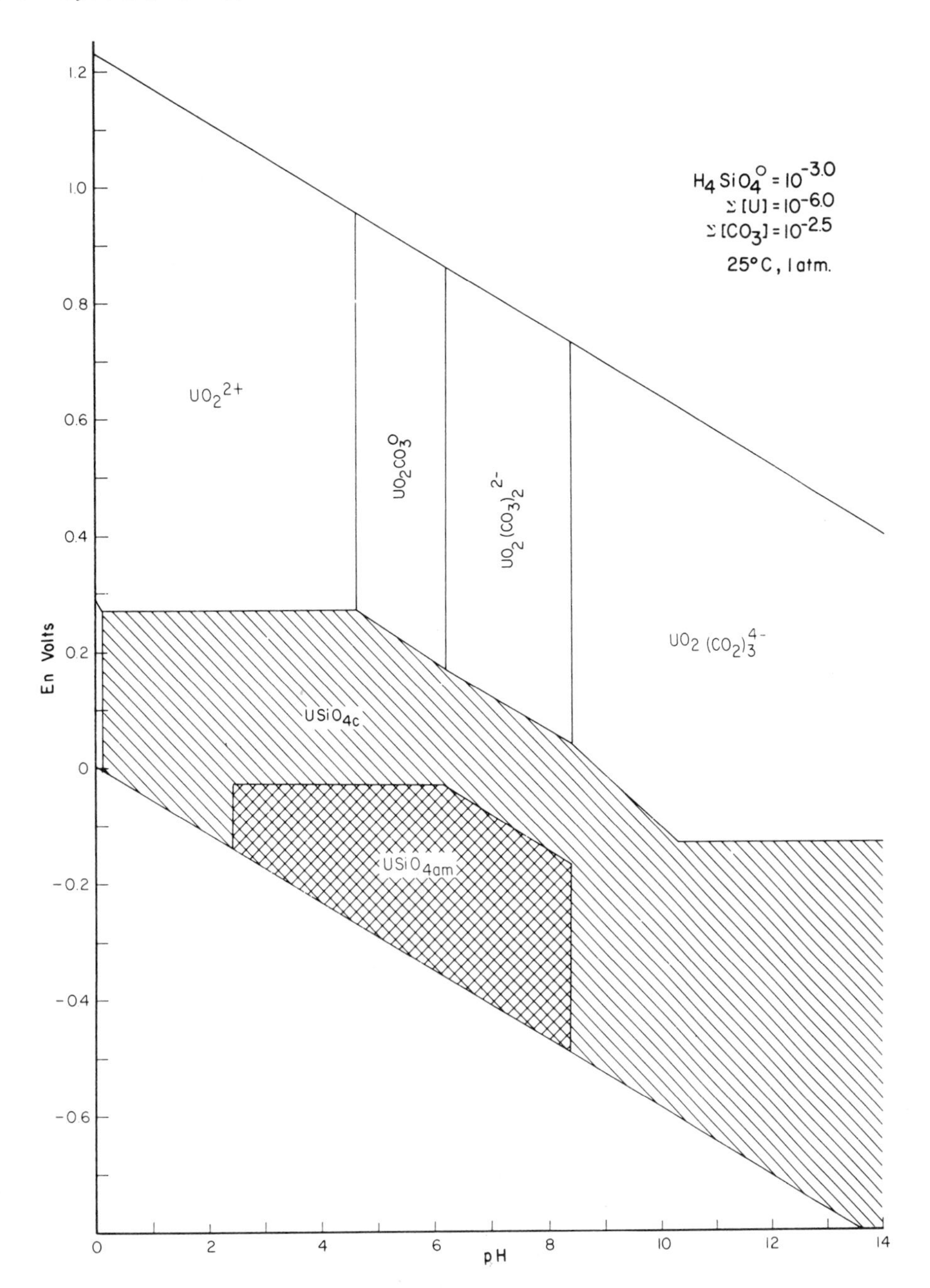

Figure 13-2. Eh–pH diagram showing the stability fields of crystalline and amorphous coffinite ($USiO_4$) in carbonate-rich ground water containing approximately 60 mg/l of H_4SiO_2, 200 mg/l total carbonate species, and 240 μg/l uranium. (From Galloway and Kaiser, 1980.)

ion exchange, or complexing with solids dispersed through the sediment or aquifer matrix may provide the critical first step in concentration of uranium from dilute solutions, and may control dissolved uranium concentrations at levels below saturation for mineral phases (Kochenov *et al.*, 1977; Langmuir, 1978).

Although specific mechanisms of sorption are complex and commonly poorly documented, numerous solids have been demonstrated to remove uranium from solution and concentrate it on or near particle surfaces. Amorphous titanium hydroxide and iron oxhydroxides can concentrate uranium by 10^4 to 10^6 (Langmuir, 1978). Organic materials, including peat and humic acid, have comparably high concentration efficiencies (Szalay, 1964; Schmidt-Collerus, 1969). However, sorptive capacity decreases with increasing thermal maturity of the organic material (Cameron and Leclair, 1975). Solid phases capable of absorbing uranium include natural zeolites such as clinoptilolite, clay minerals, particularly those such as the smectites characterized by high cation-exchange capacity, and phosphorites (Doi *et al.*, 1975; Tsunashima *et al.*, 1981). Enrichment factors for these phases are considerably lower, ranging from a few times to several hundred times.

For all sorbents, maximum concentration of uranium occurs within the pH range of 5–8.5 (Langmuir, 1978). The majority of natural sorbents adsorb positively charged uranium species, but reject negatively charged uranyl carbonate ion complexes (Lisitsin *et al.*, 1967). The boundary between neutral and anionic forms of uranium in carbonate-rich water occurs at about pH 6. At a slightly alkaline pH, only about 10 percent of the dissolved uranium is present as cationic species (Lisitsin *et al.*, 1967).

Adsorption of uranium can occur in both oxidizing or reducing waters; however, it is reversible in an oxidizing medium but irreversible under strongly reducing conditions (Kochenov *et al.*, 1965). Adsorption by inorganic compounds such as clay minerals and zeolites occurs in both oxidizing and reducing waters. Adsorption by organic matter is most likely at low Eh and pH values. Humic acids are capable of breaking down uranyl carbonate ions and adsorption of UO_2^{2+} can take place on insoluble humic acids (Szalay and Szilagi, 1969). Spectroscopic studies on the binding of uranium by organic matter point to complexing of UO_2^{2+} by carbozylate groups (Koglin *et al.*, 1978), explaining the observed decreased adsorptive capacity of higher rank coal or organic materials.

In near-surface waters containing typical dissolved uranium concentrations of a few tens of ppb, efficient adsorption by peat or amorphous iron or titanium hydroxide phases could produce concentrations in the sediments or rock matrix of up to several hundred ppm weight percent U_3O_8. However, unless subsequently reduced and precipitated, sorbed uranium is subject to remobilization if pH or composition of the fluid changes. Sorption as a prelude to reduction offers an extremely attractive mechanism for efficient removal and localized concentration of metals such as uranium from dilute solutions.

Coprecipitation of uranium may be important in certain geochemical conditions. Uranium can substitute for calcium in the crystal lattice of apatite and thus may be anomalously rich in phosphate deposits. Uranium can also be incorporated in precipitated amorphous silica, or opal.

Apparently, certain *biologic processes* are capable of concentrating uranium within simple organisms. Diatoms and coccoliths have been shown to accumulate uranium within their tests (Degens *et al.*, 1977). Enrichment by factors of 10^4 above average content in surface waters of the microorganism's environment has been suggested.

Classification of Uranium Deposits

An examination of the fundamental uranium cycle in the context of active sediment deposition and subsequent ground-water circulation suggests a straightforward genetic subdivision of sedimentary uranium deposits. However, as with most genetic classifications, application to individual deposits may lead to conflicting conclusions. Furthermore, genetic mineralization processes may overlap in both time and space. Nonetheless, recognition of related families of deposits focuses attention on the stratigraphic, geochemical, and hydrogeologic parameters that are the key to the discovery and development of additional deposits. Such attention to process is critical if exploration is to expand beyond attempts to discover geologic look-alikes, attempts which may emphasize inconsequential geologic features unrelated to actual uranium resource potential.

Three broad classes of uranium deposits are differentiated based on (1) timing of primary mineralization relative to deposition of the host sediment facies, and (2) relative dominance of the surface or ground-water system in transport of uranium to the site of concentration. Each genetic class contains two or more representative styles of mineralization, some of which appear quite disparate superficially.

Syngenetic Deposits

Syngenetic deposits are formed within the host sediments during their active deposition (Fig. 13-

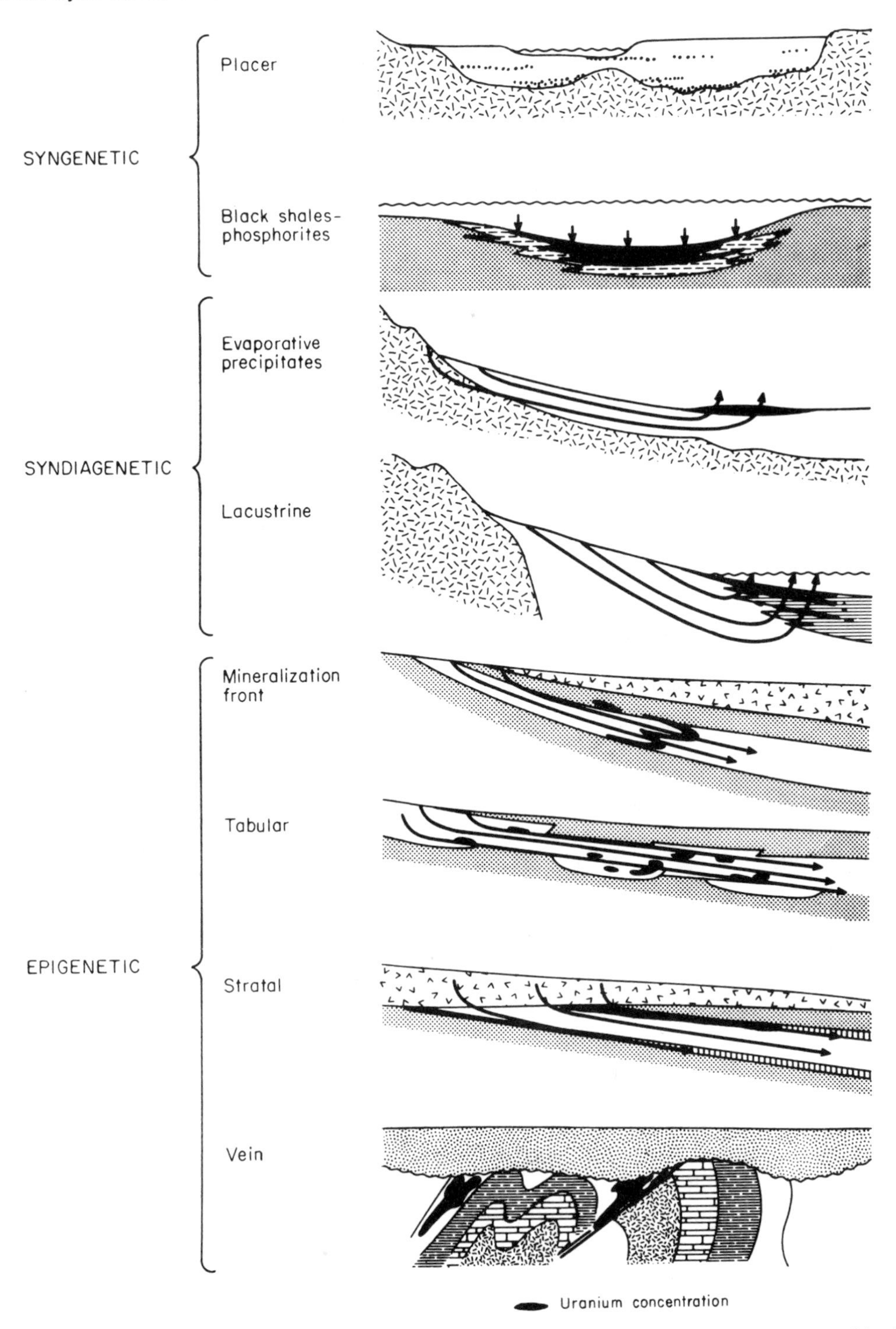

Figure 13-3. Principal genetic classes of sedimentary uranium deposits and their stratigraphic and hydrologic setting.

3). Uranium is transported by surface waters, though such waters may readily interchange with shallow portions of the sediment pile. Two very different styles of mineralization are syngenetic in origin. The Proterozoic quartz-pebble conglomerate ores originated as detrital placers of uranium minerals. Uraniferous black shales and phosphorites constitute younger syngenetic accumulations forming by extraction of uranium from solution within surface waters.

Syngenetic deposits are characteristically both stratiform and stratabound. Uranium content is directly related to specific depositional facies, and correlates closely with features or components that are also of depositional origin. Both modern (Boberg and Runnells, 1971) and best-documented ancient examples of syngenetic concentration of dissolved uranium suggest that expected grades are low, at best measuring several tens to a few hundreds of ppm. Consequently, postdepositional redistribution and grade enhancement are usually invoked to adapt syngenetic models to commercial deposits.

Syndiagenetic Deposits

Syndiagenetic processes are characterized by the presence of large amounts of interstitial water and extreme variation in Eh and pH. The term is used here in a somewhat restrictive sense in that syngenetic processes are excluded. Syndiagenetic deposits are the product of circulating shallow but partially confined meteoric ground water within the buried sediments of an active depositional system. This definition has two inherent corollaries: (1) flow direction reflects the regional topography of the host depositional system and its bounding uplands and basin; (2) mineralizing waters are modified by significant lateral flow and residence time below the water table. Obviously, uranium is transported in solution.

Two significantly different types of syndiagenetic deposits produce economic uranium reserves (Fig. 13-3). Calcrete deposits form where meteoric waters are discharged along the lower reaches of bed-rock defined fluvial systems, and uranium increased to supersaturation by evaporation and depletion of complexing agents. A different variation results from discharge of uranium-enriched waters across organic-rich sediments of lacustrine or marsh swamp origin. Resulting deposits are stratabound and locally may appear stratiform. Because lateral flow and discharge of ground water reflect the depositional morphology of the basin, mineralization commonly correlates closely to specific depositional facies. However, degree of mineralization of a particular bed or facies is a function of later exposure to circulating ground water. Thus, enrichment cuts across bedding on a regional scale, and large volumes of otherwise appropriate host facies display no uranium enrichment.

Epigenetic Deposits

Epigenetic deposits are produced by the continued or rejuvenated flow of meteoric ground water following total burial of the host depositional system. Although the mineralizing flow system may be very similar to that operating during active accumulation of the depositional system, strata of the buried system serve primarily as passive plumbing for the mineralizing ground waters. Portions of the ground-water flow system are typically confined, and considerable geochemical evolution of the waters may occur along flow paths. Additional complexities may be introduced by sequential fluxes of geochemically different waters, or by development of a dynamic interface between two reactive ground-waters (Nash and Granger, 1981). Resultant deposits are commonly stratabound, reflecting preferred conduits for fluid flow but are only locally, if at all, stratiform.

Resultant deposits assume a number of morphologies (Fig. 13-3). Sandstone uranium ore bodies, an extremely important class of epigenetic deposits, occur as dispersed pods or masses and as elongate, sinuous mineralization fronts (Fischer, 1968; Harshman, 1970). Facies that are particularly favorable for entrapment and retention of uranium, such as permeable, highly adsorptive lignite beds, may produce local stratiform deposits (Denson *et al.*, 1959). Structural features, including fault or shear zones, that provide enhanced permeability within underlying bedrock may become conduits for downward fluid movement and resultant mineralization, producing vein-like deposits. In summary, geometry and distribution of epigenetic deposits are functions of the permeability configuration, the localization of regional recharge, lateral flow, and discharge zones, and the distribution of geochemical traps for dissolved uranium.

The three classes reflect a continuum. Just as ground-water flow and depositional patterns can gradually evolve, boundaries between the classes can be transitional in any one suite of deposits. More than one period of mineralization or redistribution may be superimposed on the same host system. Thus, though many definitive examples of each class may be cited, few deposits do not show some evidence for overlap or superposition of mineralization processes. However, as in the case of stratigraphic interpretation, regional patterns should be emphasized. Primary focus should be on the timing and nature of the processes responsible for initial importation and concentration of uranium within the overall uraniferous province or district. Subsequent modification or redistribution, which overprints primary mineralization patterns, may produce nearly random "noise," or may itself be a major ongoing process responsible for development of economically important secondary trends within the larger provinces.

Syngenetic Quartz-Pebble Conglomerate Deposits

Quartz-pebble conglomerate ores form world-class deposits in the Elliot Lake uranium district, Canada, and the Witwatersrand gold-uranium district, South Africa. Elliot Lake reserves are estimated to exceed one billion pounds (450,000 m.t.) of U_3O_8. Together, these two provinces account for approximately 20 percent of known world reserves. Additional uraniferous Proterozoic conglomerates occur in Australia, North and South America, and Eurasia. The geology of quartz-pebble conglomerate ores has been reviewed in detail by Button and Adams (1981), who interpreted their genesis as a part of a distinct, time-bounded metallogenic epoch produced by three key factors:

1. The evolution of sialic continental crust and formation of emergent continental nuclei upon which fluvial depositional systems were established;
2. The development and exposure of uranium-enriched granitic rocks that provided both a source of coarse sediment and of uranium; and
3. The presence of an early oxygen-deficient atmosphere that allowed survival of reduced uranium minerals during long periods of intense mechanical reworking at the earth's surface.

Geologic Setting

Depositional basins of Proterozoic quartz-pebble conglomerates lie exclusively on the ancient Archean cratons, within Precambrian shield areas, rather than on post-Archean mobile belts. Basement rocks consist of high-grade gneisses, low-grade metasediments, granitoids, and greenstone belts, which typically comprise an older mafic volcanic sequence and younger sedimentary sequence. The most favorable basins occur on crust dominated by granite-gneiss terranes.

Basins typically formed as broad, regional downwarps, partly bounded by faults. Typical dimensions range from 10^4 to 10^5 km^2 (10^3 to 10^5 mi^2). The basins were long-lived features, persisting for millions to hundreds of millions of years (Tankard *et al.*, 1982; Button and Adams, 1981). The Witwatersrand basin, which is nearly completely preserved, exhibits an asymmetric, yoked geometry, with maximum thicknesses of fill approaching 7500 m (25,000 ft) along its fault-bounded northern and western periphery (Pretorius, 1976). Younger strata typically onlap eroded basement crust along margins of the basin. Basin fill was metamorphosed only to lower greenschist facies, and is gently deformed.

During subsidence, basins were filled by terrigenous clastic sediment along with some carbonate, iron formation, and volcanics. Clastic sediments are commonly coarse, particularly around the margins of the basin. Three facies assemblages characterize uraniferous depositional episodes (Button and Adams, 1981). (1) A quartzarenite assemblage consists of beds of medium to coarse, texturally and mineralogically mature sand with sericite matrix. (2) A subarkose-quartz wacke assemblage is dominated by coarse to very coarse sandstone containing minor amounts of conglomerate. In addition to quartz, potassium feldspar, sericite, and chlorite are common. (3) An extrusive basalt association consists of dominantly subaerial flow units and interbedded volcaniclastic sediments.

As emphasized by Button and Adams (1981), the basin fills are laced with unconformities and nonconformities of variable type and temporal and areal extent. Interformational disconformities bound major depositional episodes, extending over 10^4 to 10^5 km^2 (10^3 to 10^5 mi^2), and separate rocks of substantially different ages. Erosion surfaces commonly extend onto Archean basement around the periphery of the basin. Intraformational disconformities are more local, typically lying between facies sequences and separating depositional events of a single system. Areas range from 10^2 to 10^3 km^2 (10 to 10^3 mi^2) and the time span represented is inferred to extend from a few hundred to tens of thousands of years. Localized scour surfaces and residual lags occur within individual facies sequences.

Paleotopography of the unconformity surfaces,

particularly where Proterozoic sediments lay directly on basement, exerted considerable influence on both depositional and mineralization patterns. Paleotopographic features were products of syndepositional structural deformation as well as differential erosion. For example, subcrops of greenstone belts or synclinal axes commonly produced paleotopographic lows. Dipping, bedded, shaly strata developed a series of strike valleys (Button and Tyler, 1979). Coarse, uranium-bearing conglomerates typically are restricted to or concentrated within such paleotopographic depressions or valleys.

Studies of weathering profiles and paleosol horizons within quartz-pebble conglomerate-bearing depositional complexes and their contemporary source terranes reflect limited chemical weathering in the absence of an oxygen-rich atmosphere and land plants (Button and Adams, 1981). Iron is significantly leached from weathered zones, indicating mobilization in the ferrous state. Thick, residual soils of granitic terranes contain quartz, potassium feldspar, sericite, and heavy minerals, including pyrite grains. However, demonstrated leaching of uranium, as well as the presence of much of the remaining dispersed iron in the ferric state, suggests that the weathering zone was slightly oxidizing. Although deposition of the uraniferous conglomerates in a glacial climatic regime has been suggested, Proterozoic strata containing diagnostic evidence of glaciation are of generally different age than the quartz-pebble bearing systems (Button and Adams, 1981), and the significance of contemporaneous glaciation is problematical.

Host Depositional Systems

Uraniferous conglomerates of both the Matinenda Formation (Elliot Lake) and Central Rand Group (Upper Witwatersrand) were deposited in various depositional elements of braided bed-load fluvial and fan systems (Vos, 1975; Pretorius, 1976; Minter, 1978; Smith and Minter, 1980; Button and Adams, 1981). Configuration of fluvial systems, as outlined by the coarse, conglomeratic framework deposits, was complex and locally included: coalescing "wet" alluvial fans that issued from point sources around the faulted basin margins (Fig. 13-4B) (Pretorius, 1976), bedrock-confined valley fills deposited within structural or erosional lows (Fig. 13-4A, B and F) (Pienaar, 1963; Button and Tyler, 1979), and widespread unconfined fluvial braidplains (Fig. 13-4D) (Minter, 1978). Associated depositional facies include lacustrine or tidally influenced marine basinal shales and silts, which are separated from fluvial and fan systems by reworked fan margin and strandline facies, and thin, transgressive strandline sands. The well-preserved and extensively studied Central Rand Group illustrates features of these Proterozoic depositional systems (Tankard *et al.*, 1982).

The Central Rand Group accumulated during a depositional episode of basin filling that is bounded in part below and above by regional basin-margin angular unconformities. It pinches out abruptly against the bedrock basin margin, and expands to a thickness of almost 3,000 m (10,000 ft) basinward. Sand/shale ratios range between 10 and 5 to 1. About 8 percent of the sequence consists of conglomerate (Pretorius, 1976). Coarsest deposits are concentrated around the basin-margin uplands, and are dominantly of the subarkose quartz wacke facies assemblage. The Central Rand Group constitutes a wet alluvial-fan delta system (Pretorius, 1976; Minter, 1978). The system is bounded landward by confined tributary valleys and uplands, and grades basinward into mudrocks of a large lake or marine basin system.

Component facies of the Central Rand fan delta system include the following:

Conglomeratic, braided channel fill, characterized by abundant clast-supported conglomerate, occurs as lags, sheets, lenses, and scour fills exhibiting bar-accretion foreset bedding, trough cross bedding, and common cross-cutting scour surfaces. Interpreted depositional units include longitudinal gravel bars, braid channel fills and scour lags, and deflation lags (Minter, 1978; Smith and Minter, 1980).

Sandy, braided channel fill consists of sand, locally containing floating pebbles or matrix-supported conglomerate and minor thin, interbedded mud drapes and mud clasts. Primary structures include trough and tabular cross stratification, planar stratification, and common local scour and channeling. Channel-filling cycles are less than 3 m (10 ft) thick (Fig. 13-5).

Fan delta margin quartzarenite occurs within basin margin sequences dominated by channel fill as thin but widespread tongues of mature,

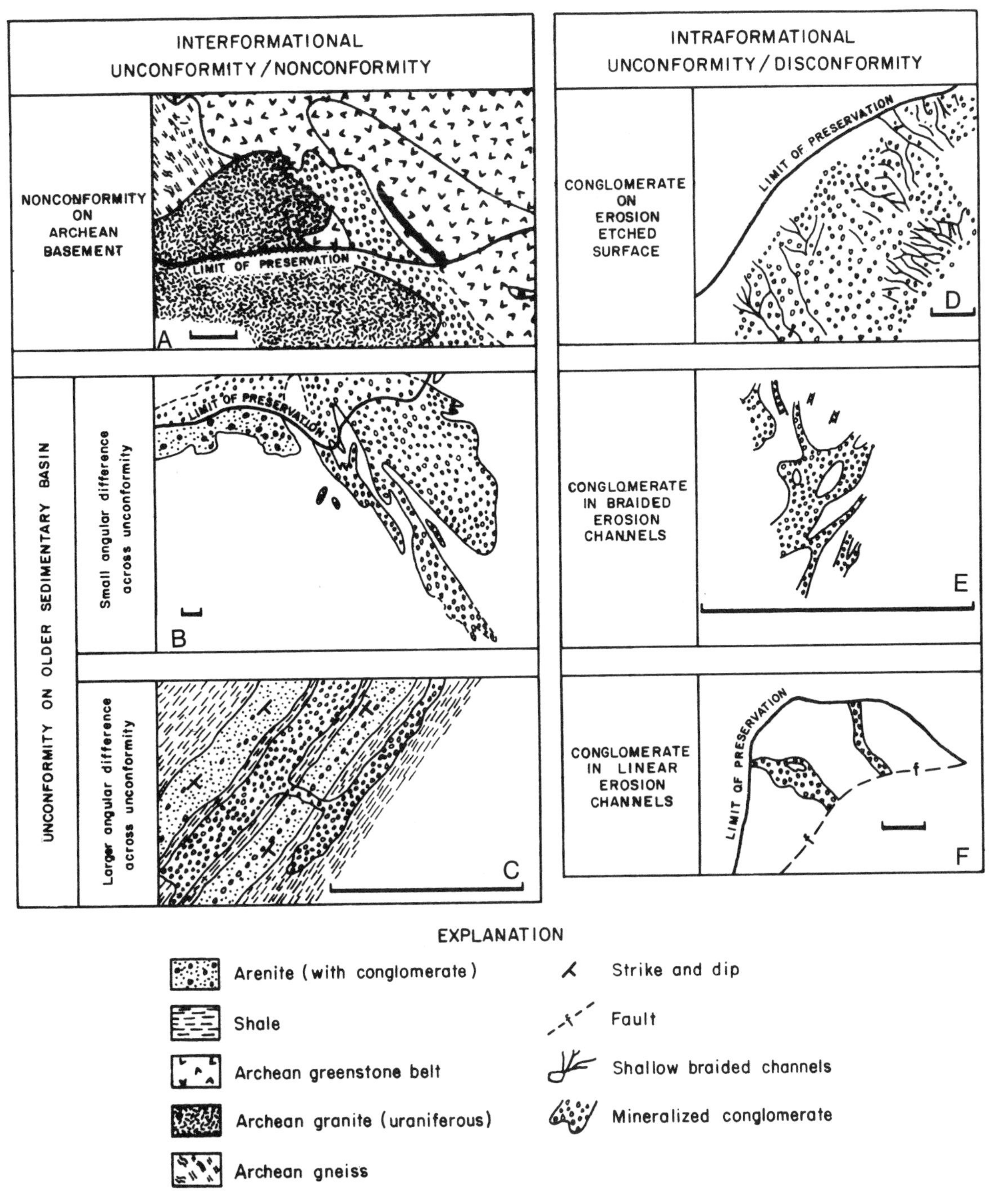

Figure 13-4. Schematic map views of uraniferous quartz pebble conglomerate units deposited on regional interformational and intraformational erosion surfaces: (A) Alluvial belt confined within erosional low, or valley on basement, surface. (B) Unconfined "wet" alluvial fan along basin margin. (C) Strike valley drainage pattern on high-angle unconformity. (D) Wide, unconfined fluvial braidplain. (E) Local braided erosion-confined channel segments. (F) Linear erosion channel segments. Length of bar is 1 mi (1.6 km). (Modified from Button and Adams, 1981.)

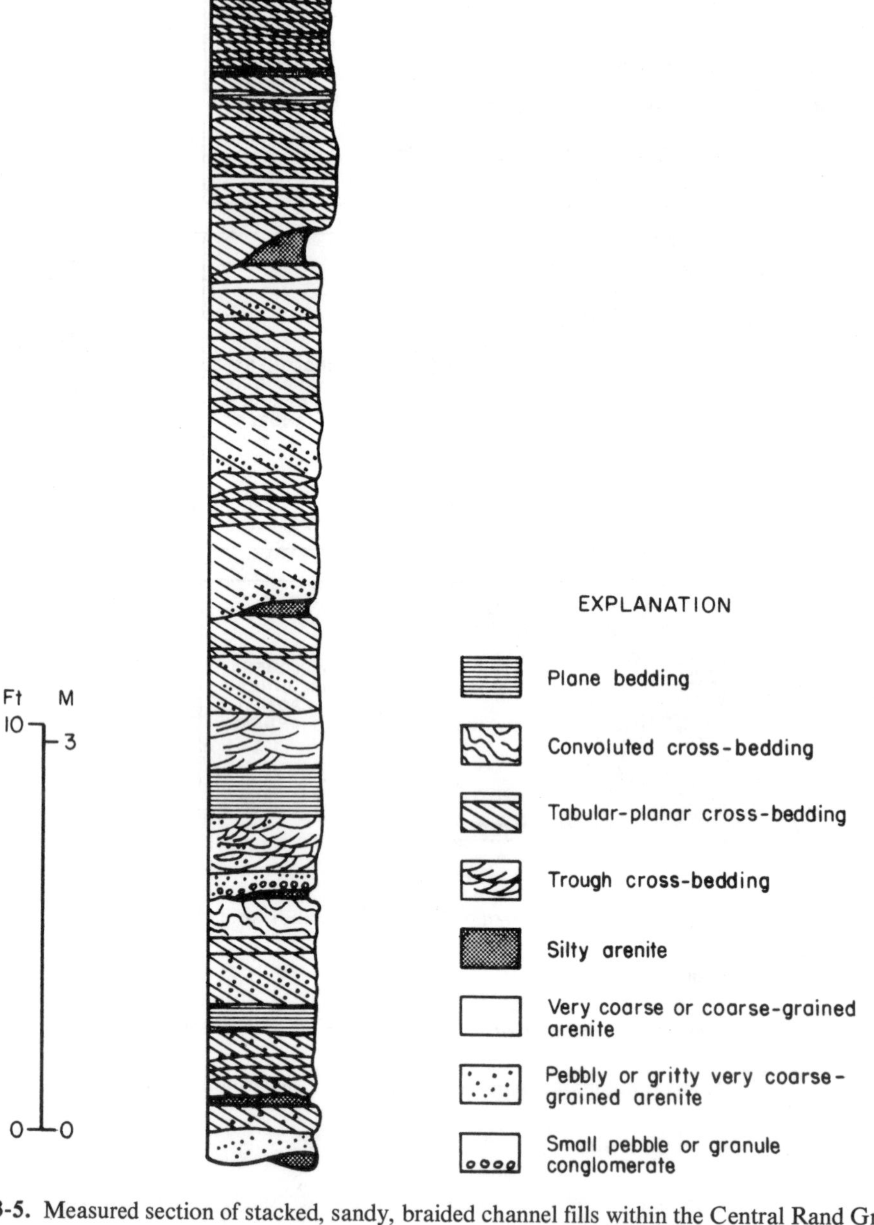

Figure 13-5. Measured section of stacked, sandy, braided channel fills within the Central Rand Group (South Africa). Dominant sedimentary structures include tabular and trough cross stratification and planar stratification. (From Button and Adams, 1981.)

medium to coarse, trough crossbedded, planar or ripple laminated sand. Oscillation and interference ripples, as well as mud cracks indicate intermittent, shallow, standing water. Basinward, sequences of quartzarenite coalesce to form thick successions that are interbedded with basinal mudstone. They grade up into shale, limestone, and iron formation of the basinal system.

Although most authors propose a lacustrine origin for the Witwatersrand basin deposits,

Button and Adams (1981) and Tankard *et al.* (1982) argue for significant tidal influence, thus implying a connection to the world ocean.

The Central Rand Group succession displays large-scale coarsening- and fining-upward cycles hundreds of meters thick. These, along with the associated intraformational disconformable surfaces, are interpreted to reflect local or basinwide changes in base level associated with regional tectonism (Vos, 1975). Within laterally equivalent deposits of a single cycle, conglomeratic channel-fill facies grade downflow into sandy channel-fill facies over distances of tens of kilometers. Deposits of the conglomeratic channel facies form readily mapped skeletal elements of the fan delta system. As shown in Figure 13-4, they display elongate, dip-oriented geometries and commonly occur within erosionally defined channels or on extensive scour surfaces. Mappable channel-fill units have high width/depth ratios (commonly ranging between 200 and 1,000), are quite shallow, and display low sinuosity. Larger erosional valleys may be hundreds to thousands of feet wide and tens of feet deep (Minter, 1978). Deposits of both confined and unconfined channel fills are characterized by complex interfingering of thin, texturally diverse beds and lenses.

In summary, the facies of the uraniferous portion of the Central Rand Group exhibit all the features of bed-load dominated, wet fluvial fans. Interfingering along their distal margins with wave or tide-reworked, texturally mature quartzarenites and associated basinal deposits completes the facies tract from fault-bounded source to a permanent marine or lacustrine reservoir characteristic of fan delta deposition. Abundant disconformable surfaces and associated truncation, valley incision, and planation are all features typical of large fans within tectonically active terranes. The near absence of fine sediment reflects the coarse grain size of material derived under pre-Devonian weathering conditions from coarsely crystalline source granite and gneiss. Lacking appreciable suspended load or land vegetation to stabilize banks, unconfined channels were broad and rapidly shifting; sheet-flow was likely during floods. Perhaps the closest modern analogs for Witwatersrand systems are the braided outwash plains and fans of glacial areas. The Sandur plains of Iceland (Boothroyd and Nummedal, 1978) may offer models comparable in scale, dominance of mechanical weathering processes in the source area, and depositional processes (Tankard *et al.*, 1982).

Uranium Mineralization

The primary uranium within quartz-pebble conglomerate ores occurs as discrete, rounded grains of thorium uraninite dispersed within coarse, typically conglomeratic sediment. Uraninite is directly associated with other detrital heavy minerals, including abundant pyrite, brannerite, and (in the Witwatersrand basin) gold. Associated heavy-mineral suites typically show hydraulic equivalency, which along with rounded shapes and texturally controlled distribution, supports the generally accepted detrital origin for the uraninite. Grain sizes are small, ranging between a few hundredths to two tenths of a millimeter in diameter (Theis, 1979; Pretorius, 1976). Highest concentrations occur in the matrix of clast-supported conglomerate, along sandy partings and lenses within conglomerate, and along bedding and cross-bedding surfaces that are rich in pyrite. Richest concentration occurs in conglomerates or lags on erosion surfaces. In plan view, highest grades of uranium parallel trends established by conglomerate distribution or erosional morphology (Fig. 13-6).

The broad zone of uranium distribution within a fan, valley fill, or other fluvial element occurs as a strike-parallel band, up to several kilometers in width, that represents the portion of the drainage element that had the correct hydraulic properties to concentrate heavy mineral grains of the appropriate grain size and density. Typically, highest concentrations of uranium occur down flow from maximum gold concentrations in the Witwatersrand (Minter, 1978). This pattern, however, reflects the source-controlled grain sizes and thus relative hydraulic equivalency of uraninite and gold particles, and would not necessarily hold for other provinces. In contrast, heavy minerals in the Late Proterozoic (?) Van Horn wet alluvial-fan system of West Texas, are concentrated in distal fan sandstone facies rather than in coarser, conglomeratic channel fill deposits (McGowen and Groat, 1971). The combination of grain size and shape, determined by source rocks and the hydraulic properties of the host fluvial-fan system combine to determine the preferred site of accumulation of a specific heavy mineral within a particular stratigraphic unit.

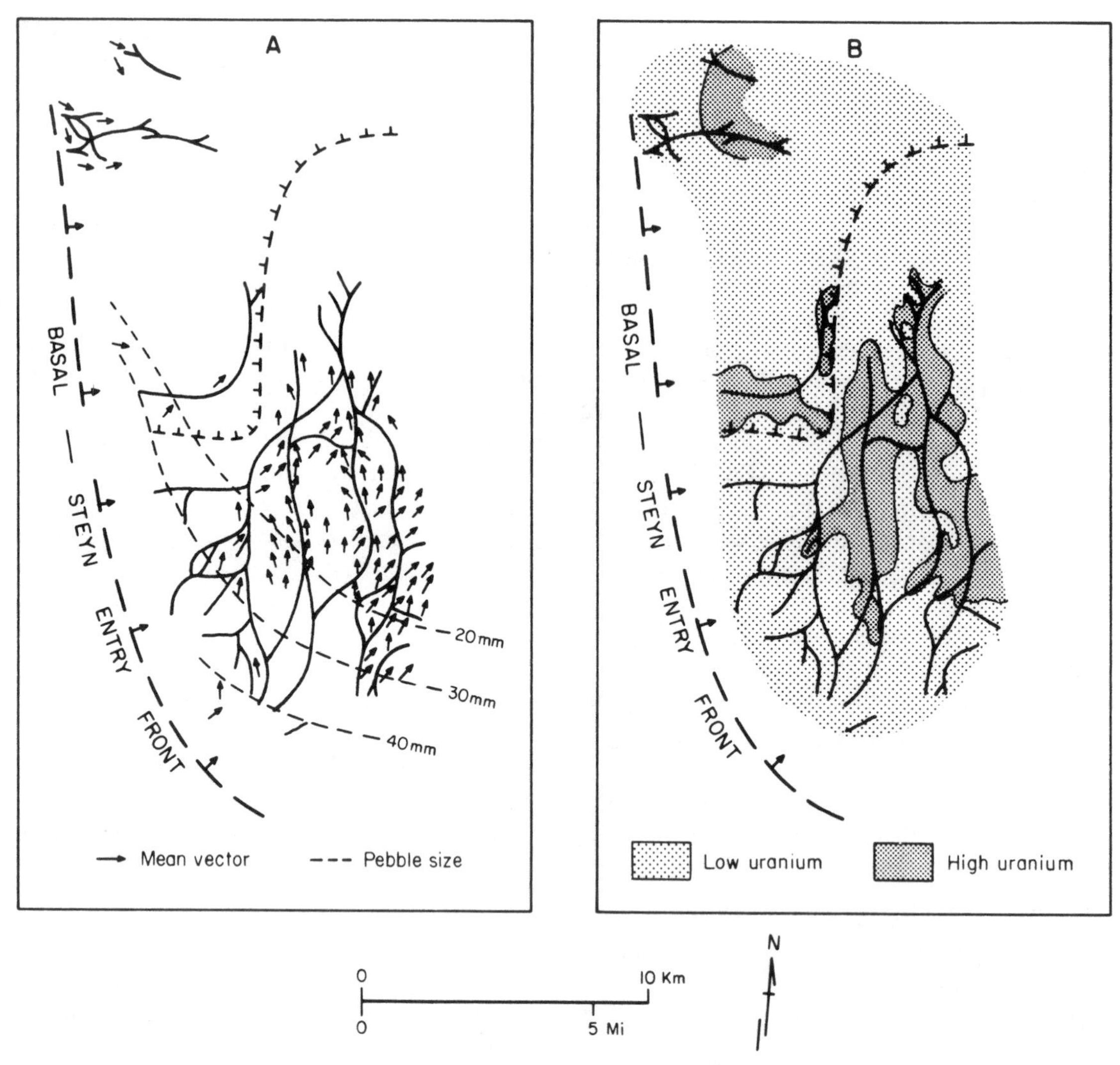

Figure 13-6. Maps of the Basal-Steyn placer deposit, Central Rand Group, showing: (A) paleocurrent vector means and down-flow decrease in maximum pebble size, and (B) distribution of uranium within the placer. In both maps, combined gold-uranium mineralization trends are shown by the anastomosing solid lines, and the "T" pattern outlines the junction of the Basal and Steyn placers. (Modified from Minter, 1978.)

Secondary, or redistributed uranium is concentrated in the organic films and laminae (thucolite) that occur within the sandy braided channel facies of the Central Rand Group fans. In addition, considerable uranium occurs at Elliot Lake in titanium-rich grains that were produced by postdepositional alteration of detrital iron-titanium oxides. Such intrastratal redistribution of some primary uranium both by syndiagenetic and epigenetic processes is to be expected in the highly transmissive host sediments. Adsorptive solids, such as organic material or amorphous titanium oxide would effectively trap uranium leached by postdepositional ground water flow.

Although uranium is concentrated in small amounts in many mineralogically mature Proterozoic conglomerates, as typical of syngenetic deposits, economic grades are extremely rare. Rich Witwatersrand ore bodies, for example, average only a few hundreds of ppm uranium. Nearly all uranium from the basin has been produced as a byproduct of gold extraction. The Elliot Lake province remains unique in that ore grade of large deposits averages above 0.1 percent U_3O_8 (Button and Adams, 1981).

In summary, ore deposits of the quartz-pebble type reflect a variation of the uranium cycle unique to a specific period of crustal and atmospheric evolution. Mechanical rather than geochemical weathering processes were primarily responsible for releasing uranium from its source. However, as in most applications of the cycle concept, an unusually enriched source seems necessary for large-scale economic mineralization. Potential host bed-load fluvial and wet fan systems dominated by coarse, granite/gneiss-derived sediment are common in Proterozoic rocks. Few such systems yield economic concentrations of uranium, however. Following mobilization, uranium was transported and concentrated by repeated recycling and mechanical sorting within a high energy, surface hydrologic environment. Close association of most large deposits with regional unconformities documents the importance of erosion and sediment recycling in the concentration process (Pretorius, 1976; Button and Adams, 1981). Final concentration of the heavy mineral-enriched fraction occurred in localized environments characterized by sediment bypass and high turbulence rather than in sites of diverging flow and comparatively rapid deposition (Smith and Minter, 1980). Ore bodies are distinctly stratabound and typically thin, multilateral, and dip elongate, all reflections of the geometry of the host braided bed-load channels.

Syndiagenetic Uraniferous Lacustrine and Swamp Deposits

Uranium concentration, locally attaining economic grade and volume, occurs in basin-center lacustrine playa and swamp depositional facies ranging from late Quaternary lowmoor bogs of Russia to Miocene fill of the Date Creek Basin of Arizona (U.S.A.) and the Permian Lodeve Basin of France. Various hypotheses for the origin of potentially economic deposits of the Lodeve and Date Creek basins have been proposed. Syngenetic or a combination of syngenetic and subsequent epigenetic processes involving compactional expulsion of pore waters have been favored. Stratigraphic and basin setting, as well as the nature and distribution of mineralization, suggest that such deposits are likely to be the product of syndiagenetic processes as described in this chapter.

The lowmoor uraniferous peat deposits have been well described (Kochenov *et al.*, 1965; Lisitsin *et al.*, 1967), and provide an example of active syndiagenetic mineralization. Lowmoor bogs and interspersed lakes occupy the center of ground-water basins; plant growth is a product of abundant ground-water discharge and the associated shallow water table. Uranium is concentrated by the organic-rich bog sediments from discharging or laterally inflowing ground water. Concentration processes, which efficiently remove uranium from extremely dilute waters, are believed to involve both adsorption and direct reduction and precipitation of reduced uranium minerals (Lisitsin *et al.*, 1967). The hydrologic and geochemical processes responsible for mineralization are reflected in several predictable ways:

1. Areal distribution of uranium-enriched peat is irregular. Some portions of the deposits contain no excess uranium, whereas adjacent deposits that are subjected to intense ground-water flow are uraniferous.
2. Uranium enrichment is greatest toward the base of peat beds, along the margins of the moor, and above buried permeable, oxidized fluvial sediments. Again, richness is directly a function of the potential for contact with upward or laterally flowing ground water.
3. Concentrations of uranium increase in dry climates, where abundant, circulating, oxidized and mineralized ground water encounters the slightly acidic, reducing peat beds.

Comparison of uraniferous lowmoor peat deposits with uranium ores in ancient lacustrine systems reveals many similarities, suggesting a comparable syndiagenetic origin.

Lacustrine Deposits of the Date Creek Basin

The geology of the uraniferous lacustrine fill of the northern Date Creek Basin has been described by Sherborne *et al.* (1979). The depositional basin was produced by Early Miocene block faulting of the Basin and Range type. A fault bounded, restricted embayment at the northern end of the basin was filled by sediments of the Anderson Mine Formation (Fig. 13-7), which is bounded above and below by incised, erosional unconformities. Up to 400 ft (120 m) of relief was present on the lower erosional surface. Basin filling began with deposition of fluvial sediments, consisting of basal arkosic sands and gravels. With rising lake level, a widespread unit con-

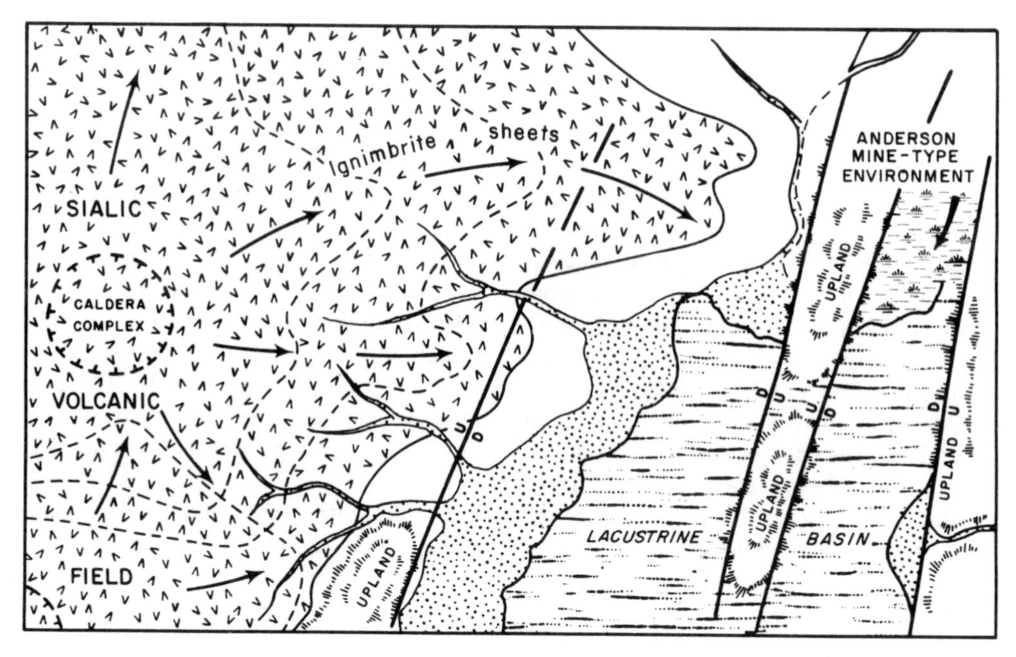

Figure 13-7. Geologic setting of the down-faulted Date Creek Basin, Arizona during deposition of the uraniferous Anderson Mine Formation. (From Sherborne *et al.*, 1979.)

sisting of interbedded tuffaceous mudstone and siltstone, water-reworked tuff, marlstone, limestone, carbonaceous mudstone, and fine-grained arkosic and tuffaceous sandstone onlapped the lower fluvial deposits and surrounding basement highs. Contemporaneous volcanic centers to the west filled the main portion of the basin with volcaniclastic sand and rhyolitic tuffs (Fig. 13-7). Palynological evidence indicates surrounding mountainous areas characterized by a temperate climate, whereas the depositional basin was semiarid.

Major facies assemblages within the lacustrine depositional system of the restricted Anderson Mine arm of the basin include uraniferous carbonaceous mudstone, calcareous tuff, fossiliferous limestone, and reworked, tuffaceous siltstone and sandstone. The carbonaceous mudstone facies include thin, discontinuous interbeds of mudstone, siltstone, tuffaceous, gastropod-rich marlstone, and local impure lignite. The facies assemblage is best developed in flooded inlets and depressions along the lake margin and interfingers basinward with fossiliferous limestone beds, which contain various freshwater snails and bivalves, as well as macerated carbonaceous materials. Calcareous tuffs are massive to burrowed subaqueous deposits. Tuffaceous siltstone and fine to medium sandstone are scattered through lake margin facies. The feldspathic and volcaniclastic sand bodies display both sharp and progradational bases. The arenites are green to gray and are interbedded with micritic limestone and carbonaceous mudstone. Throughout, the lacustrine sequence is distinguished by its textural and compositional variability as well as thin, laterally discontinuous bedding. Diagenetic modifications to the mineralogically complex lithologic suite include abundant silicification, calcification, argillation, and zeolitization.

Facies variability exhibited by the Anderson Mine Formation is a product of the environmental mosaic of a lake margin. Vertical and lateral facies changes reflect fluctuating lake levels, irregular bottom topography, and variable geochemistry typical of a closed, tectonic lake basin. The diversity of diagenetic products is similarly characteristic of alkaline saline lakes containing abundant, highly reactive volcanic debris (Hay, 1977).

Uranium mineralization consists of dispersed, poorly crystalline coffinite forming low-grade, stacked stratiform ore bodies that individually are less than 10 ft (3 m) thick, and cover areas of about one square mile (2.6 km^2). Several aspects of mineralization are directly comparable to mineralization within the lowmoor peats. (1) Uranium is closely associated with lake-margin carbonaceous mudstones, although it is not restricted to lignitic units. (2) Lower portions of lacustrine facies assemblages are commonly more abundantly mineralized, suggesting an upward component to migration of mineralizing fluids. (3) Ore-grade mineralization is localized within otherwise laterally uniform strata, and occurs within a variety of fine-grained sediment types of the carbonaceous mudstone assemblage. Thus, syndepositional concentration in a particular depositional environment is not indicated. (4) Interbedded and subjacent fluvial sandstone and conglomerate deposits are commonly oxidized and nonuraniferous.

In summary, both the lowmoor bogs as well as lacustrine systems exhibit similar styles of uranium mineralization that can be most simply explained by a mineralization cycle operative during deposition of the host lacustrine system. The cycle involves leaching and transport of uranium from source sediments or rocks of the subaerial basin margin by oxidizing meteoric ground water, and concentration in localized areas of ground-water discharge through reduced but unconsolidated lake-margin sediments. The Date Creek deposits closely parallel the observations and criteria established by studies of young, uraniferous playa lakes (Leach *et al.*, 1980). Postdepositional, epigenetic mineralization by expelled pore waters of the finer-grained lake sediments is not precluded, but seems unnecessary. The efficiency of such compactional expulsion may be further questioned because the supply of interstitial water is limited and because it is likely to be reducing (and thus a poor carrier of dissolved uranium).

Such a model predicts partial facies control of concentration in lake margin sediments, which are characteristically deposited within a zone of regional ground-water discharge. Localization of enrichment in geochemically favorable facies reflects topographic, basement, and shallow formational features that focus ground-water flow and discharge into restricted areas of the basin margin. Organic material is important both as a feedstock for bacterial generation of syngenetic sulfide, which reduces introduced uranium, and as a potential adsorber of uranium.

Yeelirrie Calcrete Uranium Deposits

Calcrete uranium ores, which are best known for the major deposits of the Yeelirrie districts of Australia, offer a somewhat different example of syndiagenetic mineralization produced by genetically interrelated depositional and ground-water processes (Mann and Deutscher, 1978a and b). Chemical processes dominate erosion and deposition, as well as uranium transport and concentration. Interestingly, the general geologic setting of these calcrete ores differs little from that of the quartz-pebble conglomerate ores. The Yeelirrie calcretes occur within a closed cratonic basin floored by granitoids and greenstones of an Archean basement complex (Fig. 13-8). Mineralized calcretes are a major facies component of fluvial and lacustrine-margin delta systems. However, in contrast to the Proterozoic ores, mineralization is the product of intense chemical weathering of a low relief terrane under an increasingly arid climatic regime in late Tertiary and Quaternary time. Weathering products include abundant kaolinite, silicified crusts, and thick colluvial mantles.

Uraniferous calcrete occurs along the surface drainage axis of ephemeral streams and at the margins of playa lakes. Uranium is in the oxidized form as the mineral carnotite (Fig. 13-8). Precipitation of diagenetic phases, including the abundant Ca and Mg carbonates of the calcrete itself, is such an important depositional process that the lake-margin deposits are called "chemical deltas." The associated lacustrine systems contain bedded evaporite deposits.

The prevailing hydrogeologic system, though apparently only active in redistributing carnotite already in place, provides considerable insight into the hydrodynamics and geochemistry of primary mineralization (Mann and Deutscher, 1978b). Uranium is slowly leached from the intensely weathered bedrock highlands and colluvial mantle by recharging meteoric ground water. The source of vanadium, a necessary ingredient for carnotite precipitation, is less certain, but is likely to be the greenstones of the basement complex. Ground-water flow is focused into the topographically low valley axes where the water table is comparatively shallow. Within regional flow cells, ground waters show increasing contents of calcium, potassium, chloride, and uranium. Carbonate content is also high. Along

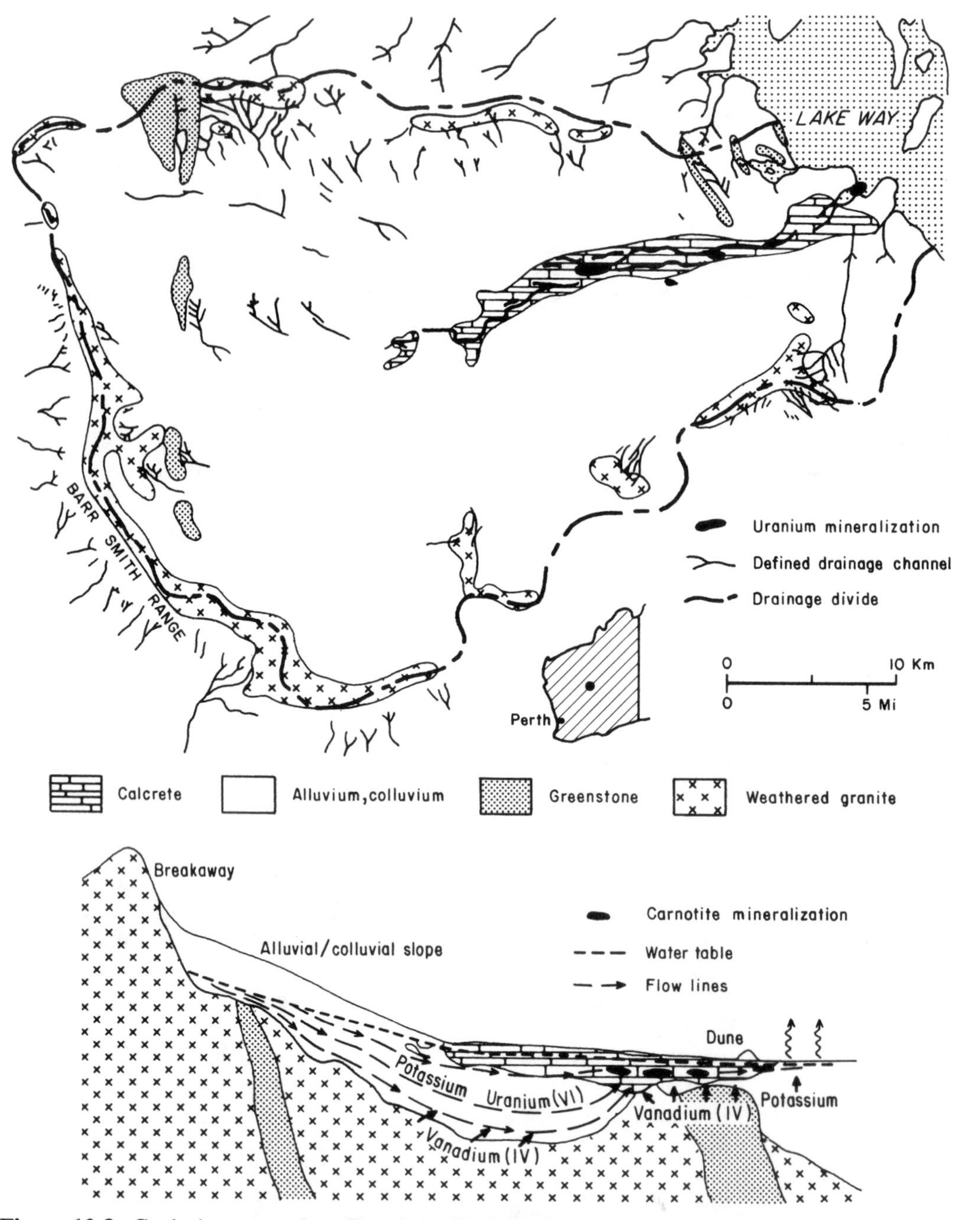

Figure 13-8. Geologic map and profile of the Yeelirrie drainage basin, Western Australia, showing the distribution of calcrete and its contained uranium deposits. Mineralized ground-water converges on the drainage axis, flows down the channel trend, and discharges along lower reaches of the channel in response to bedrock ridges or the flattened topography of the lake margin. (Modified from Mann and Deutscher, 1978b.)

the lower reaches of the alluvial valley fill and at their deltas, discharge and evapotranspiration results in precipitation of calcium and magnesium carbonate (Fig. 13-8), and in increasing concentration of other dissolved solids, including uranium. Together, evaporative concentration, decomplexing of uranium and dicarbonate due to precipitation of carbonate minerals or mixing of water masses at the playa margin, and increasing Eh and oxidation of dissolved vanadium (IV) combine to favor direct carnotite precipitation within regional or local zones of ground-water discharge (Fig. 13-8).

The deposits are thus the product of long-lived, contemporaneous depositional and hydrologic systems. Localization of sites of active mineralization reflect both the depositional and hydrogeologic environments. An analogous syndiagenetic model for mineralization in evaporative basin margins, or sabhkas, that was initially developed for sedimentary metal deposits, has been suggested for other uranium occurrences (Rawson, 1980).

Epigenetic Sandstone Uranium Deposits

Sandstone-hosted epigenetic ore deposits are the dominant type of uranium occurrence in the United States. In addition, major reserves are known in Russia and Niger, and ore bodies have been found in many countries. Approximately 40 percent of world reserves lie within sandstone ores (Robertson *et al.*, 1978). Size of individual deposits is small. About 99 percent of all United States ore bodies contain less than 3 million pounds (1,360 m.t.) of U_3O_8; larger districts rarely contain more than 100 million pounds (45,500 m.t.) in a series of contiguous ore bodies (Robertson *et al.*, 1978). Ore bodies assume a variety of morphologies that reflect the geometry as well as internal sedimentary features of the host sand unit. A common form, particularly for Tertiary deposits, is called a roll or mineralization front. In cross section, the uranium-enriched roll is stratabound, displays an irregular "C"-shaped outline, and lies at the boundary of the oxidation-alteration tongue (Fig. 13-9). In plan view, the front is elongate and sinuous. Remnants or islands of mineralization may lie behind the mineralization front within unaltered portions of the sand aquifer. In some geochemical settings continuous mineralization fronts break up into isolated pods or tabs within strata characterized by poorly defined iron-alteration zonation.

Most sandstone-hosted ore occurs within facies of terrestrial systems deposited in closed, continental basins (Stokes, 1967; Gabelman, 1971). Marginal marine coastal plains are a secondary, but important, setting for sandstone deposits. Thus, sandstone uranium host systems provide good examples of fluvial, alluvial fan, and strandline depositional facies, as well as of ore deposit morphologies and mineralization histories typical of epigenetic deposits.

Terrestrial Systems: Wyoming Tertiary Basins

Large uranium districts in late Paleocene and Eocene units of Wyoming include the Fort Union, Wind River, Wasatch, and Battle Spring Formations. Geology of the province has been synthesized by Galloway *et al.* (1979b) and Harshman and Adams (1981). The description by Harshman (1972) of the Shirley Basin district remains one of the most thorough descriptions of Wyoming roll-type mineralization.

Major Tertiary uranium districts of the central Wyoming intermontane basins occur in or along oxidation-alteration tongues coincident with bedload fluvial and fan axes of two major integrated late Paleocene–Eocene fluvial drainage systems (Fig. 13-10). The largest system drained along the Wind River Basin across the Casper Arch, and then northward along the axis of the Powder River Basin. Tributaries drained portions of the Precambrian granite cores of the Owl Creek, Sweetwater (Granite Mountains), Laramie, Wind River, and Hartville Uplifts (Fig. 13-10). The Gas Hills and Shirley Basin districts, and the small Copper Mountain deposits occur in fan and proximal fluvial deposits of this system. Regional alteration tongues all indicate groundwater movement roughly parallel to the drainage direction. In the Powder River Basin, a broad belt composed of multiple alteration tongues, which tend to climb section northward along the basin, occupies the central axis of the system and is bounded on the east by several districts (Fig. 13-10). Crooks Gap is the only major producing district within the southwestern fluvial system, which drained into the Green River Basin (Fig. 13-10). Uranium occurs in proximal tributary-fan deposits. Post-depositional deformation reversed local dip before the north-trending epigenetic alteration tongue formed.

The regional similarities displayed by these districts in terms of facies associations, timing of alteration tongue development (as defined by structural growth, flow reconstruction, and isotopic age dating), potential for recharge through overlying tuffaceous Oligocene–Miocene units,

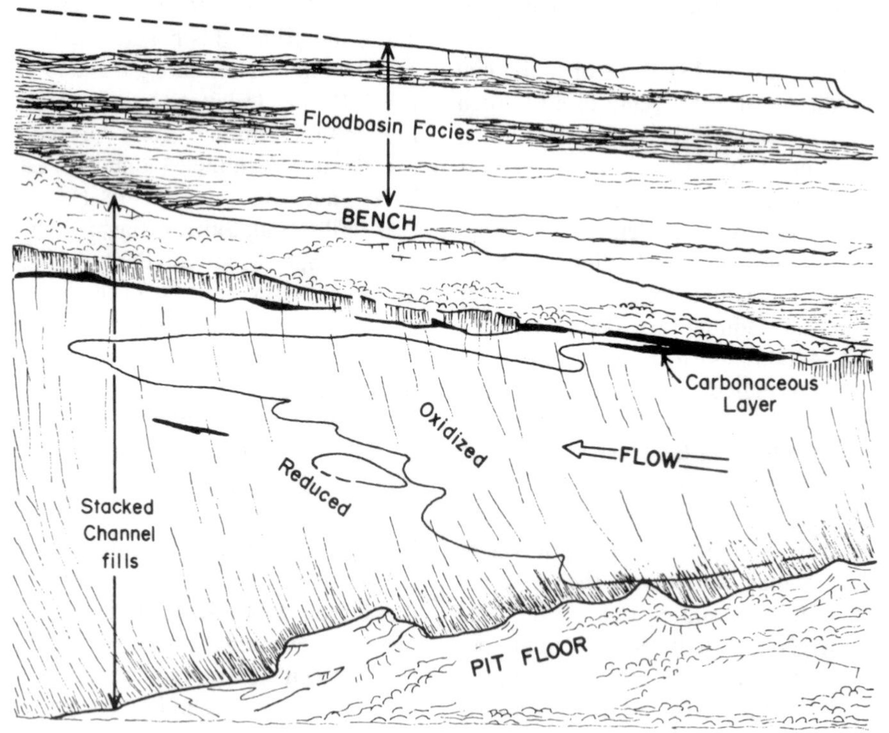

Figure 13-9. Mineralization front, or roll front, exposed by mining, Shirley Basin, Wyoming.

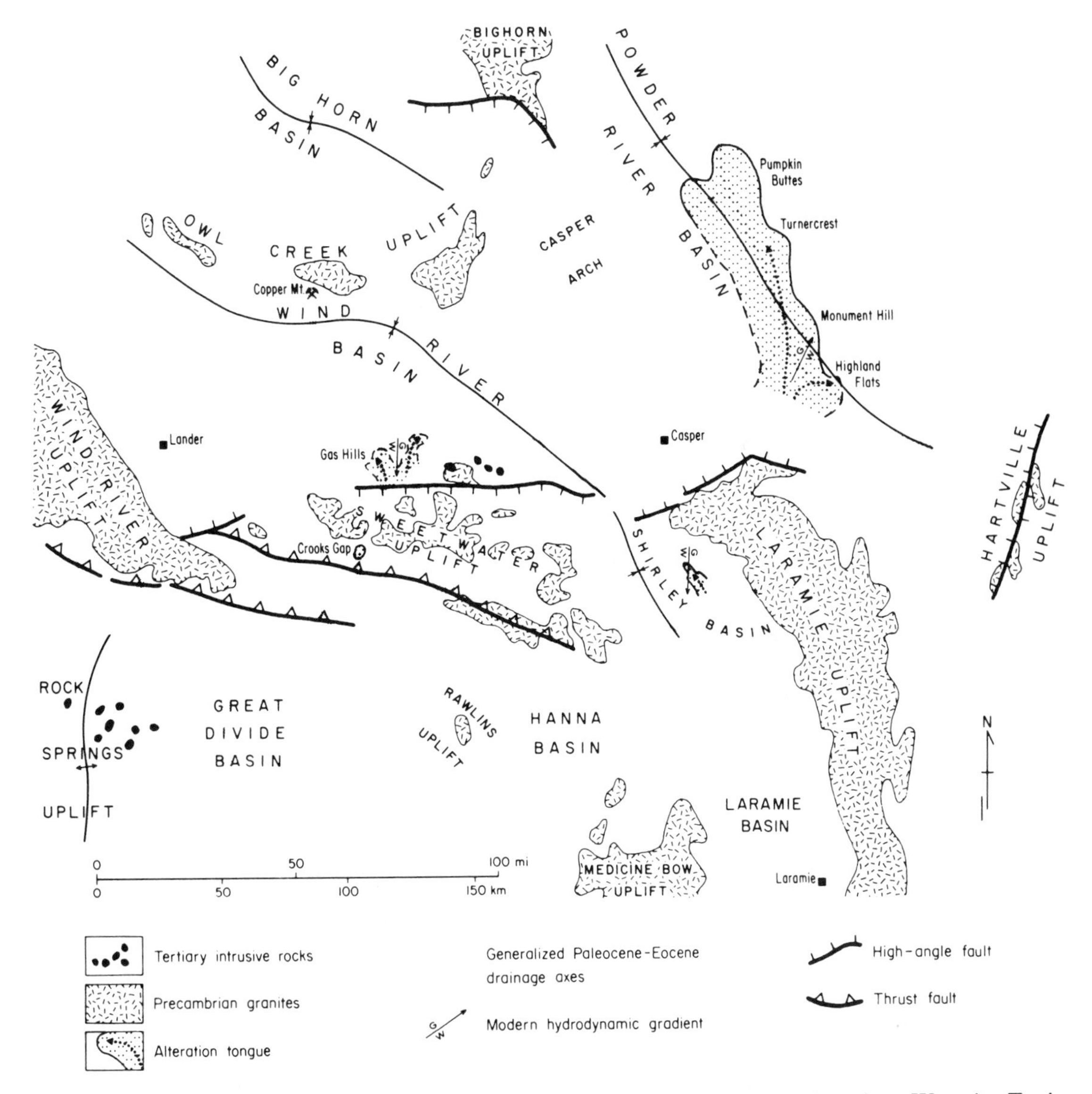

Figure 13-10. Regional setting of the principal uranium districts of central and northern Wyoming Tertiary intermontane basins. Major alteration tongues and their associated uranium-producing districts occur along tributary alluvial fan or channel axes of two large, integrated late Paleocene/Eocene fluvial systems, which drained eroding, granitic uplands. (From Galloway *et al.*, 1979b.)

and nonconcurrence of preserved alteration patterns with modern ground-water flow systems suggest a common history:

1. In limited areas of the Tertiary basins coarse-grained fan and fluvial facies were deposited and became highly transmissive aquifers upon burial. Granitic uplands, which were faulted, uplifted, and unroofed in late Paleocene time, provided the major sources of gravel and coarse sand-sized detritus to the active fluvial systems. Consequently, highly transmissive aquifers naturally developed on the periphery of granitic uplands, and were deposited only during early Tertiary uplift and erosion of these uplands.

2. Regional tectonic activity during the late Eocene disrupted the drainage network, placing portions of the axial fan and channel-fill deposits in structural and topographic positions suitable for active meteoric recharge along basin-margin outcrop belts.

3. Burial of basins and intervening uplifts by volcanic debris in Oligocene and Miocene time resulted in recharge of the highly transmissive aquifers by waters that had percolated through freshly deposited siliceous, uraniferous, volcanic debris. Geologic relationships of alteration tongues of the major districts indicate that this period was the most likely time of alteration and ore formation.

4. Late Tertiary and Quaternary regional uplift and exhumation of the basins initiated a period of active erosion and surficial modification of the primary mineralization trends. Modern ground-water flow directions are commonly discordant to inferred flow direction during the formation of the oxidation tongues and associated mineralization (Fig. 13-10).

Puddle Springs Wet Alluvial Fan System: Gas Hills District. The Puddle Springs Arkose Member of the Wind River Formation (Eocene) was deposited along the southeastern margin of the precursor Wind River basin on a northward-sloping, incised erosion surface. Uplift of the Granite Mountain block resulted in progressive unroofing of the Mesozoic and Paleozoic sections, recorded by deposition of fine-grained lower Wind River clastics, and erosion of considerable Precambrian granite, recorded by the coarse wedge of the Puddle Springs fan system (Soister, 1968; Love, 1970).

The Puddle Springs fan system consists of two north–northeast-trending belts of massive sandstone and conglomerate embedded within a lobate wedge of sandstone and subordinate sandy mudstone and shale (Fig. 13-11). The coarse belts consist of mid-fan facies, including tabular sheetflood and braided-channel deposits. Interbedded massive, but relatively gravel-poor, sand units are distal fan and interfan braided-stream deposits. These coarse facies grade laterally into interbedded sand and mudstone deposits of distal fan and fluvial origin. Width of the Puddle Springs fan system is about 12–15 miles (19–24 km), and the system is elongated north–south. Thickness of the unit is partially controlled by buried bedrock topography (granite knobs and breached anticlinal ridges project well up into the Puddle Springs) and ranges from 400 to 800 ft (120–240 m).

Internal features of the sand bodies suggest deposition by multiple, laterally migrating, shallow, flashy, low-sinuosity, braided bed-load streams. Transitional flow regime structures, including planar stratification and low-angle trough cross stratification, are abundant, and small scour channels and gravel bars are locally preserved. Textural properties vary abruptly, both laterally and vertically. Local muddy, organic-rich channel plugs and muddy sand beds and lenses were deposited on the distal fan plain or in braided channels during waning flow.

Uranium deposits occur along three subparallel trends. About one-half of the total uranium reserves are evenly divided between the outer flanks of the eastern and western lobes (Fig. 13-11). A third trend, containing about one-half of the district reserves, lies in the center of the fan system, flanking the western lobe. Regional patterns and local front trends suggest a rather simple first-order picture of two north-trending oxidation-alteration tongues centered around the transmissive coarse conglomeratic sand lobes, and enclosed within the main mass of the fan system. Ore-grade mineralization occurs as classic C-shaped rolls that follow alteration fronts; ore bodies are commonly concentrated within muddy sand intervals along the margins of the coarsest facies. Conglomerates also contain uranium mineralization, but host few major ore bodies.

Upper Fort Union and Wasatch Fluvial System: Southern Powder River Basin. Continental fluvial and lacustrine deposition prevailed in the Southern Powder River Basin from late Cretaceous through middle Paleocene time. The drainage system was low-gradient, dominantly suspended-load, and poorly integrated. Uplift of the flanking highlands in late Paleocene time produced an influx of coarse, arkosic debris through a well-integrated, axially trending bed-load to mixed-load fluvial system (Fig. 13-10) that is preserved as parts of both upper Fort Union and Wasatch Formations (Galloway *et al.*, 1979b). The fluvial deposits can be subdivided into several mappable sandstone packages, which illustrate depositional style and ground-water flow evolution (Fig. 13-12). Axes of the sand packages initially trend along the structural axis of the basin, but northward the system diverges to the east, suggesting offset of drainage by tributary alluvial fans derived from the Big Horn Mountains.

Late Eocene thrusting along the flanks of the Wind River Basin terminated the influx of sediment across the Casper Arch (Love, 1970), and

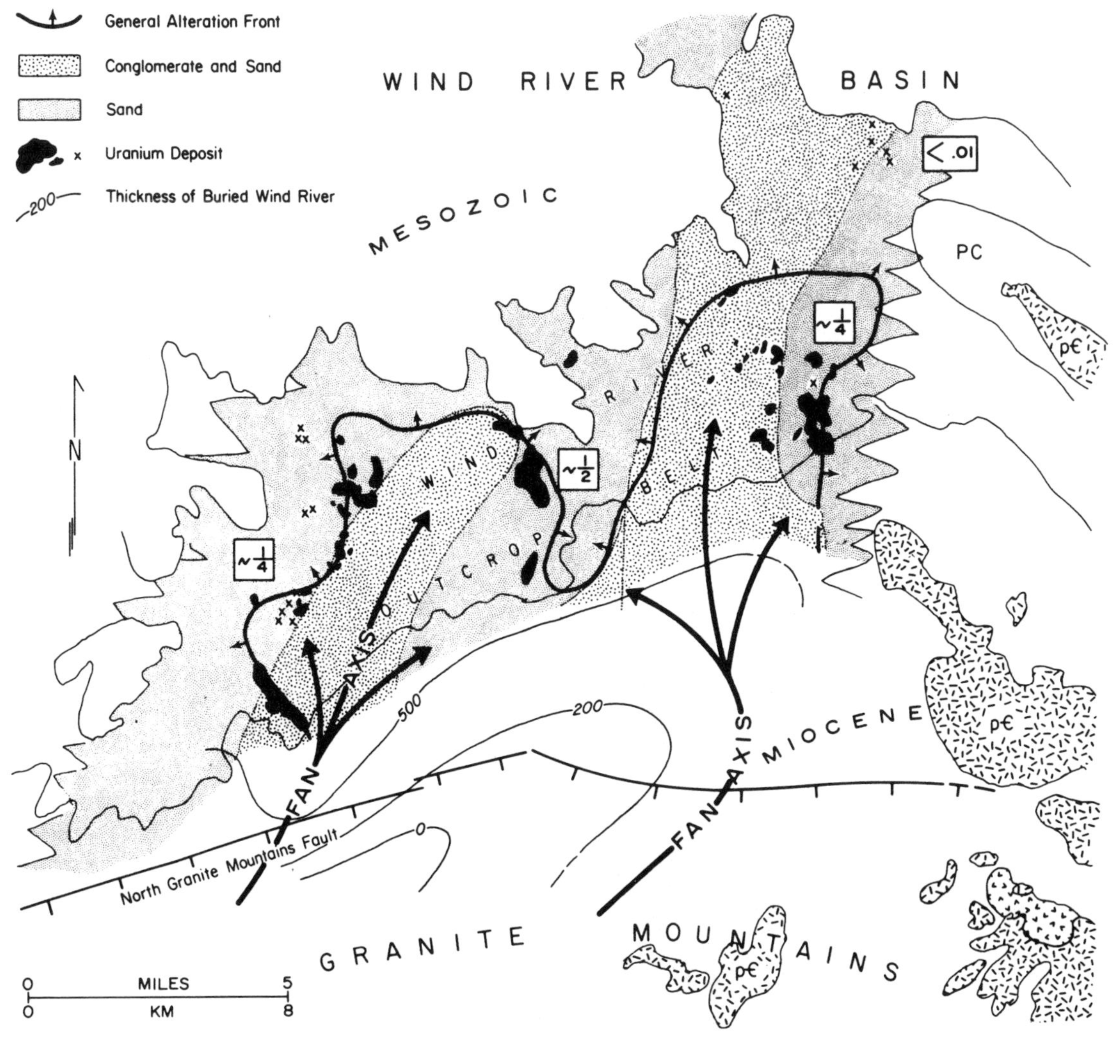

Figure 13-11. Depositional patterns and distribution of alteration and associated uranium mineralization in the Puddle Springs alluvial fan system (Gas Hills uranium district). Two major alteration tongues extend through the fan axes, separated by a reentrant within the intervening, less conglomeratic interfan area. The relative proportion of total district reserves contained in east, west, and central portions of the district are indicated in the boxes. Outcrop mapping and location of fronts in part after Soister (1968) and DeNault (1974). (From Galloway *et al.*, 1979b.)

regional incision resulted in an extensive unconformity between the Wasatch and overlying Oligocene White River deposits. Further uplift along the basin margin produced dips into the basin within the fluvial system ranging from 2 degrees or less along the southern outcrop margin to as steep as 20 degrees along the western margin. The structural picture has changed little since Eocene disturbance.

Framework facies of the upper Fort Union fluvial system consist of a core of mixed-load to bed-load channel-fill sands that interfinger laterally and downslope into fine mixed-load channel fills.

The mixed-load/bed-load channel facies was deposited by broad, sinuous to straight, laterally migrating channels of shallow to moderate depth. The facies is characterized by medium to coarse sand containing beds and lenses of sandy conglomerate, conglomeratic sandstone, pebbly sandstone, and granule conglomerate (Fig. 13-13). In general, sand units contain more than 50 percent medium to coarse sand. Local claystone, mudstone, and muddy sandstone drapes and

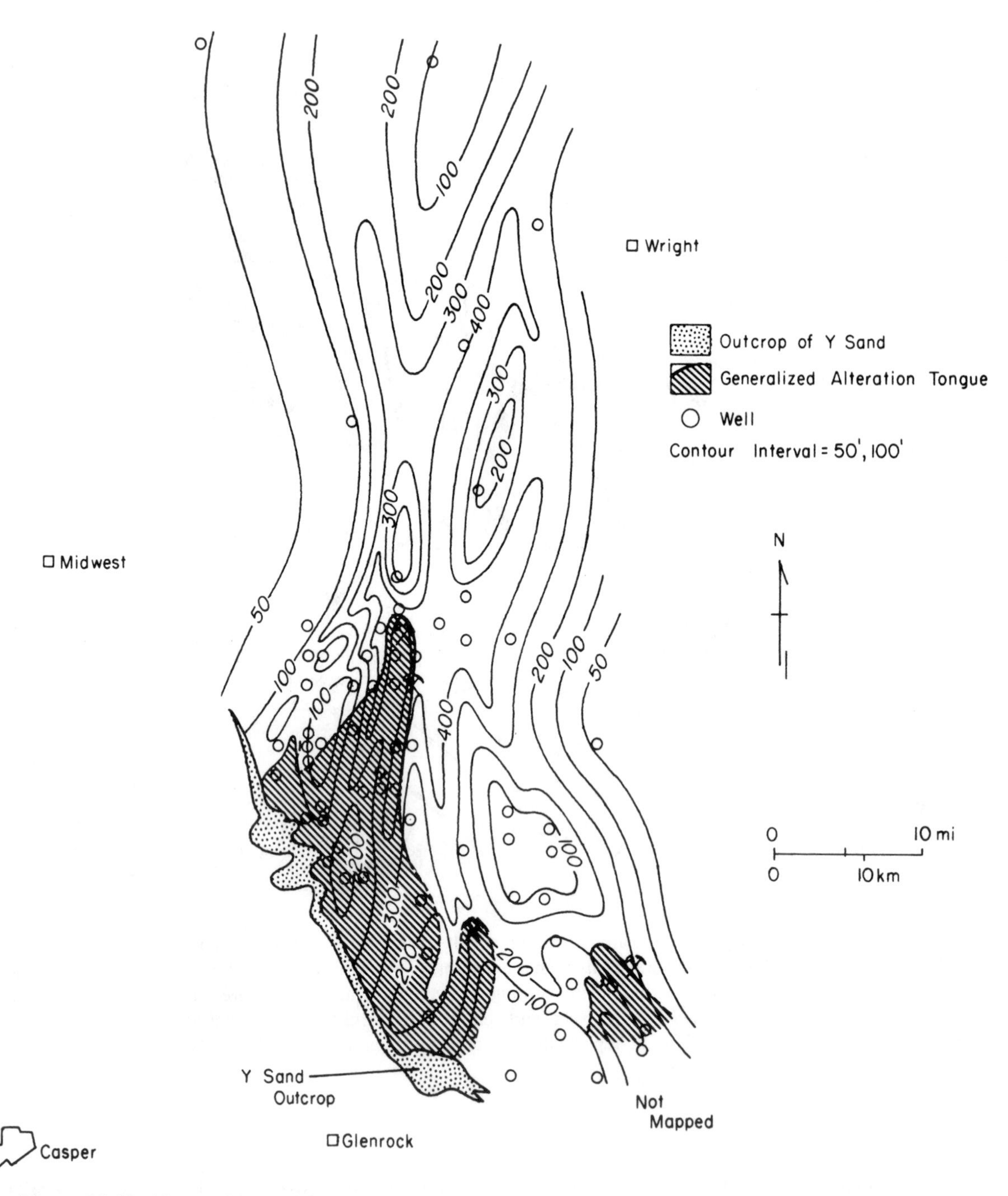

Figure 13-12. Net sand isopach map of a sand package within the upper Fort Union Formation, southern Powder River Basin. The interpreted alteration tongue geometry reflects the sand distribution pattern. Large uranium deposits lie along the tongue. (From Galloway *et al.*, 1979b.)

lenses occur within the sand body. Basal contacts are sharp and irregular. Upper contacts may be sharp or gradational over a few feet. Sand bodies are lenticular to tabular with a high width to thickness (w/t) ratio; close-spaced drilling indicates consistent w/t values ranging between approximately 170 and 230. Sedimentary structures, vertical sequence, and log response of mixed-to-bed-load channel fills are illustrated graphically in Figure 13-13. The suite of internal structures, abundance of mud clasts and mud drapes, and common deformation structures all

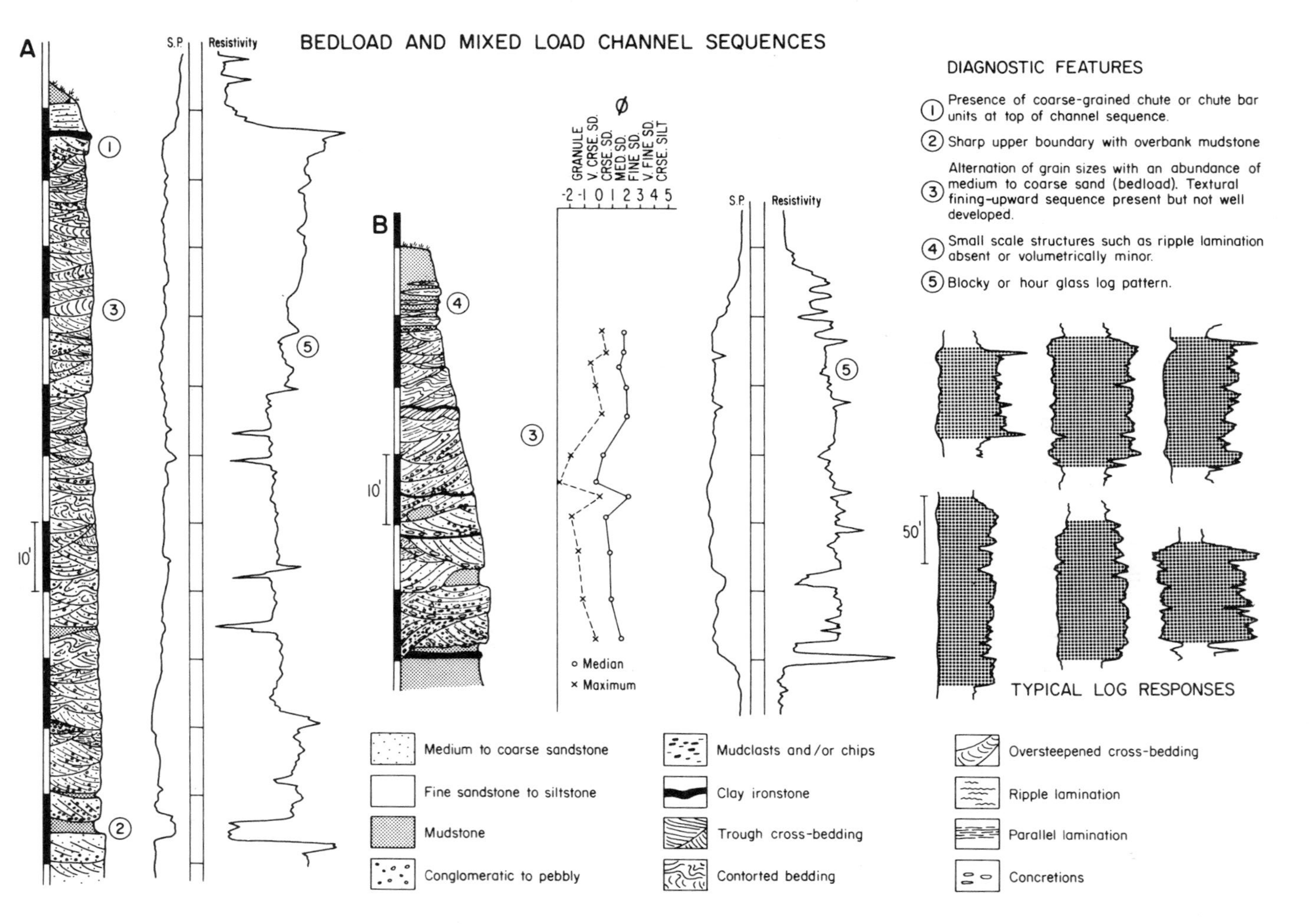

Figure 13-13. Sedimentary features of bed-load and coarse mixed-load channel fills that occupy the axis of the upper Fort Union fluvial system in the southern Powder River Basin. (From Galloway *et al.*, 1979b.)

indicate extreme variation in stream discharge and flow depth.

The fine mixed-load channel facies was deposited by sinuous streams, which in turn resulted in better preservation of upper point-bar and overbank deposits, greater variation in depth of scour at the base of channel fills, and formation of large-scale lateral accretion bedding within the channel fill. Textural parameters are quite variable, but channel fills contain a significant percentage of very fine sand and silt. Median sand size ranges from very fine to medium; lenses of conglomeratic sand occur near the base of the channel. Suspended-load channel sands typically grade upward into overlying silts and muds. Sand bodies are lenticular with abundant overbank mudstone.

Bounding facies of the Fort Union and Wasatch fluvial systems consist of fine-grained, heterogeneous sediments of tributary channel, well-drained floodplain, crevasse splay, lacustrine, and swamp origin (Ethridge *et al.*, 1981). Units are discontinuous and show both burrowing and root disturbance. To the north, backswamp and lacustrine conditions became more prevalent as the gradient and coarseness of channels decreased, and extensive coal beds surrounded the remnants of the fine-grained mixed-load fluvial axes.

Uranium ore is concentrated within and along the flanks of major sandstone axes containing coarse mixed-load to bed-load channel facies. Mineralization is found primarily along laterally persistent alteration fronts, and forms both roll and tabular ore bodies. Details of the geometry and mineralogy of the Highland deposits [estimated at 35 to 50 million pounds (17,000 to 23,000 m.t.) of U_3O_8] were given by Langen and Kidwell (1974) and Dahl and Hagmaier (1974). A contrasting style of mineralization (Childers, 1970) occurs in fine-grained mixed-load and suspended-load channel facies along the western bank of the basin and which predate the integrated axial drainage. Here, ore bodies appear to be small and discontinuous and, at present, are not credited with large reserves.

As is typical of many districts, ore distribution and the extent of alteration locally correlate with sedimentary features such as channel margins, but on a regional scale no consistent correspondence between specific genetic facies and mineralization has been documented.

Ground-Water Flow History. Evolution of the Fort Union Wind River/Wasatch aquifers of the Powder and Wind River basins was influenced by four major geologic events (Table 13-1).

1. Deposition of the tributary fan and fluvial systems within open intermontane basins established the three-dimensional permeability distribution of the aquifer system. Flow evolved from unconfined to semiconfined as successive fluvial units were deposited. The regional hydrodynamic gradient was toward the topographic axes of the basins, and then parallel to the main fluvial axes toward the open ends of the basins. Recharge occurred in tributary fan and fluvial facies; ground water moved down regional gradient parallel to the trend of the principal fluvial axes and discharged in the topographically low basin centers. Thick accumulations of lignite in the Powder River Basin document the presence of a shallow, permanent water table or lake indicative of regional discharge. Widespread preservation of detrital plant debris within channel-fill deposits also indicates a shallow water table and surplus of groundwater.

2. Tectonism in Late Eocene time segmented the regionally integrated drainage system. Basin margins were uplifted, and deposits of the fluvial system were tilted. Recharge was shifted to the newly exposed basin-margin outcrop belts of the aquifer. Discharge occurred along the structurally defined basin centers.

3. During Oligocene and Miocene time, thick sequences of ash and reworked volcaniclastic sediments filled the basins, covering the exposed margins of the fluvial system. Recharge to these confined aquifers, which remained highly porous and permeable, percolated through the overlying tuffaceous sediments, and was collected along the subcrop belt of coarse, sand-rich bed-load and mixed-load channel facies. Flow within these aquifers was most likely eastward or northward toward open margins of hydrologic basins. The climate became increasingly arid.

4. With regional uplift in the Pliocene, the Powder River and Wind River basins were exhumed. The Fort Union, Wind River, and Wasatch were again exposed to direct recharge along their outcrop belts. Modern ground-water flow moves in response to gradients established by the marginal highlands and the valleys of major rivers such as the Cheyenne (Dahl and Hagmaier, 1974). Holocene flow systems are
the Powder River and Wind River basins were

Table 13–1. Summary geologic and hydrologic history, Powder River and eastern Wind River basins.

	Structural Event	Depositional Event	Hydrologic Event
Paleocene/Kr		Deposition of Lower Fort Union Fm.	
Paleocene/Kr	Uplift of positive areas; tilting of basin flanks.	Deposition of Upper Fort Union, Wind River, and Wasatch Fluvial systems ↓	Drainage across Casper Arch and along axis of Powder River Basin (basic aquifer geometry and transmissivity determined)
Eocene	Thrusting along west Casper Arch		Drainage across Casper Arch blocked
Eocene	Regional subsidence & incision	Widespread erosion	Regional flow cells established within structural basins.
Oligocene	Extensive volcanism to west ↓	White River Fm, deposited ↓	Recharge through ash-rich sediments ↓
Miocene			Age dates for primary uranium mineralization ↓
Miocene		Local deposition of eolian and lacustrine volcaniclastic sediments	
Pliocene	Regional uplift	Basins exhumed ↓	
Pleistocene Recent			Modern incised drainage & groundwater system established. Redistribution of shallow uranium ores.

thus controlled in part by a modern integrated, erosional drainage pattern that markedly differs from the early Tertiary fluvial systems.

Mineralization History. Recently published age dates for uranium ores in Wyoming Basins (Ludwig, 1979) suggest the primary mineralization episode occurred between 20 and 35 million years ago during early Oligocene through early Miocene time. This corresponds to the period of confinement of the aquifer system by volcanic-rich Oligocene sediments (Table 13-1). Radiometric disequilibrium between uranium and its daughters indicates that some remobilization still continues, but geologic evidence and age dates support a regional primary mineralization episode concurrent with deposition of fresh, acidic volcanic debris in the then extant recharge area of the Fort Union/Wasatch aquifer system.

Regional oxidation-alteration patterns (Figs. 13-11 and 13-12) suggest that during the primary mineralization phase, waters generally moved downdip from the ash-blanketed subcrop belt of the major fluvial and fan axes. Flow was concentrated into thick sand belts that trended subparallel to the regional gradients. Where regional hydrodynamic gradient or change in channel trend forced cross-stratal flow into low-transmissivity overbank or suspended-load channel facies, alteration fronts are preserved within the zone of interfingering massive sand and mudstone. Where flow direction paralleled fluvial or fan trends, alteration tongues nose out within the belt of thickest sand. When supplies of chemically aggressive, oxidizing water were depleted, the front stabilized at the extant geochemical boundary. Stacked, massive, sand-rich channel-fill facies produced a regionally integrated aquifer. Vertical interconnection and consequent stacking of individual alteration fronts is, therefore, to be expected and is commonly observed.

Modification Phase. During uplift and denudation of the overlying Oligocene and younger volcaniclastics, the lower Tertiary section was exposed to renewed outcrop recession and direct recharge by meteoric waters. Many shallow uranium deposits have been and are being oxidized and reworked. The shallowest portions of the original alteration fronts have probably been destroyed by erosion or surface oxidation, as have most small alteration tongues that developed within less transmissive facies of the fan and fluvial systems.

Coastal Plain Systems: South Texas Uranium Province

The Wyoming roll-front model of sandstone uranium mineralization presumes a relatively uniform distribution of intrinsic reductants, primarily organic material and early diagenetic iron disulfide, throughout the mineralized aquifer. Reducing capacity is thus an inherent characteristic of the rock. This inference does not hold in many continental sedimentary sequences that were deposited in subarid to arid climates and thus subjected to intensive syndepositional oxidation. Uranium concentration in such oxidized aquifers requires earlier alteration by extrinsic reductants.

The obvious mobility of hydrocarbon and hydrogen sulfide gases has made them principal suspects as agents of epigenetic reduction where extrinsic reductants have been interpreted. However, although upward migration of a reducing fluid is commonly indicated by geologic relationships of the resultant alteration, evidence for buoyant segregation of a free gas phase has rarely been documented. In fact, characteristics of sulfidic alteration in South Texas aquifers contradict a model dominated by free gas migration. However, appropriate reducing and sulfidic geochemistries characterize deep-basin ground waters, and the existing pressure head drives these waters upward where they may interact with the supergene environment.

Uranium occurs within several formations that comprise the early to middle Tertiary clastic wedge of the Gulf Coastal Plain. Productive units include the Jackson Group barrier-lagoon and deltaic systems (Late Eocene), Catahoula Formation fluvial systems (Oligocene to Early Miocene), Oakville Formation fluvial systems (Early Miocene), and Goliad Formation fluvial systems (Miocene). General geology of the deposits and their host stratigraphic units are reviewed by Eargle and Weeks (1973), Eargle *et al.* (1975), and Galloway *et al.* (1979a).

The coastal plain forms a one-sided geomorphic basin. Consequently, regional groundwater flow is gulfward. Deposits of successive depositional episodes dip gently basinward. Similarly, genetic sequences thicken gradually basinward until syndepositional structural features produced by shale or salt deformation are encountered. Although the uraniferous fluvial and strandline systems are not caught up in the tremendous interval thickening associated with growth fault and diapir zones, details of their depositional patterns were influenced locally by long-lived structures of both local and regional scale (Galloway *et al.*, 1979b). Concentration of major fluvial and associated delta systems within diffuse, long-lived, dip-oriented troughs, such as the Rio Grande embayment of South Texas, characterized most Tertiary depositional episodes. The strike-parallel growth-fault zones constitute zones of structural weakness that propagate vertically. Thus structural patterns initiated largely in Eocene time affected both structure and facies patterns in overlying uraniferous fluvial systems (Galloway, 1977). Just as Holocene coastal plain drainage patterns reflect deep structure, paleochannel trends and resultant facies patterns commonly were deflected along strike, parallel to deep-seated fault zones. The continuing effect of the fault zones on both shallow and deep fluid migration is reflected in the suite of unusual diagenetic features, including uranium mineralization, commonly found near fault zones.

Meteoric ground-water flow within the coastal-plain aquifers has a relatively simple history, but the ongoing deposition of a thick wedge of sediment, including thousands of feet of mud-dominated continental slope facies, produces an extremely dynamic ground-water basin in which the compactional and thermobaric flow regimes also play major roles. The uranium deposits of the fluvial systems in particular provide examples of sandstone-type deposits in which multiple epigenetic alteration events in the host aquifer have substantially modified the simple oxidation-alteration mineralization front model. The Oakville Formation has been examined in detail (Galloway *et al.*, 1979a; Galloway *et al.*, 1982; Galloway, 1982), and illustrates the depositional framework and hydrologic evolution of this variation of epigenetic mineralization front style.

Oakville Fluvial System. The Oakville Formation was deposited by contemporaneous, large-to-small coastal-plain rivers, which together form the Oakville bed-load fluvial system. Sites of fluvial sediment input into and across the aggrading South Texas coastal plain are readily delineated by the regional Oakville sand isolith map (Fig. 13-14), and are outlined on the interpretive regional facies map shown in Figure 13-15. The George West and New Davy fluvial axes are thick, fan-like dispersive sand depocenters that form the core of the Oakville system. Another bed-load fluvial axis lies to the south beyond the southern edge of the facies map. A broad area containing little sand and thick sequences of calcareous mud characterizes the southernmost outcrop belt; sparse lithologic data suggest a broad interfluvial playa-floodplain depositional environment. To the northeast, distinctive calclithic conglomeratic sands mark local fluvial channel sequences deposited by small, flashy, mixed-load streams that drained a primarily Lower Tertiary and Upper Cretaceous terrane to the northwest. This interval has been informally named the Moulton streamplain (Galloway *et al.*, 1982b).

Permeable framework elements of the Oakville system consist of bed-load and mixed-load channel fills and associated sheet-flood splay sands. Floodplain muds and silts are the principal bounding facies. Framework sands consist of vertically and laterally amalgamated, braided, bed-load channel fill and associated crevasse splay units (Fig. 13-16). Splay units are laterally extensive, medium to fine sand sheets deposited along channel margins during short-lived, high-velocity floods. The extensive development of "sheet splays," which are similar to flood deposits of modern ephemeral streams, suggests flashy flow in trunk channels. Channel-fill units consist of broad, scour-based, tabular, medium to coarse grained, locally conglomerate sand units. Preserved sedimentary structures indicate that linguoid bars were a principal depositional element within sand-filled channels. Longitudinal bar deposits occur in local conglomeratic channel fills. Amalgamated channel and sheet splay units produce a cumulative Oakville sand section more than 200 ft (60 m) thick across the broad George West and New Davy axes (Fig. 13-14). Sand bodies may be up to 100 ft (30 m) thick where several individual channel fill units (which average 15 to 25 ft or 4 to 8 m thick) stack along drainage axes. Along these major fluvial axes, sand content commonly exceeds 50 percent of the total Oakville interval and is locally much greater. Less permeable bounding facies include: (1) interbedded mud-clast conglomerate, sand, mud, and clay sequences deposited in distal crevasse splays, abandoned channels, or poorly developed channel-margin levees; and (2) massive, syndepositionally oxidized, calcareous floodplain muds containing abundant pedogenic micrite nodules.

The bed-load, flashy channels of the Oakville fluvial system migrated laterally in the course of coastal-plain aggradation. Resultant sand bodies are a complex mosaic of channel fill and sheet splay facies that form broad dip-oriented belts within bounding floodplain muds and silts (Fig. 13-17). Coarse, conglomeratic, braided channel-fill deposits appear to constitute a core within the broader sand belt.

The Miocene climate in South Texas was arid to semiarid. Confirming evidence includes: (1) abundance of reworked sand-sized limestone grains in Oakville sands; (2) sparsity of vegetal debris or indirect indicators of extensive plant cover, such as root traces or stabilized channel levees; (3) abundance of pedogenic carbonate and probable syngenetic oxidation within overbank facies; and (4) sedimentologic evidence for flashy to ephemeral stream discharge. Although the underlying Catahoula Formation records a peak in deposition of tephra across the Gulf Coastal Plain, continued volcanic activity introduced new airfall ash that was reworked by Miocene

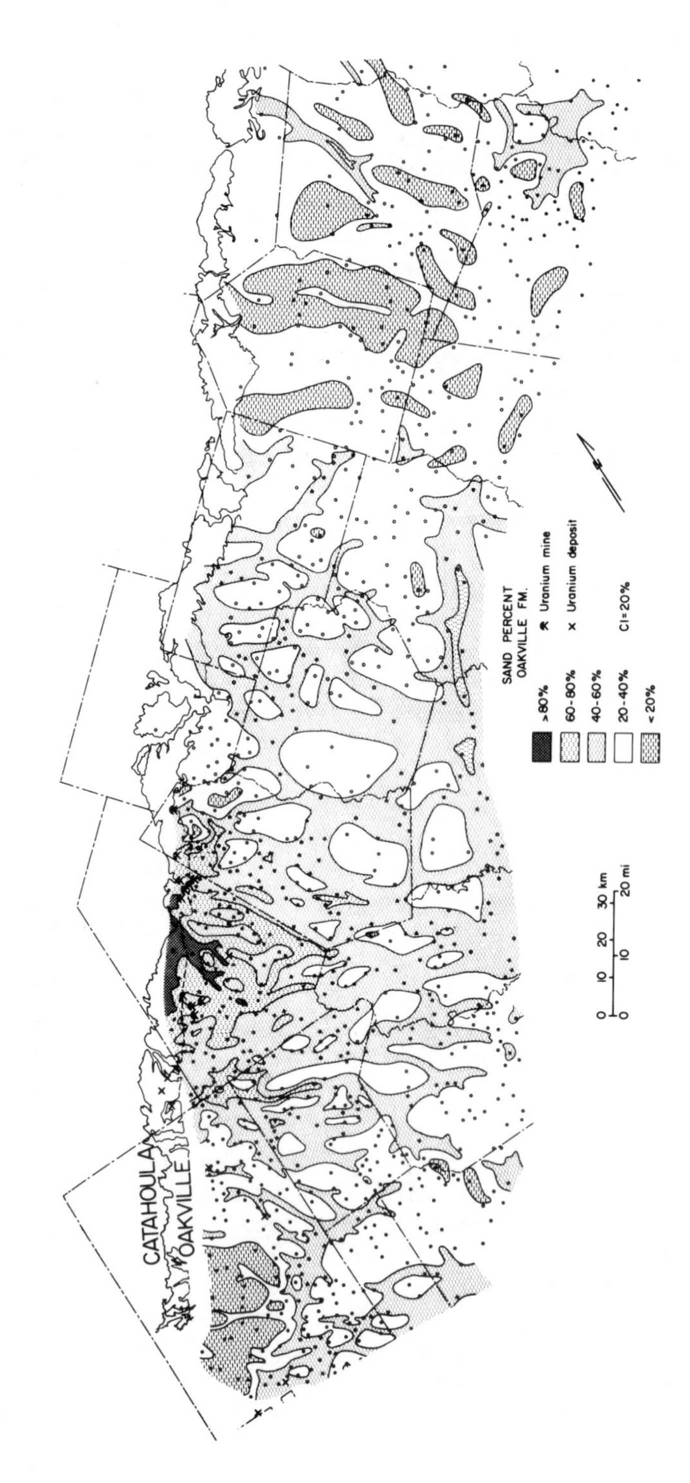

Figure 13-14. Sandstone percentage map of the Oakville (Miocene) bed-load fluvial system, South Texas Coastal Plain. (Modified from Galloway *et al.*, 1982b.)

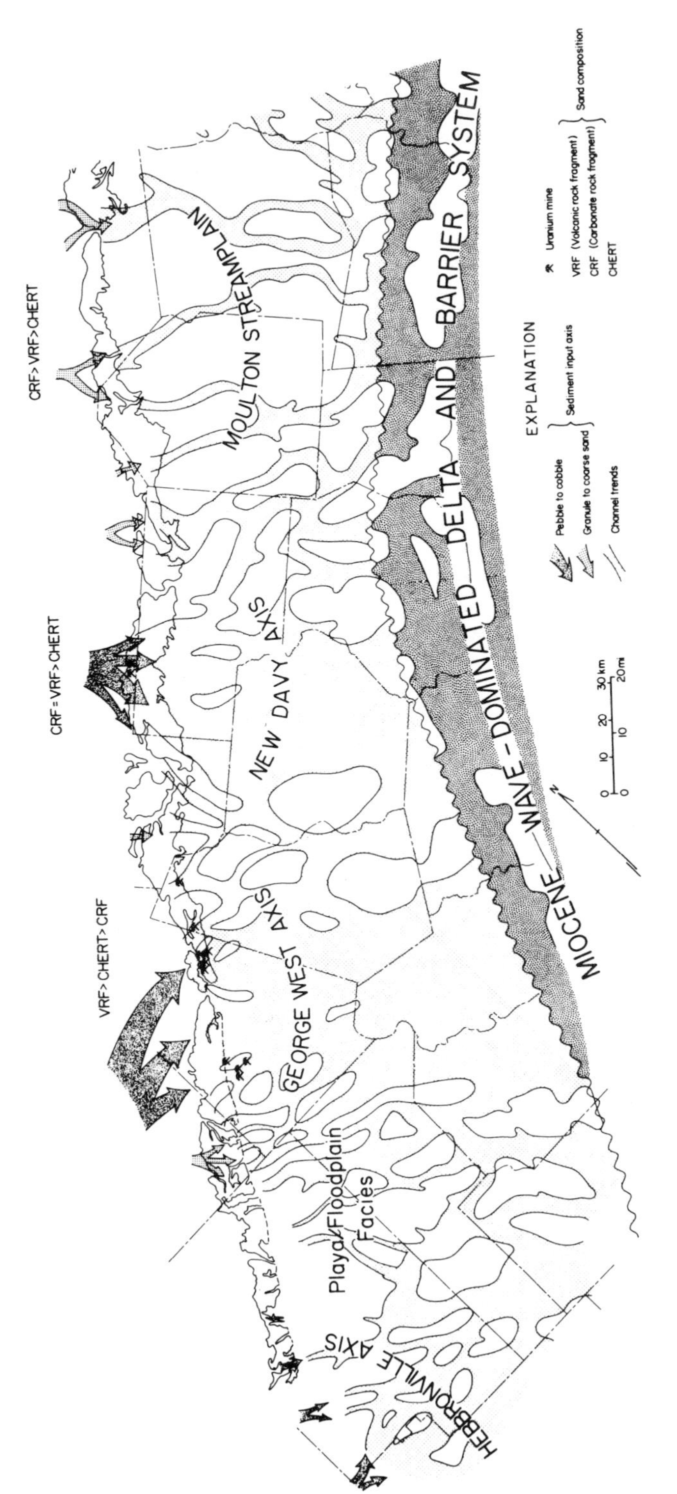

Figure 13-15. Interpreted depositional elements of the Oakville fluvial system (from Galloway *et al.*, 1982).

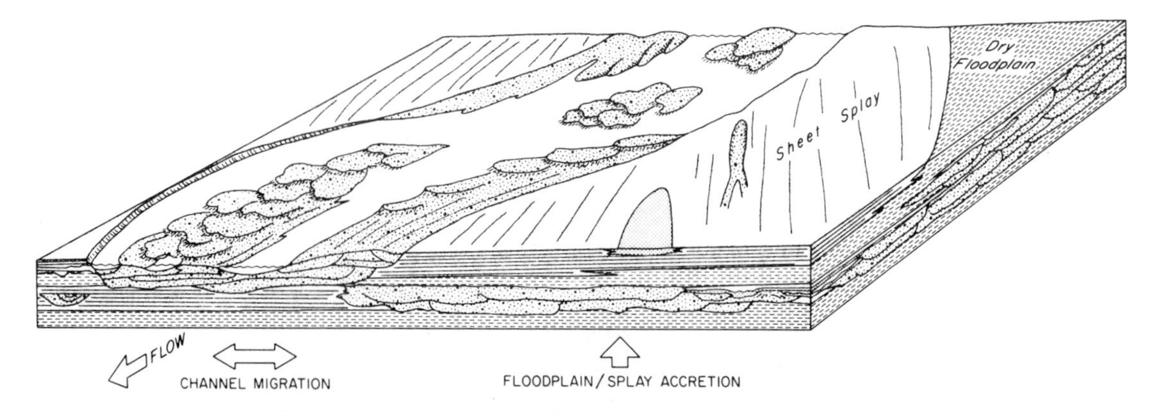

Figure 13-16. Reconstruction of Oakville channel morphology and associated environments typical of the George West axis. Sandy, braided bed-load channel fill and associated sheet splay deposits host the major uranium deposits. (From Galloway *et al.*, 1979a.)

streams, mixed with detrital mud, and deposited within the Oakville Formation.

The physical hydrogeology of the Oakville fluvial system has been systematically analyzed by Smith *et al.* (1980) and Galloway *et al.* (1982b). The Oakville bed-load system corresponds closely to a regional hydrostratigraphic unit, the Jasper aquifer system. Analysis of water-well pump-test results within the genetic stratigraphic framework demonstrated a correlation between fluvial channel facies type and aquifer permeability (Fig. 13-18). Updip (proximal) bed-load channel facies exhibit the highest average permeability. More deeply buried (distal) channel-fill facies of the bed-load axes have reduced permeabilities, and mixed-load channel fill facies exhibit the lowest average permeability. Significant reduction in average permeability is consistently noted in fault-associated portions of the aquifer. Thus, both depositional facies and structural features effect overall aquifer permeability.

Using the generalized calibration between framework depositional facies and average permeability, the net-sand isolith map of the Oakville fluvial system was utilized to prepare the semi-quantitative aquifer transmissivity map shown in Figure 13-19. The map shows the bed-load channel-dominated George West fluvial axis to be the most transmissive element within the Oakville. The New Davy axis, in comparison, is de-emphasized because it contains abundant mixed-load channel deposits. Low sand content as well as dominance of mixed-load channel units results in comparatively low transmissivity for the entire northeastern Oakville section. It is important to remember that the Oakville is everywhere an aquifer. The map simply emphasizes transmissivity variation within a unit that is generally more transmissive than underlying or overlying confining layers. The potential effect of subjacent growth fault zones is indicated by the narrow, strike-oriented ribbons of possible low transmissivity. Facies-controlled transmissive belts are dip-oriented, parallel to the direction of regional meteoric ground-water flow in a coastal plain setting. The structurally controlled ribbons of lessened transmissivity may locally affect the basinward flow patterns. In fact, growth-fault zones are commonly found to act as flow barriers in modern ground-water systems (Galloway and Kaiser, 1980).

Uranium Mineralization. Uranium occurs within the Oakville across most of South Texas. Two areas of active mining contain the largest known reserves (Fig. 13-16). Epigenetic Oakville uranium deposits cluster along laterally continuous, sinuous mineralization fronts that developed at or near margins of fluvial axes. Major districts lie near shallow projections of deep-seated faults, but faulting is not an absolute prerequisite, and mineralization is found along fault zones rooted in both Cretaceous carbonates and geopressured Tertiary muds. District reserves (Fig. 13-16) are directly proportional to the size and relative transmissivity of the host fluvial axis.

Integration of both regional and district alteration characteristics provides the basis for a

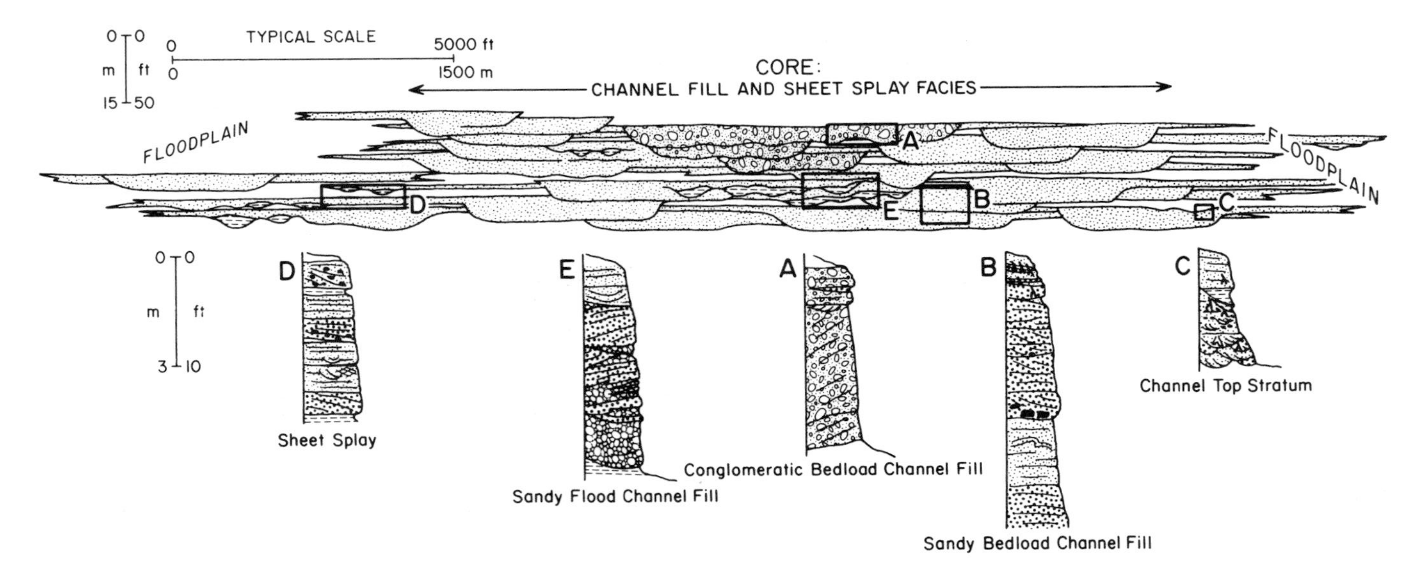

Figure 13-17. Schematic reconstruction of a section across a typical Oakville bed-load channel belt. The composite sand body consists of amalgamated, tabular, sandy, braided channel-fill units and associated sheet/splay beds. A core of conglomeratic braided channel fill may occur within proximal portions of the channel belt. Sections (A) through (E) are based on typical measured sections, and illustrate features of facies within the belt.

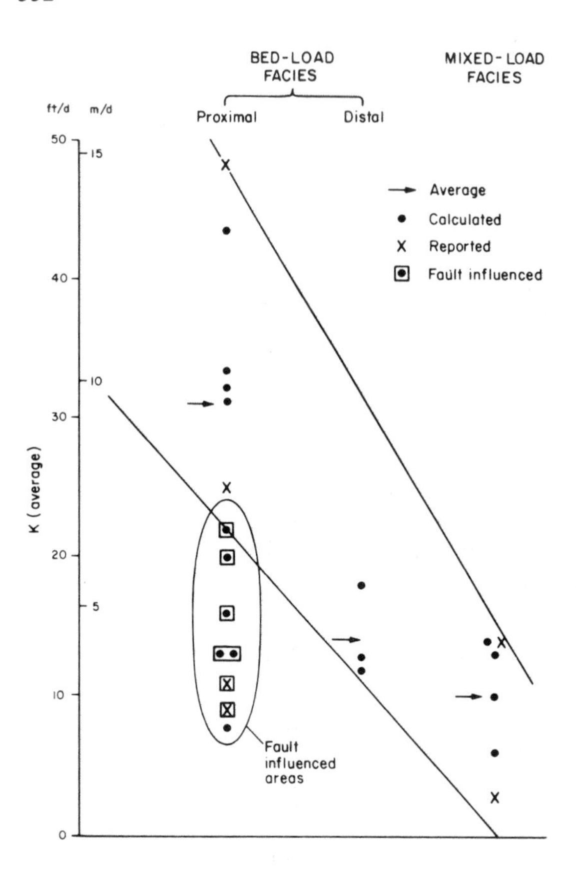

Figure 13-18. Average permeabilities for channel-fill facies of the Oakville fluvial system calculated from well pump tests. Bounding floodplain deposits contribute little water in relation to channel and splay facies, and can be ignored. Proximal (updip) bed-load channel fill sand bodies have the greatest average permeability, but may exhibit considerably lower values near shallow fault zones. (From Galloway *et al.*, 1982b.)

generalized alteration model for the Oakville fluvial system that, unlike the simple Wyoming roll-front model, requires at least two geochemically opposed alteration processes. (1) Epigenetic sulfidization of dominantly syndepositionally oxidized sediments extends into aquifers along segments of major fault zones (Fig. 13-20A). (2) Epigenetic oxidation of the sulfidized sediment intrudes the aquifer from the upflow direction (Fig. 13-20A). Specific alteration zones are defined by the oxidation state of contained iron and by textural features of the oxidized or reduced iron mineral phases present. To a lesser degree, abundance and morphology of calcite and distribution of uranium, molybdenum, and selenium reflect the zonation (Galloway, 1982). The typical alteration zonation of middle-Tertiary Texas coastal-plain uranium deposits (Fig. 13-20B) records the superposition of several such episodes of oxidation and sulfidization.

The general mineralization-alteration model includes at least three discrete epigenetic alteration events—two episodes of sulfidization interrupted by an episode of oxidation and metallogenesis. Combined mineralogic, geochemical, and isotopic description of the alteration products and waters of the meteoric, compactional, and thermobaric regimes leads to the interpretation that uranium mineralization of the South Texas fluvial aquifer systems records the sequential interplay of meteoric and thermobaric waters (Galloway, 1982). Only oxidizing meteoric waters can contain significant amounts of dissolved metals and are capable of producing limonitic alteration. Sulfide-rich thermobaric waters produced both replacement and authigenic pyrite within invaded parts of the aquifer system, leaving reduced sands and bounding mudstones. The complex, polycyclic structure of some alteration fronts preserves the evidence for recurrent flushing by waters of both regimes.

Epigenetic events and flow phenomena that are recorded by alteration and metallogenesis in typical ore deposits of the George West axis of the Oakville fluvial system reflect this interplay.

1. During and immediately after deposition of fluvial sands, shallow, unconfined to slightly confined, oxidizing meteoric ground water flushed portions of the aquifer. Such waters might have contained uranium and associated metals,

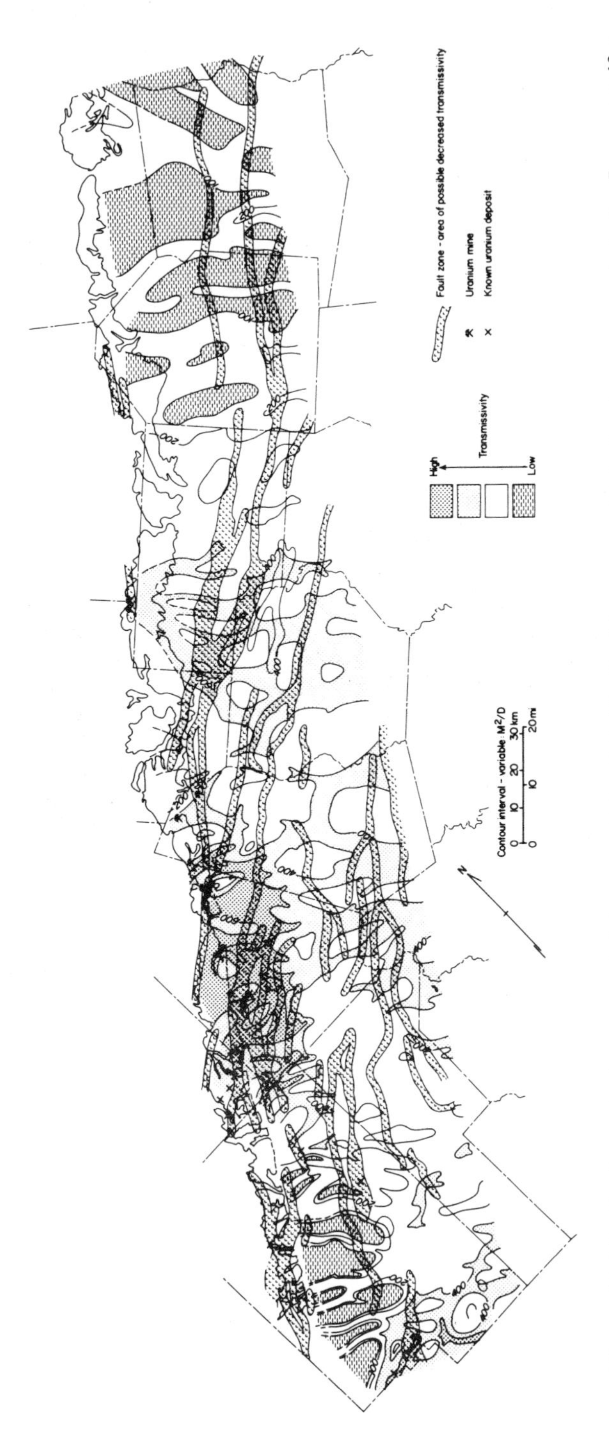

Figure 13-19. Semiquantitative transmissivity map of the Oakville fluvial system, which closely corresponds to the Jasper aquifer. (From Galloway *et al.*, 1982b.)

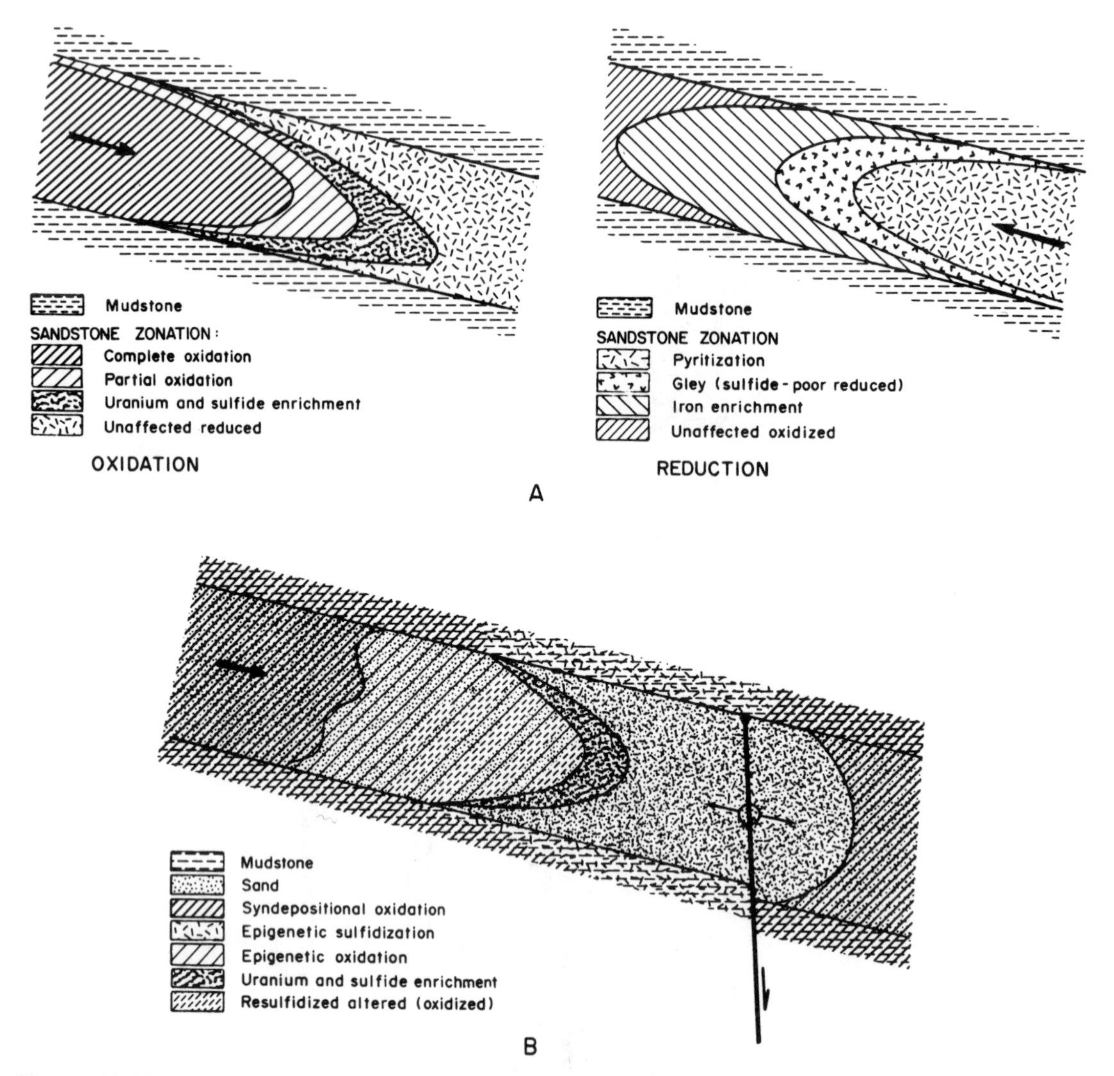

Figure 13-20. Geochemical zonations produced by epigenetic oxidation and reduction of an aquifer. (A) Idealized zones produced in a reduced aquifer by circulating, oxidizing, uranium-bearing meteoric water and in an oxidized aquifer by intrusion of sulfide-bearing ground water (modified from Shmariovich, 1973). (B) Idealized geochemical zonation typical of mineralized parts of the Oakville aquifer of the South Texas uranium province. (From Galloway, 1982.)

but syngenetic geochemical barriers were poorly developed.

2. The first of a series of pulses of reducing thermobaric waters invaded the aquifer, moving upward in response to the extreme pressure head generated within the faulted, geopressured Mesozoic carbonate section. Sulfide reacted with iron, producing isotopically heavy epigenetic pyrite in decreasing abundance away from the feeder fault zone or zones (Goldhaber *et al.*, 1978, 1979). This flushing and associated reduction was a critical precursor for subsequent large-scale uranium concentration.

3. After intrusion of the aquifer by thermobaric waters, normal coastward meteoric circulation was reestablished. Reduced, thermobaric waters were flushed from the aquifer, but reducing conditions remained because of the epigenetic pyrite within the aquifer matrix. Parts of the sulfidized aquifer were epigenetically oxidized as meteoric ground waters containing dissolved oxygen moved coastward down regional hydrodynamic gradient. Early oxidizing waters were enriched in dissolved uranium, molybdenum, and selenium, which were concentrated where flow traversed the geochemical interface between

limonitic and pyritic sediment. Oxidation of the abundant pyrite produced a sharp pH drop, which favored the authigenesis of marcasite as the ore-stage iron sulfide and the leaching and redistribution of calcium carbonate along the advancing front.

4. Discharge of thermobaric fluids from the fault again invaded the aquifer. Additional sulfide precipitated in parts of the aquifer not affected by epigenetic oxidation, and a new generation of iron disulfide formed within affected parts of the epigenetic oxidation tongue.

5. Multiple episodes of oxidation by meteoric waters and sulfidization by thermobaric waters ensued. As one result of this see-saw process, multiple "nested" mineralization fronts locally occur within the same sand body.

6. Active epigenetic sulfidization by shallow intrusion of thermobaric waters appears to have slowed or ceased. Age dates indicate that active metallogenesis ended with a final resulfidization event about 5 m.y.b.p. (Reynolds *et al.*, 1980).

Subsequent supergene redistribution of sulfide and uranium is occurring where modern meteoric waters are again intruding the aquifer. However, modern waters are incapable of producing significant new mineralization fronts (Galloway, 1982).

Incised Valley Fill Deposits: Lake Frome

A third important class of epigenetic sandstone ores occurs in paleovalley fill deposits commonly incised in granite and other crystalline basement terranes. In such a highly confined aquifer, tabular geometry and dispersed distribution of ore bodies result. Mineralization style is illustrated by uranium deposits of the Lake Frome area, South Australia. Although recognized as roll-type deposits (Ellis, 1980; Harshman and Adams, 1981), ore bodies occur as isolated masses within buried paleovalleys, such as the Lower Tertiary Yarramba channel (Fig. 13-21).

The valley fill consists of up to 175 ft (50 m) of interbedded sand and mud and grades from braided fluvial channel deposits at the base to finer grained, meandering stream deposits at the top (Harshman and Adams, 1981). Cross sections suggest that the uranium ore lies dominantly within the lower part of the valley fill sequence. Most of the valley fill has been oxidized. Uranium occurs in irregular masses and pods, poorly defined and stacked rolls, and tabular beds. The position and geometry of deposits are commonly determined by the geometry of unoxidized, less permeable remnants within the valley fill (Fig. 13-21, cross sections). Patterns of mineralization are quite reminiscent of the Colorado Plateau tabular deposits. Superposition of reducing alteration by later flux of connate or thermobaric fluids, such as in South Texas Coastal Plain, would obscure the origin of the Yarramba deposits.

Sandstone-Hosted Tabular Deposits

In many sandstone uranium deposits, ore occurs in pod-shaped, tabular, or irregular masses showing no obvious relationship to existing oxidation/reduction patterns. Such ore bodies, though sometimes described as "plums suspended in a pudding," rarely occur in random patterns. Rather, clusters of deposits commonly display ordered distributions or trends that reflect facies or structural grain.

Deposits of the Westwater Canyon and Salt Wash Members of the Morrison Formation (Jurassic) of the Colorado Plateau (southwestern U.S.A.) are classic examples and contain world-class reserves. Host facies for these deposits consist of channel and channel-margin facies of large-scale bed-load fluvial and wet alluvial fan systems that grade distally into lacustrine facies (Craig *et al.*, 1955; Galloway, 1980; Peterson, 1980; Tyler and Ethridge, 1983). Largest deposits occur in association with the most highly transmissive facies assemblages within the stratigraphic section. Both organic material as well as vanadiferous clays provided potential reductants capable of concentrating uranium from oxidizing ground waters. The different geometry and sometimes enigmatic geochemistry of the tabular deposits has led many authors to consider them as a distinct style of mineralization. Recent research indicates some tabular deposits formed at an interface between meteoric and saline ground waters. Like South Texas, some occurrences may record the sequential or contemporaneous interaction of oxidizing meteoric waters with fluids of the deeper ground-water regimes. Alteration histories were further complicated by the variety of postmineralization hydrologic systems that have developed within these geologically old aquifers. As illustrated by the deposits in the

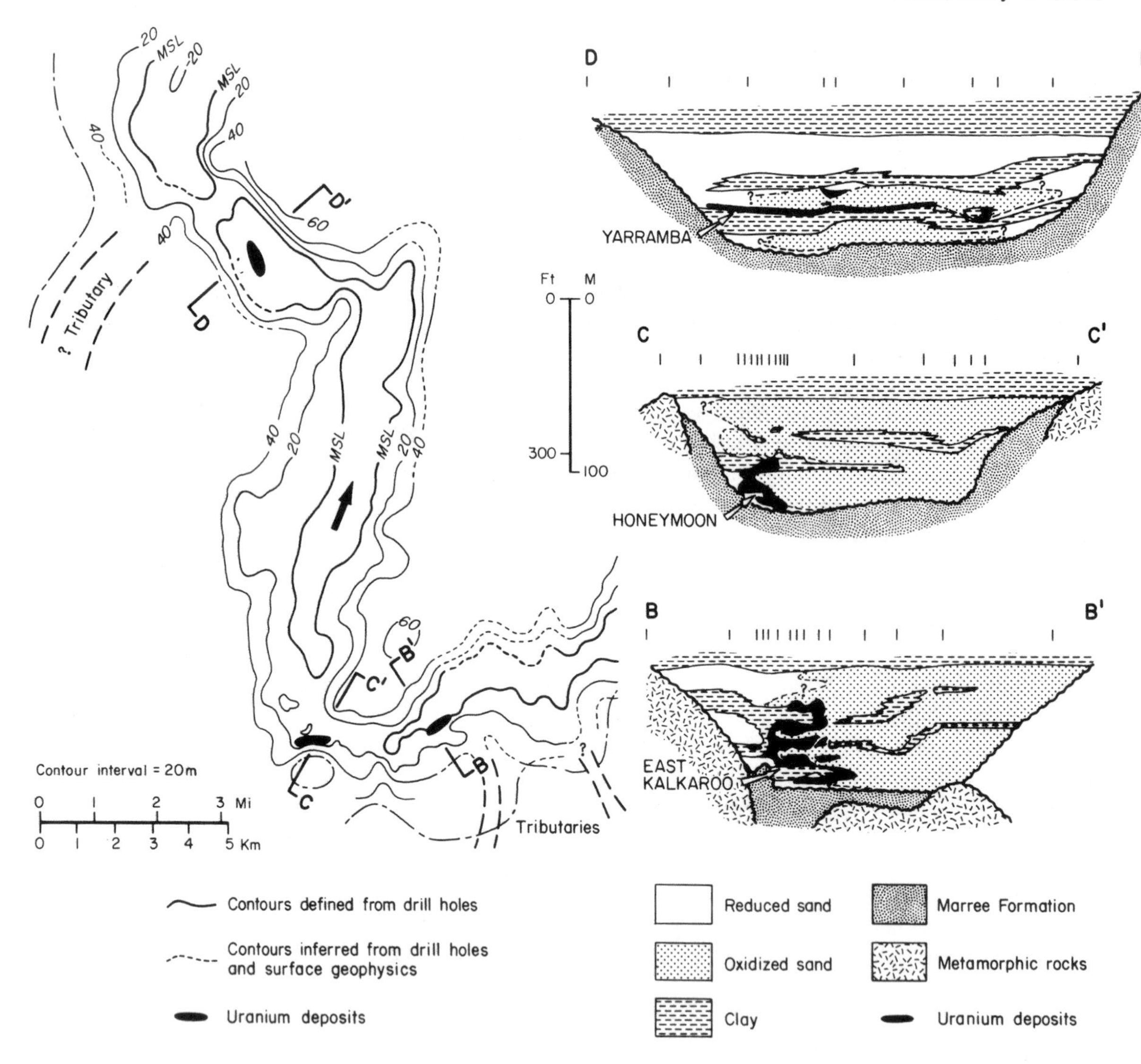

Figure 13-21. Geologic setting of the Yarramba paleovalley and its contained uranium deposits, Lake Frome area, Australia. The map shows the elevation of the bedrock valley walls and floor. Cross sections of the valley-fill show the dominance of oxidized sediment and localization of ore deposits adjacent to reduced remnants. (Modified from Harshman and Adams, 1981; original from Brunt, 1978.)

Westwater Canyon Formation of the southern San Juan Basin, large-scale redistribution of Mesozoic primary mineralization trends can be related to multiple phases of meteoric circulation in Tertiary time (Saucier, 1980).

Genesis of Sandstone Uranium Deposits

Analysis of several major sandstone-type uranium districts and their depositional setting and hydrodynamic evolution suggests a recurrent, generalized cycle of uranium mobilization, transport, and accumulation. A variety of geochemical trapping mechanisms could operate within the context of the general cycle, provided that uranium is redistributed primarily by mass transfer in ground water. The cycle consists of two principal phases (Fig. 13-22): (1) a primary or constructional mineralization phase during which uranium migration and concentration is most active, and regional mineralization patterns are established; and (2) a modification phase, in which all or parts of the primary trends are redistributed or further

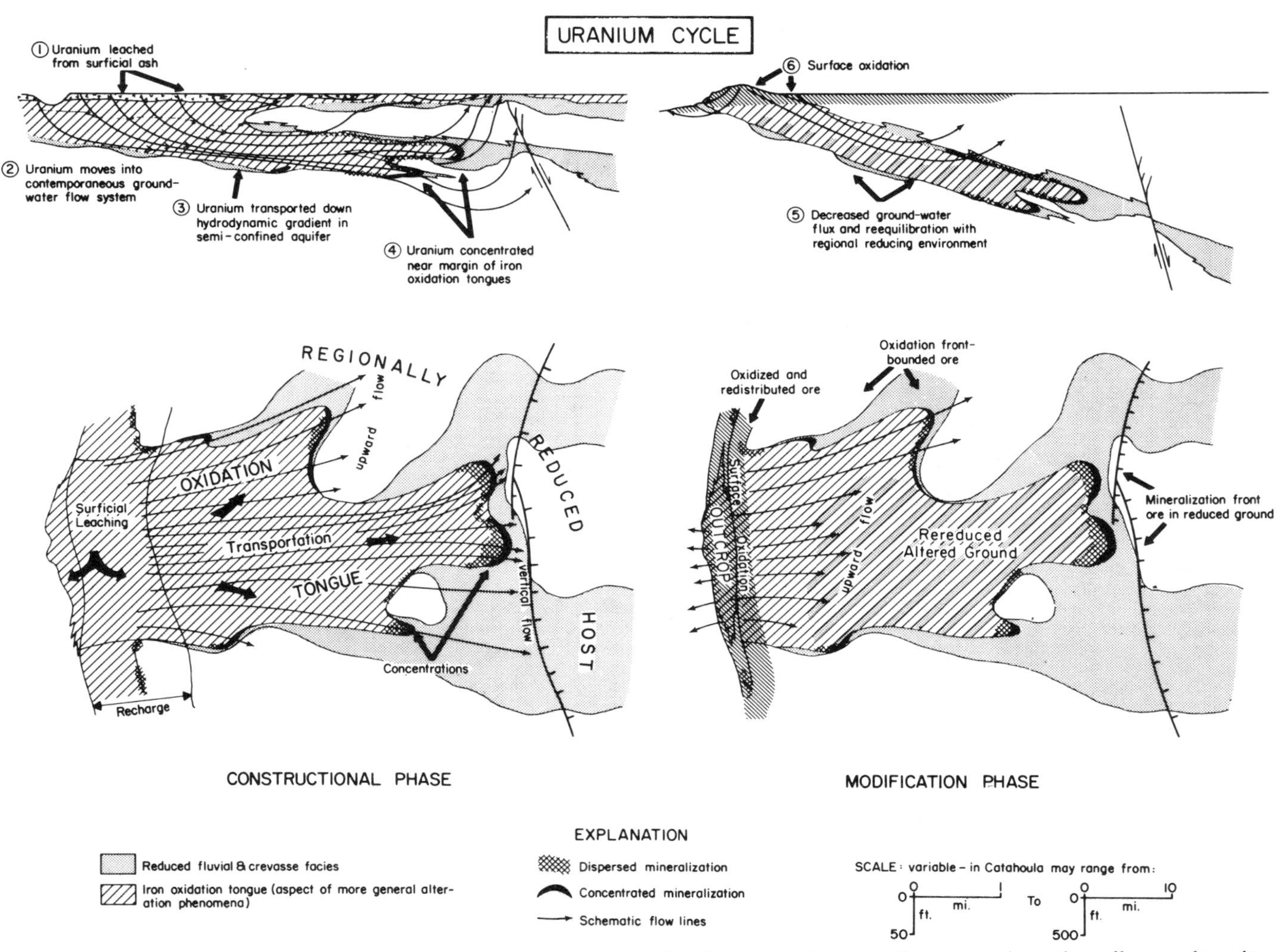

Figure 13-22. Two principal phases of the uranium mineralization cycle typical of many epigenetic roll-type deposits. Constructional events include primary mobilization, migration, and concentration of uranium within a semiconfined aquifer system during or soon after deposition of volcanic ash (or alternative uranium source) in the regional recharge area. Possible modifications of the primary mineralization trends include rereduction of parts of the alteration tongue and local remobilization or destruction of shallow deposits by surface oxidation at or above the ambient water table. Scale of the cycle is a function of the size of the aquifer system. (From Galloway, 1977.)

altered. Some deposits also experienced a period of entombment during deep burial and exposure to compactional or thermobaric regimes.

Constructional Phase events include:

1. Uranium release from interbedded or overlying source materials. Although debate about possible source rocks persists, it is significant that the primary mineralization epoch most often corresponds to a period of deposition of fresh volcanic ash *in the recharge area* of the mineralized aquifer. Although uranium can be rapidly released from siliceous ash in several ways, pedogenesis demonstrably causes extensive argillation of glass and consequent uranium release (Walton *et al.*, 1981).
2. Uranium mobilization into the ground-water flow system in areas of regional recharge. In areas of regional discharge, dissolved uranium moves into the surface drainage and is lost to the epigenetic system.
3. Entry of uraniferous, oxidizing ground waters into the regional flow system producing well-defined salients of altered matrix within regionally reduced portions of the aquifer (Granger and Warren, 1978). Potential reductants include intrinsic organic debris and sulfides, or extrinsic reductants and their diagenetic products, and possibly density stratified reducing brines. Regional alteration patterns are determined by original oxidation state of the host depositional system, extant hydrodynamic gradients and boundaries, and transmissivity, orientation, and interconnection of permeable elements within the depositional system.
4. Concentration of uranium and associated metals where flow crosses from oxidized to reduced portions of the aquifer. Such geochemical boundaries occur as elongate, linear fronts along the margins of reductant-rich pods or islands within pervasively oxidized ground, or at local sites of introduction or collection of extrinsic reducing solids or fluids. Reducing boundaries commonly develop and persist where flow crosses from massive sand facies into interbedded or finer grained facies. However, a geochemical gradient exists at all margins of the oxidation tongue, and uranium accumulation is not restricted to areas of facies change. The actual concentrating mechanism might be chemical precipitation or adsorption.

Modification Phase events that can affect primary mineralization trends established in the constructional phase of the uranium cycle include:

1. Postdepositional changes in the flow system caused by compaction and sealing of bounding aquitards, structural modification, or diagenetic reduction of permeability in transmissive sands.
2. Outcrop recession, exposing shallow ore deposits to oxidation above the water table, or to erosion.
3. Geomorphic changes that induce local or regional changes in hydrodynamic gradient, but producing flow patterns that differ significantly from patterns extant during the mineralization phase or by rejuvenating the meteoric system.
4. Invasion of portions of the aquifer by chemically reactive waters derived from deeper ground-water regimes.

Operation of such a generalized mineralization cycle is consistent with the known distribution and geologic relationships of most sandstone-type uranium ores. Application of the conceptual framework allows prediction of the extent and probable nature of potential mineralization in other systems, provided the depositional framework and ground-water flow history of the system can be reconstructed.

Applications to Resource Evaluation, Exploration, and Development

Large sedimentary uranium deposits can be syngenetic, syndiagenetic, or epigenetic. The occurrence of syngenetic ore depends primarily on the depositional environment and surface hydrology of the host, and is closely facies controlled. Syndiagenetic and epigenetic ores have more complex origins dependent on the depositional environment and facies associations and on the history of ground-water flow through those facies. Post-depositional ores are most likely to be found within or adjacent to facies possessing high transmissivity and high likelihood of ground-water flushing; these characteristics are determined in part by the depositional system. Therefore, the largest uranium-producing districts are hosted by a relatively limited suite of depositional systems, including lacustrine margin, wet alluvial fan, and bed-load or coarse mixed-load fluvial sytems. The reason for this associ-

Table 13–2. Depositional Setting and Order-of-Magnitude Resources of Some Major Uranium Occurrences

Host Depositional System	District/Formation	Resource (lbs (m.t.))
Syngenetic		
Bed-load fan delta	Witwatersand (S. Africa)	10^8 (10^5)
Bed-load fluvial (fan?)	Elliot Lake (Canada)	10^9 (10^6)
Mud-rich marine shelf/basin	Chattanooga Sh. (USA)	10^{10} (10)[a]
	Alum Sh. (Sweden)	10^{10} (10)[a]
Syndiagenetic		
Lacustrine	Lodève Basin (France)	10^7 (10^4)
	Date Creek Basin (USA)	10^7 (10^4)
Fluvial/lacustrine	Yeelirrie (Australia)	10^7–10^8 (10^4–10^5)
Epigenetic		
Bed-load wet alluvial fan or apron	Gas Hills (USA)	10^8 (10^5)
	Grants Mineral Belt (USA)	10^8–10^9 (10^5–10^6)
	Salt Wash Member (USA)	10^8 (10^5)
Bed-load fluvial system (increasing mixed load component)	Oakville Ss. (USA)	10^7–10^8 (10^4–10^5)
	Gueydan (Catahousa) Ss. (USA)	10^7–10^8 (10^4–10^5)
	Shirley Basin (USA)	10^8 (10^5)
	Powder River Basin (USA)	10^8 (10^5)
Valley fill	Tallahassee Creek (USA)	10^7 (10^4)
	Lake Frome Area (Australia)	10^7 (10^4)
Eolian	Browns Park Fm. (USA)	10^7 (10^4)
Fan delta/lacustrine	Chinle Fm. (USA)	10^7 (10^4)
Deltaic (increasing interdeltaic strandline component)	Fall River Ss. (USA)	10^6 (10^3)
	Fox Hills Ss. (USA)	10^6 (10^3)
	Jackson Group (USA)	10^7 (10^4)

[a]Grade of 60 ppm or better.

ation is straightforward. Typical marginal and deep-marine systems are rapidly buried and are not exposed to extensive erosional recycling or to meteoric circulation during their early history. Though tectonic uplift may later expose portions of marine systems to ground-water recharge, the component sands and muds are most likely consolidated so that their transmissivities are low. In contrast, subaerial systems can extensively recycle sediment and are typically exposed to meteoric circulation both during and immediately following deposition. Fan and fluvial systems have an initial topographic gradient that parallels depositional trends and persists until major tectonic activity disrupts the configuration of the basin.

Wet alluvial-fan and bed-load dominated fluvial systems are extremely transmissive because of their coarse grain size and high sand content. Eolian systems are also highly permeable, but typically lack the intrinsic reductants necessary to trap migrating uranium. Mixed-load fluvial systems and various strandline systems possess highly transmissive facies elements, but the framework sands are commonly vertically and laterally bounded by fine sediments, isolating permeable members, and restricting total ground-water flux. Suspended-load fluvial systems are typically products of low-energy hydraulic regimes and contain few laterally continuous, highly transmissive elements. Thus, they are poor hosts for either syngenetic or epigenetic mineralization.

The correlation between depositional processes and mineralization favorability is documented by Table 13-2, which lists several major uranium provinces and their host depositional systems. It is quite obvious that alluvial fan, fan-delta, and bed-load fluvial systems are the preferred targets for exploration for both syngenetic placer and epigenetic sandstone-type deposits. Valley-fill, eolian, and lacustrine systems locally contain significant reserves. Marginal marine deltaic and barrier systems are productive where stratigraphic relationships result in postdepositional recharge of their most permeable facies. With the exception of the low-grade, subcommercial syngenetic black-shale accumulations, marine-shelf and basin systems remain nonprospective, as would be expected from their depositional and hydrologic settings. Intermontane lacustrine basins are prospective targets for syndiagenetic and low-grade syngenetic classes of mineralization; they are commonly associated with uraniferous fluvial and fan systems.

Thus, at the outset of exploration or resource analysis, depositional systems that offer the greatest potential for hosting large, commercial reserves can be targeted. Exploration can then be directed at appropriate facies within such systems.

Regional Exploration Criteria

Within a favorable depositional basin or large depositional system, relative prospectivity of particular stratigraphic intervals or facies may be interpreted by consideration of the basic mineralization processes. The epigenetic, roll-type uranium cycle within a depositional basin can be compared to an ore-processing mill (Galloway *et al.*, 1979b). Both extract a dispersed element from a large volume of rock and concentrate it into usable form. The amount of enriched product (U_{tot}) is equal to the product of the volume of uranium transporting fluid moved through the system per unit time (Q) multiplied by the duration of the flux (T), the concentration of the desired element in the fluid (ppm), and the efficiency of the extraction process (E):

$$U_{tot} = Q \times T \times \text{ppm} \times E$$

Precise quantification of this expression is impossible in natural systems. Nevertheless, it is obvious that any geologic factor that results in a positive change in Q, ppm, E, or T constitutes a positive exploration criterion. For example:

1. Thick sequences rich in air-fall volcanic ash should be interbedded with or should overlie potential host sands in the area of regional recharge of the aquifer system if epigenetic or syndiagenetic deposits are anticipated. Volcanic ash is a preferred source in all but the most stable, geologically long-lived ground-water systems because uranium is readily released from vitric material (Walton *et al.*, 1981). Measured concentrations of uranium in ground waters found in uraniferous granitic terranes average about 1 ppb and rarely exceed 5 ppb (DeNault, 1974; Doi *et al.*, 1975). A few deeply circulating, old waters in granite have yielded dissolved uranium values as high as 10 ppb (Fritz *et al.*, 1979). In contrast, the dissolved uranium content in many tuffaceous aquifers averages tens of ppb, and ranges up to 60 ppb (DeNault, 1974), providing a ten-fold increase in "ppm."

2. Coarser crystalline igneous rocks such as granite provide optimal sources where physical reworking is responsible for uranium concentration. Here, as in the calcrete deposits of Australia, low concentration of uranium in the transporting fluid is compensated by the very long time interval (T) of mobilization and recycling found in the stable repository basins. Extensional tectonics or erosional unroofing and attendant fracturing also favor uranium release from plutonic sources.

3. Dry climates result in recharge through a thick, aerated, phreatic zone and thus are the most favorable for mineralization by solution-transport processes. Humid climates are less favorable because they produce reducing organic-rich, biologically active soils and excess ground water that is rejected by regional aquifers. Uranium that is leached from soils is partially lost by discharge into surface drainage, in effect decreasing concentration within the aquifer.

4. Regional reduction or development of dispersed concentration reducing centers is necessary for the efficient trapping (E) of dissolved uranium in most aquifer systems and must predate concentration processes. Reductants could be intrinsic to the depositional system and thus be facies-related, or may have permeated the system after deposition. Here, greatest reduction characterizes permeable facies.

5. The depositional system should contain one or more thick, vertically integrated, highly transmissive, semiconfined aquifer units that are distributed so as to allow efficient recharge, down-gradient flow, and discharge of meteoric water. Syngenetic placer deposits similarly occur within the high energy, permeable facies. The three-dimensional distribution of such units reflects the volume of surface water discharge through the active system and determines the volume of ground water (Q) that can later be transmitted through different parts of the buried depositional system.

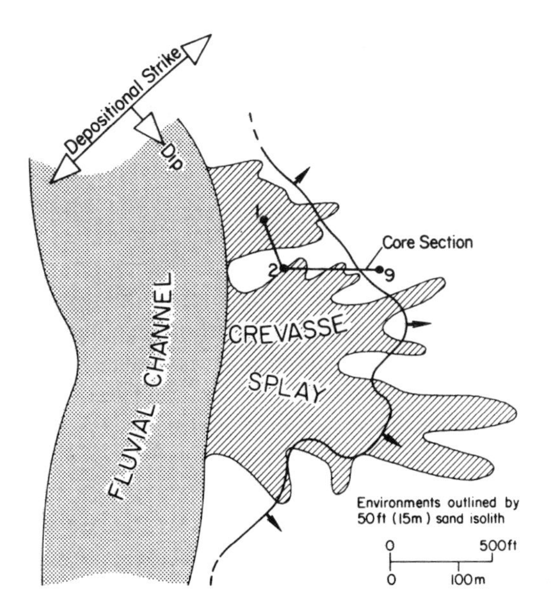

Figure 13-23. Localized expansion of a mineralization tongue from its host mixed-load channel fill unit into the adjacent permeable crevasse splay deposits shown in Figure 4-10. The crevasse sandstones are dominantly oxidized in wells 1 and 2, but are entirely reduced in well 9. The diverging splay channels offer secondary axes of higher transmissivity producing minor down-dip salients along the front. (From Galloway and Kaiser, 1980.)

Local Exploration Criteria

The localization of epigenetic uranium within a single depositional system is primarily a function of spatial variations in surface or ground-water flux, Q. A high ground-water flux produces well-developed, laterally extensive alteration tongues and associated large concentrations of uranium for the prevailing ppm, T and E of the system. Placer uranium deposits form in facies or on surfaces that are produced by long-lived, extensive flow and sediment reworking. Thus, within a particular depositional system the distribution and nature of ore deposits tends to follow predictable patterns that can serve as exploration guides.

1. In axial portions of major sand belts that are oriented parallel to the regional hydrodynamic gradient extant during the mineralization phase, mineralization fronts or clusters of mineralized pods may form salients into unaltered portions of the system. Largest deposits are likely to occur along the nose or distal margins of such salients, which define areas of maximum Q (and therefore maximum importation of oxidant and uranium). Channel axes similarly define sites of highest depositional energy and sediment bypass.

2. In both placer and epigenetic mineralization processes, numerous but commonly small- to-medium-sized deposits tend to form where local stratigraphic or structural features disperse flow.

3. Structures, particularly fault zones, and bedrock topographic features may form flow boundaries or discontinuities that modify flow patterns of both surface and ground waters. In addition, syndepositional structural features commonly localize facies changes. For example, vertical flux of ground water at a fault zone may distort the ideal roll geometry of a mineralization front by accentuating the upper or lower wing or may lead to syndiagenetic mineralization in shallow reduced or evaporative facies. Fault zones are likely conduits for discharge of compactional or thermobaric waters into the shallow aquifers, and thus produce geochemical traps within shallower aquifers.

4. The geometry of framework, high-energy, permeable facies is commonly reflected in the geometry of mineralization. Ore bodies tend to parallel mapped facies boundaries, scour surfaces, interpreted transport directions, or isolith contours. For example, Figure 13-23 illustrates the geometry of an epigenetic alteration front associated with uranium mineralization. The front extends from the channel into the body of the crevasse splay. Local fingers over 200 ft (60 m) long extend along the diverging splay channel sands. Early recognition of such facies-related patterns or trends in alteration or distribution of

uranium concentration allow interpretation or extrapolation of ore bodies using limited data.

Applications in Uranium Extraction

Careful analysis of host depositional facies before and during mining operations can significantly improve recovery as well as reduce cost. Enrichment of uranium in lag deposits, either by physical entrapment within large intergranular pores or by the increased adsorption on organic material or clay, may provide localized but high-grade ore that can be mined selectively. Association of organic debris or reactive clays with certain facies, such as the crevasse-splay deposits (Fig. 13-23), may guide mining to specific but localized facies, and improve anticipation of the vertical and lateral irregularities in ore body distribution.

Understanding of the depositional and hydrologic framework can also anticipate problems in mining. For example, shallow sand bodies that are oriented parallel to hydraulic gradient commonly present greater problems for mine dewatering than do strike-parallel sands. In the South Texas Coastal Plain, open-pit mines in fluvial or deltaic channel facies predictably produce much greater volumes of water than do pits in the strike-parallel barrier bar sand units that grade up and downdip into mudstone facies. *In situ* leaching of uranium is effected by the spatial variability of permeability within the host sand body. Recognition of permeability anisotropy, aquifer compartmentalization, and facies localization of higher grade mineralization can lead to improved leaching methods, well field design, and recovery factors.

Chapter 14

Petroleum

Introduction

For the industrial nations of the world, it may be reasonably stated that petroleum has fueled the twentieth century. Despite accelerating depletion of the resource, oil and gas will remain major contributors to the world energy supply into the next century. Because the discovery of hydrocarbon reserves is an extremely geology-intensive process, the petroleum industry has directly and indirectly initiated, supported, and incorporated much of the accumulated research on sedimentary basins and their contained deposits. Similarly, the industry has been the single largest world-wide employer of earth scientists. This tremendous expenditure of manpower and research on a particular energy resource and its geologic environs has led to considerable understanding of its origins and distribution patterns. Still, answers to many questions remain elusive, and petroleum liquids are described in legal terminology as "fugacious," a term which most petroleum geologists can readily support.

Distribution of Petroleum in Time and Space

Oil and gas are generated by the thermal maturation of organic matter incorporated within the basin fill. Initial diagenesis of plant and animal debris produces kerogen. With further increase in temperature, or more correctly, with time-integrated thermal flux, kerogen, in turn, yields hydrocarbon liquids and then gas (Tissot *et al.*, 1971; Connan, 1974; Waples, 1980). Additional quantities of dry gas, consisting almost entirely of methane, form in organic-rich sediments at low temperature and shallow burial by bacterial action. Such biogenic gas constitutes the dominant hydrocarbon in several large basins (Rice, 1980; Rice and Claypool, 1981). Following generation, petroleum moves from fine-grained source rocks into adjacent, permeable stratigraphic or structural conduits. These primary migration processes remain poorly understood. However, once separate-phase oil and gas are in permeable conduits, they migrate upward because of their buoyancy in water until trapped beneath or against a seal. Size, depth, and composition of the hydrocarbon occurrence, as well as matrix properties of the reservoir, all affect commerciality of the resultant accumulation.

No world-wide statistics for all known oil and gas occurrences exist. However, the task of statistical description of petroleum distribution is simplified by the important observation that giant fields (defined as those containing over 500 million barrels of oil or BTU-equivalent gas) hold 75 percent of discovered reserves, although they account for only about one percent of the total number of known fields (Klemme, 1980). Compilation of data on the giants shows that sediments of the Late Mesozoic and Cenozoic eras are the most prolific hosts of petroleum (Barker, 1977). A subsidiary reserve peak occurs within Carboniferous strata. As pointed out by Barker (1977), a cross plot of known reserves vs. reservoir age closely resembles a similar cross plot of age vs. average organic carbon content within shales. The parallelism exhibited by the curves evidences the importance of source facies distribution in controlling the stratigraphic distribution of hydrocarbons. Younger basins and strata are logically the most productive, given comparable source capacity, because of the lesser time for post-migration loss of oil and gas by trap breaching or long-term diffusion.

Geographic and geologic distribution of hydrocarbon-rich basins has been discussed by several authors, including Bally and Snelson (1980) and Klemme (1980), who distinguished eight basin types on the basis of tectonic and depositional styles.

1. *Simple intracratonic basins* form wide, shallow depressions. The Illinois and Paris basins are well known, petroliferous examples. Such basins contain significant volumes of mixed terrigenous clastic and carbonate sediment, but produce proportionally minimal amounts of hydrocarbons (Table 14-1). Thin, laterally exten-

Table 14–1. Proportion of Total World Hydrocarbon Endowment and Volume of Basin Fill (Buried at least 4,500 ft or 1,400 m) within the Major Types of Basins[a]

Basin Type	Percentage of Total Proven and Produced Reserve	Percentage of Total Basin Volume
Simple intracratonic	1.5	6
Composite intracratonic	25.0	25
Rift	10.0	6
Extra-continental downwarp	47.0	26
Passive margin	1.0	19
Subduction	7.5	19
Median	2.5	4
Deltaic	6.0	5

[a]Data from Klemme (1980).

sive fluvial, deltaic, shelf and lacustrine systems are typical of the basin fill. They are commonly mature hydrologic basins, flushed throughout by meteoric ground water.

2. *Composite intracratonic or foreland basins*, such as the Alberta and southern North Sea basins, are large, linear to elliptical, and asymmetric. Block faulting may characterize the deeper basin core. Sediment fill is derived from the orogenic craton margin and includes mixed or dominantly siliciclastic deposits. Moderately thick fluvial, deltaic, interdeltaic shorezone, and shelf systems are major components of basin fill. Subaerial fan and slope systems may be significant. Meteoric invasion is typically extensive, and a deep, overpressured basin core may be produced by hydrocarbon generation. Such basins contain about 25 percent of the total basin volume, and hydrocarbon productivity is proportionally large.

3. *Rift basins* are small, linear, deep depressions, such as the Sirte or Rhine basins, formed on or near the craton. Basin fill is dominated by siliciclastics of continental to subaqueous fan and fan delta systems deposited in marine or lacustrine settings. Early formed bedded evaporite deposits commonly develop diapiric intrusions and related traps. Oil productivity is disproportionally great (Table 14-1).

4. *Extracontinental downwarps* along small oceanic basins, including the northern Gulf Coast basin, are large, elongate depressions that fill asymmetrically from the landward side. This simple geometry may be modified by continental collision, producing an irregular trough. Partial to extensive meteoric circulation penetrates the mixed carbonate-siliciclastic basin fill; an overpressured core is common in younger examples. Fluvial, deltaic, shorezone, and slope systems are thick, extensive, and well developed. Nearly 50 percent of world hydrocarbon reserves occur in these basins, which include many of the highly prolific basins of the Tethys seaway.

5. *Passive continental margin depressions* are elongate, asymmetrically filled basins formed along divergent plate margins. Sediments are dominantly fluvial, deltaic, shelf, and slope siliciclastics and may be extensively flushed by meteoric flow. Examples include the North Western Shelf of Australia and East Coast basins of the United States. Hydrocarbon productivity has thus far been generally low (Table 14-1).

6. *Subduction basins*, such as the Los Angeles and Central Sumatra basins, form at convergent plate margins and are either arc or wrench-fault related. The basins are small, linear, exhibit an irregular profile, and are filled largely with siliciclastics deposited in alluvial fan, fan delta, and marine slope systems. They are oil prone, and individual basin productivities are extremely variable.

7. *Median, or intermontane basins* form small, linear troughs within mountainous, folded, and intruded orogenic belts that lie marginal to convergent plate boundaries. Fill consists largely of deposits of fluvial, alluvial fan, and lacustrine depositional systems, and hydrocarbon productivity is relatively low (Table 14-1).

8. *Late Tertiary deltaic basins* include the thick extracontinental deposits of the Mississippi, Niger, McKenzie, and Mahakam delta systems. These basins occur where major rivers discharge

along axes of structural weakness into oceanic basins. The distinction between certain other basin types, particularly along passive margins, is somewhat arbitrary. Gas-prone hydrocarbon production is dominated by syndepositional growth-fault and diapiric traps. Deeply buried basin fill is characterized by the overpressured compactional and thermobaric hydrologic regimes.

Although contained depositional systems vary greatly within each of the basin types, depositional style is directly related to specific basin configuration and tectonic history. For example, delta systems are major components of both composite intracratonic and extracontinental downwarp basins (which together contain over three-fourths of discovered reserves), but style of deltaic sedimentation differs depending upon whether progradation is onto oceanic or continental crust. This contrast is well illustrated by comparison of depositional and production styles of the Gulf Coast Tertiary and Midland basins at the end of this chapter.

Depositional Systems and Hydrocarbon Exploration and Production

The long-recognized critical elements for formation of hydrocarbon accumulations include a source, reservoir, trap, and seal. More recent recognition of the important dynamic aspects of oil and gas occurrence has added burial history (evolving temperature/pressure regime) and migration pathway to the original list. Depositional system analysis answers questions about reservoir volume and distribution, and probable nature and extent of source and sealing facies. Trapping requires three-dimensional isolation of all or part of an elevated portion of the reservoir facies. Traps may be produced by structural flexures or discontinuities, by facies distribution patterns, or by a combination of both. Consequently, depositional systems analysis may also provide useful information about trapping potential or style.

Recognition of the dynamic component of petroleum generation and subsurface fluid migration indicates additional, though less commonly utilized, applications of genetic facies studies. In combination with structural features, framework sand bodies of major basin-filling systems provide the three-dimensional stratigraphic plumbing of the basin, thus defining potential petroleum migration pathways. Integration with concepts of basin hydrology may allow prediction or early recognition of potential hydrodynamic entrapment, reservoir drive mechanisms, and diagenetic features, all of which may affect oil and gas production.

Recently, the effects of internal heterogeneity and permeability anisotropy, which are inherent products of depositional processes, are being recognized and quantified as a part of reservoir exploitation and advanced recovery programs (Harris and Hewitt, 1977).

Source-Rock Recognition

Prediction, recognition, and delineation of petroleum source units are primary tasks in basin evaluation and subsequent exploration (Tissot and Welte, 1978; Hunt, 1979). Potential source rocks contain adequate quantities of organic matter to generate significant volumes of petroleum. Functional source rocks, in addition to their initial endowment of organic matter, have been subjected to sufficient burial and consequent thermal maturation to generate oil or gas. Obviously, presence of an adequate volume of functional source-rock facies is critical to the ultimate productivity of a sedimentary basin.

Empirical as well as experimental and theoretical considerations have led to the use of a variety of indices of source potential of a sediment or rock. A relatively simple measurement of total organic carbon content (TOC) is widely applied to screen for potential source rocks (Ronov, 1959; Dow, 1978). A commonly accepted minimal TOC content for a potential source rock is 0.4 percent; values of one percent or more are preferable. TOC is, however, only a crude index of source potential. Organic carbon may be recycled from older sediments, possessing little capacity for further release of liquids. Consequently, more sophisticated analytical procedures determine the content and composition of extractable hydrocarbons within the rock matrix. Because expulsion of generated hydrocarbons from functional source rocks is an inefficient process, the effect of active generation and loss on measured extractable hydrocarbon values is considered by most geochemists to be negligible (Hunt, 1979, p. 271).

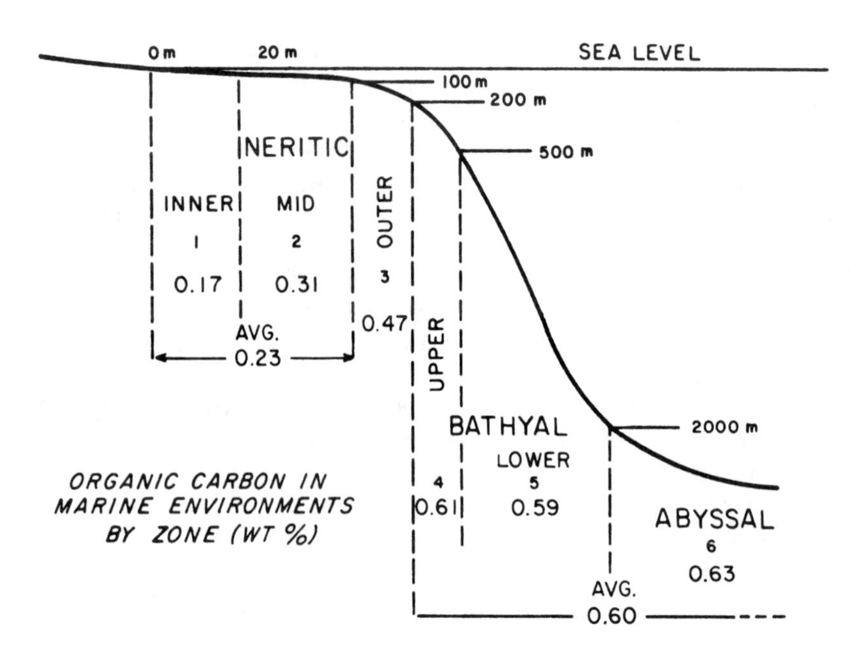

Figure 14-1. Mean total organic carbon (TOC) content of environmental depth zones in the Gulf Coast Tertiary section of Louisiana. (From Dow, 1978.)

The amount of organic matter in a sediment is a function of three variables: (1) the rate of organic productivity of the system; (2) the rate of destruction by biologic or inorganic processes; and (3) the rate of dilution by detrital sediment. Significantly, TOC of a sediment is commonly limited by destruction and dilution rather than by biologic productivity of the overlying water column. Greatest proportions of organic material occur in fine, muddy, terrigenous marine sediments deposited under restricted, reducing bottom conditions characterized by moderate rates of deposition (Ronov, 1959; Ibach, 1982). Low depositional energy and minimal reworking favor preservation of organic material; reworking by currents allows additional time for oxidation by bacteria or benthonic organisms. High terrigenous sedimentation rates dilute indigenous organic matter but may also introduce a significant proportion of land-derived organic material.

The interaction of various factors combine to produce a common pattern of organic richness in continental margins (Dow, 1978). As shown in Figure 14-1, continental shelf muds are characterized by low TOC, whereas continental slope deposits are enriched. Low TOC values of the shelf reflect the abundant biogenic reworking of the bottom sediments beneath an oxidizing water column. Greater, though still modest, TOC values in the bathyal and abyssal setting result from the more rapid accumulation of mud on the slope, combined with the lesser current activity and oxidizing capacity of the environment.

Burial and diagenesis of deposited organic matter produces kerogen, the precursor of oil and gas. Kerogen is classified on the basis of chemical composition into three types, commonly designated I, II, and III (Tissot *et al.*, 1973). Type I is an oil-prone kerogen that consists of algal or amorphous organic material. Algal-rich varieties typically produce waxy, paraffinic crude oils. Type II kerogen contains a mix of amorphous and herbaceous organic material and is also an oil-rich precursor. Type III kerogen consists of woody and coaly material; it yields largely gas upon thermal maturation. Exinite-rich coals of Australia have also been found to be adequate sources of hydrocarbon liquids (Thomas, 1982). Interpretation of the depositional facies of potential source mudstones allows logical predictions of kerogen types and thus of probable attributes and chemical composition of any generated petroleum. For example, Hedberg (1968) showed that waxy, paraffinic crude oils, which are the products of Type I or degraded Type III kerogen, are commonly associated with land-derived shallow marine or lacustrine organic matter. Similarly, sulfur content is lower in terrestrial than in marine organic debris and their resultant oils.

In summary, interpretation of depositional systems and their component facies provides the basis for estimation of the probable distribution, richness, and oil- or gas-prone nature of hydrocarbon source rocks. However, the nature of the depositional basin, degree of bottom water restriction, or organic productivity can vary greatly within any depositional complex. Quantitative description of representative potential source facies is required for adequate evaluation of the hydrocarbon resource potential or source-rock characterization.

Trap Prediction

Hydrocarbon traps are produced by geometric configurations of reservoirs and sealing units that prevent further upward migration or escape of gas or oil. Traps have been discussed and classified by Levorsen (1967) and Rittenhouse (1972). Several classes of traps are products of post-depositional deformation or reservoir truncation that is unrelated to depositional processes. However, structures such as growth faults are inherent features of some depositional systems. Geometry and distribution of such syndepositional structures are thus closely related to facies architecture. Growth faults, slumps, diapirs, differential compaction and drape structures, and intraformational folds all may be large enough to form economic traps. In some basins such structures control the distribution of oil and gas fields. Large-scale growth faults, which are rooted within thick, over-pressured prodelta and continental slope mudstone sequences, are a characteristic structural and producing style of northwestern Gulf Coast depositional systems ranging in age from Cretaceous through Miocene (Fig. 14-2). Repeated development of growth fault zones within successive offlapping systems produces multiple, coast-parallel series of fields that trap both gas and oil.

Productive structural configurations within growth fault belts include roll-over anticlines on the down-thrown side (Fig. 14-2D), updip sealing against the fault (Fig. 14-2B), coastward pinchout of sands (Fig. 14-2A), or deep-seated salt domes and associated withdrawal sags (Fig. 14-2C). At a smaller scale, drape structures of sufficient size to form commercial fields occur where sand bodies are superimposed on elongate channel or bar sand bodies encased in compactable mud. Carbonate reefs or other large, relatively rigid bulwarks surrounded by mudstone may produce large-scale drape structures in overlying sand bodies. In the Midland Basin, for example, Pennsylvanian submarine fan deposits draped against an underlying reef (Bloomer, 1977) have produced nearly 40 million barrels of oil.

Stratigraphic traps seal hydrocarbons by means of facies changes that truncate or reduce the permeability of the reservoir. A variety of specific types exist (MacKenzie, 1972), though not all are strictly features of depositional processes. For this discussion, it is useful to recognize three possible stratigraphic trap styles that are primarily products of the depositional system.

(1) Completely isolated sand bodies called isolani by Silver (1973) are ideal stratigraphic traps. Once filled with hydrocarbons, lateral facies changes and porosity pinch-out preclude migration and loss unless the reservoir is subsequently breached by an erosional or structural discontinuity. Discontinuous "clean," porous sandstone bodies isolated within marine mudstones in the Sussex sandstone of the Powder River Basin form prolific, stratigraphically-trapped oil reservoirs (Fig. 14-3). The geometry, distribution, and trapping capability of these isolani reflect their origin as a shelf depositional system (Brenner, 1978).

(2) Erosional surfaces produced by sedimentary processes, such as channel incision, within a depositional system may truncate potential reservoir sands. If the surface is sealed by fine-grained or poorly sorted sediment, oil or gas can be trapped against the erosion surface. On a small scale, mud-filled channels or plugs within fluvial sand bodies can form traps (Figs. 4-11; 4-14B). Large submarine gorges, which truncate multiple sandstone units, produce multistory stratigraphic traps in the early Tertiary fill of the Sacramento Basin.

(3) Depositional topography, though more common and preserved on a larger scale within carbonate depositional systems, may produce adequate "structure" on the top of the sand body to cause at least local oil or gas entrapment. Such topography is likely to be enhanced by lateral facies changes along the margins of the sand body, and is best revealed by mapping the elevation of the top of effective porosity. Many Pennsylvanian reservoirs of the Midcontinent reveal such topography in contour maps of the structure of the top of porous sandstone.

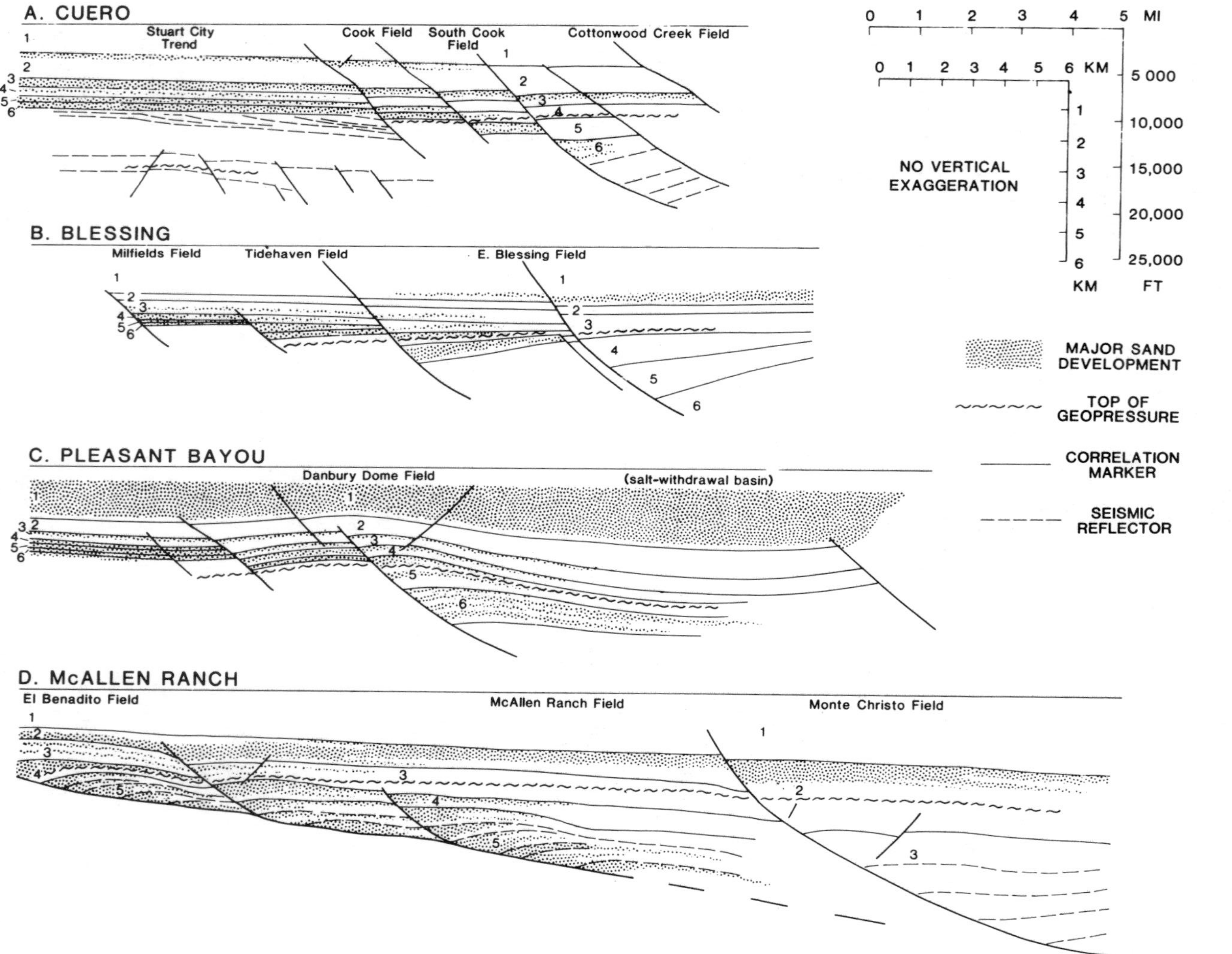

Figure 14-2. Dip cross sections showing the structural styles of four growth-faulted, shelf-margin, Tertiary deltaic sequences in the Texas Gulf Coast Basin. (A) Faulting and expansion initiated where the Eocene Wilcox Group prograded beyond the subjacent, faulted Cretaceous Stuart City reef carbonate shelf margin. (B) Repetitious faults with marked landward thickening of Frio units. (C) Combined doming and faulting in the Frio Formation reflecting deep salt mobilization. (D) Prominent décollement within prodelta and delta-front deposits of the Oligocene Vicksburg delta system of South Texas. Numbers 1–6 trace correlation markers across fault blocks. (Modified from Winker, 1982.)

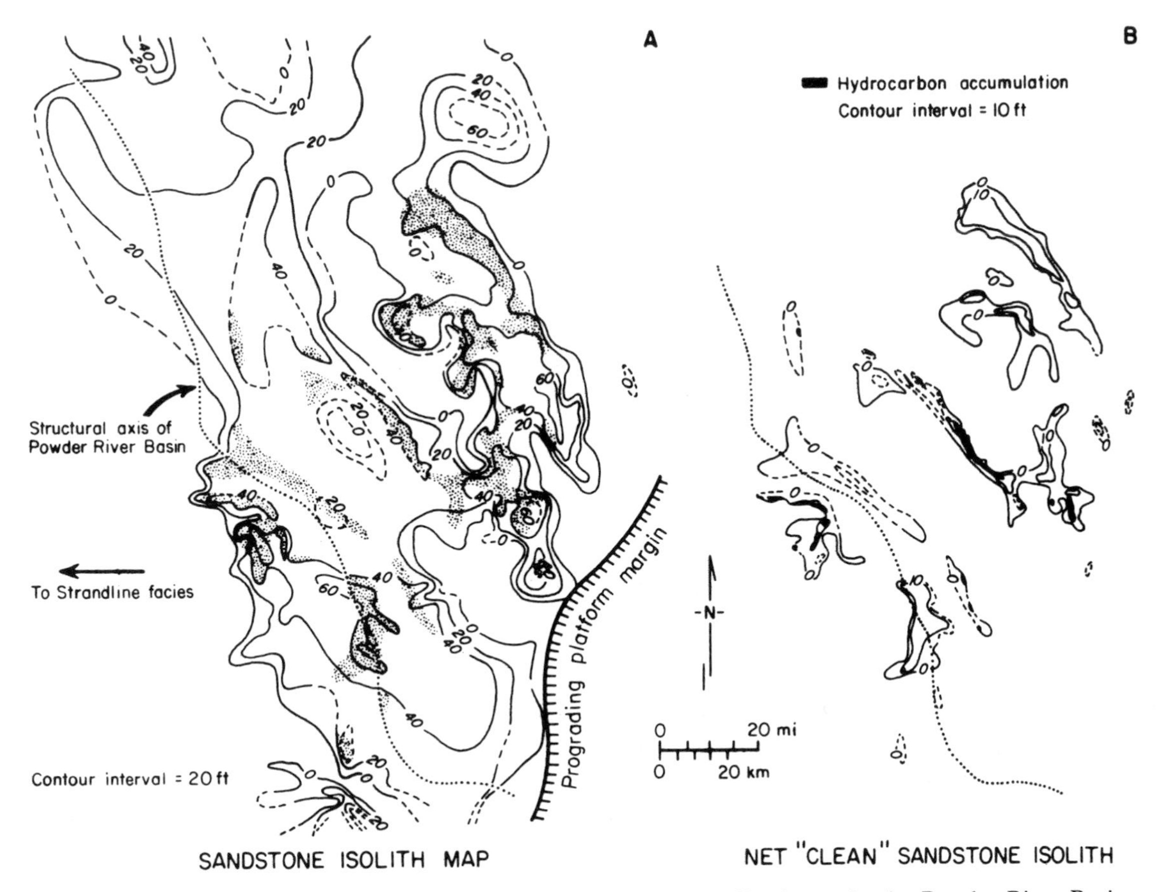

Figure 14-3. Sandstone isolith maps of the Sussex (Cretaceous) Sandstone in the Powder River Basin. (A) Regionally the Sussex forms a widespread irregular sheet. (B) Porous sandstone is distributed as a series of elongate, isolated pods forming isolani traps. (Modified from Brenner, 1978.)

Any or all of the three stratigraphic features—lateral facies gradation, erosional truncation, or preservation of sand-body topography, may occur in association with structure to produce combination traps.

In addition to static trapping mechanisms, ground-water flux within the basin may play a role in hydrocarbon accumulation. The concept of hydrodynamic entrapment was first elucidated by Hubbert (1953), and is obviously most likely within the zone of active meteoric circulation. A hybrid mechanism, involving sealing of a hydrocarbon accumulation by diagenetic cementation following initial entrapment, has been proposed (Wilson, 1977). Such entrapment by a post-accumulation, diagenetic facies change necessitates a complex but not unrealistic sequence of fluid flow events.

The inherent discontinuity of framework sand facies within some depositional systems predisposes them for development of various sorts of stratigraphic or syndepositional structural traps. In such systems, genetic facies analysis offers more than insight into reservoir distribution. Though the exception rather than the rule, some depositional systems in major petroliferous basins may constitute the complete hydrocarbon producing factory, integrating source, migration pathway, reservoir, trap, and seal within the same depositional framework.

Delineation of Fluid Migration Pathways

The framework sand facies of the depositional systems are major pathways for fluid migration. Whereas sedimentary uranium ore deposits form within the meteoric regime, the genesis and early migration of petroleum occurs within the compactional and thermobaric regimes. Sand content, trend, lateral continuity, and interconnectedness of framework sand units, and relationship of sand bodies to cross-cutting structural elements define the potential migration pathways, both for expelled pore waters and generated hydrocarbons. Because fluid migration is a basin-wide phenomenon, only regional lithostratigraphic description provides an adequate framework for analysis of this dynamic aspect of hydrocarbon entrapment.

Abnormal Pressure. Development of fluid pressures that exceed the hydrostatic gradient was discussed in Chapter 11. Deep drilling for oil and gas commonly penetrates abnormally pressured portions of basin fill. Abnormally high fluid-pressure zones require specialized drilling and production techniques. Thus, their prediction and detection are important elements of exploration within some basins (Fertl and Chilingarian, 1977).

One common mechanism for generation of overpressure—rapid deposition and burial of sand-poor sediment—directly relates to facies distribution. Expectedly, variations of fluid-pressure gradient commonly correlate with facies distribution within deltaic, slope, and interdeltaic shore-zone systems in rapidly subsiding basins of the young extracontinental downwarp and deltaic types. In the Gulf Coast Basin, deep-water prodelta and continental-slope mudstones are moderately to highly overpressured, as are interbedded distal delta-front sand facies. In contrast, well-interconnected delta-plain barrier/strandplain, and fluvial facies exhibit normal or near-normal fluid pressure (Jones and Wallace, 1974; Galloway *et al.*, 1982a). Stuart (1970) noted a close correlation between paleontologically defined paleobathymetry of mud-rich intervals and the probability and degree of overpressure.

Correlations between overpressure of other origins and depositional facies are less predictable. However, preservation of abnormally high fluid-pressure gradients inherently implies the existence of hydraulic isolation whether stratigraphic, structural, or diagenetic within the basin fill.

Oil and Gas Migration. Primary migration of petroleum from its site of generation within organic-rich mudrocks into adjacent permeable silt or sand units is a poorly understood and much debated process (Barker, 1977; Magara, 1980; Bray and Foster, 1980; Harrison, 1980). Regardless of the specific migration process, primary migration occurs mainly within the compactional or thermobaric regimes. Pore fluids, including oil or gas, are most likely to move in the direction of decreasing pressure. Several studies have suggested that the direction of migration from source mudrock sequences is determined by adjacent sedimentary or structural conduits that provide avenues for fluid flow out of the compactional regime (Hunt, 1979). Harrison (1980) correlated the distribution of gas with thermal, pressure, and salinity variations indicative of discharge of deep, geopressured, hydrocarbon-rich waters through faults into shallower, normally pressured sand bodies. Conversely, inefficient dewatering, expressed as abnormal pressure in a rapidly deposited sand-shale system, implies limitations on expulsion of generated hydrocarbons (Magara, 1980). Once in the permeable conduit, hydrocarbons tend to follow the direction of expelled waters, up and out. The extent and direction of migration depend on the lateral or vertical extent of the conduit and the distance to contemporaneous traps.

Once in the reservoir, oil or gas migrates in response to buoyancy and to hydrodynamic flux. In hydrologically mature basins, particularly of the simple and composite intracratonic type, basinal circulation of meteoric ground water is argued by Tóth (1980) to be a major control on direction of oil and gas migration. Thus, in mature basins characterized by centripetal flow, petroleum fields would be concentrated in areas of vertical, cross-formational discharge of the basinal meteoric system. Hydrodynamic arguments best explain the Athabasca and Peace River tar-sand accumulations in the Western Canadian sedimentary basin (Jones, 1980). Only the long-term circulation of meteoric water, following Tertiary uplift of the Rocky Mountains, could reasonably have localized one trillion barrels of oil derived from a dispersed, mediocre source.

The ultimate distribution of oil and gas fields reflects the structural and stratigraphic continuity of framework sand facies comprising the basin fill. Where the plumbing is interrupted laterally by facies or structural discontinuities, as is typical in large oceanic delta systems, and rift, subduction, and median basins, lateral migration is limited, but vertical migration may extend several thousand feet (Galloway *et al.*, 1982a). In contrast, in structurally simple intracratonic and passive margin basins characterized by stratigraphic continuity of sand bodies, lateral migration can be extensive. For example, sourcerock–oil correlation studies in the Denver and Williston basins (Dow, 1974; Clayton and Swetland, 1980) indicate lateral migration of oil of as much as 100 mi (160 km) from the thermally mature basin-center source rocks.

Reservoir Delineation and Characterization

First and foremost, genetic facies interpretation has been directed toward improving delineation and prediction of reservoir distribution within petroliferous basins. Sand body "models" were defined to allow early recognition of reservoir origin so that the direction and probable extent of specific oil- or gas-bearing sands could be extrapolated (Potter, 1967; LeBlanc, 1972; Conybeare, 1976). The application of facies analysis to stratigraphic trap exploration or locating offsets to discovery-wells necessitates models that predict external geometry of a sand body—its trend, lateral extent, or potential for recurrence. Despite more than 20 years of effort directed at development and application of sandbody models, site-specific prediction of an individual reservoir remains a challenging and commonly risky task.

One thesis of this book is that reservoir extrapolation, like many other applications of facies interpretation, is best carried out within a depositional system framework. Probable dimensions, trends, and recurrence intervals can then be calibrated for the system being explored. For example, lateral extent of delta-front sand bodies can vary greatly between contemporaneous delta systems prograding into the same basin, depending on the nature of the source fluvial system and basin physiography. Models constructed for one delta system may be poor predictors for another system. However, within each delta system, trend, lateral extent, and continuity of delta-front sand bodies may be generally predictable. Nonetheless, facies variability is the rule rather than the exception, and site-specific facies extrapolation remains at best an interpretive guess that attempts to accommodate all available data within the framework of the host system. Quantitative facies maps and interpretive cross sections graphically synthesize data and also may indicate subtle relationships between reservoir geometry and other, perhaps more easily delineated features such as deeper structure, regional facies patterns, or interval isopach variation. For example, the location of maximum expansion along a growth fault may also define a locus for distributary mouth bar accumulation and thus guide facies extrapolation within a delta system. Deeply rooted growth fault zones may continue to affect channel courses and resultant sand geometry in fluvial systems (Galloway, 1977).

Qualitative aspects of reservoir development within each of the major types of clastic depositional systems are reviewed in subsequent sections of this chapter. Delineation of the host system followed by recognition and qualitative description of genetic reservoir types are keys to facies-directed exploration.

A much smaller body of literature illustrates the potential use of genetic stratigraphic analysis in field development and enhanced recovery programs. In many areally extensive fields, external dimensions of the sand bodies, rather than trap size, determine the productive limits of the reservoirs. In a classic study of the Frio Sandstone (Oligocene) in Seelingson Field in South Texas, Nanz (1954) described and interpreted the complex distributary channel geometry of a major reservoir sand body. In Seelingson, reservoir dimensions are areally delimited by the sandbody geometries which, in turn, reflect deposition by upper delta plain distributary channels within a large, long-lived delta system. In a particularly useful paper, Weber (1971) illustrated the different areal and cross-sectional geometries of reservoir sand bodies deposited in distributary channel, coastal barrier, and meanderbelt environments of the Tertiary Niger delta system (Fig. 14-4). The variable trend and continuity of individual depositional units result in multiple oil-water contacts and areal configurations for productive horizons in the structurally trapped fields.

In addition to illustrating complex lateral and vertical relationships among different reservoirs, Figure 14-4 shows that single reservoirs, as defined by well log correlation and uniform fluid content, may in fact consist of a mosaic of individual genetic units. Pennsylvanian sandstones in the Elk City Field of the southern Anadarko Basin provide a graphic example of the genetic complexity inherent in many large reservoirs. Elk City is a large asymmetric anticline

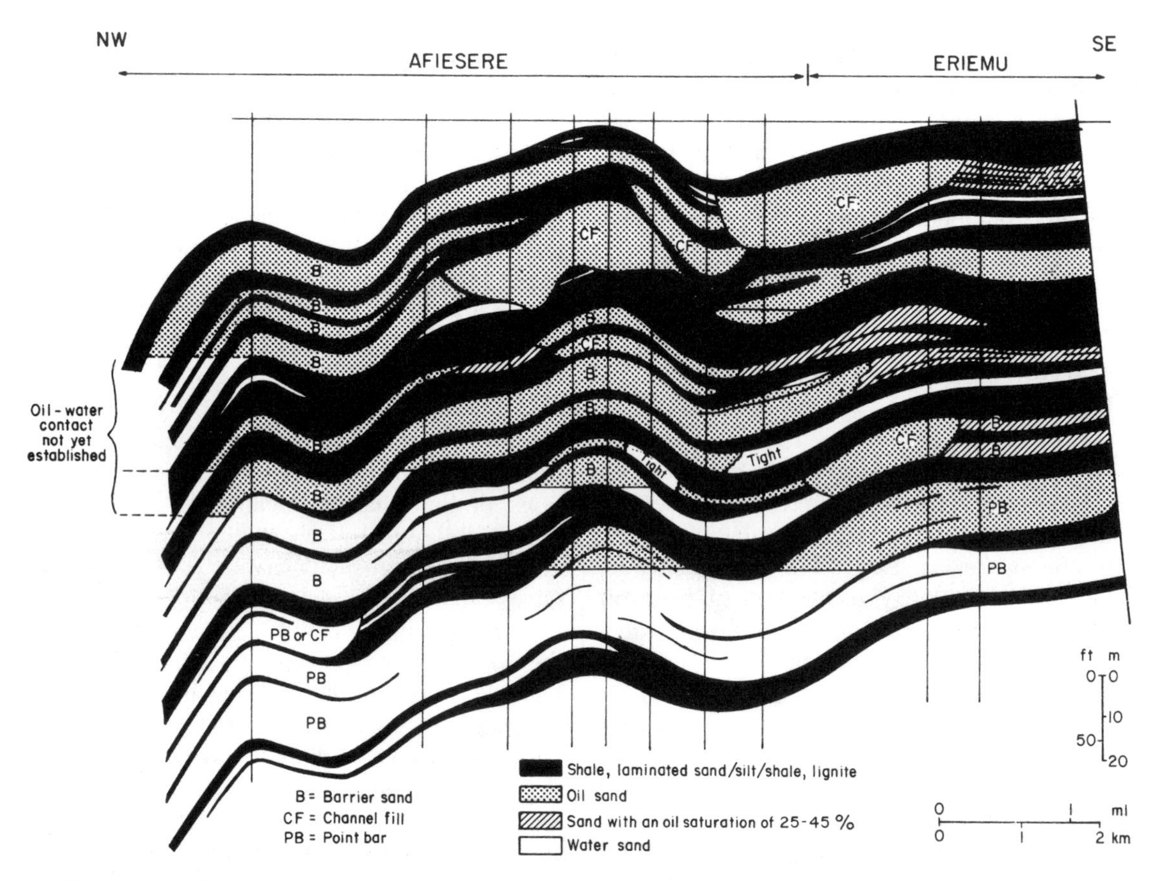

Figure 14-4. Complex genetic facies assemblage displayed by stacked reservoirs within the Afiesere and Eriemu oil fields of the Tertiary Niger delta system. (From Weber, 1971.)

covering about 25 mi^2 (65 km^2). Detailed stratigraphic analysis of one reservoir, the L_3 zone, revealed highly variable thickness and distribution patterns (Fig. 14-5A) that reflect an equally complex facies composition (Fig. 14-5B). Core, log pattern, and isolith data were combined to differentiate and map alluvial channel fill, distributary channel fill, delta margin, and barrier-bar sandstone facies. Distribution of these facies affects the efficiency of well completion and recovery practices. Hartman and Paynter (1979) describe several examples of reservoir drainage anomalies, some of which are clearly related to facies boundaries within single reservoir sand bodies. For example, wells completed within distributary channel fills were found to have poorly drained adjacent delta-margin facies. Closely spaced in-fill wells tapped essentially virgin reservoir pressure and oil–water contacts. Significantly, porosity and permeability of the geologically young Gulf Coast reservoirs are high, reflecting the unconsolidated condition of the sands.

Within a single genetic facies, macroscopic heterogeneities are introduced by bedding and spatial variability of textural parameters (Weber, 1982). Bedding produces a stratified permeability distribution that restricts cross-flow, and channels fluids within the more permeable beds (Polasek and Hutchinson, 1967; Alpay, 1972). Preliminary studies (Zeito, 1965, for example) indicated the potential for continuity of internal permeability stratification and showed that the geometry and continuity of bedding was correlative with interpreted depositional environment of the sand body. The impact of horizontal layering is well recognized in reservoir simulation studies; however, more complex bedding styles associated with lateral accretion or progradation are less commonly recognized. Shannon and Dahl (1971) demonstrated facies compartmentalization of a delta distributary mouth bar reservoir by pro-

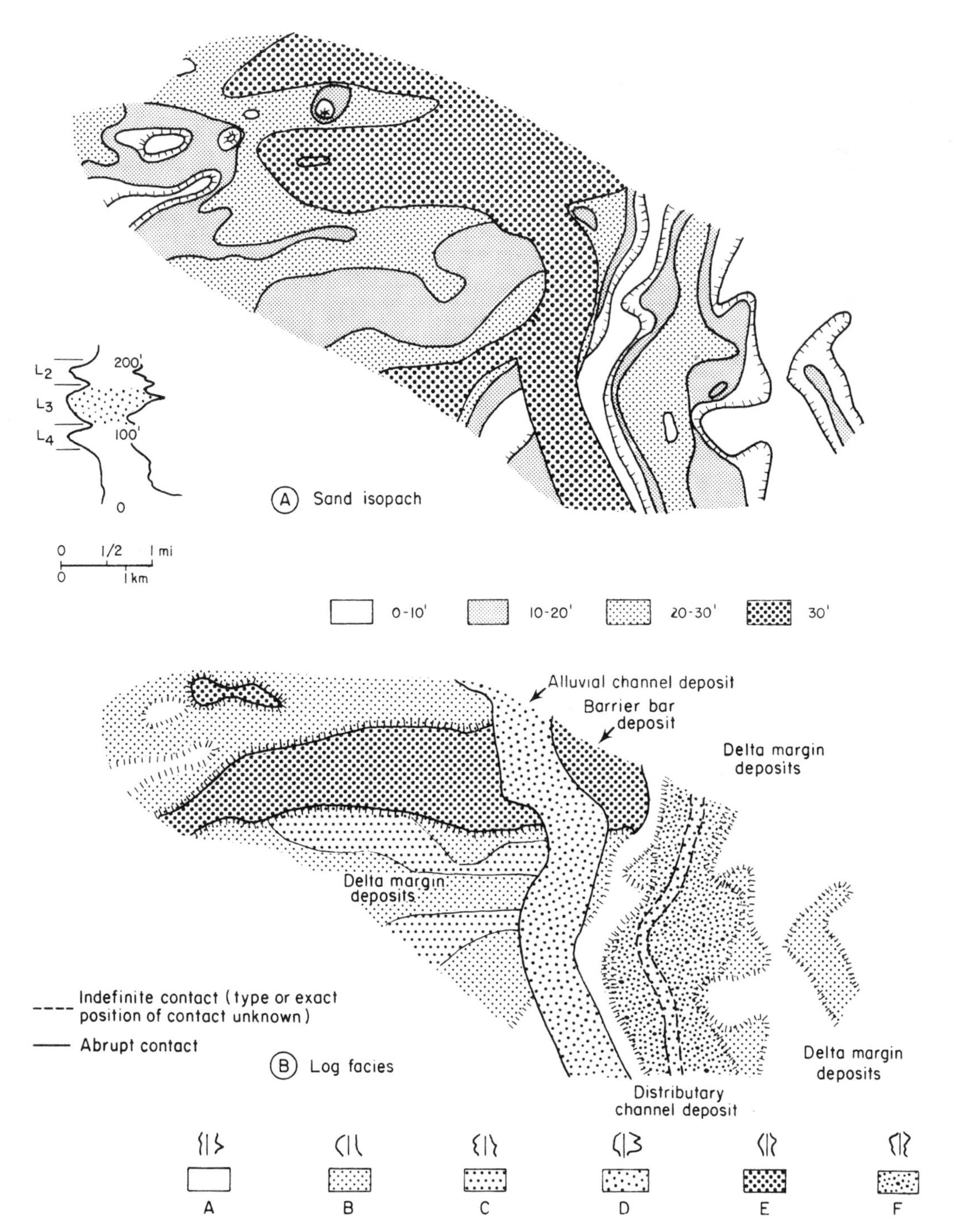

Figure 14-5. The genetic facies mosaic typical of a single, laterally continuous reservoir sand body of the Elk City Field, Oklahoma. Net sandstone isopach (A) and log facies (B) maps detail the complexity of the L_3 reservoir (inset) within the field perimeter. Interpretation of genetic units was based on thickness, vertical sequence, and lateral relationships of recognized units and reflects deposition of the reservoir as part of a fan-delta system derived from adjacent granitic uplands. (Modified from Sneider *et al.*, 1977. Copyright 1977, SPE-AIME.)

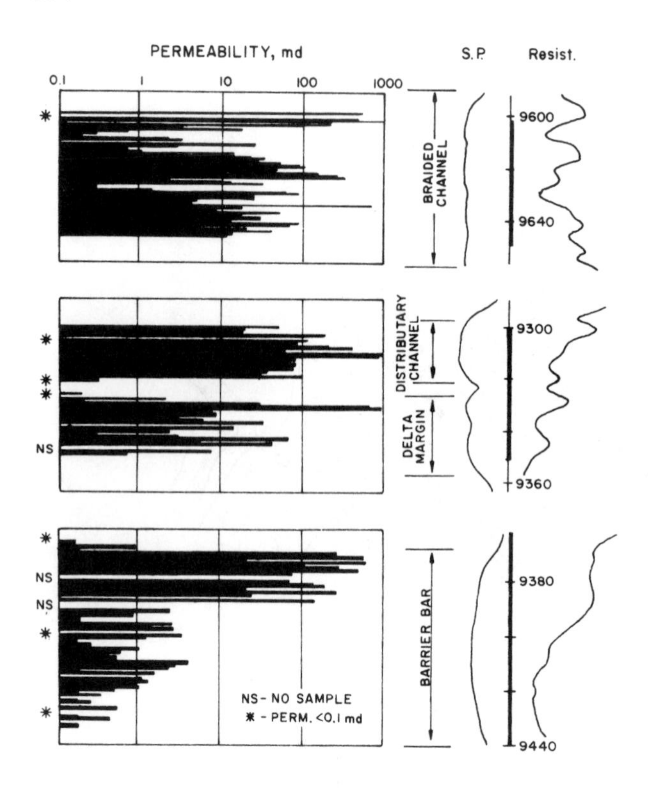

Figure 14-6. Log patterns and permeability profiles of three fan-delta facies sequences constituting the L_3 reservoir in the Elk City Field. Permeability correlates with grain size. Thus, the braided-channel sequence shows little vertical trend, the distributary channel fill sandstone has uniformly high permeability, the underlying delta margin sequence exhibits irregularly upward-increasing permeability, and the barrier bar sandstone is characterized by systematic upward-increasing permeability. (Modified from Sneider *et al.*, 1977. Copyright 1977, SPE-AIME.)

gradational bedding geometry. Recognition of the individual reservoir lenses, which reflect the deposition of frontal splays, suggested modifications to well-completion practices and improved oil recovery.

Within relatively uniform sand bodies or their component beds, permeability can vary systematically either laterally or vertically, and thus influence fluid-flow geometry. Distinctive vertical permeability trends reflect textural trends characteristic of channel fill, delta front, and barrier shore-face sequences found in the Elk City reservoirs (Fig. 14-6). From this and other studies, Sneider *et al.* (1978) suggested generalized trends of various reservoir properties for framework bar- and channel-type facies of delta systems (Table 14-2). The trends suggested are qualitative, but may be calibrated with engineering data and used to develop an accurate reservoir simulation model and improve oil recovery in deltaic reservoirs (Weber *et al.*, 1978). Spatial organization of reservoir properties of the sand facies of other types of depositional systems are less well documented. In addition, systematic lateral variation in permeability is poorly described, but is suggested by limited studies of modern sand bodies. For example, Pryor (1973) demonstrated a general down-channel decrease in average permeability in two fluvial point bars.

Finally, grain orientation and textural lamination introduce microscopic variability, which, if systematic, produces permeability anisotropy within the sand bed. Limited study of modern sand bodies (Pryor, 1973) showed maximum permeability in alluvial sands to be oriented along the channel axis. Thus, flow would be easiest along the axis of the resultant genetic unit. In contrast, upper shoreface and beach sands have maximum permeability axes that are oriented parallel to swash, producing an axis of maximum permeability that is perpendicular to the trend of the sand body.

Taken together, studies of both modern sand bodies and their ancient counterparts suggest that genetic interpretation allows prediction of a hierarchy of parameters that affect reservoir performance, ranging from external dimensions to internal compartmentalization and permeability stratification, heterogeneity, and anisotropy. Integrating and calibrating these predictions with reservoir engineering data can produce considerable improvement in the recovery efficiency.

Table 14–2. Generalized Reservoir Characteristics of Delta System Framework Sand Bodies[a]

	Grain Size	Sorting	Porosity	Pore Size	Permeability	Sw[b]	Continuity
Distributary Channels							
Top	Finest	Best	Highest	Small	Lowest	Highest	Deteriorates upward
	↑	↑	↑	↑	↑	↑	↑
Bottom	Coarsest	Poorest	Lowest	Large	Highest	Lowest	Best
Channel Mouth Bars; Delta Front Sands; Coastal Barriers							
Top	Coarsest	Best	Highest	Large	Highest	Lowest	Best
	↓	↓	↓	↓	↓	↓	↓
Bottom	Finest	Poorest	Lowest	Small	Lowest	Highest	Deteriorates downward

[a]Modified from Sneider *et al.* (1978).
[b]Sw = irreducible water saturation.

Occurrence of Petroleum in Depositional Systems

Fluvial and Alluvial Fan Systems

As would be expected from the variety of fluvial depositional styles described in Chapter 4, fluvial systems constitute diverse hosts for oil and gas. At one extreme, bed-load systems are reservoir-rich but source- and seal-poor; conversely, suspended-load systems may contain only moderate quantities of reservoir lithologies encased in abundant mudstone. However, all fluvial systems share several common attributes. (1) Principal reservoirs are the channel-fill and bar sands. Crevasse splay sands are a secondary reservoir facies. (2) General sand-body trends are parallel to depositional dip, but considerable local or subregional variation is likely. (3) Reservoir continuity is good to excellent, at least parallel to the channel. In structurally simple basins, fluvial channel deposits provide excellent conduits for regional migration of oil and gas. (4) Internally, fluvial reservoirs are extremely heterogeneous and anisotropic. (5) Mud facies of fluvial systems may contain significant quantities of herbaceous or woody organic material, and thus provide an internal gas-prone source. However, fluvial reservoirs rely on adjacent systems for oil-prone sources.

Bed-Load Systems. Permeable, framework sand bodies form abundant, well-interconnected reservoirs in bed-load systems (Fig. 14-7A). Broad sand belts tend to show minimal divergence from regional depositional dip. Although bed-load channel fills are internally complex and texturally variable, the lack of organized stratification or laterally continuous permeability barriers results in highly productive reservoirs that behave homogeneously to fluid flow at the scale of typical reservoir development.

Bounding facies are poorly developed. Floodplain muds are commonly silty or sandy and discontinuous. Thus, potential for well-developed

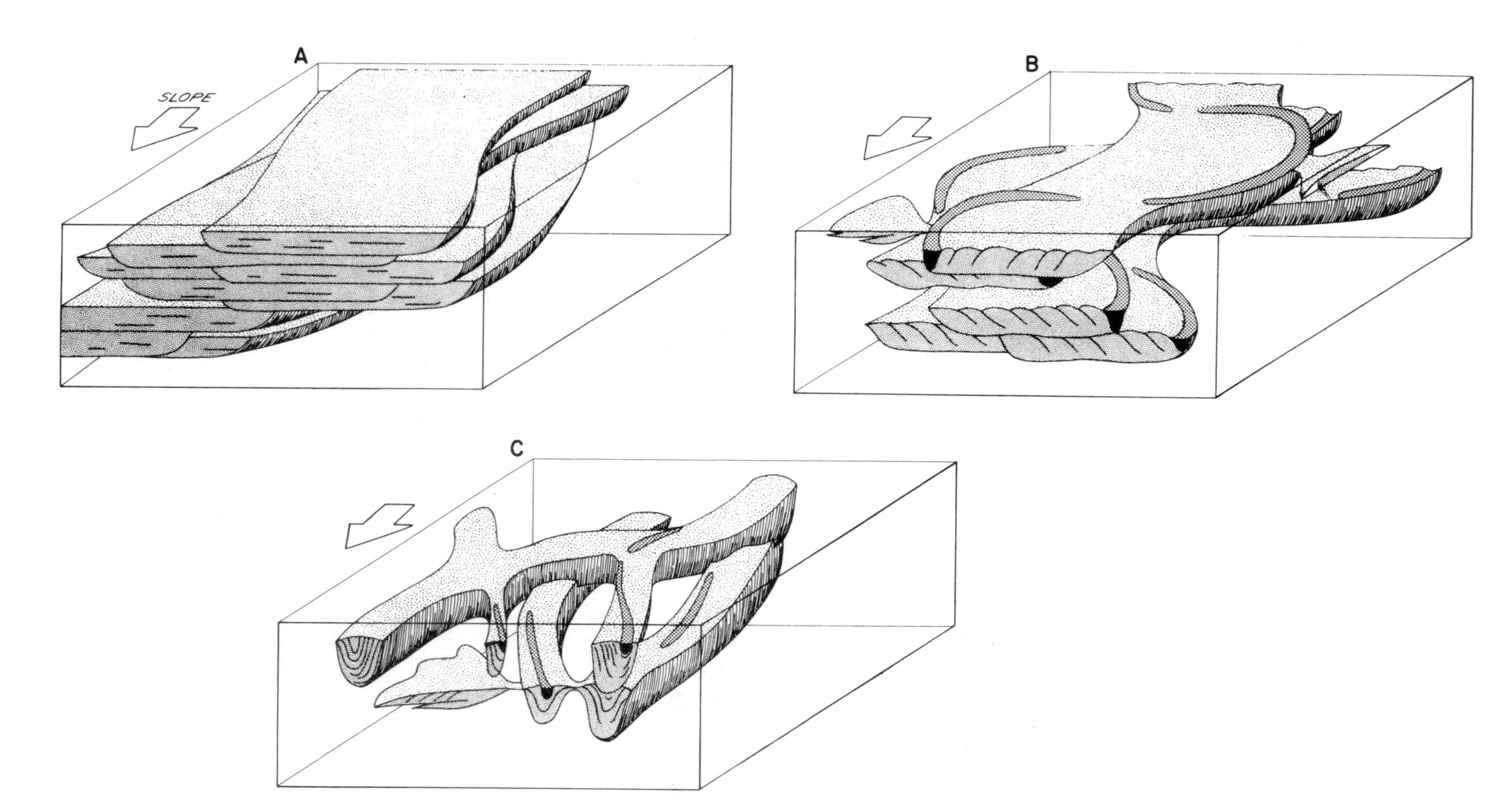

Figure 14-7. Schematic three-dimensional geometry, lateral relationships, and internal bedding architecture of reservoir sand bodies within (A) bed-load, (B) mixed-load, and (C) suspended-load fluvial systems.

sealing units or significant internal hydrocarbon sources is poor. The excellent reservoirs of bed-load systems are likely to be most prolific where overlain by sealing facies deposited by another depositional system. Similarly, source rocks likely occur as juxtaposed or interbedded facies of another system. The potential for stratigraphic traps is limited; accumulation is most common in structural or regional unconformity traps. Several giant oil fields of the Sirte Basin, Libya, including the Sarir and more recently discovered Messlah fields, produce from the updip limit of the Lower Cretaceous Sarir bed-load system. The massive, heterogeneous fluvial sandstones are up to several hundreds of feet thick, and act as a single integrated reservoir. Where truncated by a low-angle intra-Cretaceous unconformity, Sarir Group sandstone reservoirs are overlain by Upper Cretaceous marine shale that provides both a seal and probable source rock (Arabian Gulf Oil Company, 1980).

Mixed-Load Systems. Increased deposition and preservation of bounding mud facies within mixed-load fluvial systems results in greater isolation of the sandy meanderbelt sand bodies (Fig. 14-7B). Framework sands produce well-integrated, dominantly dip-oriented reservoirs. Crevasse splay deposits form isolated wedges or lobes within floodplain muds that may, in large fluvial systems, contain sufficient volume to constitute small but economic reservoirs. More importantly, splay and levee deposits together form texturally distinct "wings" along the reservoir margin that provide proximity indicators for the main belts of porous channel fill. Internally, meanderbelt sand bodies are characterized by well-developed and complex anisotropy and heterogeneity, particularly in their upper portion, where hydrocarbons preferentially accumulate. The systematic fining-upward textural trend is reflected by upward-decreasing permeability. Lateral-accretion bedding introduces permeability stratification that cuts across the sand body. The resultant permeability units are arcuate in plan view. The reservoir may be partially compartmentalized by mud plugs. In addition, the top of the permeable reservoir lithology commonly displays buried topography, reflecting ridge-and-swale and channel plugs.

Potential sealing facies within the system include levee, floodplain, backswamp, and possibly lacustrine muds. The transitional upper contact with overlying fine-grained units commonly results in poor definition of the seal, and may limit the total hydrocarbon column because of the relatively low pore-entry pressures typical of sandy and silty mudstone. Potential internal source facies include floodplain, backswamp, and lacustrine mudstone and claystone; all are predominantly gas prone. Preservation of organic material is more dependent on syndepositional ground-water setting than on depositional environment.

Mixed-load channel-fill reservoirs commonly occur in structural, stratigraphic, and combination traps. A classic production style consists of a meanderbelt that trends along structural strike. The series of convex point-bar scallops, which are bounded updip by mud plugs (Fig. 14-7B), provide a linear succession of stratigraphic traps that may contain reserves of several tens of million barrels (see example from the Fall River Sandstone shown in Figure 4-14B).

The Cretaceous Woodbine Formation, a major producing unit of the East Texas basin, contains several large structural and stratigraphic fields that produce from mixed-load fluvial meanderbelt facies of its component fluvial system (Oliver, 1971).

Suspended-Load Systems. Greater preservation of muddy facies and consequent isolation of permeable channel-fill facies make suspended load systems ideal targets for stratigraphic trap exploration. Isolated meanderbelt or anastomosed "shoe-string" sand bodies commonly display great variation in trend relative to both paleoslope and structural dip (Fig. 14-7C). Although individual traps are typically small, rarely containing more than a few tens of millions of barrels of oil or gas equivalent (b.o.e.) of producible hydrocarbons, they may be abundant. Conversely, channel-fill and associated crevasse-splay reservoirs pose great difficulty in exploitation of large structurally defined fields because of their limited dimensions, variable orientation, and erratic isolation.

Both lateral accretion or symmetric bedding produce cross-cutting permeability stratification (Fig. 14-7C). The fining-upward textural trend is reflected in upward-decreasing reservoir quality. Channel plugs, which are typically muddy, further compartmentalize the reservoir, and, as in mixed-load channels, may play a major role in defining the trap. Topography on the top of a permeable, hydrocarbon-saturated sand may exhibit substantial relief, which is usually accentuated by differential compaction and draping. Because suspended-load systems commonly occur on topographically lowest parts of the depositional basin near hydrologic base level, surrounding floodbasin deposits may be rich in herbaceous or woody debris. Bedded back-swamp coal or lignite may occur. Thus, significant potential for internal generation of biogenic and thermal gas exists within suspended-load systems. If volumetrically important, associated interchannel lacustrine muds could provide an algae-rich source for modest amounts of oil. However, mudstone facies of adjacent marine or lacustrine systems are more likely sources for large volumes of oil.

Mannville Group sandstones illustrated in Figure 4-14C display a producing style typical of mixed-load fluvial systems (Putnam and Oliver, 1980). Abundant small stratigraphic accumulations in Cisco fluvial and deltaic systems of the Midland Basin are reviewed in a subsequent section of this chapter.

Alluvial Fan Systems. Alluvial fans constitute a proximal, typically sand and gravel-rich portion of the drainage network. Fan systems, particularly wet fans, contain abundant potential reservoir facies. The inherent dearth of fine-grained sealing and potential source facies makes their overall hydrocarbon productivity dependent on the nature of surrounding systems. Lateral and vertical continuity of permeability, combined with the typical localization of fans near structurally active margins of the depositional basin, result in dominance of structural traps. Fan-delta margins, in which coarse, permeable fan facies interfinger with marine or lacustrine mudstone or limestone, are particularly favorable sites for hydrocarbon accumulation. However, structural entrapment still dominates. The relatively small dimensions of the drainage elements and the sporadic nature of fan deposition both produce internally complex reservoirs of the type described by Sneider *et al.* (1977) in the Elk City field.

Delta Systems

Like fluvial systems, the diversity of deltaic depositional styles leads, in turn, to great diversity in producing styles. Nonetheless, several important generalizations are possible.

1. Deltas have long been considered the ultimate hydrocarbon generating machines. Although some overly simplistic ideas about the mechanism of hydrocarbon generation within deltaic deposits have undergone necessary evolution and change, the fact remains that delta systems are inherently volumetrically large assemblages of potential reservoir, sealing, and source facies.

2. In all three types of delta systems, two major reservoir facies occur. Distributary channels form a branching network of irregularly dip-oriented, discontinuous reservoirs. Delta margin sands provide more extensive, commonly finer grained but better sorted reservoir facies concentrated along the basinward margin of deltaic depositional episodes. Splay, prodelta slump, and local destructional sand bodies constitute volumetrically less important but locally productive reservoirs.

3. Source facies are typically of low to moderate quality and contain largely type III kerogen. Consequently, internal sourcing of delta systems leads to gas-dominated production (Barker, 1980). However, distal deltaic deposits interfinger with shelf and slope deposits, and thus may tap more oil-prone source rocks. Source-rock richness is controlled primarily by the rapid depositional rates and by the abundance of allochthonous herbaceous organic matter in the prodelta environment.

3. Syndepositional structural traps are common to all types of delta systems. Their economic importance is determined by the scale of deltaic progradation. All of the late Tertiary deltaic basins are characterized by thousands of feet of progradational sediment containing growth-fault or diapir-related traps. Although similar features occur within delta systems of simple or composite intracratonic basins, scale is typically too small for development of commercial traps.

Delta systems are justifiably credited with being the most productive of depositional systems (Fisher *et al.*, 1972; Barker, 1980).

Fluvial-Dominated Delta Systems. Perhaps because of the popularity of the well-known Mississippi delta model, interpretations and descriptions of fluvial-dominated deltaic reservoirs are legion. Distributary channel, channel mouth bar, and delta front sands are commonly recognized reservoir facies in productive units ranging in age from Pleistocene through early Paleozoic.

As suggested in Figure 14-8A, reservoir sand bodies of the lower deltaic plain are typically multilateral, branching, isolated, irregularly oriented, and lenticular in cross section. Interdistributary crevasse splays are a volumetrically subordinate but potentially important reservoir facies, which is likely to be partially or completely isolated from adjacent sand bodies (Fig. 14-8A). In large deltas, a single splay can contain as much as several million barrels of oil in an isolani-type trap. In addition, thin local destructional bars constitute minor but commonly highly productive isolani reservoirs. Landward, the upper delta plain facies assemblage contains primarily channel-fill reservoirs, commonly of the suspended load type (Fig. 14-7C).

Both organic-rich delta-plain muds and carbonaceous prodelta deposits surround and interfinger with potential reservoir sand facies. The strong fluvial overprint, inherent in this variety of deltaic system, favors the dominance of land-derived, herbaceous, gas-prone organic matter. Local or regional destructional, transgressive shales constitute optimum seals. Many authors have noted the preferred association of oil and gas with distal and uppermost portions of deltaic depositional episodes (White, 1980, Fig. 7, for example). In both sites deltaic reservoirs are typically capped by local destructional or transgressive shelf deposits, providing an effective seal and possibly an oil-prone source.

As suggested in Figure 14-8A, delta-fringe reservoirs likely consist of nested channel fill and mouth bar or delta front sand units. Each sand facies has distinctly different reservoir properties and geometry of permeability stratification (Sneider, *et al.*, 1978). These differences may be accentuated by subsequent burial and diagenesis. In older deltaic systems the coarser channel-fill sand facies alone may retain adequate permeability to produce hydrocarbons. Conversely, in a different diagenetic system, the better sorted mouth bar sands may provide the only reservoirs, whereas the associated channel fills are tight and nonproductive.

The multiplicity of isolated or partially isolated sand bodies, which exhibit highly diverse trends on extensive, flat delta plains (Fig. 5-9), provide a multiplicity of potential stratigraphic traps within fluvially dominated delta systems.

Wave-Dominated Delta Systems. With increasing wave reworking, the isolated channel mouth bars coalesce, forming laterally extensive, interconnected beach ridge and coastal barrier sand bodies (Fig. 14-8B). The abundance of highly transmissive, well-sorted, strike-oriented delta fringe and interlaced dip-oriented distributary and fluvial channel sand bodies (Fig. 14-8B) make wave-dominated deltas optimum hosts for high quality reservoirs. Conversely, the high degree of sand-body interconnection and general funnel-like sand distribution, which opens into the feeder fluvial system in the updip direction (Fig. 5-15), limits opportunities for stratigraphic entrapment. Structural traps, including growth faults and diapirs, dominate production.

Marine domination results in slower sedimentation rates on the prodelta and adjacent shelf and slope environments. Thus, a greater proportion of marine organic material can be incorporated in these facies. This, combined with greater opportunity for bacterial breakdown of river-derived herbaceous debris, favors development of somewhat more oil-prone deltaic source rocks. However, the proportional volume of potential source mud facies is less, and the slow accumulation rates may allow complete oxidation of some organic matter. As with fluvial-dominated delta systems, highly prolific oil production likely necessitates hydrocarbon sourcing by adjacent systems, whereas gas-rich, light paraffinic oils can be generated internally.

Examples of important productive wave-dominated delta systems include portions of the Frio Formation (Oligo–Miocene) discussed below, and the Tertiary ancestral Niger Delta (Weber, 1971; Evamy *et al.*, 1978). In the Niger sheet-like delta fringe coastal barrier, sands are the most important genetic reservoir type (Fig. 14-4); associated distributary and tidal channel fills are also productive (Weber, 1971; Weber *et al.*, 1978). Traps are mainly roll-over anticlines formed by large growth faults.

Tide-Dominated Systems. Productive characteristics of tide-dominated delta systems are less well documented. However, their characteristics as hydrocarbon hosts can be reasonably inferred.

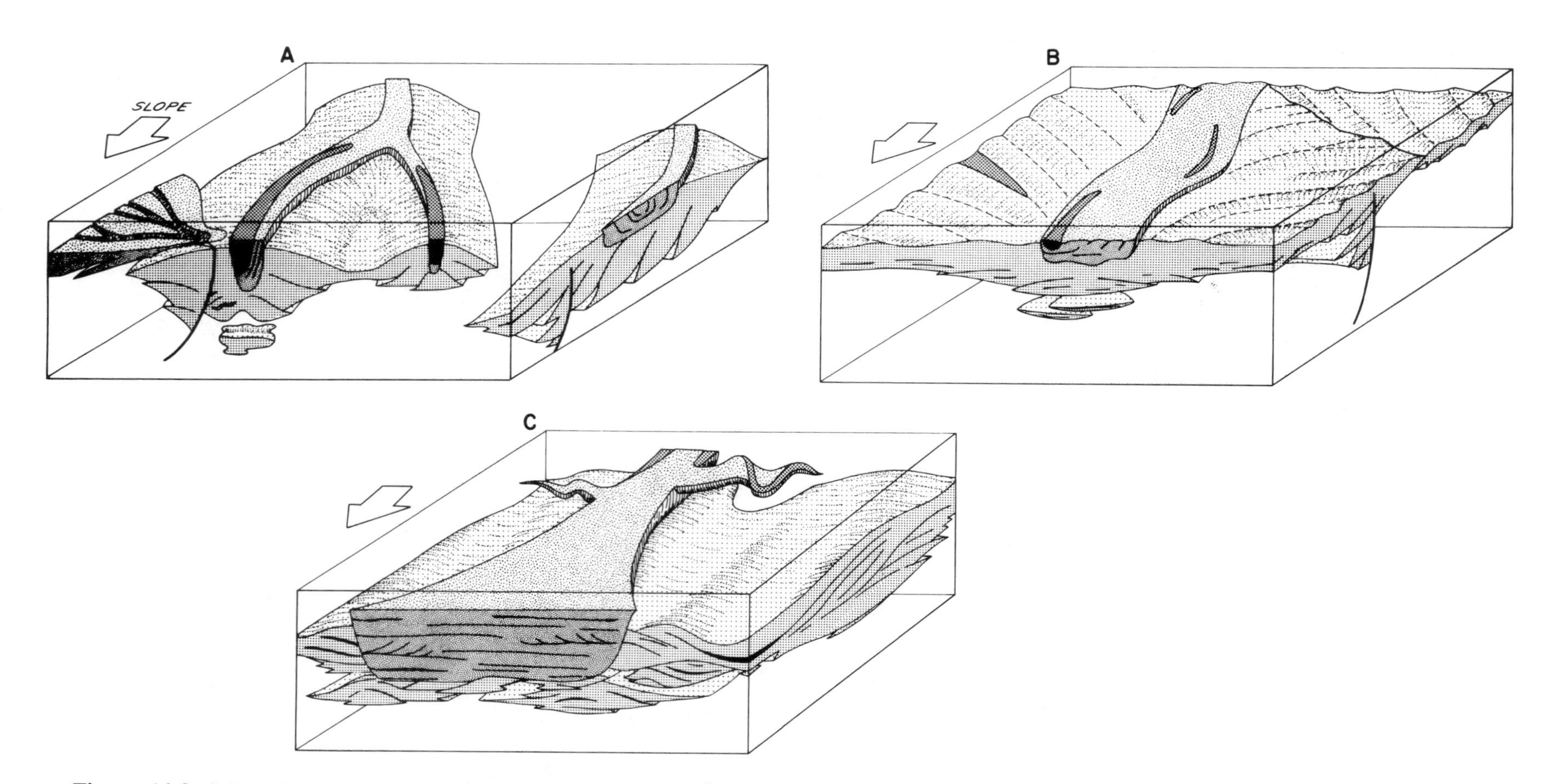

Figure 14-8. Schematic three-dimensional geometry, lateral relationships, and internal architecture of reservoir sand bodies within (A) fluvial-, (B) wave-, and (C) tide-dominated delta systems.

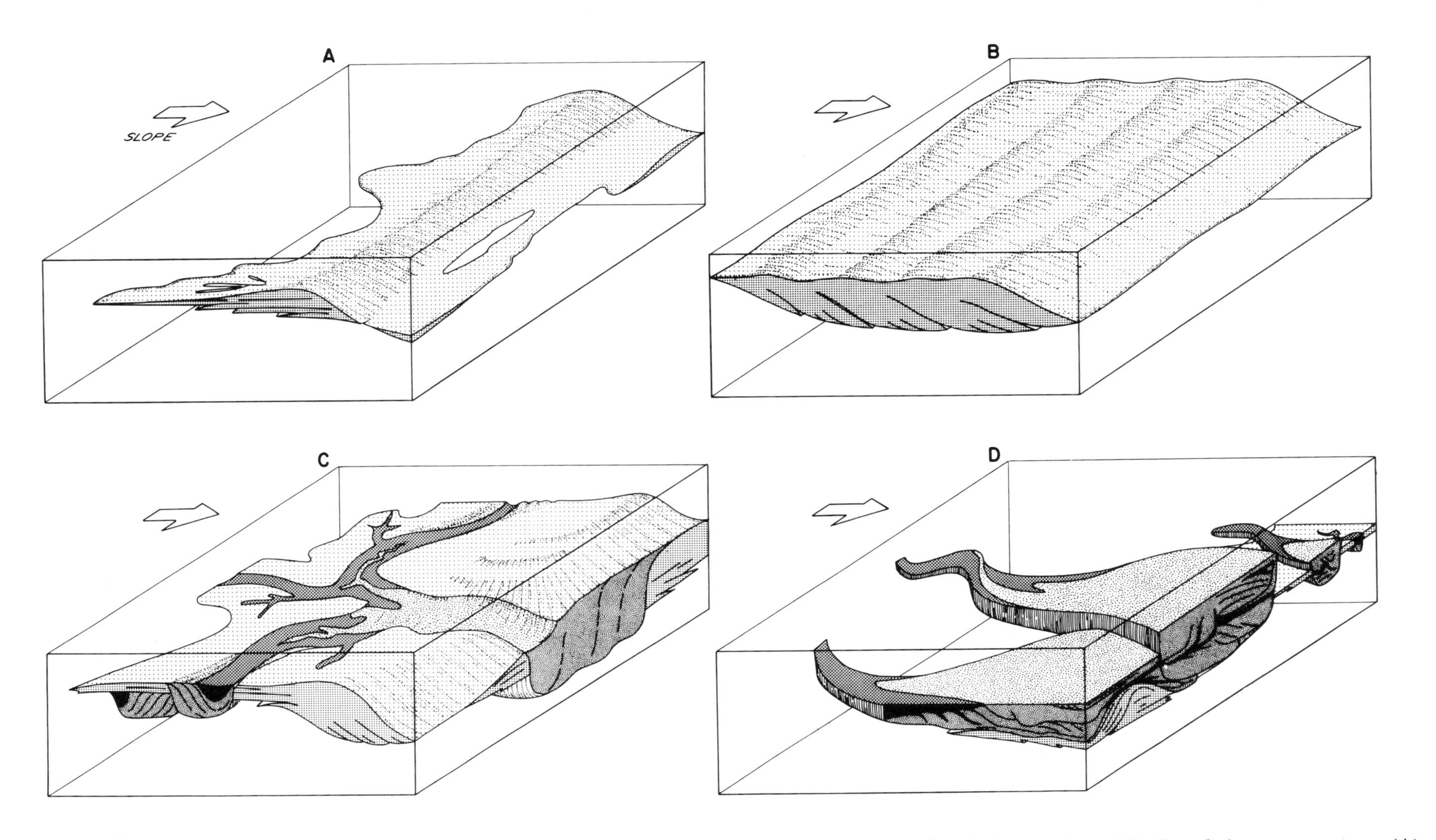

Figure 14-9. Schematic three-dimensional geometry, lateral relationships, and internal architecture of typical reservoir sand bodies of shore-zone systems. (A) Transgressive barrier bar. (B) Progradational, sand-rich strandplain. (C) Barrier bar and inlet complex of a transitional micro- to meso-tidal coast. (D) Estuary and subtidal sand-flat complex of a macrotidal shorezone.

Potential source facies include prodelta muds (gas plus some oil) and delta-plain marsh organics (gas). Although reservoirs are the product of extensive marine reworking, they are likely to be complex and discontinuous in all but the most sand-rich systems. Potential reservoir facies include (1) the large, basinward flaring distributary channel fills and associated delta-plain splays, and (2) the coalescent to isolated tidal current ridge sand bodies of the delta front (Fig. 14-8C). Both reservoir types would likely exhibit dip-orientation of both external geometry and internal compartmentalization. Potential for stratigraphic entrapment ranges from fair to poor, depending on sand content of the fluvial system and extent of wave reworking of the delta fringe. Considerable deposition and preservation of mud within tidally produced channel and bar sands would likely result in complex permeability stratification and heterogeneity (Fig. 14-8C).

The late Tertiary Mahakam delta system (Kalimantan) seems to reflect persistence through geologic time of the modern tide-dominated deltaic depositional system (Magnier *et al.*, 1975). Hydrocarbon traps are dominantly anticlines associated with gravity sliding. Reservoir geometries are described as complex. As is typical of delta systems, source rocks interbedded with productive delta-margin channel and bar reservoirs contain mostly thermally immature, Type III kerogen. More deeply buried distal prodelta and upper slope muds contain substantial portions of degraded herbaceous and marine organic constituents and are the likely source of the gas-rich paraffinic crudes produced at large fields such as Attaka and Handil (Combaz and De Matharel, 1978). An additional noteworthy feature of the Mahakam is the juxtaposition of carbonate reef deposits and prodelta shelf muds in both the Holocene and late Tertiary delta systems.

Shore-Zone Systems

Basin-margin depositional systems typically contain a mix of shallow-water marine or marine-influenced mud facies interbedded with potential reservoir sand bodies. However, sand content can range from as little as a few percent to nearly 100 percent. Sealing capacity within shore-zone systems is naturally inversely related to sand content. Because sand bodies typically parallel depositional strike in belts of variable width and complexity, up-dip pinch outs are inherent in most shore-zone systems. Thus, they have a high potential for development of at least a few stratigraphic traps. In sand-poor systems, accumulation of bed-load sediment in disconnected bars may produce isolani traps. The specific reservoir facies formed depends on the sediment supply as well as the wave and tidal energy flux of the strandline. However, marine-shelf shales can be interbedded with, underlie, or cap virtually all of the sand-body types and provide direct access to potentially oil-prone source rocks. Extensive reworking inherent in the shore-zone produces texturally mature sand which initially possesses optimum reservoir quality. However, porosity and permeability may be degraded by burrowing or by pervasive calcite cement produced by leaching and reprecipitation of contained shell material.

Transgressive Barrier Sand Bodies. Transgressive barrier and beach-ridge sand bodies tend to occur as isolated, narrow, strike-parallel stringers (Fig. 14-9A) and are ideal targets for stratigraphic exploration. Oscillation of the strandline, such as might occur down-drift of major delta systems, may stack transgressive units, producing a thick composite sand body. Sand accumulation is commonly localized at paleotopographic nick points or structural hinge lines and discontinuities. Transgressive reworking commonly collects the coarsest material available; the resultant large pore throats make such units permeability "survivors," provided shell debris or mechanically unstable clasts are minor constituents. Internal bedding, if well developed, is most likely to be horizontal (Fig. 14-9A).

Basal Niobrara sandstones of the San Juan Basin, New Mexico, are excellent examples of transgressive barrier reservoirs (McCubbin, 1969). The basal Niobrara forms a discontinuous sand sheet consisting of relatively coarse, glauconitic sand. Individual sandstone bodies were deposited on the basinward side of cuestas on the transgressed paleotopographic surface. The linear, strike-elongate sand bodies are capped and sealed by marine Niobrara shale. Hydrocarbons are trapped stratigraphically by abrupt up-dip pinch out of the reservoirs.

Sand-Rich Strandplain Sand Bodies. Progradational, sand-rich strandplain sands form excellent, widespread reservoir facies (Fig. 14-9B) that are prone to produce in structural traps. Opportunity for stratigraphic entrapment is limited to regional up- and down-dip pinch out into coastal plain or shelf muds; however, resultant stratigraphic fields may contain large reserves. Best reservoir properties occur at the top of the sand sheet in upper shoreface and beach deposits. Progradational bedding is likely to occur in the lower shoreface sequence and might influence productivity in completely saturated reservoirs.

Barrier-Bar Sand Bodies. The presence of a protected lagoon, which is necessary for true barrier bar formation, effectively isolates sandy reservoir facies in the landward direction and introduces internal facies complexities not present in the simple strandplain. Tidal inlets produce dip-oriented bulges and isolated sand thicks within the generally strike-parallel reservoir unit. Successive barrier bars are offset in the downdrift direction. The resultant genetic sand body is a strike-elongated belt encased in mud and exhibiting a moderately irregular basinward margin, and thin, highly irregular landward aprons (Fig. 14-9C). With continued basin subsidence and accentuation of depositional slope, the irregular pinchout of inlet and washover sands into lagoonal mudstone defines a succession of potential stratigraphic traps that together constitute a strike-parallel fairway. Several such fairways may be found within a barrier/lagoon system. With increased tidal influence, inlet facies are more prominent and the back-barrier is segmented by multiple tidal channel fills and plugs. Thus, while providing numerous built-in stratigraphic traps, barrier/lagoon systems tend to store the oil and gas in thin, highly interbedded and compartmentalized portions of the reservoir. The Jackson and Yegua (Eocene) Groups of the South Texas coastal plain illustrate the producing style typical of barrier/lagoon systems characterized by stratigraphic and combination traps (Fisher *et al.*, 1968).

In contrast, amalgamation and stacking of barrier sand bodies may produce massive, highly productive reservoirs such as the Greta sand (Upper Frio Formation) of South Texas. Here well-sorted strike-trending, composite sand bodies up to several hundred feet thick are prolific oil and gas reservoirs in shelf-margin, growth fault-related traps (Boyd and Dyer, 1964).

Tidal-Inlet Sand Bodies. Sand macrotidal shore-zone deposits include nested estuary channel fills and associated subaqueous sand bars and tongues. The resultant sand body is a regionally strike-elongate, irregular to discontinuous belt traversed by multiple landward-trailing tongues that grade into narrow sand, silt, and mud-filled channels (Fig. 14-9D). The estuary-fills form large funnels that open basinward into the more continuous interbedded marine muds and subaqueous bar sands and pinch out landward into muddy, impermeable marsh, tidal flat, and gully sediments. They make ideal stratigraphic traps in structurally simple basins. Landward pinch-out is less likely in tide-dominated deltas because the estuarine distributary channels connect with sand-rich fluvial channel facies (Fig. 14-8C). As in barrier systems, multiple stratigraphic traps likely occur in elongate, strike-parallel fairways.

Tidally-dominated shore zone sand bodies are internally complex, highly compartmentalized, and transected by mud and silt drapes typical of tidal deposits (Fig. 14-9D). Permeability stratification would likely be greatest in the mud-rich upper reaches of the inlet fills. Despite their probable abundance in the stratigraphic record, reservoirs in macrotidal shore-zone depositional systems have not been well described in the literature. Suggested examples, such as some Pennsylvanian sandstones of the Anadarko Basin, remain controversial, and insufficient regional data have been published for a conclusive interpretation.

Shelf Systems

Reservoir sand bodies deposited in shelf systems, though a volumetrically minor component of most basin fills, can contain significant reserves of oil and gas. The Sussex Sandstone of the North American Cretaceous seaway described by Brenner (1978) provides an excellent three-dimensional example of a shelf-sand reservoir complex.

Reservoirs tend to be porous and permeable lenses or thicks within irregular sheets composed of interlaminated or poorly sorted sand and mud (Fig. 14-10). Individual sand bodies occur in swarms and display strong preferred orientation, usually parallel or oblique to the adjacent shore line. Vertical sequences and, consequently, permeability trends include both coarsening- and

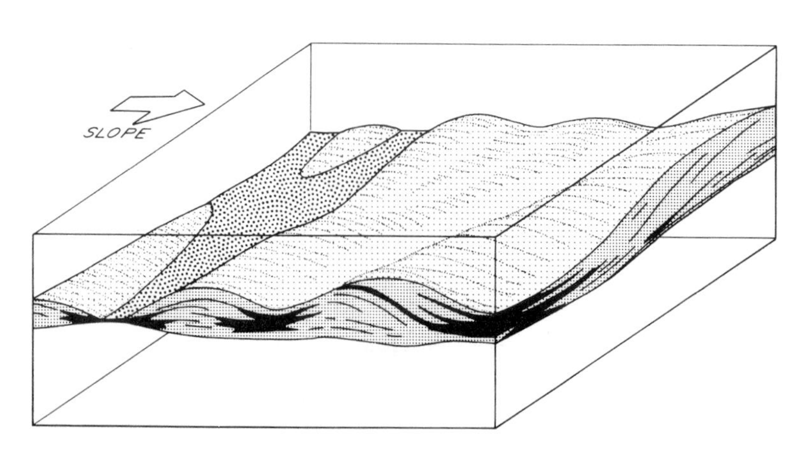

Figure 14-10. Schematic three-dimensional geometry, lateral relationships, and internal architecture of a generalized shelf sand body.

fining-upward patterns. As suggested in Figure 14-10, internal accretionary bedding and amalgamation of subsequent generations of subaqueous bars may introduce considerable permeability stratification within the sand body. Reservoir quality may be further degraded by chemical or mechanical breakdown of unstable constituents such as shell debris, glauconite, or mud pellets.

Associated shelf mudrocks provide ample reservoir seals and may also serve as adequate source rocks. Type I or II kerogen would typically dominate.

In the Sussex Sandstone of the Powder River Basin, overpressured hydrocarbons occur in several thin, but highly prolific, bar-like isolani traps (Fig. 14-3). Surrounding sediments form a broad, irregular, impermeable sandy sheet containing abundant dispersed and laminated mud. Associated shelf mudstones contain from 0.6 to 1.1 percent TOC and are the inferred source of Sussex oils (Brenner, 1978).

Marine Slope and Basin Systems

Marine slope and basin sand bodies include at least three well-described, productive reservoir configurations.

(1) The sand-rich turbidite fan association is typical of both offlapping (progradational) and uplapping (aggradational) basin fills. Reservoir facies include fan channel and suprafan deposits. Fan channel fills are lenticular, commonly multilateral and multistory, and generally oriented down depositional slope (Fig. 14-11A). Both distributary and anastomosing multichannel patterns occur. In contrast, suprafan deposits form lobate to irregular sheets and occur as well-bedded packets (Fig. 14-11A). Bedding is dominantly horizontal within packets; external geometry of individual packets or lobes reflects the basin-floor configuration. Although many of the best-known examples of marine turbidite-fan deposits occur in tectonically complex subduction basins, well-preserved examples occasionally are important oil and gas hosts in cratonic basins (Galloway and Brown, 1972).

(2) Submarine-slope and basin-floor sand bodies emplaced by nonturbid density flows have been distinguished recently (Harms, 1973) and constitute a major class of reservoirs within cratonic basins. Sand bodies are broad, shallow, amalgamated, channel-shaped ribbons to irregular aprons, displaying complex internal bedding and prominent permeability stratification (Fig. 14-11B). Examples include the numerous fine-grained basin-center reservoirs in the Permian Basin, including the Bell Canyon, Brushy Canyon, and Cherry Canyon Formations (Williamson, 1979) and the Spraberry and Dean Siltstones of the Midland Basin (Handford, 1982).

(3) Erosional canyon and gorge fills are lenticular, channel-shaped units of regional scale that parallel the paleoslope. Reservoir-quality sandstone is typically limited to the erosional gorge, forming thick, narrow, dip-oriented lenses or belts (Fig. 14-11C). In onlapping stratigraphic settings, where submarine gorge fills are common, gorge sediment may consist of abundant mud with individual sand bodies exhibiting up-channel onlap and pinch out (Fig. 14-11C). Such sands constitute internally complex reservoirs containing cross-cutting scour surfaces, horizontal and lateral accretionary permeability stratification, and shale or mudstone lenses and interbeds.

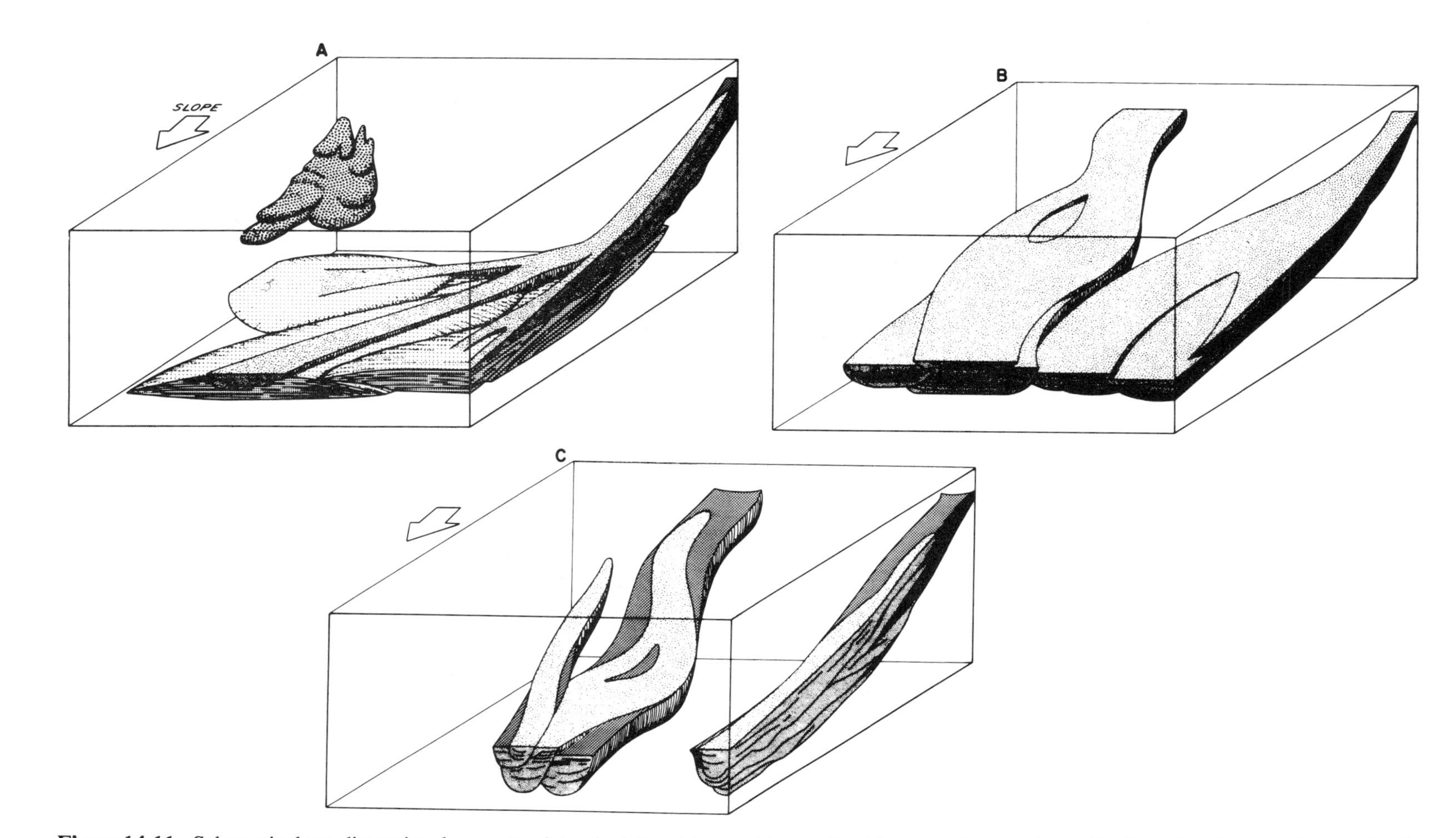

Figure 14-11. Schematic three-dimensional geometry, lateral relationships, and internal architecture of reservoir sand bodies of terrigenous slope and basin systems. (A) Offlapping, sand-rich turbidite fan and adjacent slump. (B) Nonturbid density-flow fan. (C) Onlapping submarine gorge fill.

Gorges may occur singly, as in the Tertiary of the Sacramento Valley, California (Almgren, 1978), or as part of an extensively eroded wedge. The highly prolific sandstones of the erosionally-based Hackberry Embayment (Oligocene, Louisiana) illustrate both the depositional and producing styles of sandy onlapping gorge fills within a regionally progradational, salt-floored continental slope (Paine, 1971).

Because bed-load sediment in slope and basin systems is transported by gravity-induced underflow of a dense water mass, transport paths avoid contemporaneous bathymetric highs. Actively rising structures, such as salt massifs or fault blocks, receive little if any sand across their crestal highs. Consequently, pinchout, successive wedge-out, or drastic thinning of slope reservoirs over or against structures is commonly observed and may result in a dominance of offstructure production and development of stacked stratigraphic traps (Weser, 1978; MacPherson, 1978). Additional potential for stratigraphic entrapment exists in structurally simple basins where inherent up-slope pinchout typical of all types of fan channel and gorge sand bodies (Fig. 14-11) is preserved in the regional dip. The base of the depositional slope also provides a likely site for updip wedge-out of basinal fan and channel deposits. On a larger scale, geometry and trend of slope deposits must be recognized as extremely sensitive to basin configuration. In tectonically active basins, simple fan morphologies may be highly skewed (Hsü, 1977). Axial transport in elongate subduction and rift basins is an extreme but common manifestation of the gravity-redeposition process.

Submarine slope and basin reservoirs are almost uniformly characterized by comparatively poor reservoir quality. Turbidity-current transport mechanisms are, by their nature, suspended-load rich, and a small but consequential fine fraction is typically deposited with the bed load. Nonturbid density or geostrophic current deposits, though clean and well sorted, are typically extremely fine grained. Their already low permeability is readily subject to further reduction by diagenesis. Although capable of transporting large volumes of sand onto the slope or basin floor, slumping and other forms of mass transport churn and mix sediment indiscriminately, severely degenerating original reservoir quality. Production efficiency is further decreased by the complex internal structure and intricate bedding of the texturally heterogeneous channel-fill and suprafan sand bodies.

On a positive note, however, slope and basin reservoirs are likely to be intimately interbedded with some of the most favorable oil-source facies within a depositional basin (Dow, 1978). Where traps are available and the deep basin fill remains unbreached by erosional unroofing, slope-fan reservoirs may be charged with tremendous volumes of oil. For example, Yerkes *et al.* (1965) estimated that the late Tertiary slope-fan systems of the Los Angeles basin contain over 30 million barrels of oil, perhaps one third of all that was generated from the surrounding 380 mi^2 (985 km^2) of thermally mature basinal mudstones averaging 3.3 percent TOC. In a single large complex of stratigraphic traps covering nearly 800 mi^2 (2,000 km^2), the Permian Spraberry Sandstone of the Midland Basin has produced more than 220 million barrels of oil. However, because of the very low permeability of the fine sand and siltstone, this figure represents only a small percentage of the oil in place. Source rocks are the surrounding organic-rich basinal mudstones (Dutton, 1980b).

Eolian Systems

Eolian systems produce extensive, blanket reservoir sand bodies with excellent and relatively uniform primary porosity and permeability. In rare situations low-energy transgression and burial may preserve dune topography, producing paleotopographic traps (Vincelette and Chittum, 1981). Local interdune wadi or sabkha deposits are likely to be poorly sorted and contain fine-grained material or chemical cements, further restricting permeability and leading to a generally horizontal permeability stratification.

Eolian systems must rely primarily on bounding systems for both seal and source facies. The Permian Rotliegende Sandstone of the Mid-European basin illustrates the associations that can lead to large-scale hydrocarbon accumulation in eolian systems (Lutz *et al.*, 1975; Glennie, 1972). Upper Carboniferous coal beds that underlie the dune and wadi reservoir facies of the Rotliegende eolian system provided a source for the giant reserves of methogenous gas. Overlying evaporites of the Zechstein seal the Rotliegende reservoirs. Traps are structural. The productive area is limited in part by northward gradation of dune deposits into fine-grained impermeable sabkha facies.

Lacustrine Systems

Large lacustrine basins may constitute major petroleum provinces. Shore-zone and deltaic deposits provide reservoir facies that are analogous to their marine counterparts. Subaqueous slope and basin-floor sequences include sandy turbidite and slump deposits. Interbedded evaporite and carbonate units may also play a role in petroleum occurrence. Perhaps the most significant aspect of lacustrine systems, however, is the potential for development of extremely rich source-rock mudstone facies of both deep, anoxic lakes and shallow, restricted, saline lakes. Organic matter in lake sediments tends to be rich in algal debris, with varying amounts of allochthonous vegetal matter (Ryder, 1980). Thus, many lacustrine source rocks yield abundant low sulfur, paraffinic oil, and variable amounts of gas.

The Uinta Basin of the western United States was partly filled with highly petroliferous facies of the Eocene Green River lacustrine system. Along the lake margin, strandline and deltaic sands and carbonate grainstones yield paraffinic crude at Red Wash field, a major stratigraphic trap resulting from pinchout of reservoir facies into lake basin mudstone (Ryder, 1980). The deep, structural basin center contains the giant Altamont–Bluebell trend, a complex of thin, widespread lake-margin sand and siltstone reservoirs, which pinch out up structural dip into organic-rich lacustrine mudstones (Lucas and Drexler, 1976). Production is augmented by natural fracturing of the tight, thinly interbedded sand and shale units, abnormally high fluid pressures within the low-transmissivity lacustrine sequence, and high temperatures resulting from deep burial. Oils are paraffinic and have a high pour point. In both the Red Wash and Altamont–Bluebell trend, dip reversal caused by postdepositional migration of the structural basin axis, and development of peripheral folds formed traps.

Example: Intracratonic Basin Depositional Systems and Hydrocarbon Occurrence

A network of simple to composite intracratonic basins developed throughout the Mid-Continent of the United States during late Paleozoic time. The extensive occurrence of oil and gas within these structurally simple basins provides an unusual opportunity to examine the interrelationships between genetic facies and hydrocarbon productivity.

Three generalized tectonic-environmental assemblages may be recognized within the Mid-Continent basins (Galloway and Dutton, 1979):

(1) Shallow seas marginal to block-faulted and uplifted basement highlands (Fig. 14-12A). Block-faulted uplands are most common.
(2) Extremely shallow, stable platforms (Fig. 14-12B). Average water depths varied from several feet to a few tens of feet, although the shallow basins commonly covered hundreds to thousands of square miles. Such platforms included much or all of the simple intracratonic basins such as the Illinois Basin, as well as broad, shallow shelves and platforms adjoining the deeper, moderately-subsiding basins.
(3) Deep-water cratonic seas (Fig. 14-12C). Water depths ranged from a few to several hundred feet and commonly increased through time due to sediment starvation of subsiding basin centers.

Tectonic-Environmental Assemblages and Their Reservoir Facies

In basins where faulted uplands border a cratonic sea, deposition is strongly influenced by tectonic activity. Multiple, short-headed streams produce numerous coarse-grained alluvial fan and fan-delta systems around the periphery of the uplands. These fan deltas are recognized as lobate to digitate sand thicks on regional sand isopach maps (Fig. 14-13). Thickest sand bodies are deposited in main braided channels and on the subaerial delta plain and subaqueous fan margin. These constitute major reservoir facies of the fan

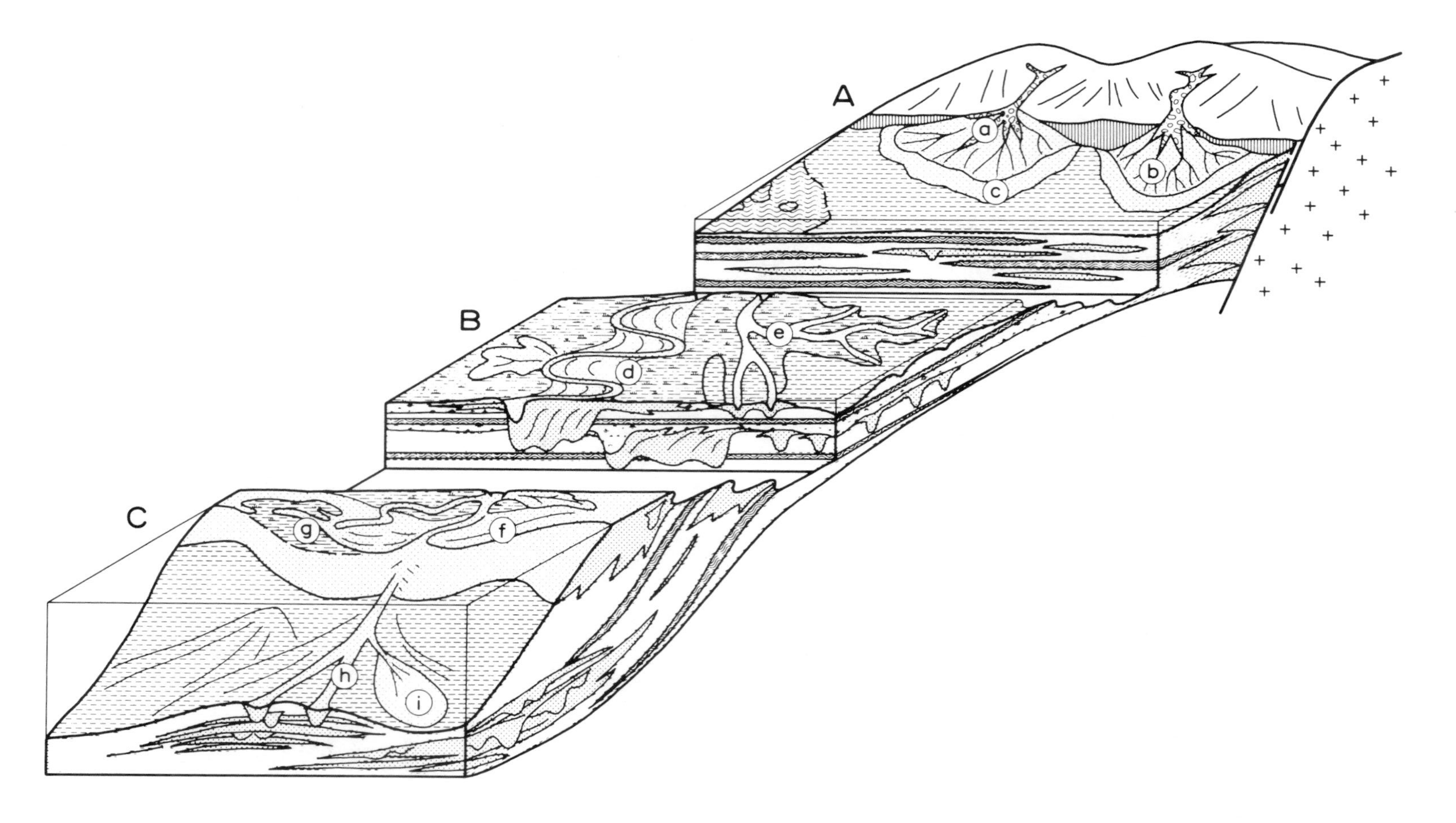

Figure 14-12. Principal producing reservoir facies of intracratonic basin depositional systems, (A) Cratonic sea marginal to actively uplifted highlands. Reservoirs include (a) braided channel fills, (b) distal fan-plain distributary channels and destructional bars, and (c) subaqueous distal fan-channel mouth bar and delta front sand bodies. (B) Shallow, stable platform. Incised fluvial channel fill and point-bar units (d) are the principal reservoirs; distributary channel fills (e) are locally significant. (C) Deep-water, stable basin. In addition to fluvial and distributary channel-fill reservoirs, delta front and channel mouth bar facies (f), destructional bars (g), and submarine fan channel (h) and suprafan lobes (i) all provide potential reservoirs (From Galloway and Dutton, 1979.)

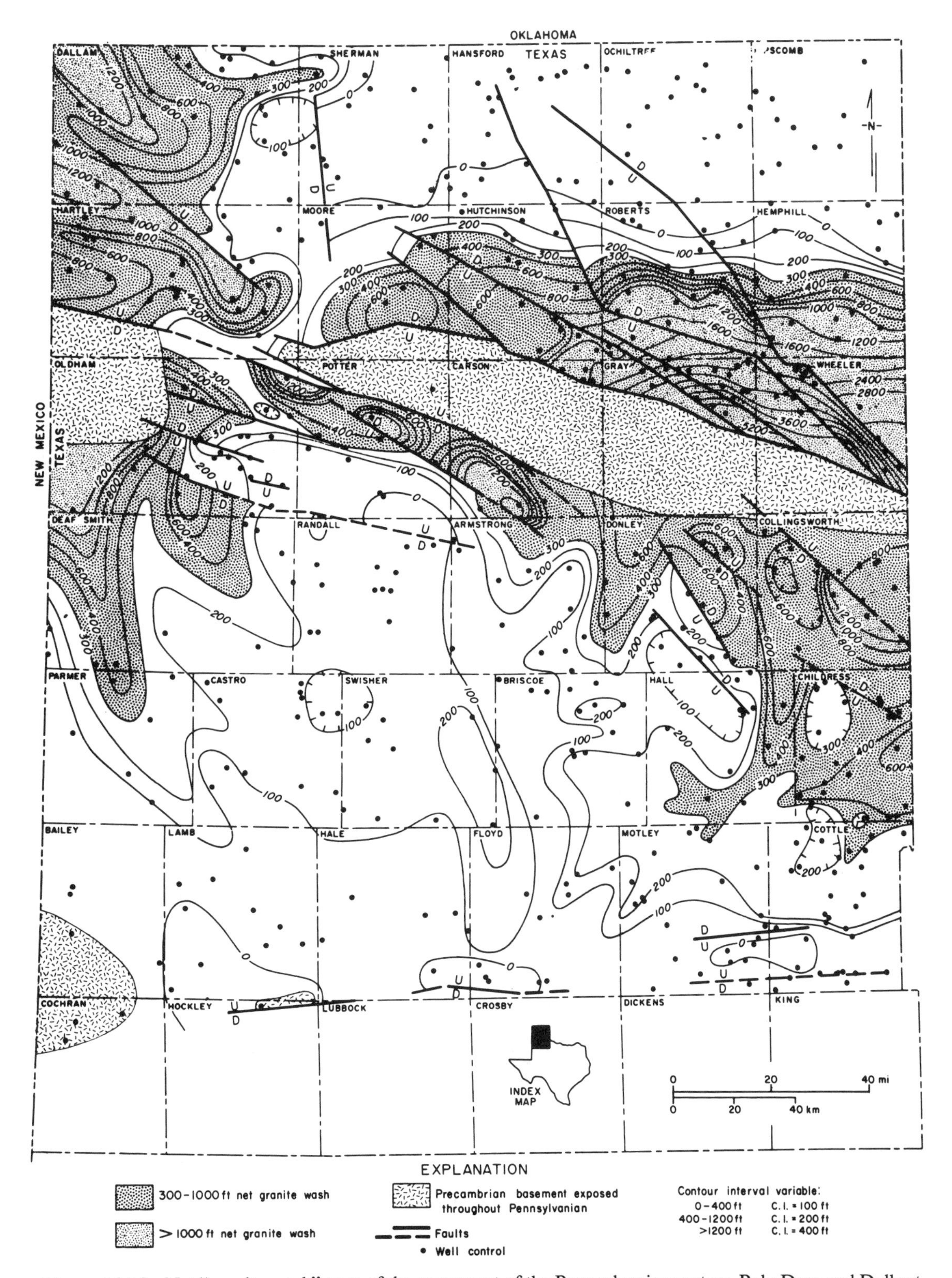

Figure 14-13. Net "granite-wash" map of the upper part of the Pennsylvanian system, Palo Duro and Dalhart basins. (From Dutton, 1980b.)

systems (Table 14-3). Reworked delta fringe and destructional sand bars are additional potential reservoir facies in some fan-delta systems (Sneider *et al.*, 1977). Reservoir facies interfinger over short distances laterally and vertically with marine shale, mudstone, and impure limestone which may provide seal and source units. Adjacent inter-fan and pro-fan shelf facies commonly include pure, open-marine limestones (Dutton, 1980a, 1982).

In the extremely shallow water of broad, stable constructional platforms, or within simple intracratonic basins, thickness of progradational sequences and marine energy flux are both limited by water depth. The subaqueous basin fills by aggradation of suspended sediment or carbonate debris and prograding distributary and alluvial channels cut into or through associated prodelta shelf deposits. Shore-zone deposition of bed-load sediment is severely limited by the low wave energy. In contrast, delta systems are laterally extensive, and their sand framework is dominated by the scour-based channel-fill facies of the superimposed fluvial channels (Fig. 14-12B). Distributary and fluvial channel-fill deposits, commonly of mixed-load and suspended-load systems, are dominant reservoir facies; valley fill deposits are locally important (Table 14-3).

Where progradation occurs in deeper water, the complete progradational sequence, including prodelta, delta-front, and delta-plain facies, is preserved. Increased wave energy may result in strike reworking and formation of sand-rich shore-zone deposits. Moderately deep, stable basins of the Mid-Continent filled from the margins by progradation of shallow shelves. A distinct morphologic break separated the crest of the depositional shelf platform from the deeper basin center. This marginal slope had a declination of only a few degrees. Where differential relief between the platform and basin floor exceeded 300 to 400 ft (90 to 120 m), gravitational remobilization of sediment became an increasingly significant process. As the upper slope became a zone of bed-load bypass, submarine fans formed at the base of the slope, providing an additional suite of potential reservoir facies. In areas of lesser terrigenous influx the platform margins were preferred sites for accumulation of carbonate reef or bank deposits. As suggested in Figure 14-12 stable platform and cratonic basin settings were commonly coeval parts of a single sediment dispersal system. In addition to the various fluvial and delta channel-fill facies, delta margin, destructional bar, fan channel and suprafan sand bodies may constitute important reservoirs (Table 14-3).

Proximal Upland Association. Within the Pennsylvanian fill of the Palo Duro and surrounding basins of northernmost Texas and western Oklahoma, thick sequences of polymictic sandstone and conglomerate rim block-faulted uplifts, which are commonly cored by Precambrian crystalline rocks. These thick, coarse-grained clastic sequences, which are locally called "granite wash," form wedges and tongues that abut against the uplifts and interfinger basinward with marine shelf and basin mudrocks and limestones (Fig. 14-13). They are interpreted by Dutton (1980a) to be fan-delta sequences deposited by locally sourced, flashy, braided streams. Granite-wash sand bodies range from a few feet to as much as 40 ft (15 m) thick and are laterally discontinuous. Stacking of multiple depositional lobes produced fan delta sequences in excess of 1,000 ft (300 m) thick. Detailed description of granite-wash sand bodies of the Elk City Field reveal the small scale of genetic facies and consequent internal complexity of individual reservoirs (Sneider *et al.*, 1977).

The small size of isolated sand bodies, heterogeneous bedding, and updip thickening and increasing amalgamation of the dip-oriented fluvial and distributary channel fills allows little opportunity for stratigraphic entrapment. However, large granite-wash fields, such as Elk City, produce from structural traps.

Stable-Platform and Deep-Basin Associations. Detailed stratigraphic studies of the Eastern Shelf of the Midland Basin by Brown (1972), Galloway and Brown (1972), and Brown *et al.* (1973) provide examples of both extremely shallow-water shelf as well as deeper-water intracratonic depositional and hydrocarbon producing styles.

During deposition of the Cisco Group (Late Pennsylvanian through Early Permian), successive deltaic progradations constructed a dominantly clastic platform behind the remnants of a fringing carbonate bank that developed during earlier Strawn and Canyon depositional intervals (Fig. 14-14). Sediments were derived from the eroded trunk of the Ouachita uplift and the gentle Bend Arch, which lay to the east of the Midland Basin. Following deposition of the Ivan and Blach

Table 14–3. Characteristics of Principal Reservoir Facies of Intracratonic Basins[a]

	Thickness	Width	Geometry	Boundary Relations	Common Trap Types
Proximal Upland Settings					
Fan plain	20 to 60 ft may be stacked	10^3 to 10^4 ft	Dendroid	Gradational basal and lateral contacts	Structural Combination Stratigraphic
Distal fan	10 to 40 ft may be stacked	10^3 to 10^4 ft	Dendroid	Gradational to sharp basal and lateral contacts	Structural Combination Stratigraphic
Main braided channels	10 to 20 ft may be stacked	10^2 to 10^3 ft	Anastomosing belt	Sharp basal and lateral contacts	Structural Combination Stratigraphic
Shallow Stable Platform Settings					
Fluvial channel fill	20 to 60 ft may be stacked	10^3 to 10^4 ft	Belt	Sharp lateral and basal contacts	Stratigraphic Combination Structural
Distributary channel	10 to 50 ft may be stacked	10^2 to 10^3 ft	Anastomosing to distributing dendroid	Sharp lateral and basal contacts	Stratigraphic Combination Intraformational drape structural
Valley fill	30 to 200 ft	10^3 to 10^4 ft	Belt	Sharp lateral and basal contacts	Structural Combination Stratigraphic
Cratonic Basin Settings					
Fluvial, distributary, and valley fill reservoirs as above					
Delta margin/ coastal barrier	10 to 50 ft	10^3 to 10^5 ft	Ameboid to corrugated sheet; ribbon	Gradational basal and lateral contacts	Stratigraphic Structural Combination
Destructional bar	5 to 25 ft	10^2 to 10^3 ft	Ribbon	Gradational to sharp basal and lateral contacts	Stratigraphic
Fan channel	10 to 50 ft	10^2 to 10^3 ft	Dendroid	Sharp basal; gradational to sharp lateral contacts	Stratigraphic
Suprafan	Amalgamated to 100 ft	10^3 to 10^5 ft	Ameboid to irregular sheet	Gradational lateral; sharp to gradational basal contacts	Stratigraphic Structural

[a]From Galloway and Dutton (1979).

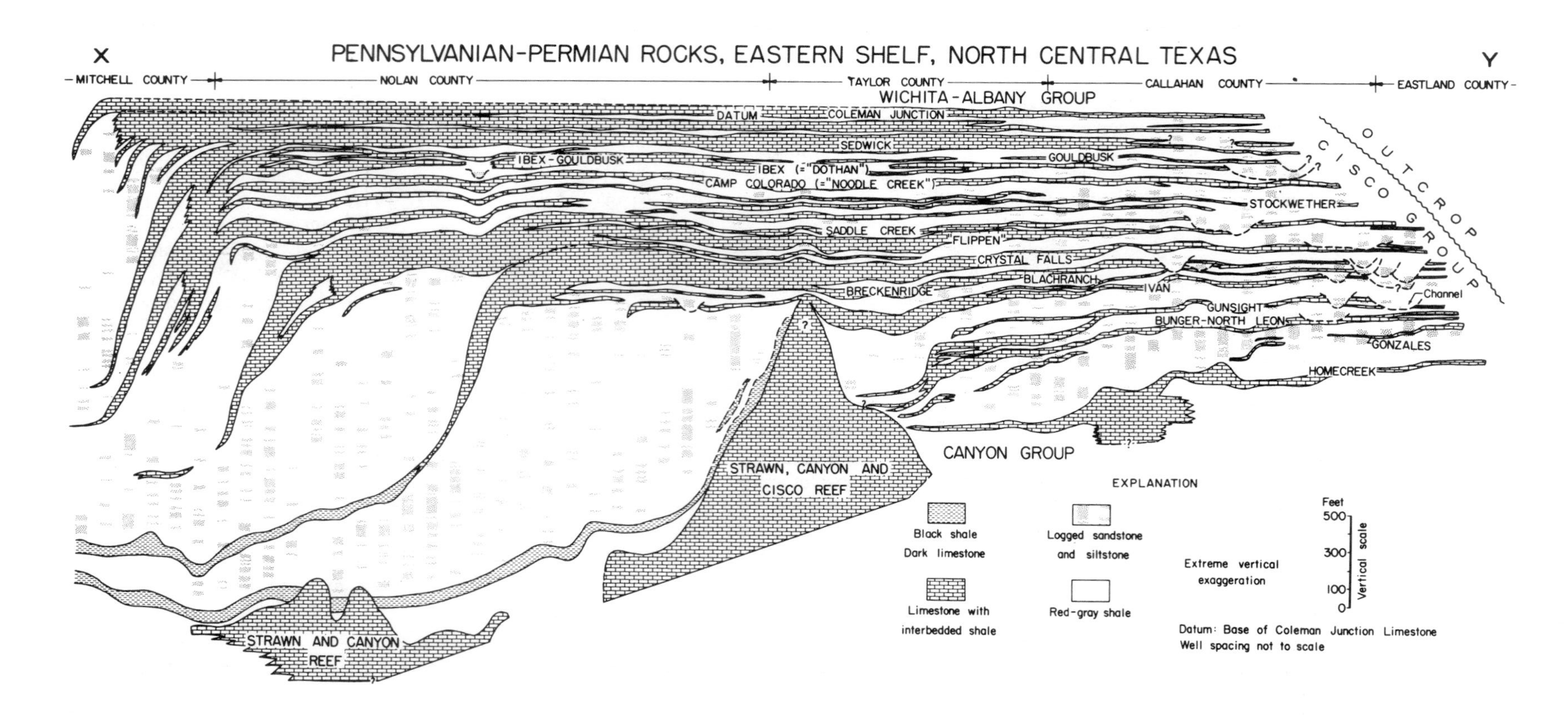

Figure 14-14. Regional dip cross section, based on about 60 wells, illustrating the offlapping depositional architecture of the Upper Paleozoic Cisco Group of the Eastern Shelf (Midland Basin, Texas). Burial of a widespread shelf-margin reef dam resulted in progradational deposition of thick deltaic and submarine-slope wedges in the previously sediment-starved basin. Section extends approximately 100 mi (160 km). (From Brown *et al.*, 1973.)

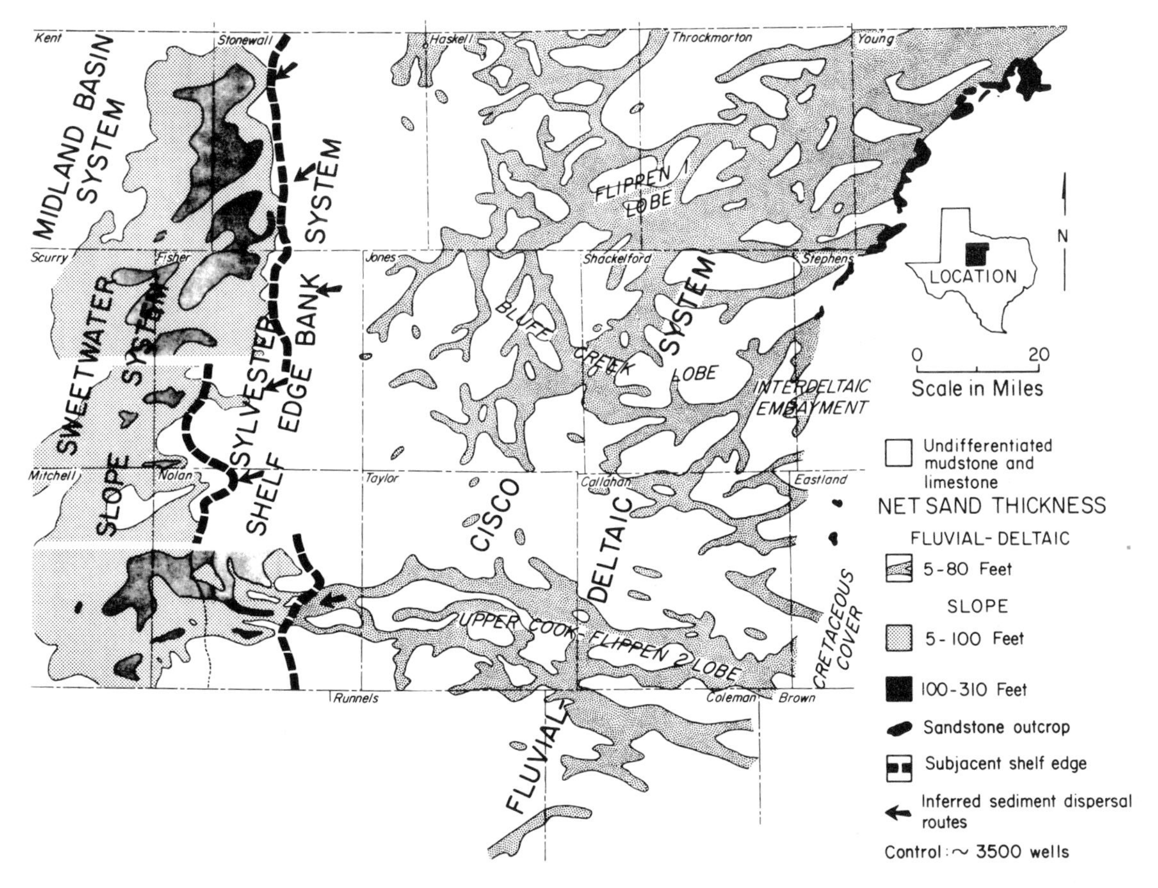

Figure 14-15. Simplified net-sandstone map of one Cisco depositional episode on the Eastern Shelf. Original map based on approximately 3500 well logs. (From Brown *et al.*, 1973.)

Ranch Limestones (Fig. 14-14), deltas buried the remnant of the bank and prograded directly into the relatively deep water of the starved basin center. Thick rhomboidal wedges of interbedded sand and mud successively filled the basin, and were periodically draped by shelf-edge carbonate sediments during periods of large-scale delta lobe shifting or regional base-level change. The resultant depositional architecture is well displayed on regional dip-oriented geophysical or lithologic cross sections (Fig. 14-14).

Bedding style and lithologic composition readily suggest the three depositional systems defined by Galloway and Brown (1972). Updip of and capping the prograded platform is the Cisco platform fluvial-deltaic system, consisting of cyclically interbedded, thin clastic sheets bounded by widespread shelf-limestone beds. The prograding shelf margin is outlined by massive limestones of the Sylvester shelf-edge bank system. Deltaic clastics can be traced only locally across this massive carbonate complex. Basinward, expanded deltaic sequences overlie the thick clastic wedges of the Sweetwater slope system. Basinal facies consisting of dark, siliceous, organic shales interfinger with lower portions of the slope system.

Regional isolith maps of successive Cisco depositional episodes reveal the areal geometry of framework facies. Channel-fill belts of the fluvial-deltaic system cluster into a number of well-defined lobes or axes that prograded from the east and northeast across the platform and locally into the deep basin (Fig. 14-15). Updip, the aggradational fluvial deposits are best preserved, and meanderbelts and local valleys cut down through underlying shelf limestones (Fig. 14-14, Callahan County). Down-dip, narrow, lenticular, distributary sand-body geometries reflect preservation of delta plain and thin, progradational mouth bar deposits (Fig. 14-15, Bluff Creek lobe in Jones County). Thicker, aggregate sandstone sequences

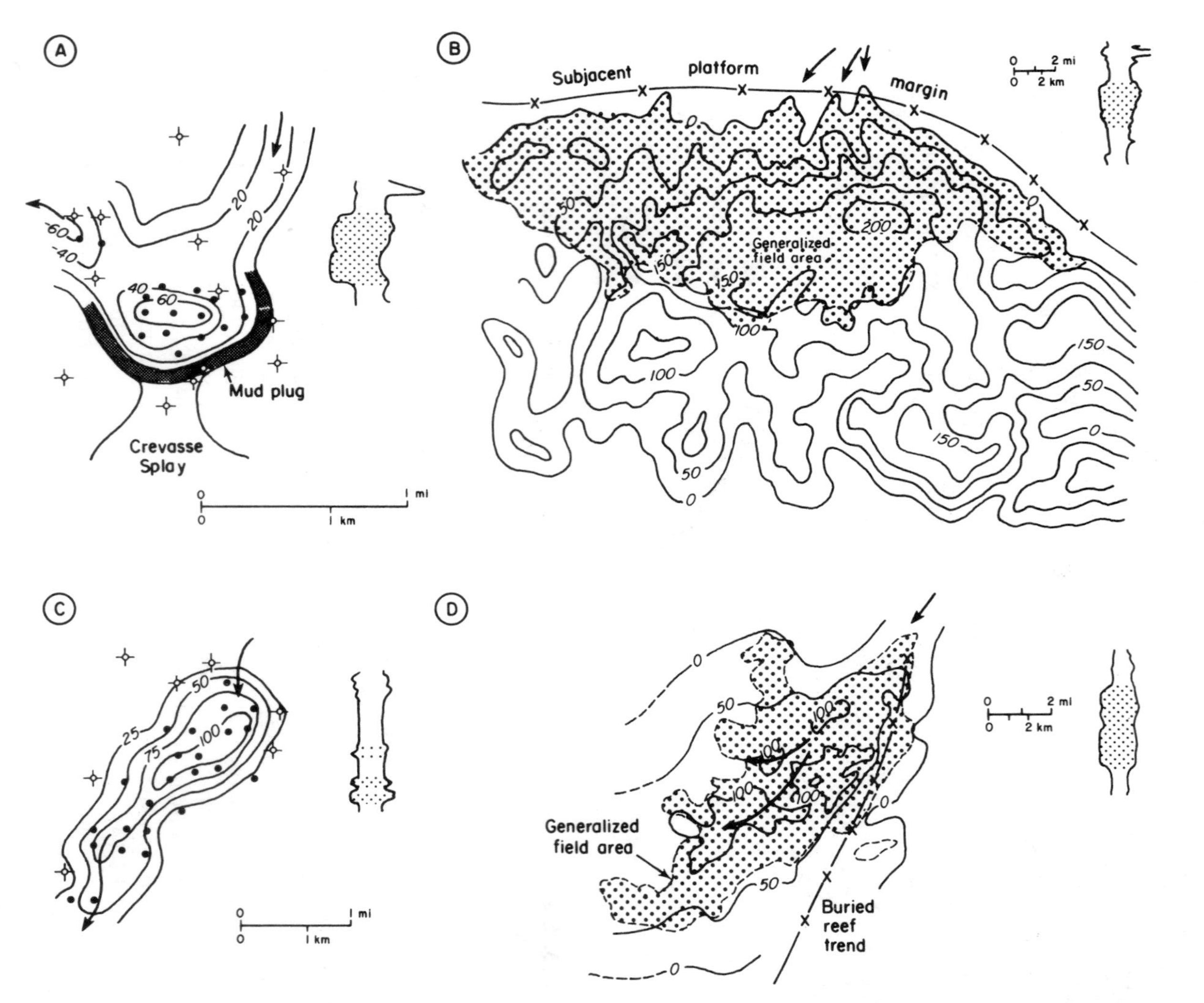

Figure 14-16. Stratigraphic traps in fluvial, deltaic, and slope systems of Mid-Continent intracratonic basins. (A) Point-bar sandstone reservoir; platform fluvial-deltaic system, eastern Midland Basin. (B) Delta-front and distributary-channel reservoir; deep-water deltaic system, northern Anadarko Basin. (C) Submarine fan-channel sand body; sand-rich fan system, eastern Midland Basin. (D) Sand-rich submarine fan channel and suprafan complex; eastern Midland Basin. (A, C, and D modified from Bloomer, 1977; B modified from Pate, 1968.)

seaward of the subjacent shelf edges include the vertical succession of submarine-fan facies, including proximal fan valley and distal suprafan deposits, and thick, progradational delta front and channel mouth bar sandstones. The bimodal vertical distribution of sand bodies within the slope wedges (Fig. 14-14) reflects this very important development of an offlapping submarine-fan system beneath the prograding deltaic system.

Together, the platform and deeper-water depositional systems of Mid-Continent intracratonic basins offer a variety of potential reservoir facies. Low-relief structural and combination traps are abundant, and opportunities for stratigraphic entrapment are numerous (Bloomer, 1977). Four representative examples are illustrated in Figure 14-16. Many small to medium-sized fields produce from point-bar sand bodies of the platform fluvial-deltaic systems (Fig. 14-16A). Sealing is commonly accomplished or aided by the mud plug. Individual reservoirs are typically at most a few tens of feet thick, and exploration demands detailed mapping of individual meanderbelts. Thicker and laterally more extensive sand bodies deposited where deltas prograded beyond the shelf-edge and into deeper water provide opportunities for less numerous but larger stratigraphic or combination traps. The Laverne area (Fig. 14-16B) is a giant gas field developed in this type of stratigraphic setting on the northern shelf of the Anadarko Basin. An analogous producing configuration occurs in several Cisco depositional episodes of the Eastern Shelf. Submarine-fan deposits of the lower portions of the slope wedges contain numerous stratigraphic accumulations, including some of the largest fields within Cisco strata. Small fields occur where individual upper-fan channels pinch out up-slope within the thick slope muds (Fig. 14-16C). Jameson Field has produced more than 36 million barrels of oil from amalgamated fan valley and suprafan sand bodies that lap against an older, buried reef and the prograding toe of the slope (Fig. 14-16D). Reservoir productivity is limited by the low porosity and permeability of the fine-grained sands.

Source-rock studies of the Palo Duro and Midland Basins reveal significant relationships between depositional systems and source potential (Dutton, 1980b). Basinal and lower-slope shales contain the highest concentrations of oil-prone organic carbon, and are commonly well above the cut-off for a quality source rock. In contrast, deltaic and fluvial mudrock facies are typically lean and, because of their commonly shallow burial, are thermally immature. Thus submarine fan reservoirs are optimally located to be charged with oil. Although not documented by geochemical study, the widespread lateral continuity of the sand-body framework of the Cisco and similar tectonic-environmental assemblages suggests extensive updip migration of oil generated in the deep basin centers into deltaic and fluvial reservoirs. Such migration would explain the laterally extensive producing fairways (noted in Galloway and Brown, 1972, Fig. 31) within specific Cisco depositional episodes surrounded by a nonproductive section that contains otherwise comparable reservoir facies and trap configurations.

Example: Frio Depositional Systems, Northern Gulf Coast Basin

The Frio Formation (Oligo–Miocene) is one of the major progradational wedges of the Gulf Coast Tertiary Basin, an extracontinental downwarp basin. Several hundred feet of aggradation of the coastal plain accompanied progradation of a continental platform prism more than 15,000 ft (4,500 m) thick and up to 50 mi (80 km) wide. In the Texas coastal plain the Frio has produced more than 16 billion b.o.e. of hydrocarbons. Using regional stratigraphic mapping such as illustrated in Figures 2-6 through 2-9, Galloway *et al.* (1982a, 1982b) synthesized the depositional framework (Fig. 14-17) and hydrocarbon distributional patterns of the Frio in the Texas coastal plain. The following discussion is derived largely from these papers.

Structural and Depositional Framework

Major progradational wedges of the Gulf Coast basin, such as the Frio Formation, consist of an updip and relatively shallow section of interbedded continental and marginal-marine facies overlying several thousand feet of abnormally pressured slope and basinal mudstone. Underlying oceanic crust rapidly subsided when subjected to sediment loading. Resultant instability within the undercompacted, water-saturated, and

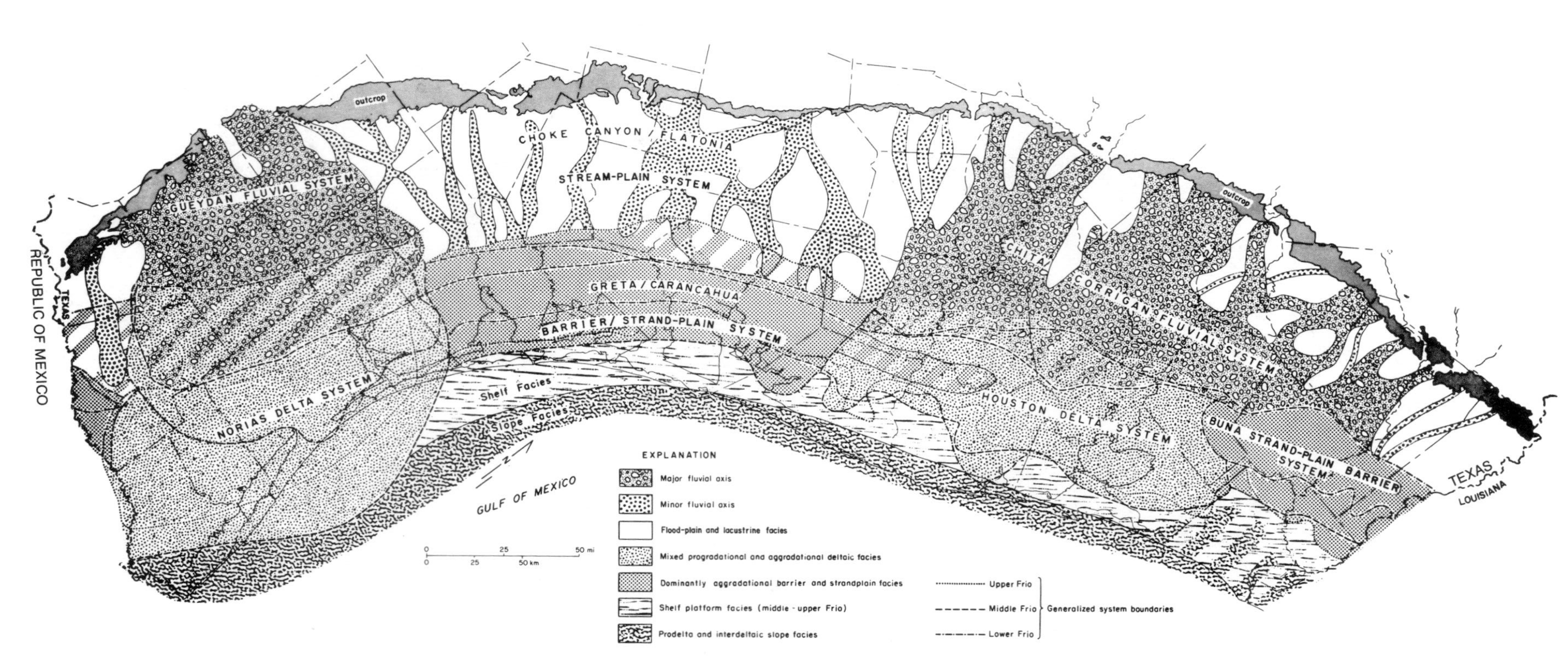

Figure 14-17. Depositional systems of the Frio (Oligo-Miocene) depositional episode, Texas Gulf Coast Tertiary basin. (From Galloway *et al.*, 1982a.)

gently inclined slope wedge resulted in large-scale syndepositional down-to-the-basin faulting and intrastratal deformation. Three structural provinces may be defined by their syndepositional deformational style and nature of the diapiric material intruded into shallower sediment. At the northeast end of the Texas coastal plain the broad, ill-defined Houston embayment was characterized by mobilization of thick Jurassic salt. Diapirs and associated faults and withdrawal basins initiated by progradation of high-constructive embanking delta systems, such as the Eocene Wilcox Rockdale delta system (Fisher and McGowen, 1967), were inherited by, and continued to deform, successor progradational systems such as the Frio. The middle coastal plain is characterized as a broad, coastward-plunging arch or platform that subsided less than surrounding embayments. In the Rio Grande embayment of the South Texas coastal plain salt is thin or absent; consequently, shale tectonics dominate, and discontinuous belts of strike-parallel growth faults and deep-seated shale ridges and massifs occur. The Vicksburg flexure, a uniquely continuous, narrow fault zone characterized by extreme vertical displacement in the deep stratigraphic section, forms the updip limit of significant deformation of the Frio Formation and coincides with the position of the buried shelf margin of a major older Eocene depositional episode (Fig. 14-18).

Division of the Frio Formation into several depositional systems was based on regional differences in sandstone distribution patterns, position within the basin, and areal and vertical log facies distributions as defined on maps and stratigraphic cross sections. The principal progradational delta systems, informally named the Houston and Norias delta systems, occupy the two embayments (Fig. 14-17). The Houston delta system is characterized by numerous small, laterally offset lobes, in contrast to more clearly defined, vertically stacked lobes of the Norias delta to the south. The Chita-Corrigan fluvial system consisted of several mixed-load rivers that fed down-dip into the Houston deltaic axes. To the south, the large Gueydan fluvial system, which is also the host of large uranium deposits of the South Texas uranium province, records deposition by a continent-scale, bed-load river (Galloway, 1977). Separating the two fluvial-deltaic facies tracts was an area spanning the middle coastal plain stable platform that was traversed by numerous minor streams. The Choke Canyon/Flatonia streamplain system was dominated by silt and mud accumulation between widely spaced, minor belts of coarse-grained stream deposits (Fig. 14-17). Strike-fed shorezone sediments derived from the adjacent deltaic headlands combined with wave-reworked sediments of the small rivers of the streamplain to build up the massive, dominantly aggradational system of Frio coastal barrier and strandplain sands (Fig. 14-19), first documented by Boyd and Dyer (1964). The strike-parallel Greta/Carancahua barrier/strandplain extends almost 150 mi (240 km) between the deltaic depocenters (Fig. 14-17). The similar but smaller Buna strandplain system lies on the eastern flank of the Houston delta system and extends across the Louisiana border. Basinward, Frio depositional elements include a shelf platform fronting the interdeltaic coastal plain and a broad, unstable, offlapping slope apron. The extensive, deepwater Hackberry embayment of Louisiana extends into the northeastern corner of the Houston delta system. This thick, erosionally based onlapping slope wedge occurs within the middle Frio section and is highly productive of oil and gas (Paine, 1971). Smaller, strike-aligned intraslope basins produced by salt or mud diapirism may have locally trapped large quantities of sediment transported by slumping and density flows. Attributes of the principal Frio depositional systems of the Texas coastal plain are summarized in Table 14-4.

Hydrocarbon Source, Maturation, and Migration

Hydrocarbon production in the Frio is geographically widespread, but can be subdivided into ten geologically defined plays (Fig. 14-20). Seven of the plays contain most of the producible oil and gas. Defining characteristics of plays include the host depositional system or predominant reservoir facies assemblage, structural style, and type of hydrocarbon produced. Although genetic stratigraphy, structural style, and production characteristics are typically interrelated, a single dominant characteristic may define a play. For example, Play III consists of fields producing from broad anticlines that lie along the Vicksburg growth fault zone. Plays V and VI are distinguished by their positions on the shelf and lagoonal sides of the axis of the Greta barrier

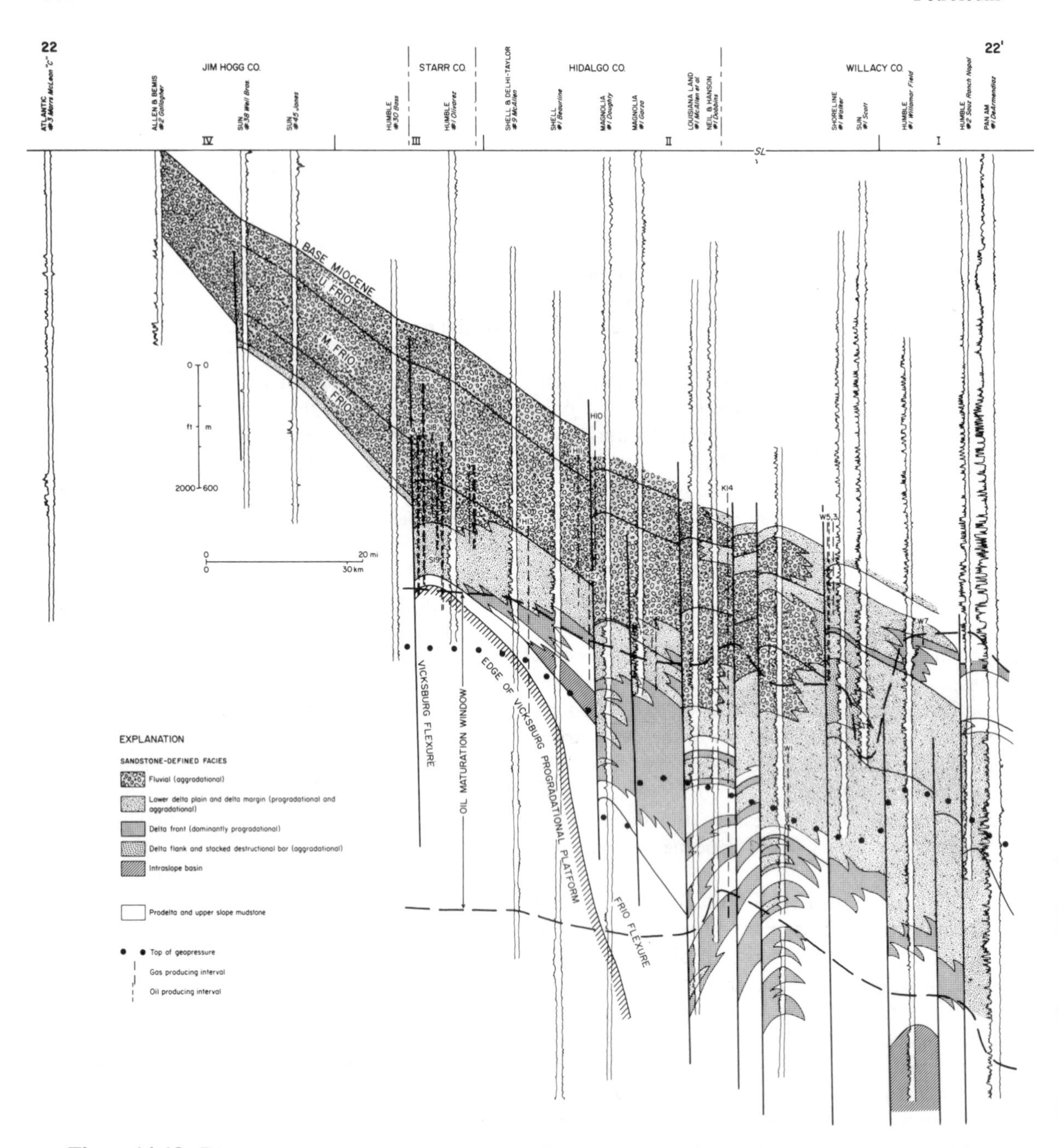

Figure 14-18. Dip cross section through the Gueydan fluvial system and Norias delta system of the South Texas Frio trend. Section illustrates relationships between genetic facies assemblages, principal growth faults, depth of the top of substantial overpressure, boundaries of the oil maturation window, and distribution of oil and gas reservoirs in fields lying near the plane of the section. (From Galloway *et al.*, 1982a.)

system (Fig. 14-20). Evaluation of production style, history, and geochemical characteristics within each play or in groups of plays reveals trends that are lost in more general compilations, but which are significant for interpretation of the source and migration history of Frio oil and gas (Galloway *et al.*, 1982a) and for projection of remaining undiscovered hydrocarbon volumes within the Frio (Galloway *et al.*, 1982b).

The geographic (Fig. 14-20) and vertical distributions of oil and gas (Figs. 14-18 and 14-19), as well as comparative productivity (Fig. 14-

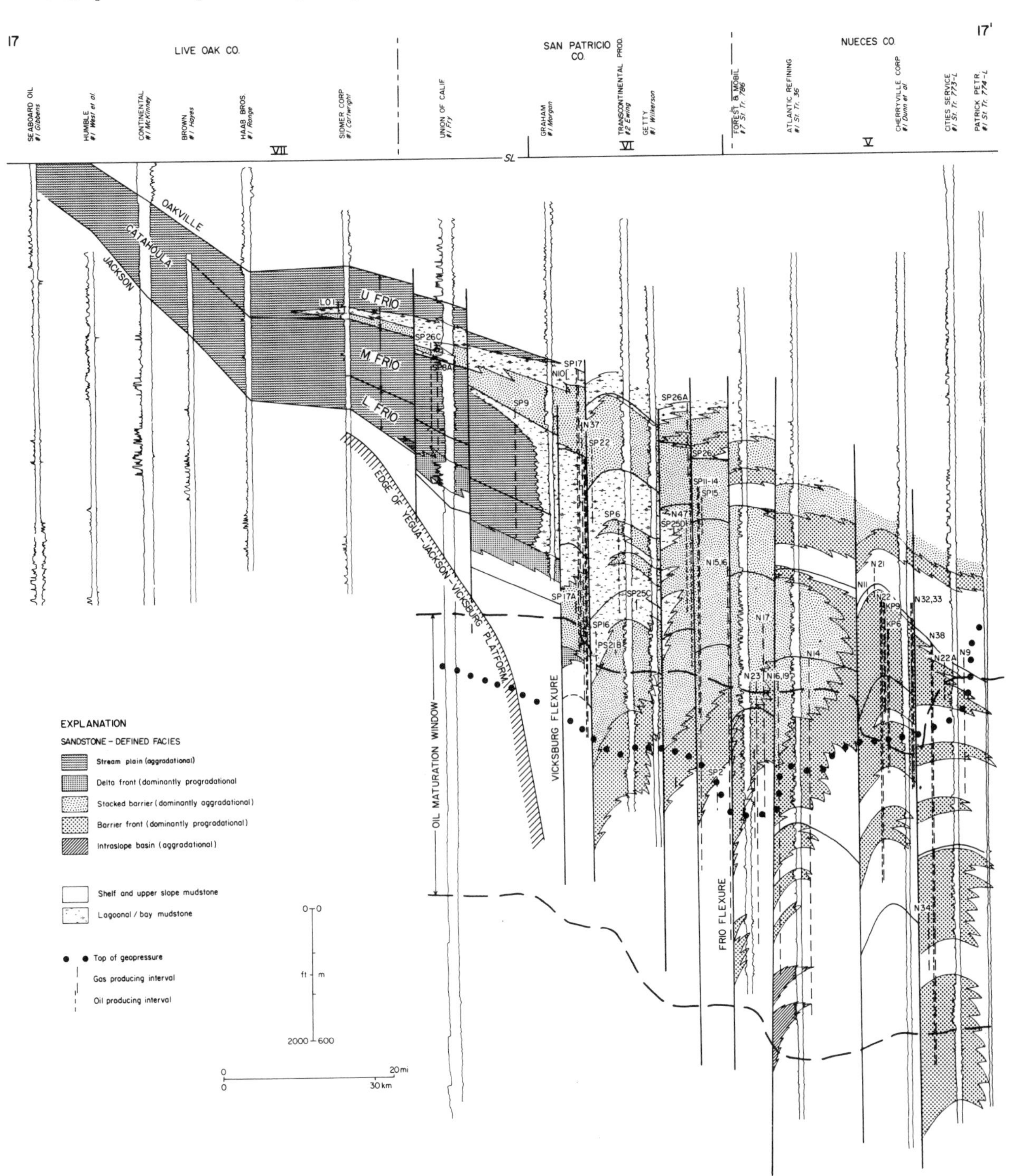

Figure 14-19. Dip cross section through the muddy streamplain facies and the sand-rich Greta barrier system of the Frio Formation in the Middle Texas Coastal Plain. Compare structural, stratigraphic, and hydrocarbon distribution patterns with those shown on Figure 14-18. (From Galloway *et al.*, 1982a.)

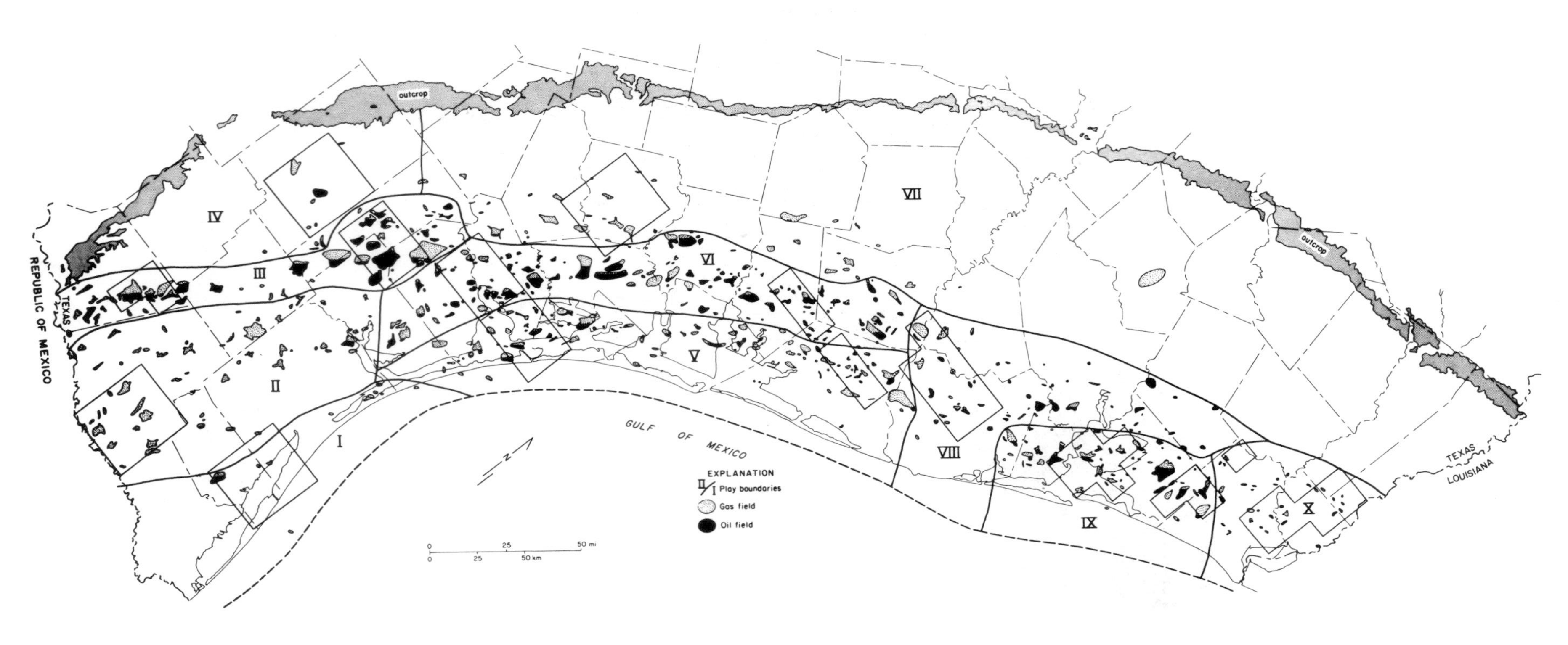

Figure 14-20. Distribution of significant Frio oil and gas fields in the Texas coastal plain. Producing plays are outlined and identified by Roman numerals. Smaller rectangles enclose particularly productive areas within each play. (From Galloway *et al.*, 1982a.)

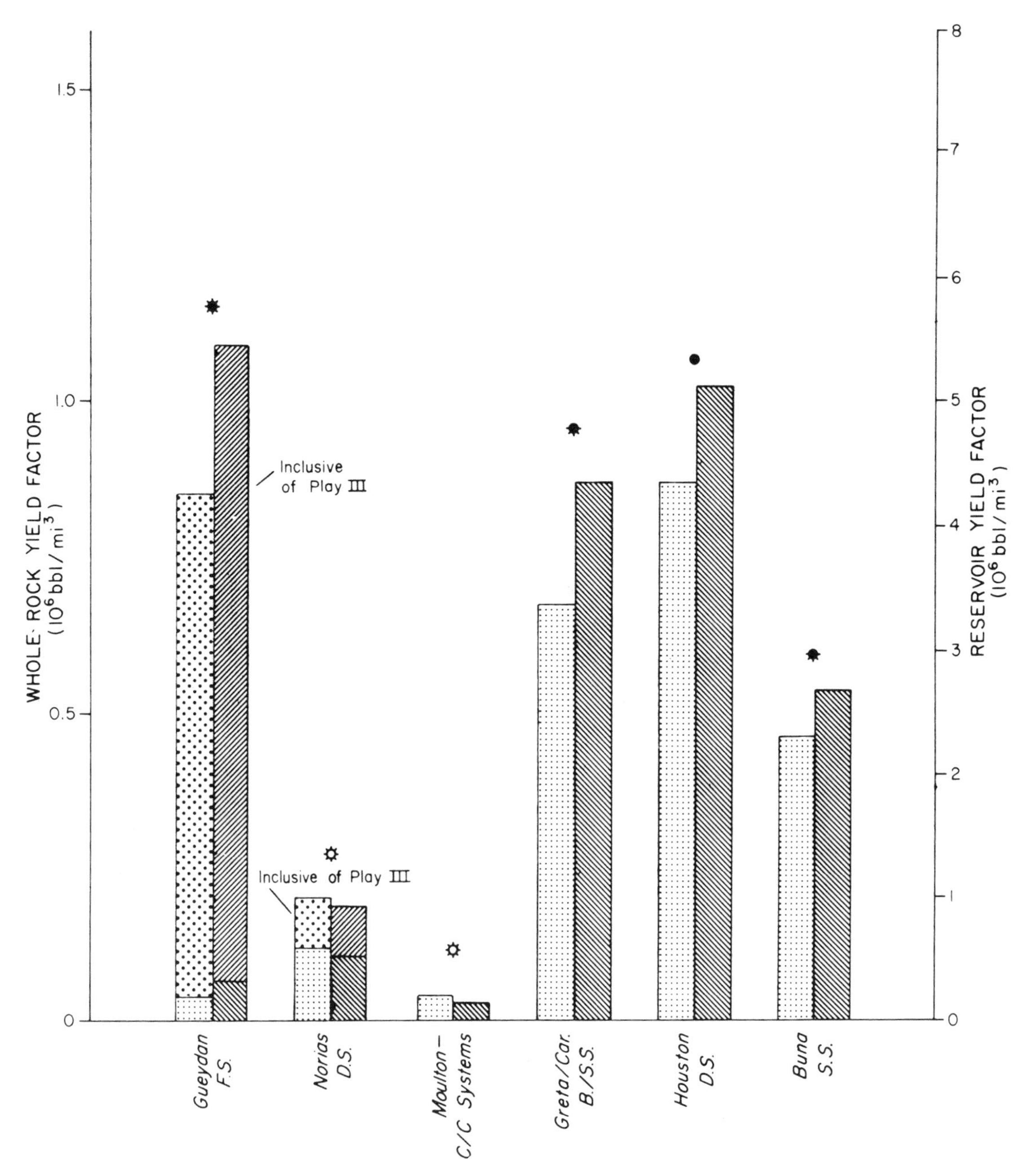

Figure 14-21. Whole rock and reservoir yield factors for each of the Frio depositional systems of the Texas Coastal Plain. Play III bridges the Gueydan and Norias systems, and its production has been proportioned accordingly. F.S., fluvial system; D.S., delta system; B.S., barrier system; S.S., strandplain system. (From Galloway *et al.*, 1982a.)

21) of each depositional system must be explained by relationships among source facies, including prodelta, slope, shelf, delta plain, lagoonal, and flood-basin muds within the Frio (as well as shelf, prodelta, and slope deposits of subjacent units), maturation history, and migration history.

1. *Source rocks.* Gulf Coast Tertiary mudstones generally have low concentrations of organic carbon. TOC contents within sampled portions of the Frio Formation average below 0.3 percent and rarely exceed 0.4 percent. Sparse analyses indicate that woody and herbaceous organic matter dominate in deltaic mudstones,

Table 14–4. Attributes of Frio Depositional Systems[a]

System	Total Volume (mi^3)	Sand Volume (mi^3)	Reservoir Facies	Reservoir Quality	Potential Source Facies	Degree of Thermal Maturity	Hydrocarbon Type[b]	Yield Factor (10^6boe/mi^3)[c]
Gueydan Fluvial system	1,700	370	Bed-to mixed-load channel fill	Good ↓	Floodplain mud (largely oxidized)	Immature	Moderately gas prone	5.5
			Crevasse splay	Fair				
Choke Canyon/ Flatonia Streamplain System	2,000±	400±	Mixed-to bed-load channel fill	Good	Floodplain & lacustrine mud	Immature	Dry gas prone	≪0.5
Chita/Corrigan Fluvial system	3,400±	900±	Meander belt and crevasse splay	Excellent	Floodplain, backswamp, and lacustrine mud	Immature	Dry gas prone	<0.5
Norias delta system	13,200+	2,800+	Distributary channel/splay Coastal barrier Proximal delta front Distal delta front	Fair ↓ Poor	Delta plain, prodelta, and upper slope mud	Mature oil to gas zone	Highly gas prone	1.0
Greta/ Carancahua Barrier/ Strandplain System	9,600	2,100	Barrier bar and strandplain association Distal shoreface/ shelf	Excellent ↓ Fair	Lagoon, shelf, and upper slope mud	Mature oil zone downdip	Mixed	4.5
Houston Delta System	4,700	1,000	Distributary channel Delta front and mouthbar Delta flank	Excellent ↓ Good	Delta plain, prodelta and upper slope mud	Mature oil to gas zone downdip	Moderately oil prone	5.0

[a]From Galloway *et al.* (1982 b).
[b]Calculated on an energy-equivalent basis.
[c]Total energy-equivalent producible hydrocarbon yield per volume of sandstone.

and amorphous organic matter is somewhat more abundant in marine shelf and slope facies.

2. *Maturation.* The Frio oil-generation window was estimated to lie approximately between the 200 and 300°F (94–150°C) isotherms (Galloway *et al.*, 1982b). Accordingly, the degree of thermal maturity of potential internal Frio source facies varies greatly (Table 14-4). Large portions of the deltaic and barrier/lagoon systems do lie within the oil generation window (Figs. 14-18 and 19). Significantly, Norias delta-plain deposits occupy the oil-generation zone, whereas prodelta and slope facies lie within the oil window in the less deeply buried Houston delta system. In contrast, the highly prolific Gueydan fluvial system is nearly devoid of mature source rocks (Fig. 14-18).

3. *Migration.* Fluid migration from the deep basin is restricted. The top of the substantially overpressured (or geopressured) section is quite sensitive to facies and typically approximates the position of the transition from (1) sandy delta plain and delta margin facies into delta front and prodelta facies (Fig. 14-18), or (2) barrier front and strandplain facies into distal barrier front and shelf facies (Fig. 14-19). Thus, a strong pressure head directs compactional fluids upward into the shallower, normally pressured section.

Within all systems the center of the mass of in-place oil and gas lies above the top of geopressure, commonly several thousand feet above the corresponding maturation window. This, combined with the high productivity of the Gueydan fluvial system, provides a strong case for upward migration of hydrocarbons as well as pore water. Structural features in the form of growth faults (such as the Vicksburg flexure) and salt domes are thus inferred to play major roles in controlling hydrocarbon distribution by acting both as principal migration pathways as well as traps. Not surprisingly, stratigraphic traps are relatively unimportant in the Frio Formation. The significant upward migration of oil and gas obscures simple facies-production associations, and results in multiple reservoirs that produce over several hundreds or even thousands of feet of section (Figs. 14-18 and 19, fields K14 and N38, for example).

Further, some of the produced Frio hydrocarbons may reflect characteristics of source facies that are much different from those of the host depositional system. For instance, although the highly gas-prone production of the Norias delta system reasonably reflects generation from Type III kerogen typical of delta plain and prodelta facies that lie within the oil and gas windows (Fig. 14-18), similar deltaic facies of the Houston delta system occur largely above the oil window and were charged from oil-prone sources in underlying slope and shelf facies of Frio and older systems. Thus, an understanding of generation and migration mechanisms is necessary to avoid misinterpretation of source rock studies of analogous depositional systems in sparsely drilled frontier basins.

In summary, analysis of the Frio Formation shows that hydrocarbon productivity varies widely among different depositional systems within the same basin or even the same stratigraphic unit (Fig. 14-21). Further, types of hydrocarbons, as well as styles of production, commonly show relationships to the genetic stratigraphic setting. However, in complex systems such as the Gulf Coast basin, the dynamic history of subsidence, thermal alteration, structural modification, and fluid expulsion and migration, all may play a part in determining the final product.

References

Ager, D.V., Wallace, P. 1970. The distribution and significance of trace fossils in the uppermost Jurassic rocks of the Boulonnais, northern France. *In* T.P. Crimes, J.C. Harper (eds.), Trace Fossils. Geol. J. Spec. Issue, pp. 1–18.

Ahlbrandt, T.S. 1979. Textural parameters of eolian deposits. *In* E.D. McKee (ed.), A study of global sand seas. U.S. Geol. Surv. Prof. Pap. No. 1052, pp. 21–51.

Ahlbrandt, T.S., Fryberger, S.G. 1981. Sedimentary features and significance of interdune deposits. *In* Soc. Econ. Paleon. Mineral. Spec. Publ. No. 31, pp. 293–314.

Ahlbrandt, T.S., Andrews, S., Gwynne, D.T. 1978. Bioturbation in eolian deposits. J. Sediment. Petrol. 48: 839–848.

Allen, G.P., Salomon, J.C., Bassoullett, P., Du Penhoat, Y., De Granpre, C. 1980. Effects of tides on mixing and suspended sediment transport in macrotidal estuaries. Sediment. Geol. 26: 69–90.

Allen, J.R.L. 1965a. A review of the origin and characteristics of Recent alluvial sediments. Sedimentology 5: 91–191.

Allen, J.R.L. 1965b. Late Quaternary Niger Delta and adjacent areas; sedimentary environments and lithofacies. Amer. Assoc. Petrol. Geol. Bull. 49: 547–600.

Allen, J.R.L. 1970. Studies in fluvial sedimentation; a comparison of fining-upward cyclothems, with special reference of coarse-member composition and interpretation. J. Sediment. Petrol. 40: 298–323.

Allen, J.R.L. 1971. Mixing at turbidity current heads and its geological implications. J. Sediment. Petrol. 41: 97–113.

Allen, J.R.L. 1978. Studies in fluviatile sedimentation: an exploratory quantitative model for the architecture of avulsion-controlled alluvial suites. Sediment. Geol. 21: 129–147.

Almgren, A.A. 1978. Timing of Tertiary submarine canyons and marine cycles of deposition in the southern Sacramento Valley, California. *In* D.J. Stanley, G. Kelling (eds.), Sedimentation in submarine canyons, fans and trenches, pp. 2176–2291. Stroudsburg: Dowden, Hutchinson and Ross.

Alpay, O.A. 1972. A practical approach to defining reservoir heterogeneity. J. Petrol. Tech. 24: 841–843.

Amiel, A.J., Freedman, G.M. 1971. Continental sabkha in Arava Valley between Dead Sea and Red Sea: significance for origin of evaporites. Amer. Assoc. Petrol. Geol. Bull. 55: 581–592.

Anderson, A.M., McLachlan, I.R., Oelofsen, B.W., 1977. The Lower Permian White Band and the Irati Formation. Geol. Soc. S. Afr. 17th Congr. Absts., pp. 4–7.

Anderton, R. 1976. Tidal-shelf sedimentation: an example from the Scottish Dalradian. Sedimentology 23: 429–458.

Andrews, P.B. 1970. Facies and genesis of a hurricane washover fan, St. Joseph Island, Central Texas coast. Bur. Econ. Geol. Univ. Texas, Austin, Rept. Invest. No. 67.

Anstey, R.L. 1966. A comparison of alluvial fans in West Pakistan and the United States. Geogr. Rev. 21: 14–20.

Arabian Gulf Oil Company, 1980. Geology of a stratigraphic giant—the Messlah Oil Field. *In* M.J. Salem, M.T. Busrewil (eds.), The Geology of Libya, Vol. 2, pp. 521–536. London: Academic.

Augustinus, P.G.E.F. 1980. Actual development of the chenier coast of Surinam (South America). Sediment. Geol. 26: 91–113.

Averitt, P. 1969. Coal resources of the United States, January 1967. U.S. Geol. Surv. Bull. 1275.

Baas-Becking, L.G.M., Kaplan, I.R. 1956. The microbiological origin of the sulphur nodules of Lake Eyre. Trans. Roy. Soc. S. Aust. 79: 52–65.

Baas-Becking, L.G.M., Kaplan, I.R., Moore, D. 1960. Limits of the natural environment in terms of pH and oxidation-reduction potentials. J. Geol. 68: 243–284.

Baganz, B.P., Horne, J.C., Ferm, J.C. 1975. Carboniferous and Recent lower delta plains—a comparison. Gulf Coast Assoc. Geol. Socs. Trans. 37: 556–591.

Bagnold, R.A. 1941. The physics of blown sand and desert dunes. London: Methuen.

Bagnold, R.A. 1953. The surface movement of blown sand in relation to meteorology. Res. Council Israel Int. Symp., Deserts.

Baker, V.R., 1978. Adjustment of fluvial systems to climate and source terrain in tropical and subtropical environments. *In* A.D. Miall (ed.), Fluvial sedimentology, pp. 211–230. Can. Soc. Petrol. Geol. Mem. 5.

Balazs, R.J., Klein, G. deV. 1972. Roundness-mineralogical changes of some intertidal sands. J. Sediment. Petrol. 42: 425–433.

Bally, A.W., Snelson, S. 1980. Realms of subsidence. *In* A.D. Miall (ed.), Facts and principles of world petroleum occurrence, pp. 9–94. Can. Soc. Petrol. Geol. Mem. 6.

Banerjee, I. 1980. A subtidal bar model for the Eze-Aku sandstones, Nigeria. Sediment. Geol. 25: 291–309.

Banks, N.L. 1973. Innerelv Member: late Precambrian marine shelf deposit, East Finnmark. Norges Geol. Unders. 288: 7–25.

Barker, C. 1972. Aquathermal pressuring—role of temperature in development of abnormal-pressure zones. Amer. Assoc. Petrol. Geol. Bull. 56: 2068–2971.

Barker, C. 1977. Aqueous solubility of petroleum as applied to its origin and primary migration: discussion. Amer. Assoc. Petrol. Geol. Bull. 61: 2146–2199.

Barker, C. 1979. Generation and accumulation of oil and gas in deltas. *In* N.J. Hyne (ed.), Pennsylvanian sandstones of the mid-continent, pp. 83–96. Tulsa Geol. Soc. Spec. Publ. No. 1.

Barwis, J.H. 1978 Sedimentology of some South Carolina tidal-creek point bars, and a comparison with their fluvial counterparts. *In* A.D. Miall (ed.), Fluvial sedimentology, pp. 129–160. Can. Soc. Petrol. Geol. Mem. 5.

Barwis, J.H., Hayes, M.O. 1979. Regional patterns of modern barrier island and tidal inlet deposits as applied to paleoenvironmental studies. *In* J.C. Ferm, J.C. Horne (eds.), Carboniferous depositional environments in the Appalachian region, pp. 472–498. Univ. South Carolina, Carolina Coal Group.

Barwis, J.H., Horne, J.C. 1979. Paleotidal-range indicators in Carboniferous barrier sequences of eastern Kentucky. Amer. Assoc. Petrol. Geol. Bull. 63: 415.

Barwis, J.H., Hubbard, D.K. 1976. The relationship of flood-tidal delta morphology to the configuration and hydraulics of tidal inlet–bay sequences. Geol. Soc. Amer. Absts. with Progs., pp. 128–129.

Barwis, J.H., Makurath, J.H. 1978. Recognition of ancient tidal inlet sequences: an example from the Upper Silurian Keyser Limestone in Virginia. Sedimentology 25: 61–82.

Bates, C.C., 1953. Rational theory of delta formation. Amer. Assoc. Petrol. Geol. Bull. 37: 2119–2162.

Bates, R.L., Jackson, J.A., eds., 1980. Glossary of geology, 2nd ed. Falls Church: Am. Geol. Inst.

Beaty, C.B. 1970. Age and estimated rate of accumulation of an alluvial fan, White Mountains, California, U.S.A. Am. J. Sci. 268: 50–77.

Beaumont, E.A. 1979. Depositional environments of Fort Union sediments (Tertiary, northwest Colorado) and their relation to coal. Amer. Assoc. Petrol. Geol. Bull. 63: 194–217.

Bein, A., Weiler, V. 1976. The Cretaceous Talme Yafe Formation: a contour-current shaped sedimentary prism of calcareous detritus at the continental margin of the Arabian craton. Sedimentology 23: 511–532.

Belderson, R.H., Stride, A.H. 1966. Tidal current fashioning of a basal bed. Mar. Geol. 4: 237–257.

Bellaiche, G., Droz, L., Aloisi, J-C. *et al.*, 1981. The Ebro and Rhone deep-sea fans: first comparative study. Mar. Geol. 43: M75–M85.

Berger, W.H. 1970. Biogenous deep-sea sediments: fractionation by deep-sea circulation. Geol. Soc. Amer. Bull. 81: 1385–1402.

Bernard, H.A., Le Blanc, R.J. 1975. Resume of the Quaternary geology of the northeastern Gulf of Mexico Province. *In* H.E. Wright, D.G. Frey (eds.), The Quaternary of the United States, pp. 137–185. Princeton, N.J.: Princeton Univ. Press.

Bernard, H.A., Major, C.F., Parrott, B.S., Le Blanc, R.J. 1970. Recent sediments of southeast Texas. A field guide to the Brazos alluvial and deltaic plains and the Galveston barrier island complex. Bur. Econ. Geol. Univ. Texas, Austin, Guidebook No. 11.

Berner, R.A. 1970. Sedimentary pyrite formation. Amer. J. Sci. 268: 1–23.

Betzer, P.R., Richardson, P.L., Zimmerman, H.B. 1974. Bottom currents, nepheloid layers, and sedimentary features under the Gulf Stream near Cape Hatteras. Mar. Geol. 16: 21–29.

Beukes, N.J. 1977. Transition from siliciclastic to carbonate sedimentation near the base of the Transvaal Supergroup, northern Cape Province, South Africa. Sediment. Geol. 18: 201–221.

Bigarella, J.J. 1972. Eolian environments—their characteristics, recognition, and importance. *In* Soc. Econ. Paleont. Mineral. Spec. Publ. No. 16, pp. 12–62.

Bigarella, J.J. 1979. Dissipation of dunes, Lagoa, Brazil. *In* E.D. McKee (ed.), A study of global sand seas. U.S. Geol. Surv. Prof. Pap. No. 1052, pp. 124–134.

Bigarella, J.J., Becker, R.D., Duarte, G.M. 1969. Coastal dune structure from Parana (Brazil). Mar. Geol. 7: 5–55.

Bird, E.C.F. 1965. A geomorphological study of the Gippsland lakes. Res. School Pacific Studies, Aust. Nat. Univ.

Bird, E.C.F. 1976. Shoreline changes during the past century. Procs. 23rd Internat. Geog. Cong., Moscow, p. 54.

Blissenbach, E. 1954. Geology of alluvial fans in semiarid regions. Geol. Soc. Amer. Bull. 65: 175–190.

Blondel, F., Lasky, S.G. 1956. Mineral reserves and mineral resources. Econ. Geol. 51: 686–697.

Bloomer, R.R. 1977. Depositional environments of a reservoir sandstone in west-central Texas. Amer. Assoc. Petrol. Geol. Bull. 61: 344–359.

Bluck, B.J. 1964. Sedimentation of an alluvial fan in southern Nevada. J. Sediment. Petrol. 34: 395–400.

Bluck, B.J. 1967. Sedimentation of gravel beaches:

examples from south Wales. J. Sediment. Petrol. 37: 128–156.

Boberg, W.W., Runnells, D.D. 1971. Reconnaissance study of uranium in the South Platte River, Colorado. Econ. Geol. 66: 435–450.

Bogomolov, G.V., Kats, D.M. 1972. Methods and results of investigations of the hydrology, climate, hydrogeology, soils and vegetation in marsh-ridden areas of the temperate zone. *In* Hydrology of marsh-ridden areas, pp. 1197–1201. Paris: UNESCO Press.

Bogomolov, Y.G., Kudelsky, A.V., Lapshin, N.N. 1978. Hydrogeology of large sedimentary basins. *In* Hydrogeology of great sedimentary basins, pp. 117–122. Int. Assoc. Hydrolog. Sci. Publ. No. 120.

Bonham, L.C. 1980. Migration of hydrocarbons in compacting basins. Amer. Assoc. Petrol. Geol. Studies in Geology 10.

Boothroyd, J.C. 1972. Coarse-grained sedimentation on a braided outwash fan, northeast Gulf of Alaska. Univ. South Carolina, Coastal Res. Div. Tech. Rept. No. 6–CRD.

Boothroyd, J.C., Ashley, G.M. 1975. Process, bar morphology and sedimentary structures on braided outwash fans, northeastern Gulf of Alaska. *In* Soc. Econ. Paleont. Mineral. Spec. Publ. No. 23, pp. 193–222.

Boothroyd, J.C., Nummedal, D. 1978. Proglacial braided outwash: a model for humid alluvial fan deposits, pp. 641–668. *In* A.D. Miall (ed.), Fluvial sedimentology. Can. Soc. Petrol. Geol. Mem. 5.

Bouma, A.H. 1962. Sedimentology of some flysch deposits. Amsterdam: Elsevier.

Bouma, A.H., Hollister, C.D. 1973. Deep ocean basin sedimentation. *In* Turbidites and deep water sedimentation, pp. 79–118. Soc. Econ. Paleont. Mineral. Short Course.

Bourgeois, J. 1980. A transgressive shelf sequence exhibiting hummocky stratification: the Cape Sebastian Sandstone (Upper Cretaceous), southwestern Oregon. J. Sediment. Petrol. 50: 681–702.

Boyd, D.R., Dyer, B.F. 1964. Frio barrier bar system of South Texas: Gulf Coast Assoc. Geol. Socs. Trans., 14: p. 309–322.

Bozanich, R.G. 1978. Fluid density current deposition of Permian Delaware Mountain Group, West Texas—subsurface evidence. Amer. Assoc. Petrol. Geol. Bull. 62: 500.

Bradley, W.H. 1925. A contribution to the origin of the Green River Formation and its oil shale. Amer. Assoc. Petrol. Geol. Bull. 9: 247–262.

Bradley, W.H. 1929. The varves and climate of the Green River epoch. U.S. Geol. Surv. Prof. Pap. No. 158–E, pp. 87–110.

Bray, E.E., Foster, W.R. 1980. A process for primary migration of petroleum. Amer. Assoc. Petrol. Geol. Bull. 64: 107–114.

Bredehoeft, J.D., Hanshaw, B.B. 1968. On the maintenance of anomalous fluid pressures. I. Thick sedimentary sequences. Geol. Soc. Amer. Bull. 79: 1097–1106.

Breed, C.S., Grow, T. 1979. Morphology and distribution of dunes in sand seas observed by remote sensing. *In* E.D. McKee (ed.), A study of global sand seas. U.S. Geol. Surv. Prof. Pap. No. 1052, pp. 253–302.

Brenchley, P.J., Newall, G., Stanistreet, I.G. 1979. A storm surge origin for sandstone beds in an epicontinental platform sequence, Ordovician, Norway. Sediment. Geol. 22: 185–217.

Brenner, R.L. 1978. Sussex sandstone of Wyoming—example of Cretaceous offshore sedimentation. Amer. Assoc. Petrol. Geol. Bull. 62: 181–200.

Brenner, R.L. 1980. Construction of process-response models for ancient epicontinental seaway depositional systems using partial analogs. Amer. Assoc. Petrol. Geol. Bull. 64: 1223–1244.

Brenner, R.L., Davies, D.K. 1973. Storm-generated coquinoid sandstone: genesis of high-energy marine sediments from the Upper Jurassic of Wyoming and Montana. Geol. Soc. Amer. Bull. 84: 1685–1697.

Bridge, J.S., Leeder, M.R. 1979. A simulation model of alluvial stratigraphy. Sedimentology 26: 617–644.

Britten, R.A. 1972. A review of the stratigraphy of the Singleton Coal Measures and its significance to coal geology and mining in the Hunter Valley region of New South Wales. Australas. Inst. Min. Metall., Newcastle Conf. Pap. No. 2, pp. 11–22.

Britten, R.A., Smyth, M., Bennett, A.R.J., Shibaoka, M. 1973. Environmental interpretations of Gondwana coal measure sequences in the Sydney Basin of New South Wales. *In* K.S.W. Campbell (ed.), Gondwana geology, pp. 233–247. Canberra: Aust. Nat. Univ. Press.

Brookfield, M.E. 1977. The origin of bounding surfaces in ancient aeolian sandstones. Sedimentology 24: 303–332.

Brookfield, M.E. 1980. Permian intermontane basin sedimentation in southern Scotland. Sediment. Geol. 27: 167–194.

Brookins, D.G. 1976. Position of uraninite and/or coffinite accumulation to the hematite-pyrite interface in sandstone-type deposits deposits. Econ. Geol. 71: 944–948.

Brown, L.F., Jr. 1969. Virgil-Lower Wolfcamp repetition of depositional environments in north-central Texas. *In* Cyclic sedimentation in the Permian Basin, pp. 115–134. West Tex. Geol. Soc.

Brown, L.F., Jr. 1979. Deltaic sandstone facies of the mid-continent. *In* N.J. Hyne (ed.), Pennsylvanian sandstones of the Mid-Continent, pp. 35–63. Tulsa Geol. Soc.

Brown, L.F., Jr., Cleaves, A.W. II, Erxleben, A.W. 1973. Pennsylvanian depositional systems in north-central Texas. Bur. Econ. Geol. Univ. Texas, Austin, Guidebook No. 16.

Brown, L.F., Fisher, W.L. 1980. Seismic stratigraphic interpretation and petroleum exploration. Amer. Assoc. Petrol. Geol. Course Notes No. 16.

Brown, L.F., Jr., McGowen, J.H., Evans, T.J., Groat, C.G., Fisher, W.L. 1977. Environmental geologic atlas of the Texas coastal zone—Kingsville area. Bur. Econ. Geol. Univ. Texas, Austin.

Bruce, D.H. 1973. Pressured shale and sediment deformation: mechanism for development of regional contemporaneous faults. Amer. Assoc. Petrol. Geol. Bull. 57: 878–886.

Brunt, D.A. 1978. Uranium in Tertiary stream channels, Lake Frome area, South Australia: Proc. Australas. Inst. Min. Metall., No. 266, pp. 79–90.

Bull, W.B. 1962. Relations of alluvial fan size and slope to drainage basin size and lithology in western Fresno County, California. U.S. Geol. Surv. Prof. Pap. No. 450–B.

Bull, W.B. 1972. Recognition of alluvial fan deposits in the stratigraphic record. *In* Soc. Econ. Paleont. Mineral. Spec. Publ. No. 16, pp. 63–83.

Bumpus, D.F. 1965. Residual drift along the bottom of the continental shelf in the Middle Atlantic Bight Area. Limnol. Oceanog. 10: R50–R53.

Burke, K. 1972. Longshore drift, submarine canyons, and submarine fans in development of Niger delta. Amer. Assoc. Petrol. Geol. Bull. 56: 1975–1983.

Burne, R.V., Bauld, J., DeDekker, P. 1980. Saline lake charophytes and their geological significance. J. Sediment. Petrol. 50: 281–293.

Burst, J.F. 1969. Diagenesis of Gulf Coast clayey sediments and its possible relation to petroleum migration. Amer. Assoc. Petrol. Geol. Bull. 53: 73–93.

Button, A., Adams, S.S. 1981. Geology and recognition criteria for uranium deposits of the quartz-pebble conglomerate type. U.S. Dept. Energy, Rept. No. GJBX-3 (81).

Button, A., Tyler, N. 1979. Precambrian paleo-weathering and erosion surfaces in southern Africa: review of their character and economic significance. Univ. Witwatersrand Econ. Geol. Res. Unit Inf. Circ. No. 135.

Byrne, J.V., LeRoy, D.O., Riley, C.M. 1959. The chenier plain and its stratigraphy, southwestern Louisiana. Gulf Coast Assoc. Geol. Socs. Trans. 9: 1–23.

Cadle, A.B., Hobday, D.K. 1977. A subsurface investigation of the Middle Ecca and Lower Beaufort in the southern Transvaal and northern Natal. Geol. Soc. S. Afr. Trans. 80: 111–115.

Cairncross, B.C. 1980. Anastomosing river deposits: paleoenvironmental control on coal quality and distribution, northern Karoo Basin. Geol. Soc. S. Afr. Trans. 8: 327–332.

Camboz, A., De Matharel, M. 1973. Organic sedimentation and genesis of petroleum in Mahakam delta, Borneo. Amer. Assoc. Petrol. Geol. Bull. 62: 1684–1695.

Cameron, A.R., Leclair, G. 1975. Extraction of uranium from aqueous solution by coal of different rank and petrographic composition. Can. Geol. Surv. Pap. No. 74–35.

Campbell, C.V. 1973. Offshore equivalents of Upper Cretaceous Gallup beach sandstones, northwestern New Mexico. Four Corners Geol. Soc. Mem., pp. 78–84.

Campbell, C.V. 1976. Reservoir geometry of a fluvial sheet sandstone. Amer. Assoc. Petrol. Geol. Bull. 60: 1009–1020.

Cant, D.J., Walker, R.G. 1978. Fluvial processes and facies sequences in the sandy braided South Saskatchewan River, Canada. Sedimentology 25: 625–648.

Carlson, P.A., Molnia, B.F. 1977. Submarine faults and slides on the continental shelf, northern Gulf of Alaska. Mar. Geotech. 2: 275–290.

Casshyap, S.M. 1970. Sedimentary cycles and environment of deposition of the Barakar coal measures of Lower Gondwana, India. J. Sediment. Petrol. 40: 1302–1317.

Castaing, P., Allen, G.P. 1981. Mechanisms controlling seaward escape of suspended sediment from the Gironde: a macrotidal estuary in France. Mar. Geol. 40: 101–118.

Caston, V.N.D. 1972. Linear sand banks in the southern North Sea. Sedimentology 18: 63–78.

Caston, V.N.D. 1976. A wind-driven near bottom current in the southern North Sea. Est. Coast. Mar. Sci. 4: 23–32.

Caughey, C.A. 1975. Pleistocene depositional trends host valuable gulf oil reserves. Oil and Gas J., Sept. 8 and 15, pp. 90–94, 240–242.

Caughey, C.A. 1977. Depositional systems in the Paluxy Formation (Lower Cretaceous), northeast Texas—oil, gas, and groundwater resources. Bur. Econ. Geol. Univ. Texas, Austin, Geol. Circ. No. 77–8.

Caughey, C.A. 1981. Deltaic and slope deposits in the Planulina trend (basal Miocene), southwestern Louisiana. *In* Gulf Coast Section, Soc. Econ. Paleont. Mineral. 2nd Ann. Res. Conf. Abstracts, pp. 20–21.

Cecil, C.B., Stanton, R.W., Dulong, F.T., Cohen, A.D. 1979a. Experimental coalification of *Taxodium* peat from the Okefenokee Swamp, Georgia. *In* A.C. Donaldson, M.W. Presley, J.J. Renton (eds.), Carboniferous coal guidebook, Vol. 3, pp. 129–141. W. Va. Geol. and Econ. Surv.

Cecil, C.B., Stanton, R.W., Dulong, F.T., Renton, J.J. 1979b. Geologic factors that control mineral matter in coal. *In* A.C. Donaldson, M.W. Presley,

J.J. Renton (eds.), Carboniferous coal guidebook, Vol. 3, pp. 43–56. W. Va. Geol. and Econ. Surv.

Chamberlain, C.K. 1978. Recognition of trace fossils in cores. *In* Soc. Econ. Paleont. Miner. Short Course No. 5, pp. 133–183.

Champ, D.R., Gulens, J., Jackson, R.E. 1979. Oxidation-reduction sequences in ground water flow systems. Can. J. Earth Sci. 16: 12–23.

Chappell, J., Eliot, I.G. 1979. Surf-beach dynamics in time and space—an Australian case study, and elements of a predictive model. Mar. Geol. 32: 231–250.

Chebotarev, I.I. 1955. Metamorphism of natural waters in the crust of weathering. Geochim. Cosmochim. 8: 22–48, 137–170, 198–212.

Childers, M.O. 1970. Uranium geology of the Kaycee area of Johnson County, Wyoming, pp. 13–20. Wyoming Geol. Assoc. Guidebook.

Chitale, S.V. 1973. Theories and relationships of river channel patterns. J. Hydrology 19: 285–308.

Chorley, R.J., Kennedy, B.A. 1971. Physical geography: a systems approach. London: Prentice-Hall.

Christopher, J.E. 1974. The Upper Jurassic Vanguard and Lower Cretaceous Mannville Groups of Southwestern Saskatchewan. Saskatchewan Geol. Surv. Rept. No. 151.

Clarke, A.M. 1978. The world's coal reserves. *In* G.M. Philip, K.L. Williams (eds.), Australia's mineral energy resources: assessment and potential, pp. 7–61. Earth Resources Foundation, Univ. Sydney, Occas. Publ. No. 1.

Clayton, J.L., Swetland, P.J. 1980. Petroleum generation and migration in Denver Basin. Amer. Assoc. Petrol. Geol. Bull, 64: 1613–1633.

Clayton, R.N., Friedman, I., Graf, D.L., Mayeda, T.K., Meents, W.F., Shimp, N.F. 1966. The origin of saline formation waters. J. Geophys. Res. 71: 3869–3882.

Cleary, W.J., Conolly, J.R. 1974. Hatteras deep-sea fan. J. Sediment. Petrol. 44: 1140–1154.

Clifton, H.E. 1969. Beach lamination—nature and origin. Mar. Geol. 7: 553–559.

Clifton, H.E. 1973. Pebble segregation and bed lenticularity in wave-worked versus alluvial gravel. Sedimentology 20: 173–187.

Clifton, H.E., Hunter, R.E., Phillips, R.L. 1971. Depositional structures and processes in the non-barred high-energy nearshore. J. Sediment. Petrol. 41: 651–670.

Clos-Arceduc, A. 1966. L'application des methodes d'interpretation des images a des problemes geographiques. Acta Symp. Int. Photointerp. 2, Paris.

Cloud, P.E. 1972. A working model of the primitive Earth. Amer. J. Sci. 272: 537–548.

Coates, E.J., Groat, C.G., Hart, G.F. 1980. Subsurface Wilcox lignite in west-central Louisiana. Gulf Coast Assoc. Geol. Socs. Trans. 30: 309–332.

Cohen, A.D. 1973. Petrology of some Holocene peat sediments from the Okefenokee swamp-marsh complex of southern Georgia. J. Sediment. Petrol. 84: 3876–3878.

Coleman, J.M. 1966. Recent coastal sedimentation: central Louisiana coast. Baton Rouge: Louisiana State Univ. Press.

Coleman, J.M. 1969. Brahmaputra River; channel processes and sedimentation. Sediment. Geol. 3: 131–239.

Coleman, J.M., Gagliano, S.M. 1964. Cyclic sedimentation in the Mississippi River deltaic plain. Gulf Coast Assoc. Geol. Socs. Trans. 14: 67–80.

Coleman, J.M., Garrison, L.E. 1977. Geological aspects of marine slope instability, northwestern Gulf of Mexico. Mar. Geotech. 2: 9–44.

Coleman, J.M., Pryor, D.B. 1980. Deltaic sand bodies. Amer. Assoc. Petrol. Geol., Contin. Ed. Course Note Ser. 15.

Coleman, J.M., Smith, W.G. 1964. Late Recent rise of sea level. Geol. Soc. Amer. Bull. 75: 833–840.

Coleman, J.M., Wright, L.D. 1975. Modern river deltas: variability of processes and sand bodies. *In* M.L. Broussard (ed.), Deltas, pp. 99–150. Houston Geol. Soc.

Coleman, J.M., Gagliano, S.M., Morgan, J.P. 1969. Mississippi River subdeltas: natural models of deltaic sedimentation. Louisiana State Univ., Coastal Studies Bull. No. 3, pp. 23–27.

Coleman, J.M., Gagliano, S.M., Smith, W.G. 1970. Sedimentation in a Malaysian high tide tropical delta. *In* Soc. Econ. Paleont. Mineral. Spec. Publ. No. 15, pp. 185–197.

Coleman, J.M., Gagliano, S.M., Webb, J.E. 1964. Minor sedimentary structures in a prograding distributary. Mar. Geol. 1: 240–258.

Coleman, J.M., Suhayda, J.N., Whelan, T., Wright, L.D. 1974. Mass movement of Mississippi River delta sediments. Gulf Coast Assoc. Geol. Socs. Trans. 24: 49–68.

Collins, A.G. 1975. Geochemistry of oilfield waters. Amsterdam: Elsevier.

Collinson, J.D. 1969. The sedimentology of the Grindslow Shales and the Kinderscout Grit: a deltaic complex in the Namurian of northern England. J. Sediment. Petrol. 39: 194–221.

Collinson, J.D. 1978a. Vertical sequence and sand body shape in alluvial sequences. *In* A.D. Miall (ed.), Fluvial sedimentology, pp. 577–586. Can. Soc. Petrol. Geol. Mem. 5.

Collinson, J.D. 1978b. Lakes. *In* H.G. Reading (ed.), Sedimentary environments and facies, pp. 61–79. Oxford: Blackwells.

Conaghan, P.J. 1982. Lacustrine (Gilbert) deltas in the Permian Coal Measures of eastern Australia and

India: implication for the widespread hydroponic origin and deep-water diagenesis of Gondwanan coals. 16th Symposium Programme and Abstracts, Advances in the study of the Sydney Basin, Newcastle, April–May 1982, pp. 6–9.

Connan, J. 1974. Time-temperature relations in oil genesis, Amer. Assoc. Petrol. Geol. Bull. 58: 2516–2521.

Conolly, J.R., Ferm, J.C. 1971. Permo-Triassic sedimentation patterns, Sydney Basin, Australia. Amer. Assoc. Petrol. Geol. Bull. 55: 2018–2032.

Conybeare, C.E.B. 1976. Geomorphology of oil and gas fields in sandstone bodies. Amsterdam: Elsevier.

Cooke, R.V., Warren, A. 1973. Geomorphology in deserts. London: Batsford.

Cotter, E. 1975. Late Cretaceous sedimentation in a low-energy coastal zone: the Ferron Sandstone of Utah. J. Sediment. Petrol. 45: 669–685.

Cotter, E. 1978. The evolution of fluvial style, with special reference to the Central Appalachian Paleozoic. *In* A.D. Miall (ed.), Fluvial sedimentology, pp. 361–384. Can. Soc. Petrol Geol. Mem. 5.

Cousteau, H., Rumeau, L., Sourisse, C. 1975. Classification hydrodynamique des bassins sedimentaires: utilisation combinee avec d'autres methodes pour rationaliser l'exploration des bassins non-productifs. Procs. 9th World Petrol. Congr. Vol. 2, pp. 105–119.

Craig, L.C., Holmes, C.N., Cadigan, R.A., Freeman, V.L., Mullens, T.E., Weir, G.W. 1955. Stratigraphy of Morrison and related formations, Colorado Plateau region, a preliminary report, U.S. Geol. Surv. Bull. 1009-E, pp. 125–168.

Crans, W., Mandl, G., Haremboure, J. 1980. On the theory of growth faulting: A geomechanical delta model based on gravity sliding. J. Petrol. Geol. 2: 265–307.

Cronin, L.E. (ed.), 1975. Estuarine research, Vol. 2. New York: Academic Press.

Crowell, J.C. 1957. Origin of pebbly mudstones. Geol. Soc. Amer. Bull. 698: 993–1010.

Crowell, J.C. 1974. Origin of late Cenozoic basins in southern California. *In* Soc. Econ. Paleont. Mineral. Spec. Publ. No. 22, pp. 190–204.

Csanady, G.T. 1976. Mean circulation in shallow seas. J. Geophys. Res. 81: 5389–5399.

Culver, S.J. 1980. Differential two-way sediment transport in the Bristol Channel and Severn Estuary, U.K. Mar. Geol. 34: M39–M43.

Curray, J.R. 1960. Sediments and history of Holocene transgression, Gulf of Mexico. *In* F.P. Shepard, F.B. Phleger (eds.), Recent sediments, northwest Gulf of Mexico, pp. 221–266. Amer. Assoc. Petrol. Geol.

Curray, J.R. 1965. Late Quaternary history, continental shelves of the United States. *In* H.E., Wright, D.G. Frey (eds.), The Quaternary of the United States, pp. 723–736. Princeton, N.J.: Princeton Univ. Press.

Curray, J.R. 1969. Shore zone sand bodies: barriers, cheniers and beach ridges. *In* D.J. Stanley (ed.), The new concepts of continental margin sedimentation, JC–11–1 to JC–11–18. Am. Geol. Inst.

Curtis, D.M. 1970. Miocene deltaic sedimentation, Louisiana Gulf Coast. *In* Soc. Econ. Paleont. Mineral. Spec. Publ. No. 12, pp. 293–308.

Dahl, A.R., Hagmaier, J.L. 1974. Genesis and characteristics of the southern Powder River Basin uranium deposits, Wyoming, USA. *In* Formation of uranium ore deposits, pp. 201–218. Atomic Ener. Agen. Panel Procs.

Dailly, G.C. 1976. A possible mechanism relating progradation, growth faulting, clay diapirism and overthrusting in a regressive sequence of sediments. Bull. Can. Petrol. Geol. 24: 92–116.

Dalrymple, R.W., Knight, R.J., Lambiase, J.J. 1978. Bedforms and their hydraulic stability relationships in a tidal environment, Bay of Fundy, Canada. Nature, 275: 100–104.

Damberger, H.H., Nelson, W.J., Krausse, H–F. 1980. Effect of geology on roof stability of room-and-pillar mines in the Herrin No. 6 coal of Illinois. *In* Procs. First Conf. on Ground Control Problems, Illinois Basin, pp. 14–32, Carbondale, Ill.

Davidson-Arnott, R.G.D., Greenwood, B. 1976. Facies relationships on a barred coast, Kouchibouguac Bay, New Brunswick, Canada. *In* Soc. Econ. Paleont. Mineral. Spec. Publ. 24, pp. 149–168.

Davies, D.K., Brenner, R.L. 1973. Role of storms in generating ancient ridge and swale systems. Amer. Assoc. Petrol. Geol. Bull. 57: 775.

Davies, W.K. 1966. Sedimentary structures and subfacies of a Mississippi River point bar. J. Geol. 74: 234–239.

Davis, G.H., Green, J.H., Olmstead, F.H., Brown, D.W. 1959. Ground-water conditions and storage capacity in the San Joaquin Valley, California. U.S. Geol. Surv. Water Supply Pap. 1468.

Davis, J.F. 1974. The Newcastle Coal Measures, models of sedimentation and sedimentary history. Unpubl. Ph.D. thesis, Sydney Univ.

Davis, S.N. 1969. Porosity and permeability of natural materials. *In* J.M. DeWiest (ed.), Flow through porous media. New York: Academic Press.

Dean, W.E. 1981. Carbonate minerals and organic matter in sediments of modern north temperate hard-water lakes. *In* Soc. Econ. Paleont. Mineral. Spec. Publ. No. 31, pp. 213–231.

Deery, J.R., Howard, J.D. 1977. Origin and character of washover fans on the Georgia coast, USA. Gulf

Coast Assoc. Geol. Socs. Trans. 27: 259–271.

Degens, E.T., Khoo, F., Michaelis, W. 1977. Uranium anomaly in Black Sea sediments. Nature 269: 566–569.

Degens, E.T., Von Herzen, R.P., Wong, H.K. 1971. Lake Tanganyika: water chemistry, sediments, geological structure. Naturwissen. 58: 224–291.

De Jong, J.D. 1977. Dutch tidal flats. Sediment. Geol. 18: 12–23.

Demaison, G.T., Moore, G.T. 1980. Anoxic environments and oil source genesis. Amer. Assoc. Petrol. Geol. Bull. 64: 1179–1209.

DeNault, K.J. 1974. Origin of sandstone-type uranium deposits in Wyoming. Unpubl. Ph.D. diss., Stanford Univ.

Denny, C.S. 1965. Alluvial fans in the Death Valley region, California and Nevada. U.S. Geol. Surv. Prof. Pap. 466.

Denson, N.M., Backman, G.O., Zeller, H.D. 1959. Uranium-bearing lignite in northwestern North Dakota and adjacent states. U.S. Geol. Surv. Bull. 1055: 11–57.

De Raaf, J.F.M., Boersma, J.R., van Felder, A. 1977. Wave generated structures and sequences from a shallow marine succession, Lower Carboniferous, Cork County, Ireland. Sedimentology 24: 451–483.

Dickinson, G. 1953. Geological aspects of abnormal reservoir pressures in Gulf Coast Louisiana. Amer. Assoc. Petrol. Geol. 37: 410–423.

Dickinson, W.R. 1974. Plate tectonics and sedimentation. *In* Soc. Econ. Paleont. Mineral. Spec. Publ. No. 22, pp. 1–27.

Diessel, C.F.K. 1970. Paralic coal seam formation. *In* The assessment of our fuel and energy requirements, Inst. of Fuel Conference, Brisbane, pp. 1–22.

Dietz, R.S. 1963. Wave-base, marine profile of equilibrium, and wave-built terraces: a critical appraisal. Geol. Soc. Amer. Bull. 74: 971–990.

Dobkins, J.E., Jr., Folk, R.L. 1970. Shape development on Tahiti–Nui. J. Sediment. Petrol. 40: 1167–1203.

Doeglas, D.J. 1962. The structure of sedimentary deposits of braided rivers. Sedimentology 1: 167–190.

Doi, K., Hirono, S., Sakamaki, Y. 1975. Uranium mineralization by ground water in sedimentary rocks, Japan. Econ. Geol. 70: 628–646.

Dolan, R. 1972. Barrier dune systems along the Outer Banks of North Carolina: a reappraisal. Science 176: 286–288.

Donaldson, A.C. 1979. Origin of coal seam discontinuities. *In* A.C. Donaldson, M.W. Presley, J.J. Renton (eds.), Carboniferous coal guidebook, Vol. 1, pp. 102–132. W. Va. Geol. and Econ. Surv.

Donaldson, A.C., Drennan, L., Hamilton, W. *et al.* 1979. Geologic factors affecting the thickness and quality of Upper Pennsylvanian coals of the Dunkard Basin. *In* A.C. Donaldson, M.W. Presley, J.J. Renton (eds.), Carboniferous coal guidebook, Vol. 1, pp. 133–188. W. Va. Geol. and Econ. Surv.

Donaldson, A.C., Martin, R.H., Kanes, W.H. 1970. Holocene Guadalupe delta of Texas Gulf Coast. *In* Soc. Econ. Paleont. Mineral. Spec. Publ. No. 12, pp. 107–137.

Dow, W.G. 1974. Application of oil-correlation and source-rock data to exploration in Williston Basin. Amer. Assoc. Petrol. Geol. Bull. 58: 1253–1262.

Dow, W.G. 1978. Petroleum source beds on continental slopes and rises. Amer. Assoc. Petrol. Geol. Bull. 62: 1584–1606.

Doyle, L.J., Pilkey, O.H., Woo, C.C. 1979. Sedimentation on the eastern United States continental slope. *In* Soc. Econ. Paleont. Mineral. Spec. Publ. No. 27, pp. 119–130.

Drake, D.E. 1976. Suspended sediment transport and mud deposition on continental shelves. *In* D.J. Stanley, D.J.P. Swift (eds.), Marine sediment transport and environmental management, pp. 127–158. New York: Wiley.

Dulhunty, J.A. 1981. Quaternary sedimentary environments in the Lake Eyre Region, South Australia. Geol. Soc. Aust. Conv. Absts., pp. 57–58.

Du Toit, A.L. 1954. The geology of South Africa, 3rd ed. Edinburgh: Oliver and Boyd.

Dutton, S.P. 1980a. Depositional systems and hydrocarbon resource potential of the Pennsylvanian System, Palo Duro and Dalhart Basins, Texas Panhandle. Bur. Econ. Geol. Univ. Texas, Austin, Geol. Circ. 80–8.

Dutton, S.P. 1980b. Petroleum source rock potential and thermal maturity, Palo Duro Basin, Texas. Bur. Econ. Geol. Univ. Texas, Austin, Geol. Circ. 80–10.

Dutton, S.P. 1982. Pennsylvanian fan-delta and carbonate deposition, Mobeetie field, Texas Panhandle. Amer. Assoc. Petrol. Geol. Bull. 66: 389–407.

Eargle, D.H., Weeks, A.M.D. 1973. Geologic relations among uranium deposits, South Texas Coastal Plain region, U.S.A. *In* G.C. Amstutz, A.J. Bernard (eds.), Ores in sediments, pp. 101–113. New York: Springer-Verlag.

Eargle, D.H., Dickinson, K.A., Davis, B.O. 1975. South Texas uranium deposits. Amer. Assoc. Petrol. Geol. Bull. 5: 766–779.

Edwards, M.B. 1976. Growth faults in Upper Triassic deltaic sediments. Amer. Assoc. Petrol. Geol. Bull. 60: 341–355.

Elliot, T. 1975. The sedimentary history of a delta lobe from a Yoredale (Carboniferous) cyclothem. Yorks. Geol. Soc. Procs. 40: 505–536.

Ellis, G.K. 1980. Distribution and genesis of sedimentary uranium near Curnamond, Lake Frome region, South Australia. Amer. Assoc. Petrol. Geol. Bull. 64: 1643–1657.
Embley, R.W. 1980. The role of mass transport in the distribution and character of deep ocean sediments with special reference to the North Atlantic. Mar. Geol. 39: 23–50.
Embley, R.W., Jacobi, R.D. 1977. Distribution and morphology of large submarine sediment slides and slumps on Atlantic continental margins. Mar. Geotech. 2: 205–228.
Emery, K.O. 1952. Continental shelf sediments off southern California. Geol. Soc. Amer. Bull. 63: 1105–1108.
Emery, K.O. 1960. The sea off southern California. New York: Wiley.
Emery, P.A., Boettcher, A.J., Snipes, R.J., McIntyre, H.J., Jr. 1971. Hydrology of the San Louis Valley, south-central Colorado. U.S. Geol. Surv. Hydrol. Atlas HA-381.
Erdelyi, M. 1972. Hydrology of deep ground-waters. *In* Hydrological methods for developing water resources management, pp. 90–157. Res. Inst. Water Res.
Eriksson, K.A. 1977. Tidal deposits from the Archaean Moodies Group, Barberton Mountain Land, South Africa. Sediment. Geol. 18: 223–244.
Ethridge, F.G., Schumm, S.A. 1978. Reconstructing paleochannel morphologic and flow characteristics: methodology, limitations, and assessment. *In* A.D. Miall (ed.), Fluvial sedimentology, pp. 703–722. Can. Soc. Petrol. Geol. Mem. 5.
Ethridge, F.G., Jackson, T.J., Youngberg, A.D. 1981. Floodbasin sequence of a fine-grained meander belt subsystem: the coal-bearing Lower Wasatch and upper Fort Union Formations, southern Powder River Basin, Wyoming. *In* Soc. Econ. Paleont. Mineral. Spec. Publ. 31, pp. 191–209.
Eugster, H.P. 1970. Chemistry and origin of the brines of Lake Magadi, Kenya. Miner. Soc. Amer. Spec. Pap. 3, pp. 213–235.
Eugster, H.P., Hardie, L.A. 1975. Sedimentation in an ancient playa-lake complex: the Wilkins Peak member of the Green River Formation of Wyoming. Geol. Soc. Amer. Bull. 86: 319–339.
Eugster, H.P., Surdam, R.C. 1973. Depositional environment of the Green River Formation of Wyoming, a preliminary report. Geol. Soc. Amer. Bull. 84: 1115–1120.
Evamy, D.D., Harembourne, J., Kamerling, P., Knapp, W.A., Molloy, F.A., Rowlands, P.H. 1978. Hydrocarbon habitat of Tertiary Niger delta. Amer. Assoc. Petrol. Geol. Bull. 62: 1–39.
Evans, G. 1965. Intertidal sediments and their environment of deposition in the Wash. Quart. J. Geol. Soc. Lond. 121: 209–245.
Falvey, D.A. 1974. The development of continental margins in plate tectonics theory. Aust. Pet. Expl. Assoc. J. 14: 95–106.
Ferentinos, G., Collins, M. 1980. Effects of shoreline irregularities on a rectilinear tidal current and their significance in sedimentation processes. J. Sediment. Petrol. 50: 1081–1094.
Ferm, J.C. 1974. Carboniferous environmental models in eastern United States and their significance. *In* Geol. Soc. Amer. Spec. Pap. 148, p. 79–95.
Ferm, J.C. 1976. Depositional models in coal exploration and development. *In* R.S. Saxena (ed.), Sedimentary environments and hydrocarbons. Amer. Assoc. Petrol. Geol. Short Course, pp. 60–78.
Ferm, J.C. 1979. Pennsylvanian cyclothems of the Appalachian region: a retrospective view. *In* J.C. Ferm, J.C. Horne (eds.), Carboniferous depositional environments in the Appalachian region, pp. 284–290. Univ. South Carolina, Carolina Coal Group.
Ferm, J.C., Cavaroc, V.V. 1968. A nonmarine sedimentary model for the Allegheny rocks of West Virginia. *In* Geol. Soc. Amer. Spec. Pap. 106, pp. 1–19.
Ferm, J.C., Cavaroc, V.V., Jr. 1969. A field guide to Allegheny deltaic deposits in the upper Ohio Valley. Pittsburg and Ohio Geol. Socs. Guidebook.
Ferm, J.C., Horne, J.C. (eds.) 1979. Carboniferous depositional environments in the Appalachian region. Univ. South Carolina, Carolina Coal Group.
Ferm, J.C., Williams, E.G. 1963. A model for cyclic sedimentation in the Appalachian Pennsylvanian. Amer. Assoc. Petrol. Geol. Bull. 47: 356–357.
Ferm, J.C., Horne, J.C., Swinchatt, J.P., Whaley, P.W. 1971. Carboniferous depositional environments in northeastern Kentucky, Geol. Soc. Kentucky Guidebook.
Ferm, J.C., Staub, J.R., Baganz, B.P. *et al.* 1979. The shape of coal bodies. *In* J.C. Ferm, J.C. Horne (eds.), Carboniferous depositional environments in the Appalachian region, pp. 605–610. Univ. South Carolina, Carolina Coal Group.
Fertl, W.H. 1976. Abnormal formation pressures. Amsterdam: Elsevier.
Fertl, W.H., Chilingarian, G.V. 1977. Importance of abnormal formation pressures. J. Petrol. Tech. 29: 347–354.
Feth, J.H. 1964. Review and annotated bibliography of ancient lake deposits (Precambrian to Pleistocene) in the Western States. U.S. Geol. Surv. Bull. 1080.
Fettweis, G.B. 1979. World coal resources: methods of assessment and results. Developments in economic geology, 10. Amsterdam: Elsevier.
Field, M.E. 1982. The awakening continental shelf: insights from a decade of intensive study in North America. Aust. Mar. Sci. Assoc., Ann. Conf., Sydney, p. 11.

Fischer, R.P. 1968. The uranium and vanadium deposits of the Colorado Plateau region. *In* J.E. Ridge (ed.), Ore deposits of the United States, 1933–1967, Vol. 1, pp. 735–746. Amer. Inst. Min. Metall. Petrol. Eng.

Fisher, R.S., 1982. Diagenetic history of Eocene Wilcox Sandstones and associated formation waters, south-central Texas: Unpubl. Ph.D. diss., Univ. Texas, Austin.

Fisher, R.V. 1971. Features of coarse-grained, high-concentration fluids and their deposits. J. Sediment. Petrol. 41: 916–927.

Fisher, W.L. 1963. Lignites of the Texas Gulf Coastal Plain. Bur. Econ. Geol. Univ. Texas, Austin, Rept. Invest. 50.

Fisher, W.L. 1968. Variations in lignite of fluvial, deltaic, and lagoonal systems, Wilcox Group (Eocene) Texas. Geol. Soc. Amer. Abst. with Prog. p. 97.

Fisher, W.L. 1969. Facies characterization of Gulf Coast Basin delta systems, with some Holocene analogues. Gulf Coast Assoc. Geol. Socs. Trans. 19: 239–261.

Fisher, W.L., Brown, L.F., Jr. 1972. Clastic depositional systems—a genetic approach to facies analysis. Bur. Econ. Geol. Univ. Texas, Austin.

Fisher, W.L., Brown, L.F., Scott, A.J., McGowen, J.H. 1969. Delta systems in the exploration for oil and gas: Bur. Econ. Geol. Univ. Texas, Austin, 78 pp.

Fisher, W.L., McGowen, J.H. 1967. Depositional systems in the Wilcox Group of Texas and their relationship to occurrence of oil and gas. Gulf Coast Assoc. Geol. Socs. Trans. 17: 105–125.

Fisher, W.L., McGowen, J.H. 1969. Lower Eocene lagoonal systems in the Texas Gulf Coast Basin. *In* Lagunas costeras, un simposio, pp. 263–274. UNAM–UNESCO, Mexico, Nov. 1967.

Fisher, W.L., Proctor, C.V., Jr., Galloway, W.E., Nagle, J.S. 1970. Depositional systems in the Jackson Group of Texas—their relationship to oil, gas, and uranium. Gulf Coast Assoc. Geol. Socs. Trans. 20: 234–261.

Fisk, H.N. 1944. Geological investigation of the alluvial valley of the lower Mississippi River. Vicksburg: Mississippi River Commission.

Fisk, H.N. 1947. Fine-grained alluvial deposits and their effects on Mississippi River activity. Vicksburg: Mississippi River Commission.

Fisk, H.N. 1952. Geological investigation of the Atchafalaya Basin and the problem of Mississippi River diversion. Vicksburg: U.S. Army Corps Engin.

Fisk, H.N. 1955. Sand facies of Recent Mississippi Delta deposits. 4th World Petroleum Congress Procs., pp. 377–398.

Fisk, H.N. 1959. Padre Island and the Laguna Madre flats, coastal south Texas. 2nd Coastal Geog. Conf., Baton Rouge, pp. 103–151.

Fisk, H.N. 1961. Bar-finger sands of the Mississippi delta. *In* J.A. Peterson, J.C. Osmond (eds.), Geometry of sandstone bodies—a symposium, pp. 29–52. Amer. Assoc. Petrol. Geol.

Fisk, H.N., McFarlan, E., Jr., Kolb, C.R., Wilbert, L.J., Jr. 1954. Sedimentary framework of the modern Mississippi delta. J. Sediment. Petrol. 24: 76–99.

Flemming, B.W. 1978. Underwater sand dunes along the southeast African continental margin—observations and implications. Mar. Geol. 26: 177–198.

Flemming, B.W. 1980. Sand transport and bedform patterns on the continental shelf between Durban and Port Elizabeth (Southeast Africa continental margin). Sediment. Geol. 26: 179–205.

Flood, R.D., Hollister, C.D., Lonsdale, P. 1979. Disruption of the Feni Sediment Drift by debris flows from Rockall Bank. Mar. Geol. 32: 311–334.

Flores, R.M. 1981. Coal deposition in fluvial environments of the Paleocene Tongue River Member of the Fort Union Formation, Powder River area, Powder River Basin, Wyoming and Montana. *In* Soc. Econ. Paleont. Miner. Spec. Publ. 31, pp. 169–190.

Folk, R.L. 1971. Longitudinal dunes of the northwestern edge of the Simpson Desert, N.T., Australia, 1. Geomorphology and grain size relationships. Sedimentology 16: 5–54.

Forgotson, J.M., Jr. 1960. Review and classification of quantitative mapping techniques. Amer. Assoc. Petrol. Geol. Bull. 44: 83–100.

Forristall, G.Z., Hamilton, R.C., Cardone, V.J. 1977. Continental shelf currents in tropical storm Delia: observations and theory. J. Phys. Oceanog. 7: 532–546.

Frazier, D.E. 1967. Recent deltaic deposits of the Mississippi River; their development and chronology. Gulf Coast Assoc. Geol. Socs. Trans. 17: 287–315.

Frazier, D.E. 1974. Depositional episodes: their relationship to the Quaternary stratigraphic framework in the northwestern portion of the Gulf Basin. Bur. Econ. Geol. Univ. Texas, Austin, Geol. Circ. 71-1.

Frazier, D.E., Osanik, A. 1961. Point-bar deposits, Old River Locksite, Louisiana. Gulf Coast Assoc. Geol. Socs. Trans. 11: 127–137.

Frazier, D.E., Osanik, A. 1969. Recent peat deposits—Louisiana coastal plain. *In* Geol. Soc. Amer. Spec. Pap. 114, pp. 63–85.

Freeze, R.A. 1969. Theoretical analysis of regional groundwater flow, Canada. Inland Waters Branch, Dept. Energy, Mines, and Resources, Sci. Ser. No. 3.

Freeze, R.A., Cherry, J.A. 1979. Groundwater. Englewood Cliffs: Prentice-Hall.

Freeze, R.A., Witherspoon, P.A. 1968. Theoretical analysis of regional groundwater flow: 2. Effect of water-table configuration and subsurface permeability variation. Water Resour. Res. 3: 623–634.

Friedman, G.M. 1968. Geology and geochemistry of reefs, carbonate sediments, and waters, Gulf of Aqaba (Elat), Red Sea. J. Sediment. Petrol. 38: 895–919.

Fritz, P., Barker, J.F., Gale, J.E. 1979. Geochemistry and isotope hydrology of groundwaters in the Stripa Granite. Univ. Calif., Lawrence Berkeley Lab., Tech. Inf. Rept. No. 12.

Fryberger, S.G., Dean, G. 1979. Dune forms and wind regime. *In* E.D. McKee (ed.), A study of global sand seas. U.S. Geol. Surv. Prof. Pap. 1052, pp. 137–169.

Fryberger, S.G., Ahlbrandt, T.S., Andrews, S. 1979. Origins, sedimentary features, and significance of low-angle eolian "sand sheet" deposits, Great Sand Dunes National Monument and vicinity, Colorado. J. Sediment. Petrol. 49: 733–746.

Gabelman, J.W. 1971. Sedimentology and uranium prospecting. Sediment. Geol. 6: 145–186.

Gabelman, J.W. 1977. Migration of uranium and thorium: exploration significance. Amer. Assoc. Petrol. Geol. Studies in Geology No. 3.

Galloway, W.E. 1968. Depositional systems in the Lower Wilcox Group, north-central Gulf Coast basin. Gulf Coast Assoc. Geol. Socs. Trans. 18: 275–289.

Galloway, W.E. 1975. Process framework for describing the morphologic and stratigraphic evolution of deltaic depositional systems. *In* M.L. Broussard (ed.), Deltas, pp. 87–98. Houston Geol. Soc.

Galloway, W.E. 1976. Sediments and stratigraphic framework of the Copper River fan delta. J. Sediment. Petrol. 46: 726–737.

Galloway, W.E. 1977. Catahoula Formation of the Texas Coastal Plain: depositional systems, composition, structural development, ground-water flow history, and uranium distribution. Bur. Econ. Geol. Univ. Texas, Austin, Rept. Invest. No. 87.

Galloway, W.E. 1980. Deposition and early hydrologic evolution of Westwater Canyon wet alluvial-fan system, *In* C.A. Rautman (ed.), Geology and mineral technology of the Grants uranium region 1979, pp. 59–69. New Mexico Bur. Mines and Mineral Resourc. Mem. 38.

Galloway, W.E. 1981. Depositional architecture of Cenozoic Gulf Coastal Plain fluvial systems. *In* Soc. Econ. Paleont. Mineral. Spec. Publ. No. 31, pp. 127–155.

Galloway, W.E. 1982. Epigenetic zonation and fluid flow history of uranium-bearing fluvial aquifer systems, South Texas uranium province. Bur. Econ. Geol. Univ. Texas, Austin, Rept. Invest. No. 119.

Galloway, W.E., Brown, L.F., Jr. 1972. Depositional systems and shelf-slope relationships in Upper Pennsylvanian rocks, North-Central Texas. Bur. Econ. Geol. Univ. Texas, Austin, Rept. Invest. No. 75.

Galloway, W.E., Dutton, S.P. 1979. Seismic stratigraphic analysis of intracratonic basin sandstone reservoirs. *In* Tulsa Geol. Soc. Spec. Pub. 1, pp. 65–81.

Galloway, W.E., Kaiser, W.R. 1980. Catahoula Formation of the Texas coastal plain: origin, geochemical evolution, and characteristics of uranium deposits. Bur. Econ. Geol. Univ. Texas, Austin, Rept. Invest. No. 100.

Galloway, W.E., Finley, R.J., Henry, C.D. 1979a. South Texas uranium province, geologic perspective. Bur. Econ. Geol. Univ. Texas, Austin, Guidebook 18.

Galloway, W.E., Kreitler, C.W., McGowen, J.H. 1979b. Depositional and ground-water flow systems in the exploration for uranium. Bur. Econ. Geol. Univ. Texas, Austin, Research Colloquium Notes.

Galloway, W.E., Henry, C.D., Smith, G.E. 1982. Depositional framework, hydrostratigraphy and uranium mineralization of the Oakville Sandstone (Miocene), Texas Coastal Plain. Bur. Econ. Geol. Univ. Texas, Austin, Rept. Invest. No. 113.

Galloway, W.E., Hobday, D.K., Magara, K. 1982a. Frio Formation of Texas Gulf Coastal Plain: depositional systems, structural framework, and hydrocarbon distribution. Amer. Assoc. Petrol. Geol. Bull. 6: 649–688.

Galloway, W.E., Hobday, D.K., Magara, K. 1982b. Frio Formation of Texas Gulf Coastal Plain—depositional systems, structural framework, and hydrocarbon origin, migration, distribution and exploration potential. Bur. Econ. Geol. Univ. Texas, Austin, Rept. Invest. No. 122.

Galloway, W. E., Hobday, D.K., Magara, K. 1982b. Seismic stratigraphic model of depositional platform margin, eastern Anadarko Basin, Oklahoma. *In* C.E. Payton (ed.), Seismic stratigraphy—applications to exploration, Amer. Assoc. Petrol. Geol. Mem. 26: 439–449.

Galvin, C.J., Jr. 1968. Breaker type classification on three laboratory beaches. J. Geophys. Res. 73: 3651–3659.

Gerard, J., Oesterle, H.J. 1973. Facies study of the offshore Mahakam area. Procs. Indones. Petrol. Assoc., pp. 187–194.

Gersib, G.A., McCabe, P.J. 1981. Continental coal-bearing sediments of the Fort Hood Formation (Carboniferous), Cape Linzee, Nova Scotia, Canada. *In* Soc. Econ. Paleont. Mineral. Spec. Publ. 31, pp. 95–108.

Gienapp, H. 1973. Stromungen wahrend der Strumflut vom 2 November 1965 in der Deutschen Bucht und ihre Bedeutung fur den Sedimenttransport. Senck. Marit 5: 135–151.
Glaeser, J.D. 1978. Global distribution of barrier islands in terms of tectonic setting. J. Geol. 86: 283–297.
Glennie, K.W. 1970. Desert sedimentary environments. Developments in sedimentology 14, Amsterdam: Elsevier.
Glennie, K.W. 1972. Permian Rotliegendes of northwest Europe interpreted in light of modern desert sedimentation studies. Amer. Assoc. Petrol. Geol. Bull. 56: 1048–1071.
Glennie, K.W., Evamy, B.D. 1968. Dikaka: plants and plant-root structures associated with aeolian sand. Palaeogeogr. Palaeoclim. Palaeoecol. 23: 77087.
Gloppen, T.G., Steel, R.J. 1981. The deposits, internal structure, and geometry in six alluvial fan–fan delta bodies (Devonian–Norway)–a study in the significance of bedding sequence in conglomerates. *In* Soc. Econ. Paleont. Mineral. Spec. Publ. No. 31, pp. 49–69.
Goldhaber, M. B., Reynolds, R.L., Rye, R.O. 1978. Origin of a South Texas roll-type uranium deposit: II. Petrology and sulfur isotope studies. Econ. Geol. 73: 1690–1705.
Goldhaber, M.B., Reynolds, R.L., Rye, R.O. 1979. Formation and resulfidization of a South Texas roll-type uranium deposit. U.S. Geol. Surv. Open-file Rept. 79–1651.
Goldring, R. 1964. Trace fossils and the sedimentary surface in shallow marine sediments. *In* L.M.J.U. Van Straaten (ed.), Deltaic and shallow marine deposits. Developments in sedimentology, pp. 136–143. Amsterdam: Elsevier.
Goldring, R., Bridges, P., 1973. Sublittoral sheet sandstones. J. Sediment. Petrol. 43: 736–747.
Gole, C.V., Chitale, S.V. 1966. Inland delta building activity of the Kosi River. Amer. Soc. Civ. Eng. Procs., J. Hydraul. Div., 92: 111–126.
Gorsline, D.S. 1980. Deep-water sedimentological conditions and models 1980. Mar. Geol. 38: 1–21.
Gould, H.R. 1970. The Mississippi delta complex. *In* Soc. Econ. Paleont. Mineral. Spec. Publ. No. 12, pp. 3–30.
Granat, M., Ludwick, J.C. 1980. Perpetual shoals at the entrance to Chesapeake Bay: flow-substrate interactions and mutually evasive net currents. Mar. Geol. 36: 307–323.
Grandstaff, D.E. 1976. A kinetic study of the dissolution of uraninite. Econ. Geol. 71: 1493–1505.
Granger, H.C., Warren, C.G. 1978. Some speculations on the genetic geochemistry and hydrology of roll-type uranium deposits, pp. 349–361. Wyoming Geol. Assoc. Guidebook.
Green, D., Crockett, R.N., Jones, M.T. 1980. Tectonic control of Karoo sedimentation in mid-eastern Botswana. Geol. Soc. S. Afr. Trans. 83: 213–219.
Greenwood, B. 1978. Spatial variability of texture over a beach-dune complex, North Devon, England. Sediment Geol. 21: 27–44.
Greenwood, B., Mittler, P.R. 1979. Structural indices of sediment transport in a straight, wave-formed nearshore bar. Mar. Geol. 32: 191–203.
Griffin, J.J., Goldberg, E.D. 1963. Clay mineral distribution in the Pacific Ocean. *In* M.N. Hill (ed.), The sea. New York: Wiley.
Grivetti, E.J. 1981. A changing environment. Aust. Petrol. Expl. Assoc. J. 12: 5–8.
Grove, A.T. 1970. The rise and fall of Lake Chad. Geogr. Mag. 42: 432–439.
Grutt, E.W., Jr. 1972. Prospecting criteria for sandstone-type uranium deposits. *In* Uranium prospecting handbook, pp. 47–76. Inst. Min. Metall.
Gustavson, T.C. 1978. Bed forms and stratification types of modern gravel meander lobes, Nueces River, Texas. Sedimentology 25: 401–426.
Gustavson, T.C., Kreitler, C.W. 1977. Geothermal resources of the Texas Gulf Coast—environmental concerns arising from the production and disposal of geothermal waters. *In* M.D. Campbell (ed.), Geology of alternate energy resources in the South-Central United States, pp. 247–355. Houston Geol. Soc.
Halbouty, M.T. 1980. Methods used, and experience gained, in exploration for new oil and gas fields in highly explored (mature) areas. Amer. Assoc. Petrol. Geol. Bull. 64: 1210–1222.
Hamblin, A.P., Walker, R.G. 1979. Storm-dominated shelf deposits: the Fernie-Kootenay (Jurassic) transition, southern Rocky Mountains. Can. J. Earth Sci. 16: 1673–1690.
Hampton, M.A. 1975. Competence of fine-grained debris flows. J. Sediment. Petrol. 45: 834–844.
Hampton, M.A. 1979. Buoyancy in debris flows. J. Sediment. Petrol. 49: 753–758.
Handford, C.R. 1980. Lower Permian facies of the Palo Duro Basin, Texas: depositional systems, shelf-margin evolution, paleogeography, and petroleum potential. Bur. Econ. Geol. Univ. Texas, Austin, Rept. Invest. No. 102.
Handford, C.R. 1982. Sedimentology and evaporite genesis in a Holocene continental sabkha playa basin, Bristol Dry Lake, California. Sedimentology, 29: 239–254.
Handford, C.R. 1981. Sedimentology and genetic stratigraphy of Dean and Spraberry Formations (Permian), Midland Basin, Texas. Amer. Assoc. Petrol. Geol. Bull. 65: 1602–1616.
Happ, S.C. 1971. Genetic classification of valley sediment deposits. Amer. Soc. Civ. Eng. J. Hydraul.

Div., 97: 43–53.

Hardie, L.A., Smoot, J.P., Eugster, H.P. 1978. Saline lakes and their deposits: a sedimentological approach. *In* Int. Assoc. Sediment. Spec. Publ. 2, pp. 7–41.

Harms, J.C. 1966. Stratigraphic traps in a valley fill. Amer. Assoc. Petrol. Geol. Bull. 50: 2119–2149.

Harms, J.C. 1969. Hydraulic significance of some wave ripples. Geol. Soc. Amer. Bull. 80: 363–396.

Harms, J.C. 1974. Brushy Canyon Formation, Texas: a deep-water density current deposit. Geol. Soc. Amer. Bull. 85: 1763–1784.

Harms, J.C. 1975. Stratification produced by migrating bedforms. *In* Soc. Econ. Paleont. Mineral. Short Course No. 2, pp. 45–61.

Harms, J.C., Fahnestock, R.K. 1965. Stratification, bedforms, and flow phenomena (with examples from the Rio Grande). *In* Soc. Econ. Paleont. Mineral. Spec. Publ. 12, pp. 84–115.

Harms, J.C., MacKenzie, D.B., McCubbin, D.G. 1963. Stratification in modern sands of the Red River, Louisiana. J. Geol. 71: 566–580.

Harms, J.C., Southard, J.B., Spearing, D.R., Walker, R.G. 1975. Depositional environments as interpreted from primary sedimentary structures and stratification sequences. Soc. Econ. Paleont. Mineral. Short Course No. 2.

Harris, D.G., Hewitt, C.H. 1977. Synergism in reservoir management—the geologic perspective. J. Petrol. Tech. 29: 761–770.

Harrison, F.W., III. 1980. The role of pressure, temperature, salinity, lithology and structure in hydrocarbon accumulation in Constance Bayou, Deep Lake, and Southeast Little Pecan Lake Fields, Cameron Parish, Louisiana. Gulf Coast Assoc. Geol. Socs. Trans. 30: 113–127.

Harrison, S.C. 1975. Tidal-flat complex, Delmarva Peninsula, Virginia. *In* R.N. Ginsburg (ed.), Tidal deposits, pp. 31–38. New York: Springer-Verlag.

Harshman, E.N. 1970. Uranium ore rolls in the United States. *In* Uranium exploration geology, pp. 219–232. Internat. Atom. Energy Agency Procs. Series No. STI/PUB/277.

Harshman, E.N. 1972. Geology and uranium deposits, Shirley Basin area, Wyoming. U.S. Geol. Surv. Prof. Pap. 745.

Harshman, E.N., Adams, S.S. 1981. Geology and recognition criteria for roll-type uranium deposits in continental sandstones. U.S. Dept. Energy Rept. GJBX–1.

Hartman, J.A., Paynter, D.D. 1979. Drainage anomalies in Gulf Coast Tertiary sandstones. J. Petrol. Tech. 31: 1313–1322.

Hawley, J.W., Wilson, W.E. 1965. Quaternary geology of the Winnamucca area, Nevada. Univ. Nevada. Desert Res. Inst. Tech. Dept. No. 5.

Hay, R.L. 1977. Geology of zeolites in sedimentary rocks. *In* F.A. Mumpton (ed.), Minerology and geology of natural zeolites, pp. 53–64. Mineral. Soc. Amer. Course Notes.

Hayes, E.T. 1979. Energy resources available to the United States in 1985-2000. Science. 203: 233-239.

Hayes, M.O. 1967a. Hurricanes as geological agents: Case studies of Hurricane Carla, 1961, and Cindy, 1963. Bur. Econ. Geol. Univ. Texas, Austin, Rept. Invest. No. 61.

Hayes, M.O. 1967b. Relationship between coastal climate and bottom sediment type of the inner continental shelf. Mar. Geol. 5: 111–132.

Hayes, M.O. 1975. Morphology of sand accumulation in estuaries. *In* L.E. Cronin (ed.), Estuarine research, Vol. 2, pp. 3–22, New York: Academic Press.

Hayes, M.O. (ed.), 1969. Coastal environments, northeastern Massachusetts and New Hampshire. Eastern Section, Soc. Econ. Paleont. Mineral. Guidebook.

Hayes, M.O., Kana, T.W. 1976. Terrigenous clastic depositional environments—some modern examples. Coastal Res. Div., Univ. South Carolina, Tech. Rept. CRD–11.

Hayes, M.O., Anan, F.A., Bozeman, R.N. 1969. Sediment dispersal trends in the littoral zone. *In* M.O. Hayes (ed.), Coastal environments, northeastern Massachusetts and New Hampshire, pp. 290–315. Dept. Geol., Univ. Massachusetts, Publ. Ser. 1.

Hedberg, M.D. 1968. Significance of high-wax oils with respect to genesis of petroleum. Amer. Assoc. Petrol. Geol. Bull.52: 736–750.

Hedberg, H.D. 1974. Relation of methane generation to undercompacted shales, shale diapirs and mud volcanoes. Amer. Assoc. Petrol. Geol. Bull. 58: 717–722.

Heezen, B.C. 1963. Turbidity currents. *In* M.N. Hill (ed.), The sea, pp. 742–775. New York: Wiley.

Heezen, B.C. 1974. Atlantic-type continental margins. *In* C.A. Burk, C.L. Drake (eds.), The geology of continental margins, pp. 13–24. New York: Springer.

Hesp, P.A. 1981. The formation of shadow dunes. J. Sediment. Petrol. 51: 101–112.

Heward, A.P. 1978a. Alluvial fan sequence and megasequence models: with examples from Westphalian D–Stephanian B coalfields, northern Spain. *In* A.D. Miall (ed.), Fluvial sedimentology, pp. 669–702. Can. Soc. Petrol. Geol. Mem. 5.

Heward, A.P. 1978b. Alluvial fan and lacustrine sediments from the Stephanian A and B (La Magdalena, Cinera–Matallana and Sabero) coalfields, northern Spain. Sedimentology 25: 451–488.

Hickin, E.J. 1974. Development of meanders in

natural river channels. Amer. Jour. Sci. 274: 414–442.
High, L.R., Jr., Picard, M.D. 1969. Stratigraphic relations within upper Chugwater Group (Triassic), Wyoming. Amer. Assoc. Petrol. Geol. Bull. 53: 1091–1104.
Hiller, N., Stavrakis, N. 1979. Distal alluvial fan deposits in the Beaufort Group of the eastern Cape. Geol. Soc. S. Afr. 18th Cong. Abstract 2, pp. 113–120.
Hitchon, B. 1968. Geochemistry of natural gas in western Canada. *In* Natural Gases of North America, pp. 1995–2025. Amer. Assoc. Petrol. Geol. Mem. 9.
Hitchon, B. 1969a. Fluid flow in the western Canada sedimentary basin; 1. Effect of topography. Water Resour. Res. 5: 186–195.
Hitchon, B. 1969b. Fluid flow in the western Canada sedimentary basin; 2. Effect of geology. Water Resour. Res. 5: 460–469.
Hjulstorm, M. 1939. Transportation of detritus by moving water. *In* P.D. Trask (ed.), Recent marine sediments, pp. 5–31. Amer. Assoc. Petrol. Geol.
Hobday, D.K. 1973. Middle Ecca deltaic deposits in the Muden-Tugela Ferry area of Natal. Geol. Soc. S. Afr. Trans. 76: 309–318.
Hobday, D.K. 1974a. Beach- and barrier-island facies in the Upper Carboniferous of northern Alabama. *In* Geol. Soc. Amer. Spec. Pap. 148, pp. 209–223.
Hobday, D.K. 1974b. Interaction between fluvial and marine processes in the lower part of the Late Precambrian Vadsø Group, Finnmark. Norges Geol. Unders., 303: 39–56.
Hobday, D.K. 1976. Quaternary sedimentation and development of the lagoonal complex, Lake St. Lucia, Zululand. Ann. S. Afr. Mus., Vol. 71, pp. 93–113.
Hobday, D.K. 1978a. Fluvial deposits of the Ecca and Beaufort Groups in the eastern Karoo Basin, southern Africa. *In* A.D. Miall (ed.), Fluvial sedimentology. Can. Soc. Petrol. Geol. Mem. 5, pp. 413–429.
Hobday, D.K. 1978b. Geological evolution and geomorphology of the Zululand coastal plain. *In* B.R. Allanson (ed.), Ecology and biogeography of Lake Sibaya, pp. 1–20. Monog. Biol., The Hague: Junk.
Hobday, D.K., Horne, J.C. 1977. Tidally influenced barrier island and estuarine sedimentation in the Upper Carboniferous of southern West Virginia. Sediment. Geol. 18: 97–122.
Hobday, D.K., Jackson, M.P.A. 1979. Transgressive shore-zone sedimentation and syndepositional deformation in the Pleistocene of Zululand, South Africa. J. Sediment. Petrol. 49: 145–158.
Hobday, D.K., Mathew, D. 1975. Late Paleozoic fluvial and deltaic deposits in the northeast Karoo Basin. *In* M.L. Broussard (ed.), Deltas, pp. 457–469. Houston Geol. Soc.
Hobday, D.K., Morton, R.A. 1980. Lower Cretaceous shelf storm deposits, north Texas. Amer. Assoc. Petrol. Geol. Bull. 64: 723.
Hobday, D.K., Morton, R.A. 1983. Lower Cretaceous shelf storm deposits, northeast Texas. *In* Soc. Econ. Paleont. Mineral. Spec. Publ., in press.
Hobday, D.K., Reading, H.G. 1972. Fair weather versus storm processes in shallow marine sand bar sequences in the late Precambrian of Finnmark, North Norway. J. Sediment. Petrol. 42: 318–324.
Hobday, D.K., Tankard, A.J. 1978. Transgressive barrier and shallow-shelf interpretation of the lower Paleozoic Peninsula Formation, South Africa. Geol. Soc. Amer. Bull. 89: 1733–1744.
Hobday, D.K., Von Brunn, V. 1979. Fluvial sedimentation and paleogeography of an early Paleozoic failed rift, southeastern margin of Africa. Palaeogeogr. Palaeoclim. Palaeoecol. 328: 169–184.
Hobday, D.K., Morton, R. A., Collins, E.W. 1979. The Queen City Formation in the East Texas Embayment: a depositional record of riverine, tidal, and wave interraction. Gulf Coast Assoc. Geol. Socs. Trans. 29: 136–146.
Hooke, R.L. 1967. Processes on arid-region alluvial fans. J. Geol. 75: 438–460.
Hooke, R.L. 1972. Geomorphic evidence for Late-Wisconsin and Holocene deformation, Death Valley, California. Geol. Soc. Amer. Bull. 83: 2073–2098.
Horn, D.R., Ewing, M., Delach, M.N., Horn, B.M. 1971. Turbidites of the northeast Pacific. Sedimentology 16: 55–69.
Horne, J.C. 1979a. Estuarine deposits in the Carboniferous of the Pocahontas Basin. *In* J.C. Ferm, J.C. Horne (eds.), Carboniferous depositional environments in the Appalachian region, pp. 692–706. Univ. South Carolina, Carolina Coal Group.
Horne, J.C. 1979b. Criteria for the recognition of depositional environments. *In* J.C. Ferm, J.C. Horne (ed.), Carboniferous depositional environments in the Appalachian region, pp. 295–300. Univ. South Carolina, Carolina Coal Group.
Horne, J.C. 1979c. The effects of Carboniferous shoreline geometry on paleocurrent distribution. *In* J.C. Ferm, J.C. Horne (eds.), Carboniferous depositional environments in the Appalachian region, pp. 509–514. Univ. South Carolina, Carolina Coal Group.
Horne, J.C., Ferm, J.C. 1976. Carboniferous depositional environments in the Pocahontas Basin, eastern Kentucky and southern West Virginia. Univ. South Carolina, Carolina Coal Group.
Horne, J.C., Ferm, J.C., Caruccio, F.T., Baganz, B.P. 1978. Depositional models in coal exploration and mine planning in Appalachian region. Amer. Assoc.

Petrol Geol. Bull. 62: 2379–2411.

Horne, J.C., Ferm, J.C., Hobday, D.K., Saxena, R.S. 1976. A field guide to Carboniferous littoral deposits in the Warrior Basin. Amer. Assoc. Petrol. Geol. and Soc. Econ. Paleont. Mineral. Ann. Conv., New Orleans.

Hostetler, P.B., Garrels, R.M. 1962. Transportation and precipitation of uranium and vanadium at low temperatures, with special reference to sandstone-type uranium deposits. Econ. Geol. 57: 137–161.

Houbolt, J.J.H.C. 1968. Recent sediments in the southern bight of the North Sea. Geol. Mijn. 47: 245–273.

Houbolt, J.J.H.C., Jonker, J.B.M. 1968. Recent sediments in the eastern part of the Lake of Geneva (Lac Leman). Geol. Mijn. 47: 131–148.

Howard, J.D. 1972. Trace fossils as criteria for recognizing shorelines in the stratigraphic record. *In* Soc. Econ. Paleont. Mineral. Spec. Publ. No. 16, pp. 215–225.

Howard, J.D. 1975. Estuaries of the Georgia coast, U.S.A.: sedimentology and biology. IX Conclusions. Senck. Marit. 7: 297–305.

Howard, J.D. 1978. Sedimentology and trace fossils. *In* Trace fossil concepts, pp. 13–47. Soc. Econ. Paleont. Mineral. Short Course No. 5.

Howard, J.D., Frey, R.W. 1975. Estuaries of the Georgia coast, U.S.A.: sedimentology and biology. II Regional animal—sediment characteristics of the Georgia estuaries. Senck. Marit. 7: 33–103.

Howard, J.D., Reineck, H-E. 1981. Depositional facies of high-energy beach-to-offshore sequence: a comparison with low-energy sequence. Amer. Assoc. Petrol. Geol. Bull. 65: 807–830.

Howell, D.J., Ferm, J.C. 1980. Exploration model for Pennsylvanian upper delta plain coals, southwest West Virginia. Amer. Assoc. Petrol. Geol. Bull. 64: 938–941.

Hoyt, J.H., Henry, J.V. 1967. Influence of inlet migration on barrier island sedimentation. Geol. Soc. Amer. Bull. 78: 77–86.

Hoyt, W.V. 1959. Erosional channel in the middle Wilcox near Yoakum, Lavaca County, Texas. Gulf Coast Geol. Soc. Trans. 9: 41–50.

Hubbard, D.K., Barwis, J.H. 1976. Discussion of tidal inlet sand deposits: examples from the South Carolina coast. *In* M.O. Hayes, T.W. Kana (eds.), Terrigenous clastic depositional environments—some modern examples. Coastal Res. Div., Univ. South Carolina, Tech. Rept. CRD–11.

Hubbard, D.K., Oertel, G., Nummedal, D. 1979. The role of waves and tidal currents in the development of tidal-inlet sedimentary structures and sand body geometry: examples from North Carolina. J. Sediment. Petrol. 49: 1073–1092.

Hubbert, M.K. 1940. The theory of ground water motion. J. Geol. 48: 785–944.

Hubbert, M.K. 1953. Entrapment of petroleum under hydrodynamic conditions. Amer. Assoc. Petrol. Geol. Bull. 37: 1954–2026.

Hubbert, M.K. 1969. Energy resources. *In* Resources and man, pp. 157–242. Nat. Acad. Sci.

Hubbert, M.K., Rubey, W.W. 1959. Role of fluid pressure in mechanics of overthrust faulting. Geol. Soc. Amer. Bull. 70: 115–166.

Hubert, J.F., Butera, J.C., Price, R.F. 1972. Sedimentology of Upper Cretaceous Cody-Parkman delta, southwestern Powder River Basin, Wyoming. Geol. Soc. Amer. Bull. 83: 1649–1670.

Hubert, J.F., Reed, A.A., Carey, P.J. 1976. Paleogeography of the East Berlin Formation, Newark Group, Connecticut Valley. Amer. J. Sci. 276: 1183–1207.

Huffman, G.G., Price, W.A. 1949. Clay dune formation near Corpus Christi, Texas. J. Sediment. Petrol. 19: 118–127.

Hunt, J.M. 1979. Petroleum geochemistry and geology. San Francisco: Freeman.

Hunter, R.E. 1973. Pseudo-cross lamination formed by climbing adhesion ripples. J. Sediment. Petrol. 43: 1125–1127.

Hunter, R.E. 1977. Basic types of stratification in small eolian dunes. Sedimentology 24: 361–387.

Hunter, R.E. 1981. Stratification styles in eolian sandstones: some Pennsylvanian to Jurassic examples from the Western Interior USA. *In* Soc. Econ. Paleont. Mineral. Spec. Publ. 31, pp. 315–329.

Hsu, K.J. 1977. Studies of Ventura Field, California, 1: Facies geometry and genesis of Lower Pliocene turbidites. Amer. Assoc. Petrol. Geol. Bull. 61: 137–168.

Hsu, K.J., Kelts, K. 1978. Late Neogene chemical sedimentation in the Black Sea. *In* Int. Assoc. Sediment. Spec. Publ. 2, pp. 129–145.

Ibach, L.E.J. 1982. Relationship between sedimentation rate and total organic carbon content in ancient marine sediments. Amer. Assoc. Petrol. Geol. 66: 170–188.

Inman, D.L. 1957. Wave-generated ripples in nearshore sands. U.S. Army Corps Eng., Beach Erosion Board, Tech. Memo. No. 100.

Irving, E. 1964. Paleomagnetism and its application to geological and geophysical problems. New York: Wiley.

Jackson, M.P.A., Hobday, D.K. 1980. Gravity gliding and clay diapirism in a Pleistocene shoreline sequence in Zululand, South Africa. Amer. J. Sci. 280: 333–362.

Jackson, R.G., II. 1975. Velocity-bedform-texture patterns of meander bends in the lower Wabash River of Illinois and Indiana. Geol. Soc. Amer. Bull. 86: 1511–1522.

Jackson, R.G., II. 1976. Depositional model of point

bars in the lower Wabash River. J. Sediment. Petrol. 46: 579–594.

Jackson, R.G., II. 1978. Preliminary evaluation of lithofacies models for meandering alluvial streams. *In* A.D. Miall (ed.), Fluvial sedimentology, pp. 543-576. Can. Soc. Petrol. Geol. Mem. 5.

Jackson, R.G., II. 1981. Sedimentology of muddy fine-grained channel deposits in meandering streams of the American Middle West. J. Sediment. Petrol. 51: 1169–1192.

Jageler, A.H., Matuszak, D.R. 1972. Use of well logs and dipmeters in stratigraphic trap exploration. *In* R.E. King (ed.), Stratigraphic oil and gas fields—classification, exploration methods, and case histories, pp. 107–135. Amer. Assoc. Petrol. Geol. Mem. 16.

Jago, C.F. 1980. Contemporary accumulation of marine sand in a macrotidal estuary, southwest Wales. Sediment. Geol. 26: 21–49.

James, E.A., Evans, P.R. 1971. The stratigraphy of the offshore Gippsland Basin. Aust. Petrol. Expl. Assoc. J., pp. 71–74.

Johnson, G.A.L. 1960. Palaeogeography of the northern Pennines and part of north-eastern England during deposition of the Carboniferous cyclothemic deposits. 21st Int. Geol. Cong., Pt. 12, pp. 118–128.

Johnson, H.D. 1977a. Shallow marine sand bar sequences: an example from the late Precambrian of North Norway. Sedimentology 24: 245–270.

Johnson, H.D. 1977b. Sedimentation and water escape structures in some late Precambrian shallow-marine sandstones from Finnmark, North Norway. Sedimentology 24: 389–411.

Johnson, H.D. 1978. Shallow siliciclastic seas. *In* H.G. Reading (ed.), Sedimentary environments and facies. pp. 207–258. Oxford: Blackwells.

Johnson, M.A., Stride, A.H. 1970. Geological significance of North Sea sand transport rates. Nature 224: 1016–1017.

Jones, J.G., McDonnell, K.L. 1981. Papua New Guinea analogue for the Late Permian environment of northeastern New South Wales, Australia. Palaeogeogr. Palaeoclim. Palaeoecol. 34: 191–205.

Jones, P.H. 1969. Hydrology of Neogene deposits in the northern Gulf of Mexico basin. Louisiana Water Resour. Res. Inst. Bull. GT–2.

Jones, P.H. 1975. Geothermal and hydrocarbon regimes, northern Gulf of Mexico basin. *In* M.H. Dorfman, R.W. Deller (eds.), Procs. Inst. Geopress. Geotherm. Energy. Conf. pp. 15–90. Univ. Texas, Austin, Center for Energy Studies.

Jones, P.H., Wallace, R.H., Jr. 1974. Hydrogeologic aspects of structural deformation in the northern Gulf of the Mexico basin. U.S. Geol. Surv. J. Res. 2: 511–517.

Jones, R.W. 1980. Some mass balance and geological constraints on migration mechanisms. Amer. Assoc. Petrol. Geol. Studies in Geol. No. 10, pp. 47–68.

Kaiser, W.R. 1974. Texas lignite: near-surface and deep-basin resources. Bur. Econ. Geol. Univ. Texas, Austin, Rept. Invest. 79.

Kaiser, W.R., Johnson, J.E., Bach, W.N. 1978. Sand body geometry and the occurrence of lignite in the Eocene of Texas. Bur. Econ. Geol. Univ. Texas, Austin, Geol. Circ. 78–4.

Kaiser, W.R., Ayers, W.B., Jr., LaBrie, L.W. 1980. Lignite resources in Texas. Bur. Econ. Geol. Univ. Texas, Austin, Rept. Invest. 104.

Karl, H.A. 1980. Speculations on processes responsible for mesoscale current lineations on the continental shelf, southern California. Mar. Geol. 34: M9–M18.

Karweil, J. 1956. Die Metamorphose der Kohlen vom Standpunkt der physikalischen Chemie. Zeit. Deutsche. Geol. Ges. 107, 132–139.

Karweil, J. 1966. Inkohlung, Pyrolyse, und primair migration des Erdols. Brennst. Chem. 47: 161–169.

Keller, G.H., Richards, A.F. 1967. Sediments of the Malacca Strait, southeast Asia. J. Sediment. Petrol. 37: 102–127.

Kenyon, N.H. 1970. Sand ribbons of European tidal seas. Mar. Geol. 9: 25–39.

Kenyon, N.H., Stride, A.H. 1970. The tide-swept continental shelf between the Shetland Isles and France. Sedimentology 14: 159–173.

Kharaka, Y.K., Callender, E., Carothers, W.W. 1977. Geochemistry of geopressured geothermal waters of the northern Gulf of Mexico Basin, 1., Brazoria and Galveston Counties, Texas. *In* H. Paquet, Y. Tardy, (eds.), Proc. 2nd Internat. Symp. Water-Rock Interact., pp. 32–41. Univ. Louis Pasteur.

Kharaka, Y.K., Lico, M.S., Wright, V.A., Carothers, W.W. 1980. Geochemistry of formation waters from Pleasant Bayou No. 2 well and adjacent areas in coastal Texas. *In* M.H. Dorfman, W.L. Fisher (eds.), 4th U.S. Gulf Coast Geopress. Geotherm. Energy. Conf. Vol. 1, pp. 168–199. Univ. Texas, Austin, Center for Energy Studies.

King, D. 1960. The sand ridge deserts of South Australia and related aeolian landforms of the Quaternary arid cycles. Trans. Roy. Soc. S. Aust. 83: 99–108.

Kissin, I.G. 1978. The principal distinctive features of the hydrodynamic regime of intensive earth crust downwarping areas. *In* Int. Assoc. Hydrolog. Sci. Publ. 120.

Klein, G. deV. 1963. Bay of Fundy intertidal zone sediments. J. Sediment. Petrol. 33: 844–854.

Klein, G. deV. 1970. Depositional and dispersal dynamics of intertidal sandbars. J. Sediment. Petrol. 40: 1095–1127.

Klein, G. deV. 1975. Resedimented pelagic carbonate and volcaniclastic sediments and sedimentary structures in Leg 30 DSDP cores from the western equatorial Pacific. Geology 3: 39–42.

Klein, G. deV. 1977. Clastic tidal facies. Champaign: Continuing Education Publ. Co.

Klein, G. deV., Ryer, T.A. 1978. Tidal circulation patterns in Precambrian, Paleozoic, and Cretaceous epeiric and mioclinal shelf seas. Geol. Soc. Amer. Bull. 89: 1050–1058.

Klemme, H.D. 1980. Petroleum basins—classifications and characteristics. J. Petrol. Geol. 3: 187–207.

Knight, S.H. 1929. The Fountain and Casper Formations of the Laramie Basin—a study on genesis of sediments. Univ. Wyoming Sci. Pubs., Geology, Vol. 1, pp. 1–82.

Knight, R.J., Dalrymple, R.W. 1975. Intertidal sediments from the south shore of Cobequid Bay, Bay of Fundy, Nova Scotia. *In* R.N. Ginsburg (ed.), Tidal deposits, pp. 47–55. New York: Springer–Verlag.

Kochenov,. A.V., Zineviyev, V.V., Lovaleva, S.A. 1965. Some features of the accumulation of uranium in peat bogs. Geochem. Internat. 2: 65–70.

Kochenov, A.V., Korolev, K.G., Dubinchuk, V.T., Medvedev, Yu. L. 1977. Experimental data on the conditions of precipitation or uranium from aqueous solution. Geochem. Internat. 14: 82–87.

Koesmanidata, R.P. 1978. Sedimentary framework of Tertiary coal basins of Indonesia. *In* Asian Inst. Tech., Proc. 3rd Reg. Conf., pp. 621–640.

Koglin, E., Schenk, H.J., Schwochau, K. 1978. Spectroscopic studies on the binding of uranium by brown coal. App. Spectros. 32: 486–488.

Kolb, C.R., Van Lopik, J.R. 1966. Depositional environments of the Mississippi River deltaic plain southeastern Louisiana. *In* M.L. Shirley, J.A. Ragsdale (eds.), Deltas in their geologic framework, pp. 17–61. Houston Geol. Soc.

Komar, P.D. 1976. The transport of cohesionless sediment on continental shelves. *In* D.J. Stanley, D.J.P. Swift (eds.), Marine sediment transport and environmental management, pp. 108–126. New York: Wiley.

Komar, P.D., Neudeck, R.H., Kulm, L.D. 1972. Origin and significance of deep-water oscillatory ripple marks on the Oregon continental shelf. *In* D.J.P. Swift, D.B. Duane, O.H. Pilkey (eds.), Shelf sediment transport processes and pattern, pp. 601–619. Stroudsburg: Dowden, Hutchinson and Ross.

Komar, P.D., Kulm, L.D., Harlett, J.C. 1974. Observations and analyses of bottom nepheloid layers on the Oregon continental shelf. J. Geol. 82: 104–111.

Korotchansky, A.N., Mitchell, J. 1972. Rejets de dechets dans les nappes profondes. 24th Internat. Geol. Cong. Sec. 11, pp. 282–295.

Kraft, J.C. 1971. Sedimentary facies patterns and geologic history of a Holocene marine transgression. Geol. Soc. Amer. Bull. 82: 2131–2158.

Kraft, J.C., John, C.J. 1979. Lateral and vertical facies relations of transgressive barrier. Amer. Assoc. Petrol. Geol. Bull. 63: 2145–2163.

Krausse, H-F., Damberger, H.H., Nelson, W.J. *et al.* 1979. Roof strata of the Herrin (No. 6) Coal associated rock in Illinois. Illinois State Geol. Surv. Minerals Note 72.

Krinsley, D.H., Doornkamp, J.C. 1973. Atlas of sand surface textures. Cam. Earth Sci. Ser.

Kruger, W.C., Jr. 1968. Depositional environments of sandstones as interpreted from electrical measurements—an introduction. Gulf Coast Assoc. Geol. Socs., Trans. 18: 226–241.

Kruit, C. 1955. Sediments of the Rhone delta. Grain size and microfauna: Nederland. Geol. Mijn. Genoot. Verhand. Geol. 15: 357–514.

Kumar, N., Sanders, J.E. 1970. Are basal transgressive sands chiefly inlet-filling sands? Marit. Seds. 6: 12–14.

Kumar, N., Sanders, J.E. 1974. Inlet sequence: a vertical sequence of sedimentary structures and textures created by the lateral migration of tidal inlets. Sedimentology 21: 491–532.

Kumar, N., Sanders, J.E. 1976. Characteristics of shoreface deposits: modern and ancient. J. Sediment. Petrol. 46: 145–162.

Kuenzi, D.W., Horst, O.H., McGehee, R.V. 1979. Effect of volcanic activity on fluvial deltaic sedimentation in a modern arc-trench gap, southern Guatemala. Geol. Soc. Amer. Bull. 90: 827–838.

Lagaaij, R., Kopstein, F.P.H.W. 1964. Typical features of a fluvio-marine offlap sequence. *In* L.M.J.U. van Straaten (ed.), Deltaic and shallow marine deposits. pp. 216–226. Amsterdam: Elsevier.

Laird, M.G. 1972. Sedimentology of the Greenland Group of the Paparoa Range, west coast, South Island, New Zealand. New Zealand J. Geol. Geophys. 15: 372–393.

Langen, R.E., Kidwell, A.L. 1974. Geology and geochemistry of the Highland uranium deposit, Converse County, Wyoming. Mountain Geol. 11: 85–93.

Langhorne, D.N. 1973. A sandwave field in the outer Thames Estuary, Great Britain. Mar. Geol. 14: 129–143.

Langmuir, D. 1978. Uranium solution-mineral equilibrium at low temperatures with applications to sedimentary ore deposits. Geochim. Cosmochim. Acta 42: 547–569.

Larsen, V., Steel, R.J. 1978. The sedimentary history of a debris-flow dominated Devonian alluvial fan—a study of textural inversion. Sedimentology 25: 37–59.

Lauff, G.H. (ed.). 1967. Estuaries. Amer. Assoc. Adv. Sci. Publ. 83.

Leach, D.L., Puchlik, K.P., Glanzman, R.K. 1980. Geochemical exploration for uranium in playas. J. Geochem. Expl. 13: 251–282.

Leatherman, S.P. 1977. Assateague Island: a case study of barrier island dynamics. Procs. First Conf. on Scientific Research in National Parks, New Orleans.

Leatherman, S.P. 1979. Barrier dune systems: a reassessment. Sediment. Geol. 24: 1–16.

Leatherman, S.P., Williams, A.T. 1977. Lateral textural grading in overwash sediments. Earth Surf. Processes 2: 333–341.

Le Blanc, R.J. 1972. Geometry of sandstone reservoir bodies. *In* Amer. Assoc. Petrol. Geol. Mem. 18, pp. 133–190.

Le Blanc, R.J. 1975. Significant studies of modern and ancient deltaic sediments. *In* M.L. Broussard (ed.), Deltas, pp. 13–85. Houston Geol. Soc.

Le Blanc-Smith, G. 1980. Genetic stratigraphy for the Witbank Coalfield. Geol. Soc. S. Afr. Trans. 83: 313–326.

Le Blanc-Smith, G., Eriksson, K.A. 1979. A fluvioglacial and glaciolacustrine model for Permo-Carboniferous coals of the northeastern Karoo Basin South Africa. Palaeogogr. Palaeoclim. Palaeoecol. 27: 67–84.

Leeder, M.R. 1978. A quantitative stratigraphic model for alluvium, with special reference to channel deposit density and interconnectedness. *In* A.D. Miall (ed.), Fluvial sedimentology, pp. 587–596. Can. Soc. Petrol. Geol. Mem. 5.

Legget, R.F., Brown, R.J.E., Johson, G.H. 1966. Alluvial fan formation near Aklavik, North West Territories, Canada. Geol. Soc. Amer. Bull. 77: 15–30.

Lehner, P. 1969. Salt tectonics and Pleistocene stratigraphy on continental slope of northern Gulf of Mexico. Amer. Assoc. Petrol. Geol. Bull. 53: 2431–2479.

Leliavsky, S. 1966. An introduction to fluvial hydraulics. New York: Dover.

Leopold, L.B., Maddock, T., Jr. 1953. The hydraulic geometry of stream channels and some physiographic implications. U.S. Geol. Surv. Prof. Pap. 242.

Leopold, L.B., Wilman, M.G. Miller, J.P. 1964. Fluvial processes in geomorphology. San Francisco: Freeman.

Levey, R.A. 1978. Bedform distribution and internal stratification of coarse-grained point bars, Upper Congaree Rivers, S.C. *In* A.D. Miall (ed.), Fluvial sedimentology, pp. 105–128. Can. Soc. Petrol. Geol. Mem. 5.

Levorsen, A.I. 1967. Geology of Petroleum, 2nd ed. Freeman: San Francisco.

Lewis, K.B. 1971. Slumping on a continental slope inclined at 10–40°. Sedimentology 16: 97–110.

Lewis, K.B. 1979. A storm-dominated inner shelf, western Cook Strait, New Zealand. Mar. Geol. 31: 31–43.

Link, M.H., Osborne, R.H. 1978. Lacustrine facies in the Pliocene Ridge Basin Group: Ridge Basin, California. *In* Int. Assoc. Sediment. Spec. Publ. 2, pp. 169–187.

Lisitsin, A.K., Kruglov, A.I., Panteleev. V.M., Sidelnikova, V.D. 1967. Conditions of uranium accumulation in lowmoor oxbow peat bogs. Lith. Min. Resourc. 3: 360–370.

Loffler, E., Sullivan, M.E. 1979. Lake Dieri resurrected: an interpretation using satellite imagery. Zeit. Geomorph. 23: 233–242.

Long, D.G.F. 1981. Dextral strike slip faults in the Canadian Cordillera and depositional environments of related fresh-water intermontane coal basins. *In* A.D. Miall (ed.), Sedimentation and tectonics in alluvial basins, pp. 153–186. Can. Soc. Petrol. Geol. Spec. Pap. 23.

Lonsdale, P. 1981. Drifts and ponds of reworked pelagic sediment in part of the southwest Pacific. Mar. Geol. 43, 153–193.

Lonsdale, P., Malfait, B. 1974. Abyssal dunes of foraminiferal sand on the Carnegie Ridge. Geol. Soc. Amer. Bull. 85: 1697–1712.

Lonsdale, P., Smith S.M. 1980. "Lower Insular rise hills" shaped by a bottom boundary current in the mid-Pacific. Mar. Geol. 34, M19–M25.

Love, J.D. 1970. Cenozoic geology of the Granite Mountain area, central Wyoming. U.S. Geol. Surv. Prof. Pap. 695–C.

Lovell, J.P.B., Stow, D.A.V. 1981. Identification of ancient sandy contourites. Geology 9: 347–349.

Lucas, P.T., Drexler, J.M. 1976. Altamont-Bluebell—a major naturally fractured stratigraphic trap, Uinta Basin, Utah. *In* Amer. Assoc. Petrol. Geol. Mem. 24, pp. 121–135.

Ludwig, K.R. 1979. Age of uranium mineralization in the Gas Hills and Crooks Gap districts, Wyoming, as indicated by U-Pb isotope apparent ages. Econ. Geol. 74: 1654–1668.

Lupe, R., Ahlbrandt, T.S. 1979. Sediments of ancient eolian environments—reservoir inhomogeneity. *In* E.D. McKee (ed.), A study of global sand seas. U.S. Geol. Surv. Prof. Pap. 1052, pp. 241–251.

Lutz, M., Kaasschieter, J.P.H., Van Wijhe, D.H. 1975. Geologic factors controlling Rotliegende gas accumulations in the mid-European Basin. 9th World Petrol. Cong. Procs. Vol. 2, pp. 1–14.

MacKenzie, D.B. 1972. Primary stratigraphic traps in sandstones. *In* Amer. Assoc. Petrol. Geol. Mem. 16, pp. 47–63.

Mackin, J.H. 1948. Concept of the graded river. Geol. Soc. Amer. Bull. 59: 463–512.

Mackowsky, M.T. 1968. European Carboniferous

coalfields and Permian Gondwana coalfields. *In* D. Murchison, T.S. Westall (eds.), Coal and coal-bearing strata, pp. 325–345. London: Oliver and Boyd.

MacPherson, B.A. 1978. Sedimentation and trapping mechanism in Upper Miocene Stevens and older turbidite fans of southeastern San Joaquin Valley. Amer. Assoc. Petrol. Geol. Bull. 62: 2243–2274.

Madigan, C.T. 1946. The Simpson Desert expedition, 1939. Scientific Reports, 6. Geology: the sand formations. Trans. Roy. Socs. S. Aust. 70: 45–63.

Magara, K. 1968. Compaction and migration of fluids in Miocene mudstone, Nagaoka Plain, Japan. Amer. Assoc. Petrol. Geol. Bull. 52: 2466–2501.

Magara, K. 1976. Water expulsion from clastic sediments during compaction—directions and volumes. Amer. Assoc. Petrol. Geol. Bull. 60: 543–553.

Magara, K. 1978. Compaction and fluid migration—practical petroleum geology. Amsterdam: Elsevier.

Magara, K. 1980. Evidences of primary oil migration. Amer. Assoc. Petrol. Geol. Bull. 64: 2108–2117.

Magnier, P.H., Oki, T., Kartaadiputra, L.W. 1975. The Mahakam delta, Kalimantan, Indonesia. 9th World Petrol. Cong. Procs. Panel Discussion 4.

Malan, R.C. 1968. Relationship of uranium in the Rocky Mountains of southwestern Colorado to local and regional metallogenesis. New Mexico Geol. Surv. Guidebook, pp. 185–192.

Mallett, C.W., Flood, P.G., Ledger, P. 1983. Sedimentation models for the Baralaba Coal Measures, Moura, Queensland. Australian Coal Geol. 5.

Mann, A.W., Deutscher, R.L. 1978a. Genesis principles for the precipitation of carnotite in calcrete drainages of Western Australia. Econ. Geol. 73: 1724–1737.

Mann, A.W., Deutscher, R.L. 1978b. Hydrogeochemistry of a calcrete-containing aquifer near Lake Way, Western Australia. J. Hydrol. 38: 357–377.

Mansfield, S.P., Spackman, W. 1965. Petrographic composition and sulfur content of selected Pennsylvanian bituminous coal seams. Penn. State Univ. Spec. Res. Rept. No. SR–SO.

Marley, W.E., Flores, R.M., Cavaroc, V.V. 1979. Coal accumulation in Upper Cretaceous marginal deltaic environments of the Blackhawk Formation and Star Point Sandstone, Emery, Utah. Utah Geology 6: 25–40.

Mason, T.R., Tavener-Smith, R. 1978. A fluvio-deltaic Middle Ecca succession west of Empangeni, Zululand. Geol. Soc. S. Afr. Trans. 81: 13–22.

Mason, T.R., von Brunn, V. 1977. 3-Gr-old stromatolites from South Africa. Nature 266: 47–49.

Massari, S.F. 1978. High-constructive coarse-textured delta systems, Tortonian, Southern Alps. Evidence of lateral deposits in delta slope channels. Soc. Geol. Ital. 18: 93–124.

Matthews, R.K. 1974. Dynamic stratigraphy. Englewood Cliffs: Prentice-Hall.

McBride, E.F., Hayes, M.O. 1962. Dune cross-bedding on Mustang Island, Texas. Amer. Assoc. Petrol. Geol. Bull. 46: 546–551.

McCarthy, M.J. 1967. Stratigraphical and sedimentological evidence from the Durban region of major sea-level movements since the late Tertiary. Geol. Soc. S. Afr. Trans. 70: 135–165.

McCave, I.N. 1971a. Sand waves in the North Sea off the coast of Holland. Mar. Geol. 10: 199–225.

McCave, I.N. 1971b. Wave effectiveness at the sea bed and its relationship to bedforms and deposition of mud. J. Sediment. Petrol. 41: 89–96.

McCave, I.N. 1979. Tidal currents at the North Hinder Lightship, southern North Sea: flow directions and turbulence in relation to maintenance of sand banks. Mar. Geol. 31: 101–114.

McCubbin, D.G. 1969. Cretaceous strike-valley sandstone reservoirs, northwestern New Mexico. Amer. Assoc. Petrol. Geol. Bull. 53: 2114–2140.

McGowen, J.H. 1970. Gum Hollow fan delta, Nueces Bay, Texas. Bur. Econ. Geol. Univ. Texas, Austin, Rept. Invest. No. 69.

McGowen, J.H. 1979. Coastal plain systems. *In* Depositional and ground-water flow systems in exploration for uranium, a research colloquim, pp. 80–117. Bur. Econ. Geol. Univ. Texas, Austin.

McGowen, J.H., Garner, L.E. 1970. Physiographic features and sedimentation types of coarse-grained point bars; modern and ancient examples. Sedimentology 14: 77–111.

McGowen, J.H., Groat, C.G. 1971. Van Horn Sandstone, West Texas: an alluvial fan model for mineral exploration. Bur. Econ. Geol. Univ. Texas, Austin, Rept. Invest. No. 72.

McGowen, J.H., Granata, G.E., Seni, S.J. 1979. Depositional framework of the lower Dockum Group (Triassic), Texas Panhandle. Bur. Econ. Geol. Univ. Texas, Austin, Rept. Invest. No. 97.

McGregor, B.A. 1977. Geophysical assessment of submarine slide northeast of Wilmington Canyon. Mar. Geotech. 2: 229–244.

McKee, E.D. 1945. Small-scale structures in Coconino Sandstone of northern Arizona. J. Geol. 53: 313–325.

McKee, E.D. 1957. Primary structures in some Recent sediments. Amer. Assoc. Petrol. Geol. Bull. 41: 1704–1747.

McKee, E.D. 1966. Structures of dunes at White Sands National Monument New Mexico. Sedimentology 7: 1–61.

McKee, E.D. 1979a. Introduction to a study of global sand seas. *In* E.D. McKee (ed.), A study of global sand seas. U.S. Geol. Surv. Prof. Pap. 1052, pp. 1–19.

McKee, E.D. 1979b. Sedimentary structures in dunes. *In* E.D. McKee (ed.), A study of global sand seas. U.S. Geol. Surv. Prof. Pap. 1052, pp. 83–134.

McKee, E.D. 1979c. Ancient sandstones considered

to be eolian. *In* E.D. McKee (ed.), A study of global sand seas. U.S. Geol. Surv. Prof. Pap. 1052, pp. 187–238.

McKee, E.D., Bigarella, J.J. 1972. Deformational structures in Brazilian coastal dunes. J. Sediment. Petrol. 42: 670–681.

McKee, E.D., Moiola, R.J. 1975. Geometry and growth of the White Sands dune field, New Mexico. U.S. Geol. Surv., J. Res., Vol. 3, pp. 59–66.

McKee, E.D., Tibbits, G.C., Jr. 1964. Primary structures of a seif dune and associated deposits in Libya. J. Sediment. Petrol. 34: 5–17.

McKee, E.D., Crosby, E.T., Berryhill, H.L., Jr. 1967. Flood deposits, Bijou Creek, Colorado, June 1965. J. Sediment. Petrol. 37: 829–851.

McKee, E.D., Douglass, J.R., Rittenhouse, S. 1971. Deformation of lee-side laminae in eolian dunes. Geol. Soc. Amer. Bull. 82: 359–378.

McKinney, T.F., Stubblefield, W.L., Swift, D.J.P. 1974. Large scale current lineation on the Great Egg Shoal retreat massif, New Jersey shelf: investigations by sidescan sonar. Mar. Geol. 17: 79–102.

Meckel, L.D. 1975. Holocene sand bodies in the Colorado Delta area, Northern Gulf of California. *In* M.L. Broussard (ed.), Deltas, pp. 87–98. Houston Geol. Soc.

Meckel, L.D., Nath, A.K. 1977. Geologic considerations for stratigraphic modeling and interpretation. *In* C.E. Payton (ed.), Seismic stratigraphy—applications to exploration, pp. 417–438. Amer. Assoc. Petrol. Geol. Mem. 26.

Meinster, B. Tickell, S.J. 1975. Precambrian aeolian deposits in the Waterberg Supergroup. Geol. Soc. S. Afr. Trans. 78: 191–199.

Mettler, D.E. 1968. West Moorcroft and Wood Dakota fields, Crook County, Wyoming. Wyoming Geol. Assoc. Guidebook, pp. 89–94. Twentieth Ann. Field. Conf.

Miall, A.D. 1970. Devonian alluvial fans, Prince of Wales Island, Arctic Canada. J. Sediment. Petrol. 40: 556–571.

Miall, A.D. 1977. Fluvial sedimentology. Can. Soc. Petrol. Geol. Notes, October, 1977, Calgary.

Miall, A.D. 1978. Lithofacies types and vertical profile models in braided river deposits: a summary. *In* A.D. Miall (ed.), Fluvial sedimentology. Can. Soc. Petrol. Geol. Mem. 5, pp. 597–604.

Miall, A.D. 1980. Cyclicity and the facies model concept in fluvial deposits. Can. Petrol. Geol. Bull. 28: 59–80.

Middleton, G.V. 1970. Experimental studies related to problems of flysch sedimentation. *In* Geol. Assoc. Can. Spec. Pap. 7, pp. 253–272.

Middleton, G.V., Hampton, M.A. 1973. Sediment gravity flows, mechanics of flow, and deposition. *In* Pacific Sec., Soc. Econ. Paleont. Mineral. Short Course Notes, pp. 1–28.

Middleton, G.V., Hampton, M.A. 1976. Subaqueous sediment transport and deposition by sediment gravity flows. *In* D.J. Stanley, D.J.P. Swift (eds.), Marine sediment transport and environmental management, pp. 197–218. New York: Wiley.

Middleton, M.F. 1981. Coalification associated with the folded zone of the Bowen Basin. Geol. Soc. Aust. 5th Geol. Conv. Abstract, p. 63.

Miller, R.E., Green, J.H., Davis, G.H. 1971. Geology of the compacting deposits in the Los Banos-Kettlemen City subsidence area, California. U.S. Geol. Surv. Prof. Pap. 497–E.

Miller, R.L. 1976. Role of vortices in surf zone prediction: sedimentation and wave forces. *In* Soc. Econ. Paleont. Miner. Spec. Publ. No. 24, pp. 92–114.

Minter, W.E.L. 1976. Detrital gold, uranium, and pyrite concentrations related to sedimentology in the Precambrian Vaal Reef placer, Witwatersrand, South Africa. Econ. Geol. 71: 157–176.

Minter, W.E.L. 1978. A sedimentological synthesis of placer gold, uranium and pyrite concentrations in Proterozoic Witwatersrand sediments. *In* A.D. Miall (ed.), Fluvial sedimentology, pp. 801–829. Can. Soc. Petrol. Geol. Mem. 5.

Mitchum, R.M., Jr., Vail, P.R., Thompson, S. 1977. Seismic stratigraphy and global changes of sea level, Part 2: the depositional sequence as a basic unit for stratigraphic analysis. *In* C.E. Payton (ed.), Seismic stratigraphy—applications to exploration, pp. 53–62. Amer. Assoc. Petrol. Geol. Mem. 26.

Mofjeld, H.O. 1976. Tidal currents. *In* D.J. Stanley, D.J.P. Swift (eds.), Marine sediment transport and environmental management, pp. 53–64. New York: Wiley.

Moiola, R.J., Spencer, A.B. 1979. Differentiation of eolian deposits by discriminant analysis. *In* E.D. McKee (ed.), A study of global sand seas. U.S. Geol. Surv. Prof. Pap. 1052, pp. 53–58.

Moody, D. 1964. Coastal morphology and processes in relation to the development of submarine sand ridges off Bethany Beach, Delaware. Unpubl. Ph.D. diss., Johns Hopkins Univ.

Moody, J.D. 1978. The world hydrocarbon resource base and related problems. *In* G.M. Philip, K.L. Williams (eds.), Australia's mineral energy resources: assessment and potential, pp. 63–69. Earth Resources Foundation, Univ. Sydney, Occas. Publ. No. 1.

Moody-Stuart, M. 1968. High- and low-sinuosity stream deposits, with examples from the Devonian of Spitsbergen. J. Sediment. Petrol. 36: 1102–1117.

Mooers, C.N.K. 1976. Introduction to physical oceanography and fluid dynamics of continental margins. *In* D.J. Stanley, D.J.P. Swift (eds.), Marine sediment transport and environmental management, pp. 7–21. New York: Wiley.

Moore, G.T., Asquith, D.O. 1971. Delta: term and concept. Geol. Soc. Amer. Bull. 82: 2563–2568.

Moore, G.T., Starke, G.W., Bonham, L.C., Woodbury, H.O. 1978. Mississippi fan, Gulf of Mexico—physiography, stratigraphy, and sedimentational patterns. *In* A.H. Bouma, G.T. Moore, J.M. Coleman (eds.), Framework, facies, and oil-trapping characteristics of the upper continental margin, pp. 155–191. Amer. Assoc. Petrol. Geol. Studies in Geology No. 7.

Moore, L.R. 1968. Cannel coals, bogheads, and oil shales. *In* D.C. Murchison, T.S. Westoll (eds.), Coal and coal-bearing strata, pp. 19–29. Edinburgh: Oliver and Boyd.

Moore, P.D. Bellamy, D.J. 1974. Peatlands. London: Elek Science.

Morgan, J.P., Coleman, J.M., Gagliano, S.M. 1968. Mudlumps: diapiric structures in Mississippi delta sediments. *In* Amer. Assoc. Petrol. Geol. Mem. No. 8, pp. 145–161.

Morisawa, M. 1968. Streams, their dynamics and morphology. New York: McGraw-Hill.

Morris, K.A. 1979. A classification of Jurassic marine shale sequences: an example from the Toarcian (Lower Jurassic) of Great Britain. Palaeogeogr. Palaeoclim. Palaeoecol. 326: 117–120.

Morton, R.A. 1974. Shoreline changes on Galveston Island (Bolivar Roads to San Luis Pass). Bur. Econ. Geol. Univ. Texas, Austin, Geol. Circ. 74–2.

Morton, R.A. 1977. Historical shoreline changes and their causes. Gulf Coast Assoc. Geol. Socs. Trans. 27: 252–264.

Morton, R.A. 1979. Temporal and spatial variations in shoreline changes and their implications, examples from the Texas Gulf Coast. J. Sediment. Petrol. 49: 1101–1111.

Morton, R.A. 1981. Formation of storm deposits by wind-forced currents in the Gulf of Mexico and North Sea. *In* Int. Assoc. Sediment. Spec. Pub. 5, pp. 385–396.

Morton, R.A., Donaldson, A.C. 1978. The Guadalupe river and delta of Texas—a modern analogue for some ancient fluvial-deltaic systems. *In* A.D. Miall (ed.), Fluvial sedimentology, pp. 773–787. Can. Soc. Petrol. Geol. Mem. 5.

Morton, R.A., McGowen, J.H. 1980. Modern depositional environments of the Texas coast. Bur. Econ. Geol. Univ. Texas, Austin, Guidebook 20.

Moslow, T.F., Heron, S.D., Jr. 1981. Holocene depositional history of a microtidal cuspate foreland cape: Cape Lookout, North Carolina. Mar. Geol. 41: 251–270.

Murray, S.P. 1970. Bottom currents near the coast during Hurricane Camille. J. Geophys. Res. 75: 4579–4582.

Nagtegaal, P.J.C. 1973. Adhesion-ripple and barchan-dune sands of the Recent Namib (SW Africa) and Permian Rotliegend (NW Europe) deserts. Madoqua, Ser. 2, 2: 5–19.

Nanson, G.C. 1980. Point bar and floodplain formation of the meandering Beatton River, northeastern British Columbia, Canada. Sedimentology 27: 3–29.

Nanz, R.R., Jr. 1954. Genesis of Oligocene sandstone reservoir, Seeligson field, Jim Wells and Kleberg Counties, Texas: Amer. Assoc. Petrol. Geol. Bull. 38: 96–118.

Nash, J.T., Granger, H.C., Adams, S.S. 1981. Geology and concepts of genesis of important types of uranium deposits. Econ. Geol. 75: 63–116.

Naylor, M.A. 1980. The origin of inverse grading in muddy debris flow deposits—a review. J. Sediment. Petrol. 50: 1111–1116.

Nemec, W., Porebski, S., Steel, R.J. 1980. Texture and structure of resedimented conglomerates—examples of Ksiaz Formation (Famennian-Tournaisian), southwestern Poland. Sedimentology 27: 519–538.

Nio, S-D. 1976. Marine transgression as a factor in the formation of sand wave complexes. Geol. Mijnb. 55: 18–40.

Normark, W.R. 1970. Growth patterns of deepsea fans. Amer. Assoc. Petrol. Geol. Bull. 54: 2170–2195.

Nummedal, D., Oertel, G., Hubbard, D.K., Hine, A.C., III. 1977. Tidal inlet variability—Cape Hatteras to Cape Canaveral. *In* Coastal sediments 77, Amer. Soc. Civ. Engrs.

Nydegger, P. 1976. Stomungen in Seen. Beit. Geol. Schweiz. 66: 141–177.

Obernyer, S. 1978. Basin-margin depositional environments of the Wasatch Formation in the Buffalo-Lake de Smet area, Johnson County, Colorado. *In* H.E. Hodgson (ed.), Symposium on the geology of Rocky Mountain coal Procs., pp. 49–65. Colorado Geol. Surv. Res. Ser. 4.

Oertel, G.F. 1972. Sediment transport of estuary entrance shoals and the formation of swash platforms. J. Sediment. Petrol. 42: 857–863.

Oertel, G.F. 1974. Hydrography and suspended-matter distribution patterns at Doboy Sound Estuary, Sapelo Island, Georgia. Georgia. Mar. Sci. Cent., Tech. Rep. Ser. 74–4.

Oertel, G.F., Dunstan, W.M. 1981. Suspended-sediment distribution and certain aspects of phytoplankton production off Georgia, U.S.A. Mar. Geol. 40: 171–197.

Officer, C.B. 1981. Physical dynamics of estuarine suspended sediments. Mar. Geol. 40: 1–40.

Oliver, W.B. 1971. Depositional systems in the Woodbine Formation (Upper Cretaceous), Northeast Texas. Bur. Econ. Geol. Univ. Texas, Austin, Rept. Invest. No. 73.

Olwig, T.L. 1981. Channelled turbidites in the eastern Central Pacific Basin. Mar. Geol. 39: 33.

Oomkens, E. 1967. Depositional sequences and sand distribution in a deltaic complex. Geol. Mijn. 46: 265–278.

Oomkens, E. 1970. Depositional sequences and sand

distribution in the postglacial Rhone delta complex. *In* Soc. Econ. Paleont. Mineral. Spec. Publ. No. 15, pp. 198–212.

Oomkens, E. 1974. Lithofacies relations in the Late Quaternary Niger Delta complex. Sedimentology 21: 195–222.

Orme, A.R. 1973. Barrier and lagoon systems along the Zululand coast, South Africa. *In* D.R. Coates (ed.), Coastal geomorphology, pp. 181–217. State Univ. New York.

Otvos, E.G., Jr., Price, W.A. 1979. Problems of chenier genesis and terminology—an overview. Mar. Geol. 31: 251–263.

Packham, G.H. (ed.) 1969. The geology of New South Wales. J. Geol. Soc. Aust., Vol. 16.

Paine, W.R. 1968. Stratigraphy and sedimentation of subsurface Hackberry wedge and associated beds of southwestern Louisiana. Amer. Assoc. Petrol. Geol. Bull. 52: 322–342.

Paine, W.R. 1971. Petrology and sedimentation of the Hackberry sequence of southwest Louisiana: Gulf Coast Assoc. Geol. Socs. Trans 21: 37–55.

Parrish, J.T., Hansen, K.S., Ziegler, A.M. 1979. Atmospheric circulation and upwelling in Paleozoic, with reference to petroleum source beds. Amer. Assoc. Petrol. Geol. Bull. 63: 507–508.

Pate, J.D. 1968. Laverne gas area, Beaver and Harper Counties, Oklahoma. *In* B.W. Beebe, B.F. Curtis (eds.), Natural gases of North America, pp. 1509–1524. Amer. Assoc. Petrol. Geol. Mem. 9.

Payne, J.N. 1970. Geohydrologic significance of lithofacies of the Cockfield Formation of Louisiana and Mississippi and of the Yegua Formation of Texas. U.S. Geol. Surv. Prof. Pap. 569–B.

Payne, J.N. 1975. Geohydrologic significance of lithofacies of the Carrizo Sand of Arkansas, Louisiana and Texas, and the Meridian Sand of Mississippi: U.S. Geol. Surv. Prof. Pap. 569–D.

Perry, E., Hower, J. 1970. Burial diagenesis in Gulf Coast pelitic sediments. Clays and Clay Mineral. 18: 165–177.

Peterson, F. 1980. Sedimentology as a strategy for uranium exploration: concepts gained from analysis of a uranium-bearing depositional sequence in the Morrison Formation of south-central Utah, pp. 65–126. *In* Rocky Mountain Section, Soc. Econ. Paleont. Mineral. Short Course Notes.

Phleger, F.B. 1969. Some general features of coastal lagoons. *In* Lagunas costeras, un simposio, pp. 5–26. UNAM–UNESCO, Mexico, Nov. 1967.

Picard, M.D., High, L.R., Jr. 1968. Sedimentary cycles in the Green River Formation (Eocene), Uinta Basin, Utah. J. Sediment. Petrol. 38: 378–383.

Picard, M.D., High, L.R., Jr. 1972. Criteria for recognizing lacustrine rocks. *In* Soc. Econ. Paleont. Mineral. Spec. Publ. No. 16, pp. 108–145.

Picard, M.D., High, L.R., Jr. 1981. Physical stratigraphy of ancient lacustrine deposits. *In* Soc. Econ. Paleont. Mineral. Spec. Publ. No. 31, pp. 233–259.

Pienaar, P.J. 1963. Stratigraphy, petrology, and genesis of the Elliot Group, Blind River, Ontario, including the uraniferous conglomerate. Geol. Surv. Can. Bull. 83.

Pingree, R.D., Maddock, L. 1979. The tidal physics of headland flows and offshore tidal bank formation. Mar. Geol. 32: 269–289.

Pitman, W.C. 1978. Relationship between eustacy and stratigraphic sequences of passive margin. Geol. Soc. Amer. Bull. 89: 1389–1403.

Plumstead, E.P. 1969. Three thousand million years of plant life in Africa. Geol. Soc. S. Afr. Trans. 72 (Annex.).

Polasek, T.E., Hutchinson, C.A. 1967. Characterization of non-uniformities within a sandstone reservoir from a fluid mechanics standpoint. Procs. 7th World Petrol. Cong., Vol. 2, pp. 397–407.

Postma, H. 1981. Exchange of materials between the North Sea and Wadden Sea. Mar. Geol. 40: 199–213.

Potter, P.E. 1967. Sand bodies and sedimentary environments: a review. Amer. Assoc. Petrol. Geol. Bull. 51: 337–365.

Potter, P.E. 1978. Significance and origin of big rivers. J. Geol. 86: 13–33.

Potter, P.E. Pettijohn, F.J. 1977. Paleocurrents and basin analysis. New York: Springer-Verlag.

Powers, M.C. 1967. Fluid release mechanisms in compacting marine mudrocks and their importance in oil exploration. Amer. Assoc. Petrol. Geol. Bull. 51: 1240–1254.

Pretorius, D.A. 1976. The nature of the Witwatersrand gold-uranium deposits. *In* K.H. Wolf (ed.), Handbook of strata-bound and stratiform ore deposits, pp. 29–88. New York: Elsevier.

Prezbindowski, D.R., 1981, Carbonate rock-water diagenesis, Lower Cretaceous Stuart City Trend, South Texas. Unpubl. Ph.D. diss., Univ. Texas, Austin.

Price, W.A. 1958. Sedimentology and Quaternary geomorphology of South Texas. Gulf Coast Assoc. Geol. Socs. Trans. 8: 41–75.

Pryor, Q.A., Amaral, E.J. 1971. Large-scale cross-stratification in the St Peter Sandstone. Geol. Soc. Amer. Bull. 82: 229–244.

Pryor, W.A. 1973. Permeability-porosity patterns and variations in some Holocene sand bodies. Amer. Assoc. Petrol. Geol. Bull. 57: 162–189.

Pryor, W.A. 1975. Biogenic sedimentation and alteration of argillaceous sediments in shallow marine environments. Geol. Soc. Amer. Bull. 86: 1244–1254.

Psuty, N.P. 1966. The geomorphology of beach ridges in Tabasco, Mexico. Coastal Studies Inst., Louisiana State Univ., Tech. Rept. 30.

Purdy, E.G. 1964. Sediments as substrates. *In* J. Embrie, N.D. Newell (eds.), Approaches to paleoecology, pp. 238–271. New York: Wiley.

Putnam, P.E., Oliver, T.A. 1980. Stratigraphic traps in channel sandstones in the upper Mannville (Albian) of east-central Alberta. Bull. Can. Petrol. Geol. 28: 439–508.

Rahmani, R.A. 1981. Facies relationships and paleoenvironments of a Late Cretaceous tide-dominated delta, Drumheller, Alberta. A field guide. Edmonton Geol. Soc.

Rainone, M., Nanni, T., Ori, G.G., Ricci Lucci, F. 1981. A prograding gravel beach in Pleistocene fan delta deposits south of Ancona, Italy. Int. Assoc. Sediment. Second European Regional Meeting Abstracts, pp. 155–156.

Rawson, R.R. 1980. Uranium in the Jurassic Todilto Limestone of New Mexico—an example of a sabkha-like deposit. *In* C.E. Turner-Peterson (ed.), Uranium in sedimentary rocks: application of the facies concept to exploration, pp. 127–147. Rocky Mountain Sect., Soc. Econ. Paleont. Mineral.

Reading, H.G. 1964. A review of the factors affecting the sedimentation of the Millstone Grit (Namurian) in the Central Pennines. *In* L.M.J.U. van Straaten (ed.), Deltaic and shallow marine deposits, pp. 26–34. Amsterdam: Elsevier.

Redfield, A.C. 1958. The influence of the continental shelf on the tides of the Atlantic coast of the United States. J. Mar. Res. 17: 432–448.

Reeves, C.C., Jr. 1968. Introduction to paleolimnology. Developments in Sedimentology 11. Amsterdam: Elsevier.

Reeves, C.C., Jr. 1977. Intermontane basins of the arid western United States. *In* D.O. Doehring (ed.), Geomorphology in arid regions. Proc. 8th Ann. Geomorph. Symp., pp. 7–26. Binghamton, New York.

Reiche, P. 1938. An analysis of cross-lamination, the Coconino Sandstone. J. Geol. 46: 905–932.

Reidenouer, D., Williams, E.G., Dutcher, R.R. 1967. The relationship between paleotopography and sulfur distribution in some coals of western Pennsylvania. Econ. Geol. 62: 632–647.

Reimnitz, E., Gutierrez-Estrada, M. 1970. Rapid changes in the head of Rio Balsas submarine canyon system, Mexico. Mar. Geol. 8: 245–258.

Reineck, H-E. 1967. Layered sediments of tidal flats, beaches and shelf bottoms. *In* G.H. Lauff (ed.), Estuaries. Amer. Assoc. Adv. Sci. Publ. 83: 191–206.

Reineck, H-E. 1972. Tidal flats. *In* Soc. Econ. Paleont. Mineral. Spec. Publ. No. 16, pp. 146–154.

Reineck, H-E. 1975. German North Sea tidal flats. *In* R.N. Ginsburg (ed.), Tidal deposits, pp. 5–12. New York: Springer-Verlag.

Reineck, H-E., Singh, I.B. 1972. Genesis of laminated sand and graded rhythmites in storm-sand layers of shelf mud. Sedimentology 18: 123–128.

Reinson, G.E. 1979. Barrier island systems. *In* R.G. Walker (ed.), Facies Models. Geoscience Canada, pp. 57–74.

Renton, J.J. 1979. The mineral content of coal. *In* A.C. Donaldson, M.W. Presley, J.J. Renton (eds.), Carboniferous coal guidebook, Vol. 1, pp. 189–205. W. Va. Geol. and Econ. Surv.

Renton, J.J., Cecil, C.B. 1979. The origin of mineral matter in coal. *In* A.C. Donaldson, M.W. Presley, J.J. Renton (eds.), Carboniferous coal guidebook, Vol. 1, pp. 206–223. W. Va. Geol. and Econ. Surv.

Renton, J.J., Cecil, C.B., Stanton, R., Dulong, F.T. 1979. Compositional relationships of plants and peats from modern peat swamps in support of a chemical coal model. *In* A.C. Donaldson, M.W. Presley, J.J. Renton (eds.), Carboniferous coal guidebook, Vol. 3, pp. 57–102. W. Va. Geol. and Econ. Surv.

Reynolds, R.E., Goldhaber, M.B., Ludwig, K.R. 1980. History of sulfidization of South Texas roll-type uranium deposits. Amer. Assoc. Petrol. Geol. Bull. 64: 772.

Rhoads, D.C. 1975. The paleoecological and environmental significance of trace fossils. *In* R.W. Frey (ed.), The study of trace fossils, pp. 147–160, New York: Springer.

Ricci-Lucci, F., Colella, A., Ori, G.G., Ogliani, F. 1981. Pliocene fan deltas of the Intra-Apenninic Basin, Bologna. Int. Assoc. Sediment. Excursion Guidebook, Second European Regional Meeting.

Rice, D.D. 1980. Chemical and isotopic evidence for the origins of natural gases in offshore Gulf of Mexico. Gulf Coast Assoc. Geol. Socs. Trans. 30: 203–213.

Rice, D.D., Claypool, G.E. 1981. Generation, accumulation, and resource potential of biogenic gas. Amer. Assoc. Petrol. Geol. Bull. 65: 5–25.

Rieke, H.H., Chilingarian, G.V. 1974. Compaction of argillaceous sediments. Amsterdam: Elsevier.

Rittenhouse, G. 1972. Stratigraphic trap classification. *In* R.E. King (ed.), Stratigraphic oil and gas fields—classification, exploration methods, and case histories, pp. 14–29. Amer. Assoc. Petrol. Geol. Mem. 16.

Roberts, D.G. 1972. Slumping on the eastern margin of Rockall Bank, North Atlantic Ocean. Mar. Geol. 13: 225–235.

Roberts, H.H. 1980. Sediment characteristics of Mississippi River delta-front mudflow deposits. Gulf Coast Assoc. Geol. Socs. Trans. 30: 485–496.

Robertson, D.S., Tilsley, J.E., Hogg, G.M. 1978. The time-bound character or uranium deposits. Econ. Geol. 73: 1409–1419.

Robinson, A.H.W. 1966. Residual currents in relation to shoreline evolution of the East Anglian coast. Mar. Geol. 4: 57–58.

Robinson, A.H.W. 1980. Erosion and accretion along part of the Suffolk coast of East Anglia, England.

Mar. Geol. 37: 133–146.
Roep, T.B., Beets, D.J., Dronkert, H., Pagnier, H. 1979. A prograding coastal sequence of wave-built structures of Messinian age, Sorbas, Almeria, Spain. Sediment. Geol. 22: 135–163.
Rona, P.A. 1969. Linear "lower continental rise hills" off Cape Hatteras. J. Sediment. Petrol. 39, 1132–1141.
Ronov, A.B. 1958. Organic carbon in sedimentary rocks (in relation to the presence of petroleum). Geochem. 5: 510–536.
Roy, P.S., Thom, B.G., Wright, L.D. 1980. Holocene sequences on an embayed high-energy coast: an evolutionary model. Sediment. Geol. 26: 1–19.
Rupke, N.A. 1976. Large-scale slumping in a flysch basin, southwestern Pyrenees. J. Geol. Soc. Lond. 132: 121–130.
Rupke, N.A. 1978. Deep clastic seas. *In* H.G. Reading (ed.), Sedimentary environments and facies, New York: Elsevier, pp. 372–415.
Rupke, N.A., Stanley, D.J. 1974. Distinctive properties of turbiditic and hemipelagic mud layers in the Algero-Balearic Basin, western Mediterranean Sea. Smithsonian Cont. Earth Sci. 13.
Russell, R.J. 1936. Physiography of the lower Mississippi River delta. *In* Reports of the geology of Plaquemines and St Bernard Parishes, La. State Dept. Conser., Geol. Bull. 8.
Russell, R.J., Russell, R.D. 1939. Mississippi River Delta sedimentation. *In* Recent Marine Sediments, pp. 153–177. Amer. Assoc. Petrol. Geol.
Rust, B.R. 1972a. Structure and process in a braided river. Sedimentology 18: 221–245.
Rust, B.R. 1972b. Pebble orientation in fluvial sediments. J. Sediment. Petrol. 42: 384–388.
Rust, B.R. 1978. Depositional models for braided alluvium. *In* A.D. Miall (ed.), Fluvial sedimentology, pp. 605–625. Can. Soc. Petrol. Geol. Mem. 5.
Rust, I.C. 1973. The evolution of the Paleozoic Cape Basin, southern margin of Africa. *In* A.E.M. Nairn, F.G. Stehli (eds.), The ocean basins and margins, I. The South Atlantic, pp. 247–276. New York: Plenum.
Ryder, J.M. 1971. The stratigraphy and morphology of para-glacial alluvial fans in south-central British Columbia, Canada. Can. J. Earth Sci. 8: 279–298.
Ryder, R.T. 1980. Lacustrine sedimentation and hydrocarbon occurrences with emphasis on Uinta Basin models. Amer. Assoc. Petrol. Geol. Fall Educ. Conf. Notes.
Ryer, T.A., Phillips, R.E., Bohor, B.F., Pollastro, R.M. 1980. Use of altered volcanic ash falls in stratigraphic studies of coal-bearing sequences: an example from the Upper Cretaceous Ferron Sandstone Member of the Mancos Shale in central Utah. Geol. Soc. Amer. Bull. 91: 579–586.
Saitta, S.B., Visher, G.S. 1968. Subsurface study of the southern portion of the Bluejacket delta. *In* G.S. Visher (ed.), A guidebook to the geology of the Bluejacket-Bartlesvelle sandstone, pp. 52–65. Oklahoma City Geol. Soc. Guidebook.
Sanders, J.E., Kumar, N. 1975. Evidence of shoreface retreat and in place "drowning" during Holocene submergence of barriers, shelf off Fire Island, New York. Geol. Soc. Amer. Bull. 86: 65–76.
Sangree, J.B., Widmier, J.M. 1977. Seismic stratigraphy and global changes in sea level, Part 9: seismic interpretation of clastic depositional facies. *In* C.E. Payton (ed.), Seismic stratigraphy—applications to exploration, pp. 165–184. Amer. Assoc. Petrol. Geol. Mem. 26.
Sangree, J.S., Waylett, D.C., Frazier, D.E., Amery, G.B., Fennessy, W.J. 1978. Recognition of continental-slope seismic facies, offshore Texas-Louisiana. *In* A.H. Bouma, G.T. Moore, J.M. Coleman (eds), Framework, facies, and oil-trapping characteristics of the upper continental margin, pp. 37–116. Amer. Assoc. Petrol. Geol. Studies in Geology No. 7.
Saucier, A.E. 1980. Tertiary oxidation in Westwater Canyon Member of Morrison Formation. *In* C.A. Rautman (ed.), Geology and mineral technology of the Grants uranium region, pp. 116–121. New Mexico Bur. Min. Miner. Resourc. Mem. 38.
Saxena, R.S. 1976. Modern Mississippi Delta—depositional environments and processes. Amer. Assoc. Petrol. Geol. and Soc. Econ. Paleont. Mineral. Guidebook.
Schmidt-Collerus, J.J. 1969. Investigations of the relationship between organic matter and uranium deposits, Part II, experimental investigations. U.S. Atom. Energy Comm. Open-File Report GJO-933-2.
Schneider, E.D., Fox, P.J., Hollister, D.C., Needham, H.D., Heezen, B.C. 1967. Further evidence of contour currents in the western North Atlantic. Earth Plan. Sci. Lett. 2: 351–359.
Scholl, D.W. 1960. Pleistocene algal pinnacles at Searles Lake, Calif. J. Sediment. Petrol. 30: 414–431.
Schramm, W.E. 1981. Humid tropical alluvial fans, northwest Honduras, Central America. Unpubl. MS thesis, Louisiana State Univ.
Schubel, J.R. 1968. Turbidity maximum of the northern Chesapeake Bay. Science 161: 1013–1015.
Schumm, S.A. 1960. The effect of sediment type on the shape and stratification of some modern fluvial deposits. Amer. J. Sci. 258: 177–184.
Schumm, S.A. 1968a. River adjustment to altered hydrologic regimen, Murrumbidgee River and paleochannels, Australia. U.S. Geol. Surv. Prof. Pap. 598.
Schumm, S.A. 1968b. Speculations concerning paleohydrologic controls of terrestrial sedimentation. Geol. Soc. Amer. Bull. 79: 1573–1588.

Schumm, S.A. 1972. Fluvial paleochannels. *In* Soc. Econ. Paleont. Mineral. Spec. Publ. No. 16, pp. 98–107.

Schumm, S.A. 1977. The fluvial system. New York: Wiley.

Schumm, S.A. 1981. Evolution and response of the fluvial system, sedimentological implications. *In* Soc. Econ. Paleont. Mineral. Spec. Publ. 31, pp. 19–30.

Schwartz, F.W. 1977. Macroscopic dispersion in porous media: the controlling factors: Water Resour. Res. 13: 743–752.

Schwartz, M.L. 1971. The multiple causality of barrier islands. J. Geol. 79: 91–94.

Schwartz, R.K. 1975. Nature and genesis of some washover deposits. U.S. Army Corps of Engineers, Coastal Eng. Res. Cent., Tech. Mem. 61.

Scott, R., Laali, H., Fee, D.W. 1975. Density current strata in Lower Cretaceous Washita Group, north-central Texas. J. Sediment Petrol. 45: 562–575.

Scruton, P.C. 1960. Delta building and the deltaic sequence. *In* F.P. Shepard, T.H. van Andel (eds.), Recent sediments, northwest Gulf of Mexico, pp. 82–102. Amer. Assoc. Petrol. Geol. Tulsa.

Scruton, P.C., Moore, D.G. 1953. Distribution of surface turbidity off the Mississippi Delta. Amer. Assoc. Petrol. Geol. Bull. 37: 1067–1074.

Seilacher, A. 1967. Bathymetry of trace fossils. Mar. Geol. 5: 413–428.

Selley, R.C. 1978a. Ancient sedimentary environments. Ithaca, N.Y.: Cornell Univ. Press.

Selley, R.C. 1978b. Concepts and methods of subsurface facies analysis. Amer. Assoc. Petrol. Geol. Course Note Ser. No. 9.

Sellwood, B.W. 1971. The genesis of some sideritic beds in the Yorkshire Lias (England). J. Sediment. Petrol. 41: 854–858.

Semeniuk, V. 1981. Sedimentology and the stratigraphic sequence of a tropical tidal flat, northwestern Australia. Sediment. Geol. 29: 195–221.

Seni, S.J. 1980, Sand-body geometry and depositional systems, Ogallala Formation, Texas. Bur. Econ. Geol. Univ. Texas, Austin, Rept. Invest. No. 105.

Shannon, J.P., Dahl, A.R. 1971. Deltaic stratigraphic traps in West Tuscola Field, Taylor County, Texas. Amer. Assoc. Petrol. Geol. Bull. 55: 1194–1205.

Sharp, R.P., Nobles, L.H. 1953. Mudflows of 1941 at Wrightwood, southern California. Geol. Soc. Amer. Bull. 64: 547–560.

Shepard, F.P. 1973. Submarine geology, 3rd ed. New York: Harper.

Shepard, F.P. 1981. Submarine canyons: multiple causes and long-time persistence: Amer. Assoc. Petrol. Geol. Bull. 65: 1062–1077.

Shepard, F.P., Dill, R.F. 1966. Submarine canyons and other sea valleys. Chicago: Rand McNally.

Shepard, F.P., Reimnitz, E. 1981. Sedimentation bordering the Rio Balsas delta and canyons, western Mexico. Geol. Soc. Amer. Bull. 92: 395–403.

Shepard, F.P., Emery, K.O., La Fond, R. 1941. Rip currents, a process of geological importance. J. Geol. 49: 337–369.

Shepard, F.P., McLoughlin, P.A., Marshall, N.F., Sullivan, G.G. 1977. Current-meter recordings of low-speed turbidity currents. Geology 5: 297–301.

Sherborne, J.E., Jr., Buckovic, W.A., Dewitt, D.B., Hellinger, T.A., Pavlak, S.J. 1979. Major uranium discovery in volcaniclastic sediments, Basin and Range Province, Yavapai County, Arizona. Amer. Assoc. Petrol. Geol. Bull. 63: 621–646.

Shibaoka, M. 1972. Silica/alumina ratios of the ashes from some Australian coals. Fuel 51: 278–283.

Shmariovick, Ye. M. 1973. Identification of epigenetic oxidation and reduction as impositions on sedimentary rock. Internat. Geol. Rev. 15: 1333–1340.

Shotten, F.W. 1956. Some aspects of the New Red desert in England. Liverpool Manch. Geol. J. 1: 450–465.

Silver, C. 1973. Entrapment of petroleum in isolated porous bodies. Amer. Assoc. Petrol. Geol. Bull. 57: 726–740.

Singer, S.F. 1970. Origin of the moon by capture and its consequences. Amer. Geophys. Union Trans. 51: 637–641.

Skilbeck, C.G. 1982. Carboniferous depositional systems of the Myall lakes, northern New South Wales. *In* P.G. Flood, B. Runnegar (eds.), New England Geology, pp. 121–132, Univ. New England.

Sly, P.G. 1978. Sedimentary processes in lakes. *In* A. Lerman (ed.), Lakes, chemistry, geology, physics, pp. 65–89. New York: Springer.

Smith, A.H.V. 1968. Seam profiles and seam characters. *In* D.G. Murchison, T.S. Westoll (eds.), Coal and coal-bearing strata, pp. 31–40. Edinburgh: Oliver and Boyd.

Smith, D.G. and Putnam, P.E. 1980. Anastomosed river deposits: modern and ancient examples in Alberta, Canada. Can. J. Earth Sci. 17: 1396–1406.

Smith, D.G., Smith, N.D. 1980. Sedimentation in anastomosed river systems: examples from alluvial valleys near Banff, Alberta. J. Sediment. Petrol. 50: 157–164.

Smith, G.E., Galloway, W.E., Henry, C.D. 1980. Effects of climatic, structural, and lithologic variables on regional hydrology within the Oakville aquifer of South Texas. Soc. Min. Eng., Amer. Inst. Min. Metall. and Petrol. Eng., Fourth Ann. Uranium Seminar, pp. 3–17.

Smith, G.I. 1978. Subsurface stratigraphy and geochemistry of late Quaternary evaporites, Searles Lake, California. U.S. Geol. Surv. Prof. Pap.

Smith, J.D. 1968. The dynamics of sand waves and sand ridges. Unpubl. Ph.D. diss., Univ. Chicago, Chicago Circle.

Smith, J.D. 1969. Geomorphology of a sand ridge. J. Geol. 77: 39–55.
Smith, J.W., Robb, W.A. 1973. Aragonite and the genesis of carbonates in Mahogany Zone oil shales in Colorado's Green River Formation. U.S. Bur. Mines. Rept. 7727.
Smith, N.D. 1970. The braided stream depositional environment: comparison of the Platte River with some Silurian clastic rocks, north-central Appalachians. Geol. Soc. Amer. Bull. 81: 2993–3014.
Smith, N.D. 1974. Sedimentology and bar formation in the upper Kicking Horse River, a braided outwash stream. J. Geol. 82: 205–224.
Smith, N.D., Minter, W.E.L. 1980. Sedimentological controls of gold and uranium in two Witwatersrand paleoplacers. Econ. Geol. 66: 114–150.
Smith, P.B. 1968. Paleoenvironment of phosphate-bearing Monterey Shale in Salinas Valley. Amer. Assoc. Petrol. Geol. Bull. 52: 1785–1791.
Smoot, J.P. 1978. Origin of carbonate sediments in the Wilkins Peak Member of the lacustrine Green River Formation (Eocene), Wyoming, U.S.A. *In* Int. Assoc. Sediment. Spec. Publ. 2, pp. 109–127.
Smyth, M. 1970. Type seam sequences for some Permian. Australian coals. Proc. Aust. Inst. Min. Metall. 233: 7–16.
Smyth, M. 1979. Hydrocarbon generation in the Fly Lake-Brolga area of the Cooper Basin. Aust. Petrol. Expl. Assoc. J. 19: 1–7.
Smyth, M. 1980. Thick coal members: products of an inflationary environment? Aust. Coal Geol. 2: 53–76.
Sneh, A. 1979. Late Pleistocene fan deltas along the Dead Sea Rift. J. Sediment. Petrol. 49: 541–552.
Sneider, R.M., Richardson, F.H., Paynter, D.D., Eddy, R.E., Wyant, I.A. 1977. Predicting reservoir rock geometry and continuity in Pennsylvanian reservoirs, Elk City field, Oklahoma. J. Petrol. Technol. 29: 851–866.
Sneider, R.M., Tinker, C.N., Meckel, L.D. 1978. Deltaic environment reservoir types and their characteristics. J. Petrol. Technol. 30: 1538–1546.
Soister, P.D. 1968. Stratigraphy of the Wind River Formation in south-central Wind River Basin, Wyoming. U.S. Geol. Surv. Prof. Pap. 594–A.
Spearing, D.R. 1975. Shallow marine sands. *In* Soc. Econ. Paleont. Mineral. Short Course 2, pp. 103–132.
Spearing, D.R. 1976. Upper Cretaceous Shannon sandstone: an offshore shallow-marine sand body. Wyoming Geol. Assoc. Guidebook, 28th Field Conf., pp. 65–72.
Stach, E. 1958. Radioaktive Inkohlung. Brennst. Chem. 39: 329–331.
Stach, E. 1968. Basic principles of coal petrology: macerals, microlithotypes and some effects of coalification. *In* D.G. Murchison, T.S. Westoll (eds.), Coal and coal-bearing strata, pp. 3–17. Edinburgh: Oliver and Boyd.
Stach, E., Mackowsky, M.T., Teichmuller, M., Taylor, G.H., Chandra, D., Teichmuller, R. 1975. Coal petrology. Stuttgart: Borntraeger.
Stanley, D.J., Unrug, R. 1972. Submarine channel deposits, fluxo-turbidites and other indicators of slope and base-of-slope environments in modern and ancient marine basins. *In* Soc. Econ. Paleont. Mineral. Spec. Publ. No. 16, p. 287–340.
Stanley, D.J., Fenner, P., Kelling, G. 1972. Currents and sediment transport at Wilmington Canyon shelf-break, as observed by underwater television. *In* D.J.P. Swift, D.B. Duane, O.H. Pilkey (eds.), Shelf sediment transport: process and pattern, pp. 621–644. Stroudsburg: Dowden, Hutchinson and Ross.
Staub, J.R., Cohen, A.D. 1978. Kaolinite enrichment beneath coals: a modern analog, Snuggedy Swamp, South Carolina. J. Sediment. Petrol. 48: 203–210.
Staub, J.R., Cohen, A.D. 1979. The Snuggedy Swamp of South Carolina: a back-barrier estuarine coal-forming environment. J. Sediment Petrol. 49: 133–144.
Stear, W.M. 1978. Sedimentary structures related to fluctuating hydrodynamic conditions in flood plain deposits of the Beaufort Group near Beaufort West, Cape. Geol. Soc. S. Afr. Trans. 81: 393–399.
Steel, R.J. 1974. New Red Sandstone floodplain and piedmont sedimentation in the Hebridean Province Scotland. J. Sediment. Petrol. 44: 336–357.
Steel, R.J. 1976. Devonian basins of western Norway, sedimentary response to tectonism and varying tectonic context. Tectonophys. 36: 207–224.
Steel, R.J., Wilson, A.C. 1975. Sedimentation and tectonism (?Permo-Triassic) on the margin of the North Minch Basin, Lewis. J. Geol. Soc. Lond. 131: 183–202.
Steel, R.J., Gjelberg, J., Haarr, G. 1977a. Helvetiafjellet Formation (Barremian) at Festningen, Spitsbergen—a field guide. Norsk Polarinst. Arbok, pp. 11–128.
Steel, R.J., Maehle, S., Nilsen, H., Roe, S.L., Spinnagr, A. 1977b. Coarsening-upward cycles in alluvium of Hornelen Basin (Devonian) Norway: sedimentary response to tectonic events. Geol. Soc. Amer. Bull. 88: 1124–1134.
Stephens, C.G., Crocker, R.L. 1946. Composition and genesis of lunettes. Trans. Roy. Soc. S. Aust. 70: 302–312.
Sternberg, R.W., Larsen, L.H. 1976. Frequency of sediment movement on the Washington continental shelf: a note. Mar. Geol. 21: M37–M47.
Stoffers, P., Hecky, R.E. 1978. Late Pleistocene-Holocene evolution of the Kivu-Tanganyika Basin. *In* Int. Assoc. Sediment. Spec. Publ. 2, pp. 43–55.
Stokes, W.L. 1967. Stratigraphy and primary sedimentary features of uranium occurrences of south-

eastern Utah. *In* Uranium districts of southeastern Utah, pp. 32–52. Utah Geol. Soc. Guidebook 21.

Stokes, W.L. 1968. Multiple parallel truncation bedding planes—a feature of wind deposited sandstone. J. Sediment. Petrol. 38: 510–515.

Stopes, M.C. 1935. On the petrology of banded bituminous coals. Fuel 14: 4–13.

Strakov, N.H. 1962. Principles of lithogenesis, Vol. 2. Edinburgh: Oliver and Boyd.

Stuart, C.A. 1970. Geopressures: Supplement to Procs., 2nd. Symp. Abnormal Subsurf. Press. Louisiana State Univ.

Stumm, W., Morgan, J.J. 1970. Aquatic chemistry. New York: Wiley.

Sturm, M., Matter, A. 1978. Turbidites and varves in Lake Brienz (Switzerland): deposition of clastic detritus by density currents. *In* Int. Assoc. Sediment. Spec. Publ. 2, pp. 147–168.

Surdam, R.C., Stanley, K.O. 1979. Lacustrine sedimentation during the culminating phase of Eocene Lake Gosiute, Wyoming (Green River Formation). Geol. Soc. Amer. Bull. 90: 93–110.

Swain, F.M. 1970. Non-marine organic chemistry. New York: Cambridge Univ.

Swanson, V.E., Palacas, J.G. 1965. Humate in coastal sands of northwest Florida. U.S. Geol. Surv. Bull. 1214–B.

Swartzendruber, D. 1962. Non-Darcy flow behavior in liquid-saturated porous media. J. Geophys. Res. 67: 5205–5213.

Swett, K., Klein, G. deV., Smit, D.E. 1971. A Cambrian tidal sand—the Eriboll Sandstone of northwest Scotland: an ancient-recent analog. J. Geol. 79: 400–415.

Swift, D.J.P. 1968. Coastal erosion and transgressive stratigraphy. J. Geol. 76: 444–456.

Swift, D.J.P. 1969. Evolution of the shelf surface, and relevance of modern shelf studies to the rock record. *In* D.J. Stanley (ed.), The new concepts of continental margin sedimentation: application to the geological record, pp. DS-7-1–DS-7-19. Washington D.C.: Amer. Geol. Inst.

Swift, D.J.P. 1970. Quaternary shelves and the return to grade. Mar. Geol. 8: 5–30.

Swift, D.J.P. 1974. Continental shelf sedimentation. *In* C.A. Burk, C.L. Drake (eds.), The geology of continental margins, pp. 117–135. Berlin: Springer.

Swift, D.J.P. 1975. Barrier island genesis: evidence from the Middle Atlantic shelf of North America. Sediment. Geol. 14: 1–43.

Swift, D.J.P. 1976. Continental shelf sedimentation. *In* D.J. Stanley, D.J.P. Swift (eds.), Marine sediment transport and environmental management, pp. 311–350. New York: Wiley.

Swift, D.J.P., Duane, D.B., McKinney, R.F. 1973. Ridge and swale topography of the Middle Atlantic Bight, North America: secular response to the Holocene hydraulic regime. Mar. Geol. 15: 227–247.

Swift, D.J.P., Field, M.F. 1981. Evolution of the classic sand ridge field: Maryland sector, North America inner shelf. Sedimentology 28: 461–482.

Swift, D.J.P., Heron, S.D., Jr., Dill, C.E., Jr. 1969. The Carolina Cretaceous: petrographic reconnaissance of a graded shelf. J. Sediment. Petrol. 39: 18–33.

Swift, D.J.P., Nelson, T.A., McHone, J.F., Jr., Holliday, B.W., Palmer, H., Shideler, G.L. 1977. Holocene evolution of the inner shelf of southern Virginia. J. Sediment. Petrol. 49: 1454–1474.

Swineford, A., Frye, J.C. 1945. A mechanical analysis of wind-blown dust compared with analyses of loess. Amer. J. Sci. 243: 249–255.

Swirydczuk, K., Wilkinson, B.H., Smith, G.R. 1980. The Pliocene Glenns Ferry Oolite—III: sedimentology of oolitic lacustrine terrace deposits. J. Sediment. Petrol. 50: 1237–1248.

Szalay, A. 1964. Cation exchange properties of humic acids and their importance in the geochemical enrichment of UO^{2+} and other cations. Geochim. Cosmochim. 28: 1605–1614.

Szalay, A., Szilagyi, M. 1969. Accumulation of microelements in peat, humic acids and coal. *In* P.A. Schenck, I. Havenaar (eds.), Advances in organic geochemistry, pp. 567–578. Oxford: Pergamon.

Tanaka, H.H., Hollowell, J.R. 1966. Hydrology of the alluvium of the Arkansas River, Muskogee, Oklahoma to Fort Smith, Arkansas: U.S. Geol. Surv. Water Supply Pap. No. 1809–T.

Tankard, A.J., Hobday, D.K. 1977. Tide-dominated back-barrier sedimentation, Early Ordovician Cape Basin, Cape Peninsula, South Africa. Sediment. Geol. 18: 135–159.

Tankard, A.J., Jackson, M.P.A., Eriksson, K.A., Hobday, D.K., Hunter, D.R., Minter, W.E.L. 1982. Crustal evolution of southern Africa. New York: Springer-Verlag.

Taylor, G., Woodyer, K.D. 1978. Bank deposition in suspended-load streams. *In* A.D. Miall (ed.), Fluvial sedimentology, pp. 257–276. Can. Soc. Petrol. Geol. Mem. 5.

Teichmuller, M., Teichmuller, R. 1968a. Cainozoic and Mesozoic coal deposits of Germany. *In* D. Murchison, T.S. Westoll (eds.), Coal and coal-bearing strata, pp. 347–379. London: Oliver and Boyd.

Teichmuller, M., Teichmuller, R. 1968b. Geological aspects of coal metamorphism. *In* Coal and coal-bearing strata pp. 233–267. D. Murchison, T.S. Westoll (eds.), London: Oliver and Boyd.

Theis, N.J. 1979. Uranium-bearing and associated minerals in their geochemical and sedimentological context, Elliot Lake, Ontario. Geol. Surv. Can. Bull. 304.

Thom, B.G. 1964. Origin of sand beach ridges. Aust. J. Sci. 26: 351.

Thomas, B.M. 1982. Land-plant source rocks for oil and their significance in Australian basins. Aust.

Petrol. Expl. Assoc. J. 22: 164–178.
Thompson, D.B. 1969. Dome-shaped aeolian dunes in the Frodsham Member of the so-called "Keuper Sandstone Formation" (Scythian-?Anisian: Triassic) at Frodsham, Cheshire (England). Sediment. Geol. 3: 263–289.
Thompson, R.W. 1968. Tidal flat sedimentation on the Colorado River Delta, northwestern Gulf of California: Geol. Soc. Amer. Mem. 107.
Thompson, R.W. 1975. Tidal flat sediments of the Colorado River delta, northwestern Gulf of California. *In* R.N. Ginsburg (ed.), Tidal deposits, pp. 57–65. New York: Springer-Verlag.
Tissot, B.P., Welte, D.H. 1978. Petroleum formation and occurrence. New York: Springer-Verlag.
Tissot, B., Califet-Debyser, Y., Deroo, G., Oudin, J.L. 1971. Origin and evolution of hydrocarbons in early Toarcian shales, Paris Basin, France. Amer. Assoc. Petrol. Geol. Bull. 55: 2177–2193.
Tissot, B., Durand, B., Espitalie, J., Combaz, A. 1974. Influence of nature and diagenesis of organic matter in formation of petroleum. Amer. Assoc. Petrol. Geol. Bull. 58: 499–506.
Tóth, J. 1962. A theory of groundwater motion in small drainage basins in central Alberta, Canada. J. Geophys. Res. 67: 4375–4387.
Tóth, J. 1972. Properties and manifestations of regional groundwater movement. 24th Internat. Geol. Cong. Procs., pp. 153–163.
Toth, J. 1978. Gravity-induced cross-formational flow of formation fluids, Red Earth region, Alberta, Canada: analysis, patterns, and evolution: Water Resources Research, v. 5, pp. 805–843.
Tóth, J. 1980. Cross-formational gravity-flow of groundwater: a mechanism of the transport and accumulation of petroleum (the generalized hydraulic theory of petroleum migration), pp. 121–167. Amer. Assoc. Petrol. Geol. Studies in Geology No. 10.
Tsunashima, A., Brindley, G.W., Bastovanov, Marija. 1981. Adsorption of uranium from solutions by montmorillonite; compositions and properties of uranyl montmorillonites: Clays and Clay Min. 29: 10–16.
Tucker, M.E. 1973. The sedimentary environment of tropical African estuaries: Freetown Peninsula, Sierra Leone. Geol. Mijn. 52: 203–215.
Tucker, M.E., Burchette, T.P. 1977. Triassic dinosaur footprints from South Wales. Palaeogeogr. Palaeoclim. Palaeoecol. 22: 195–208.
Turner, B.R. 1978. Palaeohydraulics of clast transport during deposition of the Upper Triassic Molteno Formation in the main Karoo Basin of South Africa. S. Afr. J. Sci. 74: 171–173.
Tyler, N. 1979. The stratigraphy, geochemistry and correlation of the Ventersdorp Subgroup in the Derdepoort area, west-central Transvaal. Geol. Soc. S. Afr. Trans. 82: 133–147.
Tyler, N., Ethridge, F.G., 1983. Depositional setting of the Salt Wash Member of the Morrison Formation, southwest Colorado. J. Sediment. Petrol. 53: 67–82.
Uchupi, E. 1967. The continental margin south of the Cape Hatteras North Carolina: a shallow structure. Southeastern Geol. 8: 155–177.
Uchupi, E., Austin, J.A., Jr. 1979. The stratigraphy and structure of the Laurentian Cone area. Can. J. Earth Sci. 16: 1726–1752.
Vail, P.R., Mitchum, R.M. 1979. Global cycles of sea level change and their role in exploration. 10th World Petrol. Cong. Preprint. Paper 4. Bucharest, Hungary.
Vail, P.R., Todd, R G., Sangree, J.B. 1977. Chronostratigraphic significance of seismic reflections. *In* C.E. Payton (ed.), Seismic stratigraphy—applications to exploration, pp. 99–116. Amer. Assoc. Petrol. Geol. Mem. 26.
Van Andel, T.H., 1967. The Orinoco Delta. J. Sediment. Petrol. 37: 297–310.
Van Andel, T.H., Postma, H. 1954. Recent sediments of the Gulf of Paria: reports of the Orinoco Shelf expedition. Verh. Akad. Wet. Vol. 1.
Van Beek, J.L., Koster, E.A. 1972. Fluvial and estuarine sediments exposed along the Oude Maas. Sedimentology 19: 237–256.
Van Dijk, D.E., Hobday, D.K., Tankard, A.J. 1978. Permo-Triassic lacustrine deposits in the eastern Karoo Basin, South Africa. *In* Int. Assoc. Sediment. Spec. Publ. 2, pp. 225–239.
Van Houten, F.B. 1964. Cyclic lacustrine sedimentation, Upper Triassic Lockatong Formation, central New Jersey and adjacent Pennsylvania. Kansas Geol. Surv. Bull. 169, pp. 497–531.
Van Houten, F.B. 1965. Crystal casts in Upper Triassic Lockatong and Brunswick Formations. Sedimentology 4: 301–313.
Van Rensburg, W.C.J. 1980. The classification of coal resources and reserves. Bur. Econ. Geol. Univ. Texas, Austin, Min. Res. Circ. No. 65.
Van Straaten, L.M. J.U. 1959. Littoral and submarine morphology of the Rhone delta. *In* Procs. 2nd Coastal Geog. Conf., Louisiana State Univ., pp. 223–264.
Van Straaten, L.M.J.U., Kuenen, P.H. 1957. Accumulation of fine grained sediments in the Dutch Wadden Sea. Geol. Mijn. 19: 329–354.
Veevers, J.J. 1981. Morphotectonics of rifted continental margins in embryo (East Africa), youth (Africa—Arabia) and maturity (Australia). J. Geol. 89: 57–82.
Vetter, P. 1980. Le bassin de Decazeville. *In* 26th Int. Geol. Cong., Excursion Guide 091C, pp. 48–54.
Vincelette, R.R., Chitton, W.E. 1981. Exploration for oil accumulations in Entrada Sandstone, San Juan Basin, New Mexico. Amer. Assoc. Petrol. Geol. Bull. 35: 2546–2570.

Visher, G.S. 1965. Use of vertical profile in environmental reconstruction. Amer. Assoc. Petrol. Bull. 49: 41–46.

Visher, G.S. 1972. Physical characteristics of fluvial deposits. *In* Soc. Econ. Paleont. Mineral. Spec. Publ. No. 16, pp. 108–145.

Von Brunn, V., Hobday, D.K. 1976. Early Precambrian tidal sedimentation in the Pongola Supergroup of South Africa. J. Sediment. Petrol. 46: 670–679.

Von Brunn, V., Mason, T.R. 1977. Siliciclastic-carbonate tidal deposit from the 3000 m.y. Pongola Supergroup, South Africa. Sediment. Geol. 18: 245–255.

Von der Borch, C.C., Lock, D.E. 1979. Geological significance of Coorong dolomites. Sedimentology 26: 813–824.

Vormelker, R.S. 1979. Mid-Wilcox channel: deep exploration potential. South Texas Geol. Soc. Bull. 20: 10–40.

Vos, R.G. 1975. An alluvial plain and lacustrine model for the Precambrian Witwatersrand deposits of South Africa. J. Sediment. Petrol. 45: 480–493.

Vos, R.G. 1976. Observations on the formation and location of transient rip currents. Sediment. Geol. 16: 15–19.

Vos, R.G. 1977. Sedimentology of an Upper Paleozoic river, wave, and tide influenced delta system in southern Morocco. J. Sediment. Petrol. 47; 1242–1260.

Vos, R.G. 1981. Sedimentology of an Ordovician fan delta complex, western Libya. Sed. Geol. 29: 153–170.

Vos, R.G., Eriksson, K.A., 1977. An embayment model for tidal and wave swash deposits occurring within a fluvially dominated Middle Proterozoic sequence in South Africa. Sediment. Geol. 18: 161–173.

Vos, R.G., Hobday, D.K. 1977. Storm beach deposits in the Upper Paleozoic Ecca Group of South Africa. Sediment. Geol. 19: 217–232.

Vos, R.G., Tankard, A.J. 1981. Braided fluvial sedimentation in the Lower Paleozoic Cape Basin, South Africa. Sediment. Geol. 29: 171–193.

Walker, R.G. 1966. Shale Grit and Grindslow Shales: transition from turbidite to shallow water sediments in the Upper Carboniferous of northern England. J. Sediment. Petrol. 36: 90–114.

Walker, R.G. 1975. Generalized facies models for resedimented conglomerates of turbidite origin. Amer. Assoc. Petrol. Geol. Bull. 86: 737–748.

Walker, R.G. 1978. Deep water sandstone facies and ancient submarine fans: models for exploration for stratigraphic traps. Amer. Assoc. Petrol. Geol. Bull. 62: 932–966.

Walker, R.G. 1979. Shallow marine sands. *In* R.G. Walker (ed.), Facies models, pp. 75–89. Geoscience Canada.

Walker, R.G. 1980. Exploration for turbidites and other deep water sandstones. Amer. Assoc. Petrol. Geol., Fall Ed. Conf. Notes, Vol. 2.

Walker, R.G. 1981. Shelf sedimentation in the Cretaceous seaway of western Canada. Geol. Soc. Aust. 5th Geol. Conv. Abstracts, pp. 55–56.

Walker, R.G., Middleton, G.V. 1979. Eolian sands *In* R.G. Walker (ed.), Facies models, pp. 33–41. Geoscience Canada.

Walker, R.G., Mutti, E. 1973. Turbidite facies and facies associations. *In* Pacific Sec., Soc. Econ. Paleont. Mineral. Short Course, pp. 119–157.

Walker, T.R. 1967. Formation of red beds in ancient and modern deserts. Geol. Soc. Amer. Bull. 78: 353–368.

Walker, T.R. 1979. Red color in dune sand. *In* E.D. McKee (ed.), A study of global sand seas. U.S. Geol. Surv. Prof. Pap. 1052, pp. 61–81.

Walker, T.R., Harms, J.C. 1972. Eolian origin of Flagstone Beds, Lyons Sandstone (Permian), type area, Boulder County, Colorado. Mount. Geol. 9: 279–288.

Wallick, E.I., Toth. J. 1976. Methods of regional groundwater flow analysis with suggestions for the use of environmental isotopes. *In* Interpretation of environmental isotope and hydrochemical data in groundwater hydrology, pp. 37–63. Vienna: Internat. Atom. Ener. Agency.

Walton, A.W., Galloway, W.E., Henry, C.D. 1981. Release of uranium from volcanic glass in sedimentary sequences: an analysis of two systems. Econ. Geol. 76: 69–88.

Waples, D.W. 1980. Time and temperature in petroleum formation: application of Lopatin's method to petroleum exploration, Amer. Assoc. Petrol. Geol. Bull. 64; 916–926.

Ward, L, G, 1981. Suspended-material transport in marsh tidal channels. Mar. Geol. 40: 139–154.

Wasson, R.J. 1977. Last-glacial alluvial fan sedimentation in the Lower Derwent Valley, Tasmania. Sedimentology 24: 781–799.

Wasson, R.J. 1979. Sedimentation history of the Mundi Mundi alluvial fans, western New South Wales. Sediment. Geol. 22: 21–51.

Weber, K.J. 1971. Sedimentological aspects of oil fields of the Niger delta. Geol. Mijn. 50: 569–576.

Weber, K.J. 1982. Influence of common sedimentary structures on fluid flow in reservoir models. J. Petrol. Technol. 34: 665–672.

Weber, K.J., Klootwijk, P.H., Konieczek, J., van der Vlugt, W.R. 1978. Simulation of water injection in a barrier-bar-type, oil rim reservoir in Nigeria. J. Petrol. Technol. 30: 1555–1565.

Weimer, R.J. 1970. Rates of deltaic sedimentation and intrabasin deformation, Upper Cretaceous of Rocky Mountain region. *In* Soc. Econ. Paleont. Mineral. Spec. Publ. No. 15, pp. 270–292.

Weimer, R.J. 1973. A guide to uppermost Cretaceous stratigraphy, central Front Range, Colorado: deltaic sedimentation, growth faulting and early Laramide crustal movement: Mount. Geol. 10: 53–97.

Welder, F.A. 1959. Processes of deltaic sedimentation in the lower Mississippi River: Louisiana State Univ., Coastal Stud. Inst., Tech. Rept. No. 12.

Wermund, E.G., Jenkins, W.A., Jr. 1970. Recognition of deltas by fitting trend surfaces to Upper Pennsylvanian sandstone in north-central Texas. *In* Soc. Econ. Paleontol. Mineral. Spec. Publ. No. 15, pp. 256–269.

Wescott, W.A., Ethridge, F.G. 1980. Fan-delta sedimentology and tectonic setting—Yallahs Fan Delta, southeast Jamaica. Amer. Assoc. Petrol. Geol. Bull. 64: 374–399.

Weser, O.E. 1975. Exploration for deep water sandstones. New Orleans Geol. Soc., Cont. Ed. Seminar.

Weser, O.E. 1978. Oil trapping characteristics of turbidites. Amer. Assoc. Petrol. Geol. Studies in Geology No. 7, pp. 227–242.

Whateley, M.K.G. 1980. Deltaic and fluvial deposits of the Ecca Group, Nongoma Graben, northeast Zululand. Geol. Soc. S. Afr. Trans. 83: 345–351.

White, A.H., Youngs, B.C. 1980. Cambrian alkali playa-lacustrine sequences in the northeastern Officer Basin, South Australia. J. Sediment. Petrol. 50: 1279–1286.

White, D.A. 1980. Assessing oil and gas plays in facies-cycle wedges. Amer. Assoc. Petrol. Geol. Bull. 64: 1158–1178.

White, D.E. 1965. Saline waters of sedimentary rocks. *In* A. Young, J.E. Galley (eds.), Fluids in subsurface environments, pp. 342–366. Amer. Assoc. Petrol. Geol. Mem. 4.

Williams, E.G., Keith, M.L. 1963. Relationship between sulfur in coals and the occurrences of marine roof beds. Econ. Geol. 58: 720–729.

Williams, G.E. 1971. Flood deposits of the sand-bed ephemeral streams of Central Australia. Sedimentology 17: 1–40.

Williams, P.F., Rust. B.R. 1969. The sedimentology of a braided river. J. Sediment. Petrol. 39: 649–679.

Williams, R.E. 1968. Ground-water flow systems and highway pavement failure in cold mountain valleys. J. Hydrology 6: 183–193.

Williamson, C.R. 1979. Deep-sea sedimentation and stratigraphic traps, Bell Canyon Formation (Guadalupian), Delaware Basin, Texas-New Mexico. *In* Permian Basin Sec., Soc. Econ. Paleont. Mineral. Publ. 79–18, pp. 89–94.

Wilson, H.H. 1977. "Frozen-in" hydrocarbon accumulations or diagenetic traps—exploration targets. Amer. Assoc. Petrol. Geol. Bull. 61: 483–491.

Wilson, I.G. 1971. Desert sandflow basins and a model for the development of ergs. Geog. J. 137: 180–197.

Wilson, I.G. 1972. Aeolian bedforms—their development and origins. Sedimentology 19: 193–210.

Wilson, I.G. 1973. Ergs. Sediment. Geol. 10: 77–106.

Wilson, R.G. 1976. Estimating the potential of a coal basin. *In* Coal exploration, pp. 374–400. San Francisco: Miller Freeman.

Winker, C.D. 1979. Late Pleistocene fluvial-deltaic deposition, Texas Coastal Plain and shelf. Unpubl. MA thesis, Univ. Texas, Austin.

Winker, C.D. 1980. Depositional phases in late Pleistocene cyclic sedimentation. *In* B.F. Perkins, D.K. Hobday (eds.). Middle Eocene coastal plain and nearshore deposits of East Texas: a field guide to the Queen City Formation and related papers, pp. 46–66. Gulf Coast Section, Soc. Econ. Paleont. Mineral.

Winker, C.D. 1982. Cenozoic shelf margins, northwestern Gulf of Mexico: Gulf Coast Assoc. Geol. Socs. Trans. 32: 427–448.

Wolman, M.G., Leopold, L.B. 1957. River channel patterns; braided, meandering, and straight. U.S. Geol. Surv. Prof. Pap. 282–B.

Woodbury, H.O., Spotts, J.H., Akers, W.H. 1978. Gulf of Mexico continental-slope sediments and sedimentation. *In* A.H. Bouma, G.T. Moore, J.M. Coleman (eds.), Framework, facies, and oil-trapping characteristics of the upper continental margin, pp. 117–153. Amer. Assoc. Petrol. Geol. Studies in Geology No. 7.

Woodcock, N.H. 1979. Sizes of submarine slides and their significance. J. Struct. Geol. 1: 137–142.

Woollen, I.D. 1976. Three-dimensional facies distribution of the Holocene Santee Delta. *In* M.O. Hayes, T.W. Kana (eds.), Terrigenous clastic depositional environments, pp. 52–63. Univ. South Carolina.

Wright, L.D., Chappell, J., Thom, B.G., Bradshaw, M.P., Cowell, P. 1979. Morphodynamics of reflective and dissipative beach and nearshore systems: southeastern Australia. Mar. Geol. 32: 105–140.

Yerkes, R.F., McCulloh, T.H., Schoellhammer, J.E., Vedder, J.G. 1965. Geology of the Los Angeles Basin—an introduction. U.S. Geol. Surv. Prof. Pap. No. 420A.

Zeito, G.A. 1965. Interbedding of shale breaks and reservoir heterogeneities. J. Petrol. Tech. 17: 1223–1228.

Ziegler, A.M., Parrish, J.T., Humphreville, R.G. 1979. Paleogeography, upwelling and phosphorites. *In* P.J. Cook, J.H. Shergold (eds.), Proterozoic and Cambrian phosphorites, Internat. Geol. Correlation Proj. 156.

Ziegler, P.A. 1979. Factors controlling North Sea hydrocarbon accumulations. World Oil, 189: 111–124.

Index

C

D

E

F

G

H

I

N

O

P

Q

R

S

T

U

V

W

Y

Z